ABOUT THE AUTHOR

John D. Anderson, Jr., was born in Lancaster, Pennsylvania, on October 1, 1937. He attended the University of Florida, graduating in 1959 with high honors and a Bachelor of Aeronautical Engineering Degree. From 1959 to 1962, he was a lieutenant and task scientist at the Aerospace Research Laboratory at Wright-Patterson Air Force Base. From 1962 to 1966, he attended the Ohio State University under the National Science Foundation and NASA Fellowships, graduating with a Ph.D. in aeronautical and astronautical engineering. In 1966, he joined the U.S. Naval Ordnance Laboratory as Chief of the Hypersonic Group. In 1973, he became Chairman of the Department of Aerospace Engineering at the University of Maryland, and since 1980 has been professor of Aerospace Engineering at Maryland. In 1982, he was designated a Distinguished Scholar/Teacher by the University. During 1986–1987, while on sabbatical from the University, Dr. Anderson occupied the Charles Lindbergh chair at the National Air and Space Museum of the Smithsonian Institution. He continues with the Museum in a part-time appointment as curator for aerodynamics. In addition to his appointment in aerospace engineering, in 1993 he was elected to the faculty of the Committee on the History and Philosophy of Science at Maryland, and is an affiliate faculty member in the Department of History. In July, 1999 he retired from the University and is now Professor Emeritus.

Dr. Anderson has published eight books: *Gasdynamic Lasers: An Introduction*, Academic Press (1976), *A History of Aerodynamics and Its Impact on Flying Machines*, Cambridge University Press (1997), and with McGraw-Hill, *Introduction to Flight*, 3d edition (1989), *Modern Compressible Flow*, 2d Edition (1990), *Fundamentals of Aerodynamics*, 2d edition (1991), *Hypersonic and High Temperature Gas Dynamics* (1989), *Computational Fluid Dynamics: The Basics with Applications* (1995), and *Aircraft Performance and Design* (1999). He is the author of over 120 papers on radiative gasdynamics, entry aerothermodynamics, gas dynamic and chemical lasers, computational fluid dynamics, applied aerodynamics, hypersonic flow, and the history of aerodynamics. Dr. Anderson is in *Who's Who in America*, and is a Fellow of the American Institute of Aeronautics and Astronautics (AIAA), and a Fellow of The Royal Aeronautical Society. He is also a Fellow of the Washington Academy of Sciences, and a member of Tau Beta Pi, Sigma Tau, Phi Kappa Phi, Phi Eta Sigma, The American Society for Engineering Education (ASEE), The Society for the History of Technology, and the History of Science Society. He has received the Lee Atwood Award for excellence in Aerospace Engineering Education from the AIAA and the ASEE, and the Pendray Award for Aerospace Literature from the AIAA.

INTRODUCTION TO FLIGHT

FOURTH EDITION

John D. Anderson, Jr.
Curator of Aerodynamics
National Air and Space Museum
Smithsonian Institution
and
Professor Emeritus
University of Maryland

McGraw-Hill

Boston   Burr Ridge, IL   Dubuque, IA   Madison, WI
New York   San Francisco   St. Louis
Bangkok   Bogotá   Caracas   Lisbon   London   Madrid   Mexico City
Milan   New Delhi   Seoul   Singapore   Sydney   Taipei   Toronto

WCB/McGraw-Hill

A Division of The **McGraw·Hill** *Companies*

Cover photograph of the Wright Brothers © Corbis Images.

INTRODUCTION TO FLIGHT
International Editions 2000

Exclusive rights by McGraw-Hill Book Co – Singapore, for manufacture and export. This book cannot be re-exported from the country to which it is consigned by McGraw-Hill.

10   9   8   7   6
20   9   8   7   6   5   4   3   2
UPE   MPM

Library of Congress Cataloging-in-Publication Data

Anderson, John David.
    Introduction to flight / John D. Anderson, Jr. – 4th ed.
        p.   cm.
    Includes index.
    ISBN 0-07-109282-X
    1. Aerodynamics.   2. Airplanes–Design and construction.
I. Title.
TL570.A68   2000
629.1–dc21   99-32638
                                                                          CIP

www.mhhe.com

When ordering this title, use ISBN 0-07-116034-5

Printed in Singapore

To Sarah-Allen, Katherine, and Elizabeth Anderson
For All Their Love and Understanding
John D. Anderson, Jr.

CONTENTS

Chapter   5

Airfoils, Wings, and Other Aerodynamic Shapes   227

Chapter   6

Elements of Airplane Performance   349

PREFACE TO THE FOURTH EDITION

The purpose of the present edition is the same as that of the first three: to present the basic fundamentals of aerospace engineering at the introductory level in the clearest, simplest, and most motivating way possible. Since the book is meant to be enjoyed as well as understood, I have made every effort to ensure a clear and readable text. The choice of subject matter and its organization, the order in which topics are introduced, and how these ideas are explained have been carefully planned with the uninitiated reader in mind. Because the book is intended as a self-contained text at the first- and second-year levels, I avoid tedious details and massive "handbook" data. Instead, I introduce and discuss fundamental concepts in as straightforward and clear-cut a manner as possible, knowing that the book has also found favor with those who wish to learn something about this subject outside the classroom.

The overwhelmingly favorable response to the earlier editions from students, teachers, and practicing professionals both here and abroad is a source of gratification. Particularly pleasing is the fact that those using the book have enjoyed reading its treatment of the fascinating, challenging, and sometimes awesome discipline of aerospace engineering.

Thanks to this response, the contents of the third edition have been carried over into the fourth with only minor modifications. However, this edition also contains a great deal of new material, such as the following:

1. Roadmaps have been placed at the beginning of each chapter, to help guide the reader through the logical flow of the material.

2. At numerous locations throughout the book, the implications of the local fundamental discussions in the text to aerospace vehicle design have been singled out and highlighted. These appear in *design boxes*, identified by that title, and literally set apart in boxes. This adds a substantial "design flavor" to the whole book, but still at the introductory level. These design boxes enhance the value of the reading material, and help to put the reader more in the engineering frame of mind.

3. A new chapter has been added—Chapter 10, Flight Vehicle Structures and Materials.

4. A number of new sections have been added to the existing chapters, dealing with such topics as:

   a. The anatomy of airplanes and space vehicles, introducing the basic nature of the machines themselves, and explicitly identifying their major components.

   b. Special considerations of supersonic airplane performance.

   c. Lateral and directional stability.

   d. Rocket propellants.

   e. Multi-stage rockets.

5. Many added worked examples have been sprinkled throughout the text.

Included in this new material are almost 100 new figures to help illustrate the basic ideas. All told, the new material represents a substantial enhancement of *Introduction to Flight*.

What constitutes a proper introduction to aerospace engineering? There are as many answers as there are people addressing the question. My choices, based on 25 years of experience with students at the University of Maryland, have also been influenced by conversations with university faculty and practicing professionals around the globe. Special thanks are due to these many colleagues.

At the University of Maryland this text is used in a year-long introductory course for sophomores in aerospace engineering. It leads directly into a second book by the author, *Fundamentals of Aerodynamics*, 2/e, (McGraw-Hill, 1991), which is used in a two-semester junior-senior aerodynamics course. This in turn feeds into a third, *Modern Compressible Flow: With Historical Perspective*, 2/e, (McGraw-Hill, 1990), used in a course for advanced undergraduates and first-year graduate students. The complete triad is intended to give students a reasonable technical and historical perspective on aerospace engineering in general and aerodynamics in partiicular.

I am very grateful to Mrs. Susan Cunningham, who did such an excellent job typing the manuscript. I am fortunate to have such dedicated and professional help from one of the best scientific typists in the world. My gratitude also goes to my wife, Sarah-Allen, for putting up so patiently with the turmoil surrounding a book in progress until we can breath a joint sigh of relief at the end of the project.

Finally, emphasizing that the study, understanding, and practice of the profession of aerospace engineering is one of the most gratifying of human endeavors and that my purpose is to instill a sense of enthusiasm, dedication, and love of the subject, let me simply say to the reader: read, learn and *enjoy*.

John D. Anderson, Jr.

PREFACE TO THE FIRST EDITION

This book is an introduction to aerospace engineering from both the technological and historical points of view. It is written to appeal to several groups of people: (1) students of aerospace engineering in their freshman or sophomore years in college who are looking for a comprehensive introduction to their profession, (2) advanced high school seniors who simply want to learn what aerospace engineering is all about, (3) both college undergraduate and graduate students who want to obtain a wider perspective on the glories, the intellectual demands, and the technical maturity of aerospace engineering, and (4) working engineers who simply want to obtain a firmer grasp on the fundamental concepts and historical traditions that underlie their profession.

As an introduction to aerospace engineering, this book is unique in at least three ways. First, the vast majority of aerospace engineering professionals and students have little knowledge or appreciation of the historical traditions and background associated with the technology that they use almost every day. To fill this vacuum, the present book attempts to marble some history of aerospace engineering into the parallel technical discussions. For example, such questions as who was Bernoulli, where did the Pitot tube originate, how did wind tunnels evolve, who were the first true aeronautical engineers, and how did wings and airfoils develop are answered. The present author feels strongly that such material should be an integral part of the background of all aerospace engineers.

Second, this book incorporates both the SI and the English engineering system of units. Modern students of aerospace engineering must be bilingual—on one hand, they must fully understand and feel comfortable with the SI units, because most modern and all future literature will deal with the SI system; on the other hand, they must be able to read and feel comfortable with the vast bulk of existing literature, which is predominantly in engineering units. In this book, the SI system is emphasized, but an honest effort is made to give the reader a feeling for and understanding of both systems. To this end some example problems are worked out in the SI system and others in the English system.

Third, the author feels that technical books do not have to be dry and sterile in their presentation. Instead, the present book is written in a rather informal style. It attempts to *talk* to the reader. Indeed, it is intended to be almost a self-teaching, self-pacing vehicle that the reader can use to obtain a fundamental understanding of aerospace engineering.

This book is a product of several years of teaching the introductory course in aerospace engineering at the University of Maryland. Over these years, students have constantly encouraged the author to write a book on the subject, and their repeated encouragement could not be denied. The present book is dedicated in part to these students.

Writing a book of this magnitude is a total commitment of time and effort for a longer time than the author likes to remember. In this light, this book is dedicated to my wife, Sarah-Allen, and my two daughters, Katherine and Elizabeth, who relinquished untold amounts of time with their husband and father so that these pages could be created. To them I say thank you, and hello again. Also, hidden between the lines, but ever-so-much present is Edna Brothers, who typed the manuscript in such a dedicated fashion. In addition, the author wishes to thank Dr. Richard Hallion and Dr. Thomas Crouch, curators of the National Air and Space Museum of the Smithsonian Institution, for their helpful comments on the historical sections of this manuscript, and especially Dick Hallion, for opening the vast archives of the museum for the author's historical research. Also, many thanks are due to the reviewers of this manuscript, Professor J. J. Azar of the University of Tulsa, Dr. R. F. Brodsky of Iowa State University, Dr. David Caughey of Sibley School of Mechanical and Aerospace Engineering, and Professor Francis J. Hale of North Carolina State University; their comments have been most constructive, especially those of Dr. Caughey and Professor Hale. Finally, the author wishes to thank his many colleagues in the profession for stimulating discussions about what constitutes an introduction to aerospace engineering. Hopefully, this book is a reasonable answer.

John D. Anderson, Jr.

1

The First Aeronautical Engineers

Nobody will fly for a thousand years!

Wilbur Wright, 1901, in a fit of despair

SUCCESS FOUR FLIGHTS THURSDAY MORNING ALL AGAINST TWENTY
ONE MILE WIND STARTED FROM LEVEL WITH ENGINE POWER ALONE
AVERAGE SPEED THROUGH AIR THIRTY ONE MILES LONGEST 57 SEC-
ONDS INFORM PRESS HOME CHRISTMAS.

OREVELLE WRIGHT
A telegram, with the original misprints,
from Orville Wright to his father,
December 17, 1903

1.1 INTRODUCTION

The scene: Wind-swept sand dunes of Kill Devil Hills, 4 mi south of Kitty Hawk,
North Carolina. *The time:* About 10:35 AM on Thursday, December 17, 1903. *The
characters:* Orville and Wilbur Wright and five local witnesses. *The action:* Poised,
ready to make history, is a flimsy, odd-looking machine, made from spruce and
cloth in the form of two wings, one placed above the other, a horizontal elevator
mounted on struts in front of the wings, and a double vertical rudder behind the wings
(see Fig. 1.1). A 12-hp engine is mounted on the top surface of the bottom wing,
slightly right of center. To the left of this engine lies a man—Orville Wright—prone
on the bottom wing, facing into the brisk and cold December wind. Behind him
rotate two ungainly looking airscrews (propellers), driven by two chain-and-pulley
arrangements connected to the same engine. The machine begins to move along a

Figure 1.1   Three views of the *Wright Flyer I*, 1903.

60-ft launching rail on level ground. Wilbur Wright runs along the right side of the machine, supporting the wing tip so that it will not drag the sand. Near the end of the starting rail, the machine lifts into the air; at this moment, John Daniels of the Kill Devil Life Saving Station takes a photograph which preserves for all time the most historic moment in aviation history (see Fig. 1.2). The machine flies unevenly, rising suddenly to about 10 ft, then ducking quickly toward the ground. This type of erratic flight continues for 12 s, when the machine darts to the sand, 120 ft from the point where it lifted from the starting rail. Thus ends a flight which, in Orville Wright's own words, was "the first in the history of the world in which a machine carrying a man had raised itself by its own power into the air in full flight, had sailed forward without reduction of speed, and had finally landed at a point as high as that from which it started."

Figure 1.2 The first heavier-than-air flight in history: the *Wright Flyer I* with Orville Wright at the controls, December 17, 1903. *(National Air and Space Museum.)*

The machine was the *Wright Flyer I*, which is shown in Figs. 1.1 and 1.2 and which is now preserved for posterity in the Air and Space Museum of the Smithsonian Institution in Washington, District of Columbia. The flight on that cold December 17 was momentous: It brought to a realization the dreams of centuries, and it gave birth to a new way of life. It was the first genuine powered flight of a heavier-than-air machine. With it, and with the further successes to come over the next five years, came the Wright brothers' clear right to be considered the premier aeronautical engineers of history.

However, contrary to some popular belief, the Wright brothers did not truly *invent* the airplane; rather, they represent the fruition of a century's worth of prior aeronautical research and development. The time was ripe for the attainment of powered flight at the beginning of the 20th century. The Wright brothers' ingenuity, dedication, and persistence earned them the distinction of being first. The purpose of this chapter is to look back over the years which led up to successful powered flight and to single out an important few of those inventors and thinkers who can rightfully claim to be the first aeronautical engineers. In this manner, some of the traditions and heritage that underlie modern aerospace engineering will be more appreciated when we develop the technical concepts of flight in subsequent chapters.

1.2 VERY EARLY DEVELOPMENTS

Since the dawn of human intelligence, the idea of flying in the same realm as birds has possessed human minds. Witness the early Greek myth of Daedalus and his son Icarus. Imprisoned on the island of Crete in the Mediterranean Sea, Daedalus is said to have made wings fastened with wax. With these wings, they both escaped by flying through the air. However, Icarus, against his father's warnings, flew too close to the sun; the wax melted, and Icarus fell to his death in the sea.

All early thinking of human flight centered on the imitation of birds. Various unsung ancient and medieval people fashioned wings and met with sometimes disastrous and always unsuccessful consequences in leaping from towers or roofs, flapping vigorously. In time, the idea of strapping a pair of wings to arms fell out of favor. It was replaced by the concept of wings flapped up and down by various mechanical mechanisms, powered by some type of human arm, leg, or body movement. These machines are called *ornithopters*. Recent historical research has uncovered that Leonardo da Vinci himself was possessed by the idea of human flight and that he designed vast numbers of ornithopters toward the end of the 15th century. In his surviving manuscripts, more than 35,000 words and 500 sketches deal with flight. One of his ornithopter designs is shown in Fig. 1.3, which is an original da Vinci sketch made sometime between 1486 and 1490. It is not known whether da Vinci ever built or tested any of his designs. However, human-powered flight by flapping wings was always doomed to failure. In this sense, da Vinci's efforts did not make important contributions to the technical advancement of flight.

Human efforts to fly literally got off the ground on November 21, 1783, when a balloon carrying Pilatre de Rozier and the Marquis d'Arlandes ascended into the air and drifted 5 mi across Paris. The balloon was inflated and buoyed up by hot air from an open fire burning in a large wicker basket underneath. The design and construction of the balloon were due to the Montgolfier brothers, Joseph and Etienne. In 1782, Joseph Montgolfier, gazing into his fireplace, conceived the idea of using the "lifting

Figure 1.3   An ornithopter design by Leonardo da Vinci, 1486–1490.

power" of hot air rising from a flame to lift a person from the surface of the earth. The brothers instantly set to work, experimenting with bags made of paper and linen, in which hot air from a fire was trapped. After several public demonstrations of flight without human passengers, including the 8-min voyage of a balloon carrying a cage containing a sheep, a rooster, and a duck, the Montgolfiers were ready for the big step. At 1:54 PM on November 21, 1783, the first flight with human passengers rose majestically into the air and lasted for 25 min (see Fig. 1.4). It was the first time in history that a human being had been lifted off the ground for a sustained period of time. Very quickly after this, the noted French physicist J. A. C. Charles (of Charles' gas law in physics) built and flew a hydrogen-filled balloon from the Tuileries Gardens in Paris on December 1, 1783.

So people were finally off the ground! Balloons, or "aerostatic machines," as they were called by the Montgolfiers, made no real technical contributions to human heavier-than-air flight. However, they served a major purpose in triggering the public's interest in flight through the air. They were living proof that people could really leave the ground and sample the environs heretofore exclusively reserved for birds. Moreover, balloons were the only means of human flight for almost 100 years.

Figure 1.4 The first aerial voyage in history: The Montgolfier hot-air balloon lifts from the ground near Paris on November 21, 1783.

1.3   SIR GEORGE CAYLEY (1773–1857)—THE TRUE INVENTOR OF THE AIRPLANE

The modern airplane has its origin in a design set forth by George Cayley in 1799. It was the first concept to include a *fixed* wing for generating lift, another *separate* mechanism for propulsion (Cayley envisioned paddles), and a combined horizontal and vertical (cruciform) tail for stability. Cayley inscribed his idea on a silver disk (presumably for permanence), shown in Fig. 1.5. On the reverse side of the disk is a diagram of the lift and drag forces on an inclined plane (the wing). The disk is now preserved in the Science Museum in London. Before this time, thought of mechanical flight had been oriented toward the flapping wings of ornithopters, where the flapping motion was supposed to provide both lift and propulsion. (Da Vinci designed his ornithopter wings to flap simultaneously downward and backward for lift and propulsion.) However, Cayley is responsible for breaking this unsuccessful line of thought; he separated the concept of lift from propulsion and, in so doing, set into motion a century of aeronautical development that culminated in the Wright brothers' success in 1903. George Cayley is a giant in aeronautical history: He is the parent of modern aviation and is the first true aeronautical engineer. Let us look at him more closely.

Cayley was born at Scarborough in Yorkshire, England, on December 27, 1773. He was educated at York and Nottingham and later studied chemistry and electricity under several noted tutors. He was a scholarly man of some rank, a baronet who spent much of his time on the family estate, called Brompton. A portrait of Cayley is shown in Fig. 1.6. He was a well-preserved person, of extreme intellect and open mind,

Figure 1.5   The silver disk on which Cayley engraved his concept for a fixed-wing aircraft, the first in history, in 1799. The reverse side of the disk shows the resultant aerodynamic force on a wing resolved into lift and drag components, indicating Cayley's full understanding of the function of a fixed wing. The disk is presently in the Science Museum in London.

Figure 1.6 A portrait of Sir George Cayley, painted by Henry Perronet Briggs in 1841. The portrait now hangs in the National Portrait Gallery in London.

active in many pursuits over a long life of 84 years. In 1825, he invented the caterpillar tractor, forerunner of all modern tracked vehicles. In addition, he was chairman of the Whig Club of York, founded the Yorkshire Philosophical Society (1821), cofounded the British Association for the Advancement of Science (1831), was a member of Parliament, was a leading authority on land drainage, and published papers dealing with optics and railroad safety devices. Moreover, he had a social conscience: He appealed for, and donated to, the relief of industrial distress in Yorkshire.

However, by far his major and lasting contribution to humanity was in aeronautics. After experimenting with model helicopters beginning in 1796, Cayley engraved his revolutionary fixed-wing concept on the silver disk in 1799 (see Fig. 1.5). This was followed by an intensive 10-year period of aerodynamic investigation and development. In 1804, he built a whirling-arm apparatus, shown in Fig. 1.7, for testing airfoils; this was simply a lifting surface (airfoil) mounted on the end of a long rod, which was rotated at some speed to generate a flow of air over the airfoil. In modern aerospace engineering, wind tunnels now serve this function, but in Cayley's time the whirling arm was an important development, which allowed the measurement of aerodynamic forces and the center of pressure on a lifting surface. Of course, these measurements were not very accurate, because after a number of revolutions of the arm, the surrounding air would begin to rotate with the device. Nevertheless, it was a first step in aerodynamic testing. (Cayley did not invent the whirling arm; that honor belongs to the English military engineer Benjamin Robins in 1742.) Also in 1804, Cayley designed, built, and flew the small model glider shown in Fig. 1.8; this may seem trivial today, something that you may have done as a child, *but in 1804 it represented the first modern-configuration airplane of history*, with a fixed wing, and a horizontal and vertical tail that could be adjusted. (Cayley generally flew his glider

Figure 1.7   George Cayley's whirling-arm apparatus for testing airfoils.

Figure 1.8   The first modern-configuration airplane in history: Cayley's model glider, 1804.

with the tail at a positive angle of incidence, as shown in his sketch in Fig. 1.8.) A full-scale replica of this glider is on display at the Science Museum in London—the model is only about 1 m long.

Cayley's first outpouring of aeronautical results was documented in his momentous triple paper of 1809–1810. Entitled "On Aerial Navigation" and published in the November 1809, February 1810, and March 1810 issues of Nicholson's *Journal of Natural Philosophy*, this document ranks as one of the most important aeronautical works in history. (Note that the words *natural philosophy* in history are synonymous with physical science.) Cayley was prompted to write his triple paper after hearing reports that Jacob Degen had recently flown in a mechanical machine in Vienna. In reality, Degen flew in a contraption which was lifted by a balloon. It was of no significance, but Cayley did not know the details. In an effort to let people know of his activities, Cayley documented many aspects of aerodynamics in his triple paper. It was the first treatise on theoretical and applied aerodynamics in history to be published. In it, Cayley elaborates on his principle of the separation of lift and propulsion and his use of a fixed wing to generate lift. He states that the basic principle of a flying machine is "to make a surface support a given weight by the application of power to the resistance of air." He notes that a surface inclined at some angle to the direction of motion will generate lift and that a cambered (curved) surface will do this more efficiently than a flat surface. He also states for the first time in history that lift is generated by a region of low pressure on the upper surface of the wing. The modern technical aspects of these phenomena will be developed and explained in Chaps. 4 and 5; however, stated by Cayley in 1809–1810, these phenomena were new and unique. His triple paper also addressed the matter of flight control and was the first document to discuss the role of the horizontal and vertical tail planes in airplane stability. Interestingly enough, Cayley goes off on a tangent in discussing the use of flappers for propulsion. Note that on the silver disk (see Fig. 1.5) Cayley shows some paddles just behind the wing. From 1799 until his death in 1857, Cayley was obsessed with such flappers for aeronautical propulsion. He gave little attention to the propeller (airscrew); indeed, he seemed to have an aversion to rotating machinery of any type. However, this should not detract from his numerous positive contributions. Also in his triple paper, Cayley tells us of the first successful full-size glider of history, built and flown without passengers by him at Brompton in 1809. However, there is no clue as to its configuration.

Curiously, the period from 1810 to 1843 was a lull in Cayley's life in regard to aeronautics. Presumably, he was busy with his myriad other interests and activities. During this period, he showed interest in airships (controlled balloons), as opposed to heavier-than-air machines. He made the prophetic statement that "balloon aerial navigation can be done readily, and will probably, in the order of things, come into use before mechanical flight can be rendered sufficiently safe and efficient for ordinary use." He was correct; the first successful airship, propelled by a steam engine, was built and flown by the French engineer Henri Giffard in Paris in 1852, some 51 years before the first successful airplane.

Cayley's second outpouring of aeronautical results occurred in the period from 1848 to 1854. In 1849, he built and tested a full-size airplane. During some of the flight tests, a 10-year-old boy was carried along and was lifted several meters off the ground while gliding down a hill. Cayley's own sketch of this machine, called the *boy carrier*, is shown in Fig. 1.9. Note that it is a triplane (three wings mounted on top of one another). Cayley was the first to suggest such multiplanes (i.e., biplanes and triplanes), mainly because he was concerned with the possible structural failure of a single large wing (a monoplane). Stacking smaller, more compact, wings on top of one another made more sense to him, and his concept was perpetuated into the 20th century. It was not until the late 1930s that the monoplane became the dominant airplane configuration. Also note from Fig. 1.9 that, strictly speaking, this was a "powered" airplane; that is, it was equipped with propulsive flappers.

One of Cayley's most important papers was published in *Mechanics' Magazine* on September 25, 1852. By this time he was 79 years old! The article was entitled "Sir George Cayley's Governable Parachutes." It gave a full description of a large human-carrying glider which incorporated almost all the features of the modern airplane. This design is shown in Fig. 1.10, which is a facsimile of the illustration that appeared in the original issue of *Mechanics' Magazine*. This airplane had (1) a main wing at an angle of incidence for lift, with a dihedral for lateral stability; (2) an adjustable cruciform tail for longitudinal and directional stability; (3) a pilot-operated elevator and rudder; (4) a fuselage in the form of a car, with a pilot's seat and three-wheel undercarriage; and (5) a tubular beam and box beam construction. These combined features were not to be seen again until the Wright brothers' designs at the beginning of the 20th century. Incredibly, this 1852 paper by Cayley went virtually unnoticed,

Figure 1.9   Cayley's triplane from 1849—the boy carrier. Note the vertical and horizontal tail surfaces and the flapperlike propulsive mechanism.

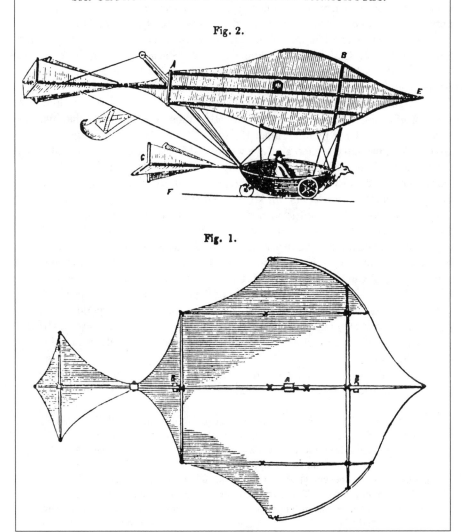

Mechanics' Magazine,

MUSEUM, REGISTER, JOURNAL, AND GAZETTE.

No. 1520.]   SATURDAY, SEPTEMBER 25, 1852.   [Price 3*d*., Stamped 4*d*.
Edited by J. C. Robertson, 166, Fleet-street.

SIR GEORGE CAYLEY'S GOVERNABLE PARACHUTES.

Fig. 2.

Fig. 1.

Figure 1.10   George Cayley's human-carrying glider, from *Mechanics' Magazine*, 1852.

even though *Mechanics' Magazine* had a large circulation. It was rediscovered by the eminent British aviation historian Charles H. Gibbs-Smith in 1960 and republished by him in the June 13, 1960, issue of *The Times*.

Sometime in 1853—the precise date is unknown—George Cayley built and flew the world's first human-carrying glider. Its configuration is not known, but Gibbs-Smith states that it was most likely a triplane on the order of the earlier boy carrier (see Fig. 1.9) and that the planform (top view) of the wings was probably shaped much as the glider in Fig. 1.10. According to several eyewitness accounts, a gliding flight of several hundred yards was made across a dale at Brompton with Cayley's coachman aboard. The glider landed rather abruptly, and after struggling clear of the vehicle, the shaken coachman is quoted as saying: "Please, Sir George, I wish to give notice. I was hired to drive, and not to fly." Very recently, this flight of Cayley's coachman was reenacted for the public in a special British Broadcasting Corporation television show on Cayley's life. While visiting the Science Museum in London in August of 1975, the present author was impressed to find the television replica of Cayley's glider (minus the coachman) hanging in the entranceway.

George Cayley died at Brompton on December 15, 1857. During his almost 84 years of life, he laid the basis for all practical aviation. He was called the *father of aerial navigation* by William Samuel Henson in 1846. However, for reasons that are not clear, the name of George Cayley retreated to the background soon after his death. His works became obscure to virtually all later aviation enthusiasts in the latter half of the 19th century. This is incredible, indeed unforgivable, considering that his published papers were available in known journals. Obviously, many subsequent inventors did not make the effort to examine the literature before forging ahead with their own ideas. (This is certainly a problem for engineers today, with the virtual explosion of written technical papers since World War II.) However, Cayley's work has been brought to light by the research of several modern historians in the 20th century. Notable among them is C. H. Gibbs-Smith, from whose book entitled *Sir George Cayley's Aeronautics* (1962) much of the above material has been gleaned. Gibbs-Smith states that had Cayley's work been extended directly by other aviation pioneers, and had they digested ideas espoused in his triple paper of 1809–1810 and in his 1852 paper, successful powered flight would have most likely occurred in the 1890s. Probably so!

As a final tribute to George Cayley, we note that the French aviation historian Charles Dollfus said the following in 1923:

> The aeroplane is a British invention: it was conceived in all essentials by George Cayley, the great English engineer who worked in the first half of last century. The name of Cayley is little known, even in his own country, and there are very few who know the work of this admirable man, the greatest genius of aviation. A study of his publications fills one with absolute admiration both for his inventiveness, and for his logic and common sense. This great engineer, during the Second Empire, did in fact not only invent the aeroplane entire, as it now exists, but he realized that the problem of aviation had to be divided between theoretical research—Cayley made the first aerodynamic experiments for aeronautical purposes—and practical tests, equally in the case of the glider as of the powered aeroplane.

1.4 THE INTERREGNUM—FROM 1853 TO 1891

For the next 50 years after Cayley's success with the coachman-carrying glider, there were no major advances in aeronautical technology comparable to those of the previous 50 years. Indeed, as stated above, much of Cayley's work became obscure to all but a few dedicated investigators. However, there was considerable activity, with numerous people striking out (sometimes blindly) in various uncoordinated directions to conquer the air. Some of these efforts are noted below, just to establish the flavor of the period.

William Samuel Henson (1812–1888) was a contemporary of Cayley. In April 1843, he published in England a design for a fixed-wing airplane powered by a steam engine driving two propellers. Called the *aerial steam carriage*, this design received wide publicity throughout the 19th century, owing mainly to a series of illustrative engravings which were reproduced and sold around the world. This was a publicity campaign of which Madison Avenue would have been proud; one of these pictures is shown in Fig. 1.11. Note some of the qualities of modern aircraft in Fig. 1.11: the engine inside a closed fuselage, driving two propellers; tricycle landing gear; and a single rectangular wing of relatively high aspect ratio. (We will discuss the aerodynamic characteristics of such wings in Chap. 5.) Henson's design was a direct product of George Cayley's ideas and research in aeronautics. The aerial steam carriage was never built, but the design, along with its widely published pictures, served to engrave George Cayley's fixed-wing concept on the minds of virtually all subsequent workers. Thus, even though Cayley's published papers fell into obscurity after his death, his major concepts were partly absorbed and perpetuated by following generations of inventors, even though most of these inventors did not know the true

Figure 1.11 Henson's aerial steam carriage, 1842–1843. *(National Air and Space Museum.)*

source of the ideas. In this manner, Henson's aerial steam carriage was one of the most influential airplanes in history, even though it never flew!

John Stringfellow, a friend of Henson, made several efforts to bring Henson's design to fruition. Stringfellow built several small steam engines and attempted to power some model monoplanes off the ground. He was close, but unsuccessful. However, his most recognized work appeared in the form of a steam-powered triplane, a model of which was shown at the 1868 aeronautical exhibition sponsored by the Aeronautical Society at the Crystal Palace in London. A photograph of Stringfellow's triplane is shown in Fig. 1.12. This airplane was also unsuccessful, but again it was extremely influential because of worldwide publicity. Illustrations of this triplane appeared throughout the end of the 19th century. Gibbs-Smith, in his book *Aviation: An Historical Survey from Its Origins to the End of World War II* (1970), states that these illustrations were later a strong influence on Octave Chanute, and through him the Wright brothers, and strengthened the concept of superimposed wings. Stringfellow's triplane was the main bridge between George Cayley's aeronautics and the modern biplane.

During this period, the first powered airplanes actually hopped off the ground, but for only hops. In 1857–1858, the French naval officer and engineer Felix Du Temple flew the first successful powered model airplane in history; it was a monoplane with swept-forward wings and was powered by clockwork! Then, in 1874, Du Temple achieved the world's first powered takeoff by a piloted, full-size airplane. Again, the airplane had swept-forward wings, but this time it was powered by some type of hot-air engine (the precise type is unknown). A sketch of Du Temple's full-size airplane is shown in Fig. 1.13. The machine, piloted by a young sailor, was launched down an inclined plane at Brest, France; it left the ground for a moment but did not come close to anything resembling sustained flight. In the same vein, the second powered airplane with a pilot left the ground near Leningrad (then St. Petersburg), Russia, in July 1884.

Figure 1.12   Stringfellow's model triplane exhibited at the first aeronautical exhibition in London, 1868.

Designed by Alexander F. Mozhaiski, this machine was a steam-powered monoplane, shown in Fig. 1.14. Mozhaiski's design was a direct descendant from Henson's aerial steam carriage—it was even powered by an English steam engine! With I. N. Golubev as pilot, this airplane was launched down a ski ramp and flew for a few seconds. As with Du Temple's airplane, no sustained flight was achieved. At various times, the Russians have credited Mozhaiski with the first powered flight in history, but of course

Figure 1.13 Du Temple's airplane: the first aircraft to make a powered but assisted takeoff, 1874.

Figure 1.14 The second airplane to make an assisted takeoff: Mozhaiski's aircraft, Russia, 1884.

it did not satisfy the necessary criteria to be called such. Du Temple and Mozhaiski achieved the first and second *assisted* powered takeoffs, respectively, in history, but neither experienced sustained flight. In his book *The World's First Aeroplane Flights* (1965), C. H. Gibbs-Smith states the following criteria used by aviation historians to judge a successful powered flight:

> In order to qualify for having made a simple powered and sustained flight, a conventional aeroplane should have sustained itself freely in a horizontal or rising flight path—without loss of airspeed—beyond a point where it could be influenced by any momentum built up before it left the ground: otherwise its performance can only be rated as a powered leap, i.e., it will not have made a fully self-propelled flight, but will only have followed a ballistic trajectory modified by the thrust of its propeller and by the aerodynamic forces acting upon its aerofoils. Furthermore, it must be shown that the machine can be kept in satisfactory equilibrium. Simple sustained flight obviously need not include full controllability, but the maintenance of adequate equilibrium in flight is part and parcel of sustention.

Under these criteria, there is no doubt in the mind of any major aviation historian that the first powered flight was made by the Wright brothers in 1903. However, the assisted "hops" described above put two more rungs in the ladder of aeronautical development in the 19th century.

Of particular note during this period is the creation in London in 1866 of the Aeronautical Society of Great Britain. Before this time, work on "aerial navigation" (a phrase coined by George Cayley) was looked upon with some disdain by many scientists and engineers. It was too out of the ordinary and was not to be taken seriously. However, the Aeronautical Society soon attracted scientists of stature and vision, people who shouldered the task of solving the problems of mechanical flight in a more orderly and logical fashion. In turn, aeronautics took on a more serious and meaningful atmosphere. The society, through its regular meetings and technical journals, provided a cohesive scientific outlet for the presentation and digestion of aeronautical engineering results. The society is still flourishing today in the form of the highly respected Royal Aeronautical Society. Moreover, it served as a model for the creation of both the American Rocket Society and the Institute of Aeronautical Sciences in the United States in this century; both of these societies merged in 1964 to form the American Institute of Aeronautics and Astronautics (AIAA), one of the most influential channels for aerospace engineering information exchange today.

In conjunction with the Aeronautical Society of Great Britain, at its first meeting on June 27, 1866, Francis H. Wenham read a paper entitled "Aerial Locomotion," one of the classics in the aeronautical engineering literature. Wenham was a marine engineer who later was to play a prominent role in the society and who later designed and built the first wind tunnel in history (see Chap. 4). His paper, which was also published in the first annual report of the society, was the first to point out that most of the lift of a wing was obtained from the portion near the leading edge. He also established that a wing with high aspect ratio was the most efficient for producing lift. (We will see why in Chap. 5.)

As noted in our previous discussion about Stringfellow, the Aeronautical Society started out in style: When it was only two years old, in 1868, it put on the first

aeronautical exhibition in history at the Crystal Palace. It attracted an assortment
of machines and balloons and for the first time offered the general public a first-
hand overview of the efforts being made to conquer the air. Stringfellow's triplane
(discussed earlier) was of particular interest. Zipping over the heads of the enthralled
onlookers, the triplane moved through the air along an inclined cable strung below
the roof of the exhibition hall (see Fig. 1.12). However, it did not achieve sustained
flight on its own. In fact, the 1868 exhibition did nothing to advance the technical
aspects of aviation, but it was a masterstroke of good public relations.

1.5 OTTO LILIENTHAL (1848–1896)—THE GLIDER MAN

With all the efforts that had been made in the past, it was still not until 1891 that
a human literally jumped into the air and flew with wings in any type of controlled
fashion. This person was Otto Lilienthal, one of the giants in aeronautical engineering
(and in aviation in general). Lilienthal designed and flew the first successful controlled
gliders in history. He was a man of aeronautical stature comparable to Cayley and
the Wright brothers. Let us examine the man and his contributions more closely.

Lilienthal was born on May 23, 1848, at Anklam, Prussia (Germany). He obtained a good technical education at trade schools in Potsdam and Berlin, the latter at
the Berlin Technical Academy, graduating with a degree in mechanical engineering
in 1870. After a one-year stint in the army during the Franco-Prussian War, Lilienthal
went to work designing machinery in his own factory. However, from early childhood
he was interested in flight and performed some youthful experiments on ornithopters
of his own design. Toward the late 1880s, his work and interests took a more mature
turn, which ultimately led to fixed-wing gliders.

In 1889, Lilienthal published a book entitled *Der Vogelflug als Grundlage der
Fliegekunst* (Bird Flight as the Basis of Aviation). This is another of the early classics
in aeronautical engineering, because not only did he study the structure and types
of birds' wings, but also he applied the resulting aerodynamic information to the
design of mechanical flight. Lilienthal's book contained some of the most detailed
aerodynamic data available at that time. Translated sections were later read by the
Wright brothers, who incorporated some of his data in their first glider designs in
1900 and 1901. (However, as discussed by Anderson in *History of Aerodynamics
and its Impact on Flying Machines*, the Wright brothers misinterpreted Lilienthal's
aerodynamic data, which caused problems with their 1900 and 1901 gliders. This
will be discussed in a subsequent section.)

By 1889, Lilienthal had also come to a philosophical conclusion which was to
have a major impact on the next two decades of aeronautical development. He concluded that to learn practical aerodynamics, he had to get up in the air and experience
it himself. In his own words,

> One can get a proper insight into the practice of flying only by actual flying experi-
> ments.... The manner in which we have to meet the irregularities of the wind, when
> soaring in the air, can only be learnt by being in the air itself.... The only way which

leads us to a quick development in human flight is a systematic and energetic practice in actual flying experiments.

To put this philosophy into practice, Lilienthal designed a glider in 1889, and another in 1890—both were unsuccessful. However, in 1891, Lilienthal's first successful glider flew from a natural hill at Derwitz, Germany. (Later, he was to build an artificial hill about 50 ft high near Lichterfelde, a suburb of Berlin; this conically shaped hill allowed glider flights to be made into the wind, no matter what the direction.) The general configuration of his monoplane gliders is shown in Fig. 1.15, which is a photograph showing Lilienthal as the pilot. Note the rather birdlike planform of the wing. Lilienthal used cambered (curved) airfoil shapes on the wing and incorporated vertical and horizontal tail planes in the back for stability. These machines were hang gliders, the grandparents of the sporting vehicles of today. Flight control was exercised by one's shifting one's center of gravity under the glider.

Contrast Lilienthal's flying philosophy with those of previous would-be aviators before him. During most of the 19th century, powered flight was looked upon in a brute-force manner: Build an engine strong enough to drive an airplane, slap it on an airframe strong enough to withstand the forces and to generate the lift, and presumably you could get into the air. What would happen *after* you got into the air would be just a simple matter of steering the airplane around the sky like a carriage or automobile on the ground—at least this was the general feeling. Gibbs-Smith called the people taking this approach the *chauffeurs*. In contrast were the *airmen*—Lilienthal was the first—who recognized the need to get up in the air, fly around in gliders, and obtain the "feel" of an airplane *before* an engine was used for powered flight. The chauffeurs were mainly interested in thrust and lift, whereas the airmen were firstly concerned with flight control in the air. The airmen's philosophy ultimately led to successful powered flight; the chauffeurs were singularly unsuccessful.

Figure 1.15   A monoplane hang glider by Lilienthal, 1894.

Lilienthal made over 2000 successful glider flights. The aerodynamic data he obtained were published in papers circulated throughout the world. In fact, his work was timed perfectly with the rise of photography and the printing industry. In 1871 the dry-plate negative was invented, which by 1890 could "freeze" a moving object without a blur. Also, the successful halftone method of printing photographs in books and journals had been developed. As a result, photographs of Lilienthal's flights were widely distributed; indeed, Lilienthal was the first human to be photographed in an airplane (see, e.g., Fig. 1.15). Such widespread dissemination of his results inspired other pioneers in aviation. The Wright brothers' interest in flight did not crystallize until Wilbur first read some of Lilienthal's papers in about 1894.

On Sunday, August 9, 1896, Lilienthal was gliding from the Gollenberg hill near Stollen in Germany. It was a fine summer's day. However, a temporary gust of wind brought Lilienthal's monoplane glider to a standstill; he stalled and crashed to the ground. Only the wing was crumpled; the rest of the glider was undamaged. However, Lilienthal was carried away with a broken spine. He died the next day in the Bergmann Clinic in Berlin. During the course of his life, Lilienthal remarked several times that "sacrifices must be made." This epitaph is carved on his gravestone in Lichterfelde cemetery.

There is some feeling that had Lilienthal lived, he would have beaten the Wright brothers to the punch. In 1893 he built a powered machine; however, the prime mover was a carbonic acid gas motor which twisted six slats at each wing tip, obviously an ornithopter-type idea to mimic the natural mode of propulsion for birds. In the spring of 1895 he built a second, but larger, powered machine of the same type. Neither one of these airplanes was ever flown with the engine operating. It seems to this author that this mode of propulsion was doomed to failure. If Lilienthal had lived, would he have turned to the gasoline engine driving a propeller and thus achieved powered flight before 1903? It is a good question for conversation.

1.6 PERCY PILCHER (1867–1899)—EXTENDING THE GLIDER TRADITION

In June 1895, Otto Lilienthal received a relatively young and very enthusiastic visitor in Berlin—Percy Pilcher, a Scot who lived in Glasgow and who had already built his first glider. Under Lilienthal's guidance Pilcher made several glides from the artificial hill. This visit added extra fuel to Pilcher's interest in aviation; he returned to the British Isles and over the next four years built a series of successful gliders. His most noted machine was the *Hawk*, built in 1896 (see Fig. 1.16). Pilcher's experiments with his hang gliders made him the most distinguished British aeronautical engineer since George Cayley. Pilcher was an airman, and along with Lilienthal he underscored the importance of learning the practical nature of flight in the air before lashing an engine to the machine.

However, Pilcher's sights were firmly set on powered flight. In 1897, he calculated that an engine of 4 hp weighing no more than 40 lb, driving a 5-ft-diameter

Figure 1.16   Pilcher's hang glider, the *Hawk*, 1896.

propeller, would be necessary to power his *Hawk* off the ground. Since no such engine was available commercially, Pilcher (who was a marine engineer by training) spent most of 1898 designing and constructing one. It was completed and bench-tested by the middle of 1899. Then, in one of those quirks of fate that dot many aspects of history, Pilcher was killed while demonstrating his *Hawk* glider at the estate of Lord Braye in Leicestershire, England. The weather was bad, and on his first flight the glider was thoroughly water-soaked. On his second flight, the heavily sodden tail assembly collapsed, and Pilcher crashed to the ground. Like Lilienthal, Pilcher died one day after this disaster. Hence, England and the world also lost the only man other than Lilienthal who might have achieved successful powered flight before the Wright brothers.

1.7   AERONAUTICS COMES TO AMERICA

Look at the geographic distribution of the early developments in aeronautics as portrayed in the previous sections. After the advent of ballooning, due to the Montgolfiers' success in France, progress in heavier-than-air machines was focused in England until the 1850s: Witness the contributions of Cayley, Henson, and Stringfellow. This is entirely consistent with the fact that England also gave birth to the industrial revolution during this time. Then the spotlight moved to the European continent with Du Temple, Mozhaiski, Lilienthal, and others. There were some brief flashes again in Britain, such as those due to Wenham and the Aeronautical Society. In contrast, throughout this time virtually no important progress was being made in the United States. The fledgling nation was busy consolidating a new government and expanding its frontiers. There was not much interest or time for serious aeronautical endeavors.

However, this vacuum was broken by Octave Chanute (1832–1910), a French-born naturalized citizen who lived in Chicago. Chanute was a civil engineer who became interested in mechanical flight in about 1875. For the next 35 years, he collected, absorbed, and assimilated every piece of aeronautical information he could find. This culminated in 1894 with the publishing of his book entitled *Progress in Flying Machines*, a work that ranks with Lilienthal's *Der Vogelflug* as one of the great classics in aeronautics. Chanute's book summarized all the important progress in aviation up to that date; in this sense, he was the first serious aviation historian. In addition, Chanute made positive suggestions as to the future directions necessary to achieve success in powered flight. The Wright brothers avidly read *Progress in Flying Machines* and subsequently sought out Chanute in 1900. A close relationship and interchange of ideas developed between them. A friendship developed which was to last in various degrees until Chanute's death in 1910.

Chanute was an airman. Following this position, he began to design hang gliders, in the manner of Lilienthal, in 1896. His major specific contribution to aviation was the successful biplane glider shown in Fig. 1.17, which introduced the effective Pratt truss method of structural rigging. The Wright brothers were directly influenced by this biplane glider, and in this sense Chanute provided the natural bridge between Stringfellow's triplane (1868) and the first successful powered flight (1903).

About 500 mi to the east, in Washington, District of Columbia, the United States' second noted pre-Wright aeronautical engineer was hard at work. Samuel Pierpont Langley (1834–1906), secretary of the Smithsonian Institution, was tirelessly designing and building a series of powered aircraft, which finally culminated in two attempted piloted flights, both in 1903, just weeks before the Wrights' success on December 17.

Figure 1.17   Chanute's hang glider, 1896. *(National Air and Space Museum.)*

Langley was born in Roxbury, Massachusetts, on August 22, 1834. He received no formal education beyond high school, but his childhood interest in astronomy spurred him to a lifelong program of self-education. Early in his career, he worked for 13 years as an engineer and architect. Then, after making a tour of European observatories, Langley became an assistant at Harvard Observatory in 1865. He went on to become a mathematics professor at the U.S. Naval Academy, a physics and astronomy professor at the University of Pittsburgh, and the director of the Allegheny Observatory at Pittsburgh. By virtue of his many scientific accomplishments, Langley was appointed secretary of the Smithsonian Institution in 1887.

In this same year, Langley, who was by now a scientist of international reputation, began his studies of powered flight. Following the example of Cayley, he built a large whirling arm, powered by a steam engine, with which he made force tests on airfoils. He then built nearly 100 different types of rubber-band-powered model airplanes, graduating to steam-powered models in 1892. However, it was not until 1896 that Langley achieved any success with his powered models; on May 6 one of his aircraft made a free flight of 3300 ft, and on November 28 another flew for more than $\frac{3}{4}$ mi. These *Aerodromes* (a term due to Langley) were tandem-winged vehicles, driven by two propellers between the wings, powered by a 1-hp steam engine of Langley's own design. (However, Langley was influenced by one of John Stringfellow's small aerosteam engines, which was presented to the Smithsonian in 1889. After studying this historic piece of machinery, Langley set out to design a better engine.)

Langley was somewhat satisfied with his success in 1896. Recognizing that further work toward a piloted aircraft would be expensive in both time and money, he "made the firm resolution not to undertake the construction of a large man-carrying machine." (Note that it was in this year that the Wright brothers became interested in powered flight, another example of the flow and continuity of ideas and developments in physical science and engineering. Indeed, Wilbur and Orville were directly influenced and encouraged by Langley's success with powered aircraft. After all, here was a well-respected scientist who believed in the eventual attainment of mechanical flight and who was doing something about it.)

Consequently, there was a lull in Langley's aeronautical work until December 1898. Then, motivated by the Spanish-American War, the War Department, with the personal backing of President McKinley himself, invited Langley to build a machine for passengers. It backed up its invitation with $50,000. Langley accepted.

Departing from his earlier use of steam, Langley correctly decided that the gasoline-fueled engine was the proper prime mover for aircraft. He first commissioned Stephan Balzer of New York to produce such an engine, but dissatisfied with the results, Langley eventually had his assistant, Charles Manly, redesign the power plant. The resulting engine produced 52.4 hp and yet weighed only 208 lb, a spectacular achievement for that time. Using a smaller, 3.2-hp, gasoline-fueled engine, Langley made a successful flight with a quarter-scale model aircraft in August 1903.

Encouraged by this success, Langley stepped directly to the full-size airplane, top and side views of which are shown in Fig. 1.18. He mounted this tandem-winged aircraft on a catapult in order to provide an assisted takeoff. In turn, the airplane and catapult were placed on top of a houseboat on the Potomac River (see Fig. 1.19).

Figure 1.18 Drawing of the Langley full-size *Aerodrome*. (National Air and Space Museum.)

On October 7, 1903, with Manly at the controls, the airplane was ready for its first attempt. The launching was given wide advance publicity, and the press was present to watch what might be the first successful powered flight in history. A photograph of the *Aerodrome* a moment after launch is shown in Fig. 1.20. Here is the resulting report from the *Washington Post* the next day:

> A few yards from the houseboat were the boats of the reporters, who for three months had been stationed at Widewater. The newspapermen waved their hands. Manly looked down and smiled. Then his face hardened as he braced himself for the flight, which might have in store for him fame or death. The propeller wheels, a foot from his head, whirred around him one thousand times to the minute. A man forward fired

Figure 1.19   Langley's full-size *Aerodrome* on the houseboat launching catapult, 1903. *(National Air and Space Museum.)*

Figure 1.20   Langley's first launch of the full-size *Aerodrome*, October 7, 1903. *(National Air and Space Museum.)*

two skyrockets. There came an answering "toot, toot," from the tugs. A mechanic stooped, cut the cable holding the catapult; there was a roaring, grinding noise—and the Langley airship tumbled over the edge of the houseboat and disappeared in the river, sixteen feet below. It simply slid into the water like a handful of mortar....

Manly was unhurt. Langley believed the airplane was fouled by the launching mechanism, and he tried again on December 8, 1903. Figure 1.21, a photograph taken

Figure 1.21   Langley's second launch of the full-size *Aerodrome*, December 8, 1903. *(National Air and Space Museum.)*

moments after launch, shows the rear wings in total collapse and the *Aerodrome* going through a 90° angle of attack. Again, the *Aerodrome* fell into the river, and again Manly was fished out, unhurt. It is not entirely certain what happened this time; again the fouling of the catapult was blamed, but some experts maintain that the tail boom cracked due to structural weakness. At any rate, that was the end of Langley's attempts. The War Department gave up, stating that "we are still far from the ultimate goal (of human flight)." Members of Congress and the press leveled vicious and unjustified attacks on Langley (human flight was still looked upon with much derision by most people). In the face of this ridicule, Langley retired from the aeronautical scene. He died on February 27, 1906, a man in despair.

In contrast to Chanute and the Wright brothers, Langley was a chauffeur. Most modern experts feel that his *Aerodrome* would not have been capable of sustained, equilibrium flight, had it been successfully launched. Langley made no experiments with gliders with passengers to get the feel of the air. He ignored completely the important aspects of flight control. He attempted to launch Manly into the air on a powered machine without Manly's having one second of flight experience. Nevertheless, Langley's aeronautical work was of some importance because he lent the power of his respected technical reputation to the cause of mechanical flight, and his *Aerodromes* were to provide encouragement to others.

Nine days after Langley's second failure, the *Wright Flyer I* rose from the sands of Kill Devil Hills.

1.8   WILBUR (1867–1912) AND ORVILLE (1871–1948) WRIGHT—INVENTORS OF THE FIRST PRACTICAL AIRPLANE

The scene now shifts to the Wright brothers, the premier aeronautical engineers of history. Only George Cayley may be considered comparable. In Sec. 1.1, it was stated that the time was ripe for the attainment of powered flight at the beginning of the 20th century. The ensuing sections then provided numerous historical brushstrokes to emphasize this statement. Thus, the Wright brothers drew on an existing heritage that is part of every aerospace engineer today.

Wilbur Wright was born on April 16, 1867 (two years after the Civil War), on a small farm in Millville, Indiana. Four years later, Orville was born on August 19, 1871, in Dayton, Ohio. The Wrights were descendants of an old Massachusetts family, and their father was a bishop of the United Brethren Church. The two brothers benefited greatly from the intellectual atmosphere of their family. Their mother was three months short of a college degree. She had considerable mechanical ability, enhanced by spending time in her father's carriage shop. She later designed and built simple household appliances and made toys for her children. In the words of Tom Crouch, the definitive biographer of the Wright brothers: "When the boys wanted mechanical advice or assistance, they came to their mother." Their father, Crouch says, "was one of those men who had difficulty driving a nail straight." (See T.

Crouch, *The Bishop's Boys*, Norton, New York, 1989.) Interestingly enough, neither Wilbur nor Orville officially received a high school diploma; Wilbur did not bother to go to the commencement services, and Orville took a special series of courses in his junior year that did not lead to a prescribed degree, and he did not attend his senior year. Afterward, the brothers immediately sampled the business world. In 1889, they first published a weekly four-page newspaper on a printing press of their own design. However, Orville had talent as a prize-winning cyclist, and this prompted the brothers to set up a bicycle sales and repair shop in Dayton in 1892. Three years later they began to manufacture their own bicycle designs, using homemade tools. These enterprises were profitable and helped to provide the financial resources for their later work in aeronautics.

In 1896, Otto Lilienthal was accidently killed during a glider flight (see Sec. 1.5). In the wake of the publicity, the Wright brothers' interest in aviation, which had been apparent since childhood, was given much impetus. Wilbur and Orville had been following Lilienthal's progress intently; recall that Lilienthal's gliders were shown in flight by photographs distributed around the world. In fact, an article on Lilienthal in an issue of *McClure's Magazine* in 1894 was apparently the first to trigger Wilbur's mature interest; but it was not until 1896 that Wilbur really became a serious thinker about human flight.

Like several pioneers before him, Wilbur took up the study of bird flight as a guide on the path toward mechanical flight. This led him to conclude in 1899 that birds "regain their lateral balance when partly overturned by a gust of wind, by a torsion of the tips of the wings." Thus emerged one of the most important developments in aviation history: the use of wing twist to control airplanes in lateral (rolling) motion. Ailerons are used on modern airplanes for this purpose, but the idea is the same. (The aerodynamic fundamentals associated with wing twist or ailerons are discussed in Chaps. 5 and 7.) In 1903, Chanute, in describing the work of the Wright brothers, coined the term *wing warping* for this idea, a term that was to become accepted but which was to cause some legal confusion later.

Anxious to pursue and experiment with the concept of wing warping, Wilbur wrote to the Smithsonian Institution in May 1899 for papers and books on aeronautics; in turn he received a brief bibliography of flying, including works by Chanute and Langley. Most important among these was Chanute's *Progress in Flying Machines* (see Sec. 1.7). Also at this time, Orville became as enthusiastic as his brother, and they both digested all the aeronautical literature they could find. This led to their first aircraft, a biplane kite with a wingspan of 5 ft, in August 1899. This machine was designed to test the concept of wing warping, which was accomplished by means of four controlling strings from the ground. The concept worked!

Encouraged by this success, Wilbur wrote to Chanute in 1900, informing him of their initial, but fruitful, progress. This letter began a close friendship between the Wright brothers and Chanute, a friendship which was to benefit both parties in the future. Also, following the true airman philosophy, the Wrights were convinced they had to gain experience in the air before applying power to an aircraft. By writing to the U.S. Weather Bureau, they found an ideal spot for glider experiments, the area around Kitty Hawk, North Carolina, where there were strong and constant winds. A full-size biplane glider was ready by September 1900 and was flown in October of that year at Kitty Hawk. Figure 1.22 shows a photograph of the Wrights' number 1 glider. It had a 17-ft wingspan and a horizontal elevator in front of the wings and was usually flown on strings from the ground; only a few brief piloted flights were made.

With some success behind them, Wilbur and Orville proceeded to build their number 2 glider (see Fig. 1.23). Moving their base of operations to Kill Devil Hills, 4 mi south of Kitty Hawk, they tested number 2 during July and August of 1901. These were mostly manned flights, with Wilbur or Orville lying prone on the bottom wing, facing into the wind, as shown in Fig. 1.23. This new glider was somewhat larger, with a 22-ft wingspan. As with all Wright machines, it had a horizontal elevator

Figure 1.22   The Wright brothers' number 1 glider at Kitty Hawk, North Carolina, 1900. *(National Air and Space Museum.)*

Figure 1.23   The Wright brothers' number 2 glider at Kill Devil Hills, 1901. *(National Air and Space Museum.)*

in front of the wings. The Wrights felt that a forward elevator would, among other functions, protect them from the type of fatal nosedive that killed Lilienthal.

During these July and August test flights, Octave Chanute visited the Wrights' camp. He was much impressed by what he saw. This led to Chanute's invitation to Wilbur to give a lecture in Chicago. In giving this paper on September 18, 1901, Wilbur laid bare their experiences, including the design of their gliders and the concept of wing warping. Chanute described Wilbur's presentation as "a devilish good paper which will be extensively quoted." Chanute, as usual, was serving his very useful function as a collector and disseminator of aeronautical data.

However, the Wrights were not close to being satisfied with their results. When they returned to Dayton after their 1901 tests with the number 2 glider, both brothers began to suspect the existing data which appeared in the aeronautical literature. To this date, they had faithfully relied upon detailed aerodynamic information generated by Lilienthal and Langley. Now they wondered about its accuracy. Wilbur wrote that "having set out with absolute faith in the existing scientific data, we were driven to doubt one thing after another, until finally, after two years of experiment, we cast it all aside, and decided to rely entirely upon our own investigations." And investigate they did! Between September 1901 and August 1902, the Wrights undertook a major program of aeronautical research. They built a wind tunnel (see Chap. 4) in their bicycle shop in Dayton and tested more than 200 different airfoil shapes. They designed a force balance to measure accurately the lift and drag. This period of research was a high-water mark in early aviation development. The Wrights learned, and with them ultimately so did the world. This sense of learning and achievement by the brothers is apparent simply from reading through *The Papers of Wilbur and Orville Wright* (1953), edited by Marvin W. McFarland. The aeronautical research carried out during this period ultimately led to their number 3 glider, which was flown in 1902. It was so successful that Orville wrote that "our tables of air pressure which we made in our wind tunnel would enable us to calculate in advance the performance of a machine." Here is the first example in history of the major impact of wind tunnel testing on the flight development of a given machine, an impact that has been repeated for all major airplanes of the 20th century. [Very recently, it has been shown by Anderson in *A History of Aerodynamics and Its Impact on Flying Machines* (Cambridge University Press, 1997) that Lilienthal's data were reasonable, but the Wrights misinterpreted them. Applying the data incorrectly, the Wrights obtained incorrect results for their 1900 and 1901 gliders. However, this is irrelevant because the Wrights went on to discover the correct results.]

The number 3 glider was a classic. It was constructed during August and September of 1902. It first flew at Kill Devil Hills on September 20, 1902. It was a biplane glider with a 32-ft 1-in wingspan, the largest of the Wright gliders to date. This number 3 glider is shown in Fig. 1.24. Note that, after several modifications, the Wrights added a vertical rudder behind the wings. This rudder was movable, and when connected to move in unison with the wing warping, it enabled the number 3 glider to make a smooth, banked turn. This combined use of rudder with wing warping (or later, ailerons) was another major contribution of the Wright brothers to flight control in particular, and aeronautics in general.

Figure 1.24   The Wright brothers' number 3 glider, 1902. *(National Air and Space Museum.)*

So the Wrights now had the most practical and successful glider in history. During 1902, they made more than 1000 perfect flights. They set a distance record of 622.5 ft and a duration record of 26 s. In the process, both Wilbur and Orville became highly skilled and proficient pilots, something that would later be envied worldwide.

Powered flight was now just at their fingertips, and the Wrights knew it! Flushed with success, they returned to Dayton to face the last remaining problem: propulsion. As with Langley before them, they could find no commercial engine that was suitable. So they designed and built their own during the winter months of 1903. It produced 12 hp and weighed about 200 lb. Moreover, they conducted their own research, which allowed them to design an effective propeller. These accomplishments, which had eluded people for a century, gushed forth from the Wright brothers like natural spring water.

With all the major obstacles behind them, Wilbur and Orville built their *Wright Flyer I* from scratch during the summer of 1903. It closely resembled the number 3 glider but had a wingspan of 40 ft 4 in and used a double rudder behind the wings and a double elevator in front of the wings. And of course, there was the spectacular gasoline-fueled Wright engine, driving two pusher propellers by means of bicycle-type chains. A three-view diagram and a photograph of the *Wright Flyer I* are shown in Figs. 1.1 and 1.2, respectively.

From September 23 to 25, the machine was transported to Kill Devil Hills, where the Wrights found their camp in some state of disrepair. Moreover, their number 3 glider had been damaged over the winter months. They set about to make repairs and afterward spent many weeks of practice with their number 3 glider. Finally, on December 12, everything was in readiness. However, this time the elements interfered: Bad weather postponed the first test of the *Wright Flyer I* until December 14. On that day, the Wrights called witnesses to the camp and then flipped a coin to see who would be the first pilot. Wilbur won. The *Wright Flyer I* began to move along the launching rail under its own power, picking up flight speed. It lifted off the rail properly but suddenly went into a steep climb, stalled, and thumped back to the ground. It was the first recorded case of pilot error in powered flight: Wilbur admitted that he put on too much elevator and brought the nose too high.

With minor repairs made, and with the weather again favorable, the *Wright Flyer I* was again ready for flight on December 17. This time it was Orville's turn at the controls. The launching rail was again laid on level sand. A camera was adjusted to take a picture of the machine as it reached the end of the rail. The engine was put on full throttle, the holding rope was released, and the machine began to move. The rest is history, as portrayed in the opening paragraphs of this chapter.

One cannot read or write of this epoch-making event without experiencing some of the excitement of the time. Wilbur Wright was 36 years old; Orville was 32. Between them, they had done what no one before them had accomplished. By their persistent efforts, their detailed research, and their superb engineering, the Wrights had made the world's first successful heavier-than-air flight, satisfying all the necessary criteria laid down by responsible aviation historians. After Orville's first flight on that December 17, three more flights were made during the morning, the last covering 852 ft and remaining in the air for 59 s. The world of flight—and along with it the world of successful aeronautical engineering—had been born!

It is interesting to note that even though the press was informed of these events via Orville's telegram to his father (see the introduction to this chapter), virtually no notice appeared before the public; even the Dayton newspapers did not herald the story. This is a testimonial to the widespread cynicism and disbelief among the general public about flying. Recall that just nine days before, Langley had failed dismally in full view of the public. In fact, it was not until Amos I. Root observed the Wrights flying in 1904 and published his inspired account in a journal of which he was the editor, *Gleanings in Bee Culture* (January 1, 1905, issue), that the public had its first detailed account of the Wrights' success. However, the article had no impact.

The Wright brothers did not stop with the *Wright Flyer I*. In May 1904, their second powered machine, the *Wright Flyer II*, was ready. This aircraft had a smaller wing camber (airfoil curvature) and a more powerful and efficient engine. In outward appearance, it was essentially like the 1903 machine. During 1904, more than 80 brief flights were made with the *Wright Flyer II*, all at a 90-acre field called Huffman Prairie, 8 mi east of Dayton. (Huffman Prairie still exists today; it is on the huge Wright-Patterson Air Force Base, a massive aerospace development center named in honor of the Wrights.) These tests included the first circular flight—made by Wilbur on September 20. The longest flight last 5 min 4 s, traversing more than $2\frac{3}{4}$ mi.

More progress was made in 1905. The *Wright Flyer III* was ready by June. The wing area was slightly smaller than that of the *Flyer II*, the airfoil camber was increased back to what it had been in 1903, the biplane elevator was made larger and was placed farther in front of the wings, and the double rudder was also larger and placed farther back behind the wings. New, improved propellers were used. This machine, the *Flyer III*, was the first *practical* airplane in history. It made more than 40 flights during 1905, the longest being 38 min 3 s and covering 24 mi. These flights were generally terminated only after gas was used up. C. H. Gibbs-Smith writes about the *Flyer III*: "The description of this machine as the world's first practical powered aeroplane is justified by the sturdiness of its structure, which withstood constant takeoffs and landings; its ability to bank, turn, and perform figures of eight; and its reliability in remaining airborne (with no trouble) for over half an hour."

Then the Wright brothers, who heretofore had been completely open about their work, became secretive. They were not making any progress in convincing the U.S. government to buy their airplane, but at the same time various people and companies were beginning to make noises about copying the Wrights' design. A patent applied for by the Wrights in 1902 to cover their ideas of wing warping combined with rudder action was not granted until 1906. So, between October 16, 1905, and May 6, 1908, neither Wilbur nor Orville flew, nor did they allow anybody to view their machines. However, their aeronautical engineering did not stop. During this period, they built at least six new engines. They also designed a new flying machine which was to become the standard Wright type A, shown in Fig. 1.25. This airplane was similar to the *Wright Flyer III*, but it had a 40-hp engine and provided for two people seated upright between the wings. It also represented the progressive improvement of a basically successful design, a concept of airplane design carried out to present day.

The public and the Wright brothers finally had their meeting, and in a big way, in 1908. The Wrights signed contracts with the U.S. Army in February 1908, and with a French company in March of the same year. After that, the wraps were off. Wilbur traveled to France in May, picked up a crated type A which had been waiting at Le Havre since July 1907, and completed the assembly in a friend's factory at Le Mans. With supreme confidence, he announced his first public flight in advance—to take place on August 8, 1908. Aviation pioneers from all over Europe, who had heard rumors about the Wrights' successes since 1903, the press, and the general public all flocked to a small race course at Hunaudieres, 5 mi south of Le Mans. On the appointed day, Wilbur took off, made an impressive, circling flight for almost 2 min, and landed. It was like a revolution. Aeronautics, which had been languishing in Europe since Lilienthal's death in 1896, was suddenly alive. The Frenchman Louis Bleriot, soon to become famous for being first to fly across the English Channel, exclaimed: "For us in France and everywhere, a new era in mechanical flight has commenced—it is marvelous." The French press, after being skeptical for years of the Wrights' supposed accomplishments, called Wilbur's flight "one of the most exciting spectacles ever presented in the history of applied science." More deeply echoing the despair of many would-be French aviators who were in a race with the Wrights to be first with powered flight, Leon Delagrange said: "Well, we are beaten. We just don't exist." Subsequently, Wilbur made 104 flights in France before the end of 1908. The acclaim and honor due the Wright brothers since 1903 had finally arrived.

Orville was experiencing similar success in the United States. On September 3, 1908, he began a series of demonstrations for the Army at Fort Myer, near Washington,

Figure 1.25   A two-view of the Wright type A, 1908.

District of Columbia. Flying a type A machine, he made 10 flights, the longest for
1 h 14 min, before September 17. On that day, Orville experienced a propeller failure
that ultimately caused the machine to crash, seriously injuring himself and killing his
passenger, Lt. Thomas E. Selfridge. This was the first crash of a powered aircraft,
but it did not deter either Orville or the Army. Orville made a fast recovery and was
back to flying in 1909; and the Army bought the airplane.

The accomplishments of the Wright brothers were monumental. Their zenith occurred during the years 1908 to 1910; after that European aeronautics quickly caught up and went ahead in the technological race. The main reason for this was that all the Wrights' machines, from the first gliders, were statically unstable (see Chap. 7). This meant that the Wrights' airplanes would not fly "by themselves"; rather, they had to be constantly, every instant, controlled by the pilot. In contrast, European inventors believed in inherently stable aircraft. After their lessons in flight control from Wilbur in 1908, workers in France and England moved quickly to develop controllable, but stable, airplanes. These were basically safer and easier to fly. The concept of static stability has carried over to virtually all airplane designs through the present century. (It is interesting to note that the new designs for lightweight fighters, such as the General Dynamics F-16, are statically *unstable*, which represents a return to the Wrights' design philosophy. However, unlike the *Wright Flyers*, these new aircraft are flown constantly, every moment, by electrical means, by the new "fly-by-wire" concept.)

To round out the story of the Wright brothers, Wilbur died in an untimely fashion of typhoid fever on May 30, 1912. In a fitting epitaph, his father said: "This morning, at 3:15 Wilbur passed away, aged 45 years, 1 month, and 14 days. A short life full of consequences. An unfailing intellect, imperturbable temper, great self-reliance and as great modesty. Seeing the right clearly, pursuing it steadily, he lived and died."

Orville lived on until January 30, 1948. During World War I, he was commissioned a major in the Signal Corps Aviation Service. Although he sold all his interest in the Wright company and "retired" in 1915, he afterward performed research in his own shop. In 1920, he invented the split flap for wings, and he continued to be productive for many years.

As a final footnote to this story of two great men, there occurred a dispute between Orville and the Smithsonian Institution concerning the proper historical claims on powered flight. As a result, Orville sent the historic *Wright Flyer I*, the original, to the Science Museum in London in 1928. It resided there, through the bombs of World War II, until 1948, when the museum sent it to the Smithsonian. It is now part of the National Air and Space Museum and occupies a central position in the gallery.

1.9 THE AERONAUTICAL TRIANGLE—LANGLEY, THE WRIGHTS, AND GLENN CURTISS

In 1903—a milestone year for the Wright brothers, with their first successful powered flight—Orville and Wilbur faced serious competition from Samuel P. Langley. As portrayed in Sec. 1.7, Langley was the secretary of the Smithsonian Institution and was one of the most respected scientists in the United States at that time. Beginning in 1886, Langley mounted an intensive aerodynamic research and development program, bringing to bear the resources of the Smithsonian and later the War Department. He carried out this program with a dedicated zeal that matched the fervor that the Wrights themselves demonstrated later. Langley's efforts culminated in the full-scale

Aerodrome shown in Figs. 1.18, 1.19, and 1.20. In October 1903, this *Aerodrome* was ready for its first attempted flight, in the full glare of publicity in the national press.

The Wright brothers were fully aware of Langley's progress. During their preparations with the *Wright Flyer* at Kill Devil Hills in the summer and fall of 1903, Orville and Wilbur kept in touch with Langley's progress via the newspapers. They felt this competition keenly, and the correspondence of the Wright brothers at this time indicates an uneasiness that Langley might become the first to successfully achieve powered flight, before they would have a chance to test the *Wright Flyer.* In contrast, Langley felt no competition at all from the Wrights. Although the aeronautical activity of the Wright brothers was generally known throughout the small circle of aviation enthusiasts in the United States and Europe—thanks mainly to reports on their work by Octave Chanute—this activity was not taken seriously. At the time of Langley's first attempted flight on October 7, 1903, there is no recorded evidence that Langley was even aware of the Wrights' powered machine sitting on the sand dunes of Kill Devil Hills, and certainly no appreciation by Langley of the degree of aeronautical sophistication achieved by the Wrights. As it turned out, as was related in Sec. 1.7, Langley's attempts at manned powered flight, first on October 7 and again on December 8, resulted in total failure. A photograph of Langley's *Aerodrome*, lying severely damaged in the Potomac River on October 7, is shown in Fig. 1.26. In hindsight, the Wrights had nothing to fear from competition with Langley.

Such was not the case in their competition with another aviation pioneer—Glenn H. Curtiss—beginning five years later. In 1908—another milestone year for the

Figure 1.26   Langley's *Aerodrome* resting in the Potomac River after its first unsuccessful flight on October 7, 1903. Charles Manly, the pilot, was fished out of the river, fortunately unhurt.

Wrights, with their glorious first public flights in France and the United States—
Orville and Wilbur faced a serious challenge and competition from Curtiss, which
was to lead to acrimony and a flurry of lawsuits that left a smudge on the Wrights'
image and resulted in a general inhibition of the development of early aviation in the
United States. By 1910, the name of Glenn Curtiss was as well known throughout
the world as Orville and Wilbur Wright, and indeed, Curtiss-built airplanes were
more popular and easier to fly than those produced by the Wrights. How did these
circumstances arise? Who was Glenn Curtiss, and what was his relationship with the
Wrights? What impact did Curtiss have on the early development of aviation, and
how did his work compare and intermesh with that of Langley and that of the Wrights?
Indeed, the historical development of aviation in the United States can be compared
to a triangle, with the Wrights on one apex, Langley at another, and Curtiss at the
third. This "aeronautical triangle" is shown in Fig. 1.27. What was the nature of this
triangular relationship? These and other questions are addressed in this section. They
make a fitting conclusion to the overall early historical development of aeronautical
engineering as portrayed in this chapter.

Let us first look at Glenn Curtiss, the man. Curtiss was born in Hammondsport, New York, on May 21, 1878. Hammondsport at that time was a small town—
population less than 1000—bordering on Keuka Lake, one of the Finger Lakes in
upstate New York. (Later, Curtiss was to make good use of Keuka Lake for the
development of amphibian aircraft—one of his hallmarks.) The son of a harness
maker who died when Curtiss was 5 years old, Curtiss was raised by his mother
and grandmother. Their modest financial support came from a small vineyard which
grew in their front yard. His formal education ceased with the eighth grade, after
which he moved to Rochester, where he went to work for Eastman Dry Plate and
Film Company (later to become Kodak), stenciling numbers on the paper backing of
film. In 1900, he returned to Hammondsport, where he took over a bicycle repair
shop (shades of the Wright brothers). At this time, Glenn Curtiss began to show a
passion that would consume him for his lifetime—a passion for *speed*. He became
active in bicycle racing and quickly earned a reputation as a winner. In 1901 he
incorporated an engine on his bicycles and became an avid motorcycle racer. By
1902, his fame was spreading, and he was receiving numerous orders for motorcycles
with engines of his own design. By 1903 Curtiss had established a motorcycle factory
at Hammondsport, and he was designing and building the best (highest horsepower-
to-weight ratio) engines available anywhere. In January 1904, at Ormond Beach,
Florida, Curtiss established a new world's speed record for a ground vehicle—67 mi/h
over a 10-mi straightaway—a record that was to stand for 7 years.

Curtiss "backed into" aviation. In the summer of 1904, he received an order
from Thomas Baldwin, a California balloonist, for a two-cylinder engine. Baldwin
was developing a powered balloon—a dirigible. The Baldwin dirigibles, with the
highly successful Curtiss engines, soon became famous around the country. In 1906
Baldwin moved his manufacturing facilities to Hammondsport, to be close to the
source of his engines. A lifelong friendship and cooperation developed between
Baldwin and Curtiss and provided Curtiss with his first experience in aviation—as a
pilot of some of Baldwin's powered balloons.

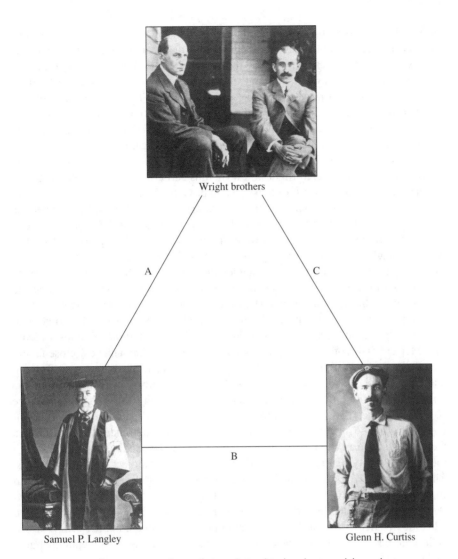

Wright brothers

A

C

Samuel P. Langley

B

Glenn H. Curtiss

Figure 1.27 The "aeronautical triangle," a relationship that dominated the early development of aeronautics in the United States during the period from 1886 to 1916.

In August 1906, Baldwin traveled to the Dayton Fair in Ohio for a week of dirigible flight demonstrations; he brought Curtiss along to personally maintain the engines. The Wright brothers also attended the fair—specifically to watch Thomas Baldwin perform. They even lent a hand in retrieving the dirigible when it strayed too far afield. This was the first face-to-face encounter between Curtiss and the Wrights. During that week, Baldwin and Curtiss visited the Wrights at the brothers' bicycle shop and entered into long discussions on powered flight. Recall from Sec. 1.8 that the Wrights had discontinued flying one year earlier, and at the time of their meeting with Curtiss, Orville and Wilbur were actively trying to interest the United States, as well as England and France, in buying their airplane. The Wrights had become very secretive about their airplane and allowed no one to view it. Curtiss and Baldwin were no exceptions. However, that week in Dayton, the Wrights were relatively free with Curtiss, giving him information and technical suggestions about powered flight. Years later, these conversations became the crux of the Wrights' claim that Curtiss had stolen some of their ideas and used them for his own gain.

This claim was probably not entirely unjustified, for by that time Curtiss had a vested interest in powered flight; a few months earlier he had supplied Alexander Graham Bell with a 15-hp motor to be used in propeller experiments, looking toward eventual application to a manned, heavier-than-air, powered aircraft. The connection between Bell and Curtiss is important. Bell, renowned as the inventor of the telephone, had an intense interest in powered flight. He was a close personal friend of Samuel Langley and, indeed, was present for Langley's successful unmanned *Aerodrome* flights in 1896. By the time Langley died in 1906, Bell was actively carrying out kite experiments and was testing air propellers on a catamaran at his Nova Scotia coastal home. In the summer of 1907, Bell formed the Aerial Experiment Association, a group of five men whose officially avowed purpose was simply "to get into the air." The Aerial Experiment Association (AEA) consisted of Bell himself, Douglas McCurdy (son of Bell's personal secretary, photographer, and very close family friend), Frederick W. Baldwin (a freshly graduated mechanical engineer from Toronto and close friend of McCurdy), Thomas E. Selfridge (an Army lieutenant with an extensive engineering knowledge of aeronautics), and Glenn Curtiss. The importance of Curtiss to the AEA is attested by the stipends that Bell paid to each member of the association—Curtiss was paid five times more than the others. Bell had asked Curtiss to join the association because of Curtiss's excellent engine design and superb mechanical ability. Curtiss was soon doing much more than just designing engines. The plan of the AEA was to conduct intensive research and development on powered flight and to build five airplanes—one for each member. The first aircraft, the *Red Wing*, was constructed by the AEA with Selfridge as the chief designer. On March 12, 1908, the *Red Wing* was flown at Hammondsport for the first time, with Baldwin at the controls. It covered a distance of 318 ft and was billed as "the first public flight" in the United States.

Recall that the tremendous success of the Wright brothers from 1903 to 1905 was not known by the general public, mainly because of indifference in the press as well as the Wrights' growing tendency to be secretive about their airplane design until they could sell an airplane to the U.S. government. However, the Wrights' growing apprehension about the publicized activities of the AEA is reflected in a letter from Wilbur to the editor of the *Scientific American* after the flight of the *Red Wing*. In this letter, Wilbur states:

> In 1904 and 1905, we were flying every few days in a field alongside the main wagon road and electric trolley line from Dayton to Springfield, and hundreds of travelers and inhabitants saw the machine in flight. Anyone who wished could look. We merely did not advertise the flights in the newspapers.

On March 17, 1908, the second flight of the *Red Wing* resulted in a crash which severely damaged the aircraft. Work on the *Red Wing* was subsequently abandoned in lieu of a new design of the AEA, the *White Wing*, with Baldwin as the chief designer. The *White Wing* was equipped with ailerons—the first in the United States. Members of the AEA were acutely aware of the Wrights' patent on wing warping for lateral control, and Bell was particularly sensitive to making certain that his association did not infringe upon this patent. Therefore, instead of using wing warping, the *White Wing* utilized triangular movable ailerons at the wing tips of both wings of the biplane. Beginning on May 18, 1908, the *White Wing* successfully made a series of flights piloted by various members of the AEA. One of these flights, with Glenn Curtiss at the controls, was reported by Selfridge to the Associated Press as follows:

> G. H. Curtiss of the Curtiss Manufacturing Company made a flight of 339 yards in two jumps in Baldwin's White Wing this afternoon at 6:47PM. In the first jump he covered 205 yards then touched, rose immediately and flew 134 yards further when the flight ended on the edge of a ploughed field. The machine was in perfect control at all times and was steered first to the right and then to the left before landing. The 339 yards was covered in 19 seconds or 37 miles per hour.

Two days later, with an inexperienced McCurdy at the controls, the *White Wing* crashed and never flew again.

However, by this time, the Wright brothers' apprehension about the AEA was growing into bitterness toward its members. Wilbur and Orville genuinely felt that the AEA had pirated their ideas and was going to use them for commercial gain. For example, on June 7, 1908, Orville wrote to Wilbur (who was in France preparing for his spectacular first public flights that summer at Le Mans—see Sec. 1.8): "I see by one of the papers that the Bell outfit is offering Red Wings for sale at $5,000 each. They have some nerve." On June 28, he related to Wilbur: "Curtiss et al. are using our patents, I understand, and are now offering machines for sale at $5,000 each, according to the *Scientific American*. They have got good cheek."

The strained relations between the Wrights and the AEA—particularly Curtiss—were exacerbated on July 4, 1908, when the AEA achieved its crowning success. A new airplane had been constructed—the *June Bug*—with Glenn Curtiss as the chief designer. In the previous year, the *Scientific American* had offered a trophy, through the Aero Club of America, worth more than $3000 to the first aviator making a straight flight of 1 km (3281 ft). On Independence Day in 1908, at Hammondsport, New York, Glenn Curtiss at the controls of his *June Bug* was ready for an attempt at the trophy. A delegation of 22 members of the Aero Club was present, and the official starter was none other than Charles Manly, Langley's dedicated assistant and pilot of the ill-fated *Aerodrome* (see Sec. 1.7 and Fig. 1.26). Late in the day, at 7:30 PM, Curtiss took off and in 1 min 40 s had covered a distance of more than 1 mi, easily winning the *Scientific American* prize. A photograph of the *June Bug* during this historic flight is shown in Fig. 1.28.

The Wright brothers could have easily won the *Scientific American* prize long before Curtiss; they simply chose not to. Indeed, the publisher of the *Scientific American*, Charles A. Munn, wrote to Orville on June 4, inviting him to make the first attempt at the trophy, offering to delay Curtiss's request for an attempt. On June

Figure 1.28   Glenn Curtiss flying *June Bug* on July 4, 1908, on his way to the *Scientific American* prize for the first public flight of greater than 1 km.

30, the Wrights responded negatively—they were too involved with preparations for their upcoming flight trials in France and at Fort Myer in the United States. However, Curtiss's success galvanized the Wrights' opposition. Remembering their earlier conversations with Curtiss in 1906, Orville wrote to Wilbur on July 19:

> I had been thinking of writing to Curtiss. I also intended to call attention of the *Scientific American* to the fact that the Curtiss machine was a poor copy of ours; that we had furnished them the information as to how our older machines were constructed, and that they had followed this construction very closely, but have failed to mention the fact in any of their writings.

Curtiss's publicity in July was totally eclipsed by the stunning success of Wilbur during his public flights in France beginning August 8, 1908, and by Orville's Army trials at Fort Myer beginning on September 3, 1908. During the trials at Fort Myer, the relationship between the Wrights and the AEA took an ironic twist. One member of the evaluation board assigned by the Army to observe Orville's flights was Lt. Thomas Selfridge. Selfridge had been officially detailed to the AEA by the Army for a year and was now back at his duties of being the Army's main aeronautical expert. As part of the official evaluation, Orville was required to take Selfridge on a flight as a passenger. During this flight, on September 17, one propeller blade cracked and changed its shape, thus losing thrust. This imbalanced the second propeller, which cut a control cable to the tail. The cable subsequently wrapped around the propeller and snapped it off. The Wright type A went out of control and crashed. Selfridge was killed, and Orville was severely injured; he was in the hospital for $1\frac{1}{2}$ months.

For the rest of his life, Orville would walk with a limp as a result of this accident. Badly shaken by Selfridge's death, and somewhat overtaken by the rapid growth of aviation after the events of 1908, the Aerial Experiment Association dissolved itself on March 31, 1909. In the written words of Alexander Graham Bell, "The A.E.A. is now a thing of the past. It has made its mark upon the history of aviation and *its work will live*."

After this, Glenn Curtiss struck out in the aviation world on his own. Forming an aircraft factory at Hammondsport, Curtiss designed and built a new airplane, improved over the *June Bug* and named the *Golden Flyer*. In August 1909, a massive air show was held in Reims, France, attracting huge crowds and the crown princes of Europe. For the first time in history, the Gordon Bennett trophy was offered for the fastest flight. Glenn Curtiss won this trophy with his *Golden Flyer*, averaging a speed of 75.7 km/h (47.09 mi/h) over a 20-km course and defeating a number of pilots flying the Wrights' airplanes. This launched Curtiss on a meteoric career as a daredevil pilot and a successful airplane manufacturer. His motorcycle factory at Hammondsport was converted entirely to the manufacture of airplanes. His airplanes were popular with other pilots of that day because they were statically stable and hence easier and safer to fly than the Wrights' airplanes, which had been intentionally designed by the Wright brothers to be statically unstable (see Chap. 7). By 1910, aviation circles and the general public held Curtiss and the Wrights in essentially equal esteem. At the lower right of Fig. 1.27 is a photograph of Curtiss at this time; the propeller ornament in his cap was a good luck charm which he took on his flights. By 1911, a Curtiss airplane had taken off from and landed on a ship. Also in that year, Curtiss developed the first successful seaplanes and forged a lasting relationship with the U.S. Navy. In June 1911, the Aero Club of America issued its first official pilot's license to Curtiss in view of the fact that he had made the first public flight in the United States, an honor which otherwise would have gone to the Wrights.

In September 1909, the Wright brothers filed suit against Curtiss for patent infringements. They argued that their wing warping patent of 1906, liberally interpreted, covered all forms of lateral control, including the ailerons used by Curtiss. This triggered five years of intensive legal maneuvering, which dissipated much of the energies of all the parties. Curtiss was not alone in this regard. The Wrights brought suit against a number of fledgling airplane designers during this period, both in the United States and in Europe. Such litigation consumed Wilbur's attention, in particular, and effectively removed him from being a productive worker toward technical aeronautical improvements. It is generally agreed by aviation historians that this was not the Wrights' finest hour. Their legal actions not only hurt their own design and manufacturing efforts, but also effectively discouraged the early development of aeronautics by others, particularly in the United States. (It is quite clear that when World War I began in 1914, the United States—birthplace of aviation—was far behind Europe in aviation technology.) Finally, in January 1914, the courts ruled in favor of the Wrights, and Curtiss was forced to pay royalties to the Wright family. (By this time, Wilbur was dead, having succumbed to typhoid fever in 1912.)

In defense of the Wright brothers, their actions against Curtiss grew from a genuine belief on their part that Curtiss had wronged them and had consciously stolen their ideas, which Curtiss had subsequently parlayed into massive economic gains.

This went strongly against the grain of the Wrights' staunchly ethical upbringing. In contrast, Curtiss bent over backward to avoid infringing on the letter of the Wrights' patent, and there is numerous evidence that Curtiss was consistently trying to mend relations with the Wrights. It is this author's opinion that both sides became entangled in a complicated course of events that followed those heady days after 1908, when aviation burst on the world scene, and that neither Curtiss nor the Wrights should be totally faulted for their actions. These events simply go down in history as a less-than-glorious, but nevertheless important, chapter in the early development of aviation.

An important postscript should be added here regarding the triangular relationship between Langley, the Wrights, and Curtiss, as shown in Fig. 1.27. In Secs. 1.7 and 1.8, we have already seen the relationship between Langley and the Wrights and the circumstances leading up to the race for the first flight in 1903. This constitutes side A in Fig. 1.27. In this section, we have seen the strong connection between Curtiss and the work of Langley, via Alexander Graham Bell—a close friend and follower of Langley and creator of the Aerial Experiment Association, which gave Curtiss a start in aviation. We have even noted that Charles Manly, Langley's assistant, was the official starter for Curtiss's successful competition for the *Scientific American* trophy. Such relationships form side B of the triangle in Fig. 1.27. Finally, we have seen the relationship, although somewhat acrimonious, between the Wrights and Curtiss, which forms side C in Fig. 1.27.

In 1914 an event occurred which simultaneously involved all three sides of the triangle in Fig. 1.27. When the Langley *Aerodrome* failed for the second time in 1903 (see Fig. 1.26), the wreckage was simply stored away in an unused room in the back of the Smithsonian Institution. When Langley died in 1906, he was replaced as secretary of the Smithsonian by Dr. Charles D. Walcott. Over the ensuing years, Secretary Walcott felt that the Langley *Aerodrome* should be given a third chance. Finally, in 1914, the Smithsonian awarded a grant of $2000 for the repair and flight of the Langley *Aerodrome* to none other than Glenn Curtiss. The *Aerodrome* was shipped to Curtiss's factory in Hammondsport, where not only was it repaired, but also 93 separate technical modifications were made, aerodynamically, structurally, and to the engine. For help during this restoration and modification, Curtiss hired Charles Manly. Curtiss added pontoons to the Langley *Aerodrome* and on May 28, 1914, personally flew the modified aircraft for a distance of 150 ft over Keuka Lake. Figure 1.29 shows a photograph of the Langley Aerodrome in graceful flight over the waters of the lake. Later, the *Aerodrome* was shipped back to the Smithsonian, where it was carefully restored to its original configuration and in 1918 was placed on display in the old Arts and Industries Building. Underneath the *Aerodrome* was placed a plaque reading: "Original Langley flying machine, 1903. The first man-carrying aeroplane in the history of the world capable of sustained free flight." The plaque did *not* mention that the *Aerodrome* demonstrated its sustained-flight capability only after the 93 modifications made by Curtiss in 1914. It is no surprise that Orville Wright was deeply upset by this state of affairs, and this is the principal reason why the original 1903 *Wright Flyer* was not given to the Smithsonian until 1948, the year of Orville's death. Instead, from 1928 to 1948, the *Flyer* resided in the Science Museum in London.

Figure 1.29 The modified Langley *Aerodrome* in flight over Keuka Lake in 1914.

This section ends with two ironies. In 1915, Orville sold the Wright Aeronautical Corporation to a group of New York businesspeople. During the 1920s, this corporation became a losing competitor in aviation. Finally, on June 26, 1929, in a New York office, the Wright Aeronautical Corporation was officially merged with the successful Curtiss Aeroplane and Motor Corporation, forming the Curtiss-Wright Corporation. Thus, ironically, the names of Curtiss and Wright finally came together after all those earlier turbulent years. The Curtiss-Wright Corporation went on to produce numerous famous aircraft, perhaps the most notable being the P-40 of World War II fame. Unfortunately, the company could not survive the lean years immediately after World War II, and its aircraft development and manufacturing ceased in 1948. This leads to the second irony. Although the very foundations of powered flight rest on the work of Orville and Wilbur Wright and Glenn Curtiss, there is not an airplane either produced or in standard operation today that bears the name of either Wright or Curtiss.

1.10 THE PROBLEM OF PROPULSION

During the 19th century, numerous visionaries predicted that manned heavier-than-air flight was inevitable once a suitable power plant could be developed to lift the aircraft off the ground. It was just a matter of developing an engine having enough horsepower while at the same time not weighing too much, that is, an engine with a high horsepower-to-weight ratio. This indeed was the main stumbling block to such people as Stringfellow, Du Temple, and Mozhaiski—the steam engine simply did not fit the bill. Then, in 1860, the Frenchman Jean Joseph Etienne Lenoir built the first practical gas engine. It was a single-cylinder engine, burning ordinary street-lighting gas for fuel. By 1865, 400 of Lenoir's engines were doing odd jobs around Paris. Further improvements in such internal combustion engines came rapidly. In 1876, N. A. Otto and E. Langen of Germany developed the four-cycle engine (the ancestor of all modern automobile engines), which also used gas as a fuel. This led to the simultaneous but separate development in 1885 of the four-cycle gasoline-burning engine by Gottlieb Daimler and Karl Benz, both in Germany. Both Benz and Daimler put their engines in motor cars, and the automobile industry was quickly born. After these "horseless carriages" were given legal freedom of the roads in 1896 in France and Britain, the automobile industry expanded rapidly. Later, this industry was to provide much of the technology and many of the trained mechanics for the future development of aviation.

This development of the gasoline-fueled internal combustion engine was a god-send to aeronautics, which was beginning to gain momentum in the 1890s. In the final analysis, it was the Wright brothers' custom-designed and -constructed gasoline engine that was responsible for lifting their *Flyer I* off the sands of Kill Devil Hills that fateful day in December 1903. A proper aeronautical propulsion device had finally been found.

It is interesting to note that the brotherhood between the automobile and the aircraft industries persists to present day. For example, in June 1926, Ford intro-duced a very successful three-engine, high-wing transport airplane—the Ford 4-AT Trimotor. During World War II, virtually all the major automobile companies built airplane engines and airframes. General Motors still maintains an airplane engine division—the Allison Division in Indianapolis, Indiana—noted for its turboprop de-signs. More recently, automobile designers are turning to aerodynamic streamlining and wind tunnel testing to reduce drag, hence increase fuel economy. Thus, the par-allel development of the airplane and the automobile over the past 100 years has been mutually beneficial.

It can be argued that propulsion has paced every major advancement in the speed of airplanes. Certainly, the advent of the gasoline engine opened the doors to the first successful flight. Then as the power of these engines increased from the 12-hp, Wrights-designed engine of 1903 to the 2200-hp, radial engines of 1945, airplane speeds correspondingly increased from 28 to more than 500 mi/h. Finally, jet and rocket engines today provide enough thrust to propel aircraft at thousands of miles per hour—many times the speed of sound. So throughout the history of manned flight, propulsion has been the key that has opened the doors to flying faster and higher.

1.11   FASTER AND HIGHER

The development of aeronautics in general, and aeronautical engineering in particular, was exponential after the Wrights' major public demonstrations in 1908 and has continued to be so to present day. It is beyond the scope of this book to go into all the details. However, marbled into the engineering text in the following chapters are various historical highlights of technical importance. It is hoped that the following parallel presentations of the fundamentals of aerospace engineering and some of their historical origins will be synergistic and that, in combination with the present chapter, they will give the reader a certain appreciation for the heritage of this profession.

As a final note, the driving philosophy of many advancements in aeronautics since 1903 has been to fly *faster and higher*. This is dramatically evident from Fig. 1.30, which gives the flight speeds for typical aircraft as a function of chronological time. Note the continued push for increased speed over the years and the particular increase in recent years made possible by the jet engine. Singled out in Fig. 1.30 are the winners

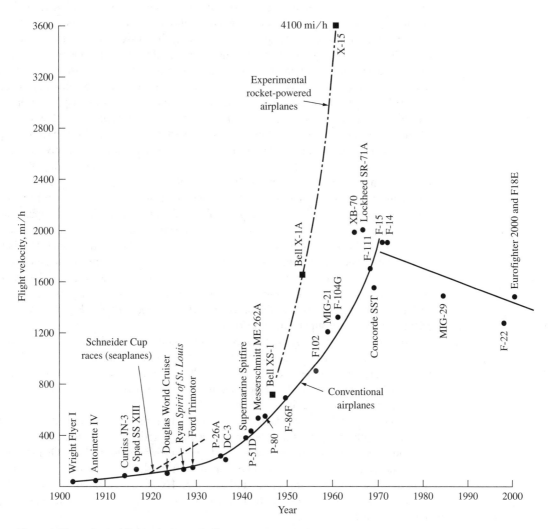

Figure 1.30   Typical flight velocities over the years.

of the Schneider Cup races between 1913 and 1931 (with a moratorium during World War I). The Schneider Cup races were started in 1913 by Jacques Schneider of France as a stimulus to the development of high-speed float planes. They prompted some early but advanced development of high-speed aircraft. The winners are shown by the dashed line in Fig. 1.30, for comparison with standard aircraft of the day. Indeed, the winner of the last Schneider race in 1931 was the *Supermarine S.6B*, a forerunner of the famous Spitfire of World War II. Of course, today the maximum speed of flight has been pushed to the extreme value of 36,000 ft/s, which is the escape velocity from the earth, by the Apollo lunar spacecraft.

Note that the almost exponential increase in speed that occurred from 1903 to 1970 has not continued in recent years. Indeed, the maximum speed of modern military fighters has actually been decreasing since 1970, as shown in Fig. 1.30. This is not due to a degradation in technology, but rather is a reflection that other airplane performance parameters (not speed) are dictating the design. For example, air-to-air combat between opposing fighter airplanes capable of high supersonic speeds quickly degenerates to flying at subsonic or near sonic speeds because of enhanced maneuverability at these lower speeds. Today, fighter airplanes are being optimized for this lower-speed combat arena. On the commercial side, most transport airplanes are subsonic, even the newest (at the time of this writing) such as the Boeing 777. There is only one type of supersonic transport in service, the Anglo-French Concorde. The Concorde was designed with 1960s' technology, and carries a relatively small number of passengers. Hence, it is not profitable. At the time of this writing, there is no commitment from any country to build a second-generation supersonic transport, although in the United States NASA has been carrying out an extensive research program to develop the basic technology for an economical high-speed supersonic transport. Even if an economically viable supersonic transport could be designed, its speed would be limited to about Mach 2.2 or less. Above this Mach number, aerodynamic heating becomes severe enough that titanium rather than aluminum would have to be used for the aircraft skin and for some internal structure. Titanium is expensive and hard to machine; it is not a preferred choice for a new supersonic transport. For these reasons, it is unlikely that the speed curve in Fig. 1.30 will be pushed up by a new supersonic transport.

As a companion to speed, the maximum altitudes of typical manned aircraft are shown in Fig. 1.31 as a function of chronological time. The same push to higher values in the decades between 1903 and 1970 is evident; so far, the record is the moon in 1969. However, the same tendency to plateau after 1970, as in the speed data, can be seen in the altitude data in Fig. 1.31.

Hence, the philosophy of *faster and higher* that has driven aeronautics throughout most of the 20th century is now being mitigated by practical constraints. To this we must add *safer*, *cheaper*, *more reliable*, *and more environmentally clean*. On the other hand, the eventual prospect of hypersonic aircraft (with Mach number greater than 5) in the 21st century is intriguing and exciting. Hypersonic airplanes may well be a new frontier in aeronautics in the future century. See Chap. 11 for a discussion of hypersonic aircraft.

In this chapter, we have only been able to note briefly several important events and people in the historical development of aeronautics. Moreover, there are many

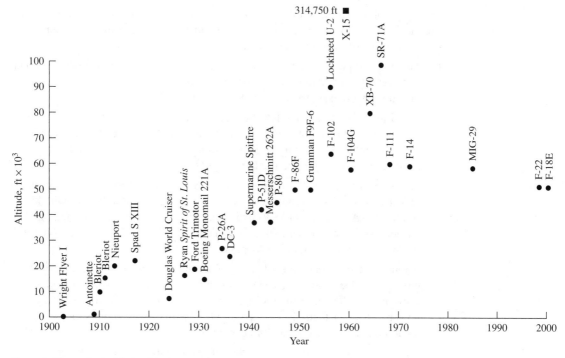

Figure 1.31 Typical flight altitudes over the years.

other places, people, and accomplishments which we simply could not mention in the interest of brevity. Therefore, the reader is urged to consult the short Bibliography at the end of this chapter for additional modern reading on the history of aeronautics.

1.12 SUMMARY

You are about to embark on a study of aerospace engineering. This chapter has presented a short historical sketch of some of the heritage underlying modern aerospace engineering. The major stepping-stones to controlled, heavier-than-air, powered flight with a human pilot are summarized as follows:

1. Leonardo da Vinci conceives the ornithopter and leaves more than 500 sketches of his design, drawn from 1486 to 1490. However, this approach to flight proves to be unsuccessful over the ensuing centuries.

2. The Montgolfier hot-air balloon floats over Paris on November 21, 1783. For the first time in history, a human being is lifted and carried through the air for a sustained period.

3. A red-letter date in the progress of aeronautics is 1799. In that year, Sir George Cayley in England engraves on a silver disk his concept of a fuselage, a fixed

wing, and horizontal and vertical tails. He is the first person to propose separate mechanisms for the generation of lift and propulsion. He is the grandparent of the concept of the modern airplane.

4. The first two powered hops in history are achieved by the Frenchman Felix Du Temple in 1874 and the Russian Alexander F. Mozhaiski in 1884. However, they do not represent truly controlled, sustained flight.

5. Otto Lilienthal designs the first fully successful gliders in history. During the period from 1891 to 1896, he achieves more than 2000 successful glider flights. If he had not been killed in a glider crash in 1896, Lilienthal might have achieved powered flight before the Wright brothers.

6. Samuel Pierpont Langley, secretary of the Smithsonian Institution, achieves the first sustained heavier-than-air, *unmanned*, powered flight in history with his small-scale *Aerodrome* in 1896. However, his attempts at manned flight are unsuccessful, the last one failing on December 8, 1903—just nine days before the Wright brothers' stunning success.

7. Another red-letter date in the history of aeronautics, indeed in the history of humanity, is December 17, 1903. On that day, at Kill Devil Hills in North Carolina, Orville and Wilbur Wright achieve the first controlled, sustained, powered, heavier-than-air, manned flight in history. This flight is to revolutionize life during the 20th century.

8. The development of aeronautics takes off exponentially after the Wright brothers' public demonstrations in Europe and the United States in 1908. The ongoing work of Glenn Curtiss and the Wrights and the continued influence of Langley's early work form an important aeronautical triangle in the development of aeronautics before World War I.

Throughout the remainder of this book, various historical notes will appear, to continue to describe the heritage of aerospace engineering as its technology advances over the 20th century. It is hoped that such historical notes will add a new dimension to your developing understanding of this technology.

BIBLIOGRAPHY

Anderson, John D., Jr.: *A History of Aerodynamics and Its Impact on Flying Machines*, Cambridge University Press, New York, 1997.

Angelucci, E.: *Airplanes from the Dawn of Flight to the Present Day*, McGraw-Hill, New York, 1973.

Combs, H.: *Kill Devil Hill*, Houghton Mifflin, Boston, 1979.

Crouch, T. D.: *The Bishop's Boys*, Norton, New York, 1989.

_____: *A Dream of Wings*, Norton, New York, 1981.

Gibbs-Smith, C. H.: *Sir George Cayley's Aeronautics 1796–1855*, Her Majesty's Stationery Office, London, 1962.

_____: *The Invention of the Aeroplane (1799–1909)*, Faber, London, 1966.

_____: *Aviation: An Historical Survey from Its Origins to the End of World War II*, Her Majesty's Stationery Office, London, 1970.

_____: *Flight Through the Ages*, Crowell, New York, 1974.

The following are a series of small booklets prepared for the British Science Museum by C. H. Gibbs-Smith, published by Her Majesty's Stationery Office, London:

The Wright Brothers, 1963

The World's First Aeroplane Flights, 1965

Leonardo da Vinci's Aeronautics, 1967

A Brief History of Flying, 1967

Sir George Cayley, 1968

Jakab, Peter L.: *Visions of a Flying Machine*, Smithsonian Institution Press, Washington, 1990.
Josephy, A. M., and A. Gordon: *The American Heritage History of Flight*, Simon and Schuster, New York, 1962.
McFarland, Marvin W. (ed.): *The Papers of Wilbur and Orville Wright*, McGraw-Hill, New York.
Roseberry, C. R.: *Glenn Curtiss: Pioneer of Flight*, Doubleday, Garden City, NY, 1972.
Taylor, J. W. R., and K. Munson: *History of Aviation*, Crown, New York, 1972.

chapter

2

Fundamental Thoughts

Engineering: "The application of scientific principles to practical ends." From the Latin word "ingenium," meaning inborn talent and skill, ingenious.

The American Heritage Dictionary of the English Language, 1969

The language of engineering and physical science is a logical collection and assimilation of symbols, definitions, formulas, and concepts. To the average person in the street, this language is frequently esoteric and incomprehensible. In fact, when you become a practicing engineer, do not expect to converse with your spouse across the dinner table about your great technical accomplishments of the day. Chances are that he or she will not understand what you are talking about. The language is intended to convey physical thoughts. It is our way of describing the phenomena of nature as observed in the world around us. It is a language that has evolved over at least 2500 years. For example, in 400 BC, the Greek philosopher Democritus introduced the word and concept of the *atom*, the smallest bit of matter that could not be cut. The purpose of this chapter is to introduce some of the everyday language used by aerospace engineers; in turn, this language will be extended and applied throughout the remainder of this book.

Throughout this book, you will be provided with road maps to guide you through the thoughts and intellectual development that constitute this introduction to flight. Please use these road maps frequently. They will tell you where you are in our discussions, where you have been, and where you are going. For example, Fig. 2.1 is an overall road map for the complete book. Examining this road map, we can obtain an overall perspective for our introduction to flight as presented in this book. First we start out with some necessary preliminaries—some fundamental thoughts that are used

51

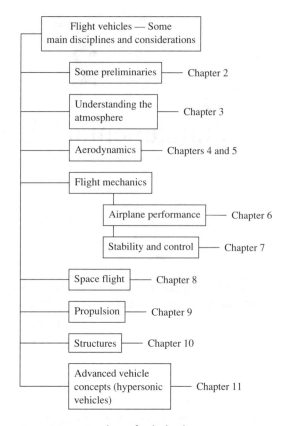

Figure 2.1   Road map for the book.

throughout the remainder of the book. This is the subject of this chapter. Since flight vehicles spend all, or at least some of, their time operating in the atmosphere, next we have to consider the properties of the atmosphere, as discussed in Chap. 3. (Airplanes spend all their time in the atmosphere. Space vehicles have to ascend through the atmosphere to get out to space, and if they carry humans or other payloads that we wish to recover on earth, space vehicles have to descend—at very high speeds—back through the atmosphere.) Now imagine a vehicle flying through the atmosphere. One of the first thoughts that comes to mind is that there is a rush of air over the vehicle. This rush of air generates a force—an aerodynamic force—on the vehicle. A study of *aerodynamics* is the subject of Chaps. 4 and 5. The vehicle itself feels not only this aerodynamic force, but also the force of gravity—its own weight. If the vehicle is powered in some fashion, it will also feel the force (called *thrust*) from the power plant. The vehicle moves under the influence of these forces. The study of the motion of the flight vehicle is labeled *flight dynamics*, which is further divided into considerations of airplane performance (Chap. 6) and stability and control (Chap. 7). In contrast, a space vehicle moving in space will, for all practical purposes, feel only the force of gravity (except when some on-board propulsion device is turned on for trajectory adjustment). The motion of a vehicle in space due to gravitational force is the subject of Chap. 8. Considering again a flight vehicle moving through the atmosphere, there needs to be something to push it along—something to keep it going. This is the function of the engine, which generates thrust to keep the vehicle going. Space vehicles also need engines—to accelerate them into orbit or deep space, and for midcourse trajectory corrections. Engines and how they generate thrust represent the discipline of *propulsion*, the subject of Chap. 9. Additionally, as the flight vehicle moves and responds to the forces acting on it, the physical structure of the vehicle is under a lot of stress and strain. You want this structure to be strong enough to not fall apart under these stresses and strains, but at the same time not to be so heavy as to render the flight vehicle inefficient. We address some aspects of *flight structures* in Chap. 10. All these major disciplines—aerodynamics, flight dynamics, propulsion, and structures—are integrated into the *design* of a flight vehicle. Such design is indeed the final objective of most aerospace research and development. Throughout this book, at appropriate places, we address pertinent aspects of vehicle design. We highlight these aspects by placing them in accented *design boxes*. You cannot miss them in your reading. Finally, looking toward the future, we discuss some advanced vehicle concepts in Chap. 11. All the above discussion is diagrammed in Fig. 2.1. This is the road map for our excursions throughout this book. From time to time, as you proceed through this book, flip back to Fig. 2.1 for a reminder of how the material you are reading fits into the whole scheme.

Returning to our considerations at hand, we look at the road map for this chapter in Fig. 2.2. We treat two avenues of thought in this chapter. As shown in the left column of Fig. 2.2, we examine some basic ideas and definitions that are rooted in

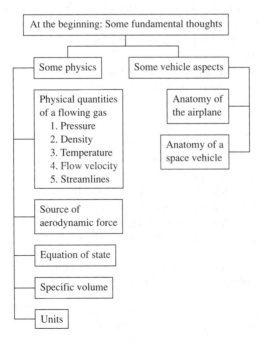

Figure 2.2 Road map for Chap. 2.

physics. These include definitions of the physical quantities of a flowing gas, that is, the language we use to talk about aerodynamics and propulsion. We discuss the fundamental sources of aerodynamic force—how aerodynamic force is exerted on a vehicle. We look at some equations that relate the physical quantities, and we also discuss the mundane (but essential) consideration of units for these physical quantities. We then move to the right column in Fig. 2.2, and discuss some fundamental aspects of flight vehicles themselves, taking a look at the anatomy of typical airplanes and space vehicles.

2.1   FUNDAMENTAL PHYSICAL QUANTITIES OF A FLOWING GAS

As you read through this book, you will soon begin to appreciate that the flow of air over the surface of an airplane is the basic source of the lifting or sustaining force that allows a heavier-than-air machine to fly. In fact, the shape of an airplane is designed to encourage the airflow over the surface to produce a lifting force in the most efficient manner possible. (You will also begin to appreciate that the design of an airplane is in reality a *compromise* between many different requirements, the production of aerodynamic lift being just one.) The science that deals with the flow of air (or, for that matter, the flow of any gas) is called *aerodynamics*, and the person who practices this science is called an *aerodynamicist*. The study of the flow of gases is important in many other aerospace applications, for example, the design of rocket and jet engines, propellers, vehicles entering planetary atmospheres from space, wind tunnels, and rocket and projectile configurations. Even the motion of the global atmosphere and the flow of effluents through smokestacks fall within the realm of aerodynamics. The applications are almost limitless.

Four fundamental quantities in the language of aerodynamics are pressure, density, temperature, and velocity. Let us look at each one.

2.1.1   Pressure

When you hold your hand outside the window of a moving automobile, with your palm perpendicular to the incoming airstream, you can feel the air pressure exerting a force and tending to push your hand rearward, in the direction of the airflow. The *force per unit area* on your palm is defined as the *pressure*. The pressure exists basically because air molecules (oxygen and nitrogen molecules) are striking the surface of your hand and transferring some of their *momentum* to the surface. More precisely,

> Pressure is the *normal* force per unit area exerted on a surface due to the time rate of change of momentum of the gas molecules impacting on that surface.

It is important to note that even though pressure is defined as force per unit area, for example, newtons per square meter or pounds per square foot, you do not need a surface that is actually 1 m² or 1 ft² to talk about pressure. In fact, pressure is usually defined at a point in the gas or a point on a surface and can vary from one point to another. We can use the language of differential calculus to see this more clearly. Referring to Fig. 2.3, we consider a point B in a volume of gas. Let

$$dA = \text{an incremental area around } B$$

$$dF = \text{force on one side of } dA \text{ due to pressure}$$

Then the pressure p at point B in the gas is defined as

$$p = \lim \left(\frac{dF}{dA} \right) \qquad dA \to 0 \qquad \text{[2.1]}$$

Equation (2.1) says that, in reality, the pressure p is the limiting form of the force per unit area where the area of interest has shrunk to zero around point B. In this formalism, it is easy to see that p is a point property and can have a different value from one point to another in the gas.

Pressure is one of the most fundamental and important variables in aerodynamics, as we shall soon see. Common units of pressure are newtons per square meter, dynes per square centimeter, pounds per square foot, and atmospheres. Abbreviations for these quantities are N/m², dyn/cm², lb/ft², and atm, respectively. See App. C for a list of common abbreviations for physical units.

2.1.2 Density

The density of a substance (including a gas) is the mass of that substance per unit volume.

Density will be designated by the symbol ρ. For example, consider air in a room that has a volume of 250 m³. If the mass of the air in the room is 306.25 kg and is evenly

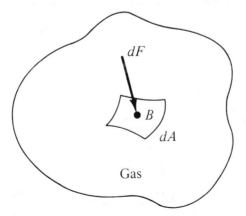

Figure 2.3 Definition of pressure.

distributed throughout the space, then $\rho = 306.25 \text{ kg}/250 \text{ m}^3 = 1.225 \text{ kg/m}^3$ and is the same at every point in the room.

Analogous to the previous discussion of pressure, the definition of density does not require an actual volume of 1 m³ or 1 ft³. Rather, ρ is a point property and can be defined as follows. Referring to Fig. 2.4, we consider point B inside a volume of gas. Let

$$dv = \text{elemental volume around point } B$$
$$dm = \text{mass of gas inside } dv$$

Then ρ at point B is

$$\rho = \lim \left(\frac{dm}{dv} \right) \qquad dv \to 0 \qquad \textbf{[2.2]}$$

Therefore, ρ is the mass per unit volume where the volume of interest has shrunk to zero around point B. The value of ρ can vary from point to point in the gas. Common abbreviated units of density are kg/m³, slug/ft³, g/cm³, and lb_m/ft^3. (The pound mass, lb_m will be discussed in a subsequent section.)

2.1.3 Temperature

Consider a gas as a collection of molecules and atoms. These particles are in constant motion, moving through space and occasionally colliding with one another. Since each particle has motion, it also has kinetic energy. If we watch the motion of a single particle over a long time during which it experiences numerous collisions with its neighboring particles, then we can meaningfully define the average kinetic energy of the particle over this long duration. If the particle is moving rapidly, it has a higher average kinetic energy than if it were moving slowly. The temperature T of the gas is directly proportional to the average molecular kinetic energy. In fact, we can define T as follows:

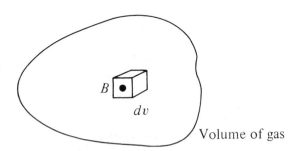

Figure 2.4   Definition of density.

Temperature is a measure of the average kinetic energy of the particles in the gas. If KE is the mean molecular kinetic energy, then temperature is given by $KE = \frac{3}{2}kT$, where k is the Boltzmann constant.

The value of k is 1.38×10^{-23} J/K, where J is an abbreviation for joule.

Hence, we can qualitatively visualize a high-temperature gas as one in which the particles are randomly rattling about at high speeds, whereas in a low-temperature gas, the random motion of the particles is relatively slow. Temperature is an important quantity in dealing with the aerodynamics of supersonic and hypersonic flight, as we shall soon see. Common units of temperature are the kelvin (K), degree Celsius (°C), degree Rankine (°R), and degree Fahrenheit (°F).

2.1.4 Flow Velocity and Streamlines

The concept of speed is commonplace: It represents the distance traveled by some object per unit time. For example, we all know what is meant by traveling at a speed of 55 mi/h down the highway. However, the concept of the velocity of a flowing gas is somewhat more subtle. First, velocity connotes *direction* as well as speed. The automobile is moving at a velocity of 55 mi/h *due north in a horizontal plane*. To designate velocity, we must quote both speed and direction. For a flowing gas, we must further recognize that each region of the gas does not necessarily have the same velocity; that is, the speed and direction of the gas may vary from point to point in the flow. Hence, flow velocity, along with p, ρ, and T, is a point property.

To see this more clearly, consider the flow of air over an airfoil or the flow of combustion gases through a rocket engine, as sketched in Fig. 2.5. To orient yourself, lock your eyes on a specific, infinitesimally small element of mass in the gas, and watch this element move with time. Both the speed and direction of this element (usually called a fluid element) can vary as it moves from point to point in the gas. Now, fix your eyes on a specific fixed point in the gas flow, say, point B in Fig. 2.5. We can now define flow velocity as follows:

> The velocity at any fixed point B in a flowing gas is the velocity of an infinitesimally small fluid element as it sweeps through B.

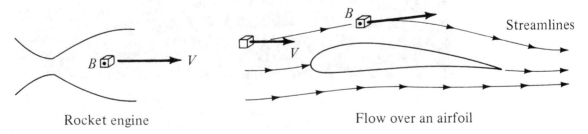

Rocket engine   Flow over an airfoil

Figure 2.5 Flow velocity and streamlines.

Again, we emphasize that velocity is a point property and can vary from point to point in the flow.

Referring again to Fig. 2.5, we note that as long as the flow is steady (as long as it does not fluctuate with time), a moving fluid element is seen to trace out a fixed *path* in space. This path taken by a moving fluid element is called a *streamline* of the flow. Drawing the streamlines of the flow field is an important way of visualizing the motion of the gas; we will frequently sketch the streamlines of the flow about various objects. For example, the streamlines of the flow about an airfoil are sketched in Fig. 2.5 and clearly show the direction of motion of the gas. Figure 2.6 is an actual photograph of streamlines over an airfoil model in a low-speed subsonic wind tunnel. The streamlines are made visible by injection of filaments of smoke upstream of the model; these smoke filaments follow the streamlines in the flow. Using another flow field visualization technique, Fig. 2.7 shows a photograph of a flow where the surface streamlines are made visible by coating the model with a mixture of white pigment in mineral oil. Clearly, the visualization of flow streamlines is a useful aid in the study of aerodynamics.

Figure 2.6   Smoke photograph of the low-speed flow over a Lissaman 7769 airfoil at 10° angle of attack. The Reynolds number based on chord is 150,000. This is the airfoil used on the Gossamer Condor human-powered aircraft. *(The photograph was taken in one of the Notre Dame University smoke tunnels by Dr. T. J. Mueller, Professor of Aerospace Engineering at Notre Dame, and is shown here through his courtesy.)*

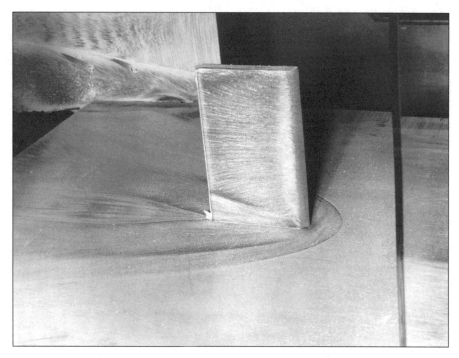

Figure 2.7 An oil streak photograph showing the surface streamline pattern for a fin mounted on a flat plate in supersonic flow. The parabolic curve in front of the fin is due to the bow shock wave and flow separation ahead of the fin. Note how clearly the streamlines can be seen in this complex flow pattern. Flow is from right to left. The Mach number is 5, and the Reynolds number is 6.7×10^6. (Courtesy of Allen E. Winkelmann, University of Maryland, and the Naval Surface Weapons Center.)

2.2 THE SOURCE OF ALL AERODYNAMIC FORCES

We have just discussed the four basic aerodynamic flow quantities: p, ρ, T, and $\mathbf{V}$, where $\mathbf{V}$ is velocity which has both magnitude and direction; that is, velocity is a vector quantity. A knowledge of p, ρ, T, and $\mathbf{V}$ at each point of a flow fully defines the *flow field*. For example, if we were concerned with the flow about a sharp-pointed cone, as shown in Fig. 2.8, we could imagine a cartesian xyz three-dimensional space, where the velocity far ahead of the cone V_∞ is in the x direction and the cone axis is also along the x direction. The specification of the following quantities then fully defines the *flow field*:

$$p = p(x, y, z)$$
$$\rho = \rho(x, y, z)$$
$$T = T(x, y, z)$$
$$\mathbf{V} = \mathbf{V}(x, y, z)$$

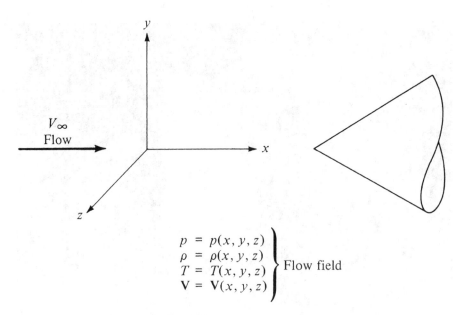

$$p = p(x, y, z)$$
$$\rho = \rho(x, y, z)$$
$$T = T(x, y, z)$$
$$\mathbf{V} = \mathbf{V}(x, y, z)$$

Flow field

Figure 2.8   Specifications of a flow field.

(In practice, the flow field about a right circular cone is more conveniently described in terms of cylindrical coordinates, but we are concerned only with the general ideas here.)

Theoretical and experimental aerodynamicists labor to calculate and measure flow fields of many types. Why? What practical information does knowledge of the flow field yield with regard to airplane design or to the shape of a rocket engine? A substantial part of the first five chapters of this book endeavors to answer these questions. However, the roots of the answers lie in the following discussion.

Probably the most practical consequence of the flow of air over an object is that the object experiences a force, an aerodynamic force, such as your hand feels outside the open window of a moving car. Subsequent chapters discuss the nature and consequences of such aerodynamic forces. The purpose here is to state that the aerodynamic force exerted by the airflow on the surface of an airplane, missile, etc., stems from only two simple natural sources:

1. Pressure distribution on the surface

2. Shear stress (friction) on the surface

We have already discussed pressure. Referring to Fig. 2.9, we see that pressure exerted by the gas on the solid surface of an object always acts *normal* to the surface, as shown by the directions of the arrows. The lengths of the arrows denote the magnitude of the pressure at each local point on the surface. Note that the surface pressure varies with location. The net *unbalance* of the varying pressure distribution over the surface creates a force, an aerodynamic force. The second source, shear stress acting on the

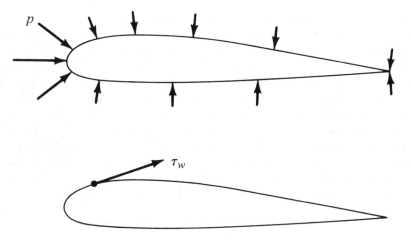

Figure 2.9 Pressure and shear-stress distributions.

surface, is due to the frictional effect of the flow "rubbing" against the surface as it moves around the body. The shear stress τ_w is defined as the force per unit area acting *tangentially* on the surface due to friction, as shown in Fig. 2.9. It is also a point property; it varies along the surface; and the net unbalance of the surface shear stress distribution creates an aerodynamic force on the body. *No matter how complex the flow field, and no matter how complex the shape of the body, the only way nature has of communicating an aerodynamic force to a solid object or surface is through the pressure and shear stress distributions which exist on the surface.* These are the basic fundamental sources of all aerodynamic forces. The pressure and shear-stress distributions are the two hands of nature that reach out and grab the body, exerting a force on the body—the aerodynamic force.

Finally, we can state that a primary function of theoretical and experimental aerodynamics is to predict and measure the aerodynamic forces on a body. In many cases, this implies prediction and measurement of p and τ_w along a given surface. Furthermore, a prediction of p and τ_w on the surface frequently requires knowledge of the complete flow field around the body. This helps to answer our earlier question as to what practical information is yielded by knowledge of the flow field.

2.3 EQUATION OF STATE FOR A PERFECT GAS

Air under normal conditions of temperature and pressure, such as those encountered in subsonic and supersonic flight through the atmosphere, behaves very much as a *perfect gas*. The definition of a perfect gas can best be seen by returning to the molecular picture. A gas is a collection of particles (molecules, atoms, electrons, etc.) in random motion, where each particle is, on average, a long distance away from its neighboring particles. Each molecule has an *intermolecular force field* about it, a ramification of the complex interaction of the electromagnetic properties of the electrons and nucleus. The intermolecular force field of a given particle extends a comparatively long distance through space and changes from a strong repulsive force at close range to a weak attractive force at long range. The intermolecular force field of a given particle reaches out and is felt by the neighboring particles. One one hand, if the neighboring particles are far away, they feel only the tail of the weak attractive force; hence the motion of the neighboring particles is only negligibly affected. On the other hand, if they are close, their motion can be greatly affected by the intermolecular force field. Since the pressure and temperature of a gas are tangible quantities derived from the motion of the particles, then p and T are directly influenced by intermolecular forces, especially when the molecules are packed closely together (i.e., at high densities). This leads to the definition of a perfect gas:

A perfect gas is one in which intermolecular forces are negligible.

Clearly, from the above discussion, a gas in nature in which the particles are widely separated (low densities) approaches the definition of a perfect gas. The air in the room about you is one such case; each particle is separated, on average, by more than 10 molecular diameters from any other. Hence, air at standard conditions can be readily approximated by a perfect gas. Such is also the case for the flow of air about ordinary flight vehicles at subsonic and supersonic speeds. Therefore, in this book, we always deal with a perfect gas for our aerodynamic calculations.

The relation among p, ρ, and T for a gas is called the *equation of state*. For a perfect gas, the equation of state is

$$\boxed{p = \rho R T} \qquad \text{[2.3]}$$

where R is the specific gas constant, the value of which varies from one type of gas to another. For normal air we have

$$R = 287 \frac{\text{J}}{\text{(kg)(K)}} = 1716 \frac{\text{ft} \cdot \text{lb}}{\text{(slug)(°R)}}$$

It is interesting that the *deviation* of an actual gas in nature from perfect gas behavior can be expressed approximately by the modified Berthelot equation of state:

$$\frac{p}{\rho R T} = 1 + \frac{ap}{T} - \frac{bp}{T^3}$$

where a and b are constants of the gas. Thus, the deviations become smaller as p decreases and T increases. This makes sense, because if p is high, the molecules are packed closely together, intermolecular forces become important, and the gas behaves less as a perfect gas. On the other hand, if T is high, the molecules move faster. Thus their average separation is larger, intermolecular forces become less significant in comparison to the inertia forces of each molecule, and the gas behaves more as a perfect gas.

Also, note that when the air in the room around you is heated to temperatures above 2500 K, the oxygen molecules begin to dissociate (tear apart) into oxygen atoms; at temperatures above 4000 K, the nitrogen begins to dissociate. For these temperatures, air becomes a *chemically reacting gas*, such that its chemical composition becomes a function of both p and T; that is, it is no longer normal air. As a result, R in Eq. (2.3) becomes a variable—$R = R(p, T)$—simply because the gas composition is changing. The perfect gas equation of state, Eq. (2.3), is still valid for such a case, except that R is no longer a constant. This situation is encountered in very high speed flight, for example, the atmospheric entry of the Apollo capsule, in which case the temperatures in some regions of the flow field reach 11,000 K.

Again, in this book, we always assume that air is a perfect gas, obeying Eq. (2.3), with a constant $R = 287$ J/(kg)(K) or 1716 ft $\cdot$ lb/(slug)(°R).

2.4 DISCUSSION OF UNITS

Physical units are vital to the language of engineering. In the final analysis, the end result of most engineering calculations or measurements is a number which represents some physical quantity, for example, pressure, velocity, or force. The number is given in terms of combinations of units, for example, 10^5 N/m^2, 300 m/s, or 5 N, where the newton, meter, and second are examples of *units*. (See App. C.)

Historically, various branches of engineering have evolved and favored systems of units which seemed to most conveniently fit their needs. These various sets of "engineering" units usually differ among themselves and are different from the metric system, preferred for years by physicists and chemists. In the modern world of technology, where science and engineering interface on almost all fronts, such duplicity and variety of units have become an unnecessary burden. Metric units are now the accepted norm in both science and engineering in most countries outside the United States. More importantly, in 1960 the Eleventh General Conference on Weights and Measures defined and officially established the *Système International d'Unités*—the SI units—which was adopted as the preferred system of units by 36 participating countries, including the United States. Since then, the United States has made progress toward the voluntary implementation of SI units in engineering. For example, several NASA (National Aeronautics and Space Administration) laboratories have made SI units virtually mandatory for all results reported in technical reports, although engineering units can be shown as a duplicate set. The AIAA (American Institute of Aeronautics and Astronautics) has made a policy of encouraging SI units for all papers reported in their technical journals. It is apparent that in a few decades, the United States, along with the rest of the world, will be using SI units almost exclusively.

For these reasons, students who prepare themselves for practicing engineering must do "double duty" with regard to familiarization with units. They must be familiar with the old engineering units in order to use the vast bulk of existing technical literature quoted in such units. At the same time, they must be intimately familiar with the SI units for future work and publications; that is, the next generation of engineers must be bilingual with regard to units.

To promote fluency in both the engineering and SI units, this book incorporates both sets. However, you should adopt the habit of working all your problems using SI units from the start, and when results are requested in engineering units, *convert* your final SI results to the desired units. This will give you every chance to develop a "feel" for SI units. It is important that you develop a natural feeling for SI units; for example, you should feel as at home with pressures in newtons per square meter as you probably already do with pounds per square inch (psi). A mark of successful experienced engineers is their feel for correct magnitudes of physical quantities in familiar units. Make the SI your familiar units.

For all practical purposes, SI is a metric system based on the meter, kilogram, second, and kelvin as basic units of length, mass, time, and temperature, respectively. It is a *coherent*, or *consistent*, set of units. Such consistent sets of units allow physical relationships to be written without the need for "conversion factors" in the basic formulas. For example, in a consistent set of units, Newton's second law can be written

$$F = m \times a$$

$$\text{Force} = \text{mass} \times \text{acceleration}$$

In SI units,

$$F = ma$$

$$1 \text{ newton} = (1 \text{ kilogram})(1 \text{ meter/second}^2)$$

[2.4]

The newton is a force defined such that it accelerates a mass of 1 kilogram by 1 meter per second squared.

The English engineering system of units is another consistent set of units. Here the basic units of mass, length, time, and temperature are the slug, foot, second, and degree Rankine, respectively. In this system,

$$F = ma$$

$$1 \text{ pound} = (1 \text{ slug})(1 \text{ foot/second}^2)$$

[2.5]

The pound is a force defined such that it accelerates a mass of 1 slug by 1 foot per second squared. Note that in both systems, Newton's second law is written simply as $F = ma$, with no conversion factor on the right-hand side.

In contrast, a nonconsistent set of units defines force and mass such that Newton's second law must be written with a conversion factor, or constant, as

$$F = \underset{\underset{\text{factor}}{\underset{\uparrow}{\text{Conversion}}}}{\frac{1}{g_c}} \times \underset{\underset{\text{Mass}}{\uparrow}}{m} \times \underset{\underset{\text{Acceleration}}{\uparrow}}{a}$$

$\underset{\underset{\text{Force}}{\uparrow}}{}$

A nonconsistent set of units frequently used in the past by mechanical engineers includes the pound force, pound mass, foot, and second, whereby

$$g_c = 32.2 \ (lb_m)(ft)/(s^2)(lb_f)$$

$$F = \frac{1}{g_c} \quad m \quad \times \quad a \qquad \qquad \textbf{[2.6]}$$

$$\uparrow \quad \uparrow \quad \uparrow \qquad \uparrow$$

$$lb_f \quad \tfrac{1}{32.2} \quad lb_m \qquad ft/s^2$$

In this nonconsistent system, the unit of mass is the pound mass lb_m. Comparing Eqs. (2.5) and (2.6), we see that 1 slug $= 32.2 \ lb_m$. A slug is a large hunk of mass, whereas the pound mass is considerably smaller, by a factor of 32.2. This is illustrated in Fig. 2.10.

Already, you can sense how confusing the various sets of units can become, especially nonconsistent units. Confusion on the use of Eq. (2.6) with the g_c factor has resulted in many heat exchangers and boilers being designed 32.2 times either too large or too small. On the other hand, the use of a consistent set of units, such as in Eqs. (2.4) and (2.5), which have no conversion factor, eliminates such confusion. This is one of the beauties of SI units.

For this reason, we will always deal with a consistent set of units in this book. We will use both the SI units from Eq. (2.4) and the English engineering units from Eq. (2.5). As stated before, you will frequently encounter the engineering units in the existing literature, whereas you will be seeing SI units with increasing frequency in the future literature; that is, you must become bilingual. To summarize, we will deal with the English engineering system units—lb, slug, ft, s, °R—and the Système International (SI) units—N, kg, m, s, K.

Therefore, returning to the equation of state, Eq. (2.3), where $p = \rho RT$, we see that the units are as follows:

	English Engineering System	SI
p	lb/ft^2	N/m^2
ρ	$slugs/ft^3$	kg/m^3
T	°R	K
R (for air)	$1716 \ ft \cdot lb/(slug)(°R)$	$287 \ J/(kg)(K)$

There are two final points about units to note. First, the units of a physical quantity can frequently be expressed in more than one combination simply by appealing to Newton's second law. From Newton's law, the relation between N, kg, m, and s is

$$F = ma$$

$$N = kg \cdot m/s^2$$

Thus, a quantity such as $R = 287 \ J/(kg)(K)$ can also be expressed in an equivalent way as

$$R = 287\frac{J}{(kg)(K)} = 287\frac{N \cdot m}{(kg)(K)} = 287\frac{kg \cdot m}{s^2}\frac{m}{(kg)(K)} = 287\frac{m^2}{(s^2)(K)}$$

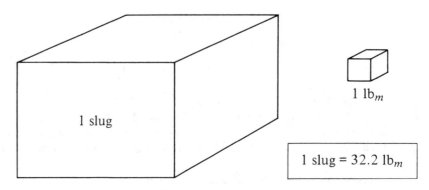

Figure 2.10   Comparison between the slug and pound mass.

Thus, R can also be expressed in the equivalent terms of velocity squared divided by temperature. In the same vein,

$$R = 1716\frac{\text{ft} \cdot \text{lb}}{(\text{slug})(^\circ\text{R})} = 1716\frac{\text{ft}^2}{(\text{s}^2)(^\circ\text{R})}$$

Second, in the equation of state, Eq. (2.3), T is always the *absolute* temperature, where zero degrees is the absolutely lowest temperature possible. Both K and °R are absolute temperature scales, where $0^\circ\text{R} = 0\text{ K} =$ temperature at which almost all molecular translational motion theoretically stops. On the other hand, the familiar Fahrenheit (°F) and Celsius (°C) scales are *not* absolute scales. Indeed,

$$0^\circ\text{F} = 460^\circ\text{R}$$
$$0^\circ\text{C} = 273\text{ K} = 32^\circ\text{F}$$

For example,   90°F is same as $460 + 90 = 550^\circ\text{R}$ **[2.7]**

and   10°C is same as $273 + 10 = 283\text{ K}$ **[2.8]**

Please remember: T in Eq. (2.3) *must* be the absolute temperature, either kelvins or degrees Rankine.

2.5 SPECIFIC VOLUME

Density ρ is the *mass per unit volume*. The inverse of this quantity is also frequently used in aerodynamics. It is called the *specific volume* v and is defined as the *volume per unit mass*. By definition,

$$v = \frac{1}{\rho}$$

Hence, from the equation of state

$$p = \rho R T = \frac{1}{v} R T$$

we also have

$$\boxed{pv = RT}$$ [2.9]

Abbreviated units for v are m³/kg and ft³/slug.

The air pressure and density at a point on the wing of a Boeing 747 are 1.10×10^5 N/m² and | **Example 2.1**
1.20 kg/m³, respectively. What is the temperature at that point?

Solution

From Eq. (2.3), $p = \rho R T$, hence $T = p/(\rho R)$, or

$$T = \frac{1.10 \times 10^5 \text{ N/m}^2}{(1.20 \text{ kg/m}^3)[287 \text{ J/(kg)(K)}]} = \boxed{319 \text{ K}}$$

The high-pressure air storage tank for a supersonic wind tunnel has a volume of 1000 ft³. If | **Example 2.2**
air is stored at a pressure of 30 atm and a temperature of 530°R, what is the mass of gas stored
in the tank in slugs? In pound mass?

Solution

$$1 \text{ atm} = 2116 \text{ lb/ft}^2$$

Hence, $p = (30)(2116)$ lb/ft² $= 6.348 \times 10^4$ lb/ft². Also, from Eq. (2.3), $p = \rho R T$, hence
$\rho = (p/RT)$, or

$$\rho = \frac{6.348 \times 10^4 \text{ lb/ft}^2}{[1716 \text{ ft} \cdot \text{lb/(slug)(}^\circ\text{R)](530}^\circ\text{R)}} = 6.98 \times 10^{-2} \text{ slug/ft}^3$$

This is the density, which is mass *per unit volume*. The total mass M in the tank of volume V
is

$$M = \rho V = (6.98 \times 10^{-2} \text{ slug/ft}^3)(100 \text{ ft}^3) = \boxed{69.8 \text{ slugs}}$$

Recall that 1 slug = 32.2 lb$_m$. Hence

$$M = (69.8)(32.2) = \boxed{2248 \text{ lb}_m}$$

Air flowing at high speed in a wind tunnel has pressure and temperature equal to 0.3 atm and | **Example 2.3**
-100°C, respectively. What is the air density? What is the specific volume?

Solution

$$1 \text{ atm} = 1.01 \times 10^5 \text{ N/m}^2$$

Hence

$$p = (0.3)(1.01 \times 10^5) = 0.303 \times 10^5 \text{ N/m}^2$$

Note that $T = -100°\text{C}$ is *not* an absolute temperature. Hence

$$T = -100 + 273 = 173 \text{ K}$$

From Eq. (2.3), $p = \rho RT$, hence $\rho = p/(RT)$, or

$$\rho = \frac{0.303 \times 10^5 \text{ N/m}^2}{[287 \text{ J/(kg)(K)}](173 \text{ K})} = \boxed{0.610 \text{ kg/m}^3}$$

$$v = \frac{1}{\rho} = \frac{1}{0.610} = \boxed{1.64 \text{ m}^3/\text{kg}}$$

Note: It is worthwhile to remember that

$$\begin{aligned} 1 \text{ atm} &= 2116 \text{ lb/ft}^2 \\ 1 \text{ atm} &= 1.01 \times 10^5 \text{ N/m}^2 \end{aligned}$$

2.6 ANATOMY OF THE AIRPLANE

In regard to fundamental thoughts, it is appropriate to discuss some basic nomenclature associated with airplanes and space vehicles—names for the machine themselves. In this section we deal with airplanes; space vehicles are discussed in the next section.

The major components of a conventional airplane are identified in Fig. 2.11. The *fuselage* is the center body, where most of the usable volume of the airplane is found. The fuselage carries people, baggage, other payload, instruments, fuel, and anything else that the airplane designer puts there. The *wings* are the main lift-producing components of the airplanes; the left and right wings are so identified as you would see them from inside the airplane, facing forward. The internal volume of the wings can be used for such items as fuel tanks and storage of the main landing gear (the wheels and supporting struts) after the gear is retracted. The horizontal and vertical *stabilizers* are located and sized so as to provide the necessary stability for the airplane in flight (we consider stability in Chap. 7). Sometimes these surfaces are called the horizontal and vertical *tails*, or *fins*. When the engines are mounted from the wings, as shown in Fig. 2.11, they are usually housed in a type of shroud called a *nacelle*. As a historical note, the French worked hard on flying machines in the late 19th and early 20th centuries; as a result, some of our conventional airplane nomenclature today comes from the French. *Fuselage* is a French word, meaning a "spindle" shape. So is the word *nacelle*, meaning a "small boat."

Figure 2.11   Basic components of an airplane.

Flaps and control surfaces are highlighted in Fig. 2.12. These are hinged surfaces, usually at the trailing edge (the back edge) of the wings and tail, that can be rotated up or down. The function of a flap is to increase the lift force on the airplane; flaps are discussed in detail in Sec. 5.17. Some aircraft are designed with flaps at the leading edge (the front edge) of the wings as well as at the trailing edge. Leading-edge flaps are not shown in Fig. 2.12. The ailerons are control surfaces that control the rolling motion of the airplane around the fuselage. For example, when the left aileron is deflected downward and the right aileron is deflected upward, lift is increased on the left wing and decreased on the right wing, causing the airplane to roll to the right. The elevators are control surfaces that control the nose up-and-down pitching motion; when the elevator is deflected downward, the lift on the tail is increased, pulling the tail up and the nose of the airplane down. The rudder is a control surface that can turn

Figure 2.12   Control surfaces and flaps.

the nose of the airplane to the right or left (called *yawing*). The nature and function of these control surfaces are discussed in greater detail in Chap. 7.

In aeronautics, it is common to convey the shape of an airplane by means of a *three-view* diagram, such as that shown in Fig. 2.13. Proceeding from the top to the bottom of Fig. 2.13, we see a front view, top view, and side view, respectively, of the North American F-86H, a famous jet fighter from the Korean war era. A three-view diagram is particularly important in the design process of a new airplane because it conveys the precise shape and dimensions of the aircraft.

DESIGN BOX

This is the first of many design boxes in this book. These design boxes highlight information pertinent to the philosophy, process, and details of flight vehicle design, as related to the local discussion at that point in the text. The purpose of these design boxes is to reflect on the design implications of various topics being discussed. This is not a book on design. But the fundamental information in this book certainly has applications to design. These design boxes are here to bring these applications to your attention. Design is a vital function, indeed usually the end product, of engineering. These design boxes can give you a better understanding of aerospace *engineering*.

This design box is associated with our discussion of the anatomy of the airplane and three-views. An example of a much more detailed three-view diagram is that in Fig. 2.14, which shows the Vought F4U Corsair, the famous Navy fighter from World War II. Figure 2.14 is an example of what, in the airplane design process, is called a *configuration layout*. In Fig. 2.14 we see not only the front view, side view, top view, and bottom view of the airplane, but also the detailed dimensions, the cross-sectional shape of the fuselage at different locations, the airfoil shape of the wing at different locations, landing gear details, and the location of various lights, radio antenna, etc. (A discussion of the role of the configuration layout in airplane design can be found in Anderson, *Aircraft Performance and Design*, McGraw-Hill, New York, 1999.)

The internal structure of an airplane is frequently illustrated by a *cutaway* drawing, such as that shown in Fig. 2.15. Here the famous Boeing B-17 bomber from World War II is shown with a portion of its skin cut away so that the internal structure is visible. Although the B-17 is a late 1930s' design, it is shown here because of its historical significance and because it represents a conventional airplane structure. Cutaway diagrams usually contain many details about the internal structure and packaging for the airplane.

Any student of the history of aeronautics knows that airplanes have been designed with a wide variety of shapes and configurations. It is generally true that *form follows function*, and airplane designers have configured their aircraft to meet specific requirements. However, airplane design is an open-ended problem—there is no single "right way" or "right configuration" to achieve the design goals. Also, airplane design is an exercise in compromise; to achieve good airplane performance in one category, other aspects of performance may have to be partly sacrificed. For example, an airplane designed for very high speed may have poor landing and takeoff

Figure 2.13 Three-view of the North American F-86H.

performance. A design feature that optimizes the aerodynamic characteristics may overly complicate the structural design. Convenient placement of the engines may disrupt the aerodynamics of the airplane. And so forth. For this reason, airplanes come in all sizes and shapes. An exhaustive listing of all the different types of airplane

Figure 2.14   Vought F4U-1D Corsair. Drawing by Paul Matt. (*Courtesy of Aviation Heritage, Inc., Destin, FL.*)
(*continued*)

(continued)

CLEAR NAV LIGHT

16'6"

32.50"

33.8 O4"

2° OFFSET

NOTE: TO PRONOUNCED INBOARD ANHEDRAL DUE TO OUTBOARD DIHEDRAL WINGS DRAWN IN PERSPECTIVE - TRUE MECHANICAL PLAN VIEW LAYOUT OF OUTER PANEL SHOWN IN DOTTED LINE.

6 feet
5
4
3
2
1
0
2 meters

SCALE BARS

GUN ACCESS PANELS

HAND GRIP

NON SKID STEP PAD

CUT OUT STEP

HAND GRIP

AMMO. BELTS ACCESS PANELS

¾" CLEARANCE, NORMAL TOLERANCE

7'6" ACTUAL

POSITION OF FORMATION LIGHT, WHEN EMPLOYED

71.4"

105"

GUN CAMERA

APPROACH LIGHT

LANDING LIGHT

4'10'

7"

R

R

R

(concluded)

Figure 2.15   Cutaway drawing of the Boeing B-17. (From Bill Gunston, *Classic World War II Aircraft Cutaways, Osprey Publishing, London, England, 1995.*)

configurations is not our purpose here. Over the course of your studies and work, you will sooner or later encounter most of these types. However, there are several general classes of airplane configurations that we do mention here.

The first is the *conventional configuration*. This is exemplified by the aircraft shown in Figs. 2.11 through 2.15. Here we see *monoplanes* (a single set of wings) with a horizontal and vertical tail at the back of the aircraft. The aircraft may have a *straight wing*, as seen in Figs. 2.11, 2.12, 2.14, and 2.15, or a *swept wing*, as seen in Fig. 2.13. Wing sweep is a design feature that reduces the aerodynamic drag at speeds near to or above the speed of sound, and that is why most high-speed aircraft today have some type of swept wing. Swept wings are discussed in greater detail in Sec. 5.16. It is an idea that goes back as far as 1935.

Figure 2.13 illustrates an airplane with a *swept-back* wing. Aerodynamically, the same benefit can be obtained by sweeping the wing forward. Figure 2.16 is a three-view of the X-29A, a research aircraft with a *swept-forward* wing. Swept-forward wings are not a new idea. However, swept-forward wings have combined aerodynamic and structural features that tend to cause the wing to twist and fail structurally. This is why most swept-wing airplanes have used swept-back wings. With the new, high-strength composite materials of today, swept-forward wings can be designed strong enough to resist this problem; the swept-forward wing of the X-29A is a composite wing. There are some advantages aerodynamically to a swept-forward wing, which are discussed in Sec. 5.16. Also note by comparing Figs. 2.13 and 2.16 that the juncture of the wing and the fuselage is farther back on the fuselage for the airplane with a swept-forward wing than for an airplane with a swept-back wing. At the wing-fuselage juncture, there is extra structure (such as a wing spar that goes through the fuselage) that can interfere with the internal packaging in the fuselage. The swept-forward wing configuration, with its more rearward fuselage-wing juncture, can allow the airplane designer greater flexibility in placing the internal

Figure 2.16   Three-view of the Grumman X-29A research aircraft.

packaging inside the fuselage. In spite of these advantages, at the time of writing, no new civilian transports or military airplanes are being designed with swept-forward wings.

The X-29A shown in Fig. 2.16 illustrates another somewhat unconventional feature—the horizontal stabilizer is mounted ahead of the wing rather than at the rear of the airplane. This is defined as a *canard* configuration, and the horizontal stabilizer in this location is called a *canard surface*. The 1903 *Wright Flyer* was a canard design, as clearly seen in Figs. 1.1 and 1.2. However, other airplane designers after the Wrights quickly placed the horizontal stabilizer at the rear of the airplane. (There is some evidence that this was done more to avoid patent difficulties with the Wrights than for technical reasons.) The rear horizontal tail location is part of the conventional aircraft configuration; it has been used on the vast majority of airplane designs since the *Wright Flyer*. One reason for this is the feeling among some designers that a canard surface has a destabilizing effect on the airplane (to call the canard a horizontal "stabilizer" might be considered by some a misnomer). However, a properly designed canard configuration can be just as stable as a conventional configuration. This is discussed in detail in Chap. 7. Indeed, there are some inherent advantages of the canard configuration, as we outline in Chap. 7. Because of this, a number of new canard airplanes have been designed in recent years, ranging from private, general aviation airplanes to military, high-performance fighters. (The word *canard* comes from the French word for duck.)

Look again at the *Wright Flyer* in Figs. 1.1 and 1.2. This aircraft has two wings mounted one above the other. The Wrights called this a *double-decker* configuration. However, within a few years such a configuration was called a *biplane*, nomenclature that persists to the present. In contrast, airplanes with just one set of wings are called *monoplanes*; Figs. 2.11 through 2.16 illustrate monoplanes, which have become the *conventional configuration*. However, this was not true through the 1930s; until about 1935, biplanes were the conventional configuration. Figure 2.17 is a three-view of the Grumman F3F-2 biplane designed in 1935. It was the U.S. Navy's last biplane fighter; it was in service as a front-line fighter with the Navy until 1940. The popularity of biplanes over monoplanes in the earlier years was mainly due to the enhanced structural strength of two shorter wings trussed together compared to that of a single, longer-span wing. However, as the cantilevered wing design, introduced by the German engineer Hugo Junkers as early as 1915, gradually became more accepted, the main technical reason for the biplane evaporated. However, old habits are sometimes hard to change, and the biplane remained in vogue far longer than any technical reason would justify. Today, biplanes still have some advantages as sport aircraft for aerobatics and as agricultural spraying aircraft. So the biplane design still lives on.

2.7 ANATOMY OF A SPACE VEHICLE

In Sec. 2.6, we discussed the conventional airplane configuration. In contrast, it is difficult to define a "conventional" spacecraft configuration. The shape, size, and arrangement of a space vehicle are determined by its particular mission, and there

Figure 2.17   Three-view of the Grumman F3F-2, the last U.S. Navy biplane fighter.

are as many (if not more) different spacecraft configurations as there are missions. In this section we discuss a few of the better-known space vehicles; although our coverage is far from complete, it does provide some perspective on the anatomy of space vehicles.

To date, all human-made space vehicles are launched into space by rocket boosters. A rather conventional booster is the Delta three-stage rocket, shown in Fig. 2.18. Built by McDonnell-Douglas (now merged with Boeing), the Delta rocket is a product of a long design and development evolution that can be traced to the Thor intermediate-range ballistic missile in the late 1950s. The spacecraft to be launched into space is housed inside a fairing at the top of the booster, which falls away af-

Figure 2.18   Delta 3914 and 3920 rocket booster configurations. (*From M. D. Griffin and J. R. French, Space Vehicle Design, AIAA: Reston, Virginia, 1991.*)

ter the booster is out of the earth's atmosphere. The rocket booster is really three rockets mounted on top of one another. The technical reasons for having such a multistage booster (as opposed to a single-stage rocket) are discussed in Sec. 9.10. Also, the fundamentals of the rocket engines that power these boosters are discussed in Chap. 9.

A not-so-conventional booster is the air-launched Pegasus, shown in Fig. 2.19. The Pegasus is a three-stage rocket that is carried aloft by an airplane. The booster is then launched from the airplane at some altitude within the sensible atmosphere. The first stage of the Pegasus has wings, which assist in boosting the rocket to higher altitudes within the sensible atmosphere.

The Delta rocket in Fig. 2.18 and the Pegasus in Fig. 2.19 are examples of *expendable launch vehicles*; no part of these boosters is recovered for reuse. There are certain economies to be realized by recovering part of (if not all) the booster and using it again. There is great interest today in such *recoverable launch vehicles*. An example of such a vehicle is the experimental X-34, shown in Fig. 2.20. This is basically a winged booster that will safely fly back to earth after it has launched its payload, to be used again for another launch.

In a sense, the Space Shuttle is partly a reusable system. The Space Shuttle is part airplane and part space vehicle. The Space Shuttle flight system is shown in Fig. 2.21. The shuttle orbiter is the airplanelike configuration that sits on the side of the rocket booster. The system is powered by two *solid rocket boosters* (SRBs) that burn out and are jettisoned after the first 2 min of flight. The SRBs are recovered and refurbished for use again. The external tank carries liquid oxygen and liquid hydrogen for the main propulsion system, which comprises the rocket engines mounted in the orbiter. The external tank is jettisoned just before the system goes into orbit; the tank falls back through the atmosphere and is destroyed. The orbiter carries on with its mission in space. When the mission is complete, the orbiter reenters the atmosphere and glides back to earth, making a horizontal landing, as a conventional unpowered airplane would.

*Optional configurations are also available.
†Optional fourth stage available for precision injection.

Figure 2.19   Orbital Sciences Pegasus, an air-launched rocket booster. (*From C. H. Eldred et al., "Future Space Transportation Systems and Launch," in Future Aeronautical and Space Systems, eds. A. K. Noor and S. L. Vennera, AIAA, Progress in Astronautics and Aeronautics, vol. 172, 1997.*)

Figure 2.20   Orbital Sciences X-34 small reusable rocket booster. (*From Eldred et al.*)

Figure 2.21   The Space Shuttle. (*From Griffin and French.*)

Let us now examine the anatomy of the payload itself—the functioning spacecraft that may be a satellite in orbit around earth or a deep-space vehicle on its way to another planet or to the sun. As mentioned earlier, these spacecraft are point designs for different specific missions, and therefore it is difficult to define a conventional configuration for spacecraft. However, let us examine the anatomy of a few of these point designs, just to obtain some idea of their nature.

A communications satellite is shown in Fig. 2.22. This is the FLTSATCOM spacecraft produced by TRW for the U.S. Navy. It is placed in a geostationary orbit—an orbit in the plane of the equator with a period (time to execute one orbit) of 24 h. Hence, a satellite in geostationary orbit appears above the same location on earth at all times—a desirable feature for a communications satellite. Orbits and trajectories for space vehicles are discussed in Chap. 8. The construction is basically aluminum. The two hexagonal compartments (buses) mounted one above the other at the center of the satellite contain all the engineering subsystems necessary for control and communications. The two antennas that project outward from the top of the bus are pointed at earth. The two solar array arms (solar panels) that project from the sides of the bus constantly rotate to remain pointed at the sun at all times. The solar panels provide power to run the equipment on the spacecraft.

The Mars Pathfinder spacecraft is sketched in Figs. 2.23 and 2.24. This spacecraft successfully landed on the surface of Mars in 1997. The package that entered the Martian atmosphere is shown in an exploded view in Fig. 2.24. The aeroshell and backshell make up the aerodynamic shape of the entry body, with the lander packaged in a folded position inside. The function of this aerodynamic entry body is to create drag to slow the vehicle as it approaches the surface of Mars, and to protect the package inside from aerodynamic heating during atmospheric entry. The dynamics of a spacecraft entering a planetary atmosphere, and entry aerodynamic heating, are discussed in Chap. 8. Figure 2.23 shows the Pathfinder lander after it has been deployed on the Martian surface. The rover, solar panel, high-gain and low-gain

NORMAL ORBIT: GEO-STATIONARY   IN ORBITAL OPERATION
LAUNCH: 1978   THE SPACECRAFT IS 3-AXIS STABILIZED
                    ETR   WITH THE BODY-FIXED ANTENNA
                    ATLAS-CENTAUR   POINTING CONSTANTLY AT THE EARTH
                    CENTAUR STANDARD FAIRING   AND THE SOLAR ARRAY ROTATED TO
                                                            POINT AT THE SUN
SPACECRAFT WEIGHT   4100 LB

Figure 2.22   The TRW communications satellite FLTSATCOM. (*From Griffin and French.*)

Figure 2.23 The Mars Pathfinder on the surface of Mars. (*From M. K. Olsen et al., "Spacecraft for Solar System Exploration," in Future Aeronautical and Space Systems, eds. A. K. Noor and S. L. Venneri, AIAA Progress in Astronautics and Aeronautics, vol. 172, 1997.*)

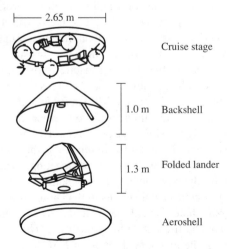

Figure 2.24 Components of the Mars Pathfinder space vehicle. (*From Olsen et al.*)

antennas, and imager for taking the transmitting pictures from the surface are shown in Fig. 2.23.

2.8 HISTORICAL NOTE: THE NACA AND NASA

Let us pick up the thread of aeronautical engineering history from Chap. 1. After Orville and Wilbur Wright's dramatic public demonstrations in the United States and Europe in 1908, there was a virtual explosion in aviation developments. In turn, this rapid progress had to be fed by new technical research in aerodynamics, propulsion, structures, and flight control. It is important to realize that then, as well as today, aeronautical research was sometimes expensive, always demanding in terms of intellectual talent, and usually in need of large testing facilities. Such research in many cases either was beyond the financial resources of, or seemed too out of the ordinary for, private industry. Thus, the fundamental research so necessary to fertilize and pace the development of aeronautics in the 20th century had to be established and nurtured by national governments. It is interesting to note that George Cayley himself (see Chap. 1) as long ago as 1817 called for "public subscription" to underwrite the expense of the development of airships. Responding about 80 years later, the British government set up a school for ballooning and military kite flying at Farnborough, England. By 1910, the Royal Aircraft Factory was in operation at Farnborough with the noted Geoffrey de Havilland as its first airplane designer and test pilot. This was the first major government aeronautical facility in history. This operation was soon to evolve into the Royal Aircraft Establishment (RAE), which today is still conducting viable aeronautical research for the British government.

In the United States, aircraft development as well as aeronautical research languished after 1910. During the next decade, the United States embarrassingly fell far behind Europe in aeronautical progress. This set the stage for the U.S. government to establish a formal mechanism for pulling itself out of its aeronautical "dark ages." On March 3, 1915, by an act of Congress, the National Advisory Committee for Aeronautics (NACA) was created, with an initial appropriation of $5000 per year for 5 years. This was at first a true committee, consisting of 12 distinguished members who were knowledgeable about aeronautics. Among the charter members in 1915 were Professor Joseph S. Ames of Johns Hopkins University (later to become president of Johns Hopkins) and Professor William F. Durand of Stanford University, both of whom were to make major impressions on aeronautical research in the first half century of powered flight. This advisory committee, NACA, was originally to meet annually in Washington, District of Columbia, on "the Thursday after the third Monday of October of each year," with any special meetings to be called by the chair. Its purpose was to advise the government on aeronautical research and development and to bring some cohesion to such activities in the United States.

The committee immediately noted that a single advisory group of 12 members was not sufficient to breathe life into U.S. aeronautics. Their insight is apparent in the letter of submittal for the first annual report of NACA in 1915, which contained the following passage:

There are many practical problems in aeronautics now in too indefinite a form to enable their solution to be undertaken. The committee is of the opinion that one of the first and most important steps to be taken in connection with the committee's work is the provision and equipment of a flying field together with aeroplanes and suitable testing gear for determining the forces acting on full-sized machines in constrained and in free flight, and to this end the estimates submitted contemplate the development of such a technical and operating staff, with the proper equipment for the conduct of full-sized experiments.

It is evident that there will ultimately be required a well-equipped laboratory specially suited to the solving of those problems which are sure to develop, but since the equipment of such a laboratory as could be laid down at this time might well prove unsuited to the needs of the early future, it is believed that such provision should be the result of gradual development.

So the first action of this advisory committee was to call for major government facilities for aeronautical research and development. The clouds of war in Europe—World War I had started a year earlier—made their recommendations even more imperative. In 1917, when the United States entered the conflict, actions followed the committee's words. We find the following entry in the third annual NACA report: "To carry on the highly scientific and special investigations contemplated in the act establishing the committee, and which have, since the outbreak of the war, assumed greater importance, and for which facilities do not already exist, or exist in only a limited degree, the committee has contracted for a research laboratory to be erected on the Signal Corps Experimental Station, Langley Field, Hampton, Virginia." The report goes on to describe a single, two-story laboratory building with physical, chemical, and structural testing laboratories. The building contract was for $80,900; actual construction began in 1917. Two wind tunnels and an engine test stand were contemplated "in the near future." The selection of a site 4 mi north of Hampton, Virginia, was based on general health conditions and the problems of accessibility to Washington and the larger industrial centers of the east, protection from naval attack, climatic conditions, and cost of the site.

Thus, the Langley Memorial Aeronautical Research Laboratory was born. It was to remain the only NACA laboratory and the only major U.S. aeronautical laboratory of any type for the next 20 years. Named after Samuel Pierpont Langley (see Sec. 1.7), it pioneered in wind tunnel and flight research. Of particular note is the airfoil and wing research performed at Langley during the 1920s and 1930s. We return to the subject of airfoils in Chap. 5, at which time the reader should note that the airfoil data included in App. D were obtained at Langley. With the work which poured out of the Langley laboratory, the United States took the lead in aeronautical development. High on the list of accomplishments, along with the systematic testing of airfoils, was the development of the NACA engine cowl (see Sec. 6.19), an aerodynamic fairing built around radial piston engines which dramatically reduced the aerodynamic drag of such engines.

In 1936, Dr. George Lewis, who was then NACA Director of Aeronautical Research (a position he held from 1924 to 1947), toured major European laboratories. He noted that NACA's lead in aeronautical research was quickly disappearing, especially in light of advances being made in Germany. As World War II drew close, NACA clearly recognized the need for two new laboratory operations: an advanced aerodynamics laboratory to probe into the mysteries of high-speed (even supersonic) flight and a major engine-testing laboratory. These needs eventually led to the construction of Ames Aeronautical Laboratory at Moffett Field, near Mountain View, California (authorized in 1939), and Lewis Engine Research Laboratory at Cleveland, Ohio (authorized in 1941). Along with Langley, these two new NACA laboratories again helped to spearhead the United States to the forefront of aeronautical research and development in the 1940s and 1950s.

The dawn of the space age occurred on October 4, 1957, when Russia launched *Sputnik I*, the first artificial satellite to orbit the earth. Swallowing its somewhat embarrassed technical pride, the United States moved quickly to compete in the race for space. On July 29, 1958, by another act of Congress (Public Law 85-568), the National Aeronautics and Space Administration (NASA) was born. At this same moment, NACA came to an end. Its programs, people, and facilities were instantly transferred to NASA, lock, stock, and barrel. However, NASA was a larger organization than just the old NACA; it absorbed in addition numerous Air Force, Navy, and Army projects for space. Within two years of its birth, NASA was authorized four new major installations: an existing Army facility at Huntsville, Alabama, renamed the George C. Marshall Space Flight Center; the Goddard Space Flight Center at Greenbelt, Maryland; the Manned Spacecraft Center (now the Johnson Spacecraft Center) in Houston, Texas; and the Launch Operations Center (now the John F. Kennedy Space Center) at Cape Canaveral, Florida. These, in addition to the existing but slightly renamed Langley, Ames, and Lewis research centers, were the backbone of NASA. Thus, the aeronautical expertise of NACA now formed the seeds for NASA, shortly thereafter to become one of the world's most important forces in space technology.

This capsule summary of the roots of NACA and NASA is included in this chapter on fundamental thoughts because it is virtually impossible for a student or practitioner of aerospace engineering in the United States not to be influenced or guided by NACA or NASA data and results. The extended discussion on airfoils in Chap. 5 is a case in point. Thus, NACA and NASA are "fundamental" to the discipline of aerospace engineering, and it is important to have some impression of the historical roots and tradition of these organizations. Hopefully, this short historical note provides such an impression. A much better impression can be obtained by taking a journey through the NACA and NASA technical reports in the library, going all the way back to the first NACA report in 1915. In so doing, a panorama of aeronautical and space research through the years will unfold in front of you.

2.9 SUMMARY

Some of the major ideas in this chapter are listed below.

1. The language of aerodynamics involves pressure, density, temperature, and velocity. In turn, the illustration of the velocity field can be enhanced by drawing streamlines for a given flow.

2. The source of all aerodynamic forces on a body is the pressure distribution and the shear-stress distribution over the surface.

3. A perfect gas is one in which intermolecular forces can be neglected. For a perfect gas, the equation of state which relates p, ρ, and T is

$$p = \rho R T \qquad \textbf{[2.3]}$$

where R is the specific gas constant.

4. To avoid confusion, errors, and a number of unnecessary "conversion factors" in the basic equations, always use consistent units. In this book, SI units (newton, kilogram, meter, second) and the English engineering system (pound, slug, foot, second) are used.

BIBLIOGRAPHY

Anderson, John D., Jr.: *Aircraft Performance and Design*, WCB/McGraw-Hill, New York, 1999.

Gray, George W.: *Frontiers of Flight*, Knopf, New York, 1948.

Griffin, Michael D., and James R. French: *Space Vehicle Design*, American Institute of Aeronautics and Astronautics, Reston, VA, 1991.

Hartman, E. P.: *Adventures in Research: A History of Ames Research Center 1940–1965*, NASA SP-4302, 1970.

Mechtly, E. A.: *The International System of Units*, NASA SP-7012, 1969.

PROBLEMS

2.1 Consider the low-speed flight of the Space Shuttle as it is nearing a landing. If the air pressure and temperature at the nose of the shuttle are 1.2 atm and 300 K, respectively, what are the density and specific volume?

2.2 Consider 1 kg of helium at 500 K. Assuming that the total internal energy of helium is due to the mean kinetic energy of each atom summed over all the atoms, calculate the internal energy of this gas. *Note*: The molecular weight of helium is 4. Recall from chemistry that the molecular weight is the mass per mole of gas; that is, 1 mol of helium contains 4 kg of mass. Also, 1 mol of any gas contains 6.02×10^{26} molecules or atoms (Avogadro's number).

2.3 Calculate the weight of air (in pounds) contained within a room 20 ft long, 15 ft wide, and 8 ft high. Assume standard atmospheric pressure and temperature of 2116 lb/ft^2 and 59°F, respectively.

2.4 Comparing with the case of Prob. 2.3, calculate the percentage change in the total weight of air in the room when the air temperature is reduced to $-10°F$ (a very cold winter day), assuming the pressure remains the same at 2116 lb/ft^2.

2.5 If 1500 lb$_m$ of air is pumped into a previously empty 900 ft^3 storage tank and the air temperature in the tank is uniformly 70°F, what is the air pressure in the tank in atmospheres?

2.6 In Problem 2.5, assume the rate at which air is being pumped into the tank is 0.5 lb_m/s. Consider the instant in time at which there is 1000 lb_m of air in the tank. Assume the air temperature is uniformly 50°F at this instant and is increasing at the rate of 1°F/min. Calculate the rate of change of pressure at this instant.

2.7 Assume that, at a point on the wing of the Concorde supersonic transport, the air temperature is −10°C and the pressure is 1.7×10^4 N/m². Calculate the density at this point.

2.8 At a point in the test section of a supersonic wind tunnel, the air pressure and temperature are 0.5×10^5 N/m² and 240 K, respectively. Calculate the specific volume.

2.9 Consider a flat surface in an aerodynamic flow (say a flat sidewall of a wind tunnel). The dimensions of this surface are 3 ft in the flow direction (the x direction) and 1 ft perpendicular to the flow direction (the y direction). Assume that the pressure distribution (in pounds per square foot) is given by $p = 2116 - 10x$ and is independent of y. Assume also that the shear-stress distribution (in pounds per square foot) is given by $\tau_w = 90/(x + 9)^{1/2}$ and is independent of y. In the above expressions, x is in feet, and $x = 0$ at the front of the surface. Calculate the magnitude and direction of the net aerodynamic force on the surface.

2.10 A pitcher throws a baseball at 85 miles per hour. The flowfield over the baseball moving through the stationary air at 85 miles per hour is the same as that over a stationary baseball in an airstream that approaches the baseball at 85 miles per hour. (This is the principle of wind tunnel testing, as will be discussed in Chapter 4.) This picture of a stationary body with the flow moving over it, is what we adopt here. Neglecting friction, the theoretical expression for the flow velocity over the surface of a sphere (hence the baseball) is $V = \frac{3}{2}V_\infty \sin \theta$. Here, V_∞ is the airstream velocity (the freestream velocity far ahead of the sphere). An arbitrary point on the surface of the sphere is located by the intersection of the radius of the sphere with the surface, and θ is the angular position of the radius measured from a line through the center in the direction of the freestream (i.e., the most forward and rearward points on the spherical surface correspond to $\theta = 0°$ and 180° respectively. The velocity V is the flow velocity at that arbitrary point on the surface. Calculate the values of the minimum and maximum velocity at the surface, and the location of the points at which these occur.

2.11 Consider an ordinary, helium-filled party balloon with a volume of 2.2 ft³. The lifting force on the balloon due to the outside air is the net resultant of the pressure distribution exerted on the exterior surface of the balloon. Using this fact, Archimedes principle can be derived, namely that the upward force on the balloon is equal to the weight of the air displaced by the balloon. Assuming the balloon is at sea level, where the air density is 0.002377 slug/ft³, calculate the maximum weight that can be lifted by the balloon. Note: The molecular weight of air is 28.8 and that of helium is 4.

3

The Standard Atmosphere

Sometimes gentle, sometimes capricious, sometimes awful, never the same for two moments together; almost human in its passions, almost spiritual in its tenderness, almost divine in its infinity.

John Ruskin, *The Sky*

Aerospace vehicles can be divided into two basic categories: atmospheric vehicles such as airplanes and helicopters, which always fly within the sensible atmosphere, and space vehicles such as satellites, the Apollo lunar vehicle, and deep-space probes, which operate outside the sensible atmosphere. However, space vehicles do encounter the earth's atmosphere during their blastoffs from the earth's surface and again during their reentries and recoveries after completion of their missions. If the vehicle is a planetary probe, then it may encounter the atmospheres of Venus, Mars, Jupiter, etc. Therefore, during the design and performance of any aerospace vehicle, the properties of the atmosphere must be taken into account.

The earth's atmosphere is a dynamically changing system, constantly in a state of flux. The pressure and temperature of the atmosphere depend on altitude, location on the globe (longitude and latitude), time of day, season, and even solar sunspot activity. To take all these variations into account when considering the design and performance of flight vehicles is impractical. Therefore, a *standard atmosphere* is defined in order to relate flight tests, wind tunnel results, and general airplane design and performance to a common reference. The standard atmosphere gives mean values of pressure, temperature, density, and other properties as functions of altitude; these values are obtained from experimental balloon and sounding-rocket measurements combined with a mathematical model of the atmosphere. To a reasonable degree, the standard atmosphere reflects average atmospheric conditions, but this is not its main importance. Rather, its main function is to provide tables of common reference conditions that can be used in an organized fashion by aerospace engineers everywhere. The purpose of this chapter is to give you some feeling for what the standard atmosphere is all about and how it can be used for aerospace vehicle analyses.

We might pose the rather glib question: *Just what is the standard atmosphere?* A rather glib answer is: *The tables in Apps. A and B at the end of this book.* Take a look at these two appendixes. They tabulate the temperature, pressure, and density for different altitudes. Appendix A is in SI units, and App. B is in English engineering units. Where do these numbers come from? Were they simply pulled out of thin air by somebody in the distant past? Absolutely not. The numbers in these tables were obtained on a rational, scientific basis. One purpose of this chapter is to develop this rational basis. Another purpose is to show you how to use these tables.

The road map for this chapter is given in Fig. 3.1. We first run down the left side of the road map, establishing some definitions and an equation from basic physics (the hydrostatic equation) which are necessary tools for constructing the numbers in the standard atmosphere tables. Then we move to the right side of the road map and discuss how the numbers in the tables are actually obtained. We go through the construction of the standard atmosphere in detail. Finally, we define some terms that are derived from the numbers in the tables—the pressure, density, and temperature altitudes—which are in almost everyday use in aeronautics.

Finally, we note that the details of this chapter are focused on the determination of the standard atmosphere for earth. The tables in Apps. A and B are for the earth's atmosphere. However, the physical principles and techniques discussed in this chapter are also applicable to constructing model atmospheres for other planets, such as Venus, Mars, and Jupiter. So the applicability of this chapter reaches far beyond the earth.

It should be mentioned that several different standard atmospheres exist, compiled by different agencies at different times, each using slightly different experimental data in the models. For all practical purposes, the differences are insignificant below 30 km (100,000 ft), which is the domain of contemporary airplanes. A standard atmosphere in common use is the 1959 ARDC model atmosphere. (ARDC stands for the U.S. Air Force's previous Air Research and Development Command, which

Figure 3.1   Road map for Chap. 3.

is now the Air Force Systems Command.) The atmospheric tables used in this book are taken from the 1959 ARDC model atmosphere.

3.1 DEFINITION OF ALTITUDE

Intuitively, we all know the meaning of altitude. We think of it as the distance above the ground. But like so many other general terms, it must be more precisely defined for quantitative use in engineering. In fact, in the following sections we define and use six different altitudes: absolute, geometric, geopotential, pressure, temperature, and density altitudes.

First, imagine that we are at Daytona Beach, Florida, where the ground is at sea level. If we could fly straight up in a helicopter and drop a tape measure to the ground, the measurement on the tape would be, by definition, the geometric altitude h_G, that is, the geometric height above sea level.

Now, if we bored a hole through the ground to the center of the earth and extended our tape measure until it hit the center, then the measurement on the tape would be, by definition, the *absolute altitude* h_a. If r is the radius of the earth, then $h_a = h_G + r$. This is illustrated in Fig. 3.2.

The absolute altitude is important, especially for space flight, because the local acceleration of gravity g varies with h_a. From Newton's law of gravitation, g varies inversely as the square of the distance from the center of the earth. By letting g_0 be

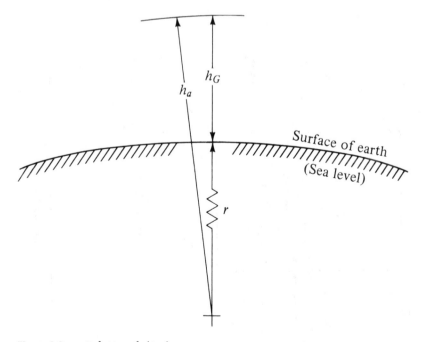

Figure 3.2   Definition of altitude.

the gravitational acceleration at *sea level*, the local gravitational acceleration g at a given absolute altitude h_a is

$$g = g_0 \left(\frac{r}{h_a} \right)^2 = g_0 \left(\frac{r}{r + h_G} \right)^2 \qquad \text{[3.1]}$$

The variation of g with altitude must be taken into account when you are dealing with mathematical models of the atmosphere, as discussed in the following sections.

3.2   HYDROSTATIC EQUATION

We will now begin to piece together a model which will allow us to calculate variations of p, ρ, and T as functions of altitude. The foundation of this model is the hydrostatic equation, which is nothing more than a force balance on an element of fluid at rest. Consider the small stationary fluid element of air shown in Fig. 3.3. We take for convenience an element with rectangular faces, where the top and bottom faces have sides of unit length and the side faces have an infinitesimally small height dh_G. On the bottom face, the pressure p is felt, which gives rise to an upward force of $p \times 1 \times 1$ exerted on the fluid element. The top face is slightly higher in altitude (by the distance dh_G), and because pressure varies with altitude, the pressure on the top face will be slightly different from that on the bottom face, differing by the infinitesimally small value dp. Hence, on the top face, the pressure $p + dp$ is felt. It gives rise to a

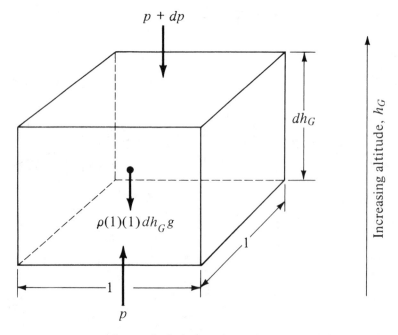

Figure 3.3   Force diagram for the hydrostatic equation.

downward force of $(p + dp)(1)(1)$ on the fluid element. Moreover, the volume of the fluid element is $(1)(1) dh_G = dh_G$, and since ρ is the mass per unit volume, then the mass of the fluid element is simply $\rho(1)(1)dh_G = \rho \, dh_G$. If the local acceleration of gravity is g, then the weight of the fluid element is $g\rho \, dh_G$, as shown in Fig. 3.3. The three forces shown in Fig. 3.3, pressure forces on the top and bottom, and the weight must balance because the fluid element is not moving. Hence

$$p = p + dp + \rho g \, dh_G$$

Thus

$$\boxed{dp = -\rho g \, dh_G} \qquad \textbf{[3.2]}$$

Equation (3.2) is the *hydrostatic equation* and applies to any fluid of density ρ, for example, water in the ocean as well as air in the atmosphere.

Strictly speaking, Eq. (3.2) is a differential equation; that is, it relates an infinitesimally small change in pressure dp to a corresponding infinitesimally small change in altitude dh_G, where in the language of differential calculus, dp and dh_G are differentials. Also note that g is a variable in Eq. (3.2); g depends on h_G as given by Eq. (3.1).

To be made useful, Eq. (3.2) should be integrated to give what we want, namely, the variation of pressure with altitude $p = p(h_G)$. To simplify the integration, we make the *assumption* that g is constant throughout the atmosphere, equal to its value at sea level g_0. This is something of a historical convention in aeronautics. At the altitudes encountered during the earlier development of human flight (less than 15 km or 50,000 ft), the variation of g is negligible. Hence, we can write Eq. (3.2) as

$$dp = -pg_0 \, dh \qquad \textbf{[3.3]}$$

However, to make Eqs. (3.2) and (3.3) numerically identical, the altitude h in Eq. (3.3) must be slightly different from h_G in Eq. (3.2), to compensate for the fact that g is slightly different from g_0. Suddenly, we have defined a new altitude h, which is called the *geopotential altitude* and which differs from the geometric altitude. For the practical mind, geopotential altitude is a "fictitious" altitude, defined by Eq. (3.3) for ease of future calculations. However, many standard atmosphere tables quote their results in terms of geopotential altitude, and care must be taken to make the distinction. Again, geopotential altitude can be thought of as that fictitious altitude which is physically compatible with the assumption of $g = \text{const} = g_0$.

3.3 RELATION BETWEEN GEOPOTENTIAL AND GEOMETRIC ALTITUDES

We still seek the variation of p with geometric altitude $p = p(h_G)$. However, our calculations using Eq. (3.3) will give, instead, $p = p(h)$. Therefore, we need to relate h to h_G, as follows. Dividing Eq. (3.3) by (3.2), we obtain

$$1 = \frac{g_0}{g} \frac{dh}{dh_G}$$

or
$$dh = \frac{g}{g_0} dh_G \qquad \text{[3.4]}$$

We substitute Eq. (3.1) into (3.4):

$$dh = \frac{r^2}{(r + h_G)^2} dh_G \qquad \text{[3.5]}$$

By convention, we set both h and h_G equal to zero at sea level. Now, consider a given point in the atmosphere. This point is at a certain geometric altitude h_G, and associated with it is a certain value of h (different from h_G). Integrating Eq. (3.5) between sea level and the given point, we have

$$\int_0^h dh = \int_0^{h_G} \frac{r^2}{(r + h_G)^2} dh_G = r^2 \int_0^{h_G} \frac{dh_G}{(r + h_G)^2}$$

$$h = r^2 \left(\frac{-1}{r + h_G} \right)_0^{h_G} = r^2 \left(\frac{-1}{r + h_G} + \frac{1}{r} \right) = r^2 \left(\frac{-r + r + h_G}{(r + h_G)r} \right)$$

Thus
$$\boxed{h = \frac{r}{r + h_G} h_G} \qquad \text{[3.6]}$$

where h is geopotential altitude and h_G is geometric altitude. This is the desired relation between the two altitudes. When we obtain relations such as $p = p(h)$, we can use Eq. (3.6) to subsequently relate p to h_G.

A quick calculation using Eq. (3.6) shows that there is little difference between h and h_G for low altitudes. For such a case, $h_G \ll r$, $r/(r + h_G) \approx 1$, hence $h \approx h_G$. Putting in numbers, $r = 6.356766 \times 10^6$ m (at a latitude of 45°), and if $h_G = 7$ km (about 23,000 ft), then the corresponding value of h is, from Eq. (3.6), $h = 6.9923$ km, about 0.1 of 1 percent difference! Only at altitudes above 65 km (213,000 ft) does the difference exceed 1 percent. (Note that 65 km is an altitude at which aerodynamic heating of NASA's Space Shuttle becomes important during reentry into the earth's atmosphere from space.)

3.4 DEFINITION OF THE STANDARD ATMOSPHERE

We are now in a position to obtain p, T, and ρ as functions of h for the standard atmosphere. The keystone of the standard atmosphere is a *defined* variation of T with altitude, based on experimental evidence. This variation is shown in Fig. 3.4.

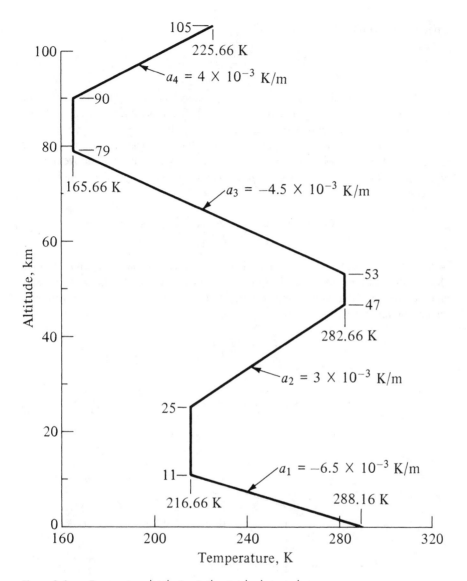

Figure 3.4 Temperature distribution in the standard atmosphere.

Note that it consists of a series of straight lines, some vertical (called the constant-temperature, or *isothermal*, regions) and others inclined (called the *gradient* regions). Given $T = T(h)$ as *defined* by Fig. 3.4, then $p = p(h)$ and $\rho = \rho(h)$ follow from the laws of physics, as shown below.

First, consider again Eq. (3.3):

$$dp = -\rho g_0 \, dh$$

Divide by the equation of state, Eq. (2.3):

$$\frac{dp}{p} = -\frac{\rho g_0\, dh}{\rho RT} = -\frac{g_0}{RT} dh \qquad\qquad \textbf{[3.7]}$$

Consider first the isothermal (constant-temperature) layers of the standard atmosphere, as given by the vertical lines in Fig. 3.4 and sketched in Fig. 3.5. The temperature, pressure, and density at the base of the isothermal layer shown in Fig. 3.5 are T_1, p_1, and ρ_1, respectively. The base is located at a given geopotential altitude h_1. Now consider a given point in the isothermal layer above the base, where the altitude is h. The pressure p at h can be obtained by integrating Eq. (3.7) between h_1 and h.

$$\int_{p_1}^{p} \frac{dp}{p} = -\frac{g_0}{RT} \int_{h_1}^{h} dh \qquad\qquad \textbf{[3.8]}$$

Note that g_0, R, and T are constants that can be taken outside the integral. (This clearly demonstrates the simplification obtained by assuming that $g = g_0 = $ const, and therefore dealing with geopotential altitude h in the analysis.) Performing the integration in Eq. (3.8), we obtain

$$\ln \frac{p}{p_1} = -\frac{g_0}{RT}(h - h_1)$$

or

$$\boxed{\frac{p}{p_1} = e^{-[g_0/(RT)](h-h_1)}} \qquad\qquad \textbf{[3.9]}$$

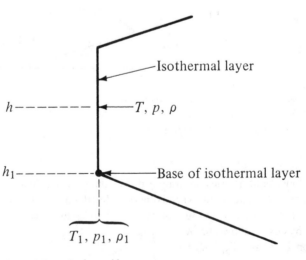

Figure 3.5   Isothermal layer.

From the equation of state

$$\frac{p}{p_1} = \frac{\rho T}{\rho_1 T_1} = \frac{\rho}{\rho_1}$$

Thus

$$\boxed{\frac{\rho}{\rho_1} = e^{-[g_0/(RT)](h-h_1)}}$$

[3.10]

Equations (3.9) and (3.10) give the variation of p and ρ versus geopotential altitude for the isothermal layers of the standard atmosphere.

Considering the gradient layers, as sketched in Fig. 3.6, we find the temperature variation is linear and is geometrically given as

$$\frac{T - T_1}{h - h_1} = \frac{dT}{dh} \equiv a$$

where a is a *specified* constant for each layer obtained from the defined temperature variation in Fig. 3.4. The value of a is sometimes called the *lapse rate* for the gradient layers.

$$a \equiv \frac{dT}{dh}$$

Thus

$$dh = \frac{1}{a}dT$$

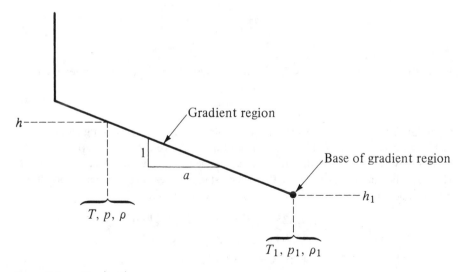

Figure 3.6 Gradient layer.

We substitute this result into Eq. (3.7):

$$\frac{dp}{p} = -\frac{g_0}{aR}\frac{dT}{T} \qquad\qquad \text{[3.11]}$$

Integrating between the base of the gradient layer (shown in Fig. 3.6) and some point at altitude h, also in the gradient layer, Eq. (3.11) yields

$$\int_{p_1}^{p}\frac{dp}{p} = -\frac{g_0}{aR}\int_{T_1}^{T}\frac{dT}{T}$$

$$\ln\frac{p}{p_1} = -\frac{g_0}{aR}\ln\frac{T}{T_1}$$

Thus

$$\frac{p}{p_1} = \left(\frac{T}{T_1}\right)^{-g_0/(aR)} \qquad\qquad \text{[3.12]}$$

From the equation of state

$$\frac{p}{p_1} = \frac{\rho T}{\rho_1 T_1}$$

Hence, Eq. (3.12) becomes

$$\frac{\rho T}{\rho_1 T_1} = \left(\frac{T}{T_1}\right)^{-g_0/(aR)}$$

$$\frac{\rho}{\rho_1} = \left(\frac{T}{T_1}\right)^{-[g_0/(aR)]-1}$$

or

$$\frac{\rho}{\rho_1} = \left(\frac{T}{T_1}\right)^{-\{[g_0/(aR)]+1\}} \qquad\qquad \text{[3.13]}$$

Recall that the variation of T is linear with altitude and is given the specified relation

$$T = T_1 + a(h - h_1) \qquad\qquad \text{[3.14]}$$

Equation (3.14) gives $T = T(h)$ for the gradient layers; when it is plugged into Eq. (3.12), we obtain $p = p(h)$; similarly from Eq. (3.13) we obtain $\rho = \rho(h)$.

Now we can see how the standard atmosphere is pieced together. Looking at Fig. 3.4, start at sea level ($h = 0$), where standard sea level values of pressure, density, and temperature—p_s, ρ_s, and T_s, respectively—are

$$p_s = 1.01325 \times 10^5 \text{ N/m}^2 = 2116.2 \text{ lb/ft}^2$$
$$\rho_s = 1.2250 \text{ kg/m}^3 = 0.002377 \text{ slug/ft}^3$$
$$T_s = 288.16 \text{ K} = 518.69°\text{R}$$

These are the base values for the first gradient region. Use Eq. (3.14) to obtain values of T as a function of h until $T = 216.66$ K, which occurs at $h = 11.0$ km. With

DESIGN BOX

The first step in the design process of a new aircraft is the determination of a set of *specifications*, or *requirements*, for the new vehicle. These specifications may include such performance aspects as a stipulated maximum velocity at a given altitude, or a stipulated maximum rate of climb at a given altitude. These performance parameters depend on the aerodynamic characteristics of the vehicle, such as lift and drag. In turn, the lift and drag depend on the properties of the atmosphere. When the specifications dictate certain performance at a given altitude, this altitude is taken to be the standard altitude in the tables. Therefore, in the preliminary design of an airplane, the designer uses the standard atmosphere tables to define the pressure, temperature, and density at the given altitude. In this fashion, many calculations made during the preliminary design of an airplane contain information from the standard altitude tables.

these values of T, use Eqs. (3.12) and (3.13) to obtain the corresponding values of p and ρ in the first gradient layer. Next, starting at $h = 11.0$ km as the base of the first isothermal region (see Fig. 3.4), use Eqs. (3.9) and (3.10) to calculate values of p and ρ versus h, until $h = 25$ km, which is the base of the next gradient region. In this manner, with Fig. 3.4 and Eqs. (3.9), (3.10), and (3.12) to (3.14), a table of values for the standard atmosphere can be constructed.

Such a table is given in App. A for SI units and App. B for English engineering units. Look at these tables carefully and become familiar with them. They *are* the standard atmosphere. The first column gives the geometric altitude, and the second column gives the corresponding geopotential altitude obtained from Eq. (3.6). The third through fifth columns give the corresponding standard values of temperature, pressure, and density, respectively, for each altitude, obtained from the discussion above.

We emphasize again that the standard atmosphere is a reference atmosphere only and certainly does not predict the actual atmospheric properties that may exist at a given time and place. For example, App. A says that at an altitude (geometric) of 3 km, $p = 0.70121 \times 10^5$ N/m^2, $T = 268.67$ K, and $\rho = 0.90926$ kg/m^3. In reality, situated where you are, if you could right now levitate yourself to 3 km above sea level, you would most likely feel a p, T, and ρ different from the above values obtained from App. A. The standard atmosphere allows us only to reduce test data and calculations to a convenient, agreed-upon reference, as will be seen in subsequent sections of this book.

Calculate the standard atmosphere values of T, p, and ρ at a geopotential altitude of 14 km. | **Example 3.1**

Solution

Remember that T is a *defined* variation for the standard atmosphere. Hence, we can immediately refer to Fig. 3.4 and find that at $h = 14$ km,

$$T = 216.66 \text{ K}$$

To obtain p and ρ, we must use Eqs. (3.9) to (3.14), piecing together the different regions from sea level up to the given altitude with which we are concerned. Beginning at sea level, the first region (from Fig. 3.4) is a gradient region from $h = 0$ to $h = 11.0$ km. The lapse rate is

$$a = \frac{dT}{dh} = \frac{216.66 - 288.16}{11.0 - 0} = -6.5 \text{ K/km}$$

or
$$a = -0.0065 \text{ K/m}$$

Therefore, using Eqs. (3.12) and (3.13), which are for a gradient region and where the base of the region is sea level (hence $p_1 = 1.01 \times 10^5$ N/m² and $\rho_1 = 1.23$ kg/m³), we find at $h = 11.0$ km

$$p = p_1 \left(\frac{T}{T_1}\right)^{-g_0/(aR)} = (1.01 \times 10^5) \left(\frac{216.66}{288.16}\right)^{-9.8/[-0.0065(287)]}$$

where $g_0 = 9.8$ m/s² in SI units. Hence, p (at $h = 11.0$ km) $= 2.26 \times 10^4$ N/m².

$$\rho = \rho_1 \left(\frac{T}{T_1}\right)^{-[g_0/(aR)+1]}$$

$$= (1.23) \left(\frac{216.66}{288.16}\right)^{-\{9.8/[-0.0065(287)]+1\}}$$

$$= 0.367 \text{ kg/m}^3 \quad \text{at } h = 11.0 \text{ km}$$

The above values of p and ρ now form the *base* values for the first isothermal region (see Fig. 3.4). The equations for the isothermal region are Eqs. (3.9) and (3.10), where now $p_1 = 2.26 \times 10^4$ N/m² and $\rho_1 = 0.367$ kg/m³. For $h = 14$ km, $h - h_1 = 14 - 11 = 3$ km $= 3000$ m. From Eq. (3.9),

$$p = p_1 e^{-[g_0/(RT)](h-h_1)} = (2.26 \times 10^4)e^{-[9.8/287(216.66)](3000)}$$

$$\boxed{p = 1.41 \times 10^4 \text{ N/m}^2}$$

From Eq. (3.10),

$$\frac{\rho}{\rho_1} = \frac{p}{p_1}$$

Hence
$$\rho = \rho_1 \frac{p}{p_1} = 0.367\frac{1.41 \times 10^4}{2.26 \times 10^4} = \boxed{0.23 \text{ kg/m}^3}$$

These values check, within roundoff error, with the values given in App. A. *Note*: This example demonstrates how the numbers in Apps. A and B are obtained!

3.5 PRESSURE, TEMPERATURE, AND DENSITY ALTITUDES

With the tables of Apps. A and B in hand, we can now define three new "altitudes"— pressure, temperature, and density altitudes. This is best done by example. Imagine that you are in an airplane flying at some real, geometric altitude. The value of your actual altitude is immaterial for this discussion. However, at this altitude, you measure the actual outside air pressure to be 6.16×10^4 N/m^2. From App. A, you find that the standard altitude that corresponds to a pressure of 6.16×10^4 N/m^2 is 4 km. Therefore, by *definition*, you say that you are flying at a *pressure altitude* of 4 km. Simultaneously, you measure the actual outside air temperature to be 265.4 K. From App. A, you find that the standard altitude that corresponds to a temperature of 265.4 K is 3.5 km. Therefore, by definition, you say that you are flying at a *temperature altitude* of 3.5 km. Thus, you are simultaneously flying at a pressure altitude of 4 km and a temperature altitude of 3.5 km while your actual geometric altitude is yet a different value. The definition of *density altitude* is made in the same vein. These quantities—pressure, temperature, and density altitudes—are just convenient numbers that, via App. A or B, are related to the actual p, T, and ρ for the actual altitude at which you are flying.

If an airplane is flying at an altitude where the actual pressure and temperature are 4.72×10^4 N/m^2 and 255.7 K, respectively, what are the pressure, temperature, and density altitudes? | **Example 3.2**

Solution

For the pressure altitude, look in App. A for the standard altitude value corresponding to $p = 4.72 \times 10^4$ N/m^2. This is 6000 m. Hence

Pressure altitude = 6000 m = 6 km

For the temperature altitude, look in App. A for the standard altitude value corresponding to $T = 255.7$ K. This is 5000 m. Hence

Temperature altitude = 5000 m = 5km

For the density altitude, we must first calculate ρ from the equation of state:

$$\rho = \frac{p}{RT} = \frac{4.72 \times 10^4}{287(255.7)} = 0.643 \text{ kg/m}^3$$

Looking in App. A and interpolating between 6.2 and 6.3 km, we find that the standard altitude value corresponding to $\rho = 0.643$ kg/m^3 is about 6.240 m. Hence

Density altitude = 6240 m = 6.24 km

Note that temperature altitude is not a unique value. The above answer for temperature altitude could equally well be 5.0, 38.2, or 59.5 km because of the multivalued nature of the altitude-versus-temperature function. In this section, only the lowest value of temperature altitude is used.

The flight test data for a given airplane refer to a level-flight maximum-velocity run made at an altitude which simultaneously corresponded to a pressure altitude of 30,000 ft and density altitude of 28,500 ft. Calculate the temperature of the air at the altitude at which the airplane was flying for the test. | **Example 3.3**

Solution

From App. B:

For pressure altitude = 30,000 ft:

$$p = 629.66 \text{ lb/ft}^2$$

For density altitude = 28,500 ft:

$$\rho = 0.9408 \times 10^{-3} \text{ slug/ft}^3$$

These are the values of p and ρ that simultaneously existed at the altitude at which the airplane was flying. Therefore, from the equation of state

$$T = \frac{p}{\rho R} = \frac{629.66}{(0.94082 \times 10^{-3})(1716)} = \boxed{390°\text{R}}$$

Example 3.4

Consider an airplane flying at some real geometric altitude. The outside (ambient) pressure and temperature are 5.3×10^4 N/m² and 253 K, respectively. Calculate the pressure and density altitudes at which this airplane is flying.

Solution

Consider the ambient pressure of 5.3×10^4 N/m². In App. A, there is not a precise entry for this number. It lies between the value $p_1 = 5.331 \times 10^4$ N/m² at altitude $h_{G,1} = 5100$ m and $p_2 = 5.2621 \times 10^4$ N/m² at altitude $h_{G,2} = 5200$ m. We have at least two choices. We could simply use the nearest entry in the table, which is for an altitude $h_{G,2} = 5100$ m, and say that the answer for pressure altitude is 5100 m. This is acceptable if we are making only approximate calculations. However, if we need greater accuracy, we can *interpolate* between entries. Using linear interpolation, the value of h_G corresponding to $p = 5.3 \times 10^4$ N/m² is

$$h_G = h_{G,1} + (h_{G,2} - h_{G,1}) \left(\frac{p_1 - p}{p_1 - p_2} \right)$$

$$h_G = 5100 + (5200 - 5100) \left(\frac{5.331 - 5.3}{5.331 - 5.2621} \right) = 5100 + 100(0.4662) = \boxed{5146.6 \text{ m}}$$

The pressure altitude at which the airplane is flying is 5146.6 m. (Note that in this example, and in Examples 3.2 and 3.3, we are interpreting the word *altitude* in the tables to be the geometric altitude h_G rather than the geopotential altitude h. This is for convenience, because h_G is tabulated in round numbers, in contrast to the column for h. Again, at the altitudes for conventional flight, the difference between h_G and h is not significant.)

To obtain the density altitude, calculate the density from the equation of state.

$$\rho = \frac{p}{RT} = \frac{5.3 \times 10^4}{(287)(253)} = 0.72992 \text{ kg/m}^3$$

Once again we note that this value of ρ falls between two entries in the table. It falls between $h_{G,1} = 5000$ m where $\rho_1 = 0.73643$ kg/m³ and $h_{G,2} = 5100$ m where $\rho_2 = 0.72851$ kg/m³. (Note that these subscripts denote different lines in the table from those used in the first part of this example. It is good never to become a slave to subscripts and symbols. Just always keep in mind the significance of what you are doing.) We could take the nearest entry, which is for an altitude $h_G = 5100$ m, and say that the answer for the density altitude is 5100 m. However, for greater accuracy, let us linearly interpolate between the two entries.

$$h_G = h_{G,1} + (h_{G,2} - h_{G,1})\left(\frac{\rho_1 - \rho}{\rho_1 - \rho_2}\right)$$

$$= 5000 + (5100 - 5000)\left(\frac{0.73643 - 0.72992}{0.73643 - 0.72851}\right)$$

$$= 5000 + 100(0.82197) = \boxed{5082.2 \text{ m}}$$

The density altitude at which the airplane is flying is 5082.2 m.

3.6 HISTORICAL NOTE: THE STANDARD ATMOSPHERE

With the advent of ballooning in 1783 (see Chap. 1), people suddenly became interested in acquiring a greater understanding of the properties of the atmosphere above ground level. However, a compelling reason for such knowledge did not arise until the coming of heavier-than-air flight in the 20th century. As we shall see in subsequent chapters, the flight performance of aircraft is dependent upon such properties as the pressure and density of the air. Thus, a knowledge of these properties, or at least some agreed-upon standard for worldwide reference, is absolutely necessary for intelligent aeronautical engineering.

The situation in 1915 was summarized by C. F. Marvin, Chief of the U.S. Weather Bureau and chairman of an NACA subcommittee to investigate and report upon the existing status of atmospheric data and knowledge. In his "Preliminary Report on the Problem of the Atmosphere in Relation to Aeronautics," NACA Report No. 4, 1915, Marvin writes:

> The Weather Bureau is already in possession of an immense amount of data concerning atmospheric conditions, including wind movements at the earth's surface. This information is no doubt of distinct value to aeronautical operations, but it needs to be collected and put in form to meet the requirements of aviation.

The following four years saw such efforts to collect and organize atmospheric data for use by aeronautical engineers. In 1920, the Frenchman A. Toussaint, director of the Aerodynamic Laboratory at Saint- Cyr-l'Ecole, France, suggested the following formula for the temperature decrease with height:

$$T = 15 - 0.0065h$$

where T is in degrees Celsius and h is the geopotential altitude in meters. Toussaint's formula was formally adopted by France and Italy with the Draft of Inter-Allied Agreement on Law Adopted for the Decrease of Temperature with Increase of Altitude, issued by the Ministere de la Guerre, Aeronautique Militaire, Section Technique, in March 1920. One year later, England followed suit. The United States was close behind. Since Marvin's report in 1915, the U.S. Weather Bureau had compiled measurements of the temperature distribution and found Toussaint's formula to be a reasonable representation of the observed mean annual values. Therefore, at its executive committee meeting of December 17, 1921, NACA adopted Toussaint's formula for airplane performance testing, with the statement: "The subcommittee on aerodynamics recommends that for the sake of uniform practice in different countries that Toussaint's formula be adopted in determining the standard atmosphere up to 10 km (33,000 ft)...."

Much of the technical data base that supported Toussaint's formula was reported in NACA Report No. 147, "Standard Atmosphere," by Willis Ray Gregg in 1922. Based on free-flight tests at McCook Field in Dayton, Ohio, and at Langley Field in Hampton, Virginia, and on the other flights at Washington, District of Columbia, as well as artillery data from Aberdeen, Maryland, and Dahlgren, Virginia, and sounding-balloon observations at Fort Omaha, Nebraska, and St. Louis, Missouri, Gregg was able to compile a table of mean annual atmospheric properties. An example of his results follows:

Altitude, m	Mean Annual Temperature in United States, K	Temperature from Toussaint's Formula, K
0	284.5	288
1,000	281.0	281.5
2,000	277.0	275.0
5,000	260.0	255.5
10,000	228.5	223.0

Clearly, Toussaint's formula provided a simple and reasonable representation of the mean annual results in the United States. This was the primary message in Gregg's report in 1922. However, the report neither gave extensive tables nor attempted to provide a document for engineering use.

Thus, it fell to Walter S. Diehl (who later became a well-known aerodynamicist and airplane designer as a captain in the Naval Bureau of Aeronautics) to provide the first practical tables for a standard atmosphere for aeronautical use. In 1925, in NACA Report No. TR 218, entitled (again) "Standard Atmosphere," Diehl presented extensive tables of standard atmospheric properties in both metric and English units. The tables were in increments of 50 m up to an altitude of 10 km and then in increments of 100 m up to 20 km. In English units, the tables were in increments of 100 ft up to 32,000 ft and then in increments of 200 ft up to a maximum altitude of 65,000 ft. Considering the aircraft of that day (see Fig. 1.31), these tables were certainly sufficient. Moreover, starting from Toussaint's formula for T up to 10,769 m, then assuming $T = \text{const} = -55°C$ above 10,769 m, Diehl obtained p and ρ in precisely the same fashion as described in the previous sections of this chapter.

The 1940s saw the beginning of serious rocket flights, with the German V-2 and the initiation of sounding rockets. Moreover, airplanes were flying higher than ever. Then, with the advent of intercontinental ballistic missiles in the 1950s and space flight in the 1960s, altitudes began to be quoted in terms of hundreds of miles rather than feet. Therefore, new tables of the standard atmosphere were created, mainly extending the old tables to higher altitudes. Popular among the various tables is the ARDC 1959 Standard Atmosphere, which is used in this book and is given in App. A and B. For all practical purposes, the old and new tables agree for altitudes of greatest interest. Indeed, it is interesting to compare values, as shown below:

Altitude, m	T from Diehl, 1925, K	T from ARDC, 1959, K
0	288	288.16
1,000	281.5	281.66
2,000	275.0	275.16
5,000	255.5	255.69
10,000	223.0	223.26
10,800	218.0	218.03
11,100	218.0	216.66
20,000	218.0	216.66

So Diehl's standard atmosphere from 1925, at least up to 20 km, is just as good as the values today.

3.7 SUMMARY

Some of the major ideas of this chapter are listed below.

1. The standard atmosphere is defined in order to relate flight tests, wind tunnel results, and general airplane design and performance to a common reference.

2. The definitions of the standard atmospheric properties are based on a given temperature variation with altitude, representing a mean of experimental data. In turn, the pressure and density variations with altitude are obtained from this empirical temperature variation by using the laws of physics. One of these laws is the hydrostatic equation:

$$dp = -\rho g \, dh_G \qquad \textbf{[3.2]}$$

3. In the isothermal regions of the standard atmosphere, the pressure and density variations are given by

$$\frac{p}{p_1} = \frac{\rho}{\rho_1} = e^{-[g_0/(RT)](h-h_1)} \qquad \textbf{[3.9] and [3.10]}$$

4. In the gradient regions of the standard atmosphere, the pressure and density variations are given by, respectively,

$$\frac{p}{p_1} = \left(\frac{T}{T_1}\right)^{-g_0/(aR)}$$ [3.12]

$$\frac{\rho}{\rho_1} = \left(\frac{T}{T_1}\right)^{-\{[(g_0/(aR)]+1\}}$$ [3.13]

where $T = T_1 + a(h - h_1)$ and a is the given lapse rate.

5. The pressure altitude is that altitude in the standard atmosphere which corresponds to the actual ambient pressure encountered in flight or laboratory experiments. For example, if the ambient pressure of a flow, no matter where it is or what it is doing, is 393.12 lb/ft^2, the flow is said to correspond to a pressure altitude of 40,000 ft (see App. B). The same idea can be used to define density and temperature altitudes.

BIBLIOGRAPHY

Minzner, R. A., K. S. W. Champion, and H. L. Pond: *The ARDC Model Atmosphere, 1959*, Air Force Cambridge Research Center Report No. TR-59-267, U.S. Air Force, Bedford, MA, 1959.

PROBLEMS

3.1 At 12 km in the standard atmosphere, the pressure, density, and temperature are 1.9399×10^4 N/m^2, 3.1194×10^{-1} kg/m^3, and 216.66 K, respectively. Using these values, calculate the standard atmospheric values of pressure, density, and temperature at an altitude of 18 km, and check with the standard altitude tables.

3.2 Consider an airplane flying at some real altitude. The outside pressure and temperature are 2.65×10^4 N/m^2 and 220 K, respectively. What are the pressure and density altitudes?

3.3 During a flight test of a new airplane, the pilot radios to the ground that she is in level flight at a standard altitude of 35,000 ft. What is the ambient air pressure far ahead of the airplane?

3.4 Consider an airplane flying at a pressure altitude of 33,500 ft and a density altitude of 32,000 ft. Calculate the outside air temperature.

3.5 At what value of the geometric altitude is the difference $h - h_G$ equal to 2 percent of the geopotential altitude, h?

3.6 Using Toussaint's formula, calculate the pressure at a geopotential altitude of 5 km.

3.7 The atmosphere of Jupiter is essentially made up of hydrogen, H_2. For H_2, the specific gas constant is 4157 J/(kg)(K). The acceleration of gravity of Jupiter is 24.9 m/s^2. Assuming an isothermal atmosphere with a temperature of 150 K, and assuming that Jupiter has a definable surface, calculate the altitude above that surface where the pressure is one-half the surface pressure.

3.8 An F-15 supersonic fighter aircraft is in a rapid climb. At the instant it passes through a standard altitude of 25,000 ft, its time rate of change of altitude is 500 ft/s, which by definition is the *rate of climb*, discussed in Chap. 6. Corresponding to this rate of climb at 25,000 ft is a time rate of change of ambient pressure. Calculate this rate of change of pressure in units of pounds per square foot per second.

3.9 Assume that you are ascending in an elevator at sea level. Your eardrums are very sensitive to minute changes in pressure. In this case, you are feeling a one-percent decrease in pressure per minute. Calculate the upward speed of the elevator in meters per minute.

3.10 Consider an airplane flying at an altitude where the pressure and temperature are 530 lb/ft^2 and 390°R respectively. Calculate the pressure and density altitudes at which the airplane is flying.

chapter

4

Basic Aerodynamics

Mathematics up to the present day have been quite useless to us in regard to flying.

From the fourteenth Annual Report
of the Aeronautical Society
of Great Britain, 1879

Mathematical theories from the happy hunting grounds of pure mathematicians are found suitable to describe the airflow produced by aircraft with such excellent accuracy that they can be applied directly to airplane design.

Theodore von Karman, 1954

Consider an airplane flying at an altitude of 3 km (9840 ft) at a velocity of 112 m/s (367 ft/s or 251 mi/h). At a given point on the wing, the pressure and airflow velocity are specific values, dictated by the laws of nature. One of the objectives of the science of aerodynamics is to decipher these laws and to give us methods to calculate the flow properties. In turn, such information allows us to calculate practical quantities, such as the lift and drag on the airplane. Another example is the flow through a rocket engine of a given size and shape. If this engine is sitting on the launch pad at Cape Canaveral and given amounts of fuel and oxidizer are ignited in the combustion chamber, the flow velocity and pressure at the nozzle exit are again specific values, dictated by the laws of nature. The basic principles of aerodynamics allow us to calculate the exit flow velocity and pressure, which in turn allow us to calculate the thrust. For reasons such as these, the study of aerodynamics is vital to the overall understanding of flight. The purpose of this chapter is to provide an introduction to the basic laws and concepts of aerodynamics and to show how they are applied to solving practical problems.

The road map for this chapter is given in Fig. 4.1. Let us walk through this road map so that we can get a better idea of what this chapter on aerodynamics is all about. First, we can identify two basic types of aerodynamic flows: (1) flow with no friction (called *inviscid* flow) and (2) flow with friction (called *viscous* flow). These two types of flow are represented by the two boxes shown near the top of the road map. This is an important distinction in aerodynamics. Any real-life aerodynamic flow has friction acting on the fluid elements moving within the flow field. However, there are many practical aerodynamic problems in which the influence of this internal friction is very small, and it can be neglected. Such flows can be *assumed* to have *no* friction and hence can be analyzed as *inviscid flows*. This is an idealization, but for many problems, a good one. By not dealing with friction, the analysis of the flow is usually simplified. However, for some flows the influence of friction is dominant, and it must be included in any analysis of such flows. The inclusion of friction usually makes the analysis of the flow more complicated.

This chapter deals with *basics*. We will start out with the statement of three fundamental physical principles from physics:

1.   Mass is conserved.

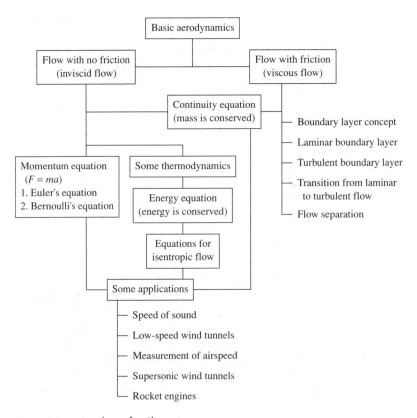

Figure 4.1   Road map for Chap. 4.

2. Newton's second law (force = mass × acceleration) holds.

3. Energy is conserved.

When these fundamental principles are applied to an aerodynamic flow, certain equations result which, in mathematical language, are statements of these principles. We will see how this can be accomplished. We will start with the physical principle that mass is conserved and obtain a governing equation labeled the continuity equation. This is represented by the center box in Fig. 4.1. The continuity equation says, in mathematical symbols, that mass is conserved in an aerodynamic flow. Mass is conserved, no matter whether the flow involves friction. Hence, the continuity equation is equally applicable to both types of flow, and that is why it is centered beneath the top two boxes in Fig. 4.1. Then we will work our way down the left side of the road map, making the assumption of an inviscid flow. We will invoke Newton's second law and obtain the momentum equation for an inviscid flow, called Euler's equation (pronounced like "oilers"). A specialized, but important form of Euler's equation is Bernoulli's famous equation. Then we will invoke the principle of conservation of energy and obtain the energy equation for a flow. However, since the science of energy is *thermodynamics*, we have to first examine some of the basic concepts of thermodynamics.

After the basic equations are in hand, we will continue down the left side of Fig. 4.1 with some applications for inviscid flows, ranging from the speed of sound to wind tunnels and rocket engines.

Finally, we will move to the right side of our road map and discuss some important aspects of viscous flows. We will introduce the idea of a viscous *boundary layer*, the region of flow immediately adjacent to a solid surface, where friction is particularly dominant. We will examine two types of viscous flows with quite different natures—*laminar* flow and *turbulent* flow—and how a laminar flow will transit to a turbulent flow. We will discuss how these flows have an impact on the aerodynamic drag on a body. Finally, we will see how a viscous aerodynamic flow can actually lift off (separate) from the surface—the phenomenon of *flow separation*.

This has been a rather long discussion of a somewhat intricate road map. However, the experience of the author has been that readers being introduced to the world of basic aerodynamics can find the subject matter sometimes bewildering. In reality, aerodynamics is a beautifully organized intellectual subject, and the road map in Fig. 4.1 is designed to prevent some of the possible bewilderment. As we progress through this chapter, it will be very important for you to frequently return to this road map for guidance and orientation.

4.1 CONTINUITY EQUATION

The laws of aerodynamics are formulated by applying to a flowing gas several basic principles from physics. For example,

Physical principle: Mass can be neither created nor destroyed.[1]

To apply this principle to a flowing gas, consider an imaginary circle drawn perpendicular to the flow direction, as shown in Fig. 4.2. Now look at all the streamlines that go through the circumference of the circle. These streamlines form a tube, called a *stream tube*. As we move along with the gas confined inside the stream tube, we see that the cross-sectional area of the tube may change, say, in moving from point 1 to point 2 in Fig. 4.2. However, as long as the flow is steady (invariant with time), the mass that flows through the cross section at point 1 must be the same as the mass that flows through the cross section at point 2, because by the definition of a streamline, there can be no flow across streamlines. The mass flowing through the stream tube is confined by the streamlines of the boundary, much as the flow of water through a flexible garden hose is confined by the wall of the hose. This is a case of "what goes in one end must come out the other end."

Let A_1 be the cross-sectional area of the stream tube at point 1. Let V_1 be the flow velocity at point 1. Now, at a given instant in time, consider all the fluid elements that are momentarily in the plane of A_1. After a lapse of time dt, these same fluid elements all move a distance $V_1\,dt$, as shown in Fig. 4.2. In so doing, the elements have swept out a volume $A_1 V_1\,dt$ downstream of point 1. The *mass* of gas dm in this volume is equal to the density times the volume, that is,

$$dm = \rho_1(A_1 V_1\,dt) \qquad\qquad \textbf{[4.1]}$$

This is the mass of gas that has *swept through* area A_1 during time interval dt.

Definition: The *mass flow* $\dot{m}$ through area A is the mass crossing A per unit time.

Therefore, from Eq. (4.1), for area A_1,

$$\text{Mass flow} = \frac{dm}{dt} \equiv \dot{m}_1 = \rho_1 A_1 V_1 \qquad \text{kg/s or slugs/s}$$

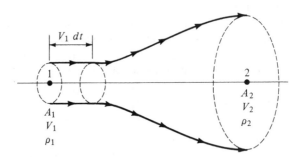

Figure 4.2   Stream tube with mass conservation.

[1] Of course, Einstein has shown that $e = mc^2$, and hence mass is truly not conserved in situations where energy is released. However, for any noticeable change in mass to occur, the energy release must be tremendous, such as occurs in a nuclear reaction. We are generally not concerned with such a case in practical aerodynamics.

Figure 4.3 A stream tube.

Also, the mass flow through A_2, bounded by the same streamlines that go through the circumference of A_1, is obtained in the same fashion, as

$$\dot{m}_2 = \rho_2 A_2 V_2$$

Since mass can be neither created nor destroyed, we have $\dot{m}_1 = \dot{m}_2$. Hence

$$\rho_1 A_1 V_1 = \rho_2 A_2 V_2 \qquad \textbf{[4.2]}$$

This is the *continuity equation* for steady fluid flow. It is a simple algebraic equation that relates the values of density, velocity, and area at one section of the stream tube to the same quantities at any other section.

There is a caveat in the above development. In Fig. 4.2, velocity V_1 is assumed to be uniform over the entire area A_1. Similarly, the density ρ_1 is assumed to be uniform over area A_1. In the same vein, V_2 and ρ_2 are assumed to be uniform over area A_2. In real life, this is an approximation; in reality, V and ρ vary across the cross-sectional area A. However, when using Eq. (4.2), we assume that ρ and V represent *mean* values of density and velocity over the cross-sectional area A. For many flow applications, this is quite reasonable. The continuity equation in the form of Eq. (4.2) is a workhorse in the calculation of flow through all types of ducts and tubes, such as wind tunnels and rocket engines.

The stream tube sketched in Fig. 4.2 does not have to be bounded by a solid wall. For example, consider the streamlines of flow over an airfoil, as sketched in Fig. 4.3. The space between two adjacent streamlines, such as the shaded space in Fig. 4.3, is a stream tube. Equation (4.2) applies to the stream tube in Fig. 4.3, where ρ_1 and V_1 are appropriate mean values over A_1, and ρ_2 and V_2 are appropriate values over A_2.

4.2 INCOMPRESSIBLE AND COMPRESSIBLE FLOW

Before we proceed, it is necessary to point out that all matter in real life is *compressible* to some greater or lesser extent. That is, if we take an element of matter and squeeze it hard enough with some pressure, the volume of the element of matter will decrease. However, its mass will stay the same. This is shown schematically in Fig. 4.4. As a result, the *density* ρ of the element changes as it is squeezed. The amount by which

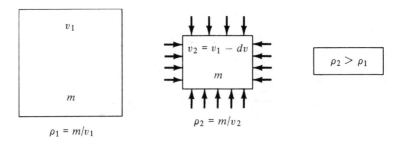

Figure 4.4   Illustration of compressibility.

ρ changes depends on the nature of the material of the element and how hard we squeeze it, that is, the magnitude of the pressure. If the material is solid, such as steel, then the change in volume is insignificantly small and ρ is constant for all practical purposes. If the material is a liquid, such as water, then the change in volume is also very small and again ρ is essentially constant. (Try pushing a tight-fitting lid into a container of liquid, and you will find out just how "solid" the liquid can be.) But, if the material is a gas, the volume can readily change and ρ can be a variable.

The preceding discussion allows us to characterize two classes of aerodynamic flow: compressible flow and incompressible flow.

1. *Compressible flow*—flow in which the density of the fluid elements can change from point to point. Referring to Eq. (4.2), we see if the flow is compressible, $\rho_1 \neq \rho_2$. The variability of density in aerodynamic flows is particularly important at high speeds, such as for high-performance subsonic aircraft, all supersonic vehicles, and rocket engines. Indeed, all real-life flows, strictly speaking, are compressible. However, there are some circumstances in which the density changes only slightly. These circumstances lead to the second definition, as follows.

2. *Incompressible flow*—flow in which the density of the fluid elements is always constant. Referring to Eq. (4.2), we see if the flow is incompressible, $\rho_1 = \rho_2$, hence

$$A_1 V_1 = A_2 V_2 \qquad \textbf{[4.3]}$$

Incompressible flow is a myth. It can never actually occur in nature, as discussed above. However, for those flows in which the actual variation of ρ is negligibly small, it is convenient to make the *assumption* that ρ is constant, to simplify our analysis. (Indeed, it is an everyday activity of engineering and physical science to make idealized assumptions about real physical systems in order to make such systems amenable to analysis. However, care must always be taken not to apply results obtained from such idealizations to those real problems in which the assumptions are grossly inaccurate or inappropriate.) The assumption of incompressible flow is an excellent approximation for the flow of liquids, such as water or oil. Moreover, the low-speed flow of air, where $V < 100$ m/s (or $V < 225$ mi/h) can also be assumed to be incompressible to a close approximation. A glance at Fig. 1.30 shows that such

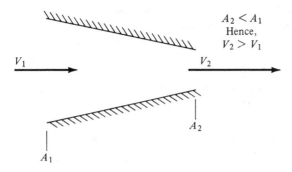

$$A_2 < A_1$$
Hence,
$$V_2 > V_1$$

Figure 4.5 Incompressible flow in a convergent duct.

velocities were the domain of almost all airplanes from the *Wright Flyer* (1903) to the late 1930s. Hence, the early development of aerodynamics always dealt with incompressible flows, and for this reason there exists a huge body of incompressible flow literature with its attendant technology. At the end of this chapter, we will be able to prove *why* airflow at velocities less than 100 m/s can be safely assumed to be incompressible.

In solving and examining aerodynamic flows, you will constantly be faced with making distinctions between incompressible and compressible flows. It is important to start that habit now, because there are some striking quantitative and qualitative differences between the two types of flow.

As a parenthetical comment, for incompressible flow, Eq. (4.3) explains why all common garden hose nozzles are convergent shapes, such as shown in Fig. 4.5. From Eq. (4.3),

$$V_2 = \frac{A_1}{A_2} V_1$$

If A_2 is less than A_1, then the velocity increases as the water flows through the nozzle, as desired. The same principle is used in the design of nozzles for subsonic wind tunnels built for aerodynamic testing, as will be discussed in a subsequent section.

Consider a convergent duct with an inlet area $A_1 = 5$ m². Air enters this duct with a velocity $V_1 = 10$ m/s and leaves the duct exit with a velocity $V_2 = 30$ m/s. What is the area of the duct exit?

Example 4.1

Solution

Since the flow velocities are less than 100 m/s, we can assume incompressible flow. From Eq. (4.3)

$$A_1 V_1 = A_2 V_2$$

$$A_2 = A_1 \frac{V_1}{V_2} = \left(5 \text{ m}^2\right) \frac{10}{30} = \boxed{1.67 \text{ m}^2}$$

Example 4.2

Consider a convergent duct with an inlet area $A_1 = 3$ ft^2 and an exit area $A_2 = 2.57$ ft^2. Air enters this duct with a velocity $V_1 = 700$ ft/s and a density $\rho_1 = 0.002$ slug/ft^3, and air leaves with an exit velocity $V_2 = 1070$ ft/s. Calculate the density of the air ρ_2 at the exit.

Solution

An inlet velocity of 700 ft/s is a high-speed flow, and we assume that the flow has to be treated as compressible. This implies that the resulting value for ρ_2 will be different from ρ_1. From Eq. (4.2),

$$\rho_1 A_1 V_1 = \rho_2 A_2 V_2$$

or

$$\rho_2 = \rho_1 \frac{A_1 V_1}{A_2 V_2} = 0.002 \frac{3(700)}{2.57(1070)} = \boxed{0.00153 \text{ slug/ft}^3}$$

Note: The value of ρ_2 is indeed different from ρ_1, which clearly indicates that the flow in this example is a compressible flow. *If* the flow were essentially incompressible, then the calculation of ρ_2 from Eq. (4.2) would have produced a value essentially equal to ρ_1. But this is not the case. Keep in mind that Eq. (4.2) is more general than Eq. (4.3). Eq. (4.2) applies for both compressible and incompressible flows; Eq. (4.3) is valid for an incompressible flow only.

4.3 MOMENTUM EQUATION

The continuity equation, Eq. (4.2), is only part of the story. For example, it says nothing about the pressure in the flow; yet, we know, just from intuition, that pressure is an important flow variable. Indeed, differences in pressure from one point to another in the flow create forces that act on the fluid elements and cause them to move. Hence, there must be some relation between pressure and velocity, and that relation is derived in this section.

Again, we first state a fundamental law of physics, namely, Newton's second law.

Physical principle: Force = mass × acceleration

or $$F = ma \qquad \text{[4.4]}$$

To apply this principle to a flowing gas, consider an infinitesimally small fluid element moving along a streamline with velocity V, as shown in Fig. 4.6. At some given instant, the element is located at point P. The element is moving in the x direction, where the x axis is oriented parallel to the streamline at point P. The y and z axes are mutually perpendicular to x. The fluid element is very small, infinitesimally small. However, looking at it through a magnifying glass, we see the picture shown at the upper right of Fig. 4.6. Question: What is the force on this element? Physically, the force is a combination of three phenomena:

1. Pressure acting in a normal direction on all six faces of the element
2. Frictional shear acting tangentially on all six faces of the element

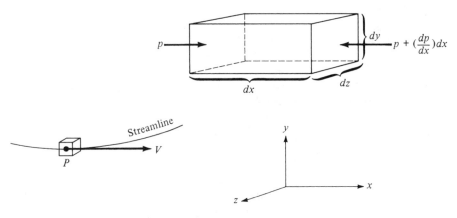

Figure 4.6   Force diagram for the momentum equation.

3.   Gravity acting on the mass inside the element

For the time being, we will ignore the presence of frictional forces; moreover, the
gravity force is generally a small contribution to the total force. Therefore, we will
assume that the only source of a force on the fluid element is pressure.

To calculate this force, let the dimensions of the fluid element be dx, dy, and dz,
as shown in Fig. 4.6. Consider the left and right faces, which are perpendicular to the
x axis. The pressure on the left face is p. The area of the left face is $dy\,dz$, hence
the force on the left face is $p(dy\,dz)$. This force is in the positive x direction. Now
recall that pressure varies from point to point in the flow. Hence, there is some change
in pressure per unit length, symbolized by the derivative dp/dx. Thus, if we move
away from the left face by a distance dx along the x axis, the *change* in pressure
is $(dp/dx)\,dx$. Consequently, the pressure on the right face is $p + (dp/dx)\,dx$.
The area of the right face is also $dy\,dz$, hence the force on the right face is $[p +
(dp/dx)\,dx](dy\,dz)$. This force acts in the negative x direction, as shown in Fig. 4.6.
The net force in the x direction F is the sum of the two:

$$F = p\,dy\,dz - \left(p + \frac{dp}{dx}dx\right)dy\,dz$$

or
$$F = -\frac{dp}{dx}(dx\,dy\,dz)$$
[4.5]

Equation (4.5) gives the force on the fluid element due to pressure. Because of the
convenience of choosing the x axis in the flow direction, the pressures on the faces
parallel to the streamlines do not affect the motion of the element along the streamline.

The mass of the fluid element is the density ρ multiplied by the volume $dx\,dy\,dz$:

$$m = \rho(dx\,dy\,dz)$$
[4.6]

Also, the acceleration a of the fluid element is, by definition of acceleration (rate of
change of velocity), $a = dV/dt$. Noting that, also by definition, $V = dx/dt$, we can

write

$$a = \frac{dV}{dt} = \frac{dV}{dx}\frac{dx}{dt} = \frac{dV}{dx}V \qquad \textbf{[4.7]}$$

Equations (4.5) to (4.7) give the force, mass, and acceleration, respectively, that go into Newton's second law, Eq. (4.4):

$$F = ma$$

$$-\frac{dp}{dx}(dx\,dy\,dz) = \rho(dx\,dy\,dz)V\frac{dV}{dx}$$

or

$$\boxed{dp = -\rho V\,dV} \qquad \textbf{[4.8]}$$

Equation (4.8) is *Euler's equation.* Basically, it relates rate of change of momentum to the force; hence it can also be designated as the *momentum equation.* It is important to keep in mind the assumptions utilized in obtaining Eq. (4.8); we neglected friction and gravity. For flow which is frictionless, aerodynamicists sometimes use another term, *inviscid flow.* Equation (4.8) is the momentum equation for inviscid (frictionless) flow. Moreover, the flow field is assumed to be steady, that is, invariant with respect to time.

Please note that Eq. (4.8) relates pressure and velocity (in reality, it relates a change in pressure dp to a change in velocity dV). Equation (4.8) is a differential equation, and hence it describes the phenomena in an infinitesimally small neighborhood around the given point P in Fig. 4.6. Now consider two points, 1 and 2, far removed from each other in the flow but on the same streamline. To relate p_1 and V_1 at point 1 to p_2 and V_2 at the other, far-removed point 2, Eq. (4.8) must be integrated between points 1 and 2. This integration is different depending on whether the flow is compressible or incompressible. Euler's equation itself, Eq. (4.8), holds for both cases. For compressible flow, ρ in Eq. (4.8) is a variable; for incompressible flow, ρ is a constant.

First, consider the case of incompressible flow. Let points 1 and 2 be located along a given streamline, such as that shown over an airfoil in Fig. 4.7. From Eq. (4.8),

$$dp + \rho V\,dV = 0$$

Figure 4.7   Two points at different locations along a streamline.

where $\rho = $ constant. Integrating between points 1 and 2, we obtain

$$\int_{p_1}^{p_2} dp + \rho \int_{V_1}^{V_2} V\, dV = 0$$

$$p_2 - p_1 + \rho \left(\frac{V_2^2}{2} - \frac{V_1^2}{2} \right) = 0$$

$$p_2 + \rho \frac{V_2^2}{2} = p_1 + \rho \frac{V_1^2}{2} \qquad\qquad \textbf{[4.9a]}$$

$$p + \rho \frac{V^2}{2} = \text{const along streamline} \qquad\qquad \textbf{[4.9b]}$$

Either form, Eq. (4.9a) or (4.9b), is called *Bernoulli's equation.* Historically, Bernoulli's equation is one of the most fundamental equations in fluid mechanics.

The following important points should be noted:

1. Equations (4.9a) and (4.9b) hold only for inviscid (frictionless), incompressible flow.

2. Equations (4.9a) and (4.9b) relate properties between different points along a streamline.

3. For a compressible flow, Eq. (4.8) must be used, with ρ treated as a variable. Bernoulli's equation *must not* be used for compressible flow.

4. Remember that Eqs. (4.8) and (4.9a) and (4.9b) say that $F = ma$ for a fluid flow. They are essentially Newton's second law applied to fluid dynamics.

To return to Fig. 4.7, if all the streamlines have the same values of p and V far upstream (far to the left in Fig. 4.7), then the constant in Bernoulli's equation is the *same for all streamlines.* This would be the case, for example, if the flow far upstream were uniform flow, such as that encountered in flight through the atmosphere and in the test sections of well-designed wind tunnels. In such cases, Eqs. (4.9a) and (4.9b) are not limited to the same streamline. Instead, points 1 and 2 can be anywhere in the flow, even on different streamlines.

For the case of compressible flow also, Euler's equation, Eq. (4.8), can be integrated between points 1 and 2; however, because ρ is a variable, we must in principle have some extra information on how ρ varies with V before the integration can be carried out. This information can be obtained; however, there is an alternate, more convenient route to treating many practical problems in compressible flow that does not explicitly require the use of the momentum equation. Hence, in this case, we will not pursue the integration of Eq. (4.8) further.

4.4   A COMMENT

It is important to make a philosophical distinction between the nature of the equation of state, Eq. (2.3), and the flow equations of continuity, Eq. (4.2), and momentum, such as Eq. (4.9a). The equation of state relates p, T, and ρ to one another at the *same* point; in contrast, the flow equations relate ρ and V (as in the continuity equation) and p and V (as in Bernoulli's equation) at one point in the flow to the same quantities at another point in the flow. There is a basic difference here, and one does well to keep it in mind when setting up the solution of aerodynamic problems.

Example 4.3

Consider an airfoil (the cross section of a wing, as shown in Fig. 4.7) in a flow of air, where far ahead (upstream) of the airfoil, the pressure, velocity, and density are 2116 lb/ft², 100 mi/h, and 0.002377 slug/ft³, respectively. At a given point A on the airfoil, the pressure is 2070 lb/ft². What is the velocity at point A?

Solution

First, we must deal in consistent units; $V_1 = 100$ mi/h is *not* in consistent units. However, a convenient relation to remember is that 60 mi/h = 88 ft/s. Hence, $V_1 = 100(88/60) = 146.7$ ft/s. This velocity is low enough that we can assume incompressible flow. Hence, Bernoulli's equation, Eq. (4.9), is valid

$$p_1 + \frac{\rho V_1^2}{2} = p_A + \frac{\rho V_A^2}{2}$$

Thus

$$V_A = \left[\frac{2(p_1 - p_A)}{\rho} + V_1^2 \right]^{1/2}$$

$$= \left[\frac{2(2116 - 2070)}{0.002377} + (146.7)^2 \right]^{1/2}$$

$$\boxed{V_A = 245.4 \text{ ft/s}}$$

Example 4.4

Consider the same convergent duct and conditions as in Example 4.1. If the air pressure and temperature at the inlet are $p_1 = 1.2 \times 10^5$ N/m² and $T_1 = 330$ K, respectively, calculate the pressure at the exit.

Solution

First, we must obtain the density. From the equation of state,

$$\rho_1 = \frac{p_1}{RT_1} = \frac{1.2 \times 10^5}{287(330)} = 1.27 \text{ kg/m}^3$$

Still assuming incompressible flow, we find from Eq. (4.9)

$$p_1 + \frac{\rho V_1^2}{2} = p_2 + \frac{\rho V_2^2}{2}$$

$$p_2 = p_1 + \tfrac{1}{2}\rho(V_1^2 - V_2^2) = 1.2 \times 10^5 + (\tfrac{1}{2})(1.27)(10^2 - 30^2)$$

$$\boxed{p_2 = 1.195 \times 10^5 \ \text{N/m}^2}$$

Note: In accelerating from 10 to 30 m/s, the air pressure decreases only a small amount, less than 0.45 percent. This is a characteristic of very low velocity airflow.

Example 4.5

Consider a long dowel with a semicircular cross section, as sketched in Fig. 4.8a. The dowel is immersed in a flow of air, with its axis perpendicular to the flow, as shown in perspective in Fig. 4.8a. The rounded section of the dowel is facing into the flow, as shown in Fig. 4.8a and 4.8b. We call this rounded section the front face of the dowel. The radius of the semicircular cross section is $R = 0.5$ ft. The velocity of the flow far ahead of the dowel (called the *freestream*) is $V_\infty = 100$ ft/s. Assume inviscid flow; that is, neglect the effect of friction. The velocity of the flow along the surface of the rounded front face of the dowel is a function of location on the surface; location is denoted by angle θ in Fig. 4.8b. Hence, along the front rounded surface, $V = V(\theta)$. This variation is given by

$$V = 2V_\infty \sin \theta \qquad \text{[E4.5.1]}$$

The pressure distribution exerted over the surface of the cross section is sketched in Fig. 4.8c. On the front face, p varies with location along the surface, where the location is denoted by the angle θ; that is, $p = p(\theta)$ on the front face. On the flat back face, the pressure, denoted by p_B, is constant. The back face pressure is given by

$$p_B = p_\infty - 0.7\rho_\infty V_\infty^2 \qquad \text{[E4.5.2]}$$

where p_∞ and ρ_∞ are the pressure and density, respectively, in the freestream, far ahead of the dowel. The freestream density is given as $\rho_\infty = 0.002378$ slug/ft^3. Calculate the aerodynamic force exerted by the surface pressure distribution (illustrated in Fig. 4.8c) on a 1-ft segment of the dowel, shown by the shaded section in Figure 4.8a.

Solution

For this solution, we appeal to the discussions in Secs. 2.2 and 4.3. Examine Fig. 4.8c. Because of the symmetry of the semicircular cross section, the pressure distribution over the upper surface is a mirror image of the pressure distribution over the lower surface; that is, $p = p(\theta)$ for $0 \le \theta \le \pi/2$ is the same as $p = p(\theta)$ for $0 \ge \theta \ge -\pi/2$. Owing to this symmetry, there is no net force on the cross section in the direction perpendicular to the freestream; that is, the force due to the pressure pushing down on the upper surface is exactly canceled by the equal and opposite force due to the pressure pushing up on the lower surface. Therefore, owing to this symmetry, the resultant aerodynamic force is parallel to the freestream direction. This resultant aerodynamic force is illustrated by the arrow labeled D in Fig. 4.8c.

Before feeding the numbers into our calculation, we obtain an analytical formula for D in terms of V_∞ and R, as follows. Consider the infinitesimal arclength segment of the surface ds and the pressure p exerted locally on this segment, as drawn in Fig. 4.8d. A magnified view of this segment is shown in Fig. 4.8e. Recall from Fig. 4.8a that we wish to calculate the aerodynamic force on a 1-ft length of the dowel, as shown by the shaded region in Fig. 4.8a. As part of the shaded region, consider a small sliver of area of width ds and length equal to 1 ft on the curved face of the dowel, as shown in Fig. 4.8f. The surface area of this sliver is

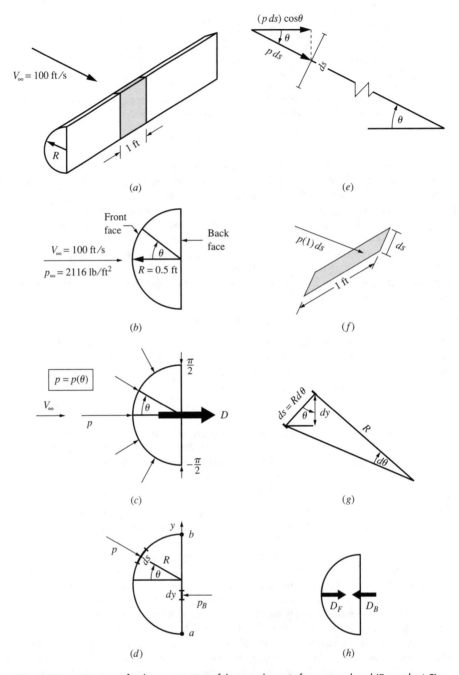

Figure 4.8   Diagrams for the construction of the aerodynamic force on a dowel (Example 4.5).

$1\,ds$. The force due to the pressure p on this area is $p(1)\,ds = p\,ds$. This force is shown in Fig. 4.8e, acting perpendicular to the segment ds. The component of this force in the horizontal directions is $(p\,ds)\cos\theta$, also shown in Fig. 4.8e. From the geometric construction shown in

Fig. 4.8g, we have

$$ds = R \, d\theta \qquad \textbf{[E4.5.3]}$$

and the vertical projection of ds, denoted by dy, is given by

$$dy = ds \, \cos \theta \qquad \textbf{[E4.5.4]}$$

Substituting Eq. (E4.5.3) into (E4.5.4), we have

$$dy = R \, \cos \theta \, d\theta \qquad \textbf{[E4.5.5]}$$

In light of Eq. (E.4.5.3), the horizontal force $(p \, ds) \cos \theta$ in Fig. 4.8e can be expressed as

$$(p \, ds) \cos \theta = pR \, \cos \theta \, d\theta \qquad \textbf{[E4.5.6]}$$

Returning to Fig. 4.8c, we see that the net horizontal force exerted by the pressure distribution on the rounded front face is the integral of Eq. (E4.5.6) over the front surface. Denote this force by D_F.

$$D_F = \int_{-\pi/2}^{\pi/2} pR \, \cos \theta \, d\theta \qquad \textbf{[E4.5.7]}$$

This force is shown in Figure 4.8h.

In Eq. (E4.5.7), p is obtained from Bernoulli's equation, Eq. (4.9), written between a point in the freestream where the pressure and velocity are p_∞ and V_∞, respectively, and the point on the body surface where the pressure and velocity are p and V, respectively.

$$p_\infty + \tfrac{1}{2}\rho V_\infty^2 = p + \tfrac{1}{2}\rho V^2$$

or

$$p = p_\infty + \tfrac{1}{2}\rho(V_\infty^2 - V^2) \qquad \textbf{[E4.5.8]}$$

Note: We can use Bernoulli's equation for this solution because the freestream velocity of $V_\infty = 100$ ft/s is low, and we can comfortably assume that the flow is incompressible. Also, because ρ is constant, the value of ρ in Eq. (E4.5.8) is the same as ρ_∞ in the freestream. Substituting Eq. (E4.5.8) into Eq. (E4.5.7), we have

$$D_F = \int_{-\pi/2}^{\pi/2} [p_\infty + \tfrac{1}{2}\rho(V_\infty^2 - V^2)]R \, \cos \theta \, d\theta \qquad \textbf{[E4.5.9]}$$

Recall that the variation of the surface velocity is given by Eq. (E4.5.1) repeated here:

$$V = 2V_\infty \, \sin \theta \qquad \textbf{[E4.5.1]}$$

Substituting (E4.5.1) into (E4.5.9), we have

$$D_F = \int_{-\pi/2}^{\pi/2} [p_\infty + \tfrac{1}{2}\rho(V_\infty^2 - 4V_\infty^2 \, \sin^2 \theta)]R \, \cos \theta \, d\theta$$

or

$$D_F = \int_{-\pi/2}^{\pi/2} [p_\infty + \tfrac{1}{2}\rho V_\infty^2(1 - 4\sin^2 \theta)]R \, \cos \theta \, d\theta \qquad \textbf{[E4.5.10]}$$

Let us put this expression for D_F on the shelf for a moment; we will come back to it shortly.

Returning to Fig. 4.8c, we now direct our attention to the pressure on the back face of the cross-section p_B. This pressure exerts a force D_B on the 1-ft length of dowel, as sketched in Fig. 4.8h. Force D_B acts toward the left, opposite to the direction of D_F. Pressure p_B is constant over the back face. The rectangular area of the 1-ft length of the back face is $(1)(2R)$. Because p_B is constant over this back face, we can directly write

$$D_B = (1)(2R)p_B \qquad\qquad \textbf{[E4.5.11]}$$

However, because the resultant aerodynamic force on the cross section is given by $D_F - D_B$, as seen in Fig. 4.8h, and because D_F is expressed in terms of an integral in Eq. (E4.5.10), it will be convenient to couch D_B in terms of an integral also, as follows. Returning to Figure 4.8d, we consider a segment of the back surface area of height dy on which p_B is exerted. Over a 1-ft length of dowel, (perpendicular to the page in Fig. 4.8d), the area of a small sliver of surface is $1\,dy$, and the force on this sliver is $p_B(1)\,dy$. The total force on the back face is obtained by integrating with respect to y from point a to point b, as noted in Fig. 4.8d.

$$D_B = \int_a^b p_B(1)\,dy \qquad\qquad \textbf{[E4.5.12]}$$

However, recall from Eq. (E4.5.5) that $dy = R\cos\theta\,d\theta$. Hence, Eq. (E4.5.12) becomes

$$D_B = \int_{-\pi/2}^{\pi/2} p_B R\cos\theta\,d\theta \qquad\qquad \textbf{[E4.5.13]}$$

Please note that Eqs. (E4.5.13) and (E4.5.11) are both valid expressions for D_B—they just look different. To see this, carry out the integration in Eq. (E4.5.13); you will obtain the result in Eq. (E4.5.11). Also recall that p_B is given by Eq. (E4.5.2), repeated here (and dropping the subscript ∞ on ρ since ρ is constant):

$$p_B = p_\infty - 0.7\rho V_\infty^2 \qquad\qquad \textbf{[E4.5.2]}$$

Hence, Eq. (E4.5.13) becomes

$$D_B = \int_{-\pi/2}^{\pi/2} (p_\infty - 0.7\rho_2 V_\infty^2)R\cos\theta\,d\theta \qquad\qquad \textbf{[E4.5.14]}$$

Returning to Fig. 4.8h, we see that the resultant aerodynamic force D is given by

$$D = D_F - D_B \qquad\qquad \textbf{[E4.5.15]}$$

Substituting Eqs. (E4.5.10) and (E4.5.14) into (E4.5.15), we have

$$D = \int_{-\pi/2}^{\pi/2} [p_\infty + \tfrac{1}{2}\rho V_\infty^2(1 - 4\sin^2\theta)]R\cos\theta\,d\theta$$
$$- \int_{-\pi/2}^{\pi/2} (p_\infty - 0.7\rho V_\infty^2)R\cos\theta\,d\theta \qquad\qquad \textbf{[E4.5.16]}$$

Combining the two integrals in Eq. (E4.5.16) and noting that the two terms involving p_∞ cancel, we have

DESIGN BOX

The results of Example 4.5 illustrate certain aspects important to the general background of airplane design:

1. It reinforces the important point made in Sect. 2.2, namely, that the resultant aerodynamic force exerted on any object immersed in a flowing fluid is due *only* to the net integration of the pressure distribution and the shear stress distribution exerted all over the body surface. In Example 4.5, we assumed the flow to be inviscid; that is, we neglected the effect of friction. So the resultant aerodynamic force was due to just the integrated effect of the pressure distribution over the body surface. Indeed, this is precisely how we calculated the force on the dowel in Example 4.5—we integrated the pressure distribution over the surface of the dowel. Instead of a dowel, if we had dealt with a Boeing 747 jumbo jet, the idea would have been the same. In airplane design, the shape of the airplane is influenced by the desire to create a surface pressure distribution that will minimize drag while at the same time creating the necessary amount of lift. We return to this basic idea several times throughout the book.

2. Equation (E4.5.17) shows that the aerodynamic force on the body is:

(a) Directly proportional to the density of the fluid ρ.

(b) Directly proportional to the *square* of the freestream velocity; $D \propto V_\infty^2$.

(c) Directly proportional to the *size* of the body, as reflected by the radius R.

These results are not just specialized to the dowel in Example 4.5, they are much more general in their application. We will see in Chap. 5 that the aerodynamic force on airfoils, wings, and whole airplanes is indeed proportional to ρ_∞, V_∞^2, and the size of the body, where size is couched in terms of a surface area. [In Eq. (E4.5.17), R really represents an area equal to $R(1)$ for the unit length of the dowel over which the aerodynamic force is calculated.] It is interesting to note that Eq. (E4.5.17) does not contain the freestream pressure p_∞. Indeed, p_∞ canceled out in our derivation of Eq. (E4.5.17). This is not just a characteristic of the dowel used in Example 4.5; in general, we will see in Chap. 5 that we do not need the explicit value of freestream pressure to calculate the aerodynamic force on a flight vehicle, in spite of the fact that the aerodynamic force fundamentally is due (in part) to the *pressure* distribution over the surface. In the final result, it is always the value of the freestream density ρ_∞ that appears in the expressions for aerodynamic force, not p_∞.

$$D = \int_{-\pi/2}^{\pi/2} [(\tfrac{1}{2}\rho + 0.7\rho)V_\infty^2 - 2\rho V_\infty^2 \sin^2 \theta]R \cos \theta \, d\theta$$

$$= 1.2\rho V_\infty^2 R \int_{-\pi/2}^{\pi/2} \cos \theta \, d\theta - 2\rho V_\infty^2 R \int_{-\pi/2}^{\pi/2} \sin^2 \theta \cos \theta \, d\theta$$

$$= 2.4\rho V_\infty^2 R - 2\rho V_\infty^2 R \left[\frac{\sin^3 \theta}{3} \right]_{-\pi/2}^{\pi/2}$$

$$= 2.4\rho V_\infty^2 R - 2\rho V_\infty^2 R(\tfrac{1}{3} + \tfrac{1}{3}) = \boxed{1.067\rho V_\infty^2 R}$$

Highlighting the above result, we have just derived an analytical expression for the aerodynamic force D, per unit length of the dowel. It is given by

$$\boxed{D = 1.067\rho V_\infty^2 R} \qquad\qquad \textbf{[E4.5.17]}$$

Putting in the numbers given in the problem, where $\rho = \rho_\infty = 0.002378$ slug/ft^3, $V_\infty = 100$ ft/s, and $R = 0.5$ ft, we obtain from Eq. (E4.5.17),

$$D = (1.067)(0.002377)(100)^2(0.5) = \boxed{12.68 \text{ lb}} \text{ per foot of length of dowel}$$

4.5   ELEMENTARY THERMODYNAMICS

As stated earlier, when the airflow velocity exceeds 100 m/s, the flow can no longer be treated as incompressible. Later we will restate this criterion in terms of the Mach number, which is the ratio of the flow velocity to the speed of sound, and we will show that the flow must be treated as compressible when the Mach number exceeds 0.3. This is the situation with the vast majority of current aerodynamic applications; hence, the study of compressible flow is of extreme importance.

A high-speed flow of gas is also a high-energy flow. The kinetic energy of the fluid elements in a high-speed flow is large and must be taken into account. When high-speed flows are slowed down, the consequent reduction in kinetic energy appears as a substantial increase in temperature. As a result, high-speed flows, compressibility, and vast energy changes are all related. Thus, to study compressible flows, we must first examine some of the fundamentals of energy changes in a gas and the consequent response of pressure and temperature to these energy changes. Such fundamentals are the essence of the science of thermodynamics.

Here the assumption is made that the reader is not familiar with thermodynamics. Therefore, the purpose of this section is to introduce those ideas and results of thermodynamics which are absolutely necessary for our further analysis of high-speed, compressible flows.

The pillar of thermodynamics is a relationship called the *first law*, which is an empirical observation of natural phenomena. It can be developed as follows. Consider a fixed mass of gas (for convenience, say a unit mass) contained within a flexible *boundary*, as shown in Fig. 4.9. This mass is called the *system*, and everything outside the boundary is the *surroundings*. Now, as in Chap. 2, consider the gas that makes up the system to be composed of individual molecules moving about with random

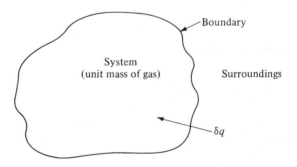

Figure 4.9   System of unit mass.

motion. The energy of this molecular motion, summed over all the molecules in the system, is called the *internal energy* of the system. Let e denote the internal energy per unit mass of gas. The *only* means by which e can be increased (or decreased) are the following:

1. Heat is added to (or taken away from) the system. This heat comes from the surroundings and is added to the system across the boundary. Let δq be an incremental amount of heat added per unit mass.

2. Work is done on (or by) the system. This work can be manifested by the boundary of the system being pushed in (work done on the system) or pushed out (work done by the system). Let δw be an incremental amount of work done on the system per unit mass.

Also, let de be the corresponding change in internal energy per unit mass. Then simply on the basis of common sense, confirmed by laboratory results, we can write

$$\delta q + \delta w = de \qquad \textbf{[4.10]}$$

Equation (4.10) is termed the *first law of thermodynamics*. It is an energy equation which states that the change in internal energy is equal to the sum of the heat added to and the work done on the system. (Note in the above that δ and d both represent infinitesimally small quantities; however, d is a "perfect differential" and δ is not.)

Equation (4.10) is very fundamental; however, it is not in a practical form for use in aerodynamics, which speaks in terms of pressures, velocities, etc. To obtain more useful forms of the first law, we must first derive an expression for δw in terms of p and v (specific volume), as follows. Consider the system sketched in Fig. 4.10. Let dA be an incremental surface area of the boundary. Assume that work ΔW is being done on the system by dA being pushed in a small distance s, as also shown in Fig. 4.10. Since work is defined as force times distance, we have

$$\Delta W = (\text{force})(\text{distance})$$
$$\Delta W = (p\,dA)s \qquad \textbf{[4.11]}$$

Figure 4.10 Work being done on the system by pressure.

Now assume that many elemental surface areas of the type shown in Fig. 4.10 are distributed over the total surface area A of the boundary. Also assume that all the elemental surfaces are being simultaneously displaced a small distance s into the system. Then the total work δw done on the unit mass of gas inside the system is the sum (integral) of each elemental surface over the whole boundary, that is, from Eq. (4.11),

$$\delta w = \int_A (p\,dA)s = \int_A ps\,dA \qquad \text{[4.12]}$$

Assume p is constant everywhere in the system (which, in thermodynamic terms, contributes to a state of thermodynamic equilibrium). Then, from Eq. (4.12),

$$\delta w = p \int_A s\,dA \qquad \text{[4.13]}$$

The integral $\int_A s\,dA$ has physical meaning. Geometrically, it is the change in volume of the unit mass of gas inside the system, created by the boundary surface being displaced inward. Let dv be the change in volume. Since the boundary is pushing in, the volume decreases (dv is a negative quantity) and work is done on the gas (hence δw is a positive quantity in our development). Thus

$$\int_A s\,dA \equiv -dv \qquad \text{[4.14]}$$

Substituting Eq. (4.14) into Eq. (4.13), we obtain

$$\boxed{\delta w = -p\,dv} \qquad \text{[4.15]}$$

Equation (4.15) gives the relation for work done strictly in terms of the thermodynamic variables p and v.

When Eq. (4.15) is substituted into Eq. (4.10), the first law becomes

$$\boxed{\delta q = de + p\,dv} \qquad \text{[4.16]}$$

Equation (4.16) is an alternate form of the first law of thermodynamics.

It is convenient to define a new quantity, called *enthalpy h*, as

$$h = e + pv = e + RT \qquad \text{[4.17]}$$

where $pv = RT$, assuming a perfect gas. Then differentiating the definition, Eq. (4.17), we find

$$dh = de + p\,dv + v\,dp \qquad \text{[4.18]}$$

Substituting Eq. (4.18) into (4.16), we obtain

$$\delta q = de + p\,dv = (dh - p\,dv - v\,dp) + p\,dv$$

$$\boxed{\delta q = dh - v\,dp} \qquad \text{[4.19]}$$

Equation (4.19) is yet another alternate form of the first law.

Before we go further, remember that a substantial part of science and engineering is simply the language. In this section, we are presenting some of the language of thermodynamics essential to our future aerodynamic applications. We continue to develop this language.

Figures 4.9 and 4.10 illustrate systems to which heat δq is added and on which work δw is done. At the same time, δq and δw may cause the pressure, temperature, and density of the system to change. The way (or means) by which changes of the thermodynamic variables (p, T, ρ, v) of a system take place is called a *process*. For example, a *constant-volume process* is illustrated at the left in Fig. 4.11. Here, the system is a gas inside a rigid boundary, such as a hollow steel sphere, and therefore the volume of the system always remains constant. If an amount of heat δq is added to this system, p and T will change. Thus, by definition, such changes take place at constant volume; this is a constant-volume process. Another example is given at the right in Fig. 4.11. Here the system is a gas inside a cylinder-piston arrangement. Consider that heat δq is added to the system, and at the same time assume the piston is moved in just exactly the right way to maintain a constant pressure inside the system. When δq is added to this system, T and v (hence ρ) will change. By definition, such changes take place at constant pressure; this is a constant-pressure process. There are many different kinds of processes treated in thermodynamics. The above are only two examples.

The last concept to be introduced in this section is that of specific heat. Consider a system to which a small amount of heat δq is added. The addition of δq will cause a small change in temperature dT of the system. By definition, specific heat is the heat added per unit change in temperature of the system. Let c denote specific heat. Thus,

$$c \equiv \frac{\delta q}{dT}$$

However, with this definition, c is multivalued. That is, for a fixed quantity δq, the resulting value of dT can be different, depending on the type of process in which δq is added. In turn, the value of c depends on the type of process. Therefore, in principle

Constant-volume process   Constant-pressure process

Figure 4.11   Illustration of constant-volume and constant-pressure processes.

we can define more precisely a different specific heat for each type of process. We will be interested in only two types of specific heat, one at constant volume and the other at constant pressure, as follows.

If the heat δq is added at *constant volume* and it causes a change in temperature dT, the *specific heat at constant volume* c_v is defined as

$$c_v \equiv \left(\frac{\delta q}{dT} \right)_{\text{constant volume}}$$

or $\delta q = c_v \, dT$ (constant volume)   **[4.20]**

On the other hand, if δq is added at constant pressure and it causes a change in temperature dT (whose value is different from the dT above), the *specific heat at constant pressure* c_p is defined as

$$c_p \equiv \left(\frac{\delta q}{dt} \right)_{\text{constant pressure}}$$

or $\delta q = c_p \, dT$ (constant pressure)   **[4.21]**

The above definitions of c_v and c_p, when combined with the first law, yield useful relations for internal energy e and enthalpy h as follows. First, consider a constant-volume process, where by definition $dv = 0$. Thus, from the alternate form of the first law, Eq. (4.16),

$$\delta q = de + p \, dv = de + 0 = de \qquad \textbf{[4.22]}$$

Substituting the definition of c_v, Eq. (4.20), into Eq. (4.22), we get

$$\boxed{de = c_v \, dT} \qquad \textbf{[4.23]}$$

By assuming that c_v is a constant, which is very reasonable for air at normal conditions, and letting $e = 0$ when $T = 0$, Eq. (4.23) may be integrated to

$$\boxed{e = c_v T} \qquad \textbf{[4.24]}$$

Next, consider a constant-pressure process, where by definition $dp = 0$. From the alternate form of the first law, Eq. (4.19),

$$\delta q = dh - v \, dp = dh - 0 = dh \qquad \textbf{[4.25]}$$

Substituting the definition of c_p, Eq. (4.21), into Eq. (4.25), we find

$$\boxed{dh = c_p \, dT} \qquad \textbf{[4.26]}$$

Again, assuming that c_p is constant and letting $h = 0$ at $T = 0$, we see that Eq. (4.26) yields

$$\boxed{h = c_p T} \qquad \textbf{[4.27]}$$

Equations (4.23) to (4.27) are very important relationships. They have been derived from the first law, into which the definitions of specific heat have been inserted. Look at them! They relate thermodynamic variables *only* (*e* to *T* and *h* to *T*); work and heat do not appear in these equations. In fact, Eqs. (4.23) to (4.27) are quite general. Even though we used examples of constant volume and constant pressure to obtain them, they hold *in general* as long as the gas is a perfect gas (no intermolecular forces). Hence, for *any* process,

$$de = c_v \, dT$$
$$dh = c_p \, dT$$
$$e = c_v T$$
$$h = c_p T$$

This generalization of Eqs. (4.23) to (4.27) to any process may not seem logical and may be hard to accept; nevertheless, it is valid, as can be shown by good thermodynamic arguments beyond the scope of this book. For the remainder of our discussions, we will make frequent use of these equations to relate internal energy and enthalpy to temperature.

4.6 ISENTROPIC FLOW

We are almost ready to return to our consideration of aerodynamics. However, there is one more concept we must introduce, a concept that bridges both thermodynamics and compressible aerodynamics, namely, that of *isentropic flow*.

First, consider three more definitions:

An *adiabatic process* is one in which no heat is added or taken away; $\delta q = 0$.

A *reversible process* is one in which no frictional or other dissipative effects occur.

An *isentropic process* is one which is both adiabatic and reversible.

Thus, an isentropic process is one in which there is neither heat exchange nor any effect due to friction. (The source of the word *isentropic* is another defined thermodynamic variable, called *entropy*. Entropy is constant for an isentropic process. A discussion of entropy is not vital to our discussion here; therefore, no further elaboration is given.)

Isentropic processes are very important in aerodynamics. For example, consider the flow of air over the airfoil shown in Fig. 4.7. Imagine a fluid element moving along one of the streamlines. There is *no* heat being added or taken away from this fluid element; heat-exchange mechanisms such as heating by a flame, cooling in a refrigerator, or intense radiation absorption are all ruled out by the nature of the physical problem we are considering. Thus, the flow of the fluid element along the streamline is *adiabatic*. At the same time, the shearing stress exerted on the surface of the fluid element due to friction is generally quite small and can be neglected (except very near the surface, as will be discussed later). Thus, the flow is also frictionless. [Recall that this same assumption was used in obtaining the momentum equation, Eq. (4.8).] Hence, the flow of the fluid element is both adiabatic and reversible (frictionless); that is, the flow is *isentropic*. Other aerodynamic flows can also be treated as isentropic, for example, the flows through wind tunnel nozzles and rocket engines.

Note that even though the flow is adiabatic, the temperature need not be constant. Indeed, the temperature of the fluid element can vary from point to point in an adiabatic, compressible flow. This is because the volume of the fluid element (of fixed mass) changes as it moves through regions of different density along the streamline; when the volume varies, work is done [Eq. (4.15)], hence the internal energy changes [Eq. (4.10)], and hence the temperature changes [Eq. (4.23)]. This argument holds for compressible flows, where the density is variable. On the other hand, for incompressible flow, where ρ = constant, the volume of the fluid element of fixed mass does not change as it moves along a streamline; hence no work is done and no change in temperature occurs. If the flow over the airfoil in Fig. 4.7 were incompressible, the entire flow field would be at constant temperature. For this reason, temperature is not an important quantity for frictionless incompressible flow. Moreover, our present discussion of isentropic flows is relevant to *compressible* flows only, as explained below.

An isentropic process is more than just another definition. It provides us with several important relationships among the thermodynamic variables T, p, and ρ at two different points (say, points 1 and 2 in Fig. 4.7) along a given streamline. These relations are obtained as follows. Since the flow is isentropic (adiabatic and reversible), $\delta q = 0$. Thus, from Eq. (4.16),

$$\delta q = de + p\, dv = 0$$
$$-p\, dv = de \tag{4.28}$$

Substitute Eq. (4.23) into (4.28):

$$-p\, dv = c_v\, dT \tag{4.29}$$

In the same manner, using the fact that $\delta q = 0$ in Eq. (4.19), we also obtain

$$\delta q = dh - v\, dp = 0$$
$$v\, dp = dh \tag{4.30}$$

Substitute Eq. (4.26) into (4.30):

$$v\, dp = c_p\, dT \tag{4.31}$$

Divide Eq. (4.29) by (4.31):

$$\frac{-p\, dv}{v\, dp} = \frac{c_v}{c_p}$$

or

$$\frac{dp}{p} = -\frac{c_p}{c_v}\frac{dv}{v} \tag{4.32}$$

The ratio of specific heats c_p/c_v appears so frequently in compressible flow equations that it is given a symbol all its own, usually γ; $c_p/c_v \equiv \gamma$. For air at normal conditions, which apply to the applications treated in this book, both c_p and c_v are constants, and hence $\gamma = \text{constant} = 1.4$ (for air). Also $c_p/c_v \equiv \gamma = 1.4$ (for air at normal conditions). Thus, Eq. (4.32) can be written as

$$\frac{dp}{p} = -\gamma \frac{dv}{v} \qquad [4.33]$$

Referring to Fig. 4.7, we integrate Eq. (4.33) between points 1 and 2:

$$\int_{p_1}^{p_2} \frac{dp}{p} = -\gamma \int_{v_1}^{v_2} \frac{dv}{v}$$

$$\ln \frac{p_2}{p_1} = -\gamma \ln \frac{v_2}{v_1}$$

$$\frac{p_2}{p_1} = \left(\frac{v_2}{v_1}\right)^{-\gamma} \qquad [4.34]$$

Since $v_1 = 1/\rho_1$ and $v_2 = 1/\rho_2$, Eq. (4.34) becomes

$$\boxed{\frac{p_2}{p_1} = \left(\frac{\rho_2}{\rho_1}\right)^{\gamma}} \quad \text{isentropic flow} \qquad [4.35]$$

From the equation of state, we have $\rho = p/(RT)$. Thus, Eq. (4.35) yields

$$\frac{p_2}{p_1} = \left(\frac{p_2}{RT_2}\frac{RT_1}{p_1}\right)^{\gamma}$$

$$\left(\frac{p_2}{p_1}\right)^{1-\gamma} = \left(\frac{T_1}{T_2}\right)^{\gamma} = \left(\frac{T_2}{T_1}\right)^{-\gamma}$$

or

$$\boxed{\frac{p_2}{p_1} = \left(\frac{T_2}{T_1}\right)^{\gamma/(\gamma-1)}} \quad \text{isentropic flow} \qquad [4.36]$$

Combining Eqs. (4.35) and (4.36), we obtain

$$\boxed{\frac{p_2}{p_1} = \left(\frac{\rho_2}{\rho_1}\right)^{\gamma} = \left(\frac{T_2}{T_1}\right)^{\gamma/(\gamma-1)}} \quad \text{isentropic flow} \qquad [4.37]$$

The relationships given in Eq. (4.37) are powerful. They provide important information for p, T, and ρ between two different points on a streamline in an isentropic flow. Moreover, if the streamlines all emanate from a uniform flow far upstream (far to the left in Fig. 4.7), then Eq. (4.37) holds for any two points in the flow, not necessarily those on the same streamline.

We emphasize again that the isentropic flow relations, Eq. (4.37), are relevant to compressible flows only. By contrast, the assumption of incompressible flow (remember, incompressible flow is a myth, anyway) is not consistent with the same physics that went into the development of Eq. (4.37). To analyze incompressible flows, we need only the continuity equation [say, Eq. (4.3)] and the momentum equation [Bernoulli's equation, Eqs. (4.9a) and (4.9b)]. To analyze compressible flows, we need the continuity equation, Eq. (4.2), the momentum equation [Euler's equation, Eq. (4.8)], and another soon-to-be-derived relation called the *energy equation*. If the compressible flow is isentropic, then Eq. (4.37) can be used to replace either the momentum or the energy equation. Since Eq. (4.37) is a simpler, more useful algebraic relation than Euler's equation, Eq. (4.8), which is a differential equation, we frequently use Eq. (4.37) in place of Eq. (4.8) for the analysis of compressible flows in this book.

As mentioned above, to complete the development of the fundamental relations for the analysis of compressible flow, we must now consider the energy equation.

Example 4.6

An airplane is flying at standard sea-level conditions. The temperature at a point on the wing is 250 K. What is the pressure at this point?

Solution

The air pressure and temperature, p_1 and T_1, far upstream of the wing correspond to standard sea level. Hence, $p_1 = 1.01 \times 10^5$ N/m^2 and $T_1 = 288.16$ K. Assume the flow is isentropic (hence compressible). Then, the relation between points 1 and 2 is obtained from Eq. (4.37):

$$\frac{p_2}{p_1} = \left(\frac{T_2}{T_1}\right)^{\gamma/(\gamma-1)}$$

$$p_2 = p_1 \left(\frac{T_2}{T_1}\right)^{\gamma/(\gamma-1)} = (1.01 \times 10^5)\left(\frac{250}{288.16}\right)^{1.4/(1.4-1)}$$

$$\boxed{p_2 = 6.14 \times 10^4 \text{ N/m}^2}$$

4.7 ENERGY EQUATION

Recall that our approach to the derivation of the fundamental equations for fluid flow is to state a fundamental principle and then to proceed to cast that principle in terms of flow variables p, T, ρ, and V. Also recall that compressible flow, high-speed flow, and massive changes in energy go hand in hand. Therefore, our last fundamental physical principle that we must take into account is as follows.

Physical principle: Energy can be neither created nor destroyed. It can only change form.

In quantitative form, this principle is nothing more than the first law of thermo-dynamics, Eq. (4.10). To apply this law to fluid flow, consider again a fluid element moving along a streamline, as shown in Fig. 4.6. Let us apply the first law of thermodynamics

$$\delta q + \delta w = de$$

to this fluid element. Recall that an alternate form of the first law is Eq. (4.19),

$$\delta q = dh - v\,dp$$

Again, we consider an adiabatic flow, where $\delta q = 0$. Hence, from Eq. (4.19),

$$dh - v\,dp = 0 \tag{4.38}$$

Recalling Euler's equation, Eq. (4.8),

$$dp = -\rho V\,dV$$

we can combine Eqs. (4.38) and (4.8) to obtain

$$dh + v\rho V\,dV = 0 \tag{4.39}$$

However, $v = 1/\rho$, hence Eq. (4.39) becomes

$$dh + V\,dV = 0 \tag{4.40}$$

Integrating Eq. (4.40) between two points along the streamline, we obtain

$$\int_{h_1}^{h_2} dh + \int_{V_1}^{V_2} V\,dV = 0$$

$$h_2 - h_1 + \frac{V_2^2}{2} - \frac{V_1^2}{2} = 0$$

$$\boxed{\begin{aligned} h_1 + \frac{V_1^2}{2} &= h_2 + \frac{V_2^2}{2} \\ h + \frac{V^2}{2} &= \text{const} \end{aligned}} \tag{4.41}$$

Equation (4.41) is the energy equation for frictionless, adiabatic flow. It can be written in terms of T by using Eq. (4.27), $h = c_p T$. Hence, Eq. (4.41) becomes

$$\boxed{\begin{aligned} c_p T_1 + \tfrac{1}{2}V_1^2 &= c_p T_2 + \tfrac{1}{2}V_2^2 \\ c_p T + \tfrac{1}{2}V^2 &= \text{const} \end{aligned}} \tag{4.42}$$

Equation (4.42) relates the temperature and velocity at two different points along a streamline. Again, if all the streamlines emanate from a uniform flow far upstream, then Eq. (4.42) holds for any two points in the flow, not necessarily on the same streamline. Moreover, Eq. (4.42) is just as powerful and necessary for the analysis of compressible flow as Eq. (4.37).

Example 4.7

A supersonic wind tunnel is sketched in Fig. 4.26. The air temperature and pressure in the reservoir of the wind tunnel are $T_0 = 1000$ K and $p_0 = 10$ atm, respectively. The static temperatures at the throat and exit are $T^* = 833$ K and $T_e = 300$ K, respectively. The mass flow through the nozzle is 0.5 kg/s. For air, $c_p = 1008$ J/(kg)(K). Calculate:
a. Velocity at the throat V^*
b. Velocity at the exit V_e
c. Area of the throat A^*
d. Area of the exit A_e

Solution

Since the problem deals with temperatures and velocities, the energy equation seems useful.
a. From Eq. (4.42), written between the reservoir and the throat,

$$c_p T_0 + \tfrac{1}{2} V_0^2 = c_p T^* + \tfrac{1}{2} V^{*2}$$

However, in the reservoir, $V_0 \approx 0$. Hence,

$$V^* = \sqrt{2 c_p (T_0 - T^*)}$$

$$= \sqrt{2(1008)(1000 - 833)} = \boxed{580 \text{ m/s}}$$

b. From Eq. (4.42) written between the reservoir and the exit,

$$c_p T_0 = c_p T_e + \tfrac{1}{2} V_e^2$$

$$V_e = \sqrt{2 c_p (T_0 - T_e)}$$

$$= \sqrt{2(1008)(1000 - 300)} = \boxed{1188 \text{ m/s}}$$

c. The basic equation dealing with mass flow and area is the continuity equation, Eq. (4.2). Note that the velocities are certainly large enough that the flow is compressible, hence Eq. (4.2), rather than Eq. (4.3), is appropriate

$$\dot{m} = \rho^* A^* V^*$$

or

$$A^* = \frac{\dot{m}}{\rho^* V^*}$$

In the above, $\dot{m}$ is given and V^* is known from part a. However, ρ^* must be obtained before we can calculate A^* as desired. To obtain ρ^*, note that, from the equation of state,

$$\rho_0 = \frac{p_0}{R T_0} = \frac{10(1.01 \times 10^5)}{287(1000)} = 3.52 \text{ kg/m}^3$$

Assuming the nozzle flow is isentropic, which is a very good approximation for the real case, from Eq. (4.37), we get

$$\left(\frac{\rho^*}{\rho_0} \right) = \left(\frac{T^*}{T_0} \right)^{1/(\gamma - 1)}$$

$$\rho^* = \rho_0 \left(\frac{T^*}{T_0} \right)^{1/(\gamma - 1)} = (3.52) \left(\frac{833}{1000} \right)^{1/(1.4 - 1)} = 2.23 \text{ kg/m}^3$$

Thus

$$A^* = \frac{\dot{m}}{\rho^* V^*} = \frac{0.5}{(2.23)(580)} = \boxed{3.87 \times 10^{-4} \text{ m}^2 = 3.87 \text{ cm}^2}$$

d. Finding A_e is similar to the above solution for A^*

$$\dot{m} = \rho_e A_e V_e$$

where, for isentropic flow,

$$\rho_e = \rho_0 \left(\frac{T_e}{T_0}\right)^{1/(\gamma-1)} = (3.52)\left(\frac{300}{1000}\right)^{1/(1.4-1)} = 0.174 \text{ kg/m}^3$$

or $\qquad A_e = \dfrac{\dot{m}}{\rho_e V_e} = \dfrac{0.5}{0.174(1188)} = \boxed{24.2 \times 10^{-4} \text{ m}^2 = 24.2 \text{ cm}^2}$

Consider an airfoil in a flow of air, where far ahead of the airfoil (the freestream), the pressure, velocity, and density are 2116 lb/ft^2, 500 mi/h, and 0.002377 slug/ft^3, respectively. At a given point A on the airfoil, the pressure is 1497 lb/ft^2. What is the velocity at point A? Assume isentropic flow. For air, $c_p = 6006$ ft·lb/(slug)(°R). | **Example 4.8**

Solution

This example is identical to Example 4.3, except here the velocity is 500 mi/h—high enough that we have to treat the flow as compressible, in contrast to Example 4.3 where we dealt with incompressible flow. Since the flow is isentropic, we can use Eq. (4.37) evaluated between the freestream and point A.

$$\frac{p_A}{p_\infty} = \left(\frac{T_A}{T_\infty}\right)^{\gamma/(\gamma-1)}$$

or $\qquad \dfrac{T_A}{T_\infty} = \left(\dfrac{p_A}{p_\infty}\right)^{(\gamma-1)/\gamma} = \left(\dfrac{1497}{2116}\right)^{0.4/1.4} = (0.7075)^{0.286} = 0.9058$

The value of T_∞ can be found from the equation of state.

$$T_\infty = \frac{p_2}{\rho_\infty R} = \frac{2116}{0.002377(1716)} = 519 \text{ °R}$$

Hence $\qquad T_A = 0.9058(519) = 470.1°\text{R}$

From the energy equation, Eq. (4.42), evaluated between the freestream and point A, and noting that $V_\infty = 500(88/60) = 733.3$ ft/s, we have

$$c_p T_\infty + \frac{V_\infty^2}{2} = c_p T_A + \frac{V_A^2}{2}$$

or $\qquad V_A = \sqrt{2c_p(T_A - T_A) + V_\infty^2}$

$$= \sqrt{2(6006)(519 - 470.1) + (733.3)^2} = \boxed{1061 \text{ ft/s}}$$

Note: The calculational procedure for this problem, where we are dealing with compressible flow, is completely different from that for Example 4.3, where we were dealing with incompressible flow. In Example 4.3, we could use Bernoulli's equation, which holds only for incompressible flow. *We cannot use Bernoulli's equation to solve the present problem, because this is a compressible flow problem and Bernoulli's equation is not valid for a compressible flow.* If we had used Bernoulli's equation to solve the present problem, following exactly the method in Example 4.3, we would have obtained a velocity of 1029 ft/s at point *A*—an *incorrect* answer. Check this yourself.

4.8   SUMMARY OF EQUATIONS

We have just finished applying some very basic physical principles to obtain equations for the analysis of flowing gases. The reader is cautioned not to be confused by the multiplicity of equations; they are useful, indeed necessary, tools to examine and solve various aerodynamic problems of interest. It is important for an engineer or scientist to look at such equations and see not just a mathematical relationship, but primarily a physical relationship. These equations talk! For example, Eq. (4.2) says that mass is conserved; Eq. (4.42) says that energy is conserved for an adiabatic, frictionless flow; etc. Never lose sight of the physical implications and limitations of these equations.

To help set these equations in your mind, here is a compact summary of our results so far:

1.   For the steady incompressible flow of a frictionless fluid in a stream tube of varying area, p and V are the meaningful flow variables; ρ and T are constants throughout the flow. To solve for p and V, use

$$A_1 V_1 = A_2 V_2 \qquad \text{continuity}$$

$$p_1 + \tfrac{1}{2}\rho V_1^2 = p_2 + \tfrac{1}{2}\rho V_2^2 \qquad \text{Bernoulli's equation}$$

2.   For steady isentropic (adiabatic and frictionless) compressible flow in a stream tube of varying area, p, ρ, T, and V are all variables. They are obtained from

$$\rho_1 A_1 V_1 = \rho_2 A_2 V_2 \qquad \text{continuity}$$

$$\frac{p_1}{p_2} = \left(\frac{\rho_1}{\rho_2}\right)^\gamma = \left(\frac{T_1}{T_2}\right)^{\gamma/(\gamma-1)} \qquad \text{isentropic relations}$$

$$c_p T_1 + \tfrac{1}{2}V_1^2 = c_p T_2 + \tfrac{1}{2}V_2^2 \qquad \text{energy}$$

$$p_1 = \rho_1 R T_1$$
$$p_2 = \rho_2 R T_2 \qquad \text{equation of state}$$

Let us now apply these relations to study some basic aerodynamic phenomena and problems.

4.9   SPEED OF SOUND

Sound waves travel through the air at a definite speed—the speed of sound. This is obvious from natural observation; a lightning bolt is observed in the distance, and thunder is heard at some later instant. In many aerodynamic problems, the speed of sound plays a pivotal role. How do we calculate the speed of sound? What does it depend on: pressure, temperature, density, or some combination thereof? Why is it so important? Answers to these questions are discussed in this section.

First let us derive a formula to calculate the speed of sound. Consider a sound wave moving into a stagnant gas, as shown in Fig. 4.12. This sound wave is created by some source, say, a small firecracker in the corner of a room. The air in the room is motionless and has density ρ, pressure p, and temperature T. If you are standing in the middle of the room, the sound wave sweeps by you at velocity a m/s, ft/s, etc. The sound wave itself is a thin region of disturbance in the air, across which the pressure, temperature, and density change slightly. (The change in pressure is what activates your eardrum and allows you to hear the sound wave.) Imagine that you now hop on the sound wave and move with it. As you are sitting on the moving wave, look to the left in Fig. 4.12, that is, look in the direction in which the wave is moving. From your vantage point on the wave, the sound wave seems to stand still and the air in front of the wave appears to be coming at you with velocity a; that is, you see the picture shown in Fig. 4.13, where the sound wave is standing still and the air ahead of the wave is moving toward the wave with velocity a. Now, return to Fig. 4.12 for a moment. Sitting on top of and riding with the moving wave, look to the right, that is, look behind the wave. From your vantage point, the air appears to be moving away from you. This appearance is sketched in Fig. 4.13, where the wave is standing still. Here, the air behind the motionless wave is moving to the right, away from the wave. However, in passing through the wave, the pressure, temperature, and density

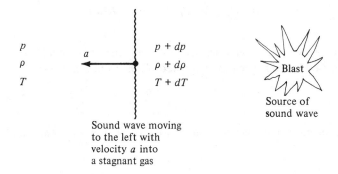

Figure 4.12 Model of a sound wave moving into a stagnant gas.

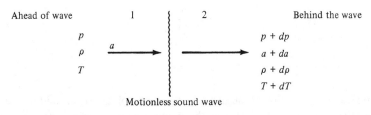

Figure 4.13 Model with the sound wave stationary.

of the air are slightly changed by the amounts dp, dT, and $d\rho$, respectively. From our previous discussions, you would then expect the airspeed a to change slightly, say, by an amount da. Thus, the air behind the wave is moving away from the wave with velocity $a + da$, as shown in Fig. 4.13. Figures 4.12 and 4.13 are completely analogous pictures; only their perspectives are different. Figure 4.12 is what you see by standing in the middle of the room and watching the wave go by; Fig. 4.13 is what you see by riding on top of the wave and watching the air go by. Both pictures are equivalent. However, Fig. 4.13 is easier to work with, so we will concentrate on it.

Let us apply our fundamental equations to the gas flow shown in Fig. 4.13. Our objective is to obtain an equation for a, where a is the speed of the sound wave, the speed of sound. Let points 1 and 2 be ahead of and behind the wave, respectively, as shown in Fig. 4.13. Applying the continuity equation, Eq. (4.2), we find

$$\rho_1 A_1 V_1 = \rho_2 A_2 V_2$$

or
$$\rho A_1 a = (\rho + d\rho) A_2 (a + da) \qquad \textbf{[4.43]}$$

Here, A_1 and A_2 are the areas of a stream tube running through the wave. Just looking at the picture shown in Fig. 4.13, we see no geometric reason why the stream tube should change area in passing through the wave. Indeed, it does not; the area of the stream tube is constant, hence $A = A_1 = A_2 = $ constant. (This is an example of a type of flow called one-dimensional, or constant-area, flow.) Thus, Eq. (4.43) becomes

$$\rho a = (\rho + d\rho)(a + da)$$

or
$$\rho a = \rho a + a\, d\rho + \rho\, da + d\rho\, da \qquad \textbf{[4.44]}$$

The product of two small quantities $d\rho\, da$ is very small in comparison to the other terms in Eq. (4.44) and hence can be ignored. Thus, from Eq. (4.44),

$$a = -\rho\, \frac{da}{d\rho} \qquad \textbf{[4.45]}$$

Now apply the momentum equation in the form of Euler's equation, Eq. (4.8):

$$dp = -\rho a\, da$$

or
$$da = -\frac{dp}{\rho a} \qquad \textbf{[4.46]}$$

Substitute Eq. (4.46) into (4.45):

$$a = \frac{\rho}{d\rho} \frac{dp}{\rho a}$$

or
$$a^2 = \frac{dp}{d\rho} \qquad \textbf{[4.47]}$$

On a physical basis, the flow through a sound wave involves no heat addition, and the effect of friction is negligible. Hence, the flow through a sound wave is isentropic.

Thus, from Eq. (4.47), the speed of sound is given by

$$a = \sqrt{\left(\frac{dp}{d\rho}\right)_{\text{isentropic}}}$$ [4.48]

Equation (4.48) is fundamental and important. However, it does not give us a straight-forward formula for computing a number for a. We must proceed further.

For isentropic flow, Eq. (4.37) gives

$$\frac{p_2}{p_1} = \left(\frac{\rho_2}{\rho_1}\right)^{\gamma}$$

or $$\frac{p_2}{\rho_2^{\gamma}} = \frac{p_1}{\rho_1^{\gamma}} = \text{const} = c$$ [4.49]

Equation (4.49) says that the ratio p/ρ^{γ} is the same constant value at every point in an isentropic flow. Thus, we can write everywhere

$$\frac{p}{\rho^{\gamma}} = c$$ [4.50]

Hence $$\left(\frac{dp}{d\rho}\right)_{\text{isentropic}} = \frac{d}{d\rho} c\rho^{\gamma} = c\gamma\rho^{\gamma-1}$$ [4.51]

Substituting for c in Eq. (4.51) the ratio of Eq. (4.50), we obtain

$$\left(\frac{dp}{d\rho}\right)_{\text{isentropic}} = \frac{p}{\rho^{\gamma}}\gamma\rho^{\gamma-1} = \frac{\gamma p}{\rho}$$ [4.52]

Substitute Eq. (4.52) into (4.48):

$$a = \sqrt{\gamma\frac{p}{\rho}}$$ [4.53]

However, for a perfect gas, p and ρ are related through the equation of state; $p = \rho RT$, hence $p/\rho = RT$. Substituting this result into Eq. (4.53) yields

$$a = \sqrt{\gamma RT}$$ [4.54]

Equations (4.48), (4.53), and (4.54) are important results for the speed of sound; however, Eq. (4.54) is the most useful. It also demonstrates a fundamental result: *The speed of sound in a perfect gas depends only on the temperature of the gas.* This simple result may appear surprising at first. However, it is to be expected on a physical basis, as follows. The propagation of a sound wave through a gas takes place via molecular collisions. For example, consider again a small firecracker in the corner of the room. When the firecracker is set off, some of its energy is transferred to the neighboring gas molecules in the air, thus increasing their kinetic energy. In turn, these energetic gas molecules are moving randomly about, colliding with some of their neighboring molecules and transferring some of their extra energy to these new molecules. Thus, the energy of a sound wave is transmitted through the air by molecules which collide with one another. Each molecule is moving at a different velocity, but if they are summed over a large number of molecules, a mean or average molecular velocity can be defined. Therefore, looking at the collection of molecules as a whole, we see that the sound energy released by the firecrackers will be transferred through the air at something approximating this mean molecular velocity. Recall from Chap. 2 that temperature is a measure of the mean molecular kinetic energy, hence of the mean molecular velocity; then temperature should also be a measure of the speed of a sound wave transmitted by molecular collisions. Equation (4.54) proves this to be a fact.

For example, consider air at standard sea-level temperature $T_s = 288.16$ K. From Eq. (4.54), the speed of sound is $a = \sqrt{\gamma R T} = \sqrt{1.4(287)(288.16)} = 340.3$ m/s. From the results of the kinetic theory of gases, the mean molecular velocity can be obtained as $\overline{V} = \sqrt{(8/\pi)RT} = \sqrt{(8/\pi)287(288.16)} = 458.9$ m/s. Thus, the speed of sound is of the same order of magnitude as the mean molecular velocity and is smaller by about 26 percent.

Again we emphasize that the speed of sound is a point property of the flow, in the same vein as T is a point property as described in Chap. 2. It is also a thermodynamic property of the gas, defined by Eqs. (4.48) to (4.54). In general, the value of the speed of sound varies from point to point in the flow.

The speed of sound leads to another, vital definition for high-speed gas flows, namely, the *Mach number*. Consider a point B in a flow field. The flow velocity at B is V, and the speed of sound is a. By definition, the Mach number M at point B is the flow velocity divided by the speed of sound:

$$M = \frac{V}{a}$$ [4.55]

We will find that M is one of the most powerful quantities in aerodynamics. We can immediately use it to define three different regimes of aerodynamic flows:

1. If $M < 1$, the flow is *subsonic*.
2. If $M = 1$, the flow is *sonic*.
3. If $M > 1$, the flow is *supersonic*.

Each of these regimes is characterized by its own special phenomena, as will be discussed in subsequent sections. In addition, two other specialized aerodynamic regimes are commonly defined, namely, *transonic* flow, where M generally ranges from slightly less than to slightly greater than 1 (for example, $0.8 \leq M \leq 1.2$), and *hypersonic* flow, where generally $M > 5$. The definitions of subsonic, sonic, and supersonic flows in terms of M as given above are precise; the definitions of transonic and hypersonic flows in terms of M are a bit more imprecise and really refer to sets of specific aerodynamic phenomena, rather than to just the value of M. This distinction will be clarified in subsequent sections.

Example 4.9 | **A** jet transport is flying at a standard altitude of 30,000 ft with a velocity of 550 mi/h. What is its Mach number?

Solution

From the standard atmosphere table, App. B, at 30,000 ft, $T_\infty = 411.86°R$. Hence, from Eq. (4.54),

$$a_\infty = \sqrt{\gamma RT} = \sqrt{1.4(1716)(411.86)} = 995 \text{ ft/s}$$

The airplane velocity is $V_\infty = 550$ mi/h; however, in consistent units, remembering that 88 ft/s = 60 mi/h, we find that

$$V_\infty = 550 \left(\frac{88}{60}\right) = 807 \text{ ft/s}$$

From Eq. (4.55)

$$M_\infty = \frac{V_\infty}{a_\infty} = \frac{807}{995} = \boxed{0.811}$$

In the nozzle flow described in Example 4.7, calculate the Mach number of the flow at the throat, M^* and at the exit M_e. | **Example 4.10**

Solution

From Example 4.7, at the throat, $V^* = 580$ m/s and $T^* = 833$ K. Hence, from Eq. (4.54),

$$a^* = \sqrt{\gamma RT^*} = \sqrt{1.4(287)(833)} = 580 \text{ m/s}$$

From Eq. (4.55)

$$M^* = \frac{V^*}{a^*} = \frac{580}{580} = \boxed{1}$$

Note: The flow is sonic at the throat. We will soon prove that the Mach number at the throat is always sonic in supersonic nozzle flows (except in special, nonequilibrium, high-temperature flows, which are beyond the scope of this book).

Also from Example 4.7, at the exit, $V_e = 1188$ m/s and $T_e = 300$ K. Hence,

$$a_e = \sqrt{\gamma RT_e} = \sqrt{1.4(287)(300)} = 347 \text{ m/s}$$

$$M_e = \frac{V_e}{a_e} = \frac{1188}{347} = \boxed{3.42}$$

4.10 LOW-SPEED SUBSONIC WIND TUNNELS

Throughout the remainder of this book, the aerodynamic fundamentals and tools (equations) developed in previous sections will be applied to specific problems of interest. The first will be a discussion of low-speed subsonic wind tunnels.

What are wind tunnels, any kind of wind tunnels? In the most basic sense, they are ground-based experimental facilities designed to produce flows of air (or sometimes other gases) which simulate natural flows occurring outside the laboratory. For most aerospace engineering applications, wind tunnels are designed to simulate flows encountered in the flight of airplanes, missiles, or space vehicles. Since these flows have ranged from the 27 mi/h speed of the early *Wright Flyer* to the 25,000 mi/h reentry velocity of the Apollo lunar spacecraft, obviously many different types of wind tunnels, from low subsonic to hypersonic, are necessary for the laboratory simulation of actual flight conditions. However, referring again to Fig. 1.30, we see that flow velocities of 300 mi/h or less were the flight regime of interest until about 1940. Hence, during the first four decades of human flight, airplanes were tested and developed in wind tunnels designed to simulate low-speed subsonic flight. Such tunnels are still in use today but are complemented by transonic, supersonic, and hypersonic wind tunnels as well.

The essence of a typical low-speed subsonic wind tunnel is sketched in Fig. 4.14. The airflow with pressure p_1 enters the nozzle at a low velocity V_1, where the area is A_1. The nozzle converges to a smaller area A_2 at the test section. Since we are dealing with low-speed flows, where M is generally less than 0.3, the flow will be assumed to be incompressible. Hence, Eq. (4.3) dictates that the flow velocity increases as the air flows through the convergent nozzle. The velocity in the test section is then, from Eq. (4.3),

$$V_2 = \frac{A_1}{A_2} V_1 \qquad \text{[4.56]}$$

After flowing over an aerodynamic model (which may be a model of a complete airplane or part of an airplane, such as a wing, tail, or engine nacelle), the air passes into a diverging duct called a diffuser, where the area increases and velocity decreases to A_3 and V_3, respectively. Again, from continuity,

$$V_3 = \frac{A_2}{A_3} V_2$$

The pressure at various locations in the wind tunnel is related to the velocity, through Bernoulli's equation, Eq. (4.9a), for incompressible flow.

$$p_1 + \tfrac{1}{2}\rho V_1^2 = p_2 + \tfrac{1}{2}\rho V_2^2 = p_3 + \tfrac{1}{2}\rho V_3^2 \qquad \text{[4.57]}$$

From Eq. (4.57), as V increases, p decreases; hence $p_2 < p_1$, that is, the test section pressure is smaller than the reservoir pressure upstream of the nozzle. In many subsonic wind tunnels, all or part of the test section is open, or vented, to the surrounding

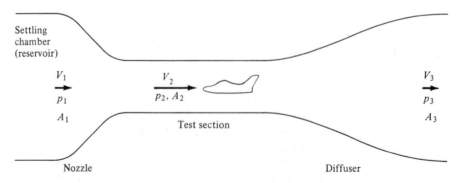

Figure 4.14   Simple schematic of a subsonic wind tunnel.

air in the laboratory. In such cases, the outside air pressure is communicated directly to the flow in the test section, and $p_2 = 1$ atm. Downstream of the test section, in the diverging area diffuser, the pressure increases as velocity decreases. Hence, $p_3 > p_2$. If $A_3 = A_1$, then from Eq. (4.56), $V_3 = V_1$; and from Eq. (4.57), $p_3 = p_1$. (*Note:* In actual wind tunnels, the aerodynamic drag created by the flow over the model in the test section causes a loss of momentum not included in the derivation of Bernoulli's equation; hence in reality, p_3 is slightly less than p_1, because of such losses.)

In practical operation of this type of wind tunnel, the test section velocity is governed by the pressure difference $p_1 - p_2$ and the area ratio of the nozzle A_2/A_1 as follows. From Eq. (4.57),

$$V_2^2 = \frac{2}{\rho}(p_1 - p_2) + V_1^2 \qquad \text{[4.58]}$$

From Eq. (4.56), $V_1 = (A_2/A_1)V_2$. Substituting this into the right-hand side of Eq. (4.58), we obtain

$$V_2^2 = \frac{2}{\rho}(p_1 - p_2) + \left(\frac{A_2}{A_1}\right)^2 V_2^2 \qquad \text{[4.59]}$$

Solving Eq. (4.59) for V_2 yields

$$\boxed{V_2 = \sqrt{\frac{2(p_1 - p_2)}{\rho[1 - (A_2/A_1)^2]}}} \qquad \text{[4.60]}$$

The area ratio A_2/A_1 is a fixed quantity for a wind tunnel of given design. Hence, the "control knob" of the wind tunnel basically controls $p_1 - p_2$, which allows the wind tunnel operator to control the value of test section velocity V_2 via Eq. (4.60).

In subsonic wind tunnels, a convenient method of measuring the pressure difference $p_1 - p_2$, hence of measuring V_2 via Eq. (4.60), is by means of a *manometer*. A basic type of manometer is the U tube shown in Fig. 4.15. Here, the left side of the tube is connected to a pressure p_1, the right side of the tube is connected to a pressure

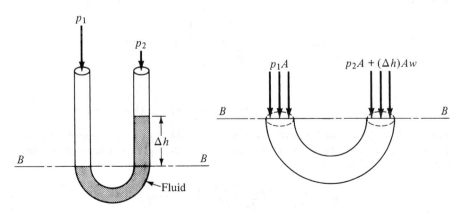

Figure 4.15   Force diagram for a manometer.

p_2, and the difference Δh in the heights of a fluid in both sides of the U tube is a measurement of the pressure difference $p_2 - p_1$. This can easily be demonstrated by considering the force balance on the liquid in the tube at the two cross sections cut by plane B-B, shown in Fig. 4.15. Plane B-B is drawn tangent to the top of the column of fluid on the left. If A is the cross-sectional area of the tube, then $p_1 A$ is the force exerted on the left column of fluid. The force on the right column at plane B-B is the sum of the weight of the fluid above plane B-B and the force due to the pressure $p_2 A$. The volume of the fluid in the right column above B-B is $A\,\Delta h$. The specific weight (weight per unit volume) of the fluid is $w = \rho_l g$, where ρ_l is the density of the fluid and g is the acceleration of gravity. Hence, the total weight of the column of fluid above B-B is the specific weight times the volume, that is, $wA\,\Delta h$. The total force on the right-hand cross section at plane B-B is then $p_2 A + wA\,\Delta h$. Since the fluid is stationary in the tube, the forces on the left- and right-hand cross sections must balance, that is, they are the same. Hence,

$$p_1 A = p_2 A + wA\,\Delta h$$

or

$$\boxed{p_1 - p_2 = w\,\Delta h} \qquad \textbf{[4.61]}$$

If the left-hand side of the U-tube manometer were connected to the reservoir in a subsonic tunnel (point 1 in Fig. 4.14) and the right-hand side were connected to the test section (point 2), then Δh of the U tube would directly measure the velocity of the airflow in the test section via Eqs. (4.61) and (4.60).

Example 4.11 | In a low-speed subsonic wind tunnel, one side of a mercury manometer is connected to the settling chamber (reservoir) and the other side is connected to the test section. The contraction ratio of the nozzle A_2/A_1 equals $\frac{1}{15}$. The reservoir pressure and temperature area $p_1 = 1.1$ atm and $T_1 = 300$ K, respectively. When the tunnel is running, the height difference between the two columns of mercury is 10 cm. The density of liquid mercury is 1.36×10^4 kg/m^3. Calculate the airflow velocity in the test section V_2.

Solution

$$\Delta h = 10\ \text{cm} = 0.1\ \text{m}$$
$$w\ (\text{for mercury}) = \rho_l g = (1.36 \times 10^4\ \text{kg/m}^3)(9.8\ \text{m/s}^2)$$
$$= 1.33 \times 10^5\ \text{N/m}^3$$

From Eq. (4.61)

$$p_1 - p_2 = w\,\Delta h = (1.33 \times 10^5\ \text{N/m}^3)(0.1\ \text{m}) = 1.33 \times 10^4\ \text{N/m}^2$$

To find the velocity V_2, use Eq. (4.60). However, in Eq. (4.60) we need a value of density ρ. This can be found from the reservoir conditions by using the equation of state. (Remember: 1 atm $= 1.01 \times 10^5$ N/m^2.)

$$\rho_1 = \frac{p_1}{RT_1} = \frac{1.1(1.01 \times 10^5)}{287(300)} = 1.29\ \text{kg/m}^3$$

Since we are dealing with a low-speed subsonic flow, assume $\rho_1 = \rho = $ constant. Hence, from Eq. (4.60)

$$V_2 = \sqrt{\frac{2(p_1 - p_2)}{\rho[1 - (A_2/A_1)^2]}} = \sqrt{\frac{2(1.33 \times 10^4)}{1.29[1 - (\frac{1}{15})^2]}} = \boxed{144 \text{ m/s}}$$

Note: This answer corresponds to approximately a Mach number of 0.4 in the test section, one slightly above the value of 0.3 that bounds incompressible flow. Hence, our assumption of $\rho = $ constant in this example is inaccurate by about 8 percent.

4.11 MEASUREMENT OF AIRSPEED

In the previous section, we demonstrated that the airflow velocity in the test section of a low-speed wind tunnel (assuming incompressible flow) can be obtained by measuring $p_1 - p_2$. However, the previous analysis implicitly assumes that the flow properties are reasonably constant over any given cross section of the flow in the tunnel (so-called quasi-one-dimensional flow). If the flow is not constant over a given cross section, for example, if the flow velocity in the middle of the test section is higher than that near the walls, then V_2 obtained from the preceding section is only a mean value of the test section velocity. For this reason and for many other aerodynamic applications, it is important to obtain a *point* measurement of velocity at a given spatial location in the flow. This measurement can be made by an instrument called a *Pitot-static tube*, as described below.

First, we must add to our inventory of aerodynamic definitions. We have been glibly talking about the pressures at points in flows, such as points 1 and 2 in Fig. 4.7. However, these pressures are of a special type, called *static*. Static pressure at a given point is the pressure we would feel if we were moving along with the flow at that point. It is the ramification of gas molecules moving about with random motion and transferring their momentum to or across surfaces, as discussed in Chap. 2. If we look more closely at the molecules in a flowing gas, we see that they have a purely random motion superimposed on a directed motion due to the velocity of the flow. Static pressure is a consequence of just the purely random motion of the molecules. When an engineer or scientist uses the word *pressure*, it always means static pressure unless otherwise identified, and we will continue such practice here. In all our previous discussions, the pressures have been static pressures.

A second type of pressure is commonly utilized in aerodynamics, namely, *total* pressure. To define and understand total pressure, consider again a fluid element moving along a streamline, as shown in Fig. 4.6. The pressure of the gas in this fluid element is the static pressure. However, now imagine that we grab this fluid element and slow it down to zero velocity. Moreover, imagine that we do this isentropically. Intuitively, the thermodynamic properties p, T, and ρ of the fluid element will change as we bring the element to rest; they will follow the conservation laws we have discussed previously in this chapter. Indeed, as the fluid element is isentropically brought to rest, p, T, and ρ would all increase above their original values when the element is moving freely along the streamline. The values of p, T, and ρ of the fluid element after it has been brought to rest are called *total* values, that is, total pressure p_0, total temperature T_0, etc. Thus, we are led to the following precise definition:

> Total pressure at a given point in a flow is the pressure that would exist if the flow were slowed down isentropically to zero velocity.

There is a perspective to be gained here. Total pressure p_0 is a property of the gas flow at a given point. It is something that is associated with the flow itself. The process of isentropically bringing the fluid element to rest is just an imaginary mental process we use to define the total pressure. It does not mean that we actually have to do it in practice. In other words, if we consider again the flow sketched in Fig. 4.7, there are *two* pressures we can consider at points 1, 2, etc., associated with each point of the flow: a static pressure p and a total pressure p_0, where $p_0 > p$.

For the special case of a gas that is not moving, that is, the fluid element has no velocity in the first place, then static and total pressures are synonymous: $p_0 = p$. This is the case in common situations such as the stagnant air in the room and gas confined in a cylinder.

The following analogy might help to further illustrate the difference between the definitions of static and total pressure. Assume that you are driving down the highway at 60 mi/h. The windows of your automobile are closed. Inside the automobile, along with you, there is a fly buzzing around in a very random fashion. Your speed is 60 mi/h, and in the mean, so is that of the fly, moving down the highway at 60 mi/h. However, the fly has its random buzzing-about motion superimposed on top of its mean directed speed of 60 mi/h. To you in the automobile, all you see is the random, buzzing-about motion of the fly. If the fly hits your skin with this random motion, you will feel a slight impact. This slight impact is analogous to the *static* pressure in a flowing gas, where the static pressure is due simply to the *random* motion of the molecules. Now assume that you open the window of your automobile, and the fly buzzes out. There is a person standing along the side of the road. If the fly that has just left your automobile hits the skin of this person, the impact will be strong (it may even really hurt) because the fly hits this person with a mean velocity of 60 mi/h plus whatever its random velocity may be. The strength of this impact is analogous to the *total* pressure of a gas.

There is an aerodynamic instrument that actually measures the total pressure at a point in the flow, namely, a *Pitot tube*. A basic sketch of a Pitot tube is shown in Fig. 4.16. It consists of a tube placed parallel to the flow and open to the flow at one end (point A). The other end of the tube (point B) is closed. Now imagine that the flow is first started. Gas will pile up inside the tube. After a few moments, there will be no motion inside the tube because the gas has nowhere to go—the gas will stagnate once steady-state conditions have been reached. In fact, the gas will be stagnant everywhere inside the tube, including at point A. As a result, the flow field sees the open end of the Pitot tube (point A) as an obstruction, and a fluid element moving along the streamline, labeled C, has no choice but to stop when it arrives at point A. Since no heat has been exchanged, and friction is negligible, this process will be

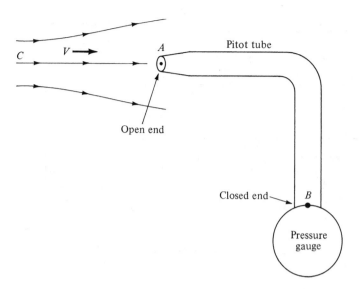

Figure 4.16   Sketch of a Pitot tube.

isentropic; that is, a fluid element moving along streamline C will be isentropically brought to rest at point A by the very presence of the Pitot tube. Therefore, the pressure at point A is, truly speaking, the total pressure p_0. This pressure will be transmitted throughout the Pitot tube, and if a pressure gauge is placed at point B, it will in actuality measure the *total pressure* of the flow. In this fashion, a Pitot tube is an instrument that measures the total pressure of a flow.

By definition, any point of a flow where $V = 0$ is called a *stagnation point*. In Fig. 4.16, point A is a stagnation point.

Consider the arrangement shown in Fig. 4.17. Here we have a uniform flow with velocity V_1 moving over a flat surface parallel to the flow. There is a small hole in the surface at point A, called a *static pressure orifice*. Since the surface is parallel to the flow, only the random motion of the gas molecules will be felt by the surface itself. In other words, the surface pressure is indeed the static pressure p. This will be the pressure at the orifice at point A. On the other hand, the Pitot tube at point B in Fig. 4.17 will feel the total pressure p_0, as discussed above. If the static pressure orifice at point A and the Pitot tube at point B are connected across a pressure gauge, as shown in Fig. 4.17, the gauge will measure the difference between total and static pressure $p_0 - p$.

Now we arrive at the main thrust of this section. The pressure difference $p_0 - p$, as measured in Fig. 4.17, gives a measure of the flow velocity V_1. A combination of a total pressure measurement and a static pressure measurement allows us to measure the velocity at a given point in a flow. These two measurements can be combined in the same instrument, a *Pitot-static probe*, as illustrated in Fig. 4.18. A Pitot-static probe measures p_0 at the nose of the probe and p at a point on the probe surface

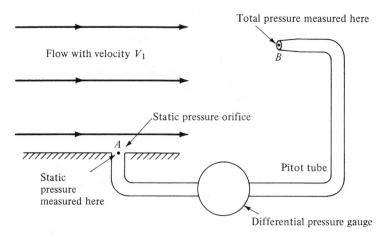

Figure 4.17   Schematic of a Pitot-static measurement.

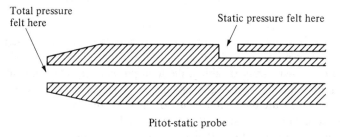

Figure 4.18   Schematic of a Pitot-static probe.

downstream of the nose. The pressure difference $p_0 - p$ yields the velocity V_1, but the quantitative formulation differs depending on whether the flow is low-speed (incompressible), high-speed subsonic, or supersonic.

4.11.1   Incompressible Flow

Consider again the sketch shown in Fig. 4.17. At point A, the pressure is p and the velocity is V_1. At point B, the pressure is p_0 and the velocity is zero. Applying Bernoulli's equation, Eq. (4.9a), at points A and B, we obtain

$$\underset{\substack{\text{Static} \\ \text{pressure}}}{p} + \underset{\substack{\text{Dynamic} \\ \text{pressure}}}{\tfrac{1}{2}\rho V_1^2} = \underset{\substack{\text{Total} \\ \text{pressure}}}{p_0}$$

[4.62]

In Eq. (4.62), for *dynamic pressure q* we have the definition

$$q \equiv \tfrac{1}{2}\rho V^2 \qquad\qquad \textbf{[4.63]}$$

which is frequently employed in aerodynamics; the grouping $\frac{1}{2}\rho V^2$ is termed the *dynamic pressure for flows of all types*, incompressible to hypersonic. From Eq. (4.62),

$$\boxed{p_0 = p + q} \qquad\qquad \textbf{[4.64]}$$

This relation holds for incompressible flow only. The total pressure equals the sum of the static and the dynamic pressure. Also from Eq. (4.62),

$$\boxed{V_1 = \sqrt{\dfrac{p_0 - p}{\rho}}} \qquad\qquad \textbf{[4.65]}$$

Equation (4.65) is the desired result; it allows the calculation of flow velocity from a measurement of $p_0 - p$, obtained from a Pitot-static tube. Again, we emphasize that Eq. (4.65) holds only for incompressible flow.

A Pitot tube can be used to measure the flow velocity at various points in the test section of a low-speed wind tunnel, as shown in Fig. 4.19. The total pressure at point B is obtained by the Pitot probe, and the static pressure, also at point B, is

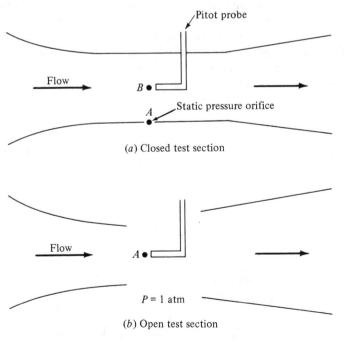

(a) Closed test section

(b) Open test section

Figure 4.19 Pressure measurements in open and closed test sections of subsonic wind tunnels.

obtained from a static pressure orifice located at point A on the wall of the closed test section, assuming that the static pressure is constant throughout the test section. This assumption of constant static pressure is fairly good for subsonic wind tunnel test sections and is commonly made. If the test section is open to the room, as also sketched in Fig. 4.19, then the static pressure at all points in the test section is $p = 1$ atm. In either case, the velocity at point A is calculated from Eq. (4.65). The density ρ in Eq. (4.65) is a constant (incompressible flow). Its value can be obtained by measurements of p and T somewhere in the tunnel, using the equation of state to calculate $\rho = p/(RT)$. These measurements are usually made in the reservoir upstream of the nozzle.

Either a Pitot tube or a Pitot-static tube can be used to measure the airspeed of airplanes. Such tubes can be seen extending from airplane wing tips, with the tube oriented in the flight direction, as shown in Fig. 4.20. If a Pitot tube is used, then the ambient static pressure in the atmosphere around the airplane is obtained from a static pressure orifice placed strategically on the airplane surface. It is placed where the surface pressure is nearly the same as the pressure of the surrounding atmosphere. Such a location is found by experience. It is generally on the fuselage somewhere between the nose and the wing. The values of p_0 obtained from the wing tip Pitot probe and p obtained from the static pressure orifice on the surface enable the calculation of the airplane's speed through the air using Eq. (4.65), *as long as the airplane's velocity is low enough to justify the assumption of incompressible flow*, that is, for velocities less than 300 ft/s. In actual practice, the measurements of p_0 and p are joined across a differential pressure gauge which is calibrated in terms of airspeed, using Eq. (4.65). This airspeed indicator is a dial in the cockpit, with units of velocity, say miles per hour, on the dial. However, in determining the calibration, that is, in determining what values of miles per hour go along with given values of $p_0 - p$, the engineer must decide what value of ρ to use in Eq. (4.65). If ρ is the true value, somehow measured in the actual air around the airplane, then Eq. (4.65) gives

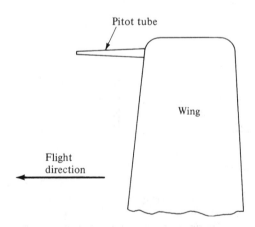

Figure 4.20   Sketch of wing-mounted Pitot probe.

the *true airspeed* of the airplane:

$$V_{true} = \sqrt{\frac{2(p_0 - p)}{\rho}} \qquad \text{[4.66]}$$

However, the measurement of atmospheric air density directly at the airplane's location is difficult. Therefore, for practical reasons, the airspeed indicators on low-speed airplanes are calibrated by using the standard sea-level value of ρ_s in Eq. (4.65). This gives a value of velocity called the *equivalent airspeed*:

$$V_e = \sqrt{\frac{2(p_0 - p)}{\rho_s}} \qquad \text{[4.67]}$$

The equivalent airspeed V_e differs slightly from V_{true}, the difference being the factor $(\rho/\rho_s)^{1/2}$. At altitudes near sea level, this difference is small.

The altimeter on a low-speed Cessna 150 private aircraft reads 5000 ft. By an independent measurement, the outside air temperature is 505°R. If a Pitot tube mounted on the wing tip measures a pressure of 1818 lb/ft², what is the true velocity of the airplane? What is the equivalent airspeed?

Example 4.12

Solution

An altimeter measures the pressure altitude (see discussion in Chap. 3). From the standard atmosphere table, App. B, at 5000 ft, $p = 1761$ lb/ft². Also, the Pitot tube measures total pressure; hence

$$p_0 - p = 1818 - 1761 = 57 \text{ lb/ft}^2$$

The true airspeed can be obtained from Eq. (4.66); however, we need ρ, which is obtained from the equation of state. For the outside, ambient air,

$$\rho = \frac{p}{RT} = \frac{1761}{1716(505)} = 2.03 \times 10^{-3} \text{ slug/ft}^3$$

From Eq. (4.66),

$$V_{true} = \sqrt{\frac{2(p_0 - p)}{\rho}} = \sqrt{\frac{2(57)}{2.03 \times 10^{-3}}} = \boxed{237 \text{ ft/s}}$$

Note: Since 88 ft/s = 60 mi/h, $V_{true} = 237(60/88) = 162$mi/h.

The equivalent airspeed (that which would be read on the airspeed indicator in the cockpit) is obtained from Eq. (4.67), where $\rho_s = 0.002377$ slug/ft³ (the standard sea-level value). Hence, from Eq. (4.67),

$$V_e = \sqrt{\frac{2(p_0 - p)}{\rho_s}} = \sqrt{\frac{2(57)}{2.377 \times 10^{-3}}} = \boxed{219 \text{ ft/s}}$$

Note that there exists a 7.6 percent difference between V_{true} and V_e.

4.11.2   Subsonic Compressible Flow

The results of the previous section are valid for airflows where $M < 0.3$, that is, where the flow can be reasonably assumed to be incompressible. This is the flight regime of, for example, small, piston-engine private aircraft. For higher-speed flows, but where the Mach number is still less than 1 (high-speed subsonic flows), other equations must be used. This is the flight regime of commercial jet transports such as the Boeing 747 and the McDonnell-Douglas DC-10 and of many military aircraft. For these cases, compressibility must be taken into account, as follows.

Consider the definition of enthalpy $h = e + pv$. Since $h = c_p T$ and $e = c_v T$, then $c_p T = c_v T + RT$, or

$$c_p - c_v = R \qquad\qquad \textbf{[4.68]}$$

Divide Eq. (4.68) by c_p:

$$1 - \frac{1}{c_p/c_v} = \frac{R}{c_p}$$

$$1 - \frac{1}{\gamma} = \frac{\gamma - 1}{\gamma} = \frac{R}{c_p}$$

or

$$\boxed{c_p = \frac{\gamma R}{\gamma - 1}} \qquad\qquad \textbf{[4.69]}$$

Equation (4.69) holds for a perfect gas with constant specific heats. It is a necessary thermodynamic relation for use in the energy equation, as follows.

Consider again a Pitot tube in a flow, as shown in Figs. 4.16 and 4.17. Assume the flow velocity V_1 is high enough that compressibility must be taken into account. As usual, the flow is isentropically compressed to zero velocity at the stagnation point on the nose of the probe. The values of the stagnation, or total, pressure and temperature at this point are p_0 and T_0, respectively. From the energy equation, Eq. (4.42), written between a point in the freestream flow where the temperature and velocity are T_1 and V_1, respectively, and the stagnation point, where the velocity is zero and the temperature is T_0,

$$c_p T_1 + \tfrac{1}{2} V_1^2 = c_p T_0$$

or

$$\frac{T_0}{T_1} = 1 + \frac{V_1^2}{2 c_p T_1} \qquad\qquad \textbf{[4.70]}$$

Substitute Eq. (4.69) for c_p in Eq. (4.70):

$$\frac{T_0}{T_1} = 1 + \frac{V_1^2}{2[\gamma R/(\gamma - 1)]T_1} = 1 + \frac{\gamma - 1}{2}\frac{V_1^2}{\gamma R T_1} \qquad\qquad \textbf{[4.71]}$$

However, from Eq. (4.54) for the speed of sound,

$$a_1^2 = \gamma R T_1$$

Thus, Eq. (4.71) becomes

$$\frac{T_0}{T_1} = 1 + \frac{\gamma - 1}{2}\frac{V_1^2}{a_1^2} \qquad \textbf{[4.72]}$$

Since the Mach number $M_1 = V_1/a_1$, Eq. (4.72) becomes

$$\boxed{\frac{T_0}{T_1} = 1 + \frac{\gamma - 1}{2}M_1^2} \qquad \textbf{[4.73]}$$

Since the gas is *isentropically* compressed at the nose of the Pitot probe in Figs. 4.16 and 4.17, Eq. (4.37) holds between the freestream and the stagnation point. That is, $p_0/p_1 = (\rho_0/\rho_1)^\gamma = (T_0/T_1)^{\gamma/(\gamma-1)}$. Therefore, from Eq. (4.73), we obtain

$$\boxed{\begin{aligned} \frac{p_0}{p_1} &= \left(1 + \frac{\gamma - 1}{2}M_1^2\right)^{\gamma/(\gamma-1)} \\ \frac{\rho_0}{\rho_1} &= \left(1 + \frac{\gamma - 1}{2}M_1^2\right)^{1/(\gamma-1)} \end{aligned}}$$

$$\textbf{[4.74]}$$
$$\textbf{[4.75]}$$

Equations (4.73) to (4.75) are fundamental and important relations for compressible, isentropic flow. They apply to many other practical problems in addition to the Pitot tube. Note that Eq. (4.73) holds for adiabatic flow, whereas Eqs. (4.74) and (4.75) contain the additional assumption of frictionless (hence isentropic) flow. Also, from a slightly different perspective, Eqs. (4.73) to (4.75) determine the total temperature, density, and pressure—T_0, ρ_0, and p_0—at any point in the flow where the static temperature, density, and pressure are T_1, ρ_1, and p_1 and where the Mach number is M_1. In other words, reflecting the earlier discussion of the definition of total conditions, Eqs. (4.73) to (4.75) give the values of p_0, T_0, and ρ_0 that are associated with a point in the flow where the pressure, temperature, density, and Mach number are p_1, T_1, ρ_1, and M_1, respectively. These equations also demonstrate the powerful influence of Mach number in aerodynamic flow calculations. It is very important to note that the ratios T_0/T_1, p_0/p_1, and ρ_0/ρ_1 are functions of M_1 only (assuming γ is known; $\gamma = 1.4$ for normal air).

Returning to our objective of measuring airspeed, and solving Eq. (4.74) for M_1, we obtain

$$M_1^2 = \frac{2}{\gamma - 1}\left[\left(\frac{p_0}{p_1}\right)^{(\gamma-1)/\gamma} - 1\right] \qquad \textbf{[4.76]}$$

Hence, for subsonic compressible flow, the ratio of total to static pressure p_0/p_1 is a direct measure of Mach number. Thus, individual measurements of p_0 and p_1 in conjunction with Eq. (4.76) can be used to calibrate an instrument in the cockpit of an airplane called a *Mach meter*, where the dial reads directly in terms of the flight Mach number of the airplane.

To obtain the actual flight velocity, recall that $M_1 = V_1/a_1$, hence Eq. (4.76) becomes

$$V_1^2 = \frac{2a_1^2}{\gamma - 1}\left[\left(\frac{p_0}{p_1}\right)^{(\gamma-1)/\gamma} - 1\right] \qquad \textbf{[4.77a]}$$

Equation (4.77) can be rearranged algebraically as

$$V_1^2 = \frac{2a_1^2}{\gamma - 1}\left[\left(\frac{p_0 - p_1}{p_1} + 1\right)^{(\gamma-1)/\gamma} - 1\right] \qquad \textbf{[4.77b]}$$

Equations (4.77a) and (4.77b) give the *true* airspeed of the airplane. However, they require a knowledge of a_1, hence T_1. The static temperature in the air surrounding the airplane is difficult to measure. Hence, all high-speed (but subsonic) airspeed indicators are calibrated from Eq. (4.77b), assuming that a_1 is equal to the standard sea-level value $a_s = 340.3$ m/s $= 1116$ ft/s. Moreover, the airspeed indicator is designed to sense the actual pressure *difference* $p_0 - p_1$, in Eq. (4.77b), not the pressure *ratio* p_0/p_1, as appears in Eq. (4.77a). Hence, the form of Eq. (4.77b) is used to define a calibrated airspeed as follows:

$$V_{cal}^2 = \frac{2a_s^2}{\gamma - 1}\left[\left(\frac{p_0 - p_1}{p_s} + 1\right)^{(\gamma-1)/\gamma} - 1\right] \qquad \textbf{[4.78]}$$

where a_s and p_s are the standard sea-level values of the speed of sound and static pressure, respectively.

Again, we emphasize that Eqs. (4.76) to (4.78) must be used to measure airspeed when $M_1 > 0.3$, that is, when the flow is compressible. Equations based on Bernoulli's equation, such as Eqs. (4.66) and (4.67), *are not valid* when $M_1 > 0.3$.

Example 4.13 | **A** high-speed subsonic McDonnell-Douglas DC-10 airliner is flying at a pressure altitude of 10 km. A Pitot tube on the wing tip measures a pressure of 4.24×10^4 N/m^2. Calculate the Mach number at which the airplane is flying. If the ambient air temperature is 230 K, calculate the true airspeed and the calibrated airspeed.

Solution
From the standard atmosphere table, App. A, at an altitude of 10,000 m, $p = 2.65 \times 10^4$ N/m^2. Hence, from Eq. (4.76),

$$M_1^2 = \frac{2}{\gamma - 1}\left[\left(\frac{p_0}{p_1}\right)^{(\gamma-1)/\gamma} - 1\right] = \frac{2}{1.4-1}\left[\left(\frac{4.24 \times 10^4}{2.65 \times 10^4}\right)^{0.286} - 1\right]$$

$$= 0.719$$

Thus $\boxed{M_1 = 0.848}$

It is given that $T_1 = 230$ K, hence

$$a_1 = \sqrt{\gamma R T_1} = \sqrt{1.4(287)(230)} = 304.0 \text{ m/s}$$

From Eq. (4.77)

$$V_1^2 = \frac{2a_1^2}{\gamma - 1}\left[\left(\frac{p_0}{p_1}\right)^{(\gamma-1)/\gamma} - 1\right] = \frac{2(304.0)^2}{1.4 - 1}\left[\left(\frac{4.24}{2.65}\right)^{0.286} - 1\right]$$

$$\boxed{V_1 = 258 \text{ m/s} \quad \text{true airspeed}}$$

Note: As a check, from the definition of Mach number,

$$V_1 = M_1 a_1 = 0.848(304.0) = 258 \text{ m/s}$$

The calibrated airspeed can be obtained from Eq. (4.78).

$$V_{\text{cal}}^2 = \frac{2a_s^2}{\gamma - 1}\left[\left(\frac{p_0 - p_1}{p_s} + 1\right)^{(\gamma-1)/\gamma} - 1\right]$$

$$= \frac{2(340.3)^2}{1.4 - 1}\left[\left(\frac{4.24 \times 10^4 - 2.65 \times 10^4}{1.01 \times 10^5} + 1\right)^{0.286} - 1\right]$$

$$\boxed{V_{\text{cal}} = 157 \text{ m/s}}$$

The difference between true and calibrated airspeeds is 39 percent. *Note:* Just out of curiosity, let us calculate V_1 the *wrong* way; that is, let us apply Eq. (4.66), which was obtained from Bernoulli's equation for incompressible flow. Equation (4.66) does *not* apply to the high-speed case of this problem, but let us see what result we get anyway.

$$\rho = \frac{p_1}{RT_1} = \frac{2.65 \times 10^4}{287(230)} = 0.4 \text{ kg/m}^3$$

From Eq. (4.66)

$$V_{\text{true}} = \sqrt{\frac{2(p_0 - p)}{\rho}} = \sqrt{\frac{2(4.24 - 2.65) \times 10^4}{0.4}} = 282 \text{ m/s} \quad \text{incorrect answer}$$

Compared with $V_1 = 258$ m/s obtained above, an error of 9.3 percent is introduced in the calculation of true airspeed by using the *incorrect* assumption of incompressible flow. This error grows very rapidly as the Mach number approaches unity, as discussed in a subsequent section.

4.11.3 Supersonic Flow

Airspeed measurements in supersonic flow, that is, for $M > 1$, are qualitatively different from those for subsonic flow. In supersonic flow, a *shock wave* will form ahead of the Pitot tube, as shown in Fig. 4.21. Shock waves are very thin regions of

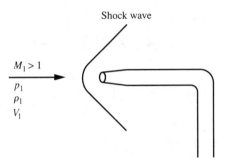

Figure 4.21   Pitot tube in supersonic flow.

the flow (for example, 10^{-4} cm), across which some very severe changes in the flow properties take place. Specifically, as a fluid element flows through a shock wave,

1. The Mach number *decreases.*
2. The static pressure *increases.*
3. The static temperature *increases.*
4. The flow velocity *decreases.*
5. The total pressure p_0 *decreases.*
6. The total temperature T_0 *stays the same* for a perfect gas.

These changes across a shock wave are shown in Fig. 4.22.

How and why does a shock wave form in supersonic flow? These are various answers with various degrees of sophistication. However, the essence is as follows. Refer to Fig. 4.16, which shows a Pitot tube in subsonic flow. The gas molecules that collide with the probe set up a disturbance in the flow. This disturbance is communicated to other regions of the flow, away from the probe, by means of weak pressure waves (essentially sound waves) propagating at the local speed of sound. If

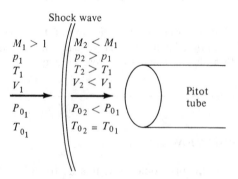

Figure 4.22   Changes across a shock wave in front of a Pitot tube in supersonic flow.

the flow velocity V_1 is less than the speed of sound, as in Fig. 4.16, then the pressure disturbances (which are traveling at the speed of sound) will work their way upstream and eventually will be felt in all regions of the flow. On the other hand, refer to Fig. 4.21, which shows a Pitot tube in supersonic flow. Here V_1 is greater than the speed of sound. Thus, pressure disturbances which are created at the probe surface and which propagate away at the speed of sound *cannot* work their way upstream. Instead, these disturbances coalesce at a finite distance from the probe and form a natural phenomenon called a *shock wave*, as shown in Figs. 4.21 and 4.22. The flow upstream of the shock wave (to the left of the shock) does not feel the pressure disturbance; that is, the presence of the Pitot tube is not communicated to the flow upstream of the shock. The presence of the Pitot tube is felt only in the regions of flow behind the shock wave. Thus, the shock wave is a thin boundary in a supersonic flow, across which major changes in flow properties take place and which divides the region of undisturbed flow upstream from the region of disturbed flow downstream.

Whenever a solid body is placed in a supersonic stream, shock waves will occur. An example is shown in Fig. 4.23, which shows photographs of the supersonic flow over several aerodynamic shapes. The shock waves, which are generally not visible to the naked eye, are made visible in Fig. 4.23 by means of a specially designed optical system, called a *schlieren system*, and a *shadow graph system*. (An example in which shock waves are sometimes visible to the naked eye is on the wing of a high-speed subsonic transport such as a Boeing 707. As we will discuss shortly, there are regions of local supersonic flow on the upper surface of the wing, and these supersonic regions are generally accompanied by weak shock waves. If the sun is

(a)

Figure 4.23   (a) Shock waves on a swept-wing airplane (left) and on a straight-wing airplane (right). Schlieren pictures taken in a supersonic wind tunnel at NASA Ames Research Center. (*continued*)

(concluded)

(b)

(b) Shock waves on a blunt body (left) and sharp-nosed body (right).

(c)

(c) Shock waves on a model of the Gemini manned space
capsule. *(Courtesy of NASA Ames Research Center.)*
Parts *b* and *c* are shadow graphs of the flow.

almost directly overhead and if you look out the window along the span of the wing,
you can sometimes see these waves dancing back and forth on the wing surface.)

Consider again the measurement of airspeed in a supersonic flow. The measurement is complicated by the presence of the shock wave in Fig. 4.21 because the flow
through a shock wave is *nonisentropic*. Within the thin structure of a shock wave
itself, very large friction and thermal conduction effects are taking place. Hence, nei-

ther adiabatic nor frictionless conditions hold, and hence the flow is *not* isentropic. As a result, Eq. (4.74) and hence Eqs. (4.76) and (4.77a) do not hold across the shock wave. A major consequence is that the total pressure p_0 is smaller behind the shock wave than in front of it. In turn, the total pressure measured at the nose of the Pitot probe in supersonic flow will *not* be the same value as that associated with the freestream, that is, as associated with M_1. Consequently, a separate shock wave theory must be applied to relate the Pitot tube measurement to the value of M_1. This theory is beyond the scope of our presentation, but the resulting formula is given here for the sake of completeness:

$$\frac{p_{0_2}}{p_1} = \left[\frac{(\gamma+1)^2 M_1^2}{4\gamma M_1^2 - 2(\gamma-1)} \right]^{\gamma/(\gamma-1)} \frac{1-\gamma+2\gamma M_1^2}{\gamma+1} \qquad \text{[4.79]}$$

This equation is called the *Rayleigh Pitot tube formula*. It relates the Pitot tube measurement of total pressure behind the shockwave, p_{0_2} and a measurement of freestream static pressure (again obtained by a static pressure orifice somewhere on the surface of the airplane) to the freestream supersonic Mach number M_1. In this fashion, measurements of p_{0_2} and p_1, along with Eq. (4.79), allow the calibration of a Mach meter for supersonic flight.

An experimental rocket-powered aircraft is flying at a velocity of 3000 mi/h at an altitude where the ambient pressure and temperature are 151 lb/ft^2 and 390°R, respectively. A Pitot tube is mounted in the nose of the aircraft. What is the pressure measured by the Pitot tube? | **Example 4.14**

Solution

First, we ask, Is the flow supersonic or subsonic, that is, what is M_1? From Eq. (4.54)

$$a_1 = \sqrt{\gamma R T_1} = \sqrt{1.4(1716)(390)} = 968.0 \text{ ft/s}$$

$$V_1 = 3000 \left(\frac{88}{60}\right) = 4400 \text{ ft/s}$$

$$M_1 = \frac{V_1}{a_1} = \frac{4400}{968.0} = 4.54$$

Hence, $M_1 > 1$; the flow is supersonic. There is a shock wave in front of the Pitot tube; therefore, Eq. (4.74) developed for isentropic flow does *not* hold. Instead, Eq. (4.79) must be used.

$$\frac{p_{0_2}}{p_1} = \left[\frac{(\gamma+1)^2 M_1^2}{4\gamma M_1^2 - 2(\gamma-1)} \right]^{\gamma/(\gamma-1)} \frac{1-\gamma+2\gamma M_1^2}{\gamma+1}$$

$$= \left[\frac{(2.4)^2(4.54)^2}{4(1.4)(4.54)^2 - 2(0.4)} \right]^{3.5} \frac{1-1.4+2(1.4)(4.54)^2}{2.4} = 27$$

Thus $\qquad p_{0_2} = 27p_1 = 27(151) = \boxed{4077 \text{ lb/ft}^2}$

Note: Again, out of curiosity, let us calculate the *wrong* answer. If we had *not* taken into account the shock wave in front of the Pitot tube at supersonic speeds, then Eq. (4.74) would give

$$\frac{p_0}{p_1} = \left(1 + \frac{\gamma - 1}{2}M_1^2\right)^{\gamma/(\gamma-1)} = \left[1 + \frac{0.4}{2}(4.54)^2\right]^{3.5} = 304.2$$

Thus $p_0 = 304.2 p_1 = 304.2(151) = 45{,}931 \ \text{lb/ft}^2$ incorrect answer

Note that the incorrect answer is off by a factor of more than 10!

4.11.4   Summary

As a summary on the measurement of airspeed, note that different results apply to different regimes of flight: low-speed (incompressible), high-speed subsonic, and supersonic. These differences are fundamental and serve as excellent examples of the application of the different laws of aerodynamics developed in previous sections. Moreover, many of the formulas developed in this section apply to other practical problems, as discussed below.

4.12   SOME ADDITIONAL CONSIDERATIONS

Section 4.11 contains information that is considerably more general than just the application to airspeed measurements. The purpose of this section is to elaborate on some of the ideas and results discussed in Sect. 4.11.

4.12.1   More on Compressible Flow

Equations (4.73) through (4.75) relating the ratios of T_0/T_1, p_0/p_1, and ρ_0/ρ_1 to the local Mach number M_1 apply in general to any isentropic flow. We state without proof that the values of T_0, p_0, and ρ_0 are constant throughout a given isentropic flow. In conjunction with Eqs. (4.73) to (4.75), this fact gives us a powerful tool for the analysis of an isentropic flow. For example, let us again consider the isentropic flow over an airfoil, which was the problem solved in Example 4.8. But now we have more information and a broader perspective from which to approach this problem.

Example 4.15   Consider the isentropic flow over the airfoil sketched in Fig. 4.24. The freestream pressure, velocity, and density are 2116 lb/ft^2, 500 mi/h, and 0.002377 slug/ft^3, respectively. At a given point A on the airfoil, the pressure is 1497 lb/ft^2. What are the Mach number and the velocity at point A?

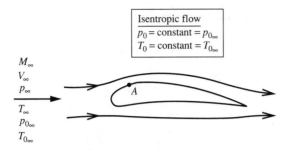

$$M_\infty$$
$$V_\infty$$
$$p_\infty$$

$$T_\infty$$
$$p_{0_\infty}$$
$$T_{0_\infty}$$

Figure 4.24 Total pressure and total temperature are constant throughout an isentropic flow.

Solution

This example is the same as Example 4.8, with the additional requirement to calculate the Mach number at point A. However, we use a different solution procedure in this example. First, we calculate the freestream Mach number, as follows.

$$T_\infty = \frac{p_\infty}{\rho_\infty R} = \frac{2116}{0.002377(1716)} = 518.8°\text{R}$$

$$a_\infty = \sqrt{\gamma R T_\infty} = \sqrt{(1.4)(1716)(518.8)} = 116.4 \text{ ft/s}$$

$$V_\infty = 500 \text{ mi/h} = 500\left(\frac{88}{60}\right) \text{ft/s} = 733.3 \text{ ft/s}$$

$$M_\infty = \frac{V_\infty}{a_\infty} = \frac{733.3}{116.4} = 0.6568$$

The freestream total temperature is, from Eq. (4.73),

$$\frac{T_{0_\infty}}{T_\infty} = 1 + \frac{\gamma - 1}{2} M_\infty^2 = 1 + 0.2(0.6568)^2 = 1.0863$$

$$T_{0_\infty} = 1.0863 T_\infty = 1.0863(518.8) = 563.6°\text{R}$$

The freestream total pressure is, from Eq. (4.74),

$$\frac{p_{0_\infty}}{p_\infty} = \left(1 + \frac{\gamma - 1}{2} M_\infty^2\right)^{\gamma/(\gamma-1)} = (1.0863)^{3.5} = 1.336$$

$$p_{0_\infty} = 1.336(2116) = 2827 \text{ lb/ft}^2$$

Since the total temperature and total pressure are constant throughout the isentropic flow over the airfoil, the total temperature and total pressure at point A are the same as the freestream values.

$$T_{0_A} = T_{0_\infty} = 563.6°\text{R}$$
$$p_{0_A} = p_{0_\infty} = 2827 \text{ lb/ft}^2$$

We can solve for the Mach number at point A by applying Eq. (4.74) at point A:

$$\frac{p_{0_A}}{p_A} = \left(1 + \frac{\gamma - 1}{2} M_A^2\right)^{\gamma/\gamma-1}$$

$$\frac{2827}{1497} = (1 + 0.2M_A^2)^{3.5}$$

or
$$1 + 0.2M_A^2 = (1.888)^{1/3.5} = (1.888)^{0.2857} = 1.1991$$

or
$$M_A = \sqrt{\frac{1.1991 - 1}{0.2}} = \boxed{0.9955}$$

Note: The Mach number at point, A is essentially 1; we have nearly sonic flow at point A. The static temperature at point A can be obtained from Eq. (4.73).

$$\frac{T_{0_A}}{T_A} = 1 + \frac{\gamma - 1}{2} M_A^2 = 1 + 0.2(0.9955)^2 = 1.1982$$

$$T_A = \frac{T_{0_A}}{1.1982} = \frac{563.6}{1.1982} = 470.4°R$$

(*Note:* The above result for $T_A = 470.4°R$ agrees well with the value of 470.1°R calculated in Example 4.8; the difference is due to roundoff error produced by carrying just four significant figures and the author's doing the calculations on a hand calculator.)

The velocity at point A can be obtained as follows.

$$a_A = \sqrt{\gamma R T_A} = \sqrt{1.4(1716)(470.4)} = 1063 \text{ ft/s}$$

$$V_A = a_A M_A = 1063(0.9955) = \boxed{1058 \text{ ft/s}}$$

(*Note:* This agrees well with the result $V_A = 1061$ ft/s calculated in Example 4.8.)

The calculation procedure used in Example 4.15 is slightly longer than that used in Example 4.8; however, it is a more fundamental approach than that used in Example 4.8. Return to Example 4.8, and note that we had to employ a value of the specific heat c_p to solve the problem. However, in the present calculation, we did not need a value of c_p. Indeed, the explicit use of c_p is not necessary in solving isentropic compressible flows. Instead, we used γ and M to solve this example. The ratio of specific heats γ and the Mach number M are both examples of *similarity parameters* in aerodynamics. The concept and power of the similarity parameters for governing fluid flows are something you will study in more advanced treatments than this book. Suffice it to say here that *Mach number* is a powerful governing parameter for compressible flow, and that the results depend on the value of γ, which is usually a fixed value for a given gas ($\gamma = 1.4$ for air, as we use here). Example 4.15 shows the power of using M and γ for solving compressible flow problems. We will continue to see the power of M and γ in some of our subsequent discussions.

4.12.2 More on Equivalent Airspeed

Equivalent airspeed was introduced in Sec. 4.11.1 and expressed by Eq. (4.67) for low-speed flight, where the flow is assumed to be incompressible. However, the concept of equivalent airspeed has a broader meaning than just a value that comes from an airspeed indicator which uses the standard sea-level density to determine its readout, as first explained in Sec. 4.11.1.

The general definition of equivalent airspeed can be introduced by the following example. Consider a Lockheed-Martin F-16 fighter cruising at a velocity of 300 m/s at an altitude of 7 km, where the freestream density is $0.59\ \text{kg/m}^3$. The velocity of 300 m/s is the airplane's true airspeed. At this speed and altitude, the dynamic pressure is $\frac{1}{2}\rho_\infty V_\infty^2 = \frac{1}{2}(0.50)(300)^2 = 2.655 \times 10^4\ \text{N/m}^2$. It is important to reinforce that dynamic pressure is a *definition*, defined by the quantity $\frac{1}{2}\rho_\infty V_\infty^2$. This definition holds no matter what the flight regime is—subsonic, supersonic, etc.—and whether the flow is incompressible or compressible. Dynamic pressure q_∞ is just the definition

$$q_\infty \equiv \tfrac{1}{2}\rho_\infty V_\infty^2$$

Now imagine the F-16 flying at standard sea level, where the freestream density is $1.23\ \text{kg/m}^3$. *Question:* What velocity would it have to have at standard sea level to experience the *same dynamic pressure* that it had when flying at 300 m/s at the altitude of 7 km? The answer is easy to calculate, as follows.

$$(q_\infty)_{\text{sea level}} = (q_\infty)_{7\,\text{km}}$$
$$(\tfrac{1}{2}\rho_\infty V_\infty^2)_{\text{sea level}} = (\tfrac{1}{2}\rho_\infty V_\infty^2)_{7\,\text{km}}$$

Dropping the subscripts ∞ for convenience, we have

$$V_{\text{sea level}} = V_{7\,\text{km}}\left(\frac{\rho}{\rho_s}\right)^{1/2}$$

where ρ is the density at 7 km and ρ_s is the standard sea-level density. Putting in the numbers, we have

$$V_{\text{sea level}} = 300\left(\frac{0.59}{1.23}\right)^{1/2} = 207.8\ \text{m/s}.$$

Hence, the F-16 flying at 300 m/s at 7-km altitude would have to fly at a velocity of 207.8 m/s at standard sea level to experience the same dynamic pressure. By definition, the F-16 flying at 300 m/s at 7-km altitude has an *equivalent airspeed* of 207.8 m/s.

This leads to the more general definition of equivalent airspeed, as follows. Consider an airplane flying at some true airspeed at some altitude. Its *equivalent airspeed* at this condition is defined as the velocity at which it would have to fly at standard sea level to *experience the same dynamic pressure*. The equation for equivalent airspeed is straightforward, as obtained above. It is

$$V_e = V\left(\frac{\rho}{\rho_s}\right)^{1/2}$$

where V_e is the equivalent airspeed, V is the true velocity at some altitude, ρ is the density at that altitude, and ρ_s is the standard sea-level density.

In retrospect, our first discussion of V_e in Sec. 4.11.1 is consistent with our discussions here; however, in Sec. 4.11.1 our discussion was focused on airspeed measurements in an incompressible flow.

The concept of equivalent airspeed is useful in studies of airplane performance which involve the aerodynamic lift and drag of airplanes. The lift and drag depend on the dynamic pressure, q_∞ as we will see in Chap. 5. Giving the *equivalent airspeed* of an airplane is the same as stating its dynamic pressure, as discussed above. Hence, equivalent airspeed is sometimes used as a convenience in reporting and analyzing airplane performance data.

4.13 SUPERSONIC WIND TUNNELS AND ROCKET ENGINES

For more than a century, projectiles such as bullets and artillery shells have been fired at supersonic velocities. However, the main aerodynamic interest in supersonic flows arose after World War II with the advent of jet aircraft and rocket-propelled guided missiles. As a result, almost every aerodynamic laboratory has an inventory of supersonic and hypersonic wind tunnels to simulate modern high-speed flight. In addition to their practical importance, supersonic wind tunnels are an excellent example of the application of the fundamental laws of aerodynamics. The flow through rocket engine nozzles is another example of the same laws. In fact, the basic aerodynamics of supersonic wind tunnels and rocket engines are essentially the same, as discussed below.

First, consider isentropic flow in a stream tube, as sketched in Fig. 4.2. From the continuity equation, Eq. (4.2),

$$\rho A V = \text{const}$$

or
$$\ln \rho + \ln A + \ln V = \ln \text{(const)}$$

Differentiating, we obtain

$$\frac{d\rho}{\rho} + \frac{dA}{A} + \frac{dV}{V} = 0 \qquad \textbf{[4.80]}$$

Recalling the momentum equation, Eq. (4.8) (Euler's), we obtain

$$dp = -\rho V \, dV$$

Hence
$$\rho = -\frac{dp}{V \, dV} \qquad \textbf{[4.81]}$$

Substitute Eq. (4.81) into (4.80):

$$-\frac{d\rho \, V \, dV}{dp} + \frac{dA}{A} + \frac{dV}{V} = 0 \qquad \textbf{[4.82]}$$

Since the flow is isentropic,

$$\frac{d\rho}{dp} = \frac{1}{dp/d\rho} \equiv \frac{1}{(dp/d\rho)_{\text{isentropic}}} \equiv \frac{1}{a^2}$$

Thus, Eq. (4.82) becomes

$$-\frac{V\,dV}{a^2} + \frac{dA}{A} + \frac{dV}{V} = 0$$

Rearranging, we get

$$\frac{dA}{A} = \frac{V\,dV}{a^2} - \frac{dV}{V} = \left(\frac{V^2}{a^2} - 1\right)\frac{dV}{V}$$

or

$$\boxed{\frac{dA}{A} = (M^2 - 1)\frac{dV}{V}}$$
[4.83]

Equation (4.83) is called the *area-velocity relation*, and it contains a wealth of information about the flow in the stream tube shown in Fig. 4.2. First, note the mathematical convention that an increasing velocity and an increasing area correspond to positive values of dV and dA, respectively, whereas a decreasing velocity and a decreasing area correspond to negative values of dV and dA. This is the normal convention for differentials from differential calculus. With this in mind, Eq. (4.83) yields the following physical phenomena:

1. If the flow is subsonic ($M < 1$), for the velocity to increase (dV positive), the area must decrease (dA negative); that is, when the flow is subsonic, the area must converge for the velocity to increase. This is sketched in Fig. 4.25a. This same result was observed in Sec. 4.2 for incompressible flow. Of course, incompressible flow is, in a sense, a singular case of subsonic flow, where $M \to 0$.

2. If the flow is supersonic ($M > 1$), for the velocity to increase (dV positive), the area must also increase (dA positive); that is, when the flow is supersonic, the area must diverge for the velocity to increase. This is sketched in Fig. 4.25b.

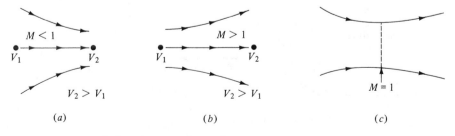

Figure 4.25 Results from the area-velocity relation.

3. If the flow is sonic ($M = 1$), then Eq. (4.83) yields for the velocity

$$\frac{dV}{V} = \frac{1}{M^2 - 1} \frac{dA}{A} = \frac{1}{0} \frac{dA}{A} \qquad \textbf{[4.84]}$$

which at first glance says that dV/V is infinitely large. However, on a physical basis, the velocity, and hence the change in velocity dV, at all times must be finite. This is only common sense. Thus, looking at Eq. (4.84), we see that the only way for dV/V to be finite is to have $dA/A = 0$, so

$$\frac{dV}{V} = \frac{1}{0} \frac{dA}{A} = \frac{0}{0} = \text{finite number}$$

that is, in the language of differential calculus, dV/V is an indeterminate form of 0/0 and hence can have a finite value. In turn, if $dA/A = 0$, the stream tube has a *minimum* area at $M = 1$. This minimum area is called a *throat* and is sketched in Fig. 4.25c.

Therefore, to expand a gas to supersonic speeds, starting with a stagnant gas in a reservoir, the above discussion says that a duct of a sufficiently converging-diverging shape must be used. This is sketched in Fig. 4.26, where typical shapes for supersonic wind tunnel nozzles and rocket engine nozzles are shown. In both cases, the flow starts out with a very low velocity $V \approx 0$ in the reservoir, expands to high subsonic speeds in the convergent section, reaches Mach 1 at the throat, and then goes supersonic in the divergent section downstream of the throat. In a supersonic wind tunnel, smooth, uniform flow at the nozzle exit is usually desired, and therefore a long, gradually converging and diverging nozzle is employed, as shown at the top of Fig. 4.26. For

(a) Supersonic wind-tunnel nozzle

(b) Rocket-engine nozzle

Figure 4.26   Supersonic nozzle shapes.

rocket engines, the flow quality at the exit is not quite as important, but the weight of the nozzle is a major concern. For the weight to be minimized, the engine's length is minimized, which gives rise to a rapidly diverging, bell-like shape for the supersonic section, as shown at the bottom of Fig. 4.26. A photograph of a typical rocket engine is shown in Fig. 4.27.

The real flow through nozzles such as those sketched in Fig. 4.26 is closely approximated by isentropic flow because little or no heat is added or taken away through the nozzle walls and a vast core of the flow is virtually frictionless. Therefore, Eqs. (4.73) to (4.75) apply to nozzle flows. Here, the total pressure and temperature p_0 and T_0 remain constant throughout the flow, and Eqs. (4.73) to (4.75) can be interpreted as relating conditions at any point in the flow to the stagnation conditions in the reservoir. For example, consider Fig. 4.26, which illustrates the reservoir conditions p_0 and T_0 where $V \approx 0$. Consider any cross section downstream of the reservoir. The static temperature, density, and pressure at this section are T_1, ρ_1, and p_1, respectively. If the Mach number M_1 is known at this point, then T_1, ρ_1, and p_1 can be found from Eqs. (4.73) to (4.75) as

$$T_1 = T_0[1 + \tfrac{1}{2}(\gamma - 1)M_1^2]^{-1} \qquad\qquad \textbf{[4.85]}$$

$$\rho_1 = \rho_0[1 + \tfrac{1}{2}(\gamma - 1)M_1^2]^{-1/(\gamma-1)} \qquad\qquad \textbf{[4.86]}$$

$$p_1 = p_0[1 + \tfrac{1}{2}(\gamma - 1)M_1^2]^{-\gamma/(\gamma-1)} \qquad\qquad \textbf{[4.87]}$$

Again, Eqs. (4.85) to (4.87) demonstrate the power of the Mach number in making aerodynamic calculations. The variation of Mach number itself through the nozzle is strictly a function of the ratio of the cross-sectional area to the throat area A/A_t. This relation can be developed from the aerodynamic fundamentals already discussed; the

Figure 4.27 A typical rocket engine. Shown is a small rocket designed by Messerschmitt-Bolkow-Blohm for European satellite launching.

resulting form is

$$\left(\frac{A}{A_t}\right)^2 = \frac{1}{M^2}\left[\frac{2}{\gamma+1}\left(1+\frac{\gamma-1}{2}M^2\right)\right]^{(\gamma+1)/(\gamma-1)} \qquad \textbf{[4.88]}$$

Therefore, the analysis of isentropic flow through a nozzle is relatively straight-forward. The procedure is summarized in Fig. 4.28. Consider that the nozzle shape, hence A/A_t, is given as shown in Fig. 4.28a. Then from Eq. (4.88), the Mach number can be obtained (implicitly). Its variation is sketched in Fig. 4.28b. Since M is now known through the nozzle, Eqs. (4.85) to (4.87) give the variations of T, ρ, and p,

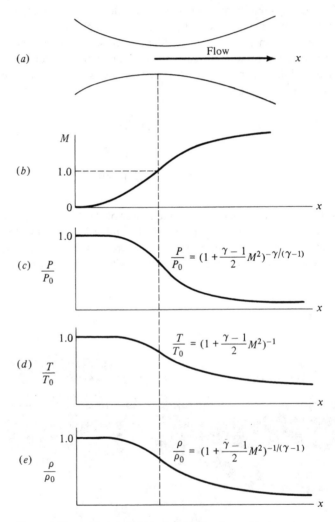

Figure 4.28 Variation of Mach number, pressure, temperature, and density through a supersonic nozzle.

which are sketched in Fig. 4.28c to e. The directions of these variations are important and should be noted. From Fig. 4.28, the Mach number continuously increases through the nozzle, going from near zero in the reservoir to $M = 1$ at the throat and to supersonic values downstream of the throat. In turn, p, T, and ρ begin with their stagnation values in the reservoir and continuously decrease to low values at the nozzle exit. Hence, a supersonic nozzle flow is an expansion process in which pressure decreases through the nozzle. In fact, it is this pressure decrease that provides the mechanical force for pushing the flow through the nozzle. If the nozzle shown in Fig. 4.28a is simply set out by itself in a laboratory, obviously nothing will happen; the air will not start to rush through the nozzle of its own accord. Instead, to establish the flow sketched in Fig. 4.28, we must provide a high-pressure source at the inlet, and/or a low-pressure source at the exit, with the pressure ratio just the right value, as prescribed by Eq. (4.87) and sketched in Fig. 4.28c.

Example 4.16

You are given the job of designing a supersonic wind tunnel that has a Mach 2 flow at standard sea-level conditions in the test section. What reservoir pressure and temperature and what area ratio A_e/A_t are required to obtain these conditions?

Solution

The static pressure $p_e = 1$ atm $= 1.01 \times 10^5$ N/m^2, and the static temperature $T_e = 288.16$ K, from conditions at standard sea level. These are the desired conditions at the exit of the nozzle (the entrance to the test section). The necessary reservoir conditions are obtained from Eqs. (4.85) and (4.87):

$$\frac{T_0}{T_e} = 1 + \frac{\gamma - 1}{2}M_e^2 = 1 + \frac{1.4 - 1}{2}(2^2) = 1.8$$

Thus

$$T_0 = 1.8T_e = 1.8(288.16) = \boxed{518.7 \text{ K}}$$

$$\frac{p_0}{p_e} = \left(1 + \frac{\gamma - 1}{2}M_e^2\right)^{\gamma/(\gamma-1)} = (1.8)^{3.5} = 7.82$$

Thus

$$p_0 = 7.82p_e = 7.82(1.01 \times 10^5) = \boxed{7.9 \times 10^5 \text{ N/m}^2}$$

The area ratio is obtained from Eq. (4.88):

$$\left(\frac{A_e}{A_t}\right)^2 = \frac{1}{M^2}\left[\frac{2}{\gamma + 1}\left(1 + \frac{\gamma - 1}{2}M^2\right)\right]^{(\gamma+1)/(\gamma-1)}$$

$$= \frac{1}{2^2}\left[\frac{2}{2.4}\left(1 + \frac{0.4}{2}2^2\right)\right]^{2.4/0.4} = 2.85$$

Hence

$$\boxed{\frac{A_e}{A_t} = 1.69}$$

Example 4.17

The reservoir temperature and pressure of a supersonic wind tunnel are 600°R and 10 atm, respectively. The Mach number of the flow in the test section is 3. A blunt-nosed model like that shown at the left in Fig. 4.23b is inserted in the test section flow. Calculate the pressure, temperature, and density at the stagnation point (at the nose of the body).

Solution

The flow conditions in the test section are the same as those at the nozzle exit. Hence, in the test section, the exit pressure is obtained from Eq. (4.87), recalling that 1 atm = 2116 lb/ft^2:

$$p_e = p_0[1 + \tfrac{1}{2}(\gamma - 1)M_e^2]^{-\gamma/(\gamma-1)}$$
$$= 10(2116)[1 + 0.5(0.4)(3)^2]^{-3.5}$$
$$= 576 \text{ lb/ft}^2$$

The pressure at the stagnation point on the model is the total pressure *behind a normal wave* because the stagnation streamline has traversed the normal portion of the curved bow shock wave in Fig. 4.23b and then has been isentropically compressed to zero velocity between the shock and the body. This is the same situation as that existing at the mouth of a Pitot tube in supersonic flow, as described in Sec. 4.11c. Hence, the stagnation pressure is given by Eq. (4.79):

$$\frac{p_{0_2}}{p_1} = \frac{p_{\text{stag}}}{p_e} = \left[\frac{(\gamma + 1)^2 M_e^2}{4\gamma M_e^2 - 2(\gamma - 1)}\right]^{\gamma/(\gamma-1)} \frac{1 - \gamma + 2\gamma M_e^2}{\gamma + 1}$$

$$\frac{p_{\text{stag}}}{p_e} = \left[\frac{2.4^2(3^2)}{4(1.4)(3^2) - 2(0.4)}\right]^{3.5} \frac{1 - 1.4 + 2(1.4)(3^2)}{2.4} = 12.06$$

$$p_{\text{stag}} = 12.06 p_e = 12.06(576) = \boxed{6947 \text{ lb/ft}^2}$$

The *total* temperature (not the static temperature) at the nozzle exit is the same as the reservoir temperature

$$T_{0,e} = T_0$$

because the flow through the nozzle is isentropic and hence adiabatic. For an adiabatic flow, the *total* temperature is constant, as demonstrated by Eq. (4.42), where at two different points in an adiabatic flow with different velocities if the flow is adiabatically slowed to zero velocity at both points, we obtain

$$c_p T_{0_1} = c_p T_{0_2}$$

hence $T_{0_1} = T_{0_2}$; that is, the total temperature at the two different points is the same. Therefore, in the present problem, the total temperature associated with the test section flow is equal to the total temperature throughout the nozzle expansion: $T_{0_e} = T_0 = 600°R$. [Note that the *static* temperature of the test section flow is 214.3°R, obtained from Eq. (4.85).] Moreover, in traversing a shock wave (see Fig. 4.22), the total temperature is unchanged; that is, the total temperature behind the shock wave on the model is also 600°R (although the static temperature behind the shock is less than 600°R). Finally, since the flow is isentropically compressed to zero velocity at the stagnation point, the stagnation point temperature is the *total temperature*, which also stays constant through the isentropic compression. Hence, the gas temperature at the stagnation point is

$$T_{\text{stag}} = T_0 = \boxed{600°R}$$

From the equation of state

$$\rho_{stag} = \frac{p_{stag}}{R T_{stag}} = \frac{6947}{1716(600)} = \boxed{0.0067 \text{ slug/ft}^3}$$

In the combustion chamber of a rocket engine, kerosene and oxygen are burned, resulting in a hot, high-pressure gas mixture in the combustion chamber with the following conditions and properties: $T_0 = 3144$ K, $p_0 = 20$ atm, $R = 378$ J/(kg)(K), and $\gamma = 1.26$. The pressure at the exit of the rocket nozzle is 1 atm, and the throat area of the nozzle is 0.1 m^2. Assuming isentropic flow through the rocket nozzle, calculate (a) the velocity at the exit and (b) the mass flow through the nozzle.

Example 4.18

Solution

a. To obtain the velocity at the exit, let us first obtain the temperature, next the speed of sound, and then the Mach number, leading to the velocity. We note that the combustion chamber conditions are the "reservoir" conditions sketched in Fig. 4.26; this is why the combustion chamber pressure and temperature have been denoted by p_0 and T_0, respectively. Since the flow is isentropic, from Eq. (3.46) we have

$$\frac{p_e}{p_o} = \left(\frac{T_e}{T_o}\right)^{\gamma/\gamma - 1}$$

$$T_e = T_o \left(\frac{p_e}{p_o}\right)^{\gamma/\gamma - 1} = (3144)\left(\frac{1}{20}\right)^{0.26/1.26} = 1694 \text{ K}$$

or

$$a_e = \sqrt{\gamma R T_e} = \sqrt{1.26(378)(1694)} = 898.2 \text{ m/s}$$

The Mach number at the exit is given by Eq. (4.73).

$$\frac{T_o}{T_e} = 1 + \frac{\gamma - 1}{2} M_e^2$$

or

$$M_e^2 = \frac{2}{\gamma - 1}\left(\frac{T_0}{T_e} - 1\right) = \frac{2}{1.26 - 1}\left(\frac{3144}{1694} - 1\right) = 6.584$$

or

$$M_e = 2.566$$

Hence

$$V_e = M_e a_e = 2.566(898.2) = \boxed{2305 \text{ m/s}}$$

b. The mass flow is given by the product $\rho A V$ evaluated at any cross section of the nozzle. Since we are given the area of the throat, the obvious location at which to evaluate $\rho A V$ is the throat, that is,

$$\dot{m} = \rho^* A^* V^*$$

where ρ^*, A^*, and V^* are the density, area, and velocity, respectively, at the throat. We will make use of the fact that the Mach number at the throat is $M^* = 1$. The pressure at the throat p^* is given by Eq. (4.74);

$$\frac{p_0}{p^*} = \left(1 + \frac{\gamma - 1}{2} M^{*2}\right)^{\gamma/(\gamma - 1)} = \left[1 + \frac{0.26}{2}(1^2)\right]^{1.26/0.26} = (1.13)^{4.846} = 1.808$$

Hence

$$p^* = \frac{p_0}{1.808} = \frac{20(1.01 \times 10^5)}{1.808} = 1.117 \times 10^6 \text{N/m}^2$$

The temperature at the throat is given by Eq. (4.73):

$$\frac{T_0}{T^*} = 1 + \frac{\gamma - 1}{\gamma} M^{*2} = 1.13$$

$$T^* = \frac{T_0}{1.13} = \frac{3144}{1.13} = 2782.3 \text{ K}$$

$$a^* = \sqrt{\gamma R T^*} = \sqrt{1.26(378)(2782.3)} = 1151 \text{ m/s}$$

$$\rho^* = \frac{p^*}{RT^*} = \frac{1.117 \times 10^6}{378(2782.3)} = 1.062 \text{ kg/m}^3$$

Since $M^* = 1$, then $V^* = a^* = 1151$ m/s. Hence

$$\dot{m} = \rho^* A^* V^* = 1.062(0.1)(1151) = \boxed{122.2 \text{ kg/s}}$$

4.14   DISCUSSION OF COMPRESSIBILITY

We have been stating all along that flows in which $M < 0.3$ can be treated as essentially incompressible and, conversely, flows where $M \geq 0.3$ should be treated as compressible. We are now in a position to prove this.

Consider a gas at rest ($V = 0$) with density ρ_0. Now accelerate this gas isentropically to some velocity V and Mach number M. Obviously, the thermodynamic properties of the gas will change, including the density. In fact, the change in density will be given by Eq. (4.75):

$$\frac{\rho_0}{\rho} = \left(1 + \frac{\gamma - 1}{2} M^2\right)^{1/(\gamma - 1)}$$

For $\gamma = 1.4$, this variation of ρ/ρ_0 is given in Fig. 4.29. Note that, for $M < 0.3$, the density change in the flow is less than 5 percent; that is, the density is essentially constant for $M < 0.3$, and for all practical purposes the flow is incompressible. Therefore, we have just demonstrated the validity of the statement

For $M < 0.3$, the flow can be treated as incompressible.

4.15   INTRODUCTION TO VISCOUS FLOW

This is a good time to look back to our road map in Fig. 4.1. We have now completed the left side of this road map—inviscid flow with some applications. Examine again the boxes on the left side, and make certain that you feel comfortable with the material represented by each box. There are a large number of aerodynamic applications in which the neglect of friction is quite reasonable and in which the assumption of inviscid flow leads to useful and reasonably accurate results.

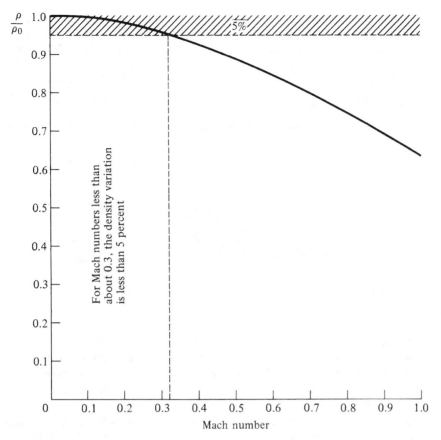

Figure 4.29   Density variation with Mach number for $\gamma = 1.4$, showing region where the density change is less than 5 percent.

However, there are numerous other practical problems in which the effect of friction is dominant, and we now turn our attention to such problems. This constitutes the right side of our road map in Fig. 4.1—viscous flow, that is, flow with friction. Indeed, there are some flows in which the *fundamental* behavior is governed by the presence of friction between the airflow and a solid surface. A classic example is sketched in Fig. 4.30, which shows the low-speed flow over a sphere. At the left is sketched the flow field that would exist if the flow were inviscid. For such an ideal, frictionless flow, the streamlines are symmetric and, amazingly, there is no aerodynamic force on the sphere. The pressure distribution over the forward surface exactly balances that over the rearward surface, and hence there is no drag (no force in the flow direction). However, this purely theoretical result is contrary to common sense; in real life there is a drag force on the sphere tending to retard the motion of the sphere. The failure of the theory to predict drag was bothersome to early 19th-century aerodynamicists and was even given a name: *d'Alembert's paradox*. The problem is caused by not including friction in the theory. The real flow over a sphere is sketched

Frictionless flow: no drag   Real flow: finite drag

Figure 4.30   Comparison between ideal frictionless flow and real flow with the effects of friction.

Figure 4.31   Frictionless flow.

on the right in Fig. 4.30. The flow separates on the rearward surface of the sphere, setting up a complicated flow in the wake and causing the pressure on the rearward surface to be less than that on the forward surface. Hence, there is a drag force exerted on the sphere, as shown by D in Fig. 4.30. The difference between the two flows in Fig. 4.30 is simply friction, but what a difference!

Consider the flow of a gas over a solid surface, such as the airfoil sketched in Fig. 4.31. According to our previous considerations of frictionless flows, we considered the flow velocity at the surface as being a finite value, such as V_2 shown in Fig. 4.31; that is, because of the lack of friction, the streamline right at the surface slips over the surface. In fact, we stated that if the flow is incompressible, V_2 can be calculated from Bernoulli's equation:

$$p_1 + \tfrac{1}{2}\rho V_1^2 = p_2 + \tfrac{1}{2}\rho V_2^2$$

However, in real life, *the flow at the surface adheres to the surface* because of friction between the gas and the solid material; that is, right at the surface, the flow velocity is zero, and there is a thin region of retarded flow in the vicinity of the surface, as sketched in Fig. 4.32. This region of viscous flow which has been retarded owing to friction at the surface is called a *boundary layer*. The inner edge of the boundary layer is the solid surface itself, such as point a in Fig. 4.32, where $V = 0$. The outer edge of the boundary layer is given by point b, where the flow velocity is essentially the value given by V_2 in Fig. 4.31. That is, point b in Fig. 4.32 is essentially equivalent to point 2 in Figure 4.31. In this fashion, the flow properties at the outer edge of the boundary layer in Fig. 4.32 can be calculated from a frictionless flow analysis, as pictured in Fig. 4.31. This leads to an important conceptual point in theoretical aerodynamics: A flow field can be split into two regions, one region in which friction

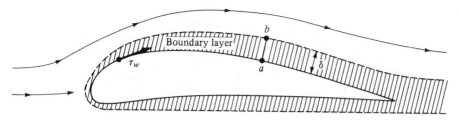

Figure 4.32   Flow in real life, with friction. The thickness of the boundary layer is greatly overemphasized for clarity.

is important, namely, in the boundary layer near the surface, and another region of frictionless flow (sometimes called *potential flow*) outside the boundary layer. This concept was first introduced by Ludwig Prandtl in 1904, and it revolutionized modern theoretical aerodynamics.

It can be shown experimentally and theoretically that the pressure through the boundary layer in a direction perpendicular to the surface is constant. That is, letting p_a and p_b be the static pressures at points a and b, respectively, in Fig. 4.32, then $p_a = p_b$. This is an important phenomenon. This is why a surface pressure distribution calculated from frictionless flow (Fig. 4.31) many times gives accurate results for the real-life surface pressures; it is because the frictionless calculations give the correct pressures at the outer edge of the boundary layer (point b), and these pressures are impressed without change through the boundary layer right down to the surface (point a). The above statements are reasonable for slender aerodynamic shapes such as the airfoil in Fig. 4.32; they do not hold for regions of separated flow over blunt bodies, as previously sketched in Fig. 4.30. Such separated flows are discussed in a subsequent section.

Refer again to Fig. 4.32. The *boundary layer thickness* δ grows as the flow moves over the body; that is, more and more of the flow is affected by friction as the distance along the surface increases. In addition, the presence of friction creates a *shear stress* at the surface τ_w. This shear stress has dimensions of force/area and acts in a direction tangential to the surface. Both δ and τ_w are important quantities, and a large part of boundary layer theory is devoted to their calculation. As we will see, τ_w gives rise to a drag force called *skin friction drag*, hence attesting to its importance. In subsequent sections, equations for the calculation of δ and τ_w will be given.

Looking more closely at the boundary layer, we see that a *velocity profile* through the boundary layer is sketched in Fig. 4.33. The velocity starts out at zero at the surface and increases continuously to its value of V_2 at the outer edge. Let us set up coordinate axes x and y such that x is parallel to the surface and y is normal to the surface, as shown in Fig. 4.33. By definition, a *velocity profile* gives the variation of velocity in the boundary layer as a function of y. In general, the velocity profiles at different x stations are different.

The slope of the velocity profile at the wall is of particular importance, because it governs the wall shear stress. Let $(dV/dy)_{y=0}$ be defined as the velocity gradient

Figure 4.33   Velocity profile through a boundary layer.

at the wall. Then the shear stress at the wall is given by

$$\tau_w = \mu \left(\frac{dV}{dy} \right)_{y=0} \qquad \text{[4.89]}$$

where μ is called the *absolute viscosity coefficient* (or simply the *viscosity*) of the gas. The viscosity coefficient has dimensions of mass/(length)(time), as can be verified from Eq. (4.89) combined with Newton's second law. It is a physical property of the fluid; μ is different for different gases and liquids. Also, μ varies with T. For liquids, μ decreases as T increases (we all know that oil gets "thinner" when the temperature is increased). But for gases, μ increases as T increases (air gets "thicker" when temperature is increased). For air at standard sea-level temperature,

$$\mu = 1.7894 \times 10^{-5} \text{ kg/(m)(s)} = 3.7373 \times 10^{-7} \text{ slug/(ft)(s)}$$

The variation of μ with temperature for air is given in Fig. 4.34.

In this section, we are simply introducing the fundamental concepts of boundary layer flows; such concepts are essential to the practical calculation of aerodynamic drag, as we will soon appreciate. In this spirit, we introduce another important dimensionless "number," a number of importance and impact on aerodynamics equal to those of the Mach number discussed earlier—the *Reynolds number*. Consider the development of a boundary layer on a surface, such as the flat plate sketched in Fig. 4.35. Let x be measured from the leading edge, that is, the front tip of the plate. Let V_∞ be the flow velocity far upstream of the plate. (The subscript ∞ is commonly used to denote conditions far upstream of an aerodynamic body, the *freestream conditions*.) The *Reynolds number* Re_x is defined as

$$\text{Re}_x = \frac{\rho_\infty V_\infty x}{\mu_\infty} \qquad \text{[4.90]}$$

Note that Re_x is dimensionless and that it varies linearly with x. For this reason, Re_x is sometimes called a *local* Reynolds number, because it is based on the local coordinate x.

Up to this point in our discussion on aerodynamics, we have always considered flow streamlines to be smooth and regular curves in space. However, in a viscous

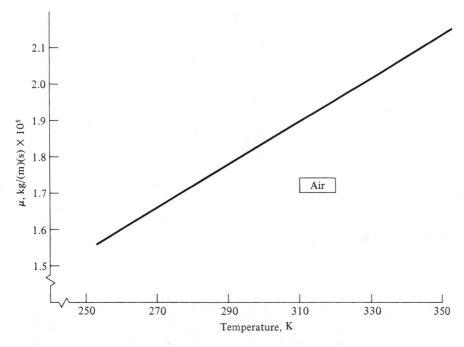

Figure 4.34 Variation of viscosity coefficient with temperature.

Figure 4.35 Growth of the boundary layer thickness.

flow, and particularly in boundary layers, life is not quite so simple. There are two basic types of viscous flow:

1. *Laminar flow*, in which the streamlines are smooth and regular and a fluid element moves smoothly along a streamline

2. *Turbulent flow*, in which the streamlines break up and a fluid element moves in a random, irregular, and tortuous fashion

If you observe smoke rising from a lit cigarette, as sketched in Fig. 4.36, you see first a region of smooth flow—laminar flow—and then a transition to irregular, mixed-up flow—turbulent flow. The differences between laminar and turbulent flow are

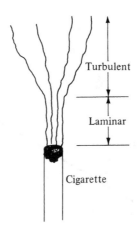

Figure 4.36   Smoke pattern from a cigarette.

dramatic, and they have a major impact on aerodynamics. For example, consider the velocity profiles through a boundary layer, as sketched in Fig. 4.37. The profiles are different, depending on whether the flow is laminar or turbulent. The turbulent profile is "fatter," or fuller, than the laminar profile. For the turbulent profile, from the outer edge to a point near the surface, the velocity remains reasonably close to the freestream velocity; it then rapidly decreases to zero at the surface. In contrast, the laminar velocity profile gradually decreases to zero from the outer edge to the surface. Now consider the velocity gradient at the wall. $(dV/dy)_{y=0}$, which is the reciprocal of the slope of the curves shown in Fig. 4.37 evaluated at $y = 0$. From Fig. 4.37, it is clear that

$$\left(\frac{dV}{dy}\right)_{y=0} \text{ for laminar flow} < \left(\frac{dV}{dy}\right)_{y=0} \text{ for turbulent flow}$$

Recalling Eq. (4.89) for τ_w leads us to the fundamental and highly important fact that *laminar shear stress is less than turbulent shear stress.*

$$\boxed{\tau_w \text{ laminar} < \tau_w \text{ turbulent}}$$

This obviously implies that the skin friction exerted on an airplane wing or body will depend on whether the boundary layer on the surface is laminar or turbulent, with laminar flow yielding the smaller skin friction drag.

It appears to be almost universal in nature that systems with the maximum amount of *disorder* are favored. For aerodynamics, this means that the vast majority of practical viscous flows are turbulent. The boundary layers on most practical airplanes, missiles, ship hulls, etc., are turbulent, with the exception of small regions near the leading edge, as we will soon see. Consequently, the skin friction on these surfaces is the higher, turbulent value. For the aerodynamicist, who is usually striving to reduce

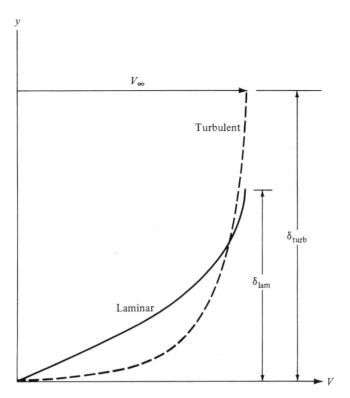

Figure 4.37 Velocity profiles for laminar and turbulent boundary layers. Note that the turbulent boundary layer thickness is larger than the laminar boundary layer thickness.

drag, this is unfortunate. However, the skin friction on slender shapes, such as wing cross sections (airfoils), can be reduced by designing the shape in such a manner as to encourage laminar flow. Figure 4.38 indicates how this can be achieved. Here two airfoils are shown; the standard airfoil on the left has a maximum thickness near the leading edge, whereas the laminar flow airfoil has its maximum thickness near the middle of the airfoil. The pressure distributions on the top surface on the airfoils are sketched above the airfoils in Fig. 4.38. Note that for the standard airfoil, the minimum pressure occurs near the leading edge, and there is a long stretch of increasing pressure from this point to the trailing edge. Turbulent boundary layers are encouraged by such increasing pressure distributions. Hence, the standard airfoil is generally bathed in long regions of turbulent flow, with the attendant high skin friction drag. However, note that for the laminar flow airfoil, the minimum pressure occurs near the trailing edge, and there is a long stretch of decreasing pressure from the leading edge to the point of minimum pressure. Laminar boundary layers are encouraged by such decreasing pressure distributions. Hence, the laminar flow airfoil can be bathed in long regions of laminar flow, thus benefiting from the reduced skin friction drag.

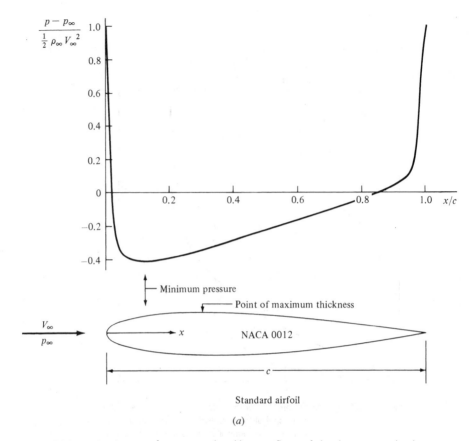

Figure 4.38   Comparison of conventional and laminar flow airfoils. The pressure distributions shown are the theoretical results obtained by NACA and are for 0° angle of attack. The airfoil shapes are drawn to scale. (*continued*)

The North American P-51 Mustang, designed at the outset of World War II, was the first production aircraft to employ a laminar flow airfoil. However, laminar flow is a sensitive phenomenon; it readily gets unstable and tries to change to turbulent flow. For example, the slightest roughness of the airfoil surface caused by such real-life effects as protruding rivets, imperfections in machining, and bug spots can cause a premature transition to turbulent flow in advance of the design condition. Therefore, most laminar flow airfoils used on production aircraft do not yield the extensive regions of laminar flow that are obtained in controlled laboratory tests using airfoil models with highly polished, smooth surfaces. From this point of view, the early laminar flow airfoils were not successful. However, they were successful from an entirely different point of view; namely, they were found to have excellent high-speed properties, postponing to a higher flight Mach number the large drag rise due to shock waves and flow separation encountered near Mach 1. (Such high-speed effects are

(concluded)

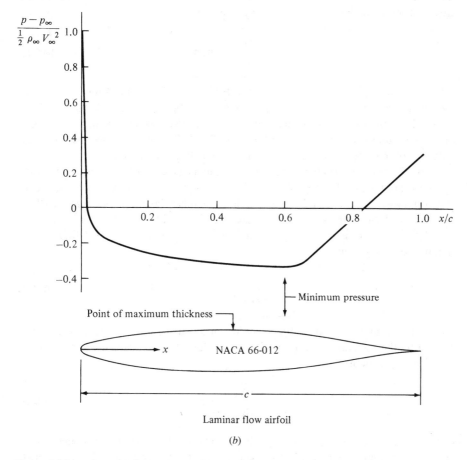

Laminar flow airfoil

(b)

Figure 4.38 (continued).

discussed in Secs. 5.9 to 5.11.) As a result, the early laminar flow airfoils were extensively used on jet-propelled airplanes during the 1950s and 1960s and are still employed today on some modern high-speed aircraft.

Given a laminar or turbulent flow over a surface, how do we actually calculate the skin friction drag? The answer is given in the following two sections.

4.16 RESULTS FOR A LAMINAR BOUNDARY LAYER

Consider again the boundary layer flow over a flat plate, as sketched in Fig. 4.35. Assume that the flow is laminar. The two physical quantities of interest are the boundary layer thickness δ and shear stress τ_w at location x. Formulas for these quantities can be obtained from laminar boundary layer theory, which is beyond the scope of this book. However, the results, which have been verified by experiment, are as follows. The laminar boundary layer thickness is

$$\delta = \frac{5.2x}{\sqrt{Re_x}} \quad \text{laminar} \tag{4.91}$$

where $Re_x = \rho_\infty V_\infty x / \mu_\infty$, as defined in Eq. (4.90). It is remarkable that a phenomenon as complex as the development of a boundary layer, which depends at least on density, velocity, viscosity, and length of the surface, should be described by a formula as simple as Eq. (4.91). In this vein, Eq. (4.91) demonstrates the powerful influence of the Reynolds number Re_x in aerodynamic calculations.

Note from Eq. (4.91) that the laminar boundary layer thickness varies inversely as the square root of the Reynolds number. Also since $Re_x = \rho_\infty V_\infty x / \mu_\infty$, then from Eq. (4.91) $\delta \propto x^{1/2}$; that is, the laminar boundary layer grows *parabolically*.

The local shear stress τ_w is also a function of x, as sketched in Fig. 4.39. Rather than deal with τ_w directly, aerodynamicists find it more convenient to define a local skin friction *coefficient* c_{f_x} as

$$c_{f_x} \equiv \frac{\tau_w}{\frac{1}{2}\rho_\infty V_\infty^2} \equiv \frac{\tau_w}{q_\infty} \tag{4.92}$$

The skin friction coefficient is dimensionless and is defined as the local shear stress divided by the dynamic pressure at the outer edge of the boundary. From laminar boundary layer theory,

$$c_{f_x} = \frac{0.664}{\sqrt{Re_x}} \quad \text{laminar} \tag{4.93}$$

where, as usual, $Re_x = \rho_\infty V_\infty x / \mu_\infty$. Equation (4.93) demonstrates the convenience of defining a dimensionless skin friction coefficient. On one hand, the dimensional shear stress τ_w (as sketched in Fig. 4.39) depends on several quantities such as ρ_∞, V_∞, and Re_x; on the other hand, from Eq. (4.93), c_{f_x} is a function of Re_x *only*. This

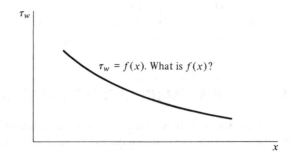

$\tau_w = f(x)$. What is $f(x)$?

Figure 4.39   Variation of shear stress with distance along the surface.

convenience, obtained from using dimensionless coefficients and numbers, reverberates throughout aerodynamics. Relations between dimensionless quantities such as those given in Eq. (4.93) can be substantiated by *dimensional analysis*, a formal procedure to be discussed later.

Combining Eqs. (4.92) and (4.93), we can obtain values of τ_w from

$$\tau_w = f(x) = \frac{0.664 q_\infty}{\sqrt{Re_x}} \qquad \textbf{[4.94]}$$

Note from Eqs. (4.93) and (4.94) that both c_{f_x} and τ_w for laminar boundary layers vary as $x^{-1/2}$; that is, c_{f_x} and τ_w decrease along the surface in the flow direction, as sketched in Fig. 4.39. The shear stress near the leading edge of a flat plate is greater than that near the trailing edge.

The variation of local shear stress τ_w along the surface allows us to calculate the total skin friction drag due to the airflow over an aerodynamic shape. Recall from Sec. 2.2 that the net aerodynamic force on any body is fundamentally due to the pressure and shear stress distributions on the surface. In many cases, it is this total aerodynamic force that is of primary interest. For example, if you mount a flat plate parallel to the airstream in a wind tunnel and measure the force exerted on the plate, by means of a balance of some sort, you are not measuring the local shear stress τ_w; rather, you are measuring the total drag due to skin friction being exerted over all the surface. This *total skin friction drag* can be obtained as follows.

Consider a flat plate of length L and unit width oriented parallel to the flow, as shown in perspective in Fig. 4.40. Consider also an infinitesimally small surface element of the plate of length dx and width unity, as shown in Fig. 4.40. The local shear stress on this element is τ_x, a function of x. Hence, the force on this element due to skin friction is $\tau_w dx(1) = \tau_w dx$. The total skin friction drag is the sum of the forces on all the infinitesimal elements from the leading to the trailing edge; that is, the total skin friction drag D_f is obtained by integrating τ_x along the surface:

$$D_f = \int_0^L \tau_w \, dx \qquad \textbf{[4.95]}$$

Combining Eqs. (4.94) and (4.95) yields

$$D_f = 0.664 q_\infty \int_0^L \frac{dx}{\sqrt{Re_x}} = \frac{0.664 q_\infty}{\sqrt{\rho_\infty V_\infty / \mu_\infty}} \int_0^L \frac{dx}{\sqrt{x}}$$

Figure 4.40 Total drag is the integral of the local shear stress over the surface.

$$D_f = \frac{1.328 q_\infty L}{\sqrt{\rho_\infty V_\infty L / \mu}}$$ [4.96]

Let us define a *total skin friction drag coefficient* C_f as

$$C_f \equiv \frac{D_f}{q_\infty S}$$ [4.97]

where S is the total area of the plate, $S = L(1)$. Thus, from Eqs. (4.96) and (4.97),

$$C_f = \frac{D_f}{q_\infty L(1)} = \frac{1.328 q_\infty L}{q_\infty L (\rho_\infty V_\infty L / \mu_\infty)^{1/2}}$$

or $$C_f = \frac{1.328}{\sqrt{\mathrm{Re}_L}}   \quad \text{laminar}$$ [4.98]

where the Reynolds number is now based on the total length L, that is, $\mathrm{Re}_L \equiv \rho_\infty V_\infty L / \mu_\infty$.

Do not confuse Eq. (4.98) with Eq. (4.93); they are different quantities. The local skin friction coefficient c_{f_x} in Eq. (4.93) is based on the local Reynolds number $\mathrm{Re}_x = \rho_\infty V_\infty x / \mu_\infty$ and is a function of x. However, the total skin friction coefficient C_f is based on Reynolds number for the plate length L: $\mathrm{Re}_L = \rho_\infty V_\infty L / \mu_\infty$.

We emphasize that Eqs. (4.91), (4.93), and (4.98) apply to laminar boundary layers only; for turbulent flow the expressions are different. Also these equations are exact only for low-speed (incompressible) flow. However, they have been shown to be reasonably accurate for high-speed subsonic flows as well. For supersonic and hypersonic flows, where the velocity gradients within the boundary layer are so extreme and where the presence of frictional dissipation creates very large temperatures within the boundary layer, the form of these equations can still be used for engineering approximations, but ρ and μ must be evaluated at some reference conditions germane to the flow inside the boundary layer. Such matters are beyond the scope of this book.

Example 4.19

Consider the flow of air over a small flat plate which is 5 cm long in the flow direction and 1 m wide. The freestream conditions correspond to standard sea level, and the flow velocity is 120 m/s. Assuming laminar flow, calculate:
a. The boundary layer thickness at the downstream edge (the trailing edge)
b. The drag force on the plate

Solution

a. At the trailing edge of the plate, where $x = 5 \text{ cm} = 0.05$ m, the Reynolds number is, from Eq. (4.90),

$$\mathrm{Re}_x = \frac{\rho_\infty V_\infty x}{\mu_\infty} = \frac{(1.225 \text{ kg/m}^3)(120 \text{ m/s})(0.05 \text{ m})}{1.789 \times 10^{-5} \text{ kg/(m)(s)}}$$

$$= 4.11 \times 10^5$$

From Eq. (4.91),

$$\delta = \frac{5.2x}{\mathrm{Re}_x^{1/2}} = \frac{5.2(0.05)}{(4.11 \times 10^5)^{1/2}} = \boxed{4.06 \times 10^{-4} \text{ m}}$$

Note how thin the boundary layer is—only 0.0406 cm at the trailing edge.

b. To obtain the skin friction drag, Eq. (4.98) gives, with $L = 0.05$ m,

$$C_f = \frac{1.328}{\mathrm{Re}_L^{1/2}} = \frac{1.328}{(4.11 \times 10^5)^{1/2}} = 2.07 \times 10^{-3}$$

The drag can be obtained from the definition of the skin friction drag coefficient, Eq. (4.97), once q_∞ and S are known.

$$q_\infty = \tfrac{1}{2}\rho_\infty V_\infty^2 = \tfrac{1}{2}(1.225)(120)^2 = 8820 \text{ N/m}^2$$
$$S = 0.05(1) = 0.05 \text{ m}^2$$

Thus, from Eq. (4.97), the drag *on one surface* of the plate (say the top surface) is

$$\text{Top} \quad D_f = q_\infty S C_f = 8820(0.05)(2.07 \times 10^{-3}) = 0.913 \text{ N}$$

Since both the top and bottom surfaces are exposed to the flow, the total friction drag will be double the above result:

$$\text{Total } D_f = 2(0.913) = \boxed{1.826 \text{ N}}$$

For the flat plate in Example 4.19, calculate and compare the local shear stress at the locations 1 and 5 cm from the front edge (the leading edge) of the plate, measured in the flow direction.

Example 4.20

Solution

The location $x = 1$ cm is near the front edge of the plate. The local Reynolds number at this location, where $x = 1$ cm $= 0.01$ m, is

$$\mathrm{Re}_x = \frac{\rho_\infty V_\infty x}{\mu_\infty} = \frac{1.225(120)(0.01)}{1.789 \times 10^{-5}} = 8.217 \times 10^4$$

From Eq. (4.93),

$$c_{f_x} = \frac{0.664}{\sqrt{\mathrm{Re}_x}} = \frac{0.664}{\sqrt{8.217 \times 10^4}} = \frac{0.664}{286.65} = 0.002316$$

From Eq. (4.92), with $q_\infty = 8820$ N/m^2 from Example 4.19,

$$\tau_w = q_\infty c_{f_x} = 8820(0.002316) = \boxed{20.43 \text{ N/m}^2}$$

At the location $x = 5$ cm $= 0.05$ m, the local Reynolds number is

$$\mathrm{Re}_x = \frac{\rho_\infty V_\infty x}{\mu_\infty} = \frac{1.225(120)(0.05)}{1.789 \times 10^{-5}} = 4.11 \times 10^5$$

(This is the same value as that calculated in Example 4.19.) From Eq. (4.93),

$$c_{f_x} = \frac{0.664}{\sqrt{\mathrm{Re}_x}} = \frac{0.664}{\sqrt{4.11 \times 10^5}} = 0.001036$$

From Eq. (4.92),

$$\tau_w = q_\infty c_{f_x} = 8820(0.001036) = \boxed{9.135 \text{ N/m}^2}$$

By comparison, note that the local shear stress at $x = 5$ cm, that is, at the back end of the plate (the trailing edge), is less than that at $x = 1$ cm near the front edge. This confirms the trend sketched in Fig. 4.39 that τ_w *decreases* with distance in the flow direction along the plate.

As a check on our calculation, we note from Eq. (4.94) that τ_w varies inversely as $x^{1/2}$. Thus, once we have calculated $\tau_w = 20.43$ N/m^2 at $x = 1$ cm, we can directly obtain τ_w at $x = 5$ cm from the ratio

$$\frac{\tau_{w_2}}{\tau_{w_1}} = \sqrt{\frac{x_1}{x_2}}$$

Setting condition 1 at $x = 1$ cm and condition 2 at $x = 5$ cm, we have

$$\tau_{w_2} = \tau_{w_1} \sqrt{\frac{x_1}{x_2}} = 20.43 \sqrt{\frac{1}{5}} = 9.135 \text{ N/m}^2$$

which verifies our original calculation of τ_w at $x = 5$ cm.

4.17   RESULTS FOR A TURBULENT BOUNDARY LAYER

Under the same flow conditions, a turbulent boundary layer will be *thicker* than a laminar boundary layer. This comparison is sketched in Fig. 4.41. Unlike in the case for laminar flows, no exact theoretical results can be presented for turbulent boundary layers. The study of turbulence is a major effort in fluid dynamics today; so far, turbulence is still an unsolved theoretical problem and is likely to remain so for an indefinite time. In fact, turbulence is one of the major unsolved problems in theoretical physics. As a result, our knowledge of δ and τ_w for turbulent boundary layers must rely on experimental results. Such results yield the following approximate formula for turbulent flow:

$$\delta = \frac{0.37x}{\text{Re}_x^{0.2}} \qquad \text{turbulent} \qquad\qquad \textbf{[4.99]}$$

Note from Eq. (4.99) that a turbulent boundary grows approximately as $x^{4/5}$. This is in contrast to the slower $x^{1/2}$ variation for a laminar boundary layer. As a result, turbulent boundary layers grow faster and are thicker than laminar boundary layers.

Figure 4.41   Turbulent boundary layers are thicker than laminar boundary layers.

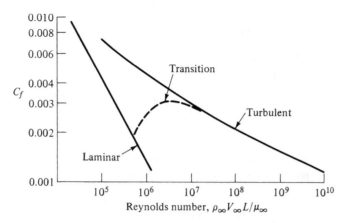

Figure 4.42 Variation of skin friction coefficient with Reynolds number for low-speed flow. Comparison of laminar and turbulent flow.

The local skin friction coefficient for turbulent flow over a flat plate can be approximated by

$$c_{f_x} = \frac{0.0592}{(\mathrm{Re}_x)^{0.2}} \qquad \text{turbulent} \qquad \textbf{[4.100]}$$

The total skin friction coefficient is given approximately as

$$C_f = \frac{0.074}{\mathrm{Re}_L^{0.2}} \qquad \text{turbulent} \qquad \textbf{[4.101]}$$

Note that for turbulent flow, C_f varies as $L^{-1/5}$; this is in contrast to the $L^{-1/2}$ variation for laminar flow. Hence C_f is larger for turbulent flow, which precisely confirms our reasoning at the end of Sec. 4.15, where we noted that τ_w (laminar) < τ_w (turbulent). Also note that C_f in Eq. (4.101) is once again a function of Re_L. Values of C_f for both laminar and turbulent flows are commonly plotted in the form shown in Fig. 4.42. Note the magnitude of the numbers involved in Fig. 4.42. The values of Re_L for actual flight situations may vary from 10^5 to 10^8 or higher; the values of C_f are generally much less than unity, on the order of 10^{-2} to 10^{-3}.

Consider the same flow over the same flat plate as in Example 4.19; however, assume that the boundary layer is now completely turbulent. Calculate the boundary layer thickness at the trailing edge and the drag force on the plate. | **Example 4.21**

Solution

From Example 4.19, $Re_x = 4.11 \times 10^5$. From Eq. (4.99), for turbulent flow,

$$\delta = \frac{0.37x}{Re_x^{0.2}} = \frac{0.37(0.05)}{(4.11 \times 10^5)^{0.2}} = \boxed{1.39 \times 10^{-3} \text{ m}}$$

Note: Compare this result with the laminar flow result from Example 4.19.

$$\frac{\delta_{turb}}{\delta_{lam}} = \frac{1.39 \times 10^{-3}}{4.06 \times 10^{-4}} = 3.42$$

Note that the turbulent boundary layer at the trailing edge is 3.42 times thicker than the laminar boundary layer—quite a sizable amount! From Eq. (4.101)

$$C_f = \frac{0.074}{Re_L^{0.2}} = \frac{0.074}{(4.11 \times 10^5)^{0.2}} = 0.00558$$

On the top surface,

$$D_f = q_\infty S C_f = 8820(0.05)(0.00558) = 2.46 \text{ N}$$

Considering both top and bottom surfaces, we have

$$\text{Total } D_f = 2(2.46) = \boxed{4.92 \text{ N}}$$

Note that the turbulent drag is 2.7 times larger than the laminar drag.

Example 4.22 | Repeat Example 4.20, except now assume that the boundary layer is completely turbulent.

Solution

From Example 4.20, at $x = 1$ cm, $Re_x = 8.217 \times 10^4$. The local turbulent skin friction coefficient at this location is, from Eq. (4.100),

$$c_{f_x} = \frac{0.0592}{Re_x^{0.2}} = \frac{0.0592}{(8.217 \times 10^4)^{0.2}} = 0.00616$$

From Example 4.20, $q_\infty = 8820$ N/m^2. Hence,

$$\tau_w = q_\infty c_{f_x} = 8820(0.00616) = \boxed{54.33 \text{ N/m}^2}$$

Note: In comparison to the laminar flow result from Example 4.20, the turbulent shear stress is $54.33/20.43 = 2.7$ times larger. By *coincidence*, this is the same ratio as the *total* drag comparison made between turbulent and laminar boundary layer cases in Example 4.20.

At $x = 5$ cm, from Example 4.20, $Re_x = 4.11 \times 10^5$. The local turbulent skin friction coefficient at this location is, from Eq. (4.100),

$$c_{f_x} = \frac{0.0592}{Re_x^{0.2}} = \frac{0.0592}{(4.11 \times 10^4)^{0.2}} = 0.00446$$

Hence $$\tau_w = q_\infty c_{f_x} = 8820(0.00446) = \boxed{39.34 \text{ N/m}^2}$$

Note: In comparison to the laminar flow result from Example 4.20, the turbulent shear stress at $x = 5$ cm is $39.34/9.135 = 4.3$ times larger.

Comparing the present results with those of Example 4.20, we see that, over a given length of plate, the *percentage* drop in shear stress for the laminar case is larger than that for the turbulent case. Specifically, the percentage drop over the 4-cm space from $x = 1$ cm to $x = 5$ cm *for the laminar case* (Example 4.20) is

$$\text{Decrease} = \frac{20.43 - 9.135}{20.43} \times 100 = 55.3\%$$

For the *turbulent case* (Example 4.22),

$$\text{Decrease} = \frac{54.33 - 39.34}{54.33} \times 100 = 27.6\%$$

4.18 COMPRESSIBILITY EFFECTS ON SKIN FRICTION

Let us examine again the expressions for laminar and turbulent skin friction coefficients given by Eqs. (4.93) and (4.100), respectively. These equations shout the important fact that c_{f_x} is a function of Reynolds number *only*, that is,

$$\text{Laminar} \quad c_{f_x} \propto \frac{1}{\sqrt{\text{Re}_x}}$$

$$\text{Turbulent} \quad c_{f_x} \propto \frac{1}{\text{Re}_x^{0.2}}$$

Once again we see the power of the Reynolds number in governing viscous flows. However, this is not the whole story. Equations (4.91), (4.93), and (4.98) give expressions for δ, c_{f_x}, and C_f, respectively, for a flat-plate boundary layer in an *incompressible* laminar flow. Similarly, Eqs. (4.99), (4.100), and (4.101) gives expressions for δ, c_{f_x}, and C_f, respectively, for a flat-plate boundary layer in an *incompressible* turbulent flow. Mainly for the benefit of simplicity, we did not emphasize in Secs. 4.16 and 4.17 that these equations apply to an incompressible flow. However, we are now bringing this to your attention. Indeed, you might want to go back to these equations and mark them in the margins as "incompressible."

This raises the question, What are the effects of compressibility on a flat-plate boundary layer? The answer lies in the *Mach number*, which, as we have already seen in Secs. 4.11 to 4.13, is *the* powerful parameter governing high-speed, compressible inviscid flows. Specifically, for a flat-plate boundary layer in a compressible flow, δ, c_{f_x}, and C_f are functions of both Mach number and Reynolds number. The effect of Mach number is not given by a nice, clean formula; rather it must be evaluated from detailed numerical solutions of the compressible boundary layer flow, which is beyond the scope of this book. It is sufficient to note that for a flat-plate compressible boundary layer, the constant 0.664 in the numerator of Eq. (4.93) is replaced by some

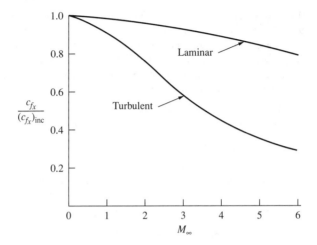

Figure 4.43   Approximate theoretical results for the compressibility effect on laminar and turbulent flat plate skin friction coefficients.

other number that depends on the value of the freestream Mach number, that is,

$$c_{f_x} = \frac{f_1(M_\infty)}{\sqrt{Re_x}} \qquad \text{laminar, compressible} \qquad \text{[4.102]}$$

Similarly, the constant 0.0592 in the numerator of Eq. (4.100) is replaced by some other number that depends on the value of M_∞, that is,

$$c_{f_x} = \frac{f_2(M_\infty)}{Re_x^{0.2}} \qquad \text{turbulent, compressible} \qquad \text{[4.103]}$$

These variations are plotted in Fig. 4.43. Here the ratio of compressible to incompressible skin friction coefficients at the same Reynolds number is plotted versus freestream Mach number for both laminar and turbulent flows. Note the following trends, shown in Figure 4.43:

1. For a constant Reynolds number, the effect of increasing M_∞ is to *decrease* c_{f_x}.
2. The decrease in c_{f_x} is much more pronounced for turbulent flow than for laminar flow.

Example 4.23 | **A** three-view of the Lockheed F-104A Starfighter is shown in Fig. 4.44. This was the first fighter aircraft designed for sustained Mach 2 flight. The airfoil section of the wing is very thin, with an extremely sharp leading edge. Assume the wing is an infinitely thin flat plate. Consider the F-104 flying at Mach 2 at a standard altitude of 35,000 ft. Assume the boundary layer over the wing is turbulent. Estimate the shear stress at a point 2 ft downstream of the leading edge.

Solution
At 35,000 ft, from App. B, $\rho_\infty = 7.382 \times 10^{-4}$ slug/ft^3 and $T_\infty = 394.08°$R. To calculate the Reynolds number, we need both V_∞ and the viscosity coefficient μ_∞. The freestream

Figure 4.44   Three-view of the Lockheed F-104 supersonic fighter.

velocity is obtained from the speed of sound as follows.

$$a_\infty = \sqrt{\gamma R T_\infty} = \sqrt{1.4(1716)(394.08)} = 973 \text{ ft/s}$$

$$V_\infty = a_\infty M_\infty = 973(2) = 1946 \text{ ft/s}$$

We obtain μ_∞ from Fig. 4.34, which shows the variation of μ with T. Note that the ambient temperature in kelvins is obtained from $394.08/1.8 = 219$ K. Extrapolating the linear curve in Fig. 4.34 to a temperature of 219 K, we find that $\mu_\infty = 1.35 \times 10^{-5}$ kg/(m)(s). Converting to English engineering units, we note that as given in Sec. 4.15 at standard sea level, $\mu = 1.7894 \times 10^{-5}$ kg/(m)(s) $= 3.7373 \times 10^{-7}$ slug/(ft)(s). The ratio of these two values gives us the conversion factor, so that at $T = 219$ K $= 394.08°$R,

$$\mu = [1.35 \times 10^{-5} \text{ kg/(m)(s)}] \frac{3.7373 \times 10^{-7} \text{ slug/(ft)(s)}}{1.7894 \times 10^{-5} \text{ kg/(m)(s)}}$$

$$= 2.82 \times 10^{-7} \text{slug/(ft)(s)}$$

Hence $$\text{Re}_x = \frac{\rho_\infty V_\infty x}{\mu_\infty} = \frac{(7.382 \times 10^{-4})(1946)(2)}{2.82 \times 10^{-7}} = 1.02 \times 10^{7}$$

From Eq. (4.100), the *incompressible* skin friction coefficient is

$$(c_{f_x})_{\text{inc}} = \frac{0.0592}{\text{Re}_x^{0.2}} = \frac{0.0592}{1.02 \times 10^{7}} = 0.00235$$

From Fig. 4.43, for a turbulent boundary layer at $M_\infty = 2$,

$$\frac{c_{f_x}}{(c_{f_x})_{\text{inc}}} = 0.74$$

Hence the value of c_{f_x} at Mach 2 is

$$c_{f_x} = 0.74(0.00235) = 0.00174$$

The dynamic pressure is

$$q_\infty = \tfrac{1}{2}\rho_\infty V_\infty^2 = \tfrac{1}{2}(7.382 \times 10^{-4})(1946^2) = 1398 \text{lb/ft}^2$$

Thus $$\tau_w = q_\infty c_{f_x} = 1398(0.00174) = \boxed{2.43 \text{ lb/ft}^2}$$

4.19 TRANSITION

In Sec. 4.16 we discussed the flow over a flat plate as if it were all laminar. Similarly, in Sec. 4.17 we assumed all-turbulent flow. In reality, the flow *always* starts out from the leading edge as laminar. Then at some point downstream of the leading edge, the laminar boundary layer become unstable and small "bursts" of turbulent flow begin to grow in the flow. Finally, over a certain region called the *transition region*, the boundary layer becomes completely turbulent. For purposes of analysis, we usually draw the picture shown in Fig. 4.45, where a laminar boundary starts out from the leading edge of a flat plate and grows parabolically downstream. Then at the *transition point*, it becomes a turbulent boundary layer growing at a faster rate, on the order of $x^{4/5}$ downstream. The value of x where transition is said to take place is the *critical value* x_{cr}. In turn, x_{cr} allows the definition of a *critical Reynolds number* for transition as

$$\text{Re}_{x_{cr}} = \frac{\rho_{\infty} V_{\infty} x_{cr}}{\mu_{\infty}} \qquad [4.104]$$

Volumes of literature have been written on the phenomenon of transition from laminar to turbulent flow. Obviously, because τ_w is different for the two flows, knowledge of where on the surface the transition occurs is vital to an accurate prediction of skin friction drag. The location of the transition point (in reality, a finite region) depends on many quantities, such as the Reynolds number, Mach number, heat transfer to or from the surface, turbulence in the freestream, surface roughness, pressure gradient. A comprehensive discussion of transition is beyond the scope of this book. However, if the critical Reynolds number is given to you (usually from experiments for a given type of flow), then the location of transition x_{cr} can be obtained directly from the definition, Eq. (4.104).

For example, assume that you have an airfoil of given surface roughness in a flow at a freestream velocity of 150 m/s and you wish to predict how far from the leading edge the transition will take place. After searching through the literature for low-speed flows over such surfaces, you may find that the critical Reynolds number determined from experience is approximately $\text{Re}_{x_{cr}} = 5 \times 10^5$. Applying this "experience" to your problem, using Eq. (4.104), and assuming the thermodynamic conditions of the

Figure 4.45   Transition from laminar to turbulent flow. The boundary layer thickness is exaggerated for clarity.

airflow correspond to standard sea level, you find

$$x_{cr} = \frac{\mu_\infty \text{Re}_{x_{cr}}}{\rho_\infty V_\infty} = \frac{[1.789 \times 10^{-5} \text{ kg/(m)(s)}](5 \times 10^5)}{(1.225 \text{ kg/m}^3)(150 \text{ m/s})} = 0.047 \text{ m}$$

Note that the region of laminar flow in this example is small—only 4.7 cm between the leading edge and the transition point. If now you double the freestream velocity to 300 m/s, the transition point is still governed by the critical Reynolds number $\text{Re}_{x_{cr}} = 5 \times 10^5$. Thus,

$$x_{cr} = \frac{(1.789 \times 10^{-5})(5 \times 10^5)}{1.225(300)} = 0.0235 \text{ m}$$

Hence, when the velocity is doubled, the transition point moves forward one-half the distance to the leading edge.

In summary, once you know the critical Reynolds number, you can find x_{cr} from Eq. (4.104). However, an accurate value of $\text{Re}_{x_{cr}}$ applicable to your problem must come from somewhere—experiment, free flight, or some semiempirical theory—and this may be difficult to obtain. This situation provides a little insight into why basic studies of transition and turbulence are needed to advance our understanding of such flows and to allow us to apply more valid reasoning to the prediction of transition in practical problems.

Example 4.24

The wingspan of the *Wright Flyer I* biplane is 40 ft 4 in, and the planform area of each wing is 255 ft^2 (see Figs. 1.1 and 1.2). Assume the wing is rectangular (obviously not quite the case, but not bad), as shown in Fig. 4.46. If the *Flyer* is moving with a velocity of 30 mi/h at standard sea-level conditions, calculate the skin friction drag on the wings. Assume the transition Reynolds number is 6.5×10^5. The areas of laminar and turbulent flow are illustrated by areas A and B, respectively, in Fig. 4.46.

Solution

The general procedure is:
a. Calculate D_f for the combined area $A + B$, assuming the flow is completely turbulent.

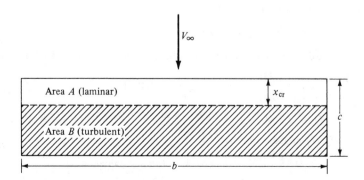

Figure 4.46   Planform view of surface experiencing transition from laminar to turbulent flow.

b. Obtain the turbulent D_f for area B only, by calculating the turbulent D_f for area A and subtracting this from the result of part (a).

c. Calculate the laminar D_f for area A.

d. Add results from parts (b) and (c) to obtain the total drag on the complete surface $A + B$. First, obtain some useful numbers in consistent units: $b = 40$ ft 4 in $= 40.33$ ft. Let $S = $ planform area $= A + B = 255$ ft^2. Hence $c = S/b = 255/40.33 = 6.32$ ft. At standard sea level, $\rho_\infty = 0.002377$ slug/ft^3 and $\mu_\infty = 3.7373 \times 10^{-7}$ slug/(ft)(s). Also $V_\infty = 30$ mi/h $= 30(88/60) = 44$ ft/s. Thus

$$\text{Re}_c = \frac{\rho_\infty V_\infty c}{\mu_\infty} = \frac{0.002377(44)(6.32)}{3.7373 \times 10^{-7}}$$

$$= 1.769 \times 10^6$$

This is the Reynolds number at the trailing edge. To find x_{cr},

$$\text{Re}_{x_{cr}} = \frac{\rho_\infty V_\infty x_{cr}}{\mu_\infty}$$

$$x_{cr} = \frac{\text{Re}_{x_{cr}} \mu_\infty}{\rho_\infty V_\infty}$$

$$= \frac{(6.5 \times 10^5)(3.7373 \times 10^{-7})}{0.002377(44)} = 2.32 \text{ ft}$$

We are now ready to calculate the drag. Assume that the wings of the *Wright Flyer I* are thin enough that the flat-plate formulas apply.

a. To calculate turbulent drag over the complete surface $S = A + B$, use Eq. (4.101):

$$C_f = \frac{0.074}{\text{Re}_L^{0.2}} = \frac{0.074}{(1.769 \times 10^6)^{0.2}} = 0.00417$$

$$q_\infty = \tfrac{1}{2}\rho_\infty V_\infty^2 = \tfrac{1}{2}(0.002377)(44^2) = 2.30 \text{ lb/ft}^2$$

$$(D_f)_s = q_\infty S C_f = 2.30(255)(0.00417) = 2.446 \text{ lb}$$

b. For area A only, assuming turbulent flow,

$$C_f = \frac{0.074}{\text{Re}_{x_{cr}}^{0.2}} = \frac{0.074}{(6.5 \times 10^5)^{0.2}} = 0.00509$$

$$(D_f)_A = q_\infty A C_f = 2.30(2.32 \times 40.33)(0.00509) = 1.095 \text{ lb}$$

Hence, the turbulent drag on area B only is

$$(D_f)_B = (D_f)_s - (D_f)_A = 2.446 - 1.095 = 1.351 \text{ lb}$$

c. Considering the drag on area A, which is in reality a laminar drag, we obtain from Eq. (4.98)

$$C_f = \frac{1.328}{\text{Re}_{x_{cr}}^{0.5}} = \frac{1.328}{(6.5 \times 10^5)^{0.5}} = 0.00165$$

$$(D_f)_A = q_\infty A C_f = 2.30(2.32 \times 40.33)(0.00165) = 0.354 \text{ lb}$$

d. The total drag D_f on the surface is

$$D_f = (\text{laminar drag on } A) + (\text{turbulent drag on } B)$$

$$= 0.354 \text{ lb} + 1.351 \text{ lb} = 1.705 \text{ lb}$$

This is the drag on one surface. Each wing has a top and bottom surface, and there are two wings. Hence, the total skin friction drag on the complete biplane wing configuration is

$$D_f = 4(1.705) = \boxed{6.820 \text{ lb}}$$

4.20 FLOW SEPARATION

We have seen that the presence of friction in the flow causes a shear stress at the surface of a body, which in turn contributes to the aerodynamic drag of the body: skin friction drag. However, friction also causes another phenomenon, called *flow separation*, which in turn creates another source of aerodynamic drag, called *pressure drag due to separation*. The real flow field about a sphere sketched in Fig. 4.30 is dominated by the separated flow on the rearward surface. Consequently, the pressure on the rearward surface is less than the pressure on the forward surface, and this imbalance of pressure forces causes a drag, hence the term *pressure drag due to separation*. In comparison, the skin friction drag on the sphere is very small.

Another example of where flow separation is important is the flow over an airfoil. Consider an airfoil at a low angle of attack (low angle of incidence) to the flow, as sketched in Fig. 4.47. The streamlines move smoothly over the airfoil. The pressure distribution over the top surface is also shown in Fig. 4.47. Note that the pressure at the leading edge is high; the leading edge is a stagnation region, and the pressure is essentially stagnation pressure. This is the highest pressure anywhere on the airfoil. As the flow expands around the top surface of the airfoil, the surface pressure decreases dramatically, dipping to a minimum pressure, which is below the freestream static pressure p_∞. Then as the flow moves farther downstream, the pressure gradually increases, reaching a value slightly above freestream pressure at the trailing edge. This region of increasing pressure is called a region of *adverse* pressure gradient, defined as a region where dp/dx is positive. This region is so identified in Fig. 4.47. The adverse pressure gradient is moderate, that is, dp/dx is small, and for all practical purposes the flow remains attached to the airfoil surface, as sketched in Fig. 4.47. The drag on this airfoil is therefore mainly skin friction drag D_f.

Now consider the same airfoil at a very high angle of attack, as shown in Fig. 4.48. First, assume that we had some magic fluid that would remain attached to the surface—purely an artificial situation. If this were the case, then the pressure distribution on the top surface would follow the dashed line in Fig. 4.48. The pressure would drop precipitously downstream of the leading edge to a value far below the freestream static pressure p_∞. Farther downstream the pressure would rapidly recover to a value above p_∞. However, in this recovery, the adverse pressure gradient would no longer be moderate, as was the case in Fig. 4.47. Instead, in Fig. 4.48, the adverse pressure gradient would be severe; that is, dp/dx would be large. In such cases, the *real* flow field tends to separate from the surface. Therefore, in Fig. 4.48, the real flow field is

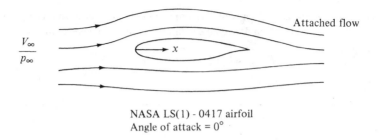

NASA LS(1) - 0417 airfoil
Angle of attack = 0°

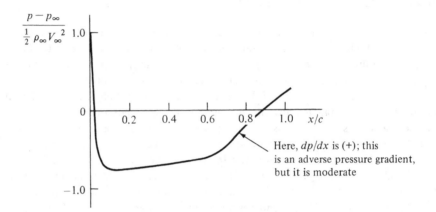

Figure 4.47   Pressure distribution over the top surface for attached flow over an airfoil.
Theoretical data for a modern NASA low-speed airfoil, from NASA
Conference Publication 2046, *Advanced Technology Airfoil Research*,
vol. II, March 1978, p. 11. *(After McGhee, Beasley, and Whitcomb.)*

sketched with a large region of separated flow over the top surface of the airfoil. In
this real separated flow, the *actual* surface pressure distribution is given by the *solid*
curve. In comparison to the dashed curve, note that the actual pressure distribution
does not dip to as low a pressure minimum and that the pressure near the trailing
edge does not recover to a value above p_∞. This has two major consequences, as
can be seen from Fig. 4.49. Here the airfoil at a large angle of attack (thus, with
flow separation) is shown with the real surface pressure distribution, symbolized by
the solid arrows. Pressure always acts normal to a surface. Hence the arrows are all
perpendicular to the local surface. The length of the arrow denotes the magnitude
of the pressure. A solid curve is drawn through the base of the arrows to form an
"envelope" to make the pressure distribution easier to visualize. However, if the flow
were *not* separated, that is, if the flow were attached, then the pressure distribution
would be that shown by the dashed arrows (and the dashed envelope). The solid and
dashed arrows in Fig. 4.49 qualitatively correspond to the solid and dashed pressure
distribution curves, respectively, in Fig. 4.48.

The solid and dashed arrows in Fig. 4.49 should be looked at carefully. They
explain the two major consequences of separated flow over the airfoil. The first

Separated flow

NASA LS(1) - 0417 airfoil
Angle of attack = 18.4°

Figure 4.48   Pressure distribution over the top surface for separated flow over an airfoil. Theoretical data for a modern NASA low-speed airfoil, from NASA Conference Publication 2045, Part 1, *Advanced Technology Airfoil Research*, vol. 1, March 1978, p. 380. *(After Zumwalt and Nack.)*

consequence is a loss of lift. The aerodynamic lift (the vertical force shown in Fig. 4.49) is derived from the net component of a pressure distribution in the vertical direction. High lift is obtained when the pressure on the bottom surface is large and the pressure on the top surface is small. Separation does not affect the bottom surface pressure distribution. However, comparing the solid and dashed arrows on the *top surface just downstream of the leading edge*, we find the solid arrows indicate a higher

Note: Length of the arrows denoting
pressure is proportional to $p - p_{ref}$,
where p_{ref} is an arbitrary reference
pressure slightly less than the minimum
pressure on the airfoil

Figure 4.49   Qualitative comparison of pressure distribution, lift, and drag for attached and
separated flows. Note that for separated flow, the lift decreases and the drag increases.

pressure when the flow is separated. This higher pressure is pushing down, hence
reducing the lift. This reduction of lift is also compounded by the geometric effect
that the portion of the top surface of the airfoil near the leading edge is approximately
horizontal in Fig. 4.49. When the flow is separated, causing a higher pressure on
this part of the airfoil surface, the direction in which the pressure is acting is closely
aligned to the vertical, and hence almost the full effect of the increased pressure is felt
by the lift. The combined effect of the increased pressure on the top surface near the
leading edge, and the fact that this portion of the surface is approximately horizontal,
leads to the rather dramatic loss of lift when the flow separates. Note in Fig. 4.49 that
the lift for separated flow (the solid vertical arrow) is smaller than the lift that would
exist if the flow were attached (the dashed vertical arrow).

Now let us concentrate on that portion of the top surface *near the trailing edge.*
On this portion of the airfoil surface, the pressure for the separated flow is now
smaller than the pressure that would exist if the flow were attached. Moreover, the
top surface near the trailing edge geometrically is inclined to the horizontal, and in
fact somewhat faces in the horizontal direction. Recall that drag is in the horizontal
direction in Fig. 4.49. Because of the inclination of the top surface near the trailing
edge, the pressure exerted on this portion of the surface has a strong component in
the horizontal direction. This component acts toward the left, tending to counter the
horizontal component of force due to the high pressure acting on the nose of the
airfoil pushing toward the right. The net pressure drag on the airfoil is the difference

between the force exerted on the front pushing toward the right, and the force exerted on the back pushing toward the left. When the flow is separated, the pressure at the back is lower than it would be if the flow were attached. Hence, for the separated flow, there is *less* force on the back pushing toward the left, and the net drag acting toward the right is therefore *increased*. Note in Fig. 4.49 that the drag for separated flow (the solid horizontal arrow) is larger than the drag that would exist if the flow were attached (the dashed horizontal arrow).

Therefore, two major consequences of the flow separating over an airfoil are

1. A drastic loss of lift (stalling)
2. A major increase in drag, caused by pressure drag due to separation

When the wing of an airplane is pitched to a high angle of attack, the wing can stall, that is, there can be a sudden loss of lift. Our discussion above gives the physical reasons for this stalling phenomenon. Additional ramifications of stalling are discussed in Chap. 5.

Before ending this discussion of separated flow, we ask the question, Why does a flow separate from a surface? The answer is combined in the concept of an adverse pressure gradient (dp/dx is positive) and the velocity profile through the boundary layer, as shown in Fig. 4.37. If dp/dx is positive, then the fluid elements moving along a streamline have to work their way "uphill" against an increasing pressure. Consequently, the fluid elements will slow down under the influence of an adverse pressure gradient. For the fluid elements moving outside the boundary layer, where the velocity (and hence kinetic energy) is high, this is not much of a problem. The fluid element keeps moving downstream. However, consider a fluid element deep inside the boundary layer. Looking at Fig. 4.37, we see its velocity is small. It has been retarded by friction forces. The fluid element still encounters the same adverse pressure gradient, but its velocity is too low to negotiate the increasing pressure. As a result, the element comes to a stop somewhere downstream and then reverses its direction. Such reversed flow causes the flow field in general to separate from the surface, as shown in Fig. 4.48. This is physically how separated flow develops.

Reflecting once again on Fig. 4.37, we note that turbulent boundary layers have fuller velocity profiles. At a given distance from the surface (a given value of y), the velocity of a fluid element in a turbulent boundary is higher than that in a laminar boundary layer. Hence, in turbulent boundary layers, there is more flow kinetic energy nearer the surface. Hence, the flow is less inclined to separate. This leads to a very fundamental fact. Laminar boundary layers separate more easily than turbulent boundary layers. *Hence, to help prevent flow field separation, we want a turbulent boundary layer.*

4.21 SUMMARY OF VISCOUS EFFECTS ON DRAG

We have seen that the presence of friction in a flow produces two sources of drag:

1. Skin friction drag D_f due to shear stress at the wall
2. Pressure drag due to flow separation D_p, sometimes identified as *form* drag

The *total* drag which is caused by viscous effects is then

$$D = D_f + D_p$$

Total drag due to viscous effects | Drag due to skin friction | Drag due to separation (pressure drag)

[4.105]

Equation (4.105) contains one of the classic compromises of aerodynamics. In previous sections, we pointed out that skin friction drag is reduced by maintaining a laminar boundary layer over a surface. However, we also pointed out at the end of Sec. 4.20 that turbulent boundary layers inhibit flow separation; hence pressure drag due to separation is reduced by establishing a turbulent boundary layer on the surface. Therefore, in Eq. (4.105) we have the following compromise:

$$D = D_f + D_p$$

Less for laminar, more for turbulent | More for laminar, less for turbulent

Consequently, as discussed at the end of Sec. 4.15, it cannot be said in general that either laminar or turbulent flow is preferable. Any preference depends on the specific application. On one hand, for a blunt body such as the sphere in Figure 4.30, the drag is mainly pressure drag due to separation; hence, turbulent boundary layers reduce the drag on spheres and are therefore preferable. (We discuss this again in Chap. 5). On the other hand, for a slender body such as a sharp, slender cone or a thin airfoil at small angles of attack to the flow, the drag is mainly skin friction drag; hence, laminar boundary layers are preferable in this case. For in-between cases, the ingenuity of the designer along with practical experience helps to determine what compromises are best.

As a final note to this section, the total drag D given by Eq. (4.105) is called *profile drag* because both skin friction and pressure drag due to separation are ramifications of the shape and size of the body, that is, the "profile" of the body. The profile drag D is the total drag on an aerodynamic shape due to viscous effects. However, it is not in general the total aerodynamic drag on the body. There is one more source of drag, induced drag, which is discussed in the next chapter.

4.22   HISTORICAL NOTE: BERNOULLI AND EULER

Equation (4.9) is one of the oldest and most powerful equations in fluid dynamics. It is credited to Daniel Bernoulli, who lived during the 18th century; little did Bernoulli know that his concept would find widespread application in the aeronautics of the 20th century. Who was Bernoulli, and how did Bernoulli's equation come about? Let us briefly look into these questions; the answers will lead us to a rather unexpected conclusion.

Daniel Bernoulli (1700–1782) was born in Groningen, Netherlands, on January 29, 1700. He was a member of a remarkable family. His father, Johann Bernoulli, was a noted mathematician who made contributions to differential and integral calculus and who later became a doctor of medicine. Jakob Bernoulli, who was Johann's brother (Daniel's uncle), was an even more accomplished mathematician; he made major contributions to the calculus; he coined the term *integral*. Sons of both Jakob and Johann, including Daniel, went on to become noted mathematicians and physicists. The entire family was Swiss and made its home in Basel, Switzerland, where they held various professorships at the University of Basel. Daniel Bernoulli was born away from Basel only because his father spent 10 years as professor of mathematics in the Netherlands. With this type of pedigree, Daniel could hardly avoid making contributions to mathematics and science himself.

And indeed, he did make contributions. For example, he had insight into the kinetic theory of gases; he theorized that a gas was a collection of individual particles moving about in an agitated fashion, and he correctly associated the increased temperature of a gas with the increased energy of the particles. These ideas, originally published in 1738, were to lead a century later to a mature understanding of the nature of gases and heat, and helped to lay the foundation for the elegant kinetic theory of gases.

Daniel's thoughts on the kinetic motion of gases were published in his book *Hydrodynamica* (1738). However, this book was to etch his name more deeply in association with fluid mechanics than with kinetic theory. The book was started in 1729, when Daniel was a professor of mathematics at Leningrad (then St. Petersburg) in Russia. By this time he was already well recognized; he had won 10 prizes offered by the Royal Academy of Sciences in Paris for his solution of various mathematical problems. In his *Hydrodynamica* (which was written entirely in Latin), Bernoulli ranged over such topics as jet propulsion, manometers, and flow in pipes. He also attempted to obtain a relationship between pressure and velocity, but his derivation was obscure. In fact, even though Bernoulli's equation, Eq. (4.9), is usually ascribed to Daniel via his *Hydrodynamica*, the precise equation is not to be found in the book! The picture is further complicated by his father, Johann, who published a book in 1743 entitled *Hydraulica*. It is clear from this latter book that the father understood Bernoulli's theorem better than the son; Daniel thought of pressure strictly in terms of the height of a manometer column, whereas Johann had the more fundamental understanding that pressure was a force acting on the fluid. However, neither of the Bernoullis understood that pressure is a point property. That was to be left to Leonhard Euler.

Leonhard Euler (1707–1783) was also a Swiss mathematician. He was born in Basel, Switzerland, on April 15, 1707, seven years after the birth of Daniel Bernoulli. Euler went on to become one of the mathematical giants of history, but his contributions to fluid dynamics are of interest here. Euler was a close friend of the Bernoullis; indeed, he was a student of Johann Bernoulli at the University of Basel. Later, Euler followed Daniel to St. Petersburg, where he became a professor of mathematics. It was here that Euler was influenced by the work of the Bernoullis in hydrodynamics, but more influenced by Johann than by Daniel. Euler originated the concept of pressure acting at a point in a gas. This quickly led to his differential equation for a fluid accelerated by gradients in pressure, the same equation we have derived as

Eq. (4.8). In turn, Euler integrated the differential equation to obtain, for the first time in history, Bernoulli's equation, just as we have obtained Eq. (4.9). Hence we see that Bernoulli's equation, Eq. (4.9), is really a historical misnomer. Credit for Bernoulli's equation is legitimately shared by Euler.

4.23   HISTORICAL NOTE: THE PITOT TUBE

The use of a Pitot tube to measure airspeed is described in Sec. 4.11; indeed, the Pitot tube today is so commonly used in aerodynamic laboratories and on aircraft that it is almost taken for granted. However, this simple little device has had a rather interesting and somewhat obscure history.

The Pitot tube is named after its inventor, Henri Pitot (1695–1771). Born in Aramon, France, in 1695, Pitot began his career as an astronomer and mathematician. He was accomplished enough to be elected to the Royal Academy of Sciences, Paris, in 1724. About this time, Pitot became interested in hydraulics and, in particular, in the flow of water in rivers and canals. However, he was not satisfied with the existing technique of measuring the flow velocity, which was to observe the speed of a floating object on the surface of the water. So he devised an instrument consisting of two tubes. One was simply a straight tube open at one end which was inserted vertically into the water (to measure static pressure), and the other was a tube with one end bent at right angles, with the open end facing directly into the flow (to measure total pressure). In 1732, between two piers of a bridge over the Seine River in Paris, he used this instrument to measure the flow velocity of the river. This invention and first use of the Pitot tube was announced by Pitot to the Academy on November 12, 1732. In his presentation, he also presented some data of major importance on the variation of water flow velocity with depth. Contemporary theory, based on experience of some Italian engineers, held that the flow velocity at a given depth was proportional to the mass above it; hence the velocity was thought to increase with depth. Pitot reported the stunning (and correct) results, measured with his instrument, that in reality the flow velocity *decreased* as the depth increased. Hence, the Pitot tube was introduced with style.

Interestingly enough, Pitot's invention soon fell into disfavor with the engineering community. A number of investigators attempted to use just the Pitot tube itself, without a local static pressure measurement. Others, using the device under uncontrolled conditions, produced spurious results. Various shapes and forms other than a simple tube were sometimes used for the mouth of the instrument. Moreover, there was no agreed-upon rational theory of the Pitot tube. Note that Pitot developed his instrument in 1732, six years *before* Daniel Bernoulli's *Hydrodynamica* and well before Euler had developed the Bernoullis' concepts into Eq. (4.9), as discussed in Sec. 4.22. Hence, Pitot used intuition, not theory, to establish that the pressure difference measured by his instrument was an indication of the square of the local flow velocity. Of course, as described in Sec. 4.11, we now clearly understand that a Pitot-static device measures the difference between total and static pressures and that for incompressible flow, this difference is related to the velocity squared through Bernoulli's equation; that is, from Eq. (4.62),

$$p_0 - p = \tfrac{1}{2}\rho V^2$$

However, for more than 150 years after Pitot's introduction of the instrument, various engineers attempted to interpret readings in terms of

$$p_0 - p = \tfrac{1}{2}K\rho V^2$$

where K was an empirical constant, generally much different from unity. Controversy was still raging as late as 1913, when John Airey, a professor of mechanical engineering from the University of Michigan, finally performed a series of well-controlled experiments in a water tow tank, using Pitot probes of six different shapes. These shapes are shown in Fig. 4.50, which is taken from Airey's paper in the April 17, 1913, issue of the *Engineering News*, entitled "Notes on the Pitot Tube." In this

Figure 4.50   Six forms of Pitot tubes tested by John Airey. (From *Engineerings News*, vol. 69, no. 16, p. 783, April 1913.)

paper, Airey states that all his measurements indicate that $K = 1.0$ within 1 percent accuracy, independent of the shape of the tube. Moreover, he presents a rational theory based on Bernoulli's equation. Further comments on these results are made in a paper entitled "Origin and Theory of the Pitot Tube" by A. E. Guy, the chief engineer of a centrifugal pump company in Pittsburgh, in a later, June 5, 1913, issue of the *Engineering News*. This paper also helped to establish the Pitot tube on firmer technical grounds.

It is interesting to note that neither of these papers in 1913 mentioned what was to become the most prevalent use of the Pitot tube, namely, the measurement of airspeed for airplanes and wind tunnels. The first practical airspeed indicator, a Venturi tube, was used on an aircraft by the French Captain A. Eteve in January 1911, more than seven years after the first powered flight. Later in 1911, British engineers at the Royal Aircraft Establishment (RAE) at Farnborough employed a Pitot tube on an airplane for the first time. This was to eventually evolve into the primary instrument for flight speed measurement.

There was still controversy over Pitot tubes, as well as the need for reliable airspeed measurements, in 1915, when the brand-new National Advisory Committee for Aeronautics (NACA) stated in its First Annual Report that "an important problem to aviation in general is the devising of accurate, reliable and durable air speed meters.... The Bureau of Standards is now engaged in investigation of such meters, and attention is invited to the report of Professor Herschel and Dr. Buckingham of the bureau on Pitot tubes." The aforementioned report was NACA Report No. 2, Part 1, "The Pitot Tube and other Anemometers for Aeroplanes," by W. H. Herschel, and Part 2, "The Theory of the Pitot and Venturi Tubes," by E. Buckingham. Part 2 is of particular interest. In clear terms, it gives a version of the theory we developed in Sec. 4.11 for the Pitot tube; moreover, it develops for the first time the theory for *compressible* subsonic flow—quite unusual for 1915! Buckingham shows that to obtain 0.5 percent accuracy with the incompressible relations, V_∞ should not exceed 148 mi/h $= 66.1$ m/s. However, he goes on to state that "since the accuracy of better than 1.0 percent can hardly be demanded of an airplane speedometer, it is evident that for all ordinary speeds of flight, no correction for compressibility is needed...." This was certainly an appropriate comment for the "ordinary" airplanes of that day; indeed, it was accurate for most aircraft until the 1930s.

In retrospect, we see that the Pitot tube was invented almost 250 years ago but that its use was controversial and obscure until the second decade of powered flight. Then between 1911 and 1915, one of those "explosions" in technical advancement occurred. Pitot tubes found a major home on airplanes, and the appropriate theory for their correct use was finally established. Since then, Pitot tubes have become commonplace. Indeed, the Pitot tube is usually the first aerodynamic instrument introduced to students of aerospace engineering in laboratory studies.

4.24 HISTORICAL NOTE: THE FIRST WIND TUNNELS

Aerospace engineering in general, and aerodynamics in particular, is an empirically based discipline. Discovery and development by experimental means have been its lifeblood, extending all the way back to George Cayley (see Chap. 1). In turn, the workhorse for such experiments has been predominantly the wind tunnel, so much so that today most aerospace industrial, government, and university laboratories have a complete spectrum of wind tunnels ranging from low subsonic to hypersonic speeds.

It is interesting to reach back briefly into history and look at the evolution of wind tunnels. Amazingly enough, this history goes back more than 400 years, because the cardinal principle of wind tunnel testing was stated by Leonardo da Vinci near the beginning of the 16th century as follows:

> For since the action of the medium upon the body is the same whether the body moves in a quiescent medium, or whether the particles of the medium impinge with the same velocity upon the quiescent body; let us consider the body as if it were quiescent and see with what force it would be impelled by the moving medium.

This is almost self-evident today, that the lift and drag of an aerodynamic body are the same whether it moves through the stagnant air at 100 mi/h or whether the air moves over the stationary body at 100 mi/h. This concept is the very foundation of wind tunnel testing.

The first actual wind tunnel in history was designed and built more than 100 years ago by Francis Wenham in Greenwich, England, in 1871. We met Wenham once before, in Sec. 1.4, where his activity in the Aeronautical Society of Great Britain was noted. Wenham's tunnel was nothing more than a 10-ft-long wooden box with a square cross section, 18 in on a side. A steam-driven fan at the front end blew air through the duct. There was no contour, hence no aerodynamic control or enhancement of flow. Plane aerodynamic surfaces were placed in the airstream at the end of the box, where Wenham measured the lift and drag on weighing beams linked to the model.

Thirteen years later, Horatio F. Phillips, also an Englishman, built the second known wind tunnel in history. Again, the flow duct was a box, but Phillips used steam ejectors (high-speed steam nozzles) downstream of the test section to suck air through the tunnel. Phillips went on to conduct some pioneering airfoil testing in his tunnel, which will be mentioned again in Chap. 5.

Other wind tunnels were built before the turning point in aviation in 1903. For example, the first wind tunnel in Russia was due to Nikolai Joukowski at the University of Moscow in 1891 (it had a 2-in diameter). A larger, 7 in $\times$ 10 in tunnel was built in Austria in 1893 by Ludwig Mach, son of the famed scientist and philosopher Ernst Mach, after whom the Mach number is named. The first tunnel in the United States was built at the Massachusetts Institute of Technology in 1896 by Alfred J. Wells, who used the machine to measure the drag on a flat plate as a check on the whirling-arm measurements of Langley (see Sec. 1.8). Another tunnel in the United States was built by Dr. A. Heb Zahm at the Catholic University of America in 1901. In light of these activities, it is obvious that at the turn of the 20th century aerodynamic testing in wind tunnels was poised and ready to burst forth with the same energy that accompanied the development of the airplane itself.

It is fitting that the same two people responsible for getting the airplane off the ground should also have been responsible for the first concentrated series of wind tunnel tests. As noted in Sec. 1.8, the Wright brothers in late 1901 concluded that a large part of the existing aerodynamic data were erroneous. This led to their construction of a 6-ft-long 16-in-square wind tunnel powered by a two-blade fan connected to a gasoline engine. A replica of the Wrights' wind tunnel is shown in Fig. 4.51. (Their original wind tunnel no longer exists.) They designed and built their own balance to measure the ratios of lift to drag. Using this apparatus, Wilbur and Orville undertook a major program of aeronautical research between September 1901 and August 1902. During this time, they tested more than 200 different airfoil shapes manufactured out of steel. The results from these tests constitute the first major impact of wind tunnel testing on the development of a successful airplane. As we quoted in Sec. 1.8, Orville said about their results: "Our tables of air pressure which we made in our wind tunnel would enable us to calculate in advance the performance of a machine." What a fantastic development! This was a turning point in the history of wind tunnel testing, and it had as much impact on that discipline as the December 17, 1903, flight had on the airplane.

The rapid growth in aviation after 1903 was paced by the rapid growth of wind tunnels, both in numbers and in technology. For example, tunnels were built at the National Physical Laboratory in London in 1903; in Rome in 1903; in Moscow in 1905; in Göttingen, Germany (by the famous Dr. Ludwig Prandtl, originator of the boundary layer concept in fluid dynamics) in 1908; in Paris in 1909 (including two built by Gustave Eiffel, of tower fame); and again at the National Physical Laboratory in 1910 and 1912.

Figure 4.51　A replica of the Wright brothers' wind tunnel in a workroom behind the Wrights' bicycle shop, now in Greenfield Village, Dearborn, Michigan.

All these tunnels, quite naturally, were low-speed facilities, but they were pioneering for their time. Then in 1915, with the creation of NACA (see Sec. 2.8), the foundation was laid for some major spurts in wind-tunnel design. The first NACA wind tunnel became operational at the Langley Memorial Aeronautical Laboratory at Hampton, Virginia, in 1920. It had a 5-ft-diameter test section which accommodated models up to 3.5 ft wide. Then in 1923, in order to simulate the higher Reynolds numbers associated with flight, NACA built the first variable-density wind tunnel, a facility that could be pressurized to 20 atm in the flow and therefore obtain a 20-fold increase in density, hence Re, in the test section. During the 1930s and 1940s, subsonic wind tunnels grew larger and larger. In 1931, an NACA wind tunnel with a 30 ft × 60 ft oval test section went into operation at Langley with a 129 mi/h maximum flow velocity. This was the first million-dollar tunnel in history. Later, in 1944, a 40 ft × 80 ft tunnel with a flow velocity of 265 mi/h was initiated at Ames Aeronautical Laboratory at Moffett Field, California. This is still the largest wind tunnel in the world today. Figure 4.52 shows the magnitude of such tunnels: whole airplanes can be mounted in the test section!

The tunnels mentioned above were low-speed, essentially incompressible flow tunnels. They were the cornerstone of aeronautical testing until the 1930s and remain an important part of the aerodynamic scene today. However, airplane speeds were progressively increasing, and new wind tunnels with higher velocity capability were needed. Indeed, the first requirement for high-speed subsonic tunnels was established by propellers—in the 1920s and 1930s the propeller diameters and rotational speeds were both increasing so as to encounter compressibility problems at the tips. This problem led NACA to build a 12-in-diameter high-speed tunnel at Langley in 1927. It could produce a test section flow of 765 mi/h. In 1936, to keep up with increasing airplane speeds, Langley built a large 8-ft high-speed wind tunnel providing 500 mi/h. This was increased to 760 mi/h in 1945. An important facility was built at Ames in 1941, a 16-ft tunnel with an airspeed of 680 mi/h. A photograph of the Ames 16-ft tunnel is shown in Fig. 4.53 just to give a feeling for the massive size involved with such a facility.

In the early 1940s, the advent of the V-2 rocket as well as the jet engine put supersonic flight in the minds of aeronautical engineers. Suddenly, the requirement for supersonic tunnels became a major factor. However, supersonic flows in the laboratory and in practice date farther back than this. The first supersonic nozzle was developed by Laval about 1880 for use with steam turbines. This is why the convergent-divergent nozzles are frequently called Laval nozzles. In 1905, Prandtl built a small Mach 1.5 tunnel at Göttingen to be used to study steam turbine flows and (of all things) the moving of sawdust around sawmills.

The first practical supersonic wind tunnel for aerodynamic testing was developed by Dr. A. Busemann at Braunschweig, Germany, in the mid-1930s. Using the "method of characteristics" technique, which he had developed in 1929, Busemann designed the first smooth supersonic nozzle contour which produced shock-free isentropic flow. He had a diffuser with a second throat downstream to decelerate the flow and to obtain efficient operation of the tunnel. A photograph of Busemann's tunnel is shown in Fig. 4.54. All supersonic tunnels today look essentially the same.

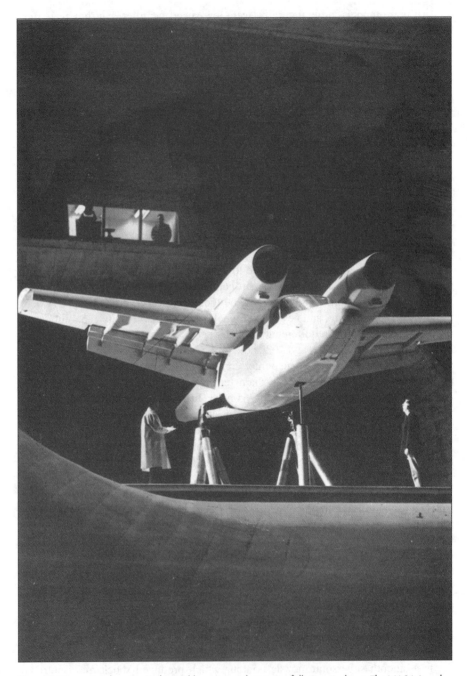

Figure 4.52   A subsonic wind tunnel large enough to test a full-size airplane. The NASA Langley Research Center 30 ft × 60 ft tunnel.

Figure 4.53 The Ames 16-ft high-speed subsonic wind tunnel, illustrating the massive size that goes along with such a wind tunnel complex. *(Courtesy NASA Ames Research Center.)*

Figure 4.54 The first practical supersonic wind tunnel, built by A. Busemann in the mid-1930s. *(Courtesy of A. Busemann.)*

Working from Busemann's example, the Germans built two major supersonic tunnels at their research complex at Peenemünde during World War II. These were used for research and development of the V-2 rocket. After the war, these tunnels were moved almost in total to the U.S. Naval Ordnance Laboratory (later, one was moved to the University of Maryland), where they are still in use today. However, the first supersonic tunnel built in the United States was designed by Theodore von Karman and his colleagues at the California Institute of Technology in 1944 and was built and operated at the Army Ballistics Research Laboratory at Aberdeen, Maryland, under contract with Cal Tech. Then the 1950s saw a virtual harvest of supersonic wind tunnels, one of the largest being the 16 ft × 16 ft continuously operated supersonic tunnel of the Air Force at the Arnold Engineering Development Center (AEDC) in Tennessee.

About this time, the development of the intercontinental ballistic missile (ICBM) was on the horizon, soon to be followed by the space program of the 1960s. Flight vehicles were soon to encounter velocities as high as 36,000 ft/s in the atmosphere—hypersonic velocities. In turn, hypersonic wind tunnels ($M > 5$) were suddenly in demand. The first hypersonic wind tunnel was operated by NACA at Langley in 1947. It had an 11-in-square test section capable of Mach 7. Three years later, another hypersonic tunnel went into operation at the Naval Ordnance Laboratory. These tunnels are distinctly different from their supersonic relatives in that, to obtain hypersonic speeds, the flow has to be expanded so far that the temperature decreases to the point of liquefying the air. To prevent this, all hypersonic tunnels, both old and new, have to have the reservoir gas heated to temperatures far above room temperature before its expansion through the nozzle. Heat transfer is a problem for high-speed flight vehicles, and such heating problems feed right down to the ground-testing facilities for such vehicles.

In summary, modern wind tunnel facilities range across the whole spectrum of flight velocities, from low subsonic to hypersonic speeds. These facilities are part of the everyday life of aerospace engineering; hopefully, this brief historical sketch has provided some insight into their tradition and development.

4.25   HISTORICAL NOTE: OSBORNE REYNOLDS AND HIS NUMBER

In Secs. 4.15 to 4.19, we observed that the Reynolds number, defined in Eq. (4.90) as $Re = \rho_\infty V_\infty x / \mu_\infty$, was the governing parameter for viscous flow. Boundary layer thickness, skin friction drag, transition to turbulent flow, and many other characteristics of viscous flow depend explicitly on the Reynolds number. Indeed, we can readily show that the Reynolds number itself has physical meaning—it is proportional to the ratio of inertia forces to viscous forces in a fluid flow. Clearly, the Reynolds number is an extremely important dimensionless parameter in fluid dynamics. Where did the Reynolds number come from? When was it first introduced, and under what circumstances? The Reynolds number is named after a man—Osborne Reynolds. Who was Reynolds? This section answers these questions.

First, let us look at Osborne Reynolds, the man. He was born on October 23, 1842, in Belfast, Ireland. He was raised in an intellectual family atmosphere; his father had been a fellow of Queens College, Cambridge, a principal of Belfast Collegiate School, headmaster of Dedham Grammar School in Essex, and finally rector at Debach-with-Boulge in Suffolk. Indeed, Anglican clerics were a tradition in the Reynolds family; in addition to his father, his grandfather and great-grandfather had been rectors at Debach. Against this background, Osborne Reynolds's early education was carried out by his father at Dedham. In his teens, he already showed an intense interest in the study of mechanics, for which he had a natural aptitude. At the age of 19, he served a short apprenticeship in mechanical engineering before attending Cambridge University a year later. Reynolds was a highly successful student at Cambridge, graduating with the highest honors in mathematics. In 1867, he was elected a fellow of Queens College, Cambridge (an honor earlier bestowed upon his father). He went on to serve one year as a practicing civil engineer in the office of John Lawson in London. However, in 1868, Owens College in Manchester (later to become the University of Manchester) established its chair of engineering—the second of its kind in any English university (the first was the chair of civil engineering established at the University College, London, in 1865). Reynolds applied for this chair, writing in his application:

> From my earliest recollection I have had an irresistible liking for mechanics and the physical laws on which mechanics as a science are based. In my boyhood I had the advantage of the constant guidance of my father, also a lover of mechanics and a man of no mean attainment in mathematics and their application to physics.

Despite his youth and relative lack of experience, Reynolds was appointed to the chair at Manchester. For the next 37 years he served as a professor at Manchester until his retirement in 1905.

During those 37 years, Reynolds distinguished himself as one of history's leading practitioners of classical mechanics. During his first years at Manchester, he worked on problems involving electricity, magnetism, and the electromagnetic properties of solar and cometary phenomena. After 1873, he focused on fluid mechanics—the area in which he made his lasting contributions. For example, he (1) developed Reynolds's analogy in 1874, a relation between heat transfer and frictional shear stress in a fluid; (2) measured the average specific heat of water between freezing and boiling, which ranks among the classic determinations of physical constants; (3) studied water currents and waves in estuaries; (4) developed turbines and pumps; and (5) studied the propagation of sound waves in fluids. However, his most important work, and the one which gave birth to the concept of the Reynolds number, was reported in 1883 in a paper entitled "An Experimental Investigation of the Circumstances which Determine whether the Motion of Water in Parallel Channels Shall Be Direct or Sinuous, and of the Law of Resistance in Parallel Channels." Published in *Proceedings of the Royal Society*, this paper was the first to demonstrate the transition from laminar to turbulent flow and to relate this transition to a critical value of a dimensionless parameter—later to become known as the Reynolds number. Reynolds studied this phenomenon in water flow through pipes. His experimental apparatus is illustrated in Fig. 4.55, taken from his original 1883 paper. (Note that before the day of modern photographic

Figure 4.55   Osborne Reynolds's apparatus for his famous pipe flow experiments. This figure is from his original paper, referenced in the text.

techniques, some technical papers contained rather elegant hand sketches of experimental apparatus, of which Fig. 4.55 is an example.) Reynolds filled a large reservoir with water, which fed into a glass pipe through a larger bell-mouth entrance. As the water flowed through the pipe, Reynolds introduced dye into the middle of the stream, at the entrance of the bell mouth. What happened to this thin filament of dye as it flowed through the pipe is illustrated in Fig. 4.56, also from Reynolds's original paper. The flow is from right to left. If the flow velocity were small, the thin dye filament would travel downstream in a smooth, neat, orderly fashion, with a clear demarcation between the dye and the rest of the water, as illustrated in Fig. 4.56*a*. However, if the flow velocity were increased beyond a certain value, the dye filament would suddenly become unstable and would fill the entire pipe with color, as shown in Fig. 4.56*b*. Reynolds clearly pointed out that the smooth dye filament in Fig. 4.56*a* corresponded to laminar flow in the pipe, whereas the agitated and totally diffused dye filament in Fig. 4.56*b* was due to turbulent flow in the pipe. Furthermore, Reynolds studied the details of this turbulent flow by visually observing the pipe flow illuminated by a momentary electric spark, much as we would use a strobe light today. He saw that the turbulent flow consisted of a large number of distinct eddies, as sketched in

(a)

(b)

(c)

Figure 4.56   Development of turbulent flow in pipes, as observed and sketched by Reynolds. This figure is from his original paper, referenced in the text.

Fig. 4.56c. The transition from laminar to turbulent flow occurred when the parameter defined by $\rho V D / \mu$ exceeded a certain critical value, where ρ was the density of the water, V was the mean flow velocity, μ was the viscosity coefficient, and D was the diameter of the pipe. This dimensionless parameter, first introduced by Reynolds, later became known as the Reynolds number. Reynolds measured the critical value of this number, above which turbulent flow occurred, as 2300. This original work of Reynolds initiated the study of transition from laminar to turbulent flow as a new field of research in fluid dynamics—a field which is still today one of the most important and insufficiently understood areas of aerodynamics.

Reynolds was a scholarly man with high standards. Engineering education was new to English universities at that time, and Reynolds had definite ideas about its proper form. He felt that all engineering students, no matter what their specialty, should have a common background based on mathematics, physics, and in particular the fundamentals of classical mechanics. At Manchester, he organized a systematic engineering curriculum covering the basics of civil and mechanical engineering. Ironically, despite his intense interest in education, as a lecturer in the classroom Reynolds left something to be desired. His lectures were hard to follow, and his topics frequently wandered with little or no connection. He was known to come up with new ideas during the course of a lecture and to spend the remainder of the lecture working out these ideas on the board, seemingly oblivious to the students in the classroom. That is, he did not "spoon-feed" his students, and many of the poorer students did not pass his courses. In contrast, the best students enjoyed his lectures and found them stimulating. Many of Reynolds's successful students went on to become distinguished engineers and scientists, the most notable being Sir J. J. Thomson, later the Cavendish Professor of Physics at Cambridge; Thomson is famous for first demonstrating the existence of the electron in 1897, for which he received the Nobel Prize in 1906.

In regard to Reynolds's interesting research approach, his student colleague, and friend Professor A. H. Gibson had this to say in his biography of Reynolds, written for the British Council in 1946:

> Reynolds' approach to a problem was essentially individualistic. He never began by reading what others thought about the matter, but first thought this out for himself. The novelty of his approach to some problems made some of his papers difficult to follow, especially those written during his later years. His more descriptive physical papers, however, make fascinating reading, and when addressing a popular audience, his talks were models of clear exposition.

At the turn of the century, Reynolds's health began to fail, and he subsequently had to retire in 1905. The last years of his life were ones of considerably diminished physical and mental capabilities, a particularly sad state for such a brilliant and successful scholar. He died at Somerset, England, in 1912. Sir Horace Lamb, one of history's most famous fluid dynamicists and a long-time colleague of Reynolds, wrote after Reynolds's death:

> The character of Reynolds was, like his writings, strongly individual. He was conscious of the value of his work, but was content to leave it to the mature judgement of the scientific world. For advertisement he had no taste, and undue pretensions on the part of others only elicited a tolerant smile. To his pupils he was most generous in the opportunities for valuable work which he put in their way, and in the share of co-operation. Somewhat reserved in serious or personal matters and occasionally combative and tenacious in debate, he was in the ordinary relations of life the most kindly and genial of companions. He had a keen sense of humor and delighted in startling paradoxes, which he would maintain, half seriously and half playfully, with astonishing ingenuity and resource. The illness which at length compelled his retirement was felt as a grievous calamity by his pupils, his colleagues and other friends throughout the country.

The purpose of this section has been to relate the historical beginnings of the Reynolds number in fluid mechanics. From now on, when you use the Reynolds number, hopefully you will view it not only as a powerful dimensionless parameter governing viscous flow, but also as a testimonial to its originator—one of the famous fluid dynamicists of the 19th century.

4.26   HISTORICAL NOTE: PRANDTL AND THE DEVELOPMENT OF THE BOUNDARY LAYER CONCEPT

The modern science of aerodynamics has its roots as far back as Isaac Newton, who devoted the entire second book of his *Principia* (1687) to fluid dynamics—especially to the formulation of "laws of resistance" (drag). He noted that drag is a function of fluid density, velocity, and shape of the body in motion. However, Newton was unable to formulate the correct equation for drag. Indeed, he derived a formula which gave the drag on an inclined object as proportional to the sine squared of the angle of attack. Later, Newton's sine-squared law was used to demonstrate the "impossibility of heavier-than-air flight" and served to hinder the intellectual advancement of flight in the 19th century. Ironically, the physical assumptions used by Newton in deriving his sine-squared law approximately reflect the conditions of hypersonic flight, and the newtonian law has been used since 1950 in the design of high-Mach-number vehicles. However, Newton correctly reasoned the mechanism of shear stress in a fluid. In section 9 of book 2 of *Principia*, Newton states the following hypothesis: "The resistance arising from want of lubricity in the parts of a fluid is ...proportional to the velocity with which the parts of the fluid are separated from each other." This is the first statement in history of the friction law for laminar flow; it is embodied in Eq. (4.89), which describes a "newtonian fluid."

Further attempts to understand fluid dynamic drag were made by the French mathematician Jean le Rond d'Alembert, who is noted for developing the calculus of partial differences (leading to the mathematics of partial differential equations). In 1768, d'Alembert applied the equations of motion for an incompressible, inviscid (frictionless) flow about a two-dimensional body in a moving fluid and found that no drag is obtained. He wrote: "I do not see then, I admit, how one can explain the resistance of fluids by the theory in a satisfactory manner. It seems to me on the contrary that this theory, dealt with and studied with profound attention gives, at least in most cases, resistance absolutely zero: a singular paradox which I leave to geometricians to explain." That this theoretical result of zero drag is truly a paradox was clearly recognized by d'Alembert, who also conducted experimental research on drag and who was among the first to discover that drag was proportional to the square of the velocity, as derived in Sec. 5.3 and given in Eq. (5.18).

D'Alembert's paradox arose due to the neglect of friction in the classical theory. It was not until a century later that the effect of friction was properly incorporated in the classical equations of motion by the work of M. Navier (1785–1836) and Sir George Stokes (1819–1903). The so-called Navier-Stokes equations stand today as the classical formulation of fluid dynamics. However, in general they are nonlinear equations and are extremely difficult to solve; indeed, only with the numerical power of modern high-speed digital computers are "exact" solutions of the Navier-Stokes equations finally being obtained for general flow fields. Also in the 19th century, the first experiments on transition from laminar to turbulent flow were carried out by Osborne Reynolds (1842–1912), as related in Sec. 4.25. In his classic paper of 1883 entitled "An Experimental Investigation of the Circumstances which Determine whether the Motion of Water in Parallel Channels Shall Be Direct or Sinuous, and of the Law of Resistance in Parallel Channels," Reynolds observed a filament of colored dye in a pipe flow and noted that transition from laminar to turbulent flow always corresponded to approximately the same value of a dimensionless number $\rho V D/\mu$, where D was the diameter of the pipe. This was the origin of the Reynolds number, defined in Sec. 4.15 and discussed at length in Sec. 4.25.

Therefore, at the beginning of the 20th century, when the Wright brothers were deeply involved in the development of the first successful airplane, the development of theoretical fluid dynamics still had not led to practical results for aerodynamic drag. It was this environment into which Ludwig Prandtl was born on February 4, 1875, at Freising, in Bavaria, Germany. Prandtl was a genius who had the talent of cutting through a maze of complex physical phenomena to extract the most salient points and putting them in simple mathematical form. Educated as a physicist, Prandtl was appointed in 1904 as professor of applied mechanics at Göttingen University in Germany, a post he occupied until his death in 1953.

In the period from 1902 to 1904, Prandtl made one of the most important contributions to fluid dynamics. Thinking about the viscous flow over a body, he reasoned that the flow velocity right at the surface was zero and that if the Reynolds number was high enough, the influence of friction was limited to a thin layer (Prandtl first called it a transition layer) near the surface. Therefore, the analysis of the flow field could be divided into two distinct regions—one close to the surface, which included friction, and the other farther away, in which friction could be neglected. In one of the most important fluid dynamics papers in history, entitled "Uber Flussigkeitsbewegung bei sehr kleiner Reibung," Prandtl reported his thoughts to the Third International Mathematical Congress at Heidelberg in 1904. In this paper, Prandtl observed:

> A very satisfactory explanation of the physical process in the boundary layer (Grenzschicht) between a fluid and a solid body could be obtained by the hypothesis of an adhesion of the fluid to the walls, that is, by the hypothesis of a zero relative velocity between fluid and wall. If the viscosity is very small and the fluid path along the wall not too long, the fluid velocity ought to resume its normal value at a very short distance from the wall. In the thin transition layer however, the sharp changes of velocity, even with small coefficient of friction, produce marked results.

In the same paper, Prandtl's theory is applied to the prediction of flow separation:

> In given cases, in certain points fully determined by external conditions, the fluid flow ought to separate from the wall. That is, there ought to be a layer of fluid which, having been set in rotation by the friction on the wall, insinuates itself into the free fluid, transforming completely the motion of the latter....

Prandtl's boundary layer hypothesis allows the Navier-Stokes equations to be reduced to a simpler form; by 1908, Prandtl and one of his students, H. Blasius, had solved these simpler boundary layer equations for laminar flow over a flat plate, yielding the equations for boundary layer thickness and skin friction drag given by Eqs. (4.91) and (4.93). Finally, after centuries of effort, the first rational resistance laws describing fluid dynamic drag due to friction had been obtained.

Prandtl's work was a stroke of genius, and it revolutionized theoretical aerodynamics. However, possibly due to the language barrier, it only slowly diffused through the worldwide technical community. Serious work on boundary layer theory did not emerge in England and the United States until the 1920s. By that time, Prandtl and his students at Göttingen had applied it to various aerodynamic shapes and were including the effects of turbulence.

Prandtl has been called the *father of aerodynamics*, and rightly so. His contributions extend far beyond boundary layer theory; for example, he pioneered the development of wing lift and drag theory, as seen in the next chapter. Moreover, he was interested in more fields than just fluid dynamics—he made several important contributions to structural mechanics as well.

As a note on Prandtl's personal life, he had the singleness of purpose which seems to drive many giants of humanity. However, his almost complete preoccupation with his work led to a somewhat naive outlook on life. Theodore von Karman, one of Prandtl's most illustrious students, relates that Prandtl would rather find fancy in the examination of children's toys than participate in social gatherings. When Prandtl was almost 40, he suddenly decided that it was time to get married, and he wrote to a friend for the hand of one of his two daughters—Prandtl did not care which one! During the 1930s and early 1940s, Prandtl had mixed emotions about the political problems of the day. He continued his research work at Göttingen under Hitler's Nazi regime but became continually confused about the course of events. Von Karman writes about Prandtl in his autobiography:

> I saw Prandtl once again for the last time right after the Nazi surrender. He was a sad figure. The roof of his house in Göttingen, he mourned, had been destroyed by an American bomb. He couldn't understand why this had been done to him! He was also deeply shaken by the collapse of Germany. He lived only a few years after that, and though he did engage in some research work in meteorology, he died, I believe, a broken man, still puzzled by the ways of mankind.

Prandtl died in Göttingen on August 15, 1953. Of any fluid dynamicist or aero-dynamicist in history, Prandtl came closest to deserving a Nobel Prize. Why he never received one is an unanswered question. However, as long as there are flight vehicles, and as long as students study the discipline of fluid dynamics, the name of Ludwig Prandtl will be enshrined for posterity.

4.27 SUMMARY

Now that you have finished this chapter, return to our road map in Fig. 4.1, and run your mind over all the items shown there. Make yourself feel comfortable with these items. Then proceed with this chapter summary, putting each equation and each concept in its proper perspective relative to our road map.

A few of the important concepts from this chapter are summarized as follows:

1. The basic equations of aerodynamics, in the form derived here, are as follows:

   Continuity $$\rho_1 A_1 V_1 = \rho_2 A_2 V_2 \qquad \textbf{[4.2]}$$

   Momentum $$dp = -\rho V \, dV \qquad \textbf{[4.8]}$$

   Energy $$c_p T_1 + \tfrac{1}{2} V_1^2 = c_p T_2 + \tfrac{1}{2} V_2^2 \qquad \textbf{[4.42]}$$

   These equations hold for a compressible flow. For an incompressible flow, we have these:

Continuity $$A_1 V_1 = A_2 V_2$$ [4.3]

Momentum $$p_1 + \rho \frac{V_1^2}{2} = p_2 + \rho \frac{V_2^2}{2}$$ [4.9a]

Equation (4.9a) is called Bernoulli's equation.

2. The change in pressure, density, and temperature between two points in an isentropic process is given by

$$\frac{p_2}{p_1} = \left(\frac{\rho_2}{\rho_1} \right)^{\gamma} = \left(\frac{T_2}{T_1} \right)^{\gamma/(\gamma-1)}$$

3. The speed of sound is given by

$$a = \sqrt{\left(\frac{dp}{d\rho} \right)_{\text{isentropic}}}$$ [4.48]

For a perfect gas, this becomes

$$a = \sqrt{\gamma R T}$$ [4.54]

4. The speed of a gas flow can be measured by a Pitot tube, which senses the total pressure p_0. For incompressible flow,

$$V_1 = \sqrt{\frac{2(p_0 - p_1)}{\rho}}$$ [4.66]

For subsonic compressible flow,

$$V_1^2 = \frac{2a_1^2}{\gamma - 1} \left[\left(\frac{p_0}{p_1} \right)^{(\gamma-1)/\gamma} - 1 \right]$$ [4.77a]

For supersonic flow, a shock wave exists in front of the Pitot tube, and Eq. (4.79) must be used in lieu of Eq. (4.77a) to find the Mach number of the flow.

5. The area-velocity relation for isentropic flow is

$$\frac{dA}{A} = (M^2 - 1) \frac{dV}{V}$$ [4.83]

From this relation, we observe that (1) for a subsonic flow, the velocity increases in a convergent duct and decreases in a divergent duct, (2) for a supersonic flow, the velocity increases in a divergent duct and decreases in a convergent duct; and (3) the flow is sonic only at the minimum area.

6. The isentropic flow of a gas is governed by

$$\frac{p_0}{p_1} = \left(1 + \frac{\gamma - 1}{2} M_1^2 \right)^{\gamma/(\gamma-1)}$$ [4.74]

$$\frac{T_0}{T_1} = 1 + \frac{\gamma - 1}{2} M_1^2$$ [4.73]

$$\frac{\rho_0}{\rho_1} = \left(1 + \frac{\gamma - 1}{2} M_1^2 \right)^{1/(\gamma-1)}$$ [4.75]

Here T_0, p_0, and ρ_0 are the total temperature, pressure, and density, respectively. For an isentropic flow, $p_0 = $ constant throughout the flow. Similarly, $\rho_0 = $ constant and $T_0 = $ constant throughout the flow.

7. Viscous effects create a boundary layer along a solid surface in a flow. In this boundary layer, the flow moves slowly and the velocity goes to zero right at the surface. The shear stress at the wall is given by

$$\tau_w = \mu \left(\frac{dV}{dy} \right)_{y=0} \qquad \textbf{[4.89]}$$

The shear stress is larger for a turbulent boundary layer than for a laminar boundary layer.

8. For a laminar incompressible boundary layer, on a flat plate,

$$\delta = \frac{5.2x}{\sqrt{\mathrm{Re}_x}} \qquad \textbf{[4.91]}$$

and
$$C_f = \frac{1.328}{\sqrt{\mathrm{Re}_L}} \qquad \textbf{[4.98]}$$

where δ is the boundary layer thickness, C_f is the total skin friction drag coefficient, and Re is the Reynolds number;

$$\mathrm{Re}_x = \frac{\rho_\infty V_\infty x}{\mu_\infty} \qquad \text{local Reynolds number}$$

$$\mathrm{Re}_L = \frac{\rho_\infty V_\infty L}{\mu_\infty} \qquad \text{plate Reynolds number}$$

Here, x is the running length along the plate, and L is the total length of the plate.

9. For a turbulent incompressible boundary layer on a flat plate,

$$\delta = \frac{0.37x}{\mathrm{Re}_x^{0.2}} \qquad \textbf{[4.99]}$$

$$C_f = \frac{0.074}{\mathrm{Re}_L^{0.2}} \qquad \textbf{[4.101]}$$

Any real flow along a surface first starts out as laminar but then changes into a turbulent flow. The point where this transition effectively occurs (in reality, transition occurs over a finite length) is designated x_{cr}. In turn, the critical Reynolds number for transition is defined as

$$\mathrm{Re}_{x_{\mathrm{cr}}} = \frac{\rho_\infty V_\infty x_{\mathrm{cr}}}{\mu_\infty} \qquad \textbf{[4.104]}$$

10. Whenever a boundary layer encounters an adverse pressure gradient (a region of increasing pressure in the flow direction), it can readily separate from the surface. On an airfoil or wing, such flow separation decreases the lift and increases the drag.

BIBLIOGRAPHY

Airey, J.: "Notes on the Pitot Tube," *Engineering News*, vol. 69, no. 16, April 17, 1913, pp. 782–783.

Anderson, J. D., Jr.: *Fundamentals of Aerodynamics*, 2nd Edition, McGraw-Hill, New York, 1991.

Anderson, J. D., Jr.: *A History of Aerodynamics and Its Impact on Flying Machines*, Cambridge University Press, New York, 1998

Goin, K. L.: "The History, Evolution, and Use of Wind Tunnels," *AIAA Student Journal*, February 1971, pp. 3–13.

Guy, A. E.: "Origin and Theory of the Pitot Tube," *Engineering News*, vol. 69, no. 23, June 5, 1913, pp. 1172–1175.

Kuethe, A. M., and C. Y. Chow: *Foundations of Aerodynamics*, 3d ed., Wiley, New York, 1976.

Pope, A: *Aerodynamics of Supersonic Flight*, Pitman, New York, 1958.

von Karman, T.: *Aerodynamics*, McGraw-Hill, New York, 1963.

PROBLEMS

4.1 Consider the incompressible flow of water through a divergent duct. The inlet velocity and area are 5 ft/s and 10 ft^2, respectively. If the exit area is 4 times the inlet area, calculate the water flow velocity at the exit.

4.2 In the above problem, calculate the pressure difference between the exit and the inlet. The density of water is 62.4 lb$_m$/ft^3.

4.3 Consider an airplane flying with a velocity of 60 m/s at a standard altitude of 3 km. At a point on the wing, the airflow velocity is 70 m/s. Calculate the pressure at this point. Assume incompressible flow.

4.4 An instrument used to measure the airspeed on many early low-speed airplanes, principally during 1919 to 1930, was the venturi tube. This simple device is a convergent-divergent duct. (The front section's cross-sectional area A decreases in the flow direction, and the back section's cross-sectional area increases in the flow direction. Somewhere in between the inlet and exit of the duct, there is a minimum area, called the throat.) Let A_1 and A_2 denote the inlet and throat areas, respectively. Let p_1 and p_2 be the pressures at the inlet and throat, respectively. The venturi tube is mounted at a specific location on the airplane (generally on the wing or near the front of the fuselage), where the inlet velocity V_1 is essentially the same as the freestream velocity, that is, the velocity of the airplane through the air. With a knowledge of the area ratio A_2/A_1 (a fixed design feature) and a measurement of the pressure difference $p_1 - p_2$, the airplane's velocity can be determined. For example, assume $A_2/A_1 = \frac{1}{4}$, and $p_1 - p_2 = 80$ lb/ft^2. If the airplane is flying at standard sea level, what is its velocity?

4.5 Consider the flow of air through a convergent-divergent duct, such as the venturi described in Prob. 4.4. The inlet, throat, and exit areas are 3, 1.5, and 2 m^2, respectively. The inlet and exit pressures are 1.02×10^5 and 1.00×10^5 N/m^2, respectively. Calculate the flow velocity at the throat. Assume incompressible flow with standard sea-level density.

An airplane is flying at a velocity of 130 mi/h at a standard altitude of 5000 ft. At a **4.6** point on the wing, the pressure is 1750.0 lb/ft^2. Calculate the velocity at that point, assuming incompressible flow.

Imagine that you have designed a low-speed airplane with a maximum velocity at sea **4.7** level of 90 m/s. For your airspeed instrument, you plan to use a venturi tube with a 1.3 : 1 area ratio. Inside the cockpit is an airspeed indicator—a dial that is connected to a pressure gauge sensing the venturi tube pressure difference $p_1 - p_2$ and properly calibrated in terms of velocity. What is the maximum pressure difference you would expect the gauge to experience?

A supersonic nozzle is also a convergent-divergent duct, which is fed by a large **4.8** reservoir at the inlet to the nozzle. In the reservoir of the nozzle, the pressure and temperature are 10 atm and 300 K, respectively. At the nozzle exit, the pressure is 1 atm. Calculate the temperature and density of the flow at the exit. Assume the flow is isentropic and, of course, compressible.

Derive an expression for the exit velocity of a supersonic nozzle in terms of the **4.9** pressure ratio between the reservoir and exit p_0/p_e and the reservoir temperature T_0.

Consider an airplane flying at a standard altitude of 5 km with a velocity of 270 m/s. **4.10** At a point on the wing of the airplane, the velocity is 330 m/s. Calculate the pressure at this point.

The mass flow of air through a supersonic nozzle is 1.5 lb$_m$/s. The exit velocity is 1500 **4.11** ft/s, and the reservoir temperature and pressure are 1000°R and 7 atm, respectively. Calculate the area of the nozzle exit. For air, $c_p = 6000$ ft · lb/(slug)(°R).

A supersonic transport is flying at a velocity of 1500 mi/h at a standard altitude of **4.12** 50,000 ft. The temperature at a point in the flow over the wing is 793.32°R. Calculate the flow velocity at that point.

For the airplane in Prob. 4.12, the total cross-sectional area of the inlet to the jet **4.13** engines is 20 ft^2. Assume that the flow properties of the air entering the inlet are those of the freestream ahead of the airplane. Fuel is injected inside the engine at a rate of 0.05 lb of fuel for every pound of air flowing through the engine (i.e., the fuel-air ratio by mass is 0.05). Calculate the mass flow (in slugs/per second) that comes out the exit of the engine.

Calculate the Mach number at the exit of the nozzle in Prob. 4.11. **4.14**

A Boeing 747 is cruising at a velocity of 250 m/s at a standard altitude of 13 km. **4.15** What is its Mach number?

A high-speed missile is traveling at Mach 3 at standard sea level. What is its velocity **4.16** in miles per hour?

Calculate the flight Mach number for the supersonic transport in Prob. 4.12. **4.17**

Consider a low-speed subsonic wind tunnel with a nozzle contraction ratio of 1 : 20. **4.18** One side of a mercury manometer is connected to the settling chamber, and the other side to the test section. The pressure and temperature in the test section are 1 atm and 300 K, respectively. What is the height difference between the two columns of mercury when the test section velocity is 80 m/s?

4.19 We wish to operate a low-speed subsonic wind tunnel so that the flow in the test section has a velocity of 200 mi/h at standard sea-level conditions. Consider two different types of wind tunnels: (a) a nozzle and a constant-area test section, where the flow at the exit of the test section simply dumps out to the surrounding atmosphere, that is, there is no diffuser, and (b) a conventional arrangement of nozzle, test section, and diffuser, where the flow at the exit of the diffuser dumps out to the surrounding atmosphere. For both wind tunnels (a) and (b) calculate the pressure differences across the entire wind tunnel required to operate them so as to have the given flow conditions in the test section. For tunnel (a) the cross-sectional area of the entrance is 20 ft^2, and the cross-sectional area of the test section is 4 ft^2. For tunnel (b) a diffuser is added to (a) with a diffuser area of 18 ft^2. After completing your calculations, examine and compare your answers for tunnels (a) and (b). Which requires the smaller overall pressure difference? What does this say about the value of a diffuser on a subsonic wind tunnel?

4.20 A Pitot tube is mounted in the test section of a low-speed subsonic wind tunnel. The flow in the test section has a velocity, static pressure, and temperature of 150 mi/h, 1 atm, and 70°F, respectively. Calculate the pressure measured by the Pitot tube.

4.21 The altimeter on a low-speed Piper Aztec reads 8000 ft. A Pitot tube mounted on the wing tip measures a pressure of 1650 lb/ft^2. If the outside air temperature is 500°R, what is the true velocity of the airplane? What is the equivalent airspeed?

4.22 The altimeter on a low-speed airplane reads 2 km. The airspeed indicator reads 50 m/s. If the outside air temperature is 280 K, what is the true velocity of the airplane?

4.23 A Pitot tube is mounted in the test section of a high-speed subsonic wind tunnel. The pressure and temperature of the airflow are 1 atm and 270 K, respectively. If the flow velocity is 250 m/s, what is the pressure measured by the Pitot tube?

4.24 A high-speed subsonic Boeing 777 airliner is flying at a pressure altitude of 12 km. A Pitot tube on the vertical tail measures a pressure of 2.96×10^4 N/m^2. At what Mach number is the airplane flying?

4.25 A high-speed subsonic airplane is flying at Mach 0.65. A Pitot tube on the wing tip measures a pressure of 2339 lb/ft^2. What is the altitude reading on the altimeter?

4.26 A high-performance F-16 fighter is flying at Mach 0.96 at sea level. What is the air temperature at the stagnation point at the leading edge of the wing?

4.27 An airplane is flying at a pressure altitude of 10 km with a velocity of 596 m/s. The outside air temperature is 220 K. What is the pressure measured by a Pitot tube mounted on the nose of the airplane?

4.28 The dynamic pressure is defined as $q = 0.5\rho V^2$. For high-speed flows, where Mach number is used frequently, it is convenient to express q in terms of pressure p and Mach number M rather than ρ and V. Derive an equation for $q = q(p, M)$.

After completing its mission in orbit around the earth, the Space Shuttle enters the **4.29** earth's atmosphere at very high Mach number and, under the influence of aerodynamic drag, slows as it penetrates more deeply into the atmosphere. (These matters are discussed in Chap. 8.) During its atmospheric entry, assume that the shuttle is flying at Mach number M corresponding to the altitudes h:

h, km	60	50	40	30	20
M	17	9.5	5.5	3	1

Calculate the corresponding values of the freestream dynamic pressure at each one of these flight path points. *Suggestion:* Use the result from Prob. 4.28. Examine and comment on the variation of q_∞ as the shuttle enters the atmosphere.

Consider a Mach 2 airstream at standard sea-level conditions. Calculate the total **4.30** pressure of this flow. Compare this result with (*a*) the stagnation pressure that would exist at the nose of a blunt body in the flow and (*b*) the erroneous result given by Bernoulli's equation, which of course does not apply here.

Consider the flow of air through a supersonic nozzle. The reservoir pressure and **4.31** temperature are 5 atm and 500 K, respectively. If the Mach number at the nozzle exit is 3, calculate the exit pressure, temperature, and density.

Consider a supersonic nozzle across which the pressure ratio is $p_e/p_0 = 0.2$. Calcu- **4.32** late the ratio of exit area to throat area.

Consider the expansion of air through a convergent-divergent supersonic nozzle. The **4.33** Mach number varies from essentially zero in the reservoir to Mach 2.0 at the exit. Plot on graph paper the variation of the ratio of dynamic pressure to total pressure as a function of Mach number; that is, plot q/p_0 versus M from $M = 0$ to $M = 2.0$.

The wing of the Fairchild Republic A-10A twin-jet close-support airplane is approx- **4.34** imately rectangular with a wingspan (the length perpendicular to the flow direction) of 17.5 m and a chord (the length parallel to the flow direction) of 3 m. The airplane is flying at standard sea level with a velocity of 200 m/s. If the flow is considered to be completely laminar, calculate the boundary layer thickness at the trailing edge and the total skin friction drag. Assume the wing is approximated by a flat plate. Assume incompressible flow.

In Prob. 4.34, assume the flow is completely turbulent. Calculate the boundary layer **4.35** thickness at the trailing edge and the total skin friction drag. Compare these turbulent results with the above laminar results.

If the critical Reynolds number for transition is 10^6, calculate the skin friction drag **4.36** for the wing in Prob. 4.34.

Let us reflect back to the fundamental equations of fluid motion dicussed in the **4.37** early sections of this chapter. Sometimes these equations were expressed in terms of differential equations, but for the most part we obtained algebraic relations by integrating the differential equations. However, it is useful to think of the differential forms as relations that govern the change in flowfield variables in an infinitesimally small region around a point in the flow. (*a*) Consider a point in an inviscid flow, where the local density is 1.1 kg/m^3. As a fluid element sweeps through this point, it is experiencing a spatial change in velocity of two percent per millimeter. Calculate the corresponding spatial change in pressure per millimeter at this point if the velocity at the point is 100 m/sec. (*b*) Repeat the calculation for the case when the velocity at the point is 1000 m/sec. What can you conclude by comparing your results for the low-speed flow in part (*a*) with the results for the high-speed flow part (*b*).

4.38 The type of calculation in Problem 4.3 is a classic one for low-speed, incompressible flow, i.e., given the freestream pressure and velocity, and the velocity at some other point in the flow, calculate the pressure at that point. In a high-speed compressible flow, Mach number is more fundamental than velocity. Consider an airplane flying at Mach 0.7 at a standard altitude of 3 km. At a point on the wing, the airflow Mach number is 1.1. Calculate the pressure at this point. Assume an isentropic flow.

4.39 Consider an airplane flying at a standard altitude of 25,000 ft at a velocity of 800 ft/sec. To experience the same dynamic pressure at sea level, how fast must the airplane be flying?

4.40 In Section 4.9, we defined hypersonic flow as that flow where the Mach number is five or greater. Wind tunnels with a test section Mach number of five or greater are called hypersonic wind tunnels. From Eq. (4.88), the exit-to-throat area ratio for supersonic exit Mach numbers increases as the exit Mach number increases. For hypersonic Mach numbers, the exit-to-throat ratio becomes extremely large, so hypersonic wind tunnels are designed with long, high-expansion ratio nozzles. In this and the following problems, let us examine some special characteristics of hypersonic wind tunnels. Assume we wish to design a Mach 10 hypersonic wind tunnel using air as the test medium. We want the static pressure and temperature in the test stream to be that for a standard altitude of 55 km. Calculate: (*a*) the exit-to-throat area ratio, (*b*) the required reservoir pressure (in atm), and (*c*) the required reservoir temperature. Examine these results. What do they tell you about the special (and sometimes severe) operating requirements for a hypersonic wind tunnel.

4.41 Calculate the exit velocity of the hypersonic tunnel in Problem 4.40.

4.42 Let us double the exit Mach number of the tunnel in Problem 4.40 simply by adding a longer nozzle section with the requisite expansion ratio. Keep the reservoir properties the same as those in Problem 4.40. Then we have a Mach 20 wind tunnel, with test section pressure and temperature considerably lower than in Problem 4.40, i.e., the test section flow no longer corresponds to conditions at a standard altitude of 55 km. Be that as it may, we have at least doubled the Mach number of the tunnel. Calculate: (*a*) the exit-to-throat area ratio of the Mach 20 nozzle, (*b*) the exit velocity. Compare these values with those for the Mach 10 tunnel in Problems 4.40 and 4.41. What can you say about the differences? In particular, note the exit velocities for the Mach 10 and Mach 20 tunnels. You will see that they are not much different. What is then giving the big increase in exit Mach number?

5

Airfoils, Wings, and Other Aerodynamic Shapes

There can be no doubt that the inclined plane is the true principle of aerial navigation by mechanical means.

Sir George Cayley, 1843

5.1 INTRODUCTION

It is remarkable that the modern airplane as we know it today, with its fixed wing and vertical and horizontal tail surfaces, was first conceived by George Cayley in 1799, more than 175 years ago. He inscribed his first concept on a silver disk (presumably for permanence), shown in Fig. 1.5. It is also remarkable that Cayley recognized that a curved surface (as shown on the silver disk) creates more lift than a flat surface. Cayley's fixed-wing concept was a true revolution in the development of heavier-than-air flight machines. Prior to his time, aviation enthusiasts had been doing their best to imitate mechanically the natural flight of birds, which led to a series of human-powered flapping-wing designs (ornithopters), which never had any real possibility of working. In fact, even Leonardo da Vinci devoted a considerable effort to the design of many types of ornithopters in the late 15th century, of course, to no avail. In such ornithopter designs, the flapping of the wings was supposed to provide simultaneously both lift (to sustain the machine in the air) and propulsion (to push it along in flight). Cayley is responsible for directing people's minds away from imitating bird flight and for separating the two principles of lift and propulsion. He proposed and demonstrated that lift can be obtained from a fixed, straight wing inclined to the airstream, while propulsion can be provided by some independent mechanism such as paddles or airscrews. For this concept and for his many other thoughts and inventions in aeronautics, Sir George Cayley is called the parent of modern aviation. A more detailed discussion of Cayley's contributions is given in Chap. 1. However, we emphasize that much of the technology discussed in the present chapter had its origins at the beginning of the 19th century—technology that came to fruition on December 17, 1903, near Kitty Hawk, North Carolina.

The following sections develop some of the terminology and basic aerodynamic fundamentals of airfoils and wings. These concepts form the heart of airplane flight, and they represent a major excursion into aeronautical engineering. The road map for this chapter is shown in Fig. 5.1. There are basically three main topics in Chap. 5, each having to do with the aerodynamic characteristics of a class of geometric shapes— airfoils, wings, and general body shapes. These three topics are shown in the three boxes at the top of our road map. We first examine the aerodynamic characteristics of airfoils and then run down the various aspects noted in the left-hand column in Fig. 5.1. This is a long list, but we will find that many thoughts on this list carry over to wings and bodies as well. We then move to the central column for a discussion of finite wings, and we will see how the aerodynamics of a wing differs from that of an airfoil. Both airfoils and wings can be classified as slender bodies. In contrast, the third column in Fig. 5.1 deals with a few examples of *blunt bodies*, namely, cylinders and spheres. We define and examine the distinctions between slender and blunt aerodynamic shapes. Finally, we discuss how aerodynamic lift is produced. Although we have alluded to this in previous chapters, it is appropriate at the end

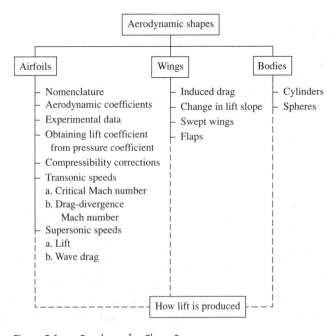

Figure 5.1   Road map for Chap. 5.

of the chapter dealing with the aerodynamics of various shapes to have a *definitive* discussion on how nature generates lift. Various physical explanations have been used in the past to explain how lift is generated, and there have been many spirited discussions in the literature as to which is proper, or more fundamental. We attempt to put all these views in perspective at the end of this chapter, as represented by the box at the bottom of Fig. 5.1. As you progress through this chapter, make certain to touch base frequently with our road map so that you can see how the details of our discussions fit into the grand scheme laid out in Fig. 5.1.

5.2 AIRFOIL NOMENCLATURE

Consider the wing of an airplane, as sketched in Fig. 5.2. The cross-sectional shape obtained by the intersection of the wing with the perpendicular plane shown in Fig. 5.2 is called an *airfoil*. Such an airfoil is sketched in Fig. 5.3, which illustrates some basic terminology. The major design feature of an airfoil is the *mean camber line*, which is the locus of points halfway between the upper and lower surfaces, as measured perpendicular to the mean camber line itself. The most forward and rearward points of the mean camber line are the *leading* and *trailing edges*, respectively. The straight line connecting the leading and trailing edges is the *chord line* of the airfoil, and the precise distance from the leading to the trailing edge measured along the chord line is simply designated the *chord* of the airfoil, given by the symbol c. The *camber* is the maximum distance between the mean camber line and the chord line, measured perpendicular to the chord line. The camber, the shape of the mean camber line, and

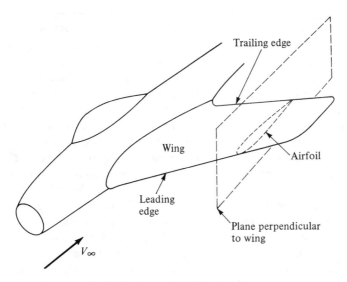

Figure 5.2 Sketch of a wing and airfoil.

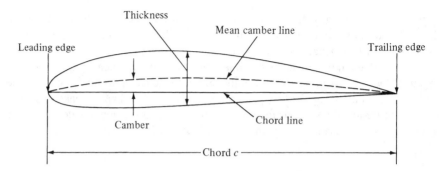

Figure 5.3   Airfoil nomenclature. The shape shown here is an NACA 4415 airfoil.

to a lesser extent, the thickness distribution of the airfoil essentially control the lift and moment characteristics of the airfoil.

More definitions are illustrated in Fig. 5.4*a*, which shows an airfoil inclined to a stream of air. The free-stream velocity V_∞ is the velocity of the air far upstream of the airfoil. The *direction* of V_∞ is defined as the *relative wind*. The angle between the relative wind and the chord line is the *angle of attack* α of the airfoil. As described in Chaps. 2 and 4, there is an aerodynamic force created by the pressure and shear stress distributions over the wing surface. This resultant force is shown by the vector R in Fig. 5.4*a*. In turn, the aerodynamic force R can be resolved into two forces, parallel and perpendicular to the relative wind. The *drag D* is always defined as the component of the aerodynamic force *parallel to the relative wind*. The *lift L* is always defined as the component of the aerodynamic force *perpendicular to the relative wind*.

In addition to lift and drag, the surface pressure and shear stress distributions create a *moment M* which tends to *rotate* the wing. To see more clearly how this moment is created, consider the surface pressure distribution over an airfoil, as sketched in Fig. 5.5 (we will ignore the shear stress for this discussion). Consider just the pressure on the top surface of the airfoil. This pressure gives rise to a net force F_1 in the general downward direction. Moreover, F_1 acts through a given point on the chord line, point 1, which can be found by integrating the pressure times distance over the surface (analogous to finding the centroid or center of pressure from integral calculus). Now consider just the pressure on the bottom surface of the airfoil. This pressure gives rise to a net force F_2 in the general upward direction, acting through point 2. The total aerodynamic force on the airfoil is the *summation* of F_1 and F_2, and lift is obtained when $F_2 > F_1$. However, note from Fig. 5.5 that F_1 and F_2 will create a moment which will tend to rotate the airfoil. Moreover, the value of this aerodynamically induced moment depends on the point about which we choose to take moments. For example, if we take moments about the leading edge, the aerodynamic moment is designated M_{LE}. It is more common in the case of subsonic airfoils to take moments about a point on the chord at a distance $c/4$ from the leading edge, the *quarter-chord point*, as illustrated in Fig. 5.4*a*. This moment about the quarter chord is designated

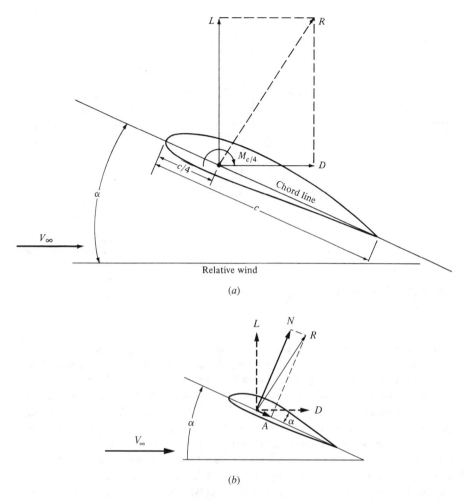

Figure 5.4   Sketch showing the definitions of (a) lift, drag, moments, angle of attack, and relative wind, (b) normal and axial force.

$M_{c/4}$. In general, $M_{\rm LE} \neq M_{c/4}$. Intuition will tell you that lift, drag, and moments on a wing will change as the angle of attack α changes. In fact, the variations of these aerodynamic quantities with α represent some of the most important information that an airplane designer needs to know. We will address this matter in the following sections. However, we point out that although $M_{\rm LE}$ and $M_{c/4}$ are both functions of α, there exists a certain point on the airfoil about which moments essentially *do not* vary with α. This point is defined as the *aerodynamic center*, and the moment about the aerodynamic center is designated $M_{\rm ac}$. By definition,

$$M_{\rm ac} = {\rm const}$$

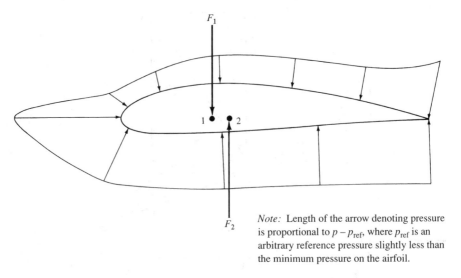

Note: Length of the arrow denoting pressure is proportional to $p - p_{ref}$, where p_{ref} is an arbitrary reference pressure slightly less than the minimum pressure on the airfoil.

Figure 5.5   The physical origin of moments on an airfoil.

independent of the angle of attack. The location of the aerodynamic center for real aerodynamic shapes can be found from experiment. For low-speed subsonic airfoils, the aerodynamic center is generally very close to the quarter-chord point.

Returning to Fig. 5.4*a*, we recall that the resultant aerodynamic force R can be resolved into components perpendicular and parallel to the relative wind, namely, the lift and drag, respectively. An alternative to this system is to resolve R into components perpendicular and parallel to the *chord line*, as shown in Fig. 5.4*b*. These components are called the *normal force* and *axial force* and are denoted by N and A, respectively, in Fig. 5.4*b*, shown by the heavy solid arrows. Also shown in Fig. 5.4*b* are the lift and drag, L and D, respectively, represented by the heavy dashed arrows. Lift and drag are easily expressed in terms of N and A from the geometry shown in Fig. 5.4*b*, as follows:

$$L = N \cos \alpha - A \sin \alpha \qquad \textbf{[5.1]}$$

$$D = N \sin \alpha + A \cos \alpha \qquad \textbf{[5.2]}$$

For airfoils and wings, the use of N and A to describe the aerodynamic force dates back as early as the work of Otto Lilienthal in 1889, as published in his book *Bird Flight as the Basis of Aviation* (see Sec. 1.5). Indeed, the famous "Lilienthal tables" which were used by the Wright brothers to design their early gliders (see Sec. 1.8) were tables dealing with normal and axial forces. The Wrights preferred to think in terms of lift and drag, and they converted Lilienthal's results by using Eqs. (5.1) and (5.2). Today, the use of N and A to describe the aerodynamic force on airfoils and wings is generally passé; L and D are almost always the system used by choice. However, N and A are still frequently used to denote the aerodynamic force on bodies of revolution, such as missiles and projectiles. Therefore, it is useful to be familiar with both systems of expressing the aerodynamic force on a body.

5.3 LIFT, DRAG, AND MOMENT COEFFICIENTS

Again appealing to intuition, we note that it makes sense that for an airplane in flight, the actual magnitudes of L, D, and M depend not only on α, but also on velocity and altitude. In fact, we can expect that the variations of L, D, and M depend at least on

1. Free-stream velocity V_∞.
2. Free-stream density ρ_∞, that is, altitude.
3. Size of the aerodynamic surface. For airplanes, we will use the *wing area S* to indicate size.
4. Angle of attack α.
5. Shape of the airfoil.
6. Viscosity coefficient μ_∞ (because the aerodynamic forces are generated in part from skin friction distributions).
7. Compressibility of the airflow. In Chap. 4 we demonstrated that compressibility effects are governed by the value of the free-stream Mach number $M_\infty = V_\infty/a_\infty$. Since V_∞ is already listed above, we can designate a_∞ as our index for compressibility.

Hence, we can write that, for a given shape airfoil at a given angle of attack,

$$L = f(V_\infty, \rho_\infty, S, \mu_\infty, a_\infty) \qquad \textbf{[5.3]}$$

and D and M are similar functions.

In principle, for a given airfoil at a given angle of attack, we could find the variation of L by performing myriad wind tunnel experiments wherein V_∞, ρ_∞, S, μ_∞, and a_∞ are individually varied, and then we could try to make sense out of the resulting huge collection of data. This is the hard way. Instead, we ask the question, Are there *groupings* of the quantities V_∞, ρ_∞, S, μ_∞, a_∞, and L such that Eq. (5.3) can be written in terms of fewer parameters? The answer is yes. In the process of developing this answer, we will gain some insight into the beauty of nature as applied to aerodynamics.

The technique we will apply is a simple example of a more general theoretical approach called *dimensional analysis*. Let us assume that Eq. (5.3) is of the functional form

$$L = Z V_\infty^a \rho_\infty^b S^d a_\infty^e \mu_\infty^f \qquad \textbf{[5.4]}$$

where Z, a, b, d, e, and f are dimensionless constants. However, no matter what the values of these constants may be, it is a physical fact that the dimensions of the left- and right-hand sides of Eq. (5.4) must match; that is, if L is a force (say, in newtons), then the net result of all the exponents and multiplication on the right-hand side must also produce a result with the dimensions of a force. This constraint will ultimately give us information on the values of a, b, etc. If we designate the basic dimensions of mass, length, and time by m, l, and t, respectively, then the dimensions of various physical quantities are as given below.

Physical Quantity	Dimensions
L	ml/t^2 (from Newton's second law)
V_∞	l/t
ρ_∞	m/l^3
S	l^2
a_∞	l/t
μ_∞	$m/(lt)$

Thus, equating the *dimensions* of the left- and right-hand sides of Eq. (5.4), we obtain

$$\frac{ml}{t^2} = \left(\frac{l}{t}\right)^a \left(\frac{m}{l^3}\right)^b (l^2)^d \left(\frac{l}{t}\right)^e \left(\frac{m}{lt}\right)^f \qquad \textbf{[5.5]}$$

Consider mass m. The exponent of m on the left-hand side is 1. Thus, the exponents of m on the right must add to 1. Hence

$$1 = b + f \qquad \textbf{[5.6]}$$

Similarly, for time t we have

$$-2 = -a - e - f \qquad \textbf{[5.7]}$$

and for length l,

$$1 = a - 3b + 2d + e - f \qquad \textbf{[5.8]}$$

Solving Eqs. (5.6) to (5.8) for a, b, and d in terms of e and f yields

$$b = 1 - f \qquad \textbf{[5.9]}$$

$$a = 2 - e - f \qquad \textbf{[5.10]}$$

$$d = 1 - \frac{f}{2} \qquad \textbf{[5.11]}$$

substituting Eqs. (5.9) to (5.11) into (5.4) gives

$$L = Z(V_\infty)^{2-e-f} \rho_\infty^{1-f} S^{1-f/2} a_\infty^e \mu_\infty^f \qquad \textbf{[5.12]}$$

Rearranging Eq. (5.12), we find

$$L = Z\rho_\infty V_\infty^2 S \left(\frac{a_\infty}{V_\infty}\right)^e \left(\frac{\mu_\infty}{\rho_\infty V_\infty S^{1/2}}\right)^f \qquad \textbf{[5.13]}$$

Note that $a_\infty/V_\infty = 1/M_\infty$, where M_∞ is the freestream Mach number. Also note that the dimensions of S are l^2; hence the dimension of $S^{1/2}$ is l, purely a length. Let us choose this length to be the chord c by convention. Hence, $\mu_\infty/(\rho_\infty V_\infty S^{1/2})$ can be replaced in our consideration by the equivalent quantity

$$\frac{\mu_\infty}{\rho_\infty V_\infty c}$$

However, $\mu_\infty/(\rho_\infty V_\infty c) \equiv 1/\text{Re}$, where Re is based on the chord length c. Hence, Eq. (5.13) becomes

$$L = Z\rho_\infty V_\infty^2 S \left(\frac{1}{M_\infty}\right)^e \left(\frac{1}{\text{Re}}\right)^f \qquad \textbf{[5.14]}$$

We now *define* a new quantity called the *lift coefficient* c_l as

$$\frac{c_l}{2} \equiv Z \left(\frac{1}{M_\infty}\right)^e \left(\frac{1}{\text{Re}}\right)^f \qquad \textbf{[5.15]}$$

Then, Eq. (5.14) becomes

$$L = \tfrac{1}{2}\rho_\infty V_\infty^2 S c_l \qquad \textbf{[5.16]}$$

Recalling from Chap. 4 that the dynamic pressure is $q_\infty \equiv \tfrac{1}{2}\rho_\infty V_\infty^2$, we transform Eq. (5.16) into

$$
\begin{array}{ccccc}
L & = & q_\infty & \times \; S \; \times & c_l \\
\uparrow & & \uparrow & \uparrow & \uparrow \\
\text{Lift} & & \text{Dynamic} & \text{Wing} & \text{Lift} \\
& & \text{pressure} & \text{area} & \text{coefficient}
\end{array}
\qquad \textbf{[5.17]}
$$

Look what has happened! Equation (5.3), written from intuition but not very useful, has cascaded to the simple, direct form of Eq. (5.17), which contains a tremendous amount of information. In fact, Eq. (5.17) is one of the most important relations in applied aerodynamics. It says that the lift is directly proportional to the dynamic pressure (hence to the square of the velocity). It is also directly proportional to the wing area S and to the lift coefficient c_l. In fact, Eq. (5.17) can be turned around and used as a *definition* for the lift coefficient:

$$c_l \equiv \frac{L}{q_\infty S} \qquad \textbf{[5.18]}$$

That is, the lift coefficient is always defined as the aerodynamic lift divided by the dynamic pressure and some reference area (for wings, the convenient reference area S, as we have been using).

The lift coefficient is a function of M_∞ and Re as reflected in Eq. (5.15). Moreover, since M_∞ and Re are dimensionless and since Z was assumed initially as a dimensionless constant, from Eq. (5.15) c_l is dimensionless. This is also consistent with Eqs. (5.17) and (5.18). Also, recall that the above derivation was carried out for an airfoil of given shape and at a given angle of attack α. If α were to vary, then c_l

would also vary. Hence, for a given airfoil,

$$c_l = f(\alpha, M_\infty, \text{Re})$$

[5.19]

This relation is important. Fix in your mind that lift coefficient is a function of angle of attack, Mach number, and Reynolds number.

To help appreciate the value of the relationship expressed by Eq. (5.19), let us assume that we are given a particular aerodynamic shape, and we wish to measure the lift and how it varies with the different parameters. So we go to the laboratory and set up a series of wind tunnel tests to measure the lift on our given shape. Reflecting on Eq. (5.3), we know the lift of the given shape at a given orientation (angle of attack) to the flow depends on the free-stream velocity, density, reference area, viscosity coefficient, and speed of sound, but we do not know precisely *how* L varies with a change in these parameters. We wish to find out how. We begin by running a set of wind tunnel tests, making measurements of L where V_∞ is varied but S, μ_∞, and a_∞ are held fixed. This gives us a stack of wind tunnel data from which we can obtain a correlation of the variation of L with V_∞. Next, we run another set of wind tunnel tests in which ρ_∞ is varied but V_∞, S, μ_∞, and a_∞ are held fixed. This gives us a second stack of wind tunnel data from which we can obtain a correlation of the variation of L with ρ_∞. Then we run a third set wind tunnel tests in which S is varied, holding everything else constant. This gives us a third stack of wind tunnel data from which we can obtain a correlation of the variation of L with S. We repeat this process two more times, alternately holding μ_∞ constant and then a_∞ constant. When we are finished, we end up with five individual stacks of wind tunnel data from which we can (in principle) obtain the precise variation of L with V_∞, ρ_∞, S, μ_∞, and a_∞, as represented by the functional relation in Eq. (5.3). As you can probably already appreciate, this represents a lot of personal effort and a lot of wind tunnel testing at great financial expense. However, if we use our knowledge obtained from our dimensional analysis, namely, Eq. (5.19), we can realize a great savings of effort, time, and expense. Instead of measuring L in five sets of wind tunnel tests as described above, let us measure the variation of *lift coefficient* [obtained from $c_l = L/(q_\infty S)$]. Keying on Eq. (5.19), for a given shape at a given angle of attack, we run a set of wind tunnel tests in which c_l is measured, with M_∞ varied but Re held constant. This gives us one stack of wind tunnel data from which we can obtain a correlation of the variation of c_l with M_∞. Then we run a second set of wind tunnel tests, varying Re and keeping M_∞ constant. This gives us a second stack of data from which we can obtain a correlation of the variation of c_l with Re. And this is all we need; we now know how c_l varies with M_∞ and Re for the given shape at the given angle of attack. With c_l, we can obtain the lift from Eq. (5.17). By dealing with the lift coefficient instead of the lift itself, and with M_∞ and Re instead of ρ_∞, V_∞, S, μ_∞, and a_∞, we have ended up with only two stacks of wind tunnel data rather than the five we had earlier. Clearly, by using the dimensionless quantities c_l, M_∞, and Re, we have achieved a great economy of effort and wind tunnel time.

But the moral to this story goes much more deeply. Dimensional analysis shows that c_l is a function of *Mach number* and *Reynolds number*, as stated in Eq. (5.19), rather than just individually of ρ_∞, V_∞, μ_∞, a_∞, and the size of the body. It is the *combination* of these physical variables in the form of M_∞ and Re that counts. The Mach number and the Reynolds number are powerful quantities in aerodynamics. They are called *similarity parameters*, for reasons that are discussed at the end of this section. We have already witnessed in Chap. 4 the power of M_∞ in governing compressible flows. For example, just look at Eqs. (4.73) through (4.75) and (4.79); only the *Mach number* and the ratio of specific heats appear on the right-hand sides of these equations.

Performing a similar dimensional analysis on drag and moments, beginning with relations analogous to Eq. (5.3), we find that

$$D = q_\infty S c_d \qquad \textbf{[5.20]}$$

where c_d is a dimensionless *drag coefficient* and

$$M = q_\infty S c c_m \qquad \textbf{[5.21]}$$

where c_m is a dimensionless *moment coefficient*. Note that Eq. (5.21) differs slightly from Eqs. (5.17) and (5.20) by the inclusion of the chord length c. This is because L and D have dimensions of a force, whereas M has dimensions of a force-length product.

The importance of Eqs. (5.17) to (5.21) cannot be over emphasized. They are fundamental to all applied aerodynamics. They are readily obtained from dimensional analysis, which essentially takes us from loosely defined functional relationships [such as Eq. (5.3)] to well-defined relations between dimensionless quantities [Eqs. (5.17) to (5.21)]. In summary, for an airfoil of given shape, the dimensionless lift, drag, and moment coefficients have been defined as

$$c_l = \frac{L}{q_\infty S} \qquad c_d = \frac{D}{q_\infty S} \qquad c_m = \frac{M}{q_\infty S c} \qquad \textbf{[5.22]}$$

where

$$c_l = f_1(\alpha, M_\infty, \text{Re}) \qquad c_d = f_2(\alpha, M_\infty, \text{Re}) \qquad c_m = f_3(\alpha, M_\infty, \text{Re}) \qquad \textbf{[5.23]}$$

Reflecting for an instant, we find there may appear to be a conflict in our aerodynamic philosophy. On one hand, Chaps. 2 and 4 emphasized that lift, drag, and moments on an aerodynamic shape stem from the detailed pressure and shear stress distributions on the surface and that measurements and/or calculations of these distributions, especially for complex configurations, are not trivial undertakings. On the other hand, Eqs. (5.22) indicate that lift, drag, and moments can be quickly obtained from simple formulas. The bridge between these two outlooks is, of course, the lift, drag, and moment coefficients. All the physical complexity of the flow field around an aerodynamic body is implicitly buried in c_l, c_d, and c_m. Before the simple Eqs. (5.22) can be used to calculate lift, drag, and moments for an airfoil, wing, and body, the appropriate aerodynamic coefficients must be known. From this point of view, the simplicity of Eqs. (5.22) is a bit deceptive. These equations simply shift the forces of aerodynamic rigor from the forces and moments themselves to the appropriate coefficients instead. So we are now led to the questions, How do we obtain values of c_l, c_d, and c_m for given configurations? and How do they vary with α, M_∞, and Re? The answers are introduced in the following sections.

However, before we leave our discussion of dimensional analysis, it is important to elaborate on why M_∞ and Re are called similarity parameters. Consider that we have two different flows, say, a red flow and a green flow, over two bodies that are geometrically similar but are different sizes for the red and green flows. The red and green flows have different values of V_∞, ρ_∞, μ_∞, and a_∞, but they both have the *same* M_∞ and Re. If M_∞ is the same for the red and green flows, and if Re is the same for the red and green flows, then from Eq. (5.23), c_l, c_d, and c_m measured in the red flow will be the *same values* as the c_l, c_d, and c_m measured in the green flow, even though the red and green flows are different flows. In this case, the red and green flows are called *dynamically similar* flows; hence, M_∞ and Re are called *similarity parameters*. The concept of dynamic flow similarity is elegant, and it goes well beyond the scope of this book. But it is mentioned here because of its importance in aerodynamics. It is the concept of dynamic similarity that allows measurements obtained in wind tunnel tests of a small-scale model of an airplane to be applied to the real airplane in free flight. If in the wind tunnel test (say, the red flow) the values of M_∞ and Re are the *same* as those for the real airplane in free flight (say, the green flow), then c_l, c_d, and c_m measured in the wind tunnel will be *precisely* the same as those values in free flight. The concept of dynamic similarity is essential to wind tunnel testing.

In most wind tunnel tests of small-scale models of real airplanes, every effort is made to simulate the values of M_∞ and Re encountered by the real airplane in free flight. Unfortunately, due to the realities of wind tunnel design and operation, this is frequently not possible. In such cases, the wind tunnel data must be "extrapolated" to the conditions of free flight. Such extrapolations are usually approximations, and they introduce a degree of error when the wind tunnel data are used to describe the conditions of full-scale free flight. The problem of not being able to simultaneously simulate free-flight values of M_∞ and Re in the same wind tunnel is still pressing today in spite of the fact that wind tunnel testing has been going on for almost 150 years. Among other reasons, this is why there are so many different wind tunnels at different laboratories around the world.

5.4 AIRFOIL DATA

A goal of theoretical aerodynamics is to predict values of c_l, c_d, and c_m from the basic equations and concepts of physical science, some of which were discussed in previous chapters. However, simplifying assumptions are usually necessary to make the mathematics tractable. Therefore, when theoretical results are obtained, they are generally not "exact." The use of high-speed digital computers to solve the governing flow equations is now bringing us much closer to the accurate calculation of aerodynamic characteristics; however, there are still limitations imposed by the numerical methods themselves, and the storage and speed capacity of current computers are still not sufficient to solve many complex aerodynamic flows. As a result, the practical aerodynamicist has to rely upon direct *experimental* measurements of c_l, c_d, and c_m for specific bodies of interest.

A large bulk of experimental airfoil data was compiled over the years by the National Advisory Committee for Aeronautics (NACA), which was absorbed in the creation of the National Aeronautics and Space Administration (NASA) in 1958. Lift, drag, and moment coefficients were systematically measured for many airfoil shapes in low-speed subsonic wind tunnels. These measurements were carried out on straight, constant-chord wings which completely spanned the tunnel test section from one side wall to the other. In this fashion, the flow essentially "saw" a wing with no wingtips, and the experimental airfoil data were thus obtained for "infinite wings." (The distinction between infinite and finite wings will be made in subsequent sections.) Some results of these airfoil measurements are given in App. D. The first page of App. D gives data for c_l and $c_{m,c/4}$ versus angle of attack for the NACA 1408 airfoil. The second page gives c_d and $c_{m,ac}$ versus c_l for the same airfoil. Since c_l is known as a function of α from the first page, the data from both pages can be cross-plotted to obtain the variations of c_d and $c_{m,ac}$ versus α. The remaining pages of App. D give the same type of data for different standard NACA airfoil shapes.

Let us examine the variation of c_l with α more closely. This variation is sketched in Fig. 5.6. The experimental data indicate that c_l varies *linearly* with α over a large range of angle of attack. Thin-airfoil theory, which is the subject of more advanced books on aerodynamics, also predicts the same type of linear variation. The slope of

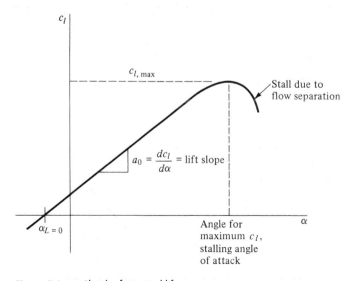

Figure 5.6   Sketch of a typical lift curve.

the linear portion of the lift curve is designated as $a_0 \equiv dc_l/d\alpha \equiv$ lift slope. Note that in Fig. 5.6, when $\alpha = 0$, there is still a positive value of c_l; that is, there is still some lift even when the airfoil is at zero angle of attack to the flow. This is due to the positive camber of the airfoil. All airfoils with such camber have to be pitched to some negative angle of attack before zero lift is obtained. The value of α when lift is zero is defined as the *zero-lift angle of attack* $\alpha_{L=0}$ and is illustrated in Fig. 5.6. This effect is further demonstrated in Fig. 5.7, where the lift curve for a cambered airfoil is compared with that for a symmetric (no camber) airfoil. Note that the lift curve for a symmetric airfoil goes through the origin. Refer again to Fig. 5.6, at the other extreme: For large values of α, the linearity of the lift curve breaks down. As α is increased beyond a certain value, c_l peaks at some maximum value $c_{l,\max}$ and then drops precipitously as α is further increased. In this situation, where the lift is rapidly decreasing at high α, the airfoil is *stalled*.

The phenomenon of airfoil stall is of critical importance in airplane design. It is caused by flow separation on the upper surface of the airfoil. This is illustrated in Fig. 5.8, which again shows the variation of c_l versus α for an airfoil. At point 1 on the linear portion of the lift curve, the flow field over the airfoil is attached to the surface, as pictured in Fig. 5.8. However, as discussed in Chap. 4, the effect of friction is to slow the airflow near the surface; in the presence of an adverse pressure gradient, there will be a tendency for the boundary layer to separate from the surface. As the angle of attack is increased, the adverse pressure gradient on the top surface of the airfoil will become stronger, and at some value of α—the stalling angle of attack—the flow becomes separated from the top surface. When separation occurs, the lift decreases drastically and the drag increases suddenly. This is the picture associated with point 2 in Fig. 5.8. (It would be well for the reader to review at this stage the discussion of flow separation and its effect on pressure distribution, lift, and drag in Sec. 4.21.)

The nature of the flow field over the wing of an airplane that is below, just beyond, and way beyond the stall is shown in Fig. 5.9a, b, and c, respectively. These

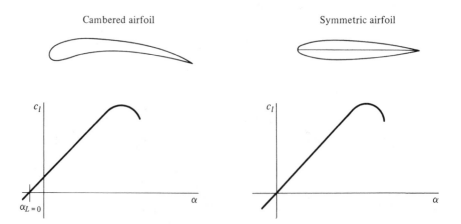

Figure 5.7   Comparison of lift curves for cambered and symmetric airfoils.

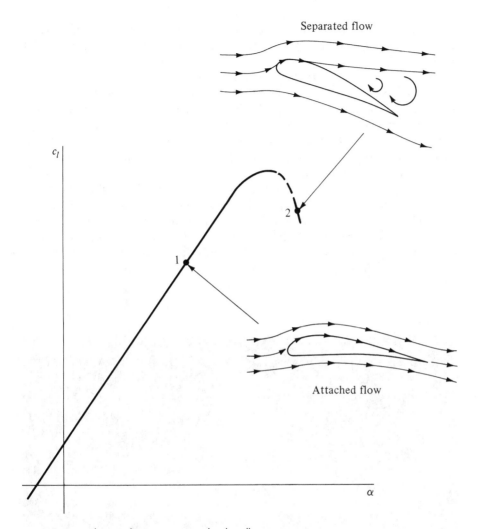

Separated flow

Attached flow

Figure 5.8 Flow mechanism associated with stalling.

figures are photographs of a wind tunnel model with a wingspan of 6 ft. The entire model has been painted with a mixture of mineral oil and a fluorescent powder, which glows under ultraviolet light. After the wind tunnel is turned on, the fluorescent oil indicates the streamline pattern on the surface of the model. In Fig. 5.9a, the angle of attack is below the stall; the flow is fully attached, as evidenced by the fact that the high surface shear stress has scrubbed most of the oil from the surface. In Fig. 5.9b, the angle of attack is slightly beyond the stall. A large, mushroom-shaped, separated flow pattern has developed over the wing, with attendant highly three-dimensional, low-energy recirculating flow. In Fig. 5.9c, the angle of attack is far beyond the stall. The flow over almost the entire wing has separated. These photographs are striking examples of different types of flow that can occur over an airplane wing at different angles of attack, and they graphically show the extent of the flow field separation that can occur.

The lift curves sketched in Figs. 5.6 to 5.8 illustrate the type of variation observed experimentally in the data of App. D. Returning to App. D, we note that the lift curves are all virtually linear up to the stall. Singling out a given airfoil, say, the NACA 2412 airfoil, also note that c_l versus α is given for three different values of the Reynolds number from 3.1×10^6 to 8.9×10^6. The lift curves for all three values of Re fall on top of one another in the linear region; that is, Re has little influence on c_l when the flow is attached. However, flow separation is a viscous effect, and as discussed in Chap. 4, Re is a governing parameter for viscous flow. Therefore, it is not surprising that the experimental data for $c_{l,\max}$ in the stalling region are affected by Re, as can be seen by the slightly different variations of c_l at high α for different values of Re. In fact, these lift curves at different Re values answer part of the question posed in Eq. (5.19): The data represent $c_l = f(\text{Re})$. Again, Re exerts little or no effect on c_l except in the stalling region.

(a)

Figure 5.9   Surface oil flow patterns on a wind tunnel model of a Grumman American Yankee, taken by Dr. Allen Winkelmann in the Glenn L. Martin Wind Tunnel at the University of Maryland. The mixture is mineral oil and a fluorescent powder, and the photographs were taken under ultraviolet light. (a) Below the stall. The wing is at $\alpha = 4°$, where the flow is attached. (*continued*)

(b)

(c)

(b) Very near the stall. The wing is at $\alpha = 11°$, where the highly three-dimensional separated flow is developing in a mushroom cell pattern. (c) Far above the stall. The wing is at $\alpha = 24°$, where the flow over almost the entire wing has separated.

On the same page as c_l versus α, the variation of $c_{m,c/4}$ versus α is also given. It has only a slight variation with α and is almost completely unaffected by Re. Also note that the values of $c_{m,c/4}$ are slightly negative. By convention, a positive moment is in a clockwise direction; it pitches the airfoil toward larger angles of attack, as shown in Fig. 5.4. Therefore, for the NACA 2412 airfoil, with $c_{m,c/4}$ negative, the moments are counterclockwise, and the airfoil tends to pitch downward. This is characteristic of all airfoils with positive camber.

On the page following c_l and $c_{m,c/4}$, the variation of c_d and $c_{m,ac}$ is given versus c_l. Because c_l varies linearly with α, the reader can visualize these curves of c_d and $c_{m,ac}$ as being plotted versus α as well; the shapes will be the same. Note that the drag curves have a "bucket" type of shape, with minimum drag occurring at small values of c_l (hence there are small angles of attack). As α goes to large negative or positive values, c_d increases. Also note that c_d is strongly affected by Re, there being a distinct drag curve for each Re. This is to be expected because the drag for a slender aerodynamic shape is mainly skin friction drag, and from Chap. 4 we have seen that Re strongly governs skin friction. With regard to $c_{m,ac}$, the definition of the aerodynamic center is clearly evident; $c_{m,ac}$ is constant with respect to α. Also, it is insensitive to Re and has a small negative value.

Refer to Eqs. (5.23); the airfoil data in App. D experimentally provide the variation of c_l, c_d, and c_m with α and Re. The effect of M_∞ on the airfoil coefficients will be discussed later. However, we emphasize that the data in App. D were measured in low-speed subsonic wind tunnels. Hence, the flow was essentially incompressible. Thus, c_l, $c_{m,c/4}$, c_d, and $c_{m,ac}$ given in App. D are incompressible flow values. It is important to keep this in mind during our subsequent discussions.

In this section we have discussed the properties of an *airfoil*. As already noted in Fig. 5.2, an airfoil is simply the shape of a wing section. The airfoils in Figs. 5.3 through 5.5 and Figs. 5.7 and 5.8 are paper-thin sections—simple drawings on a sheet of paper. So what does it mean when we talk about the lift, drag, and moments on an airfoil? How can there be a lift on an airfoil that is paper-thin? When we write Eq. (5.17) for the lift of an airfoil, what really is L? The answer is given in Fig. 5.10. Here we see a section of a wing of constant chord c. The length of the section along the span of the wing is unity (1 ft, 1 m, etc.). The lift on this wing section L, as shown in Fig. 5.10a, is the *lift per unit span*. The lift, drag, and moments on an *airfoil* are always understood to be the lift, drag, and moments *per unit span*, as sketched in Fig. 5.10. The planform area of the segment of unit span is the projected area seen by looking at the wing from above, namely, $S = c(1) = c$, as sketched in Fig. 5.10b. Hence, when we write Eq. (5.17) for an airfoil, we interpret L as the lift per unit span and S as the planform area of a unit span, that is,

$$L \text{ (per unit span)} = q_\infty c(1) c_l \qquad \textbf{[5.24]}$$

or
$$c_l = \frac{L \text{ (per unit span)}}{q_\infty c} \qquad \textbf{[5.25]}$$

Finally, return to our road map in Fig. 5.1. We have begun to work our way down the left column under airfoils. We have already accomplished a lot. We have become

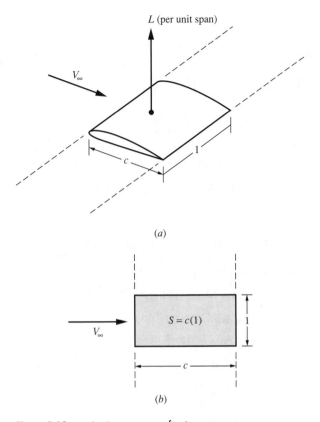

Figure 5.10 A wing segment of unit span.

familiar with airfoil nomenclature. Using dimensional analysis, we have introduced the very important concept of aerodynamic *coefficients*, and we have examined some experimental data for these coefficients. Make certain that you feel comfortable with these concepts before you continue.

A model wing of constant chord length is placed in a low-speed subsonic wind tunnel, spanning the test section. The wing has an NACA 2412 airfoil and a chord length of 1.3 m. The flow in the test section is at a velocity of 50 m/s at standard sea-level conditions. If the wing is at a 4° angle of attack, calculate (*a*) c_l, c_d, and $c_{m,c/4}$ and (*b*) the lift, drag, and moments about the quarter chord, per unit span.

Example 5.1

Solution

a. From App. D, for an NACA 2412 airfoil at a 4° angle of attack,

$$c_l = 0.63$$
$$c_{m,c/4} = -0.035$$

To obtain c_d, we must first check the value of the Reynolds number:

$$\text{Re} = \frac{\rho_\infty V_\infty c}{\mu_\infty} = \frac{(1.225 \text{ kg/m}^3)(50 \text{ m/s})(1.3 \text{ m})}{1.789 \times 10^{-5} \text{ kg/(m)(s)}} = 4.45 \times 10^6$$

For this value of Re and for $c_l = 0.63$, from App. D,

$$\boxed{c_d = 0.007}$$

b. Since the chord is 1.3 m and we want the aerodynamic forces and moments *per unit span* (a unit length along the wing, perpendicular to the flow), $S = c(1) = 1.3(1) = 1.3 \text{ m}^2$. Also

$$q_\infty = \tfrac{1}{2}\rho_\infty V_\infty^2 = \tfrac{1}{2}(1.225)(50)^2 = 1531 \text{ N/m}^2$$

From Eq. (5.22),

$$L = q_\infty S c_l = 1531(1.3)(0.63) = \boxed{1254 \text{ N}}$$

Since 1 N = 0.2248 lb, also

$$L = (1254 \text{ N})(0.2248 \text{ lb/N}) = 281.9 \text{ lb}$$

$$D = q_\infty S c_d = 1531(1.3)(0.007) = \boxed{13.9 \text{ N}}$$
$$= 13.9(0.2248) = 3.13 \text{ lb}$$

Note: The ratio of lift to drag, which is an important aerodynamic quantity, is

$$\frac{L}{D} = \frac{c_l}{c_d} = \frac{1254}{13.9} = 90.2$$

$$M_{c/4} = q_\infty S c_{m,c/4} c = 1531(1.3)(-0.035)(1.3)$$

$$\boxed{M_{c/4} = -90.6 \text{ N·m}}$$

Example 5.2 | The same wing in the same flow as in Example 5.1 is pitched to an angle of attack such that the lift per unit span is 700 N (157 lb).
a. What is the angle of attack?
b. To what angle of attack must the wing be pitched to obtain zero lift?

Solution

a. From the previous example,

$$q_\infty = 1531 \text{ N/m}^2 \qquad S = 1.3 \text{ m}^2$$

Thus

$$c_l = \frac{L}{q_\infty S} = \frac{700}{1531(1.3)} = 0.352$$

From App. D for the NACA 2412 airfoil, the angle of attack corresponding to $c_l = 0.352$ is

$$\boxed{\alpha = 1°}$$

b. Also from App. D, for zero lift, that is, $c_l = 0$,

$$\alpha_{L=0} = -2.2°$$

5.5 INFINITE VERSUS FINITE WINGS

As stated in Sec. 5.4, the airfoil data in App. D were measured in low-speed subsonic wind tunnels where the model wing spanned the test section from one sidewall to the other. In this fashion, the flow sees essentially a wing with no wing tips; that is, the wing in principle could be stretching from plus infinity to minus infinity in the spanwise direction. Such an *infinite wing* is sketched in Fig. 5.11, where the wing stretches to $\pm\infty$ in the z direction. The flow about this wing varies only in the x and y directions; for this reason the flow is called *two-dimensional*. Thus, the airfoil data in App. D apply only to such infinite (or two-dimensional) wings. This is an important point to keep in mind.

On the other hand, all real airplane wings are obviously finite, as sketched in Fig. 5.12. Here, the top view (planform view) of a finite wing is shown, where the distance between the two wing tips is defined as the *wingspan b*. The area of the wing in this planform view is designated, as before, by S. This leads to an important definition which pervades all aerodynamic wing considerations, namely, the aspect ratio AR:

$$\text{Aspect ratio} \equiv \text{AR} \equiv \frac{b^2}{S} \qquad \text{[5.26]}$$

The importance of AR will come to light in subsequent sections.

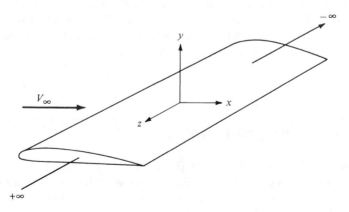

Figure 5.11　Infinite (two-dimensional) wing.

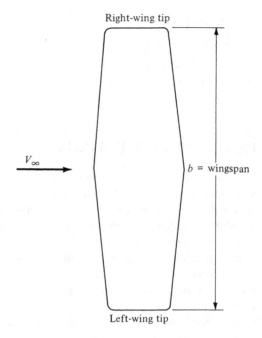

Figure 5.12   Finite wing; plan view (top).

The flow field about a finite wing is three-dimensional and is therefore inherently different from the two-dimensional flow about an infinite wing. As a result, the lift, drag, and moment coefficients for a finite wing with a given airfoil shape at a given α are different from the lift, drag, and moment coefficients for an infinite wing with the same airfoil shape at the same α. For this reason, the aerodynamic coefficients for a finite wing are designated by capital letters C_L, C_D, C_M; this is in contrast to those for an infinite wing, which we have been designating as c_l, c_d, and c_m. Note that the data in App. D are for infinite (two-dimensional) wings; that is, the data are for c_l, c_d, and c_m. In a subsequent section, we will show how to obtain the finite-wing aerodynamic coefficients from the infinite-wing data in App. D. Our purpose in this section is simply to underscore that there is a difference.

5.6   PRESSURE COEFFICIENT

We continue with our parade of aerodynamic definitions. Consider the pressure distribution over the top surface of an airfoil. Instead of plotting the actual pressure (say, in units of newtons per square meter), we define a new dimensionless quantity, called the *pressure coefficient* C_p, as

$$C_p \equiv \frac{p - p_\infty}{q_\infty} \equiv \frac{p - p_\infty}{\frac{1}{2}\rho_\infty V_\infty^2}$$ [5.27]

The pressure distribution is sketched in terms of C_p in Fig. 5.13. This figure is worth looking at closely because pressure distributions found in the aerodynamic literature are usually given in terms of the dimensionless pressure coefficient. Note from Fig. 5.13 that C_p at the leading edge is positive because $p > p_\infty$. However, as the flow expands around the top surface of the airfoil, p decreases rapidly, and C_p goes negative in those regions where $p < p_\infty$. By convention, plots of C_p for airfoils are usually shown with negative values above the abscissa, as shown in Fig. 5.13.

The pressure coefficient is an important quantity; for example, the distribution of C_p over the airfoil surface leads directly to the value of c_l, as will be discussed in a subsequent section. Moreover, considerations of C_p lead directly to the calculation of the effect of Mach number M_∞ on the lift coefficient. To set the stage for this calculation, consider C_p at a given point on an airfoil surface. The airfoil is a given shape at a fixed angle of attack. The value of C_p can be measured by testing the airfoil in a wind tunnel. Assume that, at first, V_∞ in the tunnel test section is low, say, $M_\infty < 0.3$, such that the flow is essentially incompressible. The measured value of C_p at the point on the airfoil will therefore be a low-speed value. Let us designate the low-speed (incompressible) value of C_p by $C_{p,0}$. If V_∞ is increased but M_∞ is still less than 0.3, then C_p will not change; that is, C_p is essentially constant with

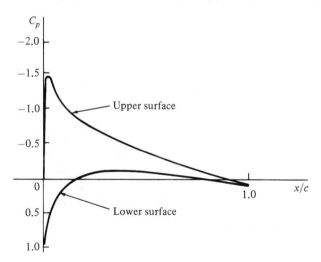

Figure 5.13 Distribution of pressure coefficient over the top and bottom surfaces of an NACA 0012 airfoil at 3.93° angle of attack. $M_\infty = 0.345$, Re $= 3.245 \times 10^6$. Experimental data from Ohio State University, in NACA Conference Publication 2045, part I, *Advanced Technology Airfoil Research*, vol. I, p. 1590. (After Freuler and Gregorek.)

velocity at low speeds. However, if we now increase V_∞ such that $M_\infty > 0.3$, then compressibility becomes a factor, and the effect of compressibility is to increase the absolute magnitude of C_p as M_∞ increases. This variation of C_p with M_∞ is shown in Fig. 5.14. Note that, at $M_\infty \approx 0$, $C_p = C_{p,0}$. As M_∞ increases to $M_\infty \approx 0.3$, essentially C_p is constant. However, as M_∞ is increased beyond 0.3, C_p increases dramatically. (That is, the absolute magnitude increases; if $C_{p,0}$ is negative, then C_p will become an increasingly more negative number as M_∞ increases, whereas if $C_{p,0}$ is positive, then C_p will become an increasingly more positive number as M_∞ increases.) The variation of C_p with M_∞ for high subsonic Mach numbers was a major focus of aerodynamic research after World War II. An approximate theoretical analysis yields

$$C_p = \frac{C_{p,0}}{\sqrt{1 - M_\infty^2}} \qquad [5.28]$$

Equation (5.28) is called the *Prandtl-Glauert rule*. It is reasonably accurate for $0.3 < M_\infty < 0.7$. For $M_\infty > 0.7$, its accuracy rapidly diminishes; indeed, Eq. (5.28) predicts that C_p becomes infinite as M_∞ goes to unity—an impossible physical situation. (It is well to note that nature abhors infinities as well as discontinuities that are sometimes predicted by mathematical, but approximate, theories in physical science.) There are more accurate, but more complicated, formulas than Eq. (5.28) for near-sonic Mach numbers. However, Eq. (5.28) will be sufficient for our purposes.

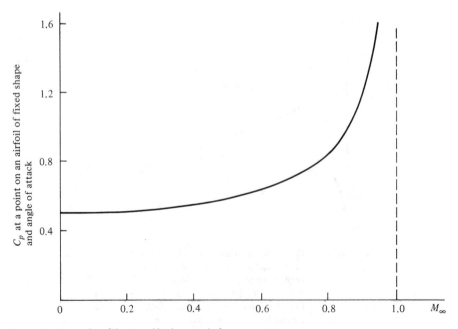

Figure 5.14   Plot of the Prandtl-Glauert rule for $C_{p,0} = 0.5$.

Formulas such as Eq. (5.28), which attempt to predict the effect of M_∞ on C_p for subsonic speeds, are called *compressibility corrections*; that is, they modify (correct) the low-speed pressure coefficient $C_{p,0}$ to take into account the effects of compressibility which are so important at high subsonic Mach numbers.

Example 5.3

The pressure at a point on the wing of an airplane is 7.58×10^4 N/m^2. The airplane is flying with a velocity of 70 m/s at conditions associated with a standard altitude of 2000 m. Calculate the pressure coefficient at this point on the wing.

Solution

For a standard altitude of 2000 m,

$$p_\infty = 7.95 \times 10^4 \text{ N/m}^2$$

$$\rho_\infty = 1.0066 \text{ kg/m}^3$$

Thus, $q_\infty = \frac{1}{2}\rho_\infty V_\infty^2 = \frac{1}{2}(1.0066)(70)^2 = 2466$ N/m^2. From Eq. (5.27),

$$C_p = \frac{p - p_\infty}{q_\infty} = \frac{(7.58 - 7.95) \times 10^4}{2466}$$

$$\boxed{C_p = -1.5}$$

Example 5.4

Consider an airfoil mounted in a low-speed subsonic wind tunnel. The flow velocity in the test section is 100 ft/s, and the conditions are standard sea level. If the pressure at a point on the airfoil is 2102 lb/ft^2, what is the pressure coefficient?

Solution

$$q_\infty = \frac{1}{2}\rho_\infty V_\infty^2 = \frac{1}{2}(0.002377 \text{ slug/ft}^3)(100 \text{ ft/s})^2 = 11.89 \text{ lb/ft}^2$$

From Eq. (5.27),

$$C_p = \frac{p - p_\infty}{q_\infty} = \frac{2102 - 2116}{11.89} = \boxed{-1.18}$$

Example 5.5

In Example 5.4, if the flow velocity is increased such that the free-stream Mach number is 0.6, what is the pressure coefficient at the same point on the airfoil?

Solution

First, what is the Mach number of the flow in Example 5.4? At standard sea level,

$$T_s = 518.69^\circ \text{ R}$$

Hence, $$a_\infty = \sqrt{\gamma R T_\infty} = \sqrt{1.4(1716)(518.69)} = 1116 \text{ ft/s}$$

Thus, in Example 5.4, $M_\infty = V_\infty/a_\infty = 100/1116 = 0.09$—a very low value. Hence, the flow in Example 5.4 is essentially incompressible, and the pressure coefficient is a low-speed value; that is, $C_{p,0} = -1.18$. Thus, if the flow Mach number is increased to 0.6, from the Prandtl-Glauert rule, Eq. (5.28),

$$C_p = \frac{C_{p,0}}{(1 - M_\infty^2)^{1/2}} = -\frac{1.18}{(1 - 0.6^2)^{1/2}}$$

$$\boxed{C_p = -1.48}$$

5.7   OBTAINING LIFT COEFFICIENT FROM C_p

If you are given the distribution of the pressure coefficient over the top and bottom surfaces of an airfoil, you can calculate c_l in a straightforward manner. Consider a segment of an infinite wing, as shown in Fig. 5.15. Assume the segment has unit span and chord c. The wing is at an angle of attack of α. Let x be the direction measured along the chord, and let s be the distance measured along the *surface* from the leading edge, as shown in Fig. 5.15. Consider the infinitesimally small sliver of surface area of length ds and unit length in the span direction, as shown by the shaded area in Fig. 5.15. The area of this surface is $1\,ds$. The dashed line ab is perpendicular to chord c. The solid line ac is locally perpendicular to the shaded area. The angle between ab and ac is θ. The aerodynamic force on the shaded area is $p(1)\,ds$, which acts in the direction of ac, normal to the surface. Its component in the direction normal to

Figure 5.15   Sketch showing how the pressure distribution can be integrated to obtain normal force per unit span, leading to lift per unit span.

the chord is $(p \cos \theta)(1) \, ds$. Adding a subscript u to designate the pressure on the upper surface of the airfoil, as well as a minus sign to indicate the force is directed downward (we use the convention that a positive force is directed upward), we see that the contribution to the normal force of the pressure on the infinitesimal strip is $-p_u \cos \theta \, ds$. If all the contributions from all the strips on the upper surface are added from the leading edge to the trailing edge, we obtain, by letting ds approach 0, the integral

$$-\int_{\mathrm{LE}}^{\mathrm{TE}} p_u \cos \theta \, ds$$

This is the force in the *normal* direction due to the pressure distribution acting on the upper surface of the wing, per unit span. Recall the definition of normal and axial forces N and A, respectively, discussed in Sec. 5.2 and sketched in Fig. 5.4a. The above integral is that part of N which is due to the pressure acting on the upper surface. A similar term is obtained that is due to the pressure distribution acting on the lower surface of the airfoil. Letting p_l denote the pressure on the lower surface, we can write for the total normal force acting on an airfoil of unit span

$$N = \int_{\mathrm{LE}}^{\mathrm{TE}} p_l \cos \theta \, ds - \int_{\mathrm{LE}}^{\mathrm{TE}} p_u \cos \theta \, ds \qquad \textbf{[5.29]}$$

From the small triangle in the box in Fig. 5.15, we see the geometric relationship $ds \cos \theta = dx$. Thus, in Eq. (5.29), the variable of integration s can be replaced by x, and at the same time the x coordinates of the leading and trailing edges become 0 and c, respectively. Thus, Eq. (5.29) becomes

$$N = \int_0^c p_l \, dx - \int_0^c p_u \, dx \qquad \textbf{[5.30]}$$

Adding and subtracting p_∞, we find that Eq. (5.30) becomes

$$N = \int_0^c (p_l - p_\infty) \, dx - \int_0^c (p_u - p_\infty) \, dx \qquad \textbf{[5.31]}$$

Putting Eq. (5.31) on the shelf for a moment, we return to the definition of normal and axial forces N and A, respectively, in Fig. 5.4a. We can define the normal and axial force coefficients for an airfoil, c_n and c_a, respectively, in the same manner as the lift and drag coefficients given by Eqs. (5.20), that is,

$$c_n = \frac{N}{q_s S} = \frac{N}{q_\infty c} \qquad \textbf{[5.32]}$$

$$c_a = \frac{A}{q_s S} = \frac{A}{q_\infty c} \qquad \textbf{[5.33]}$$

Hence, the normal force coefficient c_n can be calculated from Eqs. (5.31) and (5.32) as

$$c_n = \frac{1}{c} \int_0^c \frac{p_l - p_\infty}{q_\infty} \, dx - \frac{1}{c} \int_0^c \frac{p_u - p_\infty}{q_\infty} \, dx \qquad \textbf{[5.34]}$$

Note that

$$\frac{p_l - p_\infty}{q_\infty} \equiv C_{p,l} \equiv \text{pressure coefficient on lower surface}$$

$$\frac{p_u - p_\infty}{q_\infty} \equiv C_{p,l} \equiv \text{pressure coefficient on upper surface}$$

Hence, Eq. (5.34) becomes

$$c_n = \frac{1}{c} \int_0^c (C_{p,l} - C_{p,u})\, dx \qquad \text{[5.35]}$$

Equation (5.35) gives the normal force coefficient directly in terms of the integral of the pressure coefficient over the surface of the airfoil.

How is this related to the lift coefficient? The answer is given by Eq. (5.1), repeated here:

$$L = N \cos\alpha - A \sin\alpha \qquad \text{[5.1]}$$

Dividing Eq. (5.1) by $q_\infty S = q_\infty c$, we have

$$\frac{L}{q_\infty c} = \frac{N}{q_\infty c} \cos\alpha - \frac{A}{q_\infty c} \sin\alpha$$

or

$$c_l = c_n \cos\alpha - c_a \sin\alpha \qquad \text{[5.36]}$$

Given c_n and c_a, Eq. (5.36) allows the direct calculation of c_l. Equation (5.35) is an expression for c_n in terms of the integral of the pressure coefficients. [In Eq. (5.35), we have ignored the influence of shear stress, which contributes very little to normal force.] A similar expression can be obtained for c_a involving an integral of the pressure coefficient and an integral of the skin friction coefficient. Such an expression is derived in Chap. 1 of Anderson, *Fundamentals of Aerodynamics*, 2nd Ed., McGraw-Hill, 1991; this is beyond the scope of our discussion here.

Consider the case of small angle of attack, say, $\alpha \leq 5°$. Then in Eq. (5.36), $\cos\alpha \approx 1$ and $\sin\alpha \approx 0$, and Eq. (5.36) yields

$$c_l \approx c_n \qquad \text{[5.37]}$$

and combining Eqs. (5.37) and (5.35), we have

$$c_l \approx \frac{1}{c} \int_0^c (C_{p,l} - C_{p,u})\, dx \qquad \text{[5.38]}$$

Most conventional airplanes cruise at angles of attack of less than $5°$, so for such cases Eq. (5.38) is a reasonable representation of the lift coefficient in terms of the integral of the pressure coefficient. This leads to a useful graphical construction for c_l. Consider a combined plot of C_{p_u} and C_{p_l} as a function of x/c, as sketched in Fig. 5.16. The area between these curves is precisely the integral on the right-hand side of Eq. (5.35). Hence, this area, shown as the shaded region in Fig. 5.16, is precisely equal to the normal force coefficient. In turn, for small angles of attack, from Eq. (5.38), this area is essentially the lift coefficient, as noted in Fig. 5.16.

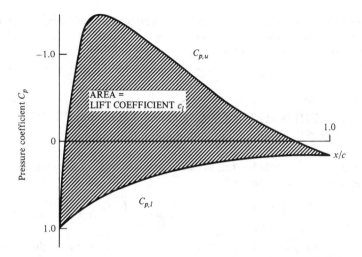

Figure 5.16 Sketch of the pressure coefficient over the upper and lower surfaces of an airfoil showing that the area between the two curves is the lift coefficient for small angles of attack.

Consider an airfoil with chord length c and the running distance x measured along the chord. The leading edge is located at $x/c = 0$ and the trailing edge at $x/c = 1$. The pressure coefficient variations over the upper and lower surfaces are given, respectively, as

Example 5.6

$$C_{p,u} = 1 - 300 \left(\frac{x}{c}\right)^2 \qquad \text{for } 0 \leq \frac{x}{c} \leq 0.1$$

$$C_{p,u} = -2.2277 + 2.2777\frac{x}{c} \qquad \text{for } 0.1 \leq \frac{x}{c} \leq 1.0$$

$$C_{p,l} = 1 - 0.95\frac{x}{c} \qquad \text{for } 0 \leq \frac{x}{c} \leq 1.0$$

Calculate the normal force coefficient.

Solution

From Eq. (5.35)

$$c_n = \frac{1}{c}\int_0^c (C_{p,l} - C_{p,u})\, dx = \int_0^1 (C_{p,l} - C_{p,u})\, d\left(\frac{x}{c}\right)$$

$$c_n = \int_0^1 \left(1 - 0.95\frac{x}{c}\right) d\left(\frac{x}{c}\right) - \int_0^{0.1}\left[1 - 300\left(\frac{x}{c}\right)^2\right] d\left(\frac{x}{c}\right)$$

$$- \int_{0.1}^{1.0}\left(-2.2277 + 2.2777\frac{x}{c}\right) d\left(\frac{x}{c}\right)$$

$$c_n = \frac{x}{c}\Big|_0^1 - 0.475\left(\frac{x}{c}\right)^2\Big|_0^1 - \frac{x}{c}\Big|_0^{0.1} + 100\left(\frac{x}{c}\right)^3\Big|_0^{0.1} + 2.2277\frac{x}{c}\Big|_{0.1}^{1.0} - 1.1388\left(\frac{x}{c}\right)^2\Big|_{0.1}^{1.0}$$

$$c_n = 1 - 0.475 - 0.1 + 0.1 + 2.2277 - 0.22277 - 1.1388 + 0.011388 = \boxed{1.40}$$

Note that the C_p variations given analytically in this problem are only crude representations of a realistic case and should not be taken too seriously, since the purpose of this example is simply to illustrate the use of Eq. (5.35).

5.8 COMPRESSIBILITY CORRECTION FOR LIFT COEFFICIENT

The pressure coefficients in Eq. (5.38) can be replaced by the compressibility correction given in Eq. (5.28) as follows

$$c_l = \frac{1}{c} \int_0^c \frac{(C_{p,l} - C_{p,u})_0}{\sqrt{1 - M_\infty^2}} \, dx = \frac{1}{\sqrt{1 - M_\infty^2}} \frac{1}{c} \int_0^c (C_{p,l} - C_{p,u})_0 \, dx \qquad \textbf{[5.39]}$$

where again the subscript 0 denotes low-speed incompressible flow values. However, referring to the form of Eq. (5.38), we see that

$$\frac{1}{c} \int_0^c (C_{p,l} - C_{p,u})_0 \, dx \equiv c_{l,0}$$

where $c_{l,0}$ is the low-speed value of the lift coefficient. Thus, Eq. (5.39) becomes

$$\boxed{c_l = \frac{c_{l,0}}{\sqrt{1 - M_\infty^2}}} \qquad \textbf{[5.40]}$$

Equation (5.40) gives the compressibility correction for the lift coefficient. It is subject to the same approximations and accuracy restrictions as the Prandtl-Glauert rule, Eq. (5.28). Also note that the airfoil data in App. D were obtained at low speeds, hence the values of lift coefficient obtained from App. D are $c_{l,0}$.

Finally, in reference to Eq. (5.19), we now have a reasonable answer to how c_l varies with Mach number. For subsonic speeds, except near Mach 1, the lift coefficient varies inversely as $(1 - M_\infty^2)^{1/2}$.

Example 5.7

Consider an NACA 4412 airfoil at an angle of attack of $4°$. If the free-stream Mach number is 0.7, what is the lift coefficient?

Solution

From App. D, for $\alpha = 4°$, $c_l = 0.83$. However, the data in App. D were obtained at low speeds, hence the lift coefficient value obtained above, namely, 0.83, is really $c_{l,0}$:

$$c_{l,0} = 0.83$$

For high Mach numbers, this must be corrected according to Eq. (5.40):

$$c_l = \frac{c_{l,0}}{(1 - M_\infty^2)^{1/2}} = \frac{0.83}{(1 - 0.7^2)^{1/2}}$$

$$\boxed{c_l = 1.16 \qquad \text{at } M_\infty = 0.7}$$

5.9 CRITICAL MACH NUMBER AND CRITICAL PRESSURE COEFFICIENT

Consider the flow of air over an airfoil. We know that as the gas expands around the top surface near the leading edge, the velocity and hence the Mach number will increase rapidly. Indeed, there are regions on the airfoil surface where the local Mach number can be greater than M_∞. Imagine that we put a given airfoil in a wind tunnel where $M_\infty = 0.3$ and that we observe the peak local Mach number on the top surface of the airfoil to be 0.435. This is sketched in Fig. 5.17a. Imagine that we now increase M_∞ to 0.5; the peak local Mach number will correspondingly increase to 0.772, as shown in Fig. 5.17b. If we further increase M_∞ to a value of 0.61, we observe that the peak local Mach number is 1.0, locally sonic flow on the surface of the airfoil. This is sketched in Fig. 5.17c. Note that the flow over an airfoil can locally be sonic (or higher), even though the free-stream Mach number is subsonic. By definition, that free-stream Mach number at which sonic flow is first obtained somewhere on the airfoil surface is called the *critical Mach number* of the airfoil. In the above example, the critical Mach number M_{cr} for the airfoil is 0.61. As we will see later, M_{cr} is an important quantity, because at some free-stream Mach number above M_{cr} the airfoil will experience a dramatic increase in drag.

Returning to Fig. 5.17, we see that the point on the airfoil where the local M is a peak value is also the point of minimum surface pressure. From the definition of the pressure coefficient, Eq. (5.27), C_p will correspondingly have its most negative value at this point. Moreover, according to the Prandtl-Glauert rule, Eq. (5.28), as M_∞ is

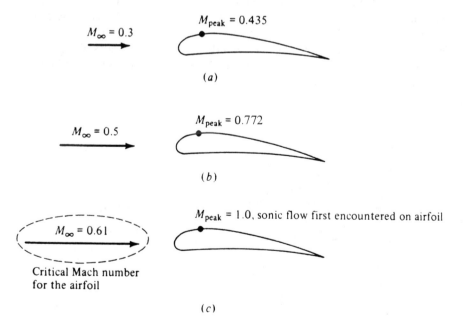

Figure 5.17 Illustration of critical Mach number.

increased from 0.3 to 0.61, the value of C_p at this point will become increasingly negative. This is sketched in Fig. 5.18. The specific value of C_p that corresponds to sonic flow is defined as the *critical pressure coefficient* $C_{p,\mathrm{cr}}$. In Fig. 5.17a and 5.17b, C_p at the minimum pressure point on the airfoil is less negative than $C_{p,\mathrm{cr}}$; however, in Fig. 5.17c, $C_p = C_{p,\mathrm{cr}}$ (by definition).

Consider now three different airfoils ranging from thin to thick, as shown in Fig. 5.19. Concentrate first on the thin airfoil. Because of the thin, streamlined profile, the flow over the thin airfoil is only slightly perturbed from its free-stream values. The expansion over the top surface is mild, the velocity increases only slightly, the pressure decreases only a relatively small amount, and hence the magnitude of C_p at the minimum pressure point is small. Thus, the variation of C_p with M_∞ is shown as the bottom curve in Fig. 5.19. For the thin airfoil, $C_{p,0}$ is small in magnitude, and the rate of increase of C_p as M_∞ increases is also relatively small. In fact, because the flow expansion over the thin airfoil surface is mild, M_∞ can be increased to a large subsonic value before sonic flow is encountered on the airfoil surface. The point corresponding to sonic flow conditions on the thin airfoil is labeled point a in Fig. 5.19. The values of C_p and M_∞ at point a are $C_{p,\mathrm{cr}}$ and M_{cr}, respectively, for the thin airfoil, by definition.

Now consider the airfoil of medium thickness. The flow expansion over the leading edge for this medium airfoil will be stronger, the velocity will increase to larger values, the pressure will decrease to lower values, and the absolute magnitude of C_p is larger. Thus, the pressure coefficient curve for the medium-thickness airfoil will lie above that for a thin airfoil, as demonstrated in Fig. 5.19. Moreover, because the flow expansion is stronger, sonic conditions will be obtained sooner (at a lower M_∞). Sonic conditions for the medium airfoil are labeled as point b in Fig. 5.19. Note that point b is to the left of point a; that is, the critical Mach number for the medium-thickness airfoil is less than M_{cr} for the thin airfoil. The same logic holds for the pressure coefficient curve for the thick airfoil, where $C_{p,\mathrm{cr}}$ and M_{cr} are given by point c. We emphasize that the thinner airfoils have higher values of M_{cr}. As we will see, this is desirable, and hence all airfoils on modern, high-speed airplanes are relatively thin.

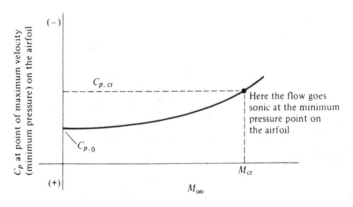

Figure 5.18   Illustration of critical pressure coefficient.

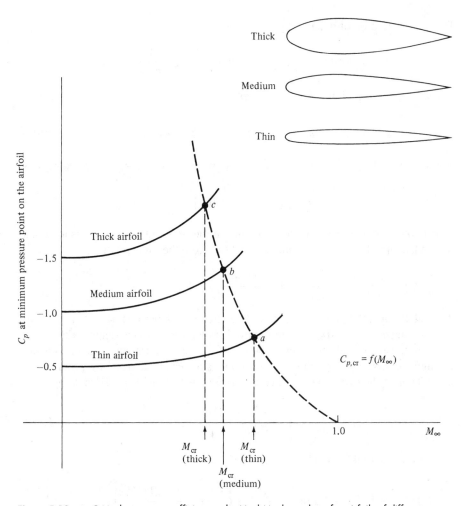

Figure 5.19 Critical pressure coefficient and critical Mach numbers for airfoils of different thicknesses.

The pressure coefficient curves in Fig. 5.19 are shown as solid curves. On these curves, only points a, b, and c are critical pressure coefficients, by definition. However, these critical points by themselves form a locus represented by the dotted curve in Fig. 5.19; that is, the critical pressure coefficients themselves are given by a curve of $C_{p,cr} = f(M_\infty)$, as labeled in Fig. 5.19. Let us proceed to derive this function. It is an important result, and it also represents an interesting application of our aerodynamic relationships developed in Chap. 4.

First, consider the definition of C_p from Eq. (5.27):

$$C_p = \frac{p - p_\infty}{q_\infty} = \frac{p_\infty}{q_\infty}\left(\frac{p}{p_\infty} - 1\right) \qquad \textbf{[5.41]}$$

From the definition of dynamic pressure,

$$q_\infty \equiv \tfrac{1}{2}\rho_\infty V_\infty^2 = \frac{1}{2}\frac{\rho_\infty}{\gamma p_\infty}(\gamma p_\infty)V_\infty^2 = \frac{1}{2}\frac{V_\infty^2}{\gamma p_\infty/\rho_\infty}(\gamma p_\infty)$$

However, from Eq. (4.53), $a_\infty^2 = \gamma p_\infty/\rho_\infty$. Thus

$$q_\infty = \frac{1}{2}\frac{V_\infty^2}{a_\infty^2}\gamma p_\infty = \frac{\gamma}{2}p_\infty M_\infty^2 \qquad\qquad \textbf{[5.42]}$$

We will return to Eq. (5.42) in a moment. Now, recall Eq. (4.74) for isentropic flow,

$$\frac{p_0}{p} = \left(1 + \frac{\gamma-1}{2}M^2\right)^{\gamma/(\gamma-1)}$$

This relates the total pressure p_0 at a point in the flow to the static pressure p and local Mach number M at the same point. Also, from the same relation,

$$\frac{p_0}{p_\infty} = \left(1 + \frac{\gamma-1}{2}M_\infty^2\right)^{\gamma/(\gamma-1)}$$

This relates the total pressure p_0 in the free-stream to the free-stream static pressure p_∞ and Mach number M_∞. For an isentropic flow, which is a close approximation to the actual, real-life, subsonic flow over an airfoil, the total pressure remains constant throughout. (We refer to more advanced books in aerodynamics for a proof of this fact.) Thus, if the two previous equations are divided, p_0 will cancel, yielding

$$\frac{p}{p_\infty} = \left[\frac{1 + \tfrac{1}{2}(\gamma-1)M_\infty^2}{1 + \tfrac{1}{2}(\gamma-1)M^2}\right]^{\gamma/(\gamma-1)} \qquad\qquad \textbf{[5.43]}$$

Substitute Eqs. (5.42) and (5.43) into (5.41):

$$C_p = \frac{p_\infty}{q_\infty}\left(\frac{p}{p_\infty}-1\right) = \frac{p_\infty}{\tfrac{1}{2}\gamma p_\infty M_\infty^2}\left\{\left[\frac{1 + \tfrac{1}{2}(\gamma-1)M_\infty^2}{1 + \tfrac{1}{2}(\gamma-1)M^2}\right]^{\gamma/(\gamma-1)}-1\right\}$$

$$\boxed{C_p = \frac{2}{\gamma M_\infty^2}\left\{\left[\frac{1 + \tfrac{1}{2}(\gamma-1)M_\infty^2}{1 + \tfrac{1}{2}(\gamma-1)M^2}\right]^{\gamma/(\gamma-1)}-1\right\}} \qquad\qquad \textbf{[5.44]}$$

For a given free-stream Mach number M_∞, Eq. (5.44) relates the local value of C_p to the local M at any given point in the flow field, hence at any given point on the airfoil surface. Let us pick that particular point on the surface where $M = 1$. Then, by definition, $C_p = C_{p,cr}$. Putting $M = 1$ into Eq. (5.44), we obtain

$$\boxed{C_{p,cr} = \frac{2}{\gamma M_\infty^2}\left\{\left[\frac{2 + (\gamma-1)M_\infty^2}{\gamma+1}\right]^{\gamma/(\gamma-1)}-1\right\}} \qquad\qquad \textbf{[5.45]}$$

Equation (5.45) gives the desired relation $C_{p,\text{cr}} = f(M_\infty)$. When numbers are fed into Eq. (5.45), the dotted curve in Fig. 5.19 results. Note that as M_∞ increases, $C_{p,\text{cr}}$ decreases.

Commentary   Pause for a moment, and let us review what all this means. In the author's experience, the concepts of critical Mach number and critical pressure coefficients are difficult for the first-time reader to fully understand. So let us elaborate. Equations (5.44) and (5.45) are strictly *aerodynamics*; they have nothing to do with the shape or angle of attack of a given airfoil. Indeed, Eq. (5.44) for a compressible flow plays a role analogous to that of Bernoulli's equation for an incompressible flow. For an incompressible flow, Bernoulli's equation, Eq. (4.9), written between the free-stream point where the pressure and velocity are p_∞ and V_∞, respectively, and another arbitrary point in the flow field where the pressure and velocity are p and V, respectively, is

$$p - p_\infty = \tfrac{1}{2}\rho(V_\infty^2 - V^2) \qquad \textbf{[5.46]}$$

For the given free-stream conditions of p_∞ and V_∞, at any other point in the incompressible flow where the local velocity is V, the pressure p at that point is obtained from Eq. (5.46). Now focus on Eq. (5.44). Here we are dealing with a *compressible flow*, where Mach number rather than velocity plays the controlling role. For the given free-stream M_∞, at any other point in the compressible flow where the local Mach number is M, the pressure coefficient at that point is obtained from Eq. (5.44). Hence the analogy with Bernoulli's equation. In turn, this reflects on Eq. (5.45). Consider a flow with a free-stream Mach number M_∞. Assume that at some local point in this flow the local Mach number is 1. Equation (5.45) gives the value of the pressure coefficient at this local point where we have Mach 1. Again, we define the value of the pressure coefficient at a point where $M = 1$ as the *critical* pressure coefficient $C_{p,\text{cr}}$. Hence, when M in Eq. (5.44) is set equal to 1, the corresponding value of the pressure coefficient at that same point where $M = 1$ is, *by definition*, the *critical* pressure coefficient. It is given by Eq. (5.45), obtained by setting $M = 1$ in Eq. (5.44). If we graph the function given in Eq. (5.45), that is, if we make a plot of $C_{p,\text{cr}}$ versus M_∞, we obtain the dashed curve in Fig. 5.19.

The fact that $C_{p,\text{cr}}$ decreases as M_∞ increases makes physical sense. For example, consider a free stream at $M_\infty = 0.5$. To expand this flow to Mach 1 requires a relatively large pressure change $p - p_\infty$, and therefore a relatively large (in magnitude) pressure coefficient because, by definition, $C_p = (p - p_\infty)/q_\infty$. However, consider a free stream at $M_\infty = 0.9$. To expand this flow to Mach 1 requires a much smaller pressure change; that is, $p - p_\infty$ is much smaller in magnitude. Hence, the pressure coefficient $C_p = (p - p_\infty)/q_\infty$ will be smaller in magnitude. As a result, $C_{p,\text{cr}}$ *decreases* with M_∞, as shown by the dashed curved in Fig. 5.19. Moreover, this dashed curve is a fixed "universal" curve—it is simply rooted in pure aerodynamics, independent of any given airfoil shape or angle of attack.

How to Estimate the Critical Mach Number for an Airfoil   Consider a given airfoil at a given angle of attack. How can we estimate the critical Mach number for this airfoil at the specified angle of attack? We will discuss two approaches to the solution: a graphical solution and an analytical solution.

The graphical solution involves several steps, as itemized below.

1. Obtain a plot of $C_{p,\text{cr}}$ versus M_∞ from Eq. (5.45). This is illustrated by curve A in Fig. 5.20. As discussed above, this curve is a fixed "universal" curve which you can use for all such problems.

2. For low speed, essentially incompressible flow, obtain the value of the *minimum* pressure coefficient on the surface of the airfoil. The minimum pressure coefficient corresponds to the point of maximum velocity on the airfoil surface. This minimum value of C_p must be given to you, from either experimental measurement or theory. This is $C_{p,0}$ shown as point B in Fig. 5.20.

3. Using Eq. (5.28), plot the variation of this minimum coefficient versus M_∞. This is illustrated by curve C in Fig. 5.20.

4. Where curve C intersects curve A, the minimum pressure coefficient on the surface of the airfoil is equal to the critical pressure coefficient. This intersection point is denoted by point D in Fig. 5.20. For the conditions associated with this point, the maximum velocity on the airfoil surface is exactly sonic. The value of M_∞ at point D is then, by definition, the critical Mach number.

The analytical solution for M_{cr} is obtained as follows. Equation (5.28), repeated here, gives the variation of C_p at a given point on the airfoil surface as a function of M_∞.

$$C_p = \frac{C_{p,0}}{\sqrt{1 - M_2^2}}$$

[5.28]

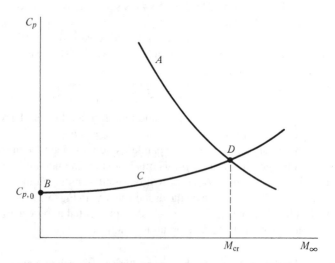

Figure 5.20   Determination of critical Mach number.

At some location on the airfoil surface, $C_{p,0}$ will be a minimum value, corresponding to the point of maximum velocity on the surface. The value of the minimum pressure coefficient will increase in absolute magnitude as M_∞ is increased, owing to the compressibility effect discussed in Sec. 5.6. Hence, Eq. (5.28) with $C_{p,0}$ being the *minimum* value on the surface of the airfoil at essentially incompressible flow conditions ($M_\infty < 0.3$) gives the value of the minimum pressure coefficient at a higher Mach number M_∞. However, at some value of M_∞, the flow velocity will become sonic at the point of minimum pressure coefficient. The value of the pressure coefficient at sonic conditions is the critical pressure coefficient, given by Eq. (5.45). When the flow becomes sonic at the point of minimum pressure, the pressure coefficient given by Eq. (5.28) is precisely the value given by Eq. (5.45). Equating these two relations, we have

$$\frac{C_{p,0}}{\sqrt{1 - M_\infty^2}} = \frac{2}{\gamma M_\infty^2} \left\{ \left[\frac{2 + (\gamma - 1)M_\infty^2}{\gamma + 1} \right]^{\gamma/(\gamma-1)} - 1 \right\} \qquad \textbf{[5.47]}$$

The value of M_∞ that satisfies Eq. (5.47) is that value when the flow becomes sonic at the point of maximum velocity (minimum pressure). That is, the value of M_∞ obtained from Eq. (5.47) *is* the critical Mach number for the airfoil. To emphasize this, we write Eq. (5.47) with M_∞ replaced by M_{cr}.

$$\boxed{\frac{C_{p,0}}{\sqrt{1 - M_{cr}^2}} = \frac{2}{\gamma M_{cr}^2} \left\{ \left[\frac{2 + (\gamma - 1)M_{cr}^2}{\gamma + 1} \right]^{\gamma/(\gamma-1)} - 1 \right\}} \qquad \textbf{[5.48]}$$

Equation (5.48) allows a direct analytical estimate for the critical Mach number of a given airfoil at a given angle of attack. Note that Eq. (5.48) must be solved implicitly for M_{cr}; for example, by trial and error, guessing at a value of M_{cr}, seeing if it satisfies Eq. (5.48), and then trying again.

Please note that Eq. (5.48) is simply an analytical representation of point D in Fig. 5.20, where curves A and C intersect.

Consider the NACA 0012 airfoil, the shape of which is shown at the top of Fig. 5.21. The pressure coefficient distribution over the surface of the airfoil at zero angle of attack is shown at the bottom of Fig. 5.21. These are low-speed values measured in a wind tunnel at Re = 3.65×10^6. From this information, estimate the critical Mach number of the NACA 0012 airfoil at zero angle of attack.

Example 5.8

Solution

First we will carry out a graphical solution, and then we will check the answer by carrying out an analytical solution.

a. Graphical solution

Let us accurately plot the curve of $C_{p,cr}$ versus M_∞, represented by curve A in Fig. 5.20. From Eq. (5.45), repeated here,

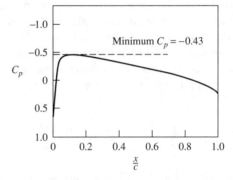

Figure 5.21   Low speed pressure coefficient distribution over the surface of an NACA 0012 airfoil at zero angle of attack. Re = 3.65 × 10⁶. After R. J. Freuler and G. M. Gregorek, "An Evaluation of Four Single Element Airfoil Analytical Methods", in *Advanced Technology Airfoil Research*, NASA CP 2045, 1978, pp. 133–162.

$$C_{p,\text{cr}} = \frac{2}{\gamma M_\infty^2} \left\{ \left[\frac{2 + (\gamma - 1)M_\infty^2}{\gamma + 1} \right]^{\gamma/(\gamma-1)} - 1 \right\}$$

for $\gamma = 1.4$, we can tabulate

M_∞	0.4	0.5	0.6	0.7	0.8	0.9	1.0
$C_{p,\text{cr}}$	−3.66	−2.13	−1.29	−0.779	−0.435	−0.188	0

The curve generated by these numbers is given in Fig. 5.22, labeled curve *A*.

Next, let us measure the minimum C_p on the surface of the airfoil from Fig. 5.21; this value is $(C_p)_{\text{min}} = -0.43$. The experimental values for pressure coefficient shown in Fig. 5.21 are for low-speed, essentially incompressible flow. Hence, in Eq. (5.28), $(C_{p,0})_{\text{min}} = -0.43$. As the Mach number is increased, the location of the point of minimum pressure stays essentially the same, but the value of the minimum pressure coefficient varies according to Eq. (5.28). Hence,

$$(C_p)_{\text{min}} = \frac{(C_{p,0})_{\text{min}}}{\sqrt{1 - M_\infty^2}} = \frac{-0.43}{\sqrt{1 - M_\infty^2}}$$

Some values of $(C_p)_{\text{min}}$ are tabulated below.

M_∞	0	0.2	0.4	0.6	0.8
$(C_p)_{\text{min}}$	−0.43	−0.439	−0.469	−0.538	−0.717

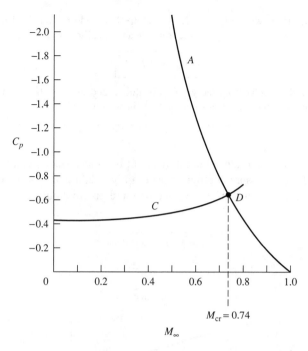

$M_{cr} = 0.74$

M_∞

Figure 5.22   Graphical solution for the critical Mach number, from Example 5.8.

The curve generated by these numbers is given in Fig. 5.22, labeled curve C. The intersection of curves A and C is at point D. The free-stream Mach number associated with point D is the critical Mach number. From Fig. 5.22, we have

$$\boxed{M_{cr} = 0.74}$$

b. Analytical solution

Solve Eq. (5.48) for M_{cr}, with $C_{p,0} = -0.43$. We can do this by trial and error. Assume different values for M_{cr}, and find by iteration the value that satisfies Eq. (5.48).

M_{cr}	$\dfrac{-0.43}{\sqrt{1 - M_{cr}^2}}$	$\dfrac{2}{\gamma M_{cr}^2}\left\{\left[\dfrac{2 + (\gamma - 1)M_{cr}^2}{\gamma + 1}\right]^{\gamma/(\gamma-1)} - 1\right\}$
0.72	−0.6196	−0.6996
0.73	−0.6292	−0.6621
0.74	−0.6393	−0.6260
0.738	−0.6372	−0.6331
0.737	−0.6362	−0.6367
0.7371	−0.6363	−0.6363

To four-place accuracy, when $M_{cr} = 0.7371$, both the left- and right-hand sides of Eq. (5.48) agree, also to four-place accuracy. Hence, from the analytical solution, we have

$$\boxed{M_{cr} = 0.7371}$$

Note: Compare the results from the graphical solution and the analytical solution. To the two-place accuracy of the graphical solution, both answers agree.

Question: How accurate is the estimate of the critical Mach number obtained in Example 5.8? The pressure coefficient data in Fig. 5.23*a* and *b* provide an answer. Wind tunnel measurements of the surface pressure distributions on an NACA 0012

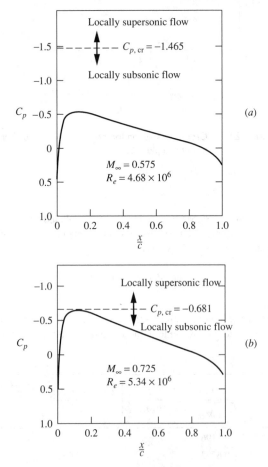

Figure 5.23   Wind tunnel measurements of surface pressure coefficient distribution for the NACA 0012 airfoil at zero angle of attack. Experimental data of Frueler and Gregorek, NASA CP 2045 (a) $M_\infty = 0.575$, (b) $M_\infty = 0.725$.

airfoil at zero angle of attack in a high-speed flow are shown in Fig. 5.23; for Fig. 5.23a, $M_\infty = 0.575$, and for Fig. 5.23b, $M_\infty = 0.725$. In Fig. 5.23a, the value of $C_{p,cr} = -1.465$ at $M_\infty = 0.575$ is shown as the dashed horizontal line. From the definition of critical pressure coefficient, any local value of C_p above this horizontal line corresponds to locally supersonic flow, and any local value below the horizontal line corresponds to locally subsonic flow. Clearly from the measured surface pressure coefficient distribution at $M_\infty = 0.575$ shown in Fig. 5.23a, the flow is locally subsonic at every point on the surface. Hence, $M_\infty = 0.575$ is *below* the critical Mach number. In Fig. 5.23b, which is for a higher Mach number, the value of $C_{p,cr} = -0.681$ at $M_\infty = 0.725$ is shown as the dashed horizontal line. Here, the local pressure coefficient is higher than $C_{p,cr}$ at every point on the surface *except* at the point of minimum pressure, where $(C_p)_{min}$ is essentially equal to $C_{p,cr}$. This means that for $M_\infty = 0.725$, the flow is locally subsonic at every point on the surface *except* the point of minimum pressure, where the flow is essentially sonic. Hence, these experimental measurements indicate that the critical Mach number of the NACA 0012 airfoil at zero angle of attack is approximately 0.73. Comparing this experimental result with the calculated value of $M_{cr} = 0.74$ from Example 5.8, we see that our calculations are amazingly accurate, to within about 1 percent.

Location of Point of Maximum Velocity (Minimum Pressure) One final observation in this section can be made from studying the pressure coefficient distributions, shown in Figs. 5.21 and 5.23, and the shape of the NACA 0012 airfoil, shown at the top of Fig. 5.21. Note that the minimum pressure (hence maximum velocity) does *not* occur at the location of maximum thickness of the airfoil. From the airfoil shape given in Fig. 5.21, the maximum thickness is at $x/c = 0.3$. From the surface pressure coefficient distributions shown in Figs. 5.21 and 5.23, the point of minimum pressure (maximum velocity) on the surface is at $x/c = 0.11$, considerably ahead of the point of maximum thickness. Your intuition might at first suggest that the point of maximum velocity (minimum pressure) might be at the point of maximum thickness, but this intuition is wrong. Nature places the maximum velocity at a point which satisfies the physics of the *whole flow field*, not just what is happening in a local region of flow. The point of maximum velocity is dictated by the *complete* shape of the airfoil, not just by the shape in a local region.

5.10 DRAG-DIVERGENCE MACH NUMBER

We now turn our attention to the airfoil drag coefficient c_d. Figure 5.24 sketches the variation of c_d with M_∞. At low Mach numbers, less than M_{cr}, c_d is virtually constant and is equal to its low-speed value given in App. D. The flow field about the airfoil for this condition (say, point a in Fig. 5.24) is noted in Fig. 5.25a, where $M < 1$ everywhere in the flow. If M_∞ is increased slightly above M_{cr}, a "bubble" of supersonic flow will occur surrounding the minimum pressure point, as shown in Fig. 5.25b. Correspondingly, c_d will still remain reasonably low, as indicated by point b in Fig. 5.24. However, if M_∞ is still further increased, a very sudden and dramatic rise in the drag coefficient will be observed, as noted by point c in Fig. 5.24. Here,

Figure 5.24   Variation of drag coefficient with Mach number.

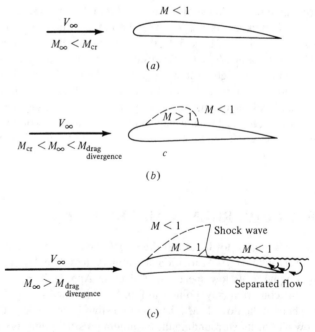

Figure 5.25   Physical mechanism of drag divergence.
(a) Flow field associated with point a in Fig. 5.19.
(b) Flow field associated with point b in Fig. 5.19.
(c) Flow field associated with point c in Fig. 5.19.

shock waves suddenly appear in the flow, as sketched in Fig. 5.25c. The effect of the shock wave on the surface pressure distribution can be seen in the experimental data given in Fig. 5.26. Here, the surface pressure coefficient is given for an NACA 0012 airfoil at zero angle of attack in a free stream with $M_\infty = 0.808$. (Figure 5.26 is a companion figure to Figs. 5.21 and 5.23.) Comparing the result of Example 5.8 and the data shown in Fig. 5.23b, we know that $M_\infty = 0.808$ is *above* the critical Mach number for the NACA 0012 airfoil at zero angle of attack. The pressure distribution in Fig. 5.26 clearly shows that fact; the shape of the pressure distribution curve is quite different from that in the previous figures. The dashed horizontal line in Fig. 5.26 corresponds to the value of $C_{p,\mathrm{cr}}$ at $M_\infty = 0.808$. Note that the flow velocity at the surface is locally supersonic in the region $0.11 < x/c < 0.45$. Recall from our discussion of shock waves in Sec. 4.11c that the pressure increases and the velocity decreases across a shock wave. We clearly see these phenomena in Fig. 5.26; the large and rather sudden increase in pressure at $x/c = 0.45$ indicates the presence of a shock wave at that location, and the flow velocity drops from supersonic in front of the shock to subsonic behind the shock. (The drop in velocity to *subsonic* behind the shock, rather than just a decrease to a smaller supersonic value, is a characteristic of shock waves that are essentially normal to the flow, as occurs here.)

The shock waves themselves are dissipative phenomena, which result in an increase in drag on the airfoil. But in addition, the sharp pressure increase across the shock waves creates a strong adverse pressure gradient, hence causing the flow to separate from the surface. As discussed in Sec. 4.20, such flow separation can create substantial increases in drag. Thus, the sharp increase in c_d shown in Fig. 5.24 is a combined effect of shock waves and flow separation. The *free-stream* Mach number

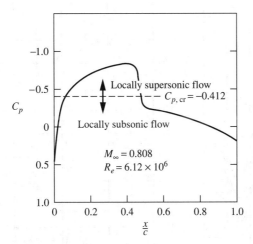

Figure 5.26 Wind tunnel measurements of the surface pressure coefficient distribution for the NACA 0012 airfoil at zero angle of attack for $M_\infty = 0.808$, which is above the critical Mach number. Experimental data are from Freuler and Gregorek, NASA CP 2045, and are a companion to the data shown in Figs. 5.21 and 5.23.

at which c_d begins to increase rapidly is defined as the *drag-divergence Mach number* and is noted in Fig. 5.24. Note that

$$M_{cr} < M_{\text{drag divergence}} < 1.0$$

The shock pattern sketched in Fig. 5.25c is characteristic of a flight regime called *transonic*. When $0.8 \leq M_\infty \leq 1.2$, the flow is generally designated as transonic flow, and it is characterized by some very complex effects only hinted at in Fig. 5.25c. To reinforce these comments, Fig. 5.27 shows the variation of both c_l and c_d as a function of Mach number with angle of attack as a parameter. The airfoil is a standard NACA 2315 airfoil. Figure 5.27, which shows actual wind tunnel data, illustrates the massive transonic-flow effects on both lift and drag coefficients. The analysis of transonic flows has been one of the major challenges in modern aerodynamics. Only in recent years, since about 1970, have computer solutions for transonic flows over airfoils

Figure 5.27   Variation of (a) lift coefficient and (b) drag coefficient versus Mach number with angle of attack as a parameter for an NACA 2315 airfoil. *(Wind tunnel measurements at the NACA Langley Memorial Laboratory.)*

come into practical use; these numerical solutions are still in a state of development and improvement. Transonic flow has been a hard nut to crack.

DESIGN BOX

The designers of transonic airplanes are frequently looking for ways to get the speed closer to Mach 1 without encountering the large transonic drag rise. These designers have two options in regard to the choice of an airfoil that will delay drag divergence to a higher Mach number: (1) Make the airfoil *thin*, and (2) adopt a specially shaped airfoil called a *supercritical airfoil*. These options can be used singly or in combination.

In regard to airfoil *thickness*, the generic trend sketched in Fig. 5.19 clearly shows that M_{cr} is increased by making the airfoil thinner. An increase in M_{cr} usually means an increase in the drag-divergence Mach number. Hence, everything else being equal, a transonic airplane with a thinner airfoil can fly at a higher Mach number before encountering drag divergence. Indeed, this knowledge was incorporated in the design of the famous Bell X-1, which was the first airplane to fly faster than sound (see Sec. 5.22). The X-1 was designed with two sets of wings, one with a 10 percent thick airfoil for more routine flights and another with an 8 percent thick airfoil for flights intended to penetrate through Mach 1. The airfoil sections were NACA 65-110 and NACA 65-108, respectively. Moreover, the horizontal tail was even thinner in both cases, being an NACA 65-008 (8 percent thickness) and an NACA 65-006 (6 percent thickness), respectively. This was done to ensure that when the wing encountered major compressibility effects, the horizontal tail and elevator would still be free of such problems and would be functional for stability and control. A three-view of the Bell X-1 is shown in Fig. 5.28.

The adverse compressibility effects which cause the dramatic increase in drag and precipitous decrease in lift, shown in Fig. 5.27, can be delayed by decreasing the airfoil thickness. The knowledge of this fact dates back as early as 1918. In that year, as World War I was coming to an end, Frank Caldwell and Elisha Fales, two engineers at the U.S. Army's McCook Field in Dayton, Ohio measured these effects in a high-speed wind tunnel capable of producing a test stream of 465 mi/h. This knowledge was reinforced by subsequent high-speed wind tunnel tests carried out by NACA in the 1920s and 1930s. (For a detailed historical treatment of the evolution of our understanding of compressibility effects during this period, see Anderson, *A History of Aerodynamics and Its Impact on Flying Machines*, Cambridge University Press, 1997. See also Anderson, "Research in Supersonic Flight and the Breaking of the Sound Barrier," chapter 3 in *From Engineering Science to Big Science*, edited by Pamela Mack, NASA SP-4219, 1998.)

Thinner airfoils are also advantageous for supersonic airplanes, for reasons to be discussed in Sec. 5.11. Indeed, in airplane design, the higher the design Mach number, usually the thinner the airfoil section. This is dramatically shown in Fig. 5.29, which is a plot of airfoil thickness versus design Mach number for a variety of high-speed airplanes since World War II. As the design Mach number of airplanes increased, thinner airfoils became a design necessity.

The *supercritical airfoil* is a different approach to the increase in drag-divergence Mach number. Here, the *shape* of the airfoil is designed with a relatively flat top surface, as shown in Fig. 5.30. When the freestream Mach number exceeds M_{cr}, a pocket of supersonic flow occurs over the top surface as usual; but because the top is relatively flat, the local supersonic Mach number is a lower value than would exist in the case of a conventional airfoil. As a result, the shock wave which terminates the pocket of supersonic flow is weaker. In turn, the supercritical airfoil can penetrate closer to Mach 1 before drag divergence occurs. In essence, the increment in Mach number (the "grace period") between M_{cr} and $M_{drag\,divergence}$ (see Fig. 5.24) is increased by the shape of the supercritical airfoil. One way to think about this is that the supercritical airfoil is "more comfortable" than conventional airfoils in the region above M_{cr}, and it can fly closer to Mach 1 before drag divergence is encountered. Because they are more comfortable in the flight regime above the critical Mach number, and because they can penetrate closer to Mach 1 after exceeding M_{cr}, these

(continued)

Figure 5.28   Three-view of the Bell X-1.

(*continued*)

Figure 5.29 Variation of thickness-to-chord ratio with Mach number for a representative sampling of different airplanes. After Ray Whitford, Design for Air Combat, Jane's Information Group, Surry, England, 1989.

airfoils are called *supercritical* airfoils. They are *designed* to cruise in the Mach number range *above* M_{cr}.

The pressure coefficient distribution over the top surface of a supercritical airfoil flying above M_{cr} but below $M_{drag\ divergence}$ is sketched in Fig. 5.30. After a sharp decrease in pressure around the leading edge, the pressure remains relatively constant over a substantial portion of the top surface. This is in contrast to the pressure coefficient distribution for a conventional airfoil flying above M_{cr}, such as shown in Fig. 5.26. Clearly, the flow over the supercritical airfoil is carefully tailored to achieve the desired results.

The early aerodynamic research on supercritical airfoils was carried out by Richard Whitcomb, an aeronautical engineer at NASA Langley Research Center, during the middle 1960s. This work by Whitcomb is described in a NASA document entitled "An Airfoil Shape for Efficient Flight at Supercritical Mach Numbers" (NASA TM X-1109, July 1965, by R. T. Whitcomb and L. R. Clark). Whitcomb's design of supercritical airfoils was pioneering; today, all modern civilian jet transports are designed with supercritical wings, incorporating custom-designed supercritical airfoil sections that have their genes in the original design by Richard Whitcomb.

The effectiveness of the supercritical airfoil was clearly established by an Air Force/NASA wind tunnel and flight test program carried out in the early 1970s called the Transonic Aircraft Technology (TACT)

(*continued*)

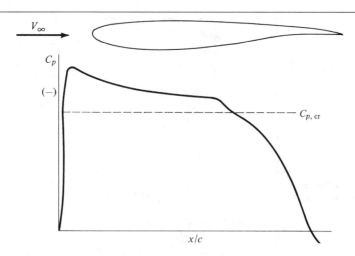

Figure 5.30   Shape of a typical supercritical airfoil and its pressure coefficient distribution over the top surface.

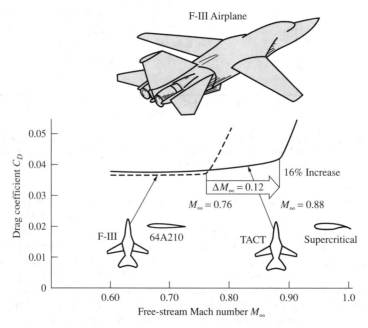

Figure 5.31   Increase in drag-divergence Mach number obtained by the TACT aircraft with a supercritical wing compared to a standard F-111. Wind tunnel data obtained at the NASA Langley Research Center. Wing sweep $=26°$. C_L held constant at 0.0465. Reported in *Symposium on Transonic Aircraft Technology* (*TACT*), AFFDL-TR-78-100, Air Force Flight Dynamics Laboratory, August 1978.

(continued)

program. A standard General Dynamics F-111 (sketched at the top of Fig. 5.31) was modified with a supercritical wing. Wind tunnel data for the variation of C_D with M_∞ for both the standard F-111 and the TACT aircraft (the F-111 modified with a supercritical wing) are shown in Fig. 5.31. The standard airfoil on the F-111 is an NACA 64-210; the supercritical airfoil on the TACT aircraft had the same 10 percent thickness. The use of the supercritical wing increased the drag-divergence Mach number from 0.76 to 0.88, a stunning 16 percent increase, as noted in Fig. 5.31.

Designers of transonic aircraft can use supercritical airfoils to accomplish one of two objectives: (1)

For a given airfoil thickness, the supercritical airfoil shape allows a higher cruise velocity; or (2) for a given lower cruise velocity, the airfoil thickness can be larger. The latter option has some design advantages. The structural design of a thicker wing is more straightforward, and actually results in a more lightweight (albeit thicker) wing. Also, a thicker wing provides more volume for an increased fuel capacity. Clearly, the use of a supercritical airfoil provides a larger "design space" for transonic airplanes.

5.11 WAVE DRAG (AT SUPERSONIC SPEEDS)

To this point, we have discussed airfoil properties at subsonic speeds, that is, for $M_\infty < 1$. When M_∞ is supersonic, a major new physical phenomenon is introduced: shock waves. We previously alluded to shock waves in Sec. 4.11.3 in conjunction with the Pitot tube measurement of supersonic airspeeds. With respect to airfoils (as well as all other aerodynamic bodies) shock waves in supersonic flow create a new source of drag, called *wave drag*. In this section, we will simply highlight some of the ideas involving shock waves and the consequent wave drag; a detailed study of shock wave phenomena is left to more advanced texts in aerodynamics.

To obtain a feel for how a shock is produced, imagine that we have a small source of sound waves, a tiny "beeper" (something like a tuning fork). At time $t = 0$, assume the beeper is at point P in Fig. 5.32. At this point, let the beeper emit a sound wave, which will propagate in all directions at the speed of sound a. Also let the beeper move with velocity V, where V is less than the speed of sound. At time t, the sound wave will have moved outward by a distance at, as shown in Fig. 5.32. At the same time t, the beeper will have moved a distance Vt, to point Q. Since $V < a$, the beeper will always stay inside the sound wave. If the beeper is constantly emitting sound waves as it moves along, these waves will constantly move outward, ahead of the beeper. As long as $V < a$, the beeper will always be inside the envelope formed by the sound waves.

On the other hand, assume the beeper is moving at supersonic speed, that is, $V > a$. At time $t = 0$, assume the beeper is at point R in Fig. 5.33. At this point, let the beeper emit a sound wave which, as before, will propagate in all directions at the speed of sound a. At time t the sound wave will have moved outward by a distance at, as shown in Fig. 5.33. At the same time t, the beeper will have moved a distance Vt, to point S. However, since $V > a$, the beeper will now be outside the sound wave. If the beeper is constantly emitting sound waves as it moves along, these waves will now pile up inside an envelope formed by a line from point S tangent to

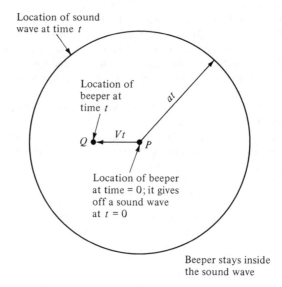

Location of sound
wave at time t

Location of
beeper at
time t

at

Q ← Vt → P

Location of beeper
at time = 0; it gives
off a sound wave
at $t = 0$

Beeper stays inside
the sound wave

Figure 5.32   Beeper moving at less than the speed of sound.

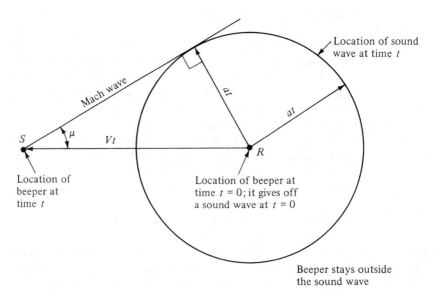

Mach wave

Location of sound
wave at time t

at

at

S μ Vt R

Location of
beeper at
time t

Location of beeper at
time $t = 0$; it gives off
a sound wave at $t = 0$

Beeper stays outside
the sound wave

Figure 5.33   The origin of Mach waves and shock waves. Beeper is moving faster than
the speed of sound.

the circle formed by the first sound wave, centered at point R. This tangent line, the
line where the pressure disturbances are piling up, is called a *Mach wave*. The vertex
of the wave is fixed to the moving beeper at point S. In supersonic flight, the air
ahead of the beeper in Fig. 5.33 has no warning of the approach of the beeper. Only
the air behind the Mach wave has felt the presence of the beeper, and this presence is

communicated by pressure (sound) waves confined inside the conical region bounded by the Mach wave. In contrast, in subsonic flight, the air ahead of the beeper in Fig. 5.32 is forewarned about the oncoming beeper by the sound waves. In this case, there is no piling up of pressure waves; there is no Mach wave.

Hence, we can begin to feel that the coalescing, or piling up, of pressure waves in supersonic flight can create sharply defined waves of some sort. In Fig. 5.33, the Mach wave that is formed makes an angle μ with the direction of movement of the beeper. This angle, defined as the *Mach angle*, is easily obtained from the geometry of Fig. 5.33, as follows:

$$\sin \mu = \frac{at}{Vt} = \frac{a}{V} = \frac{1}{M}$$

Hence

$$\boxed{\text{Mach angle} \equiv \mu \equiv \arcsin \frac{1}{M}} \qquad \textbf{[5.49]}$$

In real life, a very thin object (such as a thin needle) moving at $M_\infty > 1$ creates a very weak disturbance in the flow, limited to a Mach wave. This is sketched in Fig. 5.34. On the other hand, an object with some reasonable thickness, such as the wedge shown in Fig. 5.35, moving at supersonic speeds will create a strong disturbance, called a *shock wave*. The shock wave will be inclined at an oblique angle β, where $\beta > \mu$, as shown in Fig. 5.35. As the flow moves across the oblique shock wave, the pressure, temperature, and density increase, and the velocity and Mach number decrease.

Consider now the pressure on the surface of the wedge, as sketched in Fig. 5.36. Since p increases across the oblique shock wave, at the wedge surface $p > p_\infty$. Since the pressure acts normal to the surface and the surface itself is inclined to the relative wind, there will be a net drag produced on the wedge, as seen by simple inspection of Fig. 5.36. This drag is called *wave drag*, because it is inherently due to the pressure increase across the shock wave.

To minimize the strength of the shock wave, all supersonic airfoil profiles are thin, with relatively sharp leading edges. (The leading edge of the Lockheed F-104 supersonic fighter is almost razor-thin.) Let us approximate a thin supersonic airfoil by the flat plate illustrated in Fig. 5.37. The flat plate is inclined at a small angle of attack α to the supersonic free stream. On the top surface of the plate, the flow field is

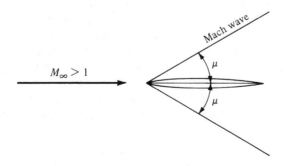

Figure 5.34   Mach waves on a needlelike body.

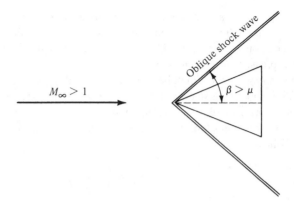

Figure 5.35   Oblique shock waves on a wedge-type body.

Figure 5.36   Pressure distribution on a wedge at supersonic speeds; origin of wave drag.

turned away from the freestream through an *expansion wave* at the leading edge; an expansion wave is a fan-shaped region through which the pressure decreases. At the trailing edge on the top side, the flow is turned back towards the free-stream direction through an oblique shock wave. On the bottom surface of the plate, the flow is turned into the free stream, causing an oblique shock wave with an increase in pressure. At the trailing edge, the flow is turned back towards the free-stream direction through an expansion wave. (Details and theory for expansion waves, as well as shock waves, are beyond the scope of this book—you will have to simply accept on faith the flow field sketched in Fig. 5.37 until your study of aerodynamics becomes more advanced.) The expansion and shock waves at the leading edge result in a surface pressure distribution in which the pressure on the top surface is less than p_∞, whereas the pressure on the bottom surface is greater than p_∞. The net effect is an aerodynamic force normal to the plate. The components of this force perpendicular and parallel to the relative wind are the lift and supersonic wave drag, respectively. Approximate relations for the lift and drag coefficients are, respectively,

$$c_l = \frac{4\alpha}{(M_\infty^2 - 1)^{1/2}}$$

[5.50]

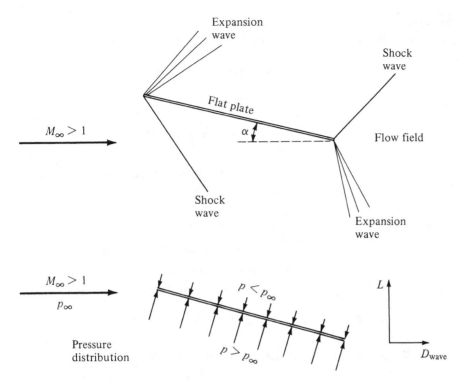

Figure 5.37 Flow field and pressure distribution for a flat plate at angle of attack in supersonic flow. There is a net lift and drag due to the pressure distribution set up by the shock and expansion waves.

and
$$c_{d,w} = \frac{4\alpha^2}{(M_\infty^2 - 1)^{1/2}}$$
[5.51]

A subscript w has been added to the drag coefficient to emphasize that it is the wave drag coefficient. Equations (5.50) and (5.51) are approximate expressions, useful for thin airfoils at small to moderate angles of attack in supersonic flow. Note that as M_∞ increases, both c_l and c_d decrease. This is not to say that the lift and drag forces themselves decrease with M_∞. Quite the contrary, for any flight regime, as the flight velocity increases, L and D usually increase because the dynamic pressure $q_\infty = \frac{1}{2}\rho V_\infty^2$ increases. In the supersonic regime, L and D increase with velocity, even though c_l and $c_{d,w}$ decrease with M_∞ according to Eqs. (5.50) and (5.51).

Example 5.9

Consider a thin supersonic airfoil with chord length $c = 5$ ft in a Mach 3 free stream at a standard altitude of 20,000 ft. The airfoil is at an angle of attack of $5°$.
a. Calculate the lift and wave drag coefficients and the lift and wave drag per unit span.
b. Compare these results with the same airfoil at the same conditions, except at Mach 2.

Solution

a. In Eqs. (5.50) and (5.51) the angle of attack α must be in radians. Hence

$$\alpha = 5° = \frac{5}{57.3} \text{ rad} = 0.0873 \text{ rad}$$

Also
$$\sqrt{M_\infty^2 - 1} = \sqrt{3^2 - 1} = 2.828$$

$$c_l = \frac{4\alpha}{\sqrt{M_\infty^2 - 1}} = \frac{4(0.0873)}{2.828} = \boxed{0.123}$$

$$c_{d,w} = \frac{4\alpha^2}{\sqrt{M_\infty^2 - 1}} = \frac{4(0.0873)^2}{2.828} = \boxed{0.0108}$$

At 20,000 ft, $\rho_\infty = 1.2673 \times 10^{-3}$ slug/ft^3 and $T = 447.43°$R. Hence

$$a_\infty = \sqrt{\gamma R T_\infty} = \sqrt{1.4(1716)(447.43)} = 1037 \text{ ft/s}$$

$$V_\infty = M_\infty a_\infty = 3(1037) = 3111 \text{ ft/s}$$

$$q_\infty = \tfrac{1}{2}\rho_\infty V_\infty^2 = \tfrac{1}{2}(1.2673 \times 10^{-3})(3111)^2 = 6133 \text{ lb/ft}^2$$

$$L \text{ (per unit span)} = q_\infty c c_l = 6133(5)(0.123) = \boxed{3772 \text{ lb}}$$

$$D_w \text{ (per unit span)} = q_\infty c c_{d,w} = 6133(5)(0.0108) = \boxed{331.2 \text{ lb}}$$

b.

$$\sqrt{M_\infty^2 - 1} = \sqrt{2^2 - 1} = 1.732$$

$$c_l = \frac{4\alpha}{\sqrt{M_\infty^2 - 1}} = \frac{4(0.0873)}{1.732} = \boxed{0.207}$$

$$c_{d,w} = \frac{4\alpha^2}{\sqrt{M_\infty^2 - 1}} = \frac{4(0.0873)^2}{1.732} = \boxed{0.0176}$$

Note: At Mach 2, c_l and $c_{d,w}$ are *higher* than at Mach 3. This is a general result; both c_l and $c_{d,w}$ decrease with increasing Mach number, as clearly seen from Eqs. (5.50) and (5.51). Does this mean that L and D_w also decrease with increasing Mach number? Intuitively this does not seem correct. Let us find out.

$$V_\infty = a_\infty M_\infty = 1037(2) = 2074 \text{ ft/s}$$

$$q_\infty = \tfrac{1}{2}\rho_\infty V_\infty^2 = \tfrac{1}{2}(1.2673 \times 10^{-3})(2074)^2 = 2726 \text{ lb/ft}^2$$

$$L \text{ (per unit span)} = q_\infty c c_l = 2726(5)(0.207) = \boxed{2821 \text{ lb}}$$

$$D_w \text{ (per unit span)} = q_\infty c c_{d,w} = 2726(5)(0.0176) = \boxed{240 \text{ lb}}$$

Hence, there is no conflict with our intuition. As the supersonic Mach numbers increase, L and D_w also increase although the lift and drag *coefficients* decrease.

5.12 SUMMARY OF AIRFOIL DRAG

Amplifying Eq. (4.105), we can write the total drag of an airfoil as the sum of three contributions:

$$D = D_f + D_p + D_w$$

where D = total drag on airfoil
D_f = skin friction drag
D_p = pressure drag due to flow separation
D_w = wave drag (present only at transonic and supersonic speeds; zero for subsonic speeds below the drag-divergence Mach number)

In terms of the drag coefficients, we can write:

$$c_d = c_{d,f} + c_{d,p} + c_{d,w}$$

where $c_d, c_{d,f}, c_{d,p}$, and $c_{d,w}$ are the total drag, skin friction drag, pressure drag, and wave drag coefficients, respectively. The sum $c_{d,f} + c_{d,p}$ is called the *profile drag coefficient*; this is the quantity that is given by the data in App. D. The profile drag coefficient is relatively constant with M_∞ at subsonic speeds.

The variation of c_d with M_∞ from incompressible to supersonic speeds is sketched in Fig. 5.38. It is important to note the qualitative variation of this curve. For M_∞ ranging from zero to drag divergence, c_d is relatively constant; it consists entirely of profile drag. For M_∞ from drag divergence to slightly above 1, the value of c_d skyrockets; indeed, the peak value of c_d around $M_\infty = 1$ can be an order of magnitude larger than the profile drag itself. This large increase in c_d is due to wave drag associated with the presence of shock waves. For supersonic Mach numbers, c_d decreases approximately as $(M_\infty^2 - 1)^{-1/2}$.

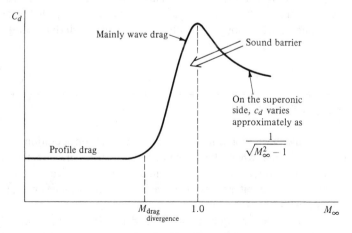

Figure 5.38   Variation of drag coefficient with Mach number for subsonic and supersonic speeds.

Good design of supersonic airplanes concentrates on minimizing wave drag. It is emphasized in Fig. 5.38 that a substantial portion of the total drag at supersonic speeds is wave drag. In turn, the way to reduce wave drag is to reduce the strength of the shock waves that occur at the nose, along the leading edges of the wing and tail, and at any other part of the aircraft that protrudes into the locally supersonic flow. The shock wave strength is reduced by having a sharp nose, slender (almost needlelike) fuselage, and very sharp wing and tail leading edges. The Lockheed F-104, shown in three-view in Fig. 4.44, is an excellent example of good supersonic airplane design. The F-104 was the first aircraft designed for

sustained speeds at Mach 2. Examining Fig. 4.44, we see an aircraft with a sharp, needlelike nose, slender fuselage, and very thin wings and tails with sharp leading edges. The wing airfoil section is a thin biconvex shape with a thickness-to-chord ratio of 0.035—3.5 percent thickness. The leading edge is almost razor-sharp, indeed sharp enough to pose a hazard to ground crew working around the airplane. Design of the F-104 began in 1953 at the famous Lockheed "Skunk Works"; it entered service with the U.S. Air Force in 1958. Now retired from the Air Force inventory, at the time of writing, F-104's are still in service with the air forces of a few other nations around the globe.

The large increase in the drag coefficient near Mach 1 gave rise to the term *sound barrier* in the 1940s. There was a camp of professionals who at that time felt that the sound barrier could not be pierced, that we could not fly faster than the speed of sound. Certainly, a glance at Eq. (5.28) for the pressure coefficient in subsonic flow, as well as Eq. (5.51) for wave drag in supersonic flow, would hint that the drag coefficient might become infinitely large as M_∞ approaches 1 from either the subsonic or supersonic side. However, such reasoning is an example of a common pitfall in science and engineering, namely, the application of equations outside their ranges of validity. Neither Eq. (5.28) nor Eq. (5.51) is valid in the transonic range near $M_\infty = 1$. Moreover, remember that nature abhors infinities. In real life, c_d does not become infinitely large. To get past the sound barrier, all that is needed (in principle) is an engine with enough thrust to overcome the high (but finite) drag.

5.13 FINITE WINGS

We now return to the discussion initiated in Sec. 5.5. Our considerations so far have dealt mainly with airfoils, where the aerodynamic properties are directly applicable to infinite wings. However, all real wings are finite, and for practical reasons we must translate our knowledge about airfoils to the case where the wing has wing tips. This is the purpose of the next two sections.

Let us pose the following questions. Consider a finite wing with a specified aspect ratio [defined by Eq. (5.26)] at an angle of attack of 6°. The airfoil section of the finite wing is an NACA 2412 section. For $\alpha = 6°$, the airfoil lift and drag coefficients, from App. D, are

$$c_l = 0.85 \qquad c_d = 0.077$$

Question: Since the finite wing is made up of the NACA 2412 airfoil section, should not the wing lift and drag coefficients be the same as those for the airfoil? That is, for the wing at $\alpha = 6°$, are the following true?

$$C_L \overset{?}{=} 0.85 \qquad C_D \overset{?}{=} 0.0077$$

(Recall from Sec. 5.5 that it is conventional to denote the aerodynamic coefficients for a finite wing with capital letters, as above.) On an intuitive basis, it may sound reasonable that C_L and C_D for the wing might be the same as c_l and c_d, respectively, for the airfoil section that makes up the wing. But intuition is not always correct. We will answer the above questions in the next few paragraphs.

The fundamental difference between flows over finite wings as opposed to infinite wings can be seen as follows. Consider the front view of a finite wing as sketched in Fig. 5.39a. If the wing has lift, then obviously the average pressure over the bottom surface is greater than that over the top surface. Consequently, there is some tendency for the air to "leak," or flow, around the wing tips from the high- to the low-pressure sides, as shown in Fig. 5.39a. This flow establishes a circulatory motion which trails

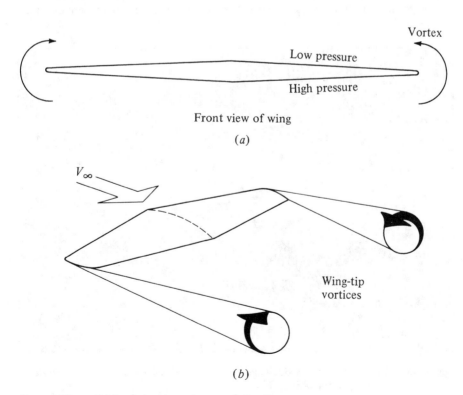

Figure 5.39   Origin of wing tip vortices on a finite wing.

downstream of the wing. The trailing circular motion is called a *vortex*. There is a major trailing vortex from each wing tip, as sketched in Fig. 5.39b and as shown in the photograph in Fig. 5.40.

These wing tip vortices downstream of the wing induce a small downward component of air velocity in the neighborhood of the wing itself. This can be seen intuitively from Fig. 5.39b; the two wing tip vortices tend to drag the surrounding air around with them, and this secondary movement induces a small velocity component in the downward direction at the wing. This downward component is called *downwash*, given the symbol w.

An effect of downwash can be seen in Fig. 5.41. As usual, V_∞ designates the relative wind. However, in the immediate vicinity of the wing, V_∞ and w add vectorally to produce a "local" relative wind which is canted downward from the original direction of V_∞. This has several consequences:

1. The angle of attack of the airfoil sections of the wing is effectively reduced in comparison to the angle of attack of the wing referenced to V_∞.

2. There is an increase in the drag. The increase is called *induced drag*, which has at least three physical interpretations. First, the wing tip vortices simply alter the flow field about the wing in such a fashion as to change the surface pressure distributions in the direction of increased drag. An alternate explanation is that because the local relative wind is canted downward (see Fig. 5.41), the lift vector itself is "tilted back." Hence it contributes a certain component of force

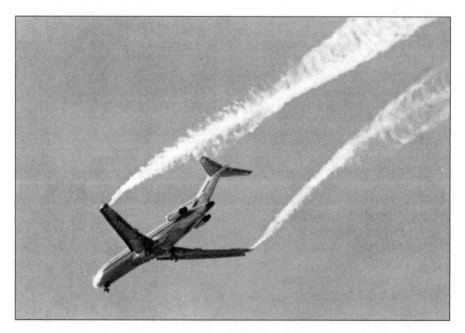

Figure 5.40   Wing tip vortices made visible by smoke ejected at the wing tips of a Boeing 727 test airplane. *(NASA.)*

Figure 5.41 The origin of downwash.

parallel to V_∞, that is, a drag force. A third physical explanation of the source of induced drag is that the wing tip vortices contain a certain amount of rotational kinetic energy. This energy has to come from somewhere; indeed, it is supplied by the aircraft propulsion system, where extra power has to be added to overcome the extra increment in drag due to induced drag. All three of these outlooks of the physical mechanism of induced drag are synonymous.

We can now answer the questions posed at the beginning of this section. Returning to the finite wing made up of the NACA 2412 airfoil section, where the wing is at $\alpha = 6°$, we now recognize that, because of the downwash, the local airfoil sections of the wing see an angle of attack *lower* than 6°. Clearly, the *local* airfoil lift coefficient will then be less than 0.85. Since the lift of the wing is an integration of the lift from each local segment, we can state that for the *finite wing*

$$C_L < 0.85$$

Also, the presence of induced drag for the finite wing, which is not present for an infinite wing, *adds* to the already existing skin friction drag and pressure drag due to flow separation which is experienced by the airfoil section itself. The value $c_d = 0.0077$ is the profile drag coefficient, which is the sum of the skin friction and pressure drag due to flow separation. For the finite wing, the induced drag must be added to the profile drag. Clearly, for the finite wing in this case:

$$C_D > 0.0077$$

So we can rest our case. The lift coefficient for a finite wing is *less* than that for its airfoil section, and the drag coefficient for a finite wing is *greater* than that for its airfoil section.

In Secs. 5.14 and 5.15, we will show how the drag coefficient and the lift coefficient, respectively, for a finite wing can be calculated. With this, we now move to the center column of our chapter road map in Fig. 5.1. Return to Fig. 5.1 for a moment, and note all the different aspects of airfoils that we have covered, as represented by the left-hand column of the road map. We are now ready to use this knowledge to examine the characteristics of finite wings, as represented by the middle column.

DESIGN BOX

For some airplane designs, the *shape* of the airfoil section *changes* along the span of the wing. For example, for the F-111 shown at the top of Fig. 5.31, the airfoil section at the root of the wing is an NACA 64A210 whereas the airfoil section at the tip of the wing is an NACA 64A209. The famous British Spitfire of World War II fame had a 13 percent thick airfoil at the root and a 7 percent thick airfoil at the tip. When a designer chooses to vary the airfoil shape along the span, it is usually for one or both of the following reasons:

1. To achieve a particular *distribution* of lift across the span of the wing which will improve the aerodynamic efficiency of the wing and/or reduce the structural weight of the wing.

2. To delay the onset of high-speed compressibility effects in the region near the wing tips. A thinner airfoil in the tip region will result in the "shock stall" pattern shown in Fig. 5.25c being delayed in that region to a higher Mach number, hence preserving aileron control effectiveness while the section of the wing closer to the root may be experiencing considerable flow separation.

In reference to our previous discussion, note that the possible variation of the airfoil shape along the span of a finite wing is yet another reason why the aerodynamic coefficients for a finite wing are different from those for an airfoil making up part of the wing itself.

5.14 CALCULATION OF INDUCED DRAG

A way of conceptualizing induced drag is shown in Fig. 5.42. Consider a finite wing as sketched in Fig. 5.42. The dashed arrow labeled $\mathbf{R}_1$ represents the resultant aerodynamic force on the wing for the *imaginary* situation of *no vortices* from the wing tips. The component of $\mathbf{R}_1$ parallel to V_∞ is the drag D_1, which in this imaginary case is due to skin friction and pressure drag due to flow separation. The solid arrow labeled $\mathbf{R}$ represents the *actual* resultant aerodynamic force, *including* the effect of wing tip vortices. The presence of the vortices changes the pressure distribution over the surface of the wing in such a fashion that $\mathbf{R}$ is tilted backward relative to $\mathbf{R}_1$. The component of $\mathbf{R}$ parallel to V_∞, denoted by D in Fig. 5.42, is the *actual* total drag, which includes the effect of the changed pressure distribution due to the wing tip vortices as well as friction drag and pressure drag due to flow separation. Since $\mathbf{R}$ is tilted backward relative to $\mathbf{R}_1$, then $D > D_1$. The *induced drag*, D_i is the difference between D and D_1: $D_i = D - D_1$. Keep in mind that induced drag is a type of *pressure drag*.

To calculate the magnitude of D_i, we will take the following perspective. Consider a section of a finite wing as shown in Fig. 5.43. The angle of attack defined between the mean chord of the wing and the direction of V_∞ (the relative wind) is called the *geometric angle of attack* α. However, in the vicinity of the wing, the local flow is (on the average) deflected downward by angle α_i because of downwash. This angle α_i, defined as the *induced angle of attack*, is the difference between the local flow direction and the free-stream direction. Hence, although the naked eye sees the wing at an angle of attack α, the airfoil section itself is seeing an *effective angle of*

Figure 5.42   Illustration of the induced drag, D_i.

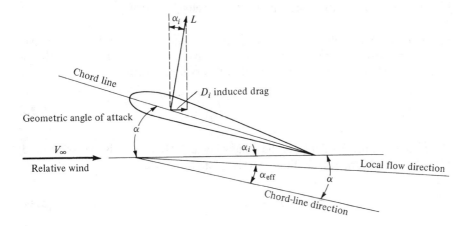

Figure 5.43   The origin of induced drag.

attack which is smaller than α. Letting α_{eff} denote the effective angle of attack, we see from Fig. 5.43 that $\alpha_{\text{eff}} = \alpha - \alpha_i$.

Let us now adopt the point of view that, because the local flow direction in the vicinity of the wing is inclined downward with respect to the free stream, the lift vector remains perpendicular to the local relative wind and is therefore tilted back through angle α_i. This is shown in Fig. 5.43. However, still considering drag to be parallel to the free-stream, we see that the tilted-lift vector contributes a certain component of drag. This drag is the *induced drag* D_i. From Fig. 5.43,

$$D_i = L \sin \alpha_i$$

Values of α_i are generally small, hence $\sin \alpha_i \approx \alpha_i$. Thus

$$D_i = L\alpha_i \tag{5.52}$$

Note that in Eq. (5.52) α_i must be in radians. Hence, D_i can be calculated from Eq. (5.52) once α_i is obtained.

The calculation of α_i is beyond the scope of this book. However, it can be shown that the value of α_i for a given section of a finite wing depends on the distribution of downwash along the span of the wing. In turn, the downwash distribution is governed by the distribution of lift over the span of the wing. To see this more clearly, consider Fig. 5.44, which shows the front view of a finite wing. The lift per unit span may vary as a function of distance along the wing because

1.   The chord may vary in length along the wing.
2.   The wing may be twisted such that each airfoil section of the wing is at a different geometric angle of attack.
3.   The shape of the airfoil section may change along the span.

Shown in Fig. 5.44 is the case of an elliptical lift distribution (the lift per unit span varies elliptically along the span), which in turn produces a uniform downwash distribution. For this case, incompressible flow theory predicts that

$$\alpha_i = \frac{C_L}{\pi \, \text{AR}} \tag{5.53}$$

where C_L is the lift coefficient of the finite wing and $\text{AR} = b^2/S$ is the aspect ratio, defined in Eq. (5.26). Substituting Eq. (5.53) into (5.52) yields

$$D_i = L\alpha_i = L\frac{C_L}{\pi \, \text{AR}} \tag{5.54}$$

Figure 5.44   Lift distribution and downwash distribution.

However, $L = q_\infty S C_L$; hence from Eq. (5.54),

$$D_i = q_\infty S \frac{C_L^2}{\pi \mathrm{AR}}$$

or
$$\frac{D_i}{q_\infty S} = \frac{C_L^2}{\pi \mathrm{AR}} \qquad [5.55]$$

Defining the *induced drag coefficient* as $C_{D,i} = D_i/(q_\infty S)$, we can write Eq. (5.55) as

$$\boxed{C_{D,i} = \frac{C_L^2}{\pi \mathrm{AR}}} \qquad [5.56]$$

This result holds for an elliptical lift distribution, as sketched in Fig. 5.44. For a wing with the same airfoil shape across the span and with no twist, an elliptical lift distribution is characteristic of an elliptical wing planform. (The famous British Spitfire of World War II was one of the few aircraft in history designed with an elliptical wing planform. Wings with straight leading and trailing edges are more economical to manufacture.)

For all wings in general, a *span efficiency factor e* can be defined such that

$$\boxed{C_{D,i} = \frac{C_L^2}{\pi e \mathrm{AR}}} \qquad [5.57]$$

For elliptical planforms, $e = 1$; for all other planforms, $e < 1$. Thus, $C_{D,i}$ and hence induced drag is a *minimum for an elliptical planform*. For typical subsonic aircraft, e ranges from 0.85 to 0.95. Equation (5.57) is an important relation. It demonstrates that induced drag varies as the square of the lift coefficient; at high lift, such as near $C_{L,\max}$, the induced drag can be a substantial portion of the total drag. Equation (5.57) also demonstrates that as AR is increased, induced drag is decreased. Hence, subsonic airplanes designed to minimize induced drag have high–aspect ratio wings (such as the long, narrow wings of the Lockheed U-2 high-altitude reconnaissance aircraft).

It is clear from Eq. (5.57) that induced drag is intimately related to lift. In fact, another expression for induced drag is *drag due to lift*. In a fundamental sense, that power provided by the engines of the airplane to overcome induced drag is the power required to sustain a heavier-than-air vehicle in the air, the power necessary to produce lift equal to the weight of the airplane in flight.

In light of Eq. (5.57), we can now write the total drag coefficient for a finite wing at subsonic speeds as

$$\boxed{C_D = \underset{\substack{\text{Total}\\\text{drag}}}{c_d} + \underset{\substack{\text{Profile}\\\text{drag}}}{} \underset{\substack{\text{Induced}\\\text{drag}}}{\frac{C_L^2}{\pi e \mathrm{AR}}}} \qquad [5.58]$$

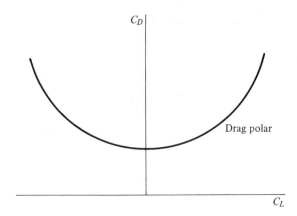

Figure 5.45   Sketch of a drag polar, that is, a plot of drag coefficient versus lift coefficient.

Keep in mind that profile drag is composed of two parts: drag due to skin friction $c_{d,f}$ and pressure drag due to separation $c_{d,p}$; that is, $c_d = c_{d,f} + c_{d,p}$. Also keep in mind that c_d can be obtained from the data in App. D. The quadratic variation of C_D with C_L given in Eq. (5.58), when plotted on a graph, leads to a curve as shown in Fig. 5.45. Such a plot of C_D versus C_L is called a *drag polar*. Much of the basic aerodynamics of an airplane is reflected in the drag polar, and such curves are essential to the design of airplanes. You should become familiar with the concept of drag polar. Note that the drag data in App. D are given in terms of drag polars for infinite wings; that is, c_d is plotted versus c_l. However, induced drag is not included in App. D because $C_{D,i}$ for an infinite wing (infinite aspect ratio) is zero.

Example 5.10   Consider the Northrop F-5 fighter airplane, which has a wing area of 170 ft². The wing is generating 18,000 lb of lift. For a flight velocity of 250 mi/h at standard sea level, calculate the lift coefficient.

Solution

The velocity in consistent units is

$$V_\infty = 250 \left(\frac{88}{60} \right) = 366.7 \text{ ft/s}$$

$$q_\infty = \tfrac{1}{2} \rho_\infty V_\infty^2 = \tfrac{1}{2}(0.002377)(366.7)^2 = 159.8 \text{ lb/ft}^2$$

Hence $$C_L = \frac{L}{q_\infty S} = \frac{18,000}{159.8(170)} = \boxed{0.6626}$$

Example 5.11   The wingspan of the Northrop F-5 is 25.25 ft. Calculate the induced drag coefficient and the induced drag itself for the conditions of Example 5.10. Assume $e = 0.8$.

Solution

The aspect ratio is AR $= b^2/S = (25.25)^2/170 = 3.75$. Since $C_L = 0.6626$ from Example 5.10, then from Eq. (5.57),

$$C_{D,i} = \frac{C_L^2}{\pi e \text{AR}} = \frac{(0.6626)^2}{\pi(0.8)(3.75)} = \boxed{0.0466}$$

From Example 5.10, $q_\infty = 159.8$ lb/ft^2. Hence

$$D_i = q_\infty S C_{D,i} = 159.8(170)(0.0466) = \boxed{1266 \text{ lb}}$$

Example 5.12

Consider a flying wing (such as the Northrop YB-49 of the early 1950s) with a wing area of 206 m^2, an aspect ratio of 10, a span effectiveness factor of 0.95, and an NACA 4412 airfoil. The weight of the airplane is 7.5×10^5 N. If the density altitude is 3 km and the flight velocity is 100 m/s, calculate the total drag on the aircraft.

Solution

First, obtain the lift coefficient. At a density altitude of 3 km = 3000 m, $\rho_\infty = 0.909$ kg/m^3 (from App. A).

$$q_\infty = \tfrac{1}{2}\rho_\infty V_\infty^2 = \tfrac{1}{2}(0.909)(100)^2 = 4545 \text{ N/m}^2$$

$$L = W = 7.5 \times 10^5 \text{ N}$$

$$C_L = \frac{L}{q_\infty S} = \frac{7.5 \times 10^5}{4545(206)} = 0.8$$

Note: This is a rather high lift coefficient, but the velocity is low—near the landing speed—hence the airplane is pitched to a rather high angle of attack to generate enough lift to keep the airplane flying.

Next, obtain the induced drag coefficient:

$$C_{D,i} = \frac{C_L^2}{\pi e \text{AR}} = \frac{0.8^2}{\pi(0.95)(10)} = 0.021$$

The profile drag coefficient must be estimated from the aerodynamic data in App. D. Assume that c_d is given by the highest Reynolds number data shown for the NACA 4412 airfoil in App. D; furthermore, assume that it is in the drag bucket. Hence, from App. D,

$$c_d \approx 0.006$$

Thus, from Eq. (5.58), the total drag coefficient is

$$C_D = c_d + C_{D,i} = 0.006 + 0.021 = 0.027$$

Note that the induced drag is about 3.5 times larger than profile drag for this case, thus underscoring the importance of induced drag.

Therefore, the total drag is

$$D = q_\infty S C_D = 4545(206)(0.027) = \boxed{2.53 \times 10^4 \text{ N}}$$

5.15   CHANGE IN THE LIFT SLOPE

The aerodynamic properties of a finite wing differ in two major respects from the data of App. D, which apply to infinite wings. The first difference has already been discussed, namely, the addition of induced drag for a finite wing. The second difference is that the lift curve for a finite wing has a smaller slope than the corresponding lift curve for an infinite wing with the same airfoil cross section. This change in the lift slope can be examined as follows. Recall that because of the presence of downwash, which is induced by the trailing wing tip vortices, the flow in the local vicinity of the wing is canted downward with respect to the free-stream relative wind. As a result, the angle of attack which the airfoil section effectively sees, called the *effective* angle of attack α_{eff}, is less than the geometric angle of attack α. This situation is sketched in Fig. 5.46. The difference between α and α_{eff} is the *induced angle of attack* α_i, first introduced in Sec 5.14, where $\alpha_i = \alpha - \alpha_{\text{eff}}$. Moreover, for an elliptical lift distribution, Eq. (5.53) gives values for the induced angle of attack $\alpha_i = C_L/(\pi \text{AR})$. Extending Eq. (5.53) to wings of any general planform, we can define a new span effectiveness factor e_1 such that

$$\alpha_i = \frac{C_L}{\pi e_1 \text{AR}} \qquad \textbf{[5.59]}$$

where e_1 and e [defined for induced drag in Eq. (5.57)] are theoretically different but are in practice approximately the same value for a given wing. Note that Eq. (5.59) gives α_i in radians. For α_i in degrees,

$$\alpha_i = \frac{57.3 C_L}{\pi e_1 \text{AR}} \qquad \textbf{[5.60]}$$

We emphasize that the flow over a finite wing at an angle of attack α is essentially the same as the flow over an infinite wing at an angle of attack α_{eff}. Keeping this in mind, assume that we plot the lift coefficient for the finite wing C_L versus the effective angle of attack $\alpha_{\text{eff}} = \alpha - \alpha_i$, as shown in Fig. 5.47a. Because we are using α_{eff}, the lift curve should correspond to that for an infinite wing; hence the lift curve slope in Fig. 5.47a is a_0, obtained from App. D for the given airfoil. However, in real life our naked eyes cannot see α_{eff}; instead, what we actually observe is a finite wing at the geometric angle of attack α (the actual angle between the free-stream relative wind

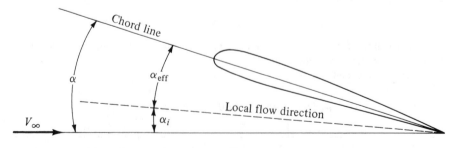

Figure 5.46   Relation between the geometric, effective, and induced angles of attack.

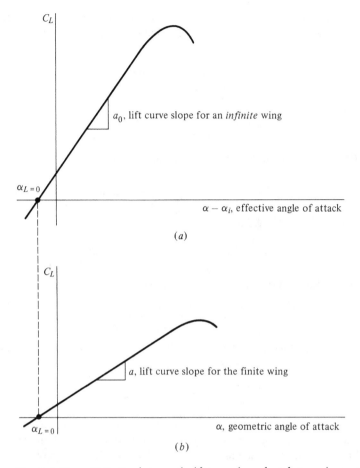

Figure 5.47 Distinction between the lift curve slopes for infinite and finite wings.

and the mean chord line). Hence, for a finite wing, it makes much more sense to plot C_L versus α, as shown in Fig. 5.47b, than C_L versus α_{eff}, as shown in Fig. 5.47a. For example, C_L versus α would be the result most directly obtained from testing a finite wing in a wind tunnel, because α (and not α_{eff}) can be measured directly. Hence, the lift curve slope for a finite wing is defined as $a \equiv dC_L/d\alpha$, where $a \neq a_0$. Noting that $\alpha > \alpha_{\text{eff}}$ from Fig. 5.46, we see that the abscissa of Fig. 5.47b is stretched out more than the abscissa of Fig. 5.47a; hence the lift curve of Fig. 5.47b is less inclined; that is, $a < a_0$. *The effect of a finite wing is to reduce the lift curve slope.* However, when the lift is zero, $C_L = 0$, and from Eq. (5.53), $\alpha_i = 0$. Thus, at zero lift, $\alpha = \alpha_{\text{eff}}$. In terms of Fig. 5.47a and 5.47b, this means that the angle of attack for zero lift $\alpha_{L=0}$ is the same for the finite and infinite wings. Thus, for finite wings, $\alpha_{L=0}$ can be obtained directly from App. D.

Question: If we know a_0 (say, from App. D), how do we find a for a finite wing with a given aspect ratio? The answer can be obtained by examining Fig. 5.47. From Fig. 5.47a,

$$\frac{dC_L}{d(\alpha - \alpha_i)} = a_0$$

Integrating, we find

$$C_L = a_0(\alpha - \alpha_i) + \text{const} \qquad [5.61]$$

Substituting Eq. (5.60) into Eq. (5.61), we obtain

$$C_L = a_0\left(\alpha - \frac{57.3C_L}{\pi e_1 \text{AR}}\right) + \text{const} \qquad [5.62]$$

Solving Eq. (5.62) for C_L yields

$$C_L = \frac{a_0\alpha}{1 + 57.3a_0/(\pi e_1 \text{AR})} + \frac{\text{const}}{1 + 57.3a_0/(\pi_1 \text{AR})} \qquad [5.63]$$

Differentiating Eq. (5.63) with respect to α, we get

$$\frac{dC_L}{d\alpha} = \frac{a_0}{1 + 57.3a_0/(\pi e_1 \text{AR})} \qquad [5.64]$$

However, from Fig. 5.47b, by definition, $dC_L/d\alpha = a$. Hence, from Eq. (5.64)

$$\boxed{a = \frac{a_0}{1 + 57.3a_0/(\pi e_1 \text{AR})}} \qquad [5.65]$$

Equation (5.65) gives the desired lift slope for a finite wing of given aspect ratio AR when we know the corresponding slope a_0 for an infinite wing. Remember: a_0 is obtained from airfoil data such as in App. D. Also note that Eq. (5.65) verifies our previous qualitative statement that $a < a_0$.

In summary, a finite wing introduces two major changes to the airfoil data in App. D:

1. Induced drag must be added to the finite wing:

$$\underset{\substack{\text{Total} \\ \text{drag}}}{C_D} = \underset{\substack{\text{Profile} \\ \text{drag}}}{c_d} + \underset{\substack{\text{Induced} \\ \text{drag}}}{\frac{C_L^2}{\pi e \text{AR}}}$$

2. The slope of the lift curve for a finite wing is less than that for an infinite wing; $a < a_0$.

Example 5.13   Consider a wing with an aspect ratio of 10 and an NACA 23012 airfoil section. Assume $\text{Re} \approx 5 \times 10^6$. The span efficiency factor is $e = e_1 = 0.95$. If the wing is at a 4° angle of attack, calculate C_L and C_D.

Solution

Since we are dealing with a finite wing but have airfoil data (App. D) for infinite wings only, the first job is to obtain the slope of this lift curve for the finite wing, modifying the data from App. D.

The infinite wing lift slope can be obtained from any two points on the linear curve. For the NACA 23012 airfoil, for example (from App. D),

$$c_l = 1.2 \quad \text{at } \alpha_{\text{eff}} = 10°$$

$$c_l = 0.14 \quad \text{at } \alpha_{\text{eff}} = 0°$$

Hence $\qquad a_0 = \dfrac{dc_l}{d\alpha} = \dfrac{1.2 - 0.14}{10 - 0} = \dfrac{1.06}{10} = 0.106 \text{ per degree}$

Also from App. D,

$$\alpha_{L=0} = -1.5 \quad \text{and} \quad c_d \approx 0.006$$

The lift slope for the finite wing can now be obtained from Eq. (5.65).

$$a = \frac{a_0}{1 + 57.3a_0/(\pi e_1 \text{AR})} = \frac{0.106}{1 + 57.3(0.106)/[\pi(0.95)(10)]} = 0.088 \text{ per degree}$$

At $\alpha = 4°$,

$$C_L = a(\alpha - \alpha_{L=0}) = 0.088[4° - (-1.5)] = 0.088(5.5)$$

$$\boxed{C_L = 0.484}$$

The total drag coefficient is given by Eq. (5.58):

$$C_D = c_d + \frac{C_L^2}{\pi e \text{AR}} = 0.006 + \frac{0.484^2}{\pi(0.95)(10)} = 0.006 + 0.0078 = \boxed{0.0138}$$

Example 5.14

In Example 4.24, we calculated the skin friction drag exerted on the biplane wings of the 1903 *Wright Flyer*. For the flight conditions given in Example 4.24, that is, $V_\infty = 30$ mi/h at sea level, calculate the induced drag exerted on the wings of the *Wright Flyer*, and compare this with the friction drag calculated earlier. For its historic first flight on December 17, 1903, the total weight of the *Flyer* including the pilot (Orville) was 750 lb. Assume the span efficiency for the wing is $e = 0.93$.

Solution

From the data given in Example 4.24, for the *Wright Flyer* the wingspan is $b = 40.33$ ft and the planform area of *each* wing is 255 ft². Hence, the aspect ratio of each wing is

$$\text{AR} = \frac{b^2}{S} = \frac{(40.33)^2}{255} = 6.38$$

For level flight, the airplane must produce a lift to counter its weight; hence for the flight of the *Wright Flyer*, the lift was equal to its weight, namely 750 lb. Also, the *Flyer* is a biplane configuration, and both wings produce lift. Let us assume that the lift is evenly divided between the two wings, hence the lift of *each* wing is $750/2 = 375$ lb. The velocity is $V_\infty = 30$ mi/h $= 44$ ft/s. The dynamic pressure is

$$q_\infty = \tfrac{1}{2}\rho_\infty V_\infty^2 = \tfrac{1}{2}(0.002377)(44^2) = 2.3 \text{ lb/ft}^2$$

Hence, the lift coefficient of each wing is

$$C_L = \frac{L}{q_\infty S} = \frac{375}{2.3(255)} = 0.639$$

From Eq. (5.57)

$$C_{D,i} = \frac{C_L^2}{\pi e \text{AR}} = \frac{(0.639)^2}{\pi(0.93)(6.38)} = 0.0219$$

The induced drag of each wing is

$$D_i = q_\infty S C_{D,i} = 2.3(255)(0.0219) = 12.84 \text{ lb}$$

The induced drag, accounting for *both* wings, is

$$D_i = 2(12.84) = \boxed{25.7 \text{ lb}}$$

Compare this with the friction drag of 6.82 lb calculated in Example 4.24. Clearly, the induced drag is much larger than the friction drag; this is because the velocity of 30 mi/h was relatively small, requiring a rather large lift coefficient to help generate the 750 lb of lift, and because the induced drag coefficient varies as the *square* of C_L, the induced drag is large compared to the friction drag at the relatively low flight speed.

Note: There is an aerodynamic interaction between the two wings of a biplane that is relatively complex; a discussion of the phenomenon is beyond the scope of this book. Because of this interaction, the induced drag of the biplane configuration is *not* equal to the sum of the induced drags acting on the single wings individually in isolation, as we have assumed in this example. Rather, the induced drag of the biplane configuration is slightly higher than the sum based on our calculations, and there is also a loss of lift. However, the above calculation is a reasonable first approximation for the biplane's induced drag.

DESIGN BOX

It is good practice to design conventional subsonic airplanes with high–aspect ratio wings. The reasons are clearly seen in Eqs. (5.57) and (5.65). The induced drag coefficient $C_{D,i}$ is inversely proportional to AR, as seen in Eqs. (5.57) and (5.58). This is a strong effect; if the aspect ratio is doubled, $C_{D,i}$ is reduced by a factor of 2. By comparison, the impact of the span efficiency factor e is minor, because changes in the wing planform and airfoil design result in only a few percent change in e, and in turn, through Eq. (5.57), result in only a few percent change in $C_{D,i}$. (Of course, when the designer is looking for every ounce of performance, the wing is designed to have a lift distribution as close to elliptical as practical, i.e., making e as close to unity as practical.) The aspect ratio is the big design feature that controls $C_{D,i}$. The

(continued)

same can be said about the lift slope. Increasing the aspect ratio increases the lift slope, as seen from Eq. (5.65). Clearly, on an aerodynamic basis, the designer of a conventional subsonic airplane would prefer to make the aspect ratio as large as possible.

However, what does *as large as possible* mean? Why do not the wings of existing airplanes look like the long and narrow slats from a venetian blind, which have very large aspect ratios? The answer is driven by structural considerations. Imagine the left and right wings on an airplane in flight; the lift acting on each wing acts to bend the wing upward, creating a bending moment where the wing joins the fuselage. The wing structure, and the structure through the fuselage, must be strong enough to resist this bending moment. Now imagine the lift acting on a venetian blind; the venetian blind slat will easily buckle under the load, unless the designer adds enough material stiffness to resist the buckling. This increase in wing stiffness can be obtained at the cost of increased wing structural weight. Consequently, the design aspect ratio for a conventional airplane is a compromise between competing values in aerodynamics and structures.

The usual outcome of this compromise is subsonic airplanes with aspect ratios on the order of 5 to 7. The following is a tabulation of wing aspect ratios for various subsonic airplane designs.

Airplane	Aspect Ratio
Wright Flyer (Fig. 1.1)	6.4
Vought F4U Corsair (Fig. 2.14)	5.35
Boeing B-17 (Fig. 2.15)	7.58
Grumman X-29 (Fig. 2.16)	3.91
Grumman F3F-2 (Fig. 2.17)	7.85
Boeing 727 (Fig. 5.40)	7.1

A dramatic example of the importance of a high aspect ratio can be seen in the Lockheed U-2 high-altitude reconnaissance airplane, shown in the three-view in Fig. 5.48. The U-2 was designed with an unusually high aspect ratio of 14.3. This was because of its mission. In 1954, the United States had an urgent need for a reconnaissance vehicle that could overfly the Soviet Union; the time was at an early stage of the cold war, and Russia had recently tested a hydrogen bomb. How-

ever, such a reconnaissance vehicle would have to fly at an altitude high enough that it could not be reached by interceptor aircraft or ground-to-air missiles; in 1954 this meant cruising at 70,000 ft or higher. The U-2 was designed by Lockheed Skunk Works, a small elite design group at Lockheed known for its innovative and advanced thinking. The airplane was essentially a *point design*; it was designed to achieve this extremely high-altitude cruise. In turn, the need for incorporating a very high–aspect ratio wing was paramount. The reason is explained below.

In steady, level flight, the airplane lift must equal its weight $L = W$. In this case, from Eq. (5.18) written for the whole airplane

$$L = W = \tfrac{1}{2}\rho_\infty V_\infty^2 S C_L \qquad \textbf{[5.66]}$$

Consider an airplane at a constant velocity V_∞. As it flies higher, ρ_∞ decreases and hence from Eq. (5.66) C_L must be increased in order to keep the lift constant, equal to the weight; that is, as ρ_∞ decreases, the angle of attack of the airplane increases in order to increase C_L. There is some maximum altitude (minimum ρ_∞) at which C_L reaches its maximum value; if the angle of attack is increased beyond this point, the airplane will stall. At its high-altitude cruise condition, the U-2 is flying at a high value of C_L with a concurrent high angle of attack, just on the verge of stalling. (This is in stark contrast to the normal cruise conditions of conventional airplanes at conventional altitudes, where the cruise lift coefficient and angle of attack are relatively small.) A high value of C_L means a high induced drag coefficient; note from Eq. (5.57) that $C_{D,i}$ varies directly as the *square* of C_L. As a result, at the design high-altitude cruise condition of the U-2, the induced drag is a major factor. To reduce the cruise value of $C_{D,i}$, the designers of the U-2 had to opt for as high an aspect ratio as possible. The wing design shown in Fig. 5.48 was the result.

It is interesting to note that at the high-altitude operating condition of the U-2, the highest velocity allowed by drag divergence and the lowest velocity allowed by stalling were almost the same; only about 7 mi/h separated these two velocities, which was not an easy situation for the pilot.

In contrast to the extreme high-altitude mission of the U-2, the opposite extreme is high-speed flight at altitudes on the order of hundreds of feet above the

(continued)

Figure 5.48   Three-view of the Lockheed U-2 high altitude reconnaissance airplane. Aspect ratio = 14.3.

ground. Consider a subsonic military aircraft designed for low-altitude, high-speed penetration of an enemy's defenses, flying close enough to the ground to avoid radar detection. The aircraft is flying at high speed in the high-density air near sea level, so it is flying at a very low C_L and very small angle of attack, as dictated by Eq. (5.66). Under these conditions, induced drag is very small compared to profile drag. At this design point, it is beneficial to have a low–aspect ratio wing with a relatively small surface area, which will reduce the profile drag. Moreover, the low aspect ratio provides another advantage under these flight conditions—it makes the aircraft less sensitive to atmospheric turbulence encountered at low altitudes. This is achieved through the effect of the aspect ratio on the lift slope, given by Eq. (5.65). The lift slope is smaller for a low–aspect ratio wing, as sketched in Fig. 5.49. Imagine the airplane encountering an atmospheric gust which momentarily perturbs its angle of attack by an amount $\Delta\alpha$, as sketched in Fig. 5.49. The lift coefficient will be correspondingly perturbed by the amount

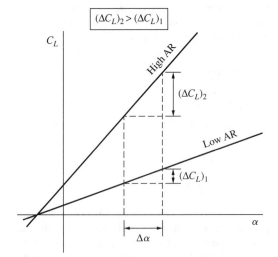

Figure 5.49   Effect of aspect ratio on the lift slope. For a given perturbation in α, the high aspect ratio wing experiences a larger perturbation in C_L than the low aspect ratio wing.

(*continued*)

ΔC_L. However, because of its larger lift slope, the high–aspect ratio wing will experience a larger perturbation $(\Delta C_L)_2$ than the low–aspect ratio wing, which experiences the smaller perturbation $(\Delta C_L)_1$. This is shown schematically in Fig. 5.49. The smaller change in C_L due to a gust for the low–aspect ratio wing results in a smoother ride, which is good for both the flight crew and the structure of the airplane.

In summary, the consideration of aspect ratio in airplane design is *not* a matter of "one size fits all."

Quite the contrary, we have discussed above two totally different flight conditions which reflect two different design points, one leading to the need for a high–aspect ratio wing and the other to a low–aspect ratio wing. It is clear that aspect ratio is one of the most important considerations for an airplane designer. The choice of what aspect ratio to use for a given airplane design depends on a number of factors and compromises. We have pointed out some of these considerations in the above discussion.

5.16 SWEPT WINGS

Almost all modern high-speed aircraft have swept-back wings, such as shown in Fig. 5.50*b*. Why? We are now in a position to answer this question.

We first consider subsonic flight. Consider the planview of a straight wing, as sketched in Fig. 5.50*a*. Assume this wing has an airfoil cross section with a critical Mach number $M_{cr} = 0.7$. (Remember from Sec. 5.10 that for M_∞ slightly greater than M_{cr}, there is a large increase in drag. Hence, it is desirable to increase M_{cr} as much as possible in high-speed subsonic airplane design.) Now assume that we sweep the wing back through an angle of, say, 30°, as shown in Fig. 5.50*b*. The airfoil, which still has a value of $M_{cr} = 0.7$, now "sees" essentially only the component of the flow normal to the leading edge of the wing; that is, the aerodynamic properties of the local section of the swept wing are governed mainly by the flow normal to the leading edge. Hence, if M_∞ is the free-stream Mach number, the airfoil in Fig. 5.50*b* is seeing effectively a smaller Mach number, $M_\infty \cos 30°$. As a result, the actual free-stream Mach number can be increased *above* 0.7 before critical phenomena on the airfoil are encountered. In fact, we could expect that the critical Mach number for the *swept wing itself* would be as high as $0.7/\cos 30° = 0.808$, as shown in Fig. 5.50*b*. This means that the large increase in drag (as sketched in Fig. 5.24) would be delayed to M_∞ much larger than M_{cr} for the airfoil—in terms of Fig. 5.50, something much larger than 0.7 and maybe even as high as 0.808. Therefore, we see the main function of a swept wing. *By sweeping the wings of subsonic aircraft, drag divergence is delayed to higher Mach numbers.*

In real life, the flow over the swept wing sketched in Fig. 5.50*b* is a fairly complex three-dimensional flow, and to say that the airfoil sees only the component normal to the leading edge is a sweeping simplification. However, it leads to a good rule of thumb. If Ω is the sweep angle, as shown in Fig. 5.50*b*, the actual critical Mach number for the swept wing is bracketed by

$$M_{cr} \text{ for airfoil} < \begin{array}{c} \text{actual } M_{cr} \\ \text{for swept wing} \end{array} < \frac{M_{cr} \text{ for airfoil}}{\cos \Omega}$$

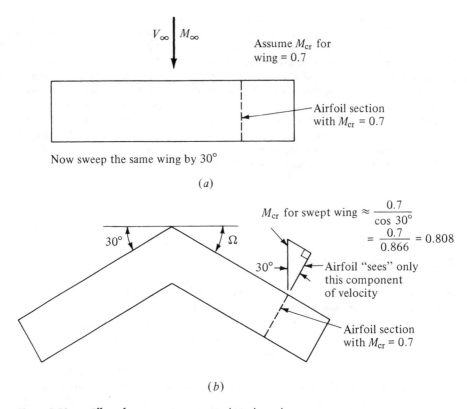

Figure 5.50   Effect of a swept wing on critical Mach number.

There is an alternate explanation of how the critical Mach number is increased by sweeping the wing. Consider the segment of a straight wing sketched in Fig. 5.51a. The airfoil section, with a thickness-to-chord ratio of $t_1/c_1 = 0.15$, is sketched at the left. The arrowed line AB represents a streamline flowing over the straight wing. This streamline "sees" the airfoil section with a 15 percent thickness. Now consider this *same wing*, but swept through the angle $\Omega = 45°$, as shown in Fig. 5.51b. The arrowed line CD represents a streamline flowing over the swept wing. (We draw streamlines AB and CD as straight lines in the free-stream direction, ignoring for simplicity any three-dimensional flow effects.) Streamline CD now travels a longer distance over the swept wing. The airfoil section that streamline CD effectively "sees" is sketched at the left in Fig. 5.51b. It has the same thickness, but a longer effective chord. Hence, the effective airfoil section that streamline CD sees is *thinner* than that seen in the case of the straight wing. Indeed, for the case of a sweep angle of 45°, the effective airfoil section seen by streamline CD has a thickness-to-chord ratio of $t_2/c_2 = 0.106$. Simply by taking the straight wing in Fig. 5.51a and sweeping it through an angle of 45°, the swept wing looks to the flow as if the effective airfoil

$$\frac{t_1}{c_1} = 0.15$$

Segment of straight wing

(a)

$$\frac{t_2}{c_2} = 0.106$$

$\Omega = 45°$

Segment of swept wing

(b)

Figure 5.51   By sweeping the wing, a streamline effectively sees a thinner airfoil.

section is almost one-third thinner than it is when the sweep angle is $0°$. From our discussion in Sec. 5.9, making the airfoil thinner increases the critical Mach number. Hence, by sweeping the wing, we can increase the critical Mach number of the wing.

Following the usual axiom that "we cannot get something for nothing," for subsonic flight increasing the wing sweep reduces the lift. Hence, although wing sweep is beneficial in terms of increasing the drag-divergence Mach number, it serves to decrease C_L. This is demonstrated in Fig. 5.52, which gives the variation of L/D with sweep angle for a representative airplane configuration at $M_\infty = 0.6$ flying at

30,000 ft. There is a considerable decrease in L/D as the sweep angle increases, mainly due to the decrease in C_L.

For supersonic flight, swept wings are also advantageous, but not quite from the same point of view as described above for subsonic flow. Consider the two swept wings sketched in Fig. 5.53. For a given $M_\infty > 1$, there is a Mach cone with vertex angle μ, equal to the Mach angle [recall Eq. (5.49)]. If the leading edge of a swept wing is *outside* the Mach cone, as shown in Fig. 5.53a, the component of the Mach number normal to the leading edge is supersonic. As a result, a fairly strong oblique shock wave will be created by the wing itself, with an attendant large wave drag. On the other hand, if the leading edge of the swept wing is *inside* the Mach cone, as shown in Fig. 5.53b, the component of the Mach number normal to the leading edge is subsonic. As a result, the wave drag produced by the wing is less. Therefore, the advantage of sweeping the wings for supersonic flight is in general to obtain a decrease in wave drag; and if the wing is swept inside the Mach cone, a considerable decrease can be obtained.

The quantitative effects of maximum thickness and wing sweep on the wave drag coefficient are shown in Fig. 5.54a and b, respectively. For all cases, the wing aspect ratio is 3.5, and the taper ratio (tip to root chord) is 0.2. Clearly, thin wings with large angles of sweepback have the smallest wave drag.

Figure 5.52   Variation of lift-to-drag ratio with wing sweep. Wind-tunnel measurements at the NASA Langley Research Center. *(From Loftin, NASA SP 468, 1985.)*

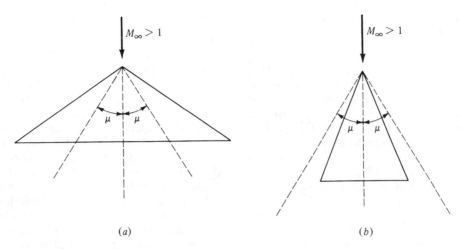

Figure 5.53 Swept wings for supersonic flow. (*a*) Wing swept outside the Mach cone. (*b*) Wing swept inside the Mach cone.

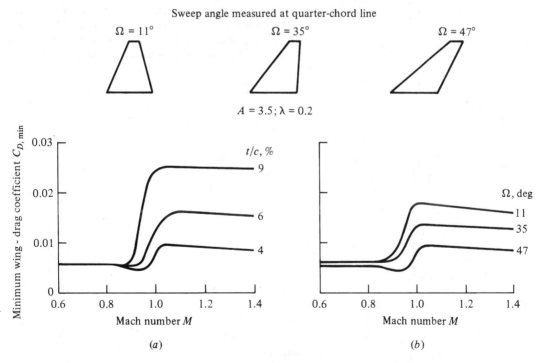

Figure 5.54 Sketch of the variation of minimum wing drag coefficient versus Mach number with (*a*) wing thickness as a parameter ($\Omega = 47°$) and (*b*) wing sweepback angle as a parameter ($t/c = 4$ percent). *(From L. Loftin, Quest for Performance, NASA SP 468, 1985.)*

DESIGN BOX

The designer of supersonic airplanes has two basic choices of wing planform: low–aspect ratio straight wing, or swept wing (including a delta wing). Both classes of wing planform result in lower wave drag compared to a high–aspect ratio straight wing. Let us examine these choices in greater detail.

First consider a low–aspect ratio straight wing at supersonic speeds. From Eq. (5.51), the wave drag coefficient for a flat plate of infinite span is

$$\frac{c_{d,w}}{\alpha^2} = \frac{4}{\sqrt{M_\infty^2 - 1}} \qquad [5.67]$$

where α is the angle of attack in *radians*. The same theory gives the wave drag coefficient for a flat plate of finite aspect ratio AR as

$$\frac{C_{D,W}}{\alpha^2} = \frac{4}{\sqrt{M_\infty^2 - 1}}\left(1 - \frac{1}{2R}\right) \qquad [5.68]$$

where

$$R \equiv AR\sqrt{M_\infty^2 - 1}$$

(See, e.g., Hilton, *High-Speed Aerodynamics*, Longman, Green and Co., 1951.) Note that Eq. (5.68) reduces to Eq. (5.67) for an aspect ratio going to infinity. Equation (5.68) is graphed in Fig. 5.55, giving $C_{D,W}/\alpha^2$ as a function of the aspect ratio for the case of $M_\infty = 2$. Note the rather dramatic drop in the wave drag coefficient at very low aspect ratios. This curve, which is for an infinitely thin flat plate, should be viewed as mainly qualitative when dealing with real wings with thickness. However, it clearly shows the advantage of low-aspect-ratio wings for supersonic flight. This is the exact opposite to the recommended practice for subsonic airplane design, as discussed earlier. However, because of the occurrence of shock waves at supersonic speeds, supersonic wave drag is usually much more important than induced drag; hence the use of low–aspect ratio wings is good practice in supersonic airplane design. A case in point is the Lockheed F-104 supersonic fighter, shown in Fig. 4.44. Return to Fig. 4.44, and study the wing planform for the F-104. This airplane was the first to be designed for sustained flight at Mach 2, and

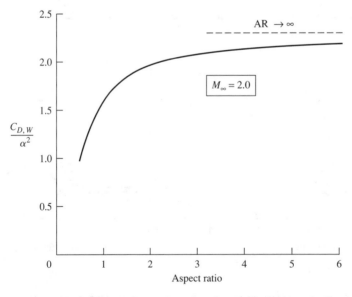

Figure 5.55   Variation of supersonic wave drag with aspect ratio for flat plates.

(continued)

the designers at Lockheed Skunk Works chose to go with a straight wing of low aspect ratio. The F-104 wing has an aspect ratio of 2.45. The airfoil section is a very thin biconvex shape; the thickness-to-chord ratio is only 0.0336. The leading edge is exceptionally sharp; the leading-edge radius of 0.016 in is so small that it poses some danger to the ground crew working around the airplane. All these features have one goal—to reduce the supersonic wave drag. They are classic examples of good supersonic airplane design.

We note that the supersonic lift coefficient is also reduced when the aspect ratio is reduced. This is illustrated in Fig. 5.56a, which gives the variation of the lift slope $dC_L/d\alpha$ as a function of aspect ratio for straight, tapered wings at $M_\infty = 1.53$. Shown here are some of the first experimental data obtained in the United States on wings at supersonic speeds. These data were obtained in the 1-ft by 3-ft supersonic tunnel at NACA Ames Laboratory by Walter Vincenti in 1947, but owing to military classification were not released until 1949. In Fig. 5.56a, the dashed triangles shown emanating from the wing leading-edge apex represent the Mach cones at $M_\infty = 1.53$. (The Mach cones are cones with a semivertex angle equal to the Mach angle μ.) Note that as AR is reduced, more of the wing is contained inside the Mach cones. The effect of decreasing AR on the lift slope at supersonic speeds is qualitatively the same as that for subsonic speeds. Recall from Sec. 5.15 that the lift slope is smaller for lower–aspect ratio wings in subsonic flight. Clearly, from Fig. 5.56a the same trend prevails for supersonic flight, even though the physical nature of the aerodynamic flow field is completely different.

The other option for a wing planform for supersonic airplanes is the swept wing. (We will consider the delta, or triangular planform, as a subset under swept wings.) In regard to Fig. 5.53, we have already discussed that supersonic wave drag can be considerably reduced by sweeping the wing inside the Mach cone, that is, by having a subsonic leading edge. This is clearly seen in the experimental data shown in Fig. 5.56b, taken from the pioneering supersonic wind tunnel work of Vincenti. In Fig. 5.56b, the minimum total drag coefficient is plotted versus wing sweep angle for $M_\infty = 1.53$. Keep in mind that the total drag coefficient is due to both pressure drag (essentially wave drag) and skin friction drag.

Positive sweep angles represent *swept-back* wings, and negative sweep angles represent *swept-forward* wings. Note the near symmetry of the data in regard to positive and negative sweep angles; the supersonic drag coefficient is essentially the same for the same degree of sweepback as it is for the same degree of sweepforward. The important message in Fig. 5.56b is the decrease in $C_{D_{\min}}$ at sweep angles greater than 49° or less than −49°. The Mach angle for $M_\infty = 1.53$ is given by $\mu = \sin^{-1}(1/M_\infty) = \sin^{-1}(1/1.53) = 41°$. Hence, wings with a sweep angle of 49° or larger will be inside the Mach cone. Note the lower drag coefficient at a sweep angle of ±60°; for this case, the wing is comfortably inside the Mach cone, with a subsonic leading edge. These data also show that when the wings are swept outside the Mach cone (supersonic leading edge), the drag coefficient is relatively flat, independent of the sweep angle. So for supersonic flight, in order to realize the drag reduction associated with a swept wing, the sweep angle must be large enough that the wing is swept *inside* the Mach cones.

A classic example of this design feature is the English Electric Lightning, a Mach 2 interceptor used by the British Royal Air Force in the 1960s and 1970s. As seen in Fig. 5.57, the Lightning has a highly swept wing, with a sweep angle $\Omega = 60°$. At Mach 2, the Mach angle is $\mu = \sin^{-1}(1/M_\infty) = \sin^{-1}\frac{1}{2} = 30°$. A swept wing, to be just inside the Mach cone at $M_\infty = 2$, must have a sweep angle of $\Omega = 60°$ or larger. Since Mach 2 was the design point, it is no surprise that the designers of the Lightning chose a sweep angle of 60°. In addition, the wing of the Lightning has a relatively low aspect ratio of 3.19, and the airfoil section is thin, with a thickness-to-chord ratio of 5 percent—both good design practices for supersonic airplanes.

Look closely at the Lightning in Fig. 5.57, and then go back and closely examine the F-104 in Fig. 4.44. Here we see classic examples of the two different wing planforms, swept wing and low–aspect ratio straight wing, from which designers of supersonic airplanes can choose.

We examined the effect of wing sweep on the *subsonic* lift coefficient (via the lift-to-drag ratio) in Fig. 5.52. *Question:* What is the effect of sweep on the *supersonic* lift coefficient? The answer is provided by the experimental data of Vincenti, shown in Fig. 5.58

(*continued*)

Figure 5.56   (a) Effect of aspect ratio on the lift curve for straight wings at supersonic speeds. $M_\infty = 1.53$. After W.G. Vincenti, "Comparison between Theory and Experiment for Wings at Supersonic Speeds", NACA TR 1033. (b) Effect of wing sweep on supersonic drag. The drag coefficient quoted is for an angle of attack that gives minimum drag. Data from Vincenti.

(*continued*)

Figure 5.57 Three-view of the English Electric Lightning supersonic fighter.

In a trend similar to that for the drag coefficient, we see from Fig. 5.58 that as long as the wing is swept outside the Mach cone (supersonic leading edge), the lift slope is relatively constant, independent of sweep angle. When the wing is swept inside the Mach cone (subsonic leading edge), the lift slope decreases with increasing sweep angle, similar to the case for subsonic flight.

The results shown in Figs. 5.56 and 5.58 clearly show a distinct change in the wing aerodynamic characteristics when the sweep angle is large enough that the wing is inside the Mach cone. This is because the pressure distribution over the wing surface changes radically when the transition is made from a supersonic to a subsonic leading edge. The nature of this change is sketched in Fig. 5.59, which shows three flat-plate wing planforms labeled A, B, and C of progressively increased sweep angle in a supersonic free stream. The shaded region shows that portion of the wing surface *not* inside the Mach cones emanating from the wing tips and the apex (midspan location). Wing A is a straight wing. The influence of the Mach cones is limited to

a small region at the tips; most of the wing is feeling the type of two-dimensional supersonic flow over a flat plate that was discussed in Sec. 5.11 and sketched in Fig. 5.37. Hence, the pressure distribution over most of the surface of wing A is the constant-pressure distribution illustrated by the vertical lines of constant-length shown near the right-hand tip of the wing. Wing B is a swept wing, but with a supersonic leading edge. A considerable portion of the wing is still outside the Mach cones; this portion is the shaded region shown on wing B. In this shaded region, the same constant-pressure distribution associated with a flat plate in supersonic flow still prevails. However, wing C is a swept wing with a subsonic leading edge; the entire wing is swept inside the Mach cone from the apex, and there is no shaded region. The pressure distribution over this wing is similar to that for subsonic flow, even though the wing is immersed in a supersonic free stream. Note that the vertical lines at the right on wing C trace out the type of subsonic pressure coefficient distribution familiar to us from our earlier discussions; for example, compare with Fig. 5.13.

(*continued*)

Figure 5.58   Effect of wing sweep on the lift slope at supersonic speed. Data from Vincenti.

--- Mach line

▭ Region of two-dimensional flow

▬ Lift distribution

Figure 5.59   Change in chordwise pressure distribution as a wing at supersonic speeds is progressively swept from outside to inside the Mach cone, i.e., as the leading edge progressively changes from supersonic to subsonic.

It is this change in the aerodynamic behavior of the flow over a wing swept inside the Mach cone that leads to the decrease in wave drag and lift coefficient associated with swept wings in supersonic flow.

There is yet another design benefit of a wing with a subsonic leading edge; namely, the leading-edge radius can be larger, similar to that for a subsonic airplane. This has benefits at low speeds, especially for landing and takeoff, for airplanes designed for supersonic flight. A wing with a sharp leading edge and a thin airfoil, such as that used on the F-104 (Fig. 4.44), experiences early flow separation at moderate angles of attack at subsonic speeds. This reduces the value of $(C_L)_{max}$ and forces the airplane to have higher landing and takeoff speeds. (For example, over its operational history, the F-104 experienced an inordinate number of accidents due to wing stall at low-speed flight conditions.) In contrast, a wing with a blunter, more rounded leading edge has much better low-speed stall characteristics. Supersonic airplanes with swept wings with subsonic leading edges can be designed with blunter, more rounded leading edges, and hence have better low-speed stalling behavior.

(continued)

Recall from Figs. 5.56 and 5.58 that the supersonic drag and lift coefficients associated with swept-forward wings are essentially the same as those for swept-back wings. Indeed, the same can be said for high-speed subsonic flight. However, airplane designers have almost always chosen sweepback rather than sweepforward. Why? The answer has to do with aeroelastic deformation of swept wings under load. For a swept-back wing, the location of the effective lift force causes the wing to twist near the tips so as to decrease the angle of attack of the outer portion of the wing. This tends to unload that portion of the wing when lift is increased—a stable situation. In contrast, for a swept-forward wing, the location of the effective lift force causes the wing to twist near the tips so as to increase the angle of attack of the outer portion of the wing, thus causing the lift to increase, which further increases the wing twist. This is an unstable situation which tends to twist the swept-forward wing right off the airplane. These aeroelastic deformation effects are even seen in the experimental data shown in Fig. 5.58. Note that the experimental data are not symmetric for swept-forward and swept-back wings. The lift slope is *smaller* for the swept-back wings, due to aeroelastic deformation of the wind tunnel models. Hence, for *structural reasons*, swept-forward wings have not been the planform of choice. However, modern advances in composite materials now allow the design of very strong, lightweight wings, and this has opened the design space for high-speed airplanes to consider the use of swept-forward wings. Indeed, swept-forward wings have certain design advantages. For example, the wing root can be placed farther back on the fuselage, allowing greater flexibility in designing the internal packaging inside the fuselage. Also, the details of the three-dimensional flow over a swept-forward wing result in flow separation occurring first near the root, preserving aileron control at the tips; in contrast, for a swept-back wing, flow separation tends to occur first near the tips, hence causing a loss of aileron control. In the 1980s, an experimental airplane, the Grumman X-29, was designed with swept-forward wings in order to examine more closely the practical aspects of swept-forward wing design. A three-view of the X-29 is shown in Fig. 5.60. The X-29 research program has been successful, but as yet there has been no rush on the part of airplane designers to go to swept-forward wings.

Return to Fig. 5.57, and examine again the highly swept wing of the English Electric Lightning. It is not much of an intellectual leap to imagine the empty notch between the wing trailing edge and the fuselage filled in with wing structure, producing a wing with a triangular planform. Such wings are called *delta*

Figure 5.60   An example of a swept-forward wing. The Grumman X-29.

(continued)

wings. Since the advent of the jet engine, there has been interest in delta wings for high-speed airplanes, both subsonic and supersonic. One design advantage of the delta wing is that by filling in that notch, as discussed above, the chord length of the wing root is considerably lengthened. For a fixed t/c ratio, this means the wing thickness at the root can be made larger, providing greater volume for structure, fuel, etc. The list of advantages and disadvantages of a delta wing is too long to discuss here. See the following book for a thorough and readable discussion of this list: Ray Whitford, *Design for Air Combat*, Janes Information Group Limited, 1989. Suffice it to say that a number of subsonic and supersonic delta wing aircraft have been designed and used extensively. An example is the French Dassault-Brequet Mirage 2000C, shown in Fig. 5.61. The Mirage 2000C is a supersonic fighter with a top speed of Mach 2.2. The leading-edge sweep angle is $\Omega = 58°$. Dassault is well known for its long line of successful delta wing airplanes since the 1950s. Note from Fig. 5.61 that the Mirage 2000C has no horizontal stabilizer; this is characteristic of many delta wing airplanes. The trailing-edge control surfaces are called *elevons* which, when deflected uniformly in the same direction (up or down), act as elevators and when deflected in opposite directions (one up and the other down) act as ailerons.

In many respects, the wing is the heart of the airplane. Great care goes into the design of the wing. Today, the design of wing shapes for supersonic airplanes is sophisticated and fine-tuned. Consider, for example, the Anglo-French Concorde supersonic transport, shown in Fig. 5.62. The Concorde is the only commercial supersonic transport in regular service today. Manufactured jointly by British Aircraft Corporation in England and Aerospatiale in France, the Concorde first flew on March 2, 1969, and went into service with British Airways and Air France in 1976. As shown in Fig. 5.62, the wing of the Concorde is a highly swept ogival delta planform with complex camber and wing droop (anhedral). The airfoil section is thin, with a thickness-to-chord ratio of 3 percent at the root and 2.15 percent from the nacelle outward. (A personal note: this author and his wife flew on the Concorde during the summer of 1998—what an exciting experience! The flight time between New York and London was only 3 h 15 min—too short to even show an in-flight movie. Unfortunately, the Concorde fare is very expensive, and by most measures the airplane is an economic failure. It will be one of the most demanding design challenges in the 21st century to design an economically and environmentally viable supersonic transport. Perhaps some of the young readers of this book will successfully rise to this challenge.)

Figure 5.61   An example of a delta wing. The French Dassault-Breguet Mirage 2000C, with an added side view (lower right) of the Mirage 2000N.

(*continued*)

Figure 5.62   The Anglo-French Aerospatiale/BAC Concorde supersonic transport.

5.17   FLAPS—A MECHANISM FOR HIGH LIFT

An airplane normally encounters its lowest flight velocities at takeoff or landing, two periods that are most critical for aircraft safety. The slowest speed at which an airplane can fly in straight and level flight is defined as the *stalling speed* V_{stall}. Hence, the calculation of V_{stall}, as well as aerodynamic methods of making V_{stall} as small as possible, is of vital importance.

The stalling velocity is readily obtained in terms of the maximum lift coefficient, as follows. From the definition of C_L,

$$L = q_\infty S C_L = \tfrac{1}{2}\rho_\infty V_\infty^2 S C_L$$

Thus
$$V_\infty = \sqrt{\frac{2L}{\rho_\infty S C_L}}$$
[5.69]

In steady, level flight, the lift is just sufficient to support the weight W of the aircraft, that is, $L = W$. Thus

$$V_\infty = \sqrt{\frac{2W}{\rho_\infty S C_L}}$$
[5.70]

Examining Eq. (5.70), for an airplane of given weight and size at a given altitude, we find the only recourse to minimize V_∞ is to maximize C_L. Hence, stalling speed corresponds to the angle of attack that produces $C_{L,max}$:

$$V_{stall} = \sqrt{\frac{2W}{\rho_\infty S C_{L,max}}}$$
[5.71]

In order to decrease V_{stall}, $C_{L,max}$ must be increased. However, for a wing with a given airfoil shape, $C_{L,max}$ is fixed by nature; that is, the lift properties of an airfoil, including maximum lift, depend on the physics of the flow over the airfoil. To assist nature, the lifting properties of a given airfoil can be greatly enhanced by the use of "artificial" high-lift devices. The most common of these devices is the flap at the trailing edge of the wing, as sketched in Fig. 5.63. When the flap is deflected downward through the angle δ, as sketched in Fig. 5.63b, the lift coefficient is increased, for the following reasons:

1. The camber of the airfoil section is effectively increased, as sketched in Fig. 5.63c. The more camber an airfoil shape has at a given angle of attack, the higher the lift coefficient.

2. When the flap is deflected, we can visualize a line connecting the leading edge of the airfoil and the trailing edge of the flap, points A and B, respectively, in Fig. 5.63d. Line AB constitutes a *virtual chord line*, rotated clockwise relative to the actual chord line of the airfoil, making the airfoil section with the

Figure 5.63 When a plain flap is deflected, the increase in lift is due to an effective increase in camber and a virtual increase in angle of attack.

deflected flap see a "virtual" increase in angle of attack. Hence, the lift coefficient is increased.

For these reasons, when the flap is deflected downward through the flap deflection angle δ, the value of $C_{L,\text{max}}$ is increased and the zero-lift angle of attack is shifted to a more negative value, as shown in Fig. 5.64. In Fig. 5.64, the lift curves for a wing with and without flaps are compared. Note that when the flaps are deflected, the lift curve shifts to the left, the value of $C_{L,\text{max}}$ increases, and the stalling angle of attack at which $C_{L,\text{max}}$ is achieved is decreased. However, the lift slope remains unchanged; trailing-edge flaps do not change the value of $dC_L/d\alpha$. Also note that for some of the airfoils given in App. D, lift curves are shown with the effect of flap deflection included.

The increase in $C_{L,\text{max}}$ due to flaps can be dramatic. If the flap is designed not only to rotate downward, but also to translate rearward so as to increase the effective wing area, then $C_{L,\text{max}}$ can be increased by approximately a factor of 2. If additional high-lift devices are used, such as slats at the leading edge, slots in the surface, or mechanical means of boundary layer control, then $C_{L,\text{max}}$ can sometimes be increased by a factor of 3 or more, as shown in Fig. 5.65. For an interesting and more detailed discussion of various high-lift devices, the reader is referred to the books by McCormick and Shevell (see the Bibliography at the end of this chapter) as well as the author's recent book; Anderson, *Aircraft Performance and Design*, McGraw-Hill, Boston, 1999.

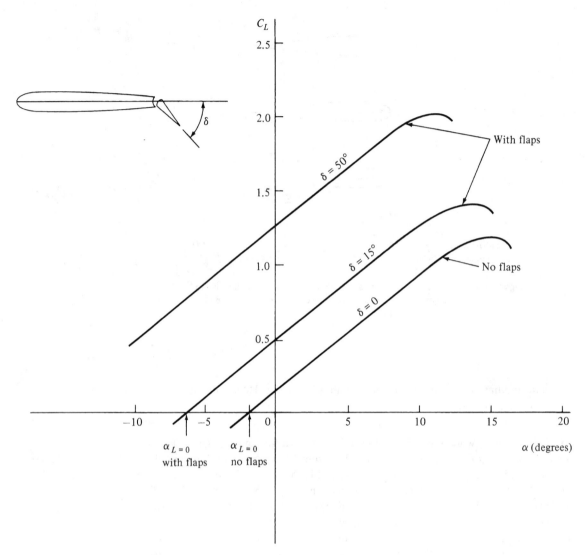

Figure 5.64   Illustration of the effect of flaps on the lift curve. The numbers shown are typical of a modern medium-range jet transport.

5.18   AERODYNAMICS OF CYLINDERS AND SPHERES

Consider the low-speed subsonic flow over a sphere or an infinite cylinder with its axis normal to the flow. If the flow were inviscid (frictionless), the theoretical flow pattern would look qualitatively as sketched in Fig. 5.66a. The streamlines would form a symmetric pattern; hence the pressure distributions over the front and rear

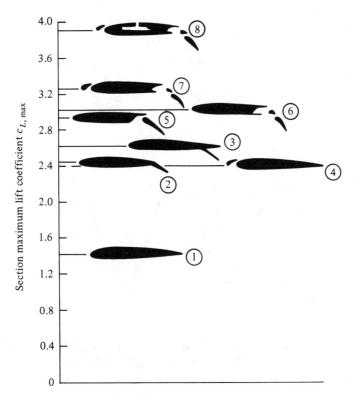

Figure 5.65   Typical values of airfoil maximum lift coefficient for various types of high-lift devices: (1) airfoil only, (2) plain flap, (3) split flap, (4) leading-edge slat, (5) single-slotted flap, (6) double-slotted flap, (7) double-slotted flap in combination with a leading-edge slat, (8) addition of boundary-layer suction at the top of the airfoil. *(From Loftin, NASA SP 468, 1985.)*

surfaces would also be symmetric, as sketched in Fig. 5.66*b*. This symmetry creates a momentous phenomenon, namely, that there is *no pressure drag on the sphere* if the flow is frictionless. This can be seen by simple inspection of Fig. 5.66*b*: The pressure distribution on the front face ($-90° \leq \theta \leq 90°$) creates a force in the drag direction, but the pressure distribution on the rear face ($90° \leq \theta \leq 270°$), which is identical to that on the front face, creates an equal and opposite force. Thus, we obtain the curious theoretical result that there is no drag on the body, quite contrary to everyday experience. This conflict between theory and experiment was well known at the end of the 19th century and is called *d'Alembert's paradox.*

The actual flow over a sphere or cylinder is sketched in Fig. 4.30; as discussed in Sec. 4.20, the presence of friction leads to separated flows in regions of adverse pressure gradients. Examining the theoretical inviscid pressure distribution shown

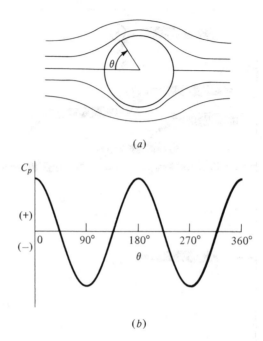

Figure 5.66   Ideal frictionless flow over a sphere. (a) Flow field. (b) Pressure coefficient distribution.

in Fig. 5.66b, we find on the rear surface ($90° \le \theta \le 270°$) the pressure increases in the flow direction; that is, an adverse pressure gradient exists. Thus, it is entirely reasonable that the real-life flow over a sphere or cylinder would be separated from the rear surface. This is indeed the case, as first shown in Fig. 4.30 and as sketched again in Fig. 5.67a. The real pressure distribution that corresponds to this separated flow is shown as the solid curve in Fig. 5.67b. Note that the average pressure is much higher on the front face ($-90° < \theta < 90°$) than on the rear face ($90° < \theta < 270°$). As a result, there is a net drag force exerted on the body. Hence, nature and experience are again reconciled, and d'Alembert's paradox is removed by a proper account of the presence of friction.

It is emphasized that the flow over a sphere or cylinder, and therefore the drag, is dominated by flow separation on the rear face. This leads to an interesting variation of C_D with the Reynolds number. Let the Reynolds number be defined in terms of the sphere diameter D: $\text{Re} = \rho_\infty V_\infty D / \mu_\infty$. If a sphere is mounted in a low-speed subsonic wind tunnel and the free-stream velocity is varied such that Re increases from 10^5 to 10^6, then a curious, almost discontinuous drop in C_D is observed at about $\text{Re} = 3 \times 10^5$. This behavior is sketched in Fig. 5.68. What causes this precipitous decrease in drag? The answer lies in the different effects of laminar and turbulent boundary layers on flow separation. At the end of Sec. 4.20 we note that laminar boundary layers separate much more readily than turbulent boundary layers. In the flow over a sphere at $\text{Re} < 3 \times 10^5$, the boundary layer is laminar. Hence, the flow is

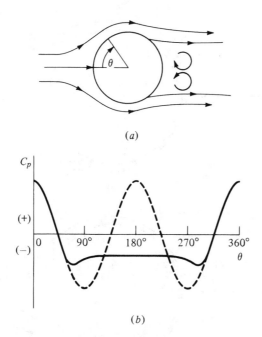

(a)

(b)

Figure 5.67 Real separated flow over a sphere; separation is due to friction. (a) Flow field. (b) Pressure coefficient distribution.

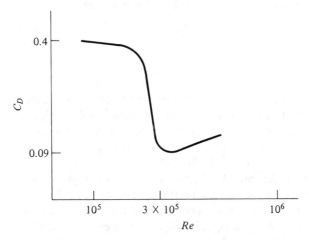

Figure 5.68 Variation of drag coefficient with Reynolds number for a sphere in low-speed flow.

totally separated from the rear face, and the wake behind the body is large, as sketched in Fig. 5.69a. In turn, the value of C_D is large, as noted at the left of Fig. 5.68 for Re $< 3 \times 10^5$. On the other hand, as Re is increased above 3×10^5, transition takes

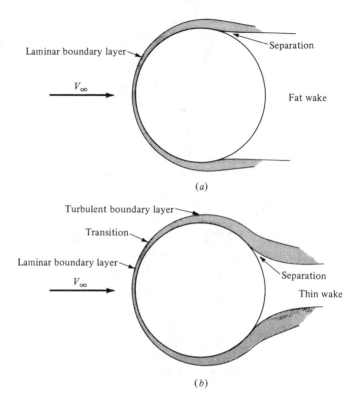

Figure 5.69   Laminar and turbulent flow over a sphere.

place on the front face, the boundary layer becomes turbulent, and the separation point moves rearward. (Turbulent boundary layers remain attached for longer distances in the face of adverse pressure gradients.) In this case, the wake behind the body is much smaller, as sketched in Fig. 5.69b. In turn, the pressure drag is less and C_D decreases, as noted as the right of Fig. 5.68.

Therefore, to decrease the drag on a sphere or cylinder, a turbulent boundary layer must be obtained on the front surface. This can be made to occur naturally by increasing Re until transition occurs on the front face. It can also be forced artificially at lower values of Re by using a rough surface to encourage early transition or by wrapping wire or other protuberances around the surface to create turbulence. (The use of such artificial devices is sometimes called *tripping the boundary layer*.)

It is interesting to note that the dimples on the surface of a golf ball are designed to promote turbulence and hence reduce the drag on the ball in flight. Indeed, some recent research has shown that polygonal dimples result in less drag than the conventional circular dimples on golf balls; but a dimple of any shape leads to less pressure drag than a smooth surface does (table tennis balls have more drag than golf balls).

DESIGN BOX

The large pressure drag associated with blunt bodies such as the cylinders and spheres discussed in this section leads to the design concept of *streamlining*. Consider a body of cylindrical cross section of diameter d, with the axis of the cylinder oriented perpendicular to the flow. There will be separated flow on the back face of the cylinder, with a relatively fat wake and with the associated high-pressure drag; this case is sketched in Fig. 5.70a. The bar to the right of Fig. 5.70a denotes the total drag of the cylinder; the shaded portion of the bar represents skin friction drag, and the open portion represents the pressure drag. Note that for the case of a blunt body, the drag is relatively large, and most of this drag is due to pressure drag. However, look at what happens when we wrap a long, mildly tapered afterbody on the back of the cylinder, creating the teardrop-shaped body sketched in Fig. 5.70b. This shape is a *streamlined body*, of the same thickness d as the cylinder. However, because of the tapered afterbody, the adverse pressure gradient along the back of the streamlined body will be much milder than that for the back surface of the cylinder, and hence flow separation on the streamlined body will be delayed until much closer to the trailing edge, as sketched in Fig. 5.70b, with an attendant, much smaller wake. As a result, the pressure drag of the streamlined body will be much smaller than that for the cylinder. Indeed, as shown by the bar to the right of Fig. 5.70b, the total drag of the streamlined body in a low-speed flow will be almost a factor or 10 smaller than that of the cylinder of same thickness. The friction drag of the streamlined body will be larger due to its increased surface area, but the pressure drag is so much less that it dominates this comparison.

This is why so much attention is placed on *streamlining* in airplane design. The value of streamlining was not totally recognized by airplane designers until the late 1920s. Jump ahead to Figs. 6.70 and 6.71. In Fig. 6.70, a typical strut-and-wire biplane from World War I, the French SPAD XIII, is shown. This airplane is definitely *not* streamlined. In contrast, by the middle 1930s, streamlined airplanes were in vogue, and the Douglas DC-3 shown in Fig. 6.71 is a classic example. The evolution of our understanding of streamlining, and how it was eventually applied in airplane design, is one of the more interesting stories in the history of aerodynamics. For this story, see Anderson, *A History of Aerodynamics and Its Impact on Flying Machines*, Cambridge University Press, New York, 1997.

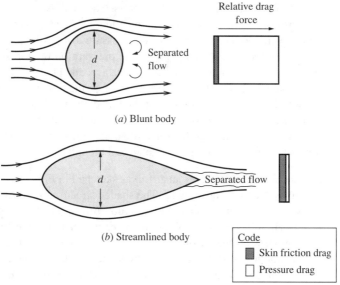

Figure 5.70   Comparison of the drag for a blunt body and a streamlined body.

5.19   HOW LIFT IS PRODUCED—SOME ALTERNATE EXPLANATIONS

Return to our road map in Fig. 5.1. We have covered all the milestones on this map except the one at the bottom labeled "How lift is produced." This is the subject of this present section.

It is amazing that today, almost 100 years after the first flight of the *Wright Flyer*, groups of engineers, scientist, pilots, and others can gather together and have a spirited debate on how an airplane wing generates lift. Various explanations are put forth, and the debate centers on which explanation is the most fundamental. The purpose of this section is to attempt to put these various explanations in perspective and to resolve the debate. Indeed, in our previous discussions in this book, we have consistently put forth one explanation as the most fundamental, and we have intentionally not burdened your thinking with any alternatives. So you may be wondering what the big deal is here. You already know how lift is produced. However, because the literature is replete with various different, and sometimes outright incorrect explanations as to how lift is produced, you need to be aware of some of the alternative thinking.

First, let us consider what this author advocates as the most fundamental explanation of lift. It is clear from our discussion in Sec. 2.2 that the two hands of nature that reach out and grab hold of a body moving through a fluid (liquid or gas) are the pressure and shear stress distributions exerted all over the exposed surface of the body. The resultant aerodynamic force on the body is the net, integrated effect of the pressure and shear stress distributions on the surface. Because the lift is the component of this resultant force perpendicular to the relative wind, and because the pressure on the surface of an airfoil at reasonable angles of attack acts mainly in the lift direction whereas the shear stress acts mainly in the drag direction, we are comfortable in saying that, for lift, the effect of shear stress is secondary and that lift is mainly due to the imbalance of the pressure distributions over the top and bottom surfaces of the airfoil. Specifically, the pressure on the top surface of the airfoil is lower than the pressure on the bottom surface, and presto—lift! However, this raises the question as to *why* the pressure is lower on the top of the airfoil and higher on the bottom. The answer is simply that the aerodynamic flow over the airfoil is obeying the laws of nature, namely, mass continuity and Newton's second law. Let us look at this more closely and see how nature applies these laws so as to produce lift on an airplane wing. There are three intellectual thoughts that follow in sequence:

1.   Consider the flow over an airfoil as sketched in Fig. 5.71a. Consider the stream tubes *A* and *B* shown here. The shaded stream tube *A* flows over the top surface, and the unshaded stream tube *B* flows over the bottom surface. Both stream tubes originate in the free stream ahead of the airfoil. As stream tube *A* flows toward the airfoil, it senses the upper portion of the airfoil as an obstruction, and stream tube *A* must move out of the way of this obstruction. In so doing, stream tube *A* is squashed to a smaller cross-sectional area as it flows over the nose of the airfoil. In turn, because of *mass continuity* ($\rho A V$ = constant), the velocity of the flow in the stream tube must *increase* in the region where the stream tube is being squashed. This higher velocity is

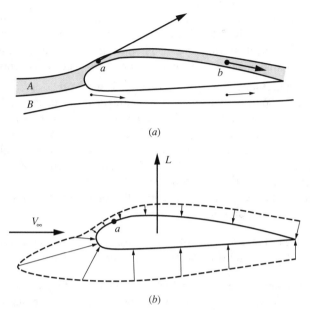

Note: The length of the arrows denoting pressure is proportional to $p - p_{ref}$, where p_{ref} is an arbitrary reference pressure slightly less than the minimum pressure on the airfoil.

Figure 5.71 (*a*) Flow velocity on upper surface is on the average higher than that on the bottom surface due to squashing of streamline A compared to streamline B. (*b*) As a result, pressure on the top surface is lower than the pressure on the bottom surface, hence creating lift in the upward direction.

shown by the long arrow at point *a* in Fig. 5.71*a*. As the stream tube flows downstream of point *a*, its cross-sectional area gradually increases and the flow velocity decreases, such as shown by the shorter arrow at point *b*. Note that stream tube *A* is squashed the most in the nose region, *ahead* of the maximum thickness of the airfoil. Hence the maximum velocity occurs *ahead* of the maximum thickness of the airfoil. Now consider stream tube *B*, which flows over the bottom surface of the airfoil. The airfoil is designed with positive camber, hence the bottom surface of the airfoil presents less of an obstruction to stream tube *B*, and so stream tube *B* is not squashed as much as stream tube *A* in flowing over the nose of the airfoil. As a result, the flow velocity in stream tube *B* remains *less* than that in stream tube *A*. Therefore, we can state the following:

Because of the law of mass continuity, that is, the continuity equation, the flow velocity increases over the top surface of the airfoil more than it does over the bottom surface.

To see the squashing of the stream tube in an actual flow, return to the smoke flow photograph in Fig. 2.6. Clearly the stream tube flowing over the top surface of the airfoil is being squashed in the region just downstream of the leading edge, and this is where the maximum flow velocity is occurring.

2. For an incompressible flow, from Bernoulli's equation $p + \frac{1}{2}\rho V^2 = $ constant, clearly where the velocity increases, the static pressure decreases. This trend is the same for compressible flow. From Euler's equation $dp = -\rho V \, dV$, when the velocity increases (dV positive), the pressure decreases (dp negative). We can label this general trend—namely, when the velocity increases, the pressure decreases—the *Bernoulli effect*. Recall that Bernoulli's equation and Euler's equation are statements of Newton's second law. Because we have shown in (1) above that the flow velocity is higher over the top surface than it is over the bottom surface, we can state the following:

> Because of the Bernoulli effect, the pressure over the top surface of the airfoil is less than the pressure over the bottom surface.

This is illustrated in Fig. 5.71b, which is a schematic of the pressure distribution over the top and bottom surfaces. Note that the minimum pressure occurs at point a.

3. Finally, it follows that

> Owing to the lower pressure over the top surface and the higher pressure over the bottom surface, the airfoil feels a lift force in the upward direction.

This lift force is shown schematically in Fig. 5.71b.

The sequence of items 1 through 3 above is the fundamental laws of nature that result in lift being produced on an airplane wing. You cannot get more fundamental than this—mass conservation and Newton's second law.

We also note that the above explanation shows why most of the lift of the wing is produced by the first 20 or 30 percent of the wing just downstream of the leading edge. This is seen in Fig. 5.71b where the largest *difference* in pressure between the top and bottom surfaces is on the front part of the airfoil. That most of the lift is generated by the forward portion of the airfoil is also seen in Figs. 5.16, 4.47, and 4.48, which demonstrate that the minimum pressure on the top surface occurs over the forward portion of the airfoil just downstream of the leading edge. In a sense, the main function of the back portion of the airfoil is to simply form a streamlined shape to avoid flow separation.

We dispel here a common *misconception* about why the flow velocity increases over the top surface of the airfoil. It is sometimes found in the literature that a fluid element which comes into the stagnation region splits into two elements, one of which flows over the top surface and the other over the bottom surface. It is then assumed that these two elements must meet up at the trailing edge, and because the running distance over the top surface of the airfoil is longer than that over the bottom surface, the element over the top surface must move faster. This is simply *not true*.

Experimental results and computational fluid dynamic calculations clearly show that a fluid element moving over the top surface of an airfoil leaves the trailing edge *long before* its companion moving over bottom surface arrives at the trailing edge. This is illustrated in Fig. 5.72. Consider a combined fluid element *CD* at time *t*, in the stagnation region at the leading edge of the airfoil, as sketched in Fig. 5.72. This element splits into element *C* moving over the top surface and element *D* moving over the bottom surface. At a later time t_2 element *C* has moved downstream of the trailing edge, and element *D* has not yet arrived at the trailing edge. The two elements simply do *not* meet at the trailing edge, and so any explanation that depends on their meeting is flawed.

There are several *alternate* explanations of the generation of lift that are in reality not *the fundamental explanation*, but rather are more of an *effect* of lift being produced, not the *cause*. Let us examine these alternate explanations.

The following alternate explanation is sometimes given: The wing deflects the airflow such that the mean velocity vector behind the wing is canted slightly downward, as sketched in Fig. 5.73. Hence, the wing imparts a downward component of momentum to the air; that is, the wing exerts a force on the air, pushing the flow downward. From Newton's third law, the equal and opposite reaction produces a lift. However, this explanation really involves the *effect* of lift, and not the cause. In

Figure 5.72 The tracking of two fluid elements in the flow over an airfoil. Element C moves over the top, and element D over the bottom.

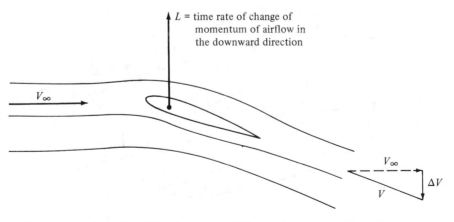

Figure 5.73 Relationship of lift to the time rate of change of momentum of the airflow.

reality, the air pressure on the surface is pushing on the surface, hence creating lift in the upward direction. As a result of the equal-and-opposite principle, the airfoil surface pushes on the air, imparting a downward force on the airflow which deflects the velocity downward. Hence, the net rate of change of downward momentum created in the airflow because of the presence of the wing can be thought of as an *effect* due to the surface pressure distribution; the pressure distribution by itself is the fundamental cause of lift.

A third argument, called the *circulation theory of lift*, is sometimes given for the source of lift. However, this turns out to be not so much an *explanation* of lift per se, but rather more of a mathematical formulation for the calculation of lift for an airfoil of given shape. Moreover, it is mainly applicable to incompressible flow. The circulation theory of lift is elegant and well developed; it is also beyond the scope of this book. However, some of its flavor is given as follows.

Consider the flow over a given airfoil, as shown in Fig. 5.74. Imagine a closed curve C drawn around the airfoil. At a point on this curve, the flow velocity is V, and the angle between V and a tangent to the curve is θ. Let ds be an incremental distance along C. A quantity called the *circulation* Γ is defined as

$$\Gamma \equiv \oint_C V \cos\theta \, ds \qquad \text{[5.72]}$$

that is, Γ is the line integral of the component of flow velocity along the closed curve C. After a value of Γ is obtained, the lift *per unit span* can be calculated from

$$\boxed{L = \rho_\infty V_\infty \Gamma_\infty} \qquad \text{[5.73]}$$

Equation (5.73) is the *Kutta-Joukowsky theorem*; it is a pivotal relation in the circulation theory of lift. The object of the theory is to (somehow) calculate Γ for a

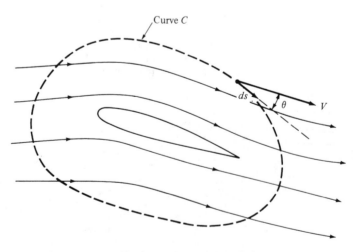

Figure 5.74   Diagram for the circulation theory of lift.

given V_∞ and airfoil shape. Then Eq. (5.73) yields the lift. A major thrust of ideal incompressible flow theory, many times called *potential flow theory*, is to calculate Γ. Such matters are discussed in more-advanced aerodynamic texts (see, e.g., Anderson, *Fundamentals of Aerodynamics*, 2d ed., McGraw-Hill, 1991).

The circulation theory of lift is compatible with the true physical nature of the flow over an airfoil, as any successful mathematical theory must be. In the simplest sense, we can visualize the true flow over an airfoil, shown at the right of Fig. 5.75, as the superposition of a uniform flow and a circulatory flow, shown at the left of Fig. 5.75. The circulatory flow is clockwise, which when added to the uniform flow, yields a higher velocity above the airfoil and a lower velocity below the airfoil. From Bernoulli's equation, this implies a lower pressure on the top surface of the airfoil and a higher pressure on the bottom surface, hence generating lift in the upward direction. The strength of the circulatory contribution, defined by Eq. (5.72), is just the precise value such that when it is added to the uniform flow contribution, the actual flow over the airfoil leaves the trailing edge smoothly, as sketched at the right of Fig. 5.75. This is called the *Kutta condition* and is one of the major facets of the circulation theory of lift.

Again, keep in mind that the actual mechanism that nature has of communicating a lift to the airfoil is the pressure distribution over the surface of the airfoil, as sketched in Fig. 5.71*b*. In turn, this pressure distribution ultimately causes a time rate of change of momentum of the airflow, as shown in Fig. 5.73, a principle which can be used as an alternate way of visualizing the generation of lift. Finally, even the circulation theory of lift stems from the pressure distribution over the surface of the airfoil, because the derivation of the Kutta-Joukowsky theorem, Eq. (5.73), involves the surface pressure distribution. Again, for more details, consult Anderson, *Fundamentals of Aerodynamics*, 2d ed., McGraw-Hill, 1991.

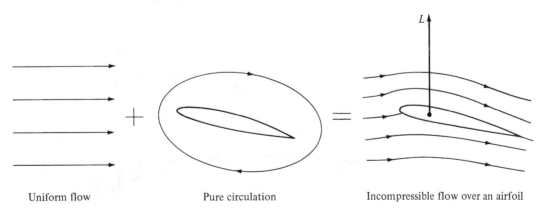

Uniform flow   Pure circulation   Incompressible flow over an airfoil

Figure 5.75   Addition of two elementary flows to synthesize a more complex flow. If one or more of the elementary flows have circulation, then the synthesized flow also has the same circulation. The lift is directly proportional to the circulation.

5.20   HISTORICAL NOTE: AIRFOILS AND WINGS

We know that George Cayley introduced the concept of a fixed-wing aircraft in 1799; this has been discussed at length in Secs. 1.3 and 5.1. Moreover, Cayley appreciated the fact that lift is produced by a region of low pressure on the top surface of the wing and high pressure on the bottom surface and that a cambered shape produces more lift than a flat surface. Indeed, Fig. 1.5 shows that Cayley was thinking of a curved surface for a wing, although the curvature was due to the wind billowing against a loosely fitting fabric surface. However, neither Cayley nor any of his immediate followers performed work even closely resembling airfoil research or development.

It was not until 1884 that the first serious airfoil developments were made. In this year Horatio F. Phillips, an Englishman, was granted a patent for a series of double-surface, cambered airfoils. Figure 5.76 shows Phillip's patent drawings for his airfoil section. Phillips was an important figure in late 19th-century aeronautical engineering; we met him before, in Sec. 4.24, in conjunction with his ejector-driven wind tunnel. In fact, the airfoil shapes in Fig. 5.76 were the result of numerous wind tunnel experiments in which Phillips examined curved wings of "every conceivable form and combination of forms." Phillips widely published his results, which had a major impact on the aeronautics community. Continuing with his work, Phillips patented more airfoil shapes in 1891. Then, moving into airplane design in 1893, he built and tested a large multiplane model, consisting of a large number of wings, each with a 19-ft span and a chord of only $1\frac{1}{2}$ in, which looked like a venetian blind! The airplane was powered by a steam engine with a 6.5-ft propeller. The vehicle ran

Figure 5.76   Double-surface airfoil sections by Phillips. The six upper shapes were patented by Phillips in 1884; the lower airfoil was patented in 1891.

on a circular track and actually lifted a few feet off the ground, momentarily. After this demonstration, Phillips gave up until 1907, when he made the first tentative hop flight in England in a similar, but gasoline-powered, machine, staying airborne for about 500 ft. This was his last contribution to aeronautics. However, his pioneering work during the 1880s and 1890s clearly designates Phillips as the grandparent of the modern airfoil.

After Phillips, the work on airfoils shifted to a search for the most efficient shapes. Indeed, work is still being done today on this very problem, although much progress has been made. This progress covers several historical periods, as described below.

5.20.1   The Wright Brothers

As noted in Secs. 1.8 and 4.24, Wilbur and Orville Wright, after their early experience with gliders, concluded in 1901 that many of the existing "air pressure" data on airfoil sections were inadequate and frequently incorrect. To rectify these deficiencies, they constructed their own wind tunnel (see Fig. 4.51), in which they tested several hundred different airfoil shapes between September 1901 and August 1902. From their experimental results, the Wright brothers chose an airfoil with a 1/20 maximum camber-to-chord ratio for their successful *Wright Flyer I* in 1903. These wind tunnel tests by the Wright brothers constituted a major advance in airfoil technology at the turn of the century.

5.20.2   British and U.S. Airfoils (1910 to 1920)

In the early days of powered flight, airfoil design was basically customized and personalized; very little concerted effort was made to find a standardized, efficient section. However, some early work was performed by the British government at the National Physical Laboratory (NPL), leading to a series of Royal Aircraft Factory (RAF) airfoils used on World War I airplanes. Figure 5.77 illustrates the shape of the RAF 6 airfoil. In the United States, most aircraft until 1915 used either an RAF section or a shape designed by the Frenchman Alexandre Gustave Eiffel. This tenuous status of airfoils led the NACA, in its First Annual Report in 1915, to emphasize the need for "the evolution of more efficient wing sections of practical form, embodying

Figure 5.77   Typical airfoils in 1917.

suitable dimensions for an economical structure, with moderate travel of the center of pressure and still affording a large angle of attack combined with efficient action." To this day, more than 85 years later, NASA is still pursuing such work.

The first NACA work on airfoils was reported in NACA Report No. 18, "Aerofoils and Aerofoil Structural Combinations," by Lt. Col. Edgar S. Gorrell and Major H. S. Martin, prepared at the Massachusetts Institute of Technology (MIT) in 1917. Gorrell and Martin summarized the contemporary airfoil status as follows:

> Mathematical theory has not, as yet, been applied to the discontinuous motion past a cambered surface. For this reason, we are able to design aerofoils only by consideration of those forms which have been successful, by applying general rules learned by experience, and by then testing the aerofoils in a reliable wind tunnel.

In NACA Report No. 18, Gorrell and Martin disclosed a series of tests on the largest single group of airfoils to that date, except for the work done at NPL and by Eiffel. They introduced the USA airfoil series and reported wind-tunnel data for the USA 1 through USA 6 sections. Figure 5.77 illustrates the shape of the USA 6 airfoil. The airfoil models were made of brass and were finite wings with a span of 18 in and a chord of 3 in, that is, AR = 6. Lift and drag coefficients were measured at a velocity of 30 mi/h in the MIT wind tunnel. These airfoils represented the first systematic series originated and studied by NACA.

5.20.3 1920 to 1930

Based on their wind tunnel observations in 1917, Gorrell and Martin stated that slight variations in airfoil design make large differences in the aerodynamic performance. This is the underlying problem of airfoil research, and it led in the 1920s to a proliferation of airfoil shapes. In fact, as late as 1929, F. A. Louden, in his NACA Report No. 331, entitled "Collection of Wind Tunnel Data on Commonly Used Wing Sections," stated that "the wing sections most commonly used in this country are the Clark Y, Clark Y-15, Gottingen G-387, G-398, G-436, N.A.C.A. M-12, Navy N-9, N-10, N-22, R.A.F.-15, Sloane, U.S.A.- 27, U.S.A.-35A, U.S.A.-35B." However, help was on its way. As noted in Sec. 4.24, the NACA built a variable-density wind tunnel at Langley Aeronautical Laboratory in 1923, a wind tunnel that was to become a workhorse in future airfoil research, as emphasized below.

5.20.4 Early NACA Four-Digit Airfoils

In a classical work in 1933, order and logic were finally brought to airfoil design in the United States. This was reported in NACA Report No. 460, "The Characteristics of 78 Related Airfoil Sections from Tests in the Variable-Density Wind Tunnel," by Eastman N. Jacobs, Kenneth E. Ward, and Robert M. Pinkerton. Their philosophy on airfoil design was as follows:

> Airfoil profiles may be considered as made up of certain profile-thickness forms disposed about certain mean lines. The major shape variables then become two,

the thickness form and the mean-line form. The thickness form is of particular importance from a structural standpoint. On the other hand, the form of the mean line determines almost independently some of the most important aerodynamic properties of the airfoil section, e.g., the angle of zero lift and the pitching-moment characteristics. The related airfoil profiles for this investigation were derived by changing systematically these shape variables.

They then proceeded to define and study for the first time in history the famous NACA four-digit airfoil series, some of which are given in App. D of this book. For example, NACA 2412 is defined as a shape that has a maximum camber of 2 percent of the chord (the first digit); the maximum camber occurs at a position of 0.4 chord from the leading edge (the second digit), and the maximum thickness is 12 percent (the last two digits). Jacobs and his colleagues tested these airfoils in the NACA variable-density tunnel using a 5-in by 30-in finite wing (again, an aspect ratio of 6). In NACA Report No. 460, they gave curves of C_L, C_D, and L/D for the finite wing. Moreover, using the same formulas developed in Sec. 5.15, they corrected their data to give results for the infinite wing case also. After this work was published, the standard NACA four-digit airfoils were widely used. Indeed, even today the NACA 2412 is used on several light aircraft.

5.20.5 Later NACA Airfoils

In the late 1930s, NACA developed a new camber line family to increase maximum lift, with the 230 camber line being the most popular. Combining with the standard NACA thickness distribution, this gave rise to the NACA five-digit airfoil series, such as the 23012, some of which are still flying today, for example, on the Cessna Citation and the Beech King Air. This work was followed by families of high-speed airfoils and laminar flow airfoils in the 1940s.

To reinforce its airfoil development, in 1939 NACA constructed a new low-turbulence two-dimensional wind tunnel at Langley exclusively for airfoil testing. This tunnel has a rectangular test section 3 ft wide and $7\frac{1}{2}$ ft high and can be pressurized up to 10 atm for high–Reynolds number testing. Most importantly, this tunnel allows airfoil models to span the test section completely, thus directly providing infinite wing data. This is in contrast to the earlier tests described above, which used a finite wing of AR = 6 and then corrected the data to correspond to infinite wing conditions. Such corrections are always compromised by tip effects. (For example, what is the *precise* span efficiency factor for a given wing?) With the new two-dimensional tunnel, vast numbers of tests were performed in the early 1940s on both old and new airfoil shapes over a Reynolds number range from 3 to 9 million and at Mach numbers less than 0.17 (incompressible flow). The airfoil models generally had a 2-ft chord and completely spanned the 3-ft width of the test section. It is interesting to note that the lift and drag are not obtained on a force balance. Rather, the lift is calculated by integrating the measured pressure distribution on the top and bottom walls of the wind tunnel, and the drag is calculated from Pitot pressure measurements made in the wake downstream of the trailing edge. However, the pitching moments are measured directly on a balance. A vast amount of airfoil data obtained in this fashion from the two-dimensional tunnel at Langley were compiled and published in a book entitled *Theory of Wing Sections Including a Summary of Airfoil Data*, by Abbott and von Doenhoff, in 1949 (see Bibliography at end of this chapter). It is important to note that all the airfoil data in App. D are obtained from this reference; that is, all the data in App. D are direct measurements for an infinite wing at essentially incompressible flow conditions.

5.20.6 Modern Airfoil Work

Priorities for supersonic and hypersonic aerodynamics put a stop to the NACA airfoil development in 1950. Over the next 15 years specialized equipment for airfoil testing was dismantled. Virtually no systematic airfoil research was done in the United States during this period.

However, in 1965 Richard T. Whitcomb made a breakthrough with the NASA supercritical airfoil. This revolutionary development, which allowed the design of wings with high critical Mach numbers (see Sec. 5.10), served to reactivate the interest in airfoils within NASA. Since that time, a healthy program in modern airfoil development has been reestablished. The low-turbulence, pressurized, two-dimensional wind tunnel at Langley is back in operation. Moreover, a new dimension has been added to airfoil research: the high-speed digital computer. In fact, computer programs for calculating the flow field around airfoils at subsonic speeds are so reliable that they shoulder some of the routine testing duties heretofore exclusively carried by wind tunnels. The same cannot yet be said about transonic cases, but current research is focusing on this problem. An interesting survey of modern airfoil activity within NASA is given by Pierpont in *Astronautics and Aeronautics* (see Bibliography).

Of special note is the modern low-speed airfoil series, designated LS(1), developed by NASA for use by general aviation on light airplanes. The shape of a typical LS(1) airfoil is contrasted with a "conventional" airfoil in Fig. 5.78. Its lifting characteristics, illustrated in Fig. 5.79, are clearly superior and should allow smaller wing areas, hence less drag, for airplanes of the type shown in Fig. 5.78.

In summary, airfoil development over the past 100 years has moved from an ad hoc individual process to a very systematic and logical engineering process. It is alive and well today, with the promise of major advancements in the future using both wind tunnels and computers.

5.20.7 Finite Wings

Some historical comments on the finite wing are in order. Francis Wenham (see Chap. 1), in his classic paper entitled *Aerial Locomotion*, given to the Aeronautical Society of Great Britain on June 27, 1866, theorized (correctly) that most of the lift of a wing occurs from the portion near the leading edge, hence a long, narrow wing would be most efficient. In this fashion, he was the first person in history to appreciate the value of high–aspect ratio wings for subsonic flight. Moreover, he suggested stacking

Figure 5.78   Shape comparison between the modern LS(1)-0417 and a conventional airfoil. The higher lift obtained with the LS(1)-0417 allows a smaller wing area and hence lower drag. *(NASA.)*

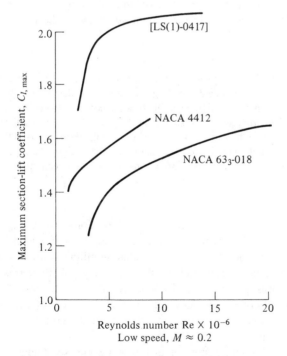

Figure 5.79   Comparison of maximum lift coefficients between the LS(1)-0417 and conventional airfoils. *(NASA.)*

a number of long, thin wings above one another to generate the required lift; hence he became an advocate of the multiplane concept. In turn, he built two full-size gliders in 1858, both with five wings each, and successfully demonstrated the validity of his ideas.

However, the true aerodynamic theory and understanding of finite wings did not come until 1907, when the Englishman Frederick W. Lanchester published his book *Aerodynamics.* In it, he outlined the circulation theory of lift, developed independently about the same time by Kutta in Germany and by Joukowsky in Russia. More importantly, Lanchester discussed for the first time the effect of wing tip vortices on finite-wing aerodynamics. Unfortunately, Lanchester was not a clear writer; his ideas were extremely difficult to understand, and hence they did not find application in this aeronautical community.

In 1908, Lanchester visited Göttingen, Germany, and fully discussed his wing theory with Ludwig Prandtl and his student, Theodore von Karman. Prandtl spoke no English, Lanchester spoke no German, and in light of Lanchester's unclear ways of explaining his ideas, there appeared to be little chance of understanding between the two parties. However, in 1914, Prandtl set forth a simple, clear, and correct theory for calculating the effect of tip vortices on the aerodynamic characteristics of finite wings. It is virtually impossible to assess how much Prandtl was influenced by Lanchester, but to Prandtl must go the credit of first establishing a practical finite wing theory, a theory which is fundamental to our discussion of finite wings in Secs. 5.13 to 5.15. Indeed, Prandtl's first published words on the subject were these:

> The lift generated by the airplane is, on account of the principle of action and reaction, necessarily connected with a descending current in all its details. It appears that the descending current is formed by a pair of vortices, the vortex filaments of which start from the airplane wingtips. The distance of the two vortices is equal to the span of the airplane, their strength is equal to the circulation of the current around the airplane, and the current in the vicinity of the airplane is fully given by the superposition of the uniform current with that of a vortex consisting of three rectilinear sections.

Prandtl's pioneering work on finite wing theory, along with his ingenious concept of the boundary layer, has earned him the title *parent of aerodynamics.* In the four years following 1914, he went on to show that an elliptical lift distribution results in the minimum induced drag. Indeed, the first coining of the terms *induced drag* and *profile drag* was made in 1918 by Max Munk in a note entitled "Contribution to the Aerodynamics of the Lifting Organs of the Airplane." Munk was a colleague of Prandtl's and the note was one of several classified German wartime reports on airplane aerodynamics.

For more details on the history of airfoils and wings, see Anderson, *A History of Aerodynamics and Its Impact on Flying Machines*, Cambridge University Press, New York, 1997.

5.21 HISTORICAL NOTE: ERNST MACH AND HIS NUMBER

Airplanes that fly at Mach 2 are commonplace today. Indeed, high-performance military aircraft such as the Lockheed SR-71 Blackbird can exceed Mach 3. As a result, the term *Mach number* has become part of our general language—the average person in the street understands that Mach 2 means twice the speed of sound. On a more technical basis, the dimensional analysis described in Sec. 5.3 demonstrated that aerodynamic lift, drag, and moments depend on two important dimensionless products—the Reynolds number and the Mach number. Indeed, in a more general treatment of fluid dynamics, the Reynolds number and Mach number can be shown as the major governing parameters for any realistic flow field; they are among a series of governing dimensionless parameters called *similarity parameters*. We already examined the historical source of the Reynolds number in Sec. 4.25; let us do the same for the Mach number in this present section.

The Mach number is named after Ernst Mach, a famous 19th-century physicist and philosopher. Mach was an illustrious figure with widely varying interests. He was the first person in history to observe supersonic flow and to understand its basic nature. Let us take a quick look at this man and his contributions to supersonic aerodynamics.

Ernst Mach was born at Turas, Moravia, in Austria, on February 18, 1838. Mach's father and mother were both extremely private and introspective intellectuals. His father was a student of philosophy and classical literature; his mother was a poet and musician. The family was voluntarily isolated on a farm, where Mach's father pursued an interest of raising silkworms—indeed pioneering the beginning of silkworm culture in Europe. At an early age, Mach was not a particularly successful student. Later, Mach described himself as a "weak pitiful child who developed very slowly." Through extensive tutoring by his father at home, Mach learned Latin, Greek, history, algebra, and geometry. After marginal performances in grade school and high school (due not to any lack of intellectual ability, but rather to a lack of interest in the material usually taught by rote), Mach entered the University of Vienna. There he blossomed, spurred by interest in mathematics, physics, philosophy, and history. In 1860, he received a Ph.D. in physics, writing a thesis entitled "On Electrical Discharge and Induction." By 1864, he was a professor of physics at the University of Graz. (The variety and depth of his intellectual interests at this time are attested by the fact that he turned down the position of a chair in *surgery* at the University of Salzburg to go to Graz.) In 1867, Mach became a professor of experimental physics at the University of Prague—a position he would occupy for the next 28 years.

In today's modern technological world, where engineers and scientists are virtually forced, out of necessity, to peak their knowledge in very narrow areas of extreme specialization, it is interesting to reflect on the personality of Mach, who was the supreme generalist. Here is only a partial list of Mach's contributions, as demonstrated in his writings: physical optics, history of science, mechanics, philosophy, origins of relativity theory, supersonic flow, physiology, thermodynamics, sugar cycle in grapes, physics of music, and classical literature. He even wrote on world affairs. (One of Mach's papers commented on the "absurdity committed by the statesman who regards the individual as existing solely for the sake of the state"; for this, Mach was severely criticized at that time by Lenin.) We can only sit back with awe and envy for Mach, who—in the words of U.S. philosopher William James—knew "everything about everything."

Mach's contributions to supersonic aerodynamics were highlighted in a paper entitled "Photographische Fixierung der durch Projektile in der Luft eingeleiten Vorgange," given to the Academy of Sciences in Vienna in 1887. Here, for the first time in history, Mach showed a photograph of a shock wave in front of a bullet moving at supersonic speeds. This historic photograph, taken from Mach's original paper, is shown in Fig. 5.80. Also visible are weaker waves at the rear of the projectile and the structure of the turbulent wake downstream of the base region. The two vertical lines are trip wires designed to time the photographic light source (or spark) with the passing of the projectile. Mach was a precise and careful experimenter; the quality of the picture shown in Fig. 5.80, along with the very fact that he was able to make the shock waves visible in the first place (he used an innovative technique called the shadowgram), attests to his exceptional experimental abilities. Note that Mach was able to carry out such experiments involving split-second timing without the benefit of electronics—indeed, the vacuum tube had not yet been invented.

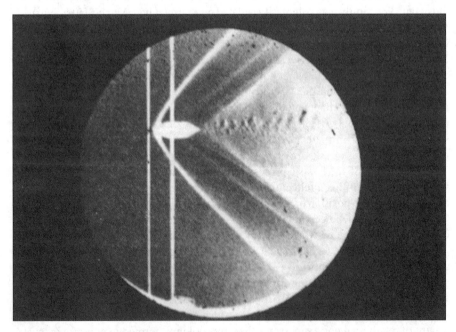

Figure 5.80   Photograph of a bullet in supersonic flight, published by Ernst Mach in 1887.

Mach was the first to understand the basic characteristics of supersonic flow. He was the first to point out the importance of the flow velocity V relative to the speed of sound a and to note the discontinuous and marked changes in a flow field as the ratio V/a changes from less than 1 to greater than 1. He did not, however, call this ratio the Mach number. The term *Mach number* was not coined until 1929, when the well-known Swiss engineer Jacob Ackeret introduced this terminology, in honor of Mach, at a lecture at the Eidgenossiche Technische Hochschule in Zurich. Hence, the term *Mach number* is of fairly recent use, not introduced into the English literature until 1932.

Mach was an active thinker, lecturer, and writer up to the time of his death on February 19, 1916, near Munich, one day after his 78th birthday. His contributions to human thought were many, and his general philosophy on epistemology—a study of knowledge itself—is still discussed in college classes in philosophy today. Aeronautical engineers know him as the originator of supersonic aerodynamics; the rest of the world knows him as a man who originated the following philosophy, as paraphrased by Richard von Mises, himself a well-known mathematician and aerodynamicist of the early 20th century:

> Mach does not start out to analyze statements, systems of sentences, or theories, but rather the world of phenomena itself. His elements are not the simplest sentences, and hence the building stones of theories, but rather—at least according to his way of speaking—the simplest facts, phenomena, and events of which the world in which we live and which we know is composed. The world open to our observation and experience consists of "colors, sounds, warmths, pressure, spaces, times, etc." and their components in greater and smaller complexes. All we make statements or assertions about, or formulate questions and answers to, are the relations in which these elements stand to each other. That is Mach's point of view.[1]

We end this section with a photograph of Mach, taken about 1910, shown in Fig. 5.81. It is a picture of a thoughtful, sensitive man; no wonder that his philosophy of life emphasized observation through the senses, as discussed by von Mises above. In honor of his memory, an entire research institute, the Ernst Mach Institute in Germany, is named. This institute deals with research in experimental gas dynamics, ballistics, high-speed photography, and cinematography. Indeed, for a much more extensive review of the technical accomplishments of Mach, see the recent paper authored by a member of the Ernst Mach Institute, H. Reichenbach, entitled "Contributions of Ernst Mach to Fluid Mechanics," in *Annual Reviews of Fluid Mechanics*, vol. 15, 1983, pp. 1–28 (published by Annual Reviews, Inc., Palo Alto, California).

---

[1] From Richard von Mises, *Positivism, A Study in Human Understanding*, Braziller, New York, 1956.

Figure 5.81   Ernst Mach (1838–1916).

5.22   HISTORICAL NOTE: THE FIRST MANNED SUPERSONIC FLIGHT

On October 14, 1947, a human being flew faster than the speed of sound for the first time in history. Imagine the magnitude of this accomplishment—just 60 years after Ernst Mach observed shock waves on supersonic projectiles (see Sec. 5.21) and a scant 44 years after the Wright brothers achieved their first successful powered flight (see Secs. 1.1 and 1.8). It is almost certain that Mach was not thinking at all about heavier-than-air manned flight of any kind, which in his day was still considered to be virtually impossible and the essence of foolish dreams. It is also almost certain that the Wright brothers had not the remotest idea that their fledgling 30 mi/h flight on December 17, 1903, would ultimately lead to a manned supersonic flight in Orville's lifetime (although Wilbur died in 1912, Orville lived an active life until his death in 1948). Compared to the total spectrum of manned flight reaching all the way back to the ideas of Leonardo da Vinci in the 15th century (see Sec. 1.2), this rapid advancement into the realm of supersonic flight is truly phenomenal. How did this advancement occur? What are the circumstances surrounding the first supersonic flight? Why was it so important? The purpose of this section is to address these questions.

Supersonic flight did not happen by chance; rather, it was an inevitable result of the progressive advancement of aeronautical technology over the years. On one hand, we have the evolution of high-speed aerodynamic theory, starting with the pioneering work of Mach, as described in Sec. 5.21. This was followed by the development of supersonic nozzles by two European engineers, Carl G. P. de Laval in Sweden and A. B. Stodola in Switzerland. In 1887 de Laval used a convergent-divergent supersonic nozzle to produce a high-velocity flow of steam to drive a turbine. In 1903 Stodola was the first person in history to definitely prove (by means of a series of laboratory experiments) that such convergent-divergent nozzles did indeed produce supersonic flow. From 1905 to 1908, Prandtl in Germany took pictures of Mach waves inside supersonic nozzles and developed the first rational theory for oblique shock waves and expansion waves. After World War I, Prandtl studied compressibility effects in high-speed subsonic flow. This work, in conjunction with independent studies by the English aerodynamicist Herman Glauert, led to the publishing of the Prandtl-Glauert rule in the late 1920s (see Sec. 5.6 for a discussion of the Prandtl-Glauert rule and its use as a compressibility correction). These milestones, among others, established a core of aerodynamic theory for high-speed flow—a core that was well established at least 20 years before the first supersonic flight. (For more historical details concerning the evolution of aerodynamic theory pertaining to supersonic flight, see Anderson, *Modern Compressible Flow: With Historical Perspective*, 2d ed., McGraw-Hill, 1990.)

On the other hand, we also have the evolution of hardware necessary for supersonic flight. The development of high-speed wind tunnels, starting with the small 12-in-diameter high-speed subsonic tunnel at NACA Langley Memorial Aeronautical Laboratory in 1927 and continuing with the first practical supersonic wind tunnels developed by Adolf Busemann in Germany in the early 1930s, is described in Sec. 4.24. The exciting developments leading to the first successful rocket engines in the late 1930s are discussed in Sec. 9.16. The concurrent invention and development of the jet engine, which would ultimately provide the thrust necessary for everyday supersonic flight, are related in Sec. 9.15. Hence, on the basis of the theory and hardware which existed at that time, the advent of manned supersonic flight in 1947 was a natural progression in the advancement of aeronautics.

However, in 1947 there was one missing link in both the theory and the hardware—the transonic regime, near Mach 1. The governing equations for transonic flow are highly nonlinear and hence are difficult to solve. No practical solution of these equations existed in 1947. This theoretical gap was compounded by a similar gap in wind tunnels. The sensitivity of a flow near Mach 1 makes the design of a proper transonic tunnel difficult. In 1947, no reliable transonic wind tunnel data were available. This gap of knowledge was of great concern to the aeronautical engineers who designed the first supersonic airplane, and it was the single most important reason for the excitement, apprehension, uncertainty, and outright bravery that surrounded the first supersonic flight.

The unanswered questions about transonic flow did nothing to dispel the myth of the "sound barrier" that arose in the 1930s and 1940s. As discussed in Sec. 5.12, the very rapid increase in drag coefficient beyond the drag-divergence Mach number led some people to believe that humans would never fly faster than the speed of sound. Grist was lent to their arguments when, on September 27, 1946, Geoffrey deHavilland, son of the famous British airplane designer, took the *D.H. 108 Swallow* up for an attack on the world's speed record. The *Swallow* was an experimental jet-propelled aircraft, with swept wings and no horizontal tail. Attempting to exceed 615 mi/h on its first high-speed, low-altitude run, the Swallow encountered major compressibility problems and broke up in the air. DeHavilland was killed instantly. The sound barrier had taken its toll. It was against this background that the first supersonic flight was attempted in 1947.

During the late 1930s, and all through World War II, some visionaries clearly saw the need for an experimental airplane designed to probe the mysteries of supersonic flight. Finally, in 1944, their efforts prevailed; the Army Air Force, in conjunction with NACA, awarded a contract to Bell Aircraft Corporation for the design, construction, and preliminary testing of a manned supersonic airplane. Designated the XS-1 (Experimental Sonic-1), this design had a fuselage shaped like a 50-caliber bullet, mated to a pair of very thin (thickness-to-chord ratio of 0.08), low–aspect ratio, straight wings, as shown in Fig. 5.82. The aircraft was powered by a four-chamber liquid-propellant rocket engine mounted in the tail. This engine, made by Reaction Motors and designated the XLR11, produced 6000 lb of thrust by burning a mixture of liquid oxygen and diluted alcohol.

The Bell XS-1 was designed to be carried aloft by a parent airplane, such as the giant Boeing B-29, and then launched at altitude; in this fashion, the extra weight of fuel that would have been necessary for takeoff and climb to altitude was saved,

Figure 5.82   The Bell XS-1, the first supersonic airplane, 1947. *(National Air and Space Museum.)*

allowing the designers to concentrate on one performance aspect—speed. Three XS-1s were ultimately built, the first one being completed just after Christmas 1945. There followed a year and a half of gliding and then powered tests, wherein the XS-1 was cautiously nudged toward the speed of sound.

Muroc Dry Lake is a large expanse of flat, hard lake bed in the Mojave Desert in California. Site of a U.S. Army high-speed flight test center during World War II, Muroc was later to become known as Edwards Air Force Base, now the site of the Air Force Test Pilots School and the home of all experimental high-speed flight testing for both the Air Force and NASA. On Tuesday, October 14, 1947, the Bell XS-1, nestled under the fuselage of a B-29, was waiting on the flight line at Muroc. After intensive preparations by a swarm of technicians, the B-29 with its cargo took off at 10 AM. On board the XS-1 was Captain Charles E. (Chuck) Yeager. That morning, Yeager was in excruciating pain from two broken ribs fractured during a horseback riding accident earlier that weekend; however, he told virtually no one. At 10:26 AM at an altitude of 20,000 ft, the Bell XS-1, with Yeager as its pilot, was dropped from the B-29. What happened next is one of the major milestones in aviation history. Let us see how Yeager himself recalled events, as stated in his written flight report:

Date: 14 October 1947
Pilot: Capt. Charles E. Yeager
Time: 14 Minutes
        9th Powered Flight

1. After normal pilot entry and the subsequent climb, the XS-1 was dropped from the B-29 at 20,000′ and at 250 MPH IAS. This was slower than desired.

2. Immediately after drop, all four cylinders were turned on in rapid sequence, their operation stabilizing at the chamber and line pressures reported in the last flight. The ensuing climb was made at .85–.88 Mach, and, as usual it was necessary to change the stabilizer setting to 2 degrees nose down from its pre-drop setting of 1 degree nose down. Two cylinders were turned off between 35,000′ and 40,000′, but speed had increased to .92 Mach as the airplane was leveled off at 42,000′. Incidentally, during the slight push-over at this altitude, the lox line pressure dropped perhaps 40 psi and the resultant rich mixture caused the chamber pressures to decrease slightly. The effect was only momentary, occurring at .6 G's, and all pressures returned to normal at 1 G.

3. In anticipation of the decrease in elevator effectiveness at all speeds above .93 Mach, longitudinal control by means of the stabilizer was tried during the climb at .83, .88, and .92 Mach. The stabilizer was moved in increments of $\frac{1}{4}-\frac{1}{3}$ degree and proved to be very effective; also, no change in effectiveness was noticed at the different speeds.

4. At 42,000′ in approximately level flight, a third cylinder was turned on. Acceleration was rapid and speed increased to .98 Mach. The needle of the machmeter fluctuated at this reading momentarily, then passed off the scale. Assuming that the off-scale reading remained linear, it is estimated that 1.05 Mach was attained at this time. Approximately 30% of fuel and lox remained when this speed was reached and the motor was turned off.

5. While the usual lift buffet and instability characteristics were encountered in the .88–.90 Mach range and elevator effectiveness was very greatly decreased at

.94 Mach, stability about all three axes was good as speed increased and elevator effectiveness was regained above .97 Mach. As speed decreased after turning off the motor, the various phenomena occurred in reverse sequence at the usual speeds, and in addition, a slight longitudinal porpoising was noticed from .98–.96 Mach which was controllable by elevators alone. Incidentally, the stabilizer setting was not changed from its 2 degrees nose down position after trial at .92 Mach.

6. After jettisoning the remaining fuel and lox at 1 G stall was performed at 45,000′. The flight was concluded by the subsequent glide and a normal landing on the lakebed.

CHARLES E. YEAGER

Capt. Air Corps

In reality, the Bell SX-1 had reached $M_\infty = 1.06$, as determined from official NACA tracking data. The duration of its supersonic flight was 20.5 s, almost twice as long as the Wright brothers' entire first flight just 44 years earlier. On that day, Chuck Yeager became the first person to fly faster than the speed of sound. It is a fitting testimonial to the aeronautical engineers at that time that the flight was smooth and without unexpected consequences. An aircraft had finally been properly designed to probe the "sound barrier," which it penetrated with relative ease. Less than a month later, Yeager reached Mach 1.35 in the same airplane. The sound barrier had not only been penetrated, it had been virtually destroyed as the myth it really was.

As a final note, the whole story of the human and engineering challenges that revolved about the quest for and eventual achievement of supersonic flight is fascinating, and it is a living testimonial to the glory of aeronautical engineering. The story is brilliantly spelled out by Dr. Richard Hallion, earlier a curator at the Air and Space Museum of the Smithsonian Institution and now chief historian of the U.S. Air Force, in his book *Supersonic Flight* (see Bibliography at the end of this chapter). The reader should make every effort to study Hallion's story of the events leading to and following Yeager's flight in 1947.

5.23 HISTORICAL NOTE: THE X-15—FIRST MANNED HYPERSONIC AIRPLANE AND STEPPING-STONE TO THE SPACE SHUTTLE

Faster and higher—for all practical purposes, this has been the driving potential behind the development of aviation since the Wrights' first successful flight in 1903. (See Sec. 1.11 and Figs. 1.30 and 1.31.) This credo was never more true than during the 15 years following Chuck Yeager's first supersonic flight in the Bell XS-1, described in Sec. 5.22. Once the sound barrier was broken, it was left far behind in the dust. The next goal became manned *hypersonic* flight—Mach 5 and beyond.

To accomplish this goal, NACA initiated a series of preliminary studies in the early 1950s for an aircraft to fly beyond Mach 5, the definition of the hypersonic flight regime. This definition is essentially a rule of thumb; unlike the severe and radical flow field changes which take place when an aircraft flies through Mach 1, nothing dramatic happens when Mach 5 is exceeded. Rather, the hypersonic regime is simply a very high–Mach number regime, where shock waves are particularly strong and the gas temperatures behind these shock waves are high. For example, consider Eq. (4.73), which gives the total temperature T_0, that is, the temperature of a gas which was initially at a Mach number M_1 and which has been adiabatically slowed to zero velocity. This is essentially the temperature at the stagnation point on a body. If $M_1 = 7$, Eq. (4.73) shows that (for $\gamma = 1.4$) $T_0/T_1 = 10.8$. If the flight altitude is, say, 100,000 ft where $T_1 = 419°R$, then $T_0 = 4525°R = 4065°F$—far above the melting point of stainless steel. Therefore, as flight velocities increase far above the speed of sound, they gradually approach a *thermal barrier*, that is, those velocities beyond which the skin temperatures become too high and structural failure can occur. As in the case of the sound barrier, the thermal barrier is only a figure of speech—it is not an inherent limitation on flight speed. With proper design to overcome the high rates of aerodynamic heating, vehicles today have flown at Mach numbers as high as 36 (e.g., the Apollo lunar return capsule). (For more details on high-speed reentry aerodynamic heating, see Sec. 8.11.)

Nevertheless, in the early 1950s, manned hypersonic flight was a goal to be achieved—an untried and questionable regime characterized by high temperatures and strong shock waves. The basic NACA studies fed into an industrywide design competition for a hypersonic airplane. In 1955, North American Aircraft Corporation was awarded a joint NACA–Air Force–Navy contract to design and construct three prototypes of a manned hypersonic research airplane capable of Mach 7 and a maximum altitude of 264,000 ft. This airplane was designated the X-15 and is shown in Fig. 5.83. The first two aircraft were powered by Reaction Motors LR11 rocket engines, with 8000 lb of thrust (essentially the same as the engine used for the Bell XS-1). Along with the third prototype, the two aircraft were later reengined with a more powerful rocket motor, the Reaction Motors XLR99, capable of 57,000 lb of thrust. The basic internal structure of the airplane was made from titanium and stainless steel, but the airplane skin was Inconel X—a nickel-alloy steel capable of withstanding temperatures up to 1200°F. (Although the theoretical stagnation temperature at Mach 7 is 4065°F, as discussed above, the actual skin temperature is cooler because of heat sink and heat dissipation effects.) The wings had a low aspect ratio of 2.5 and a thickness-to-chord ratio of 0.05—both intended to reduce supersonic wave drag.

The first X-15 was rolled out of the North American factory at Los Angeles on October 15, 1958. Then Vice President Richard M. Nixon was the guest of honor at the rollout ceremonies. The X-15 had become a political, as well as a technical, accomplishment, because the United States was attempting to heal its wounded pride after the Russians had launched the first successful unmanned satellite, *Sputnik I*, just a year earlier (see Sec. 8.15). The next day, the X-15 was transported by truck to the nearby Edwards Air Force Base (the site at Muroc which saw the first supersonic flights of the Bell XS-1).

As in the case of the XS-1, the X-15 was designed to be carried aloft by a parent airplane, this time a Boeing B-52 jet bomber. The first free flight, without power, was made by Scott Crossfield on June 8, 1959. This was soon followed by the first powered flight on September 17, 1959, when the X-15 reached Mach 2.1 in a shallow

Figure 5.83   The North American X-15, the first manned hypersonic airplane. *(North American/Rockwell Corporation.)*

climb to 52,341 ft. Powered with the smaller LR11 rocket engines, the X-15 set a speed record of Mach 3.31 on August 4, 1960, and an altitude record of 136,500 ft just eight days later. However, these records were just transitory. After November 1960, the X-15 received the more powerful XLR99 engine. Indeed, the first flight with this rocket was made on November 15, 1960; on this flight, with power adjusted to its *lowest* level and with the air brakes fully extended, the X-15 still hit 2000 mi/h. Finally, on June 23, 1961, hypersonic flight was fully achieved when U.S. Air Force test pilot Major Robert White flew the X-15 at Mach 5.3 and in so doing accomplished the first "mile-per-second" flight in an airplane, reaching a maximum velocity of 3603 mi/h. This began an illustrious series of hypersonic flight tests, peaked by a flight at Mach 6.72 on October 3, 1967, with Air Force Major Pete Knight at the controls.

Experimental aircraft are just that—vehicles designed for specific experimental purposes, which, after they are achieved, lead to the end of the program. This happened to the X-15 when on October 24, 1968, the last flight was carried out—the 199th of the entire program. A 200th flight was planned, partly for reasons of nostalgia; however, technical problems delayed this planned flight until December 20, when the X-15 was ready to go, attached to its B-52 parent plane as usual. However, of all things, a highly unusual snow squall suddenly hit Edwards, and the flight was canceled. The X-15 never flew again. In 1969, the first X-15 was given to the National

Air and Space Museum of the Smithsonian, where it now hangs with distinction in the Milestones of Flight Gallery, along with the Bell XS-1.

The X-15 opened the world of manned hypersonic flight. The next hypersonic airplane was the space shuttle. The vast bulk of aerodynamic and flight dynamic data generated during the X-15 program carried over to the space shuttle design. The pilots' experience with low-speed flights in a high-speed aircraft with low lift-to-drag ratio set the stage for flight preparations with the space shuttle. In these respects, the X-15 was clearly the major stepping-stone to the space shuttle of the 1980s.

5.24 SUMMARY

Some of the aspects of this chapter are highlighted below.

1. For an airfoil, the lift, drag, and moment coefficients are defined as

$$c_l = \frac{L}{q_\infty S} \qquad c_d = \frac{D}{q_\infty S} \qquad c_m = \frac{M}{q_\infty S c}$$

where L, D, and M are the lift, drag, and moments per unit span, respectvely, and $S = c(1)$.

For a finite wing, the lift, drag, and moment coefficients are defined as

$$C_L = \frac{L}{q_\infty S} \qquad C_D = \frac{D}{q_\infty S} \qquad C_M = \frac{M}{q_\infty S c}$$

where L, D, and M are the lift, drag, and moments, respectively, for the complete wing and S is the wing planform area.

For a given shape, these coefficients are a function of angle of attack, Mach number, and Reynolds number.

2. The pressure coefficient is defined as

$$C_p = \frac{p - p_\infty}{\frac{1}{2}\rho_\infty V_\infty^2} \qquad \textbf{[5.27]}$$

3. The Prandtl-Glauert rule is a compressibility correction for subsonic flow:

$$C_p = \frac{C_{p,0}}{\sqrt{1 - M_\infty^2}} \qquad \textbf{[5.28]}$$

where $C_{p,0}$ and C_p are the incompressible and compressible pressure coefficients, respectively. The same rule holds for the lift and moment coefficients, that is,

$$c_l = \frac{c_{l,0}}{\sqrt{1 - M_\infty^2}} \qquad \textbf{[5.40]}$$

4. The critical Mach number is that free-stream Mach number at which sonic flow is first achieved at some point on a body. The drag-divergence Mach number is that free-stream Mach number at which the drag coefficient begins to rapidly

increase due to the occurrence of transonic shock waves. For a given body, the drag-divergence Mach number is slightly higher than the critical Mach number.

5. The Mach angle is defined as

$$\mu = \arcsin \frac{1}{M} \qquad \text{[5.49]}$$

6. The total drag coefficient for a finite wing is equal to

$$C_D = c_d + \frac{C_L^2}{\pi e \mathrm{AR}} \qquad \text{[5.58]}$$

where c_d is the profile drag coefficient and $C_L^2/(\pi e \mathrm{AR})$ is the induced drag coefficient.

7. The lift slope for a finite wing a is given by

$$a = \frac{a_0}{1 + 57.3 a_0/(\pi e_1 \mathrm{AR})} \qquad \text{[5.65]}$$

where a_0 is the lift slope for the corresponding infinite wing.

BIBLIOGRAPHY

Abbott, I. H., and A. E. von Doenhoff: *Theory of Wing Sections*, McGraw-Hill, New York, 1949 (also Dover, New York, 1959).

Anderson, John D., Jr.: *Fundamentals of Aerodynamics*, 2d ed., McGraw-Hill, New York, 1991.

_____: *A History of Aerodynamics and Its Impact on Flying Machines*, Cambridge University Press, New York, 1997.

Dommasch, D. O., S. S. Sherbey, and T. F. Connolly: *Airplane Aerodynamics*, 4th ed., Pitman, New York, 1968.

Hallion, R.: *Supersonic Flight (The Story of the Bell X-1 and Douglas D-558)*, Macmillan, New York, 1972.

McCormick, B. W.: *Aerodynamics, Aeronautics and Flight Mechanics*, Wiley, New York, 1979.

Pierpont, P. K.: "Bringing Wings of Change," *Astronautics and Aeronautics*, vol. 13, no. 10, October 1975, pp. 20–27.

Shapiro, A. H.: *Shape and Flow: The Fluid Dynamics of Drag*, Anchor, Garden City, NY, 1961.

Shevell, R. S.: *Fundamentals of Flight*, Prentice-Hall, Englewood Cliffs, NJ, 1983.

von Karman, T. (with Lee Edson): *The Wind and Beyond*, Little, Brown, Boston, 1967 (an autobiography).

PROBLEMS

5.1 By the method of dimensional analysis, derive the expression $M = q_\infty S c c_m$ for the aerodynamic moment on an airfoil, where c is the chord and c_m is the moment coefficient.

5.2 Consider an infinite wing with an NACA 1412 airfoil section and a chord length of 3 ft. The wing is at an angle of attack of 5° in an airflow velocity of 100 ft/s at standard sea-level conditions. Calculate the lift, drag, and moment about the quarter chord per unit span.

Consider a rectangular wing mounted in a low-speed subsonic wing tunnel. The wing **5.3** model completely spans the test section so that the flow "sees" essentially an infinite wing. If the wing has an NACA 23012 airfoil section and a chord of 0.3 m, calculate the lift, drag, and moment about the quarter chord per unit span when the airflow pressure, temperature, and velocity are 1 atm, 303 K, and 42 m/s, respectively. The angle of attack is 8°.

The wing model in Prob. 5.3 is pitched to a new angle of attack, where the lift on the **5.4** entire wing is measured as 200 N by the wind tunnel force balance. If the wingspan is 2 m, what is the angle of attack?

Consider a rectangular wing with an NACA 0009 airfoil section spanning the test **5.5** section of a wind tunnel. The test section airflow conditions are standard sea level with a velocity of 120 mi/h. The wing is at an angle of attack of 4°, and the wind tunnel force balance measures a lift of 29.5 lb. What is the area of the wing?

The ratio of lift to drag L/D for a wing or airfoil is an important aerodynamic **5.6** parameter. Indeed, it is a direct measure of the aerodynamic efficiency of the wing. If a wing is pitched through a range of angle of attack, L/D first increases, then goes through a maximum, and then decreases. Consider an infinite wing with an NACA 2412 airfoil. Estimate the maximum value of L/D. Assume the Reynolds number is 9×10^6.

Consider an airfoil in a free stream with a velocity of 50 m/s at standard sea-level **5.7** conditions. At a point on the airfoil, the pressure is 9.5×10^4 N/m². What is the pressure coefficient at this point?

Consider a low-speed airplane flying at a velocity of 55 m/s. If the velocity at a point **5.8** on the fuselage is 62 m/s, what is the pressure coefficient at this point?

Consider a wing mounted in the test section of a subsonic wind tunnel. The velocity **5.9** of the airflow is 160 ft/s. If the velocity at a point on the wing is 195 ft/s, what is the pressure coefficient at this point?

Consider the same wing in the same wind tunnel as in Prob. 5.9. If the test section **5.10** air temperature is 510°R and if the flow velocity is increased to 700 ft/s, what is the pressure coefficient at the same point?

Consider a wing in a high-speed wind tunnel. At a point on the wing, the velocity **5.11** is 850 ft/s. If the test section flow is at a velocity of 780 ft/s, with a pressure and temperature of 1 atm and 505°R, respectively, calculate the pressure coefficient at the point.

If the test section flow velocity in Prob. 5.11 is reduced to 100 ft/s, what will the **5.12** pressure coefficient become at the same point on the wing?

Consider an NACA 1412 airfoil at an angle of attack of 4°. If the free-stream Mach **5.13** number is 0.8, calculate the lift coefficient.

An NACA 4415 airfoil is mounted in a high-speed subsonic wind tunnel. The lift **5.14** coefficient is measured as 0.85. If the test section Mach number is 0.7, at what angle of attack is the airfoil?

5.15   Consider an airfoil at a given angle of attack, say α_1. At low speeds, the minimum pressure coefficient on the top surface of the airfoil is -0.90. What is the critical Mach number of the airfoil?

5.16   Consider the airfoil in Prob. 5.15 at a smaller angle of attack, say, α_2. At low speeds, the minimum pressure coefficient is -0.65 at this lower angle of attack. What is the critical Mach number of the airfoil?

5.17   Consider a uniform flow with a Mach number of 2. What angle does a Mach wave make with respect to the flow direction?

5.18   Consider a supersonic missile flying at Mach 2.5 at an altitude of 10 km. Assume the angle of the shock wave from the nose is approximated by the Mach angle (i.e., a very weak shock). How far behind the nose of the vehicle will the shock wave impinge upon the ground? (Ignore the fact that the speed of sound, hence the Mach angle, changes with altitude.)

5.19   The wing area of the Lockheed F-104 straight-wing supersonic fighter is approximately 210 ft². If the airplane weighs 16,000 lb and is flying in level flight at Mach 2.2 at a standard altitude of 36,000 ft, estimate the wave drag on the wings.

5.20   Consider a flat plate at an angle of attack of 2° in a Mach 2.2 airflow. (Mach 2.2 is the cruising Mach number, e.g., of the Concorde supersonic transport.) The length of the plate in the flow direction is 202 ft, which is the length of the Concorde. Assume that the free-stream conditions correspond to a standard altitude of 50,000 ft. The total drag on this plate is the sum of wave drag and skin friction drag. Assume that a turbulent boundary layer exists over the entire plate. The results given in Chap. 4 for skin friction coefficients hold for incompressible flow only; there is a compressibility effect on C_f such that its value decreases with increasing Mach number. Specifically, at Mach 2.2 assume that the C_f given in Chap. 4 is reduced by 20 percent.
(a) Given all the above information, calculate the total drag coefficient for the plate.
(b) If the angle of attack is increased to 5°, assuming C_f stays the same, calculate the total drag coefficient.
(c) For these cases, what can you conclude about the relative influence of wave drag and skin friction drag?

5.21   The Cessna Cardinal, a single-engine light plane, has a wing with an area of 16.2 m² and an aspect ratio of 7.31. Assume the span efficiency factor is 0.62. If the airplane is flying at standard sea-level conditions with a velocity of 251 km/h, what is the induced drag when the total weight is 9800 N?

5.22   For the Cessna Cardinal in Prob. 5.21, calculate the induced drag when the velocity is 85.5 km/h (stalling speed at sea level with flaps down).

5.23   Consider a finite wing with an area and aspect ratio of 21.5 m² and 5, respectively (this is comparable to the wing on a Gates Learjet, a twin-jet executive transport). Assume the wing has an NACA 65-210 airfoil, a span efficiency factor of 0.9, and a profile drag coefficient of 0.004. If the wing is at a 6° angle of attack, calculate C_L and C_D.

During the 1920s and early 1930s, the NACA obtained wind tunnel data on differ- **5.24** ent airfoils by testing finite wings with an aspect ratio of 6. These data were then "corrected" to obtain infinite wing airfoil characteristics. Consider such a finite wing with an area and aspect ratio of 1.5 ft^2 and 6, respectively, mounted in a wind tunnel where the test section flow velocity is 100 ft/s at standard sea-level conditions. When the wing is pitched to $\alpha = -2°$, no lift is measured. When the wing is pitched to $\alpha = 10°$, a lift of 17.9 lb is measured. Calculate the lift slope for the airfoil (the infinite wing) if the span effectiveness factor is 0.95.

A finite wing of area 1.5 ft^2 and aspect ratio of 6 is tested in a subsonic wind tunnel at **5.25** a velocity of 130 ft/s at standard sea-level conditions. At an angle of attack of $-1°$, the measured lift and drag are 0 and 0.181 lb, respectively. At an angle of attack of $2°$, the lift and drag are measured as 5.0 and 0.23 lb, respectively. Calculate the span efficiency factor and the infinite-wing lift slope.

Consider a light, single-engine airplane such as the Piper Super Cub. If the maximum **5.26** gross weight of the airplane is 7780 N, the wing area is 16.6 m^2, and the maximum lift coefficient is 2.1 with flaps down, calculate the stalling speed at sea level.

The airfoil on the Lockheed F-104 straight-wing supersonic fighter is a thin symmetric **5.27** airfoil with a thickness ratio of 3.5 percent. Consider this airfoil in a flow at an angle of attack of 5°. The incompressible lift coefficient for the airfoil is given approximately by $c_l = 2\pi\alpha$, where α is the angle of attack in radians. Estimate the airfoil lift coefficient for: (a) $M = 0.2$, (b) $M = 0.7$, (c) $M = 2.0$.

The whirling arm test device used in 1804 by Sir George Cayley is shown in Figure **5.28** 1.7. Cayley was the first person to make measurements of the lift on inclined surfaces. In his 1804 notebook, he noted that, on a flat surface moving through the air at 21.8 ft/sec at 3 degrees angle of attack, a lift force of one ounce was measured. The flat surface was a 1 ft by 1 ft square. Calculate the lift coefficient for this condition. Compare this measured value with that predicted by the expression for lift coefficient for a flat plate airfoil in incompressible flow given by $c_l = 2\pi\alpha$, where α is in radians. What are the reasons for the differences in the two results? (See Anderson, *A History of Aerodynamics and Its Impact on Flying Machines*, Cambridge University Press, 1997, pp. 68–71 for a detailed discussion of this matter.)

Consider a finite wing at an angle of attach of 6°. The normal and axial force co- **5.29** efficients are 0.8 and 0.06 respectively. Calculate the corresponding lift and drag coefficients. What comparison can you make between the lift and normal force co-efficients?

Consider a finite wing with an aspect of ratio of 7; the airfoil section of the wing is a **5.30** symmetric airfoil with an infinite wing lift slope of 0.11 per degree. The lift-to-drag ratio for this wing is 29 when the lift coefficient is equal to 0.35. If the angle of attack remains the same, and the aspect ratio is simply increased to 10 by adding extensions to the span of the wing, what is the new value of the lift-to-drag ratio? Assume the span efficiency factors $e = e_1 = 0.9$ for both cases.

6

Elements of Airplane Performance

First Europe, and then the globe, will be linked by flight, and nations so knit together that they will grow to be next-door neighbors. This conquest of the air will prove, ultimately, to be man's greatest and most glorious triumph. What railways have done for nations, airways will do for the world.

Claude Grahame-White
British aviator, 1914

6.1 INTRODUCTION: THE DRAG POLAR

Henson's aerial steam carriage of the mid-19th century (see Fig. 1.11) was pictured by contemporary artists as flying to all corners of the world. Of course, questions about *how* it would fly to such distant locations were not considered by the designers. As with most early aeronautical engineers of that time, their main concern was simply to lift or otherwise propel the airplane from the ground; what happened once the vehicle was airborne was viewed with secondary importance. However, with the success of the Wright brothers in 1903, and with the subsequent rapid development of aviation during the pre–World War I era, the airborne performance of the airplane suddenly became of primary importance. Some obvious questions were, and still are, asked about a given design. What is the maximum speed of the airplane? How fast can it climb to a given altitude? How far can it fly on a given tank of fuel? How long can it stay in the air? Answers to these and similar questions constitute the study of *airplane performance*, which is the subject of this chapter.

In previous chapters, the physical phenomena producing lift, drag, and moments of an airplane were introduced. We emphasized that the aerodynamic forces and moments exerted on a body moving through a fluid stem from two sources:

1. The pressure distribution
2. The shear stress distribution

both acting over the body surface. The physical laws governing such phenomena were examined, with various applications to aerodynamic flows.

In this chapter, we begin a new phase of study. The airplane is considered a rigid body on which is exerted four natural forces: lift, drag, propulsive thrust, and weight. Concern is focused on the movement of the airplane as it responds to these forces. Such considerations form the core of *flight dynamics*, an important discipline of aerospace engineering. Studies of airplane performance (this chapter) and stability and control (Chap. 7) both fall under the heading of flight dynamics.

In these studies, we will no longer be concerned with aerodynamic details; rather, we will generally assume that the aerodynamicists have done their work and provided us with the pertinent aerodynamic data for a given airplane. These data are usually packaged in the form of a *drag polar* for the complete airplane, given as

$$C_D = C_{D,e} + \frac{C_L^2}{\pi e \text{AR}} \qquad \textbf{[6.1a]}$$

Equation (6.1a) is an extension of Eq. (5.58) to include the whole airplane. Here C_D is the drag coefficient for the complete airplane; C_L is the total lift coefficient, including the small contributions from the horizontal tail and fuselage; and $C_{D,e}$ is defined as the parasite drag coefficient, which contains not only the profile drag of the wing [c_d in Eq. (5.58)] but also the friction and pressure drag of the tail surfaces, fuselage, engine nacelles, landing gear, and any other component of the airplane which is exposed to the airflow. At transonic and supersonic speeds, $C_{D,e}$ also contains wave drag. Because of changes in the flow field around the airplane—especially changes in the amount of separated flow over parts of the airplane—as the angle of attack is varied, $C_{D,e}$ will change with angle of attack; that is, $C_{D,e}$ is itself a function of lift coefficient. A reasonable approximation for this function is

$$C_{D,e} = C_{D,0} + r C_L^2$$

where r is an empirically determined constant. Hence, Eq. (6.1a) can be written as

$$C_D = C_{D,0} + \left(r + \frac{1}{\pi e \text{AR}} \right) C_L^2 \qquad \textbf{[6.1b]}$$

In Eqs. (6.1a) and (6.1b), e is the familiar span efficiency factor, which takes into account the nonelliptical lift distribution on wings of general shape (see Sec. 5.14). Let us now *redefine* e such that it also includes the effect of the variation of parasite drag with lift; that is, let us write Eq. (6.1b) in the form

$$C_D = C_{D,0} + \frac{C_L^2}{\pi e \text{AR}} \qquad \textbf{[6.1c]}$$

where $C_{D,0}$ is the parasite drag coefficient at *zero lift* and the term $C_L^2/(\pi e \text{AR})$ includes both induced drag and the contribution to parasite drag due to lift. In Eq. (6.1c), our redefined e, which now includes the effect of r from Eq. (6.1b), is called the *Oswald efficiency factor* (named after W. Bailey Oswald, who first established this terminology in NACA Report No. 408 in 1932). In this chapter, the basic aerodynamic properties of the airplane are described by Eq. (6.1c), and we consider both

$C_{D,0}$ and e as known aerodynamic quantities, obtained from the aerodynamicist. We will continue to designate $C_L^2/(\pi e \text{AR})$ by $C_{D,i}$, where $C_{D,i}$ now has the expanded interpretation as the coefficient of *drag due to lift*, including both the contributions due to induced drag and the increment in parasite drag due to angle of attack different from $\alpha_L = 0$. We designate $C_{D,0}$ simply as the *zero-lift* drag coefficient, which is obvious from Eq. (6.1c) when $C_L = 0$; however, we recognize $C_{D,0}$ more precisely as the *parasite* drag coefficient at *zero lift*, that is, the value of the drag coefficient when $\alpha = \alpha_{L=0}$.

The graph of Eq. (6.1c), shown in Fig. 6.1, is also called the drag polar. With the approximations made in Eq. (6.1c), the drag polar is a parabola with its axis on the zero-lift axis, and its vertex is $C_{D,0}$. In Fig. 6.1a, C_D is plotted versus C_L; in Fig. 6.1b, C_L is plotted versus C_D. The two representations are identical; Fig. 6.1b is simply a mirror image of Fig. 6.1a rotated on its side. Both representations are found in the literature. In Fig. 6.1, negative values of C_L pertain to negative lift, which occurs when the angle of attack of the airplane is less than $\alpha_{L=0}$. This situation is not encountered frequently in the analysis of airplane performance; hence only that portion of the drag polar associated with positive C_L is usually shown.

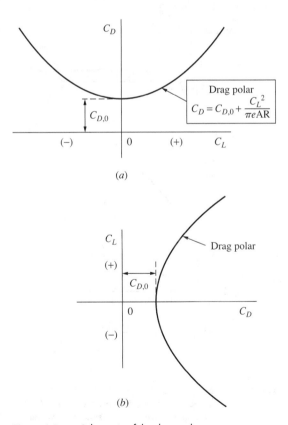

Figure 6.1 Schematic of the drag polar.

An illustration of the drag polar for a specific airplane is shown in Fig. 6.2, which gives the actual data for the Lockheed C-141A, shown in three-view at the top of the figure. Upon close examination, the drag polar for an actual airplane exhibits a subtle difference from our approximation given in Eq. (6.1c) as graphed in Fig. 6.1. Note that the zero-lift drag coefficient in Fig. 6.2 is not the minimum drag coefficient; that is, the axis of the parabolic drag polar is not the zero-lift axis, but rather is displaced slightly above the zero-lift axis. In Fig. 6.2, the minimum drag coefficient is $C_{D,\min} = 0.015$, and it occurs for a value of the lift coefficient $C_{L_{\min \text{ drag}}} = 0.16$. The zero-lift drag coefficient is $C_{D,0} = 0.017$, at $C_L = 0$. And $C_{D,0}$ is not the minimum drag coefficient because $\alpha_{L=0}$ for most airplane designs is a small but finite negative value; that is, the airplane is pitched slightly downward at this orientation, and the pressure drag due to flow separation (form drag) is slightly higher than if the airplane is at an angle of attack slightly larger, nearer a zero angle

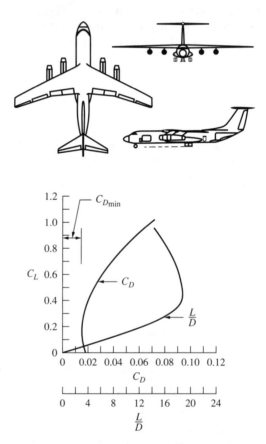

Figure 6.2   Low-speed drag polar and variation of lift-to-drag ratio for the Lockheed C-141A. The airplane is shown in a three-view above the drag polar.

of attack. The minimum drag coefficient occurs when the airplane is more aligned with the relative wind, that is, when α is slightly larger than $\alpha_{L=0}$. For this situation, the drag polar can be expressed as

$$C_D = C_{D,\min} + \frac{(C_L - C_{L_{\min \text{ drag}}})^2}{\pi e \text{AR}} \qquad [6.2]$$

The corresponding graph of the drag polar is shown in Fig. 6.3.

Now that we have made the distinction between the two generic drag polars sketched in Figs. 6.1 and 6.3, for our considerations of airplane performance in this chapter we will adopt Eq. (6.1c) and Fig. 6.1 as the representation of the drag polar. It simplifies our analysis and presentation, without loss of generality. Quantitatively, there is only a small difference between the two representations. However, for an industry standard detailed performance analysis of a particular airplane, you want to have as accurate a drag polar as you can obtain for the airplane, and you would be dealing with the more accurate representation shown in Fig. 6.3 and given by Eq. (6.2).

Return for a moment to our overall road map in Fig. 2.1. With this chapter we move to a new main discipline, namely, flight mechanics, as itemized at the left of Fig. 2.1. In particular, in this chapter we deal with airplane performance, a subheading under flight mechanics, as shown at the center of Fig. 2.1. The road map for this chapter is shown in Fig. 6.4. A study of airplane performance is frequently based on Newton's second law, which dictates the motion of the airplane through the atmosphere. We will first obtain these *equations of motion*. The remainder of the chapter is based on two forms of these equations: (1) the form associated with the assumption of *unaccelerated* flight, leading to a study of *static performance* itemized on the left side of Fig. 6.4; and (2) the form associated with acceleration of

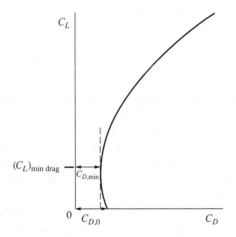

Figure 6.3 Drag polar where the zero-lift drag coefficient is not the same as the minimum drag coefficient.

Figure 6.4   Road map for Chap. 6.

the airplane, leading to a study of *dynamic performance* itemized on the right side of Fig. 6.4. (The difference between static performance and dynamic performance is analogous to taking a course in statics and another course in dynamics.) Under static performance we will examine such important aspects as how to calculate the maximum velocity of the airplane, how fast it can climb (rate of climb), how high it can fly (maximum altitude), how far it can fly (range), and how long it can stay in the air (endurance). Under dynamic performance we will examine takeoff and landing characteristics, turning flight, and accelerated rate of climb. When we arrive at the bottom of this road map, we will have toured through some of the basic aspects that dictate the design of an airplane, and will have covered some of the most important territory in aerospace engineering. So, let us hit the road.

6.2 EQUATIONS OF MOTION

To study the performance of an airplane, the fundamental equations which govern its translational motion through air must first be established. Consider an airplane in flight, as sketched in Fig. 6.5. The flight path (direction of motion of the airplane) is inclined at an angle θ with respect to the horizontal. In terms of the definitions in Chap. 5, the flight path direction and the relative wind are along the same line. The mean chord line is at a geometric angle of attack α with respect to the flight path direction. There are four physical forces acting on the airplane:

1. Lift L, which is perpendicular to the flight path direction
2. Drag D, which is parallel to the flight path direction
3. Weight W, which acts vertically toward the center of the earth (and hence is inclined at angle θ with respect to the lift direction)
4. Thrust T, which in general is inclined at the angle α_T with respect to the flight path direction

The force diagram shown in Fig. 6.5 is important. Study it carefully until you feel comfortable with it.

The flight path shown in Fig. 6.5 is drawn as a straight line. This is the picture we see by focusing locally on the airplane itself. However, if we stand back and take a wider view of the space in which the airplane is traveling, the flight path is generally curved. This is obviously true if the airplane is maneuvering; even if the airplane is flying "straight and level" with respect to the ground, it is still executing a curved flight path with a radius of curvature equal to the absolute altitude h_a (as defined in Sec. 3.1).

Figure 6.5   Force diagram for an airplane in flight.

When an object moves along a curved path, the motion is called *curvilinear*, as opposed to motion along a straight line, which is *rectilinear*. Newton's second law, which is a physical statement that force = mass × acceleration, holds in either case. Consider a curvilinear path. At a given point on the path, set up two mutually perpendicular axes, one along the direction of the flight path and the other normal to the flight path. Applying Newton's law along the flight path gives

$$\sum F_\parallel = ma = m\frac{dV}{dt} \tag{6.3}$$

where $\sum F_\parallel$ is the summation of all forces parallel to the flight path, $a = dV/dt$ is the acceleration along the flight path, and V is the instantaneous value of the airplane's flight velocity. (Velocity V is always along the flight path direction, by definition.) Now, applying Newton's law perpendicular to the flight path, we have

$$\sum F_\perp = m\frac{V^2}{r_c} \tag{6.4}$$

where $\sum F_\perp$ is the summation of all forces perpendicular to the flight path and V^2/r_c is the acceleration normal to a curved path with radius of curvature r_c. This normal acceleration V^2/r_c should be familiar from the basic physics. The right-hand side of Eq. (6.4) is nothing other than the *centrifugal force*.

Examining Fig. 6.5, we see that the forces parallel to the flight path (positive to the right, negative to the left) are

$$\sum F_\parallel = T\cos\alpha_T - D - W\sin\theta \tag{6.5}$$

and the forces perpendicular to the flight path (positive upward and negative downward) are

$$\sum F_\perp = L + T\sin\alpha_T - W\cos\theta \tag{6.6}$$

Combining Eq. (6.3) with (6.5), and Eq. (6.4) with (6.6) yields

$$T\cos\alpha_T - D - W\sin\theta = m\frac{dV}{dt} \tag{6.7}$$

$$L + T\sin\alpha_T - W\cos\theta = m\frac{V^2}{r_c} \tag{6.8}$$

Equations (6.7) and (6.8) are the *equations of motion* for an airplane in translational flight. (Note that an airplane can also rotate about its axes; this will be discussed in Chap. 7. Also note that we are not considering the possible sidewise motion of the airplane perpendicular to the page of Fig. 6.5.)

Equations (6.7) and (6.8) describe the general two-dimensional translational motion of an airplane in accelerated flight. However, in the first part of this chapter we are interested in a specialized application of these equations, namely, the case where the acceleration is zero. The performance of an airplane for such unaccelerated flight conditions is called *static performance*. This may, at first thought, seem unduly restrictive; however, static performance analyses lead to reasonable calculations of maximum velocity, maximum rate of climb, maximum range, etc.—parameters of vital interest in airplane design and operation.

With this in mind, consider level, unaccelerated flight. Referring to Fig. 6.5, level flight means that the flight path is along the horizontal; that is, $\theta = 0$. Unaccelerated flight means that the right-hand sides of Eqs. (6.7) and (6.8) are zero. Therefore, these equations reduce to

$$T \cos \alpha_T = D \qquad \qquad \textbf{[6.9]}$$

$$L + T \sin \alpha_T = W \qquad \qquad \textbf{[6.10]}$$

Furthermore, for most conventional airplanes, α_T is small enough that $\cos \alpha_T \approx 1$ and $\sin \alpha_T \approx 0$. Thus, from Eqs. (6.9) and (6.10),

$$\boxed{T = D} \qquad \qquad \textbf{[6.11]}$$

$$\boxed{L = W} \qquad \qquad \textbf{[6.12]}$$

Equations (6.11) and (6.12) are the equations of motion for level, unaccelerated flight. They can also be obtained directly from Fig. 6.5, by inspection. In level, unaccelerated flight, the aerodynamic drag is balanced by the thrust of the engine, and the aerodynamic lift is balanced by the weight of the airplane—almost trivial, but very useful, results.

Let us now apply these results to the static performance analysis of an airplane. The following sections constitute the building blocks for such an analysis, which ultimately yields answers to such questions as how fast, how far, how long, and how high a given airplane can fly. Also, the discussion in these sections relies heavily on a graphical approach to the calculation of airplane performance. In modern aerospace engineering, such calculations are made directly on high-speed digital computers. However, the graphical illustrations in the following sections are essential to the programming and understanding of such computer solutions; moreover, they help to clarify and explain the concepts being presented.

6.3 THRUST REQUIRED FOR LEVEL, UNACCELERATED FLIGHT

Consider an airplane in steady, level flight at a given altitude and a given velocity. For flight at this velocity, the airplane's power plant (e.g., turbojet engine or reciprocating engine–propeller combination) must produce a net thrust which is equal to the drag. The thrust required to obtain a certain steady velocity is easily calculated as follows. From Eqs. (6.11) and (5.20)

$$T = D = q_\infty S C_D \qquad \qquad \textbf{[6.13]}$$

and from Eqs. (6.12) and (5.17)

$$L = W = q_\infty S C_L \qquad \textbf{[6.14]}$$

Dividing Eq. (6.13) by (6.14) yields

$$\frac{T}{W} = \frac{C_D}{C_L} \qquad \textbf{[6.15]}$$

Thus, from Eq. (6.15), the thrust required for an airplane to fly at a given velocity in level, unaccelerated flight is

$$\boxed{T_R = \frac{W}{C_L/C_D} = \frac{W}{L/D}} \qquad \textbf{[6.16]}$$

(Note that a subscript R has been added to thrust to emphasize that it is thrust required.)

Thrust required T_R for a given airplane at a given altitude varies with velocity V_∞. The *thrust-required curve* is a plot of this variation and has the general shape illustrated in Fig. 6.6. To calculate a point on this curve, proceed as follows:

1. Choose a value of V_∞.
2. For this V_∞, calculate the lift coefficient from Eq. (6.14):

$$C_L = \frac{W}{\frac{1}{2}\rho_\infty V_\infty^2 S} \qquad \textbf{[6.17]}$$

Note that ρ_∞ is known from the given altitude and S is known from the given airplane. The C_L calculated from Eq. (6.17) is that value necessary for the lift to balance the known weight W of the airplane.

Figure 6.6   Thrust-required curve. The results on this and subsequent figures correspond to answers for some of the sample problems in Chap. 6.

3. Calculate C_D from the known drag polar for the airplane

$$C_D = C_{D,0} + \frac{C_L^2}{\pi e \text{AR}}$$

where C_L is the value obtained from Eq. (6.17).

4. Form the ratio C_L/C_D.

5. Calculate thrust required from Eq. (6.16).

The value of T_R obtained from step 5 is that thrust required to fly at the specific velocity chosen in step 1. In turn, the curve in Fig. 6.6 is the locus of all such points taken for all velocities in the flight range of the airplane. The reader should study Example 6.1 at the end of this section in order to become familiar with the above steps.

Note from Eq. (6.16) that T_R varies inversely as L/D. Hence, minimum thrust required will be obtained when the airplane is flying at a velocity where L/D is maximum. This condition is shown in Fig. 6.6.

The lift-to-drag ratio L/D is a measure of the aerodynamic efficiency of an airplane; it only makes sense that maximum aerodynamic efficiency should lead to minimum thrust required. Consequently, the lift-to-drag ratio is an important aerodynamic consideration in airplane design. Also note that L/D is a function of angle of attack, as sketched in Fig. 6.7. For most conventional subsonic airplanes, L/D reaches a maximum at some specific value of α, usually on the order of 2 to 5°. Hence, when an airplane is flying at the velocity for minimum T_R, as shown in Fig. 6.6, it is simultaneously flying at the angle of attack for maximum L/D, as shown in Fig. 6.7.

As a corollary to this discussion, note that different points on the thrust-required curve correspond to different angles of attack. This is emphasized in Fig. 6.8, which shows that as we move from right to left on the thrust-required curve, the airplane angle of attack increases. This also helps to explain physically why T_R goes through a minimum. Recall that $L = W = q_\infty S C_L$. At high velocities (point a in Fig. 6.8), most of the required lift is obtained from high dynamic pressure q_∞; hence C_L and

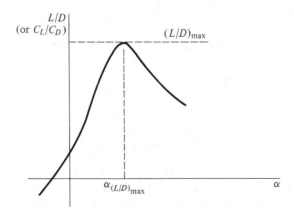

Figure 6.7 Lift-to-drag ratio versus angle of attack.

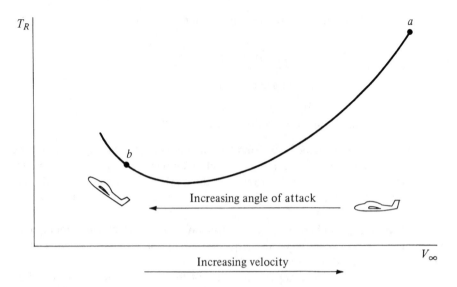

Figure 6.8   Thrust-required curve with associated angle-of-attack variation.

therefore α are small. Also, under the same conditions, drag ($D = q_\infty S C_D$) is relatively large because q_∞ is large. As we move to the left on the thrust-required curve, q_∞ decreases; hence C_L and therefore α must increase to support the given airplane weight. Because q_∞ decreases, D and hence T_R initially decrease. However, recall that drag due to lift is a component of total drag and that $C_{D,i}$ varies as C_L^2. At low velocities, such as at point b in Fig. 6.8, q_∞ is low, and hence C_L is large. At these conditions, $C_{D,i}$ increases rapidly, more rapidly than q_∞ decreases, and D and hence T_R increase. This is why, starting at point a, T_R first decreases as V_∞ decreases, then goes through a minimum and starts to increase, as shown at point b.

Recall from Eq. (6.1c) that the total drag of the airplane is the sum of the zero-lift drag and drag due to lift. The corresponding drag coefficients are $C_{D,0}$ and $C_{D,i} = C_L^2/(\pi e \text{AR})$, respectively. At the condition for minimum T_R, there exists an interesting relation between $C_{D,0}$ and $C_{D,i}$, as follows. From Eq. (6.11)

$$T_R = D = q_\infty S C_D = q_\infty S (C_{D,0} + C_{D,i})$$

$$= q_\infty S \left(C_{D,0} + \frac{C_L^2}{\pi e \text{AR}} \right)$$

$$T_R = \underset{\text{Zero-lift } T_R}{q_\infty S C_{D,0}} + \underset{\text{Lift-induced } T_R}{q_\infty S \frac{C_L^2}{\pi e \text{AR}}} \qquad \textbf{[6.18]}$$

Note that as identified in Eq. (6.18), the thrust required can be considered the sum of *zero-lift thrust required* (thrust required to balance zero-lift drag) and *lift-induced thrust required* (thrust required to balance drag due to lift). Examining Fig. 6.9, we

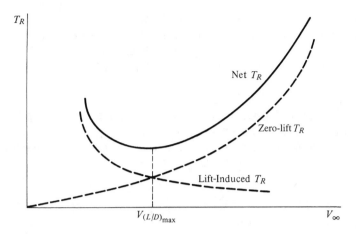

Figure 6.9   Comparison of lift-induced and zero-lift thrust required.

find lift-induced T_R decreases but zero-lift T_R increases as the velocity is increased. (Why?)

Recall that $C_L = W/(q_\infty S)$. From Eq. (6.18),

$$T_R = q_\infty S C_{D,0} + \frac{W^2}{q_\infty S \pi e \text{AR}} \qquad\qquad \textbf{[6.19]}$$

Also

$$\frac{dT_R}{dq_\infty} = \frac{dT_R}{dV_\infty}\frac{dV_\infty}{dq_\infty} \qquad\qquad \textbf{[6.20]}$$

From calculus, we find that the point of minimum T_R in Fig. 6.6 corresponds to $dT_R/dV_\infty = 0$. Hence, from Eq. (6.20), minimum T_R also corresponds to $dT_R/dq_\infty = 0$. Differentiating Eq. (6.19) with respect to q_∞ and setting the derivative equal to zero, we have

$$\frac{dT_R}{dq_\infty} = S C_{D,0} - \frac{W^2}{q_\infty^2 S \pi e \text{AR}} = 0$$

Thus

$$C_{D,0} = \frac{W^2}{q_\infty^2 S^2 \pi e \text{AR}} \qquad\qquad \textbf{[6.21]}$$

However,

$$\frac{W^2}{q_\infty^2 S^2} = \left(\frac{W}{q_\infty S}\right)^2 = C_L^2$$

Hence, Eq. (6.21) becomes

$$\boxed{C_{D,0} = \frac{C_L^2}{\pi e \text{AR}} = C_{D,i}}$$

$$\text{Zero-lift drag} = \text{drag due to lift} \qquad\qquad \textbf{[6.22]}$$

Equation (6.22) yields the interesting aerodynamic result that at minimum thrust required, zero-lift drag equals drag due to lift. Hence, the curves for zero-lift and lift-induced T_R intersect at the velocity for minimum T_R (i.e., for maximum L/D), as shown in Fig. 6.9. We will return to this result in Sec. 6.13.

Example 6.1

For all the examples given in this chapter, two types of airplanes will be considered:

a. A light, single-engine, propeller-driven, private airplane, approximately modeled after the Cessna Skylane shown in Fig. 6.10. For convenience, we will designate our hypothetical airplane as the CP-1, having the following characteristics:

Wingspan = 35.8 ft

Wing area = 174 ft^2

Normal gross weight = 2950 lb

Fuel capacity: 65 gal of aviation gasoline

Power plant: one-piston engine of 230 hp at sea level

Specific fuel consumption = 0.45 lb/(hp)(h)

Parasite drag coefficient $C_{D,0} = 0.025$

Oswald efficiency factor $e = 0.8$

Propeller efficiency = 0.8

b. A jet-powered executive aircraft, approximately modeled after the Cessna Citation 3, shown in Fig. 6.11. For convenience, we will designate our hypothetical jet as the CJ-1, having the following characteristics:

Wingspan = 53.3 ft

Wing area = 318 ft^2

Normal gross weight = 19,815 lb

Fuel capacity: 1119 gal of kerosene

Power plant: two turbofan engines of 3650-lb thrust each at sea level

Specific fuel consumption = 0.6 lb of fuel/(lb thrust)(h)

Parasite drag coefficient $C_{D,0} = 0.02$

Oswald efficiency factor $e = 0.81$

By the end of this chapter, all the examples taken together will represent a basic performance analysis of these two aircraft.

In this example, only the thrust required is considered. Calculate the T_R curves at sea level for both the CP-1 and the CJ-1.

Solution

a. For the CP-1 assume $V_\infty = 200$ ft/s $= 136.4$ mi/h. From Eq. (6.17),

$$C_L = \frac{W}{\frac{1}{2}\rho_\infty V_\infty^2 S} = \frac{2950}{\frac{1}{2}(0.002377)(200)^2(174)} = 0.357$$

The aspect ratio is

$$AR = \frac{b^2}{S} = \frac{(35.8)^2}{174} = 7.37$$

Figure 6.10 The hypothetical CP-1 studied in Chap. 6 sample problems is modeled after the Cessna Skylane shown here. *(Cessna Aircraft Corporation.)*

Figure 6.11 The hypothetical CJ-1 studied in Chap. 6 sample problems is modeled after the Cessna Citation 3 shown here. *(Cessna Aircraft Corp.)*

Thus, from Eq. (6.1c),

$$C_D = C_{D,0} + \frac{C_L^2}{\pi e \text{AR}} = 0.025 + \frac{(0.357)^2}{\pi(0.8)(7.37)} = 0.0319$$

Hence
$$\frac{L}{D} = \frac{C_L}{C_D} = \frac{0.357}{0.0319} = 11.2$$

Finally, from Eq. (6.16),

$$T_R = \frac{W}{L/D} = \frac{2950}{11.2} = \boxed{263 \text{ lb}}$$

To obtain the thrust-required curve, the above calculation is repeated for many different values of V_∞. Some sample results are tabulated below.

V_∞, ft/s	C_L	C_D	L/D	T_R, lb
100	1.43	0.135	10.6	279
150	0.634	0.047	13.6	217
250	0.228	0.028	8.21	359
300	0.159	0.026	6.01	491
350	0.116	0.026	4.53	652

The above tabulation is given so that the reader can try such calculations and compare the results. Such tabulations are given throughout this chapter. They are taken from a computer calculation in which 100 different velocities are used to generate the data. The T_R curve obtained from these calculations is given in Fig. 6.6.

b. For the CJ-1 assume $V_\infty = 500$ ft/s $= 341$ mi/h. From Eq. (6.17),

$$C_L = \frac{W}{\frac{1}{2}\rho_\infty V_\infty^2 S} = \frac{19,815}{\frac{1}{2}(0.002377)(500)^2(318)} = 0.210$$

The aspect ratio is
$$\text{AR} = \frac{b^2}{S} = \frac{(53.3)^2}{318} = 8.93$$

Thus, from Eq. (6.1c),

$$C_D = C_{D,0} + \frac{C_L^2}{\pi e \text{AR}} = 0.02 + \frac{(0.21)^2}{\pi(0.81)(8.93)} = 0.022$$

Hence
$$\frac{L}{D} = \frac{C_L}{C_D} = \frac{0.21}{0.022} = 9.55$$

Finally, from Eq. (6.16)

$$T_R = \frac{W}{L/D} = \frac{19,815}{9.55} = \boxed{2075 \text{ lb}}$$

A tabulation for a few different velocities follows.

V_∞, ft/s	C_L	C_D	L/D	T_R, lb
300	0.583	0.035	16.7	1188
600	0.146	0.021	6.96	2848
700	0.107	0.021	5.23	3797
850	0.073	0.020	3.59	5525
1000	0.052	0.020	2.61	7605

The thrust-required curve is given in Fig. 6.12.

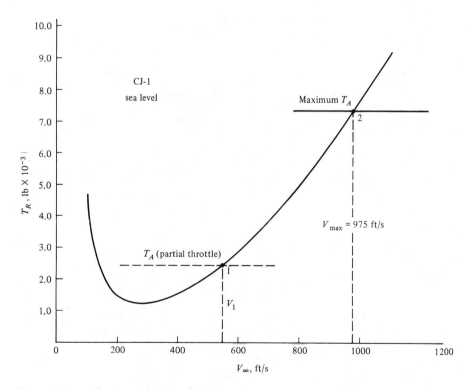

Figure 6.12   Thrust-required curve for the CJ-1.

6.4   THRUST AVAILABLE AND MAXIMUM VELOCITY

Thrust required T_R, described in the previous section, is dictated by the aerodynamics and weight of the airplane itself; it is an *airframe-associated* phenomenon. In contrast, the *thrust available* T_A is strictly associated with the engine of the airplane; it is the propulsive thrust provided by an engine-propeller combination, a turbojet, a rocket, etc. Propulsion is the subject of Chap. 9. Suffice it to say here that reciprocating piston engines with propellers exhibit a variation of thrust with velocity, as sketched in Fig. 6.13a. Thrust at zero velocity (static thrust) is a maximum and decreases with forward velocity. At near-sonic flight speeds, the tips of the propeller blades encounter the same compressibility problems discussed in Chap. 5, and the thrust available rapidly deteriorates. On the other hand, the thrust of a turbojet engine is relatively constant with velocity, as sketched in Fig. 6.13b. These two power plants are quite common in aviation today; reciprocating engine–propeller combinations power the average light, general aviation aircraft, whereas the jet engine is used by almost all large commercial transports and military combat aircraft. For these reasons, the performance analyses of this chapter consider only the above two propulsive mechanisms.

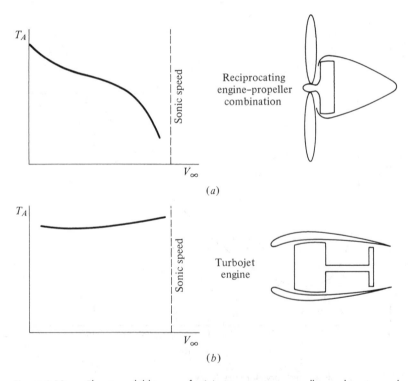

Figure 6.13   Thrust-available curves for (a) piston engine–propeller combination and (b) a turbojet engine.

Consider a jet airplane flying in level, unaccelerated flight at a given altitude and with velocity V_1, as shown in Fig. 6.12. Point 1 on the thrust-required curve gives the value of T_R for the airplane to fly at velocity V_1. The pilot has adjusted the throttle such that the jet engine provides thrust available just equal to the thrust required at this point; $T_A = T_R$. This partial-throttle T_A is illustrated by the dashed curve in Fig. 6.12. If the pilot now pushes the throttle forward and increases the engine thrust to a higher value of T_A, the airplane will accelerate to a higher velocity. If the throttle is increased to full position, maximum T_A will be produced by the jet engine. In this case, the speed of the airplane will further increase until the thrust required equals the maximum T_A (point 2 in Fig. 6.12). It is now impossible for the airplane to fly any faster than the velocity at point 2; otherwise, the thrust required would exceed the maximum thrust available from the power plant. Hence, *the intersection of the T_R curve (dependent on the airframe) and the maximum T_A curve (dependent on the engine) defines the maximum velocity V_{max} of the airplane at the given altitude*, as shown in Fig. 6.12. Calculating the maximum velocity is an important aspect of the airplane design process.

Conventional jet engines are rated in terms of thrust (usually in pounds). Hence, the thrust curves in Fig. 6.12 are useful for the performance analysis of a jet-powered aircraft. However, piston engines are rated in terms of power (usually horsepower). Hence, the concepts of T_A and T_R are inconvenient for propeller-driven aircraft. In this case, power required and power available are the more relevant quantities. Moreover, considerations of power lead to results such as rate of climb and maximum altitude for both jet and propeller-driven airplanes. Therefore, for the remainder of this chapter, emphasis is placed on power rather than thrust, as introduced in the next section.

Example 6.2

Calculate the maximum velocity of the CJ-1 at sea level (see Example 6.1).

Solution

The information given in Example 6.1 states that the power plant for the CJ-1 consists of two turbofan engines of 3650-lb thrust each at sea level. Hence,

$$T_A = 2(3650) = 7300 \text{ lb}$$

Examining the results of Example 6.1, we see that $T_R = T_A = 7300$ lb occurs when $V_\infty = 975$ ft/s (see Fig. 6.12). Hence

$$\boxed{V_{max} = 975 \text{ ft/s} = 665 \text{ mi/h}}$$

It is interesting to note that since the sea-level speed of sound is 1117 ft/s, the maximum sea-level Mach number is

$$M_{max} = \frac{V_{max}}{a} = \frac{975}{1117} = 0.87$$

In the present examples, $C_{D,0}$ is assumed constant; hence, the drag polar does not include drag-divergence effects, as discussed in Chap. 5. Because the drag-divergence Mach number for this type of airplane is normally on the order of 0.82 to 0.85, the above calculation indicates that M_{max} is greater than drag divergence, and our assumption of constant $C_{D,0}$ becomes inaccurate at this high a Mach number.

6.5 POWER REQUIRED FOR LEVEL, UNACCELERATED FLIGHT

Power is a precisely defined mechanical term; it is energy per unit time. The power associated with a moving object can be illustrated by a block moving at constant velocity V under the influence of the constant force F, as shown in Fig. 6.14. The block moves from left to right through distance d in a time interval $t_2 - t_1$. (We assume that an opposing equal force not shown in Fig. 6.14, say due to friction, keeps the block from accelerating.) *Work* is another precisely defined mechanical term; it is force multiplied by the distance through which the force moves. Moreover, work is energy, having the same units as energy. Hence

$$\text{Power} = \frac{\text{energy}}{\text{time}} = \frac{\text{force} \times \text{distance}}{\text{time}} = \text{force} \times \frac{\text{distance}}{\text{time}}$$

Applied to the moving block in Fig. 6.14, this becomes

$$\text{Power} = F\left(\frac{d}{t_2 - t_1}\right) = FV \qquad [6.23]$$

where $d/(t_2 - t_1)$ is the velocity V of the object. Thus, Eq. (6.23) demonstrates that the power associated with a force exerted on a moving object is force × velocity, an important result.

Consider an airplane in level, unaccelerated flight at a given altitude and with velocity V_∞. The thrust required is T_R. From Eq. (6.23), the *power required* P_R is therefore

$$\boxed{P_R = T_R V_\infty} \qquad [6.24]$$

The effect of the airplane aerodynamics (C_L and C_D) on P_R is readily obtained by combining Eqs. (6.16) and (6.24):

$$P_R = T_R V_\infty = \frac{W}{C_L/C_D} V_\infty \qquad [6.25]$$

Figure 6.14   Force, velocity, and power of a moving body.

From Eq. (6.12),

$$L = W = q_\infty S C_L = \tfrac{1}{2}\rho_\infty V_\infty^2 S C_L$$

Hence

$$V_\infty = \sqrt{\frac{2W}{\rho_\infty S C_L}} \qquad\qquad \textbf{[6.26]}$$

Substituting Eq. (6.26) into (6.25), we obtain

$$P_R = \frac{W}{C_L/C_D}\sqrt{\frac{2W}{\rho_\infty S C_L}}$$

$$\boxed{P_R = \sqrt{\frac{2W^3 C_D^2}{\rho_\infty S C_L^3}} \propto \frac{1}{C_L^{3/2}/C_D}} \qquad\qquad \textbf{[6.27]}$$

In contrast to thrust required, which varies inversely as C_L/C_D [see Eq. (6.16)], power required varies inversely as $C_L^{3/2}/C_D$.

The power-required curve is defined as a plot of P_R versus V_∞, as sketched in Fig. 6.15; note that it qualitatively resembles the thrust-required curve of Fig. 6.6. As the airplane velocity increases, P_R first decreases, then goes through a minimum, and

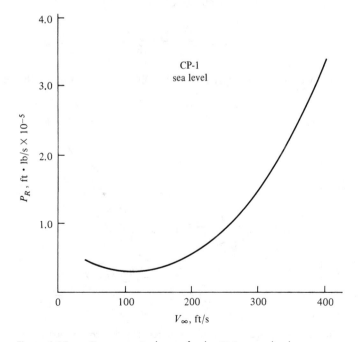

Figure 6.15 Power-required curve for the CP-1 at sea level.

finally increases. At the velocity for minimum power required, the airplane is flying at the angle of attack which corresponds to a maximum $C_L^{3/2}/C_D$.

In Sec. 6.3, we demonstrated that minimum T_R aerodynamically corresponds to equal zero-lift and lift-induced drag. An analogous but different relation holds at minimum P_R. From Eqs. (6.11) and (6.24),

$$P_R = T_R V_\infty = D V_\infty = q_\infty S \left(C_{D,0} + \frac{C_L^2}{\pi e \mathrm{AR}} \right) V_\infty$$

$$P_R = q_\infty S C_{D,0} V_\infty + q_\infty S V_\infty \frac{C_L^2}{\pi e \mathrm{AR}} \qquad \textbf{[6.28]}$$

$$\underbrace{\phantom{q_\infty S C_{D,0} V_\infty}}_{\substack{\text{Zero-lift power} \\ \text{required}}} \qquad \underbrace{\phantom{q_\infty S V_\infty \frac{C_L^2}{\pi e \mathrm{AR}}}}_{\substack{\text{Lift-induced power} \\ \text{required}}}$$

Therefore, as in the earlier case of T_R, the power required can be split into the respective contributions needed to overcome zero-lift drag and drag due to lift. These contributions are sketched in Fig. 6.16. Also as before, the aerodynamic conditions associated with minimum P_R can be obtained from Eq. (6.28) by setting $dP_R/dV_\infty = 0$. To do this, first obtain Eq. (6.28) explicitly in terms of V_∞, recalling that $q_\infty = \frac{1}{2}\rho V_\infty^2$ and $C_L = W/\left(\frac{1}{2}\rho_\infty V_\infty^2 S\right)$.

$$P_R = \frac{1}{2}\rho_\infty V_\infty^3 S C_{D,0} + \frac{1}{2}\rho_\infty V_\infty^3 S \frac{[W/(\frac{1}{2}\rho_\infty V_\infty^2 S)]^2}{\pi e \mathrm{AR}}$$

$$P_R = \frac{1}{2}\rho_\infty V_\infty^3 S C_{D,0} + \frac{W^2/(\frac{1}{2}\rho_\infty V_\infty S)}{\pi e \mathrm{AR}} \qquad \textbf{[6.29]}$$

For minimum power required, $dP_R/dV_\infty = 0$. Differentiating Eq. (6.29) yields

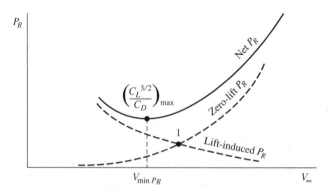

Figure 6.16   Comparison of lift-induced, zero-lift, and net power required.

$$\frac{dP_R}{dV_\infty} = \frac{3}{2}\rho_\infty V_\infty^2 S C_{D,0} - \frac{W^2/(\frac{1}{2}\rho_\infty V_\infty^2 S)}{\pi e \mathrm{AR}}$$

$$= \frac{3}{2}\rho_\infty V_\infty^2 S \left[C_{D,0} - \frac{W^2/(\frac{3}{4}\rho_\infty^2 S^2 V_\infty^4)}{\pi e \mathrm{AR}} \right]$$

$$= \frac{3}{2}\rho_\infty V_\infty^2 S \left(C_{D,0} - \frac{\frac{1}{3}C_L^2}{\pi e \mathrm{AR}} \right)$$

$$= \frac{3}{2}\rho_\infty V_\infty^2 S (C_{D,0} - \tfrac{1}{3}C_{D,i}) = 0 \qquad \text{for minimum } P_R$$

Hence, the aerodynamic conditions that holds at minimum power required is

$$\boxed{C_{D,0} = \tfrac{1}{3}C_{D,i}} \qquad\qquad \textbf{[6.30]}$$

The fact that zero-lift drag is one-third the drag due to lift at minimum P_R is reinforced by examination of Fig. 6.16. Also note that point 1 in Fig. 6.16 corresponds to $C_{D,0} = C_{D,i}$, that is, minimum T_R; hence, V_∞ for minimum P_R is less than that for minimum T_R.

The point on the power-required curve that corresponds to minimum T_R is easily obtained by drawing a line through the origin and tangent to the P_R curve, as shown in Fig. 6.17. The point of tangency corresponds to minimum T_R (hence maximum L/D). To prove this, consider any line through the origin and intersecting the P_R curve, such as the dashed line in Fig. 6.17. The slope of this line is P_R/V_∞. As we move to the right along the P_R curve, the slope of an intersecting line will first decrease, then reach a minimum (at the tangent point), and again increase. This is clearly seen simply by inspection of the geometry of Fig. 6.17. Thus, the point of tangency corresponds to a minimum slope, hence a minimum value of P_R/V_∞. In

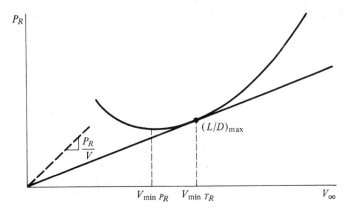

Figure 6.17 The tangent to the power-required curve locates the point of minimum thrust required (and hence the point of maximum L/D).

turn, from calculus this corresponds to

$$\frac{d(P_R/V_\infty)}{dV_\infty} = \frac{d(T_R V_\infty/V_\infty)}{dV_\infty} = \frac{dT_R}{dV_\infty} = 0$$

The above result yields $dT_R/dV_\infty = 0$ at the tangent point, which is precisely the mathematical criterion for minimum T_R. Correspondingly, L/D is maximum at the tangent point.

Example 6.3	**C**alculate the power-required curves for (*a*) the CP-1 at sea level and (*b*) the CJ-1 at an altitude of 22,000 ft.

Solution

a. For the CP-1 the values of T_R at sea level have already been tabulated and graphed in Example 6.1. Hence, from Eq. (6.24),

$$P_R = T_R V_\infty$$

we obtain the following tabulation:

V, ft/s	T_R, lb	P_R, ft·lb/s
100	279	27,860
150	217	32,580
250	359	89,860
300	491	147,200
350	652	228,100

The power-required curve is given in Fig. 6.15.

b. For the CJ-1 at 22,000 ft, $\rho_\infty = 0.001184$ slug/ft^3. The calculation of T_R is done with the same method as given in Example 6.1, and P_R is obtained from Eq. (6.24). Some results are tabulated below.

V_∞, ft/s	C_L	C_D	L/D	T_R, lb	P_R, ft·lb/s
300	1.17	0.081	14.6	1358	0.041×10^7
500	0.421	0.028	15.2	1308	0.065×10^7
600	0.292	0.024	12.3	1610	0.097×10^7
800	0.165	0.021	7.76	2553	0.204×10^7
1000	0.105	0.020	5.14	3857	0.386×10^7

The reader should attempt to reproduce these results.

The power-required curve is given in Fig. 6.18.

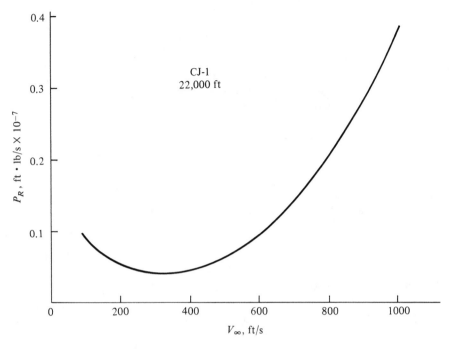

Figure 6.18 Power-required curve for the CJ-1 at 22,000 ft.

6.6 POWER AVAILABLE AND MAXIMUM VELOCITY

Note again that P_R is a characteristic of the aerodynamic design and weight of the aircraft itself. In contrast, the *power available* P_A is a characteristic of the power plant. A detailed discussion of propulsion is deferred until Chap. 9; however, the following comments are made to expedite our performance analyses.

6.6.1 Reciprocating Engine–Propeller Combination

A piston engine generates power by burning fuel in confined cylinders and using this energy to move pistons, which in turn deliver power to the rotating crankshaft, as schematically shown in Fig. 6.19. The power delivered to the propeller by the crankshaft is defined as the *shaft brake* power P (the word *brake* stems from a method of laboratory testing which measures the power of an engine by loading it with a calibrated brake mechanism). However, not all P is available to drive the airplane; some of it is dissipated by inefficiencies of the propeller itself (to be discussed in

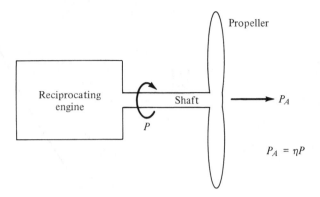

Figure 6.19   Relation between shaft brake power and power available.

Chap. 9). Hence, the power available to propel the airplane P_A is given by

$$\boxed{P_A = \eta P} \qquad \qquad \textbf{[6.31]}$$

where η is the propeller efficiency, $\eta < 1$. Propeller efficiency is an important quantity and is a direct product of the aerodynamics of the propeller. It is always less than unity. For our discussions here, both η and P are assumed to be known quantities for a given airplane.

A remark about units is necessary. In the engineering system, power is in foot-pounds per second (ft · lb/s); in SI, power is in watts [which are equivalent to Newton-meters per second (N · m/s)]. However, the historical evolution of engineering has left us with a horrendously inconsistent (but very convenient) unit of power which is widely used, namely, horsepower. All reciprocating engines are rated in terms of horsepower (hp), and it is important to note that

$$\boxed{1 \text{ hp} = 550 \text{ ft} \cdot \text{lb/s} = 746 \text{ W}}$$

Therefore, it is common to use *shaft brake horsepower* bhp in place of P, and horse-power available hp_A in place of P_A. Equation (6.31) still holds in the form

$$\text{hp}_A = (\eta)(\text{bhp}) \qquad \qquad \textbf{[6.32]}$$

However, be cautious. As always in dealing with fundamental physical relations, units must be consistent; therefore, a good habit is to immediately convert horsepower to foot-pounds per second or to watts before starting an analysis. This approach is used here.

The power-available curve for a typical piston engine–propeller combination is sketched in Fig. 6.20a.

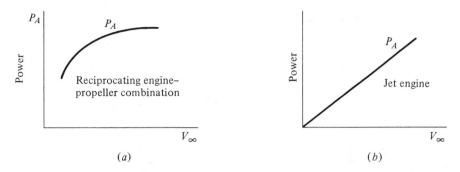

Figure 6.20 Power available for (a) piston engine–propeller combination and (b) the jet engine.

6.6.2 Jet Engine

The jet engine (see Chap. 9) derives its thrust by combustion heating an incoming stream of air and then exhausting this hot air at high velocities through a nozzle. The power available from a jet engine is obtained from Eq. (6.23) as

$$P_A = T_A V_\infty \qquad \textbf{[6.33]}$$

Recall from Fig. 6.13b that T_A for a jet engine is reasonably constant with velocity. Thus, the power-available curve varies essentially linearly with V_∞, as sketched in Fig. 6.20b.

For both the propeller- and jet-powered aircraft, the maximum flight velocity is determined by the high-speed intersection of the maximum P_A and the P_R curves. This is illustrated in Fig. 6.21. Because of their utility in determining other performance characteristics of an airplane, these power curves are essential to any performance analysis.

Calculate the maximum velocity for (a) the CP-1 at sea level and (b) the CJ-1 at 22,000 ft. | **Example 6.4**

Solution

a. For the CP-1 the information in Example 6.1 gave the horsepower rating of the power plant at sea level as 230 hp. Hence,

$$\text{hp}_A = (\eta)(\text{bhp}) = 0.80(230) = 184 \text{ hp}$$

The results of Example 6.3 for power required are replotted in Fig. 6.21a in terms of horsepower. The horsepower available is also shown, and V_{max} is determined by the intersection of the curves as

$$V_{\text{max}} = 265 \text{ ft/s} = 181 \text{ mi/h}$$

b. For the CJ-1, again from the information given in Example 6.1, the sea-level static thrust for each engine is 3650 lb. There are two engines, hence $T_A = 2(3650) = 7300$ lb. From Eq.

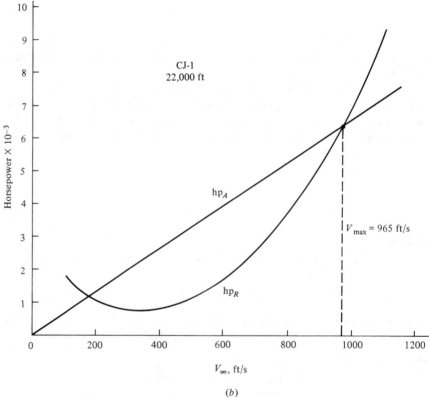

Figure 6.21   Power-available and power-required curves, and the determination of maximum velocity. (*a*) Propeller-driven airplane. (*b*) Jet-propelled airplane.

(6.33), $P_A = T_A V_\infty$, and in terms of horsepower, where T_A is in pounds and V_∞ in feet per second,

$$\text{hp}_A = \frac{T_A V_\infty}{550}$$

Let $\text{hp}_{A,0}$ be the horsepower at sea level. As we will see in Chap. 9, the thrust of a jet engine is, to a first approximation, proportional to the air density. If we make this approximation here, the thrust at altitude becomes

$$T_{A,\text{alt}} = \frac{\rho}{\rho_0} T_{A,0}$$

and hence

$$\text{hp}_{A,\text{alt}} = \frac{\rho}{\rho_0} \text{hp}_{A,0}$$

For the CJ-1 at 22,000 ft, where $\rho = 0.001184$ slug/ft^3,

$$\text{hp}_{A,\text{alt}} = \frac{(\rho/\rho_0) T_A V_\infty}{550} = \frac{(0.001184/0.002377)(7300) V_\infty}{550} = 6.61 V_\infty$$

The results of Example 6.3 for power required are replotted in Fig. 6.21b in terms of horsepower. The horsepower available, obtained from the above equation, is also shown, and V_{max} is determined by the intersection of the curves as

$$\boxed{V_{\text{max}} = 965 \text{ ft/s} = 658 \text{ mi/h}}$$

6.7 ALTITUDE EFFECTS ON POWER REQUIRED AND AVAILABLE

With regard to P_R, curves at altitude could be generated by repeating the calculations of the previous sections, with ρ_∞ appropriate to the given altitude. However, once the sea-level P_R curve is calculated by means of this process, the curves at altitude can be more quickly obtained by simple ratios, as follows. Let the subscript 0 designate sea-level conditions. From Eqs. (6.26) and (6.27),

$$V_0 = \sqrt{\frac{2W}{\rho_0 S C_L}} \qquad [6.34]$$

$$P_{R,0} = \sqrt{\frac{2W^3 C_D^2}{\rho_0 S C_L^3}} \qquad [6.35]$$

where V_0, $P_{R,0}$, and ρ_0 are velocity, power, and density, respectively, at sea level. At altitude, where the density is ρ, these relations are

$$V_{\text{alt}} = \sqrt{\frac{2W}{\rho S C_L}} \qquad [6.36]$$

$$P_{R,\text{alt}} = \sqrt{\frac{2W^3 C_D^2}{\rho S C_L^3}} \qquad \textbf{[6.37]}$$

Now, strictly for the purposes of calculation, let C_L remain fixed between sea level and altitude. Hence, because $C_D = C_{D,0} + C_L^2/(\pi e\text{AR})$, also C_D remains fixed. Dividing Eq. (6.36) by (6.34), and (6.37) by (6.35), we obtain

$$V_{\text{alt}} = V_0 \left(\frac{\rho_0}{\rho}\right)^{1/2} \qquad \textbf{[6.38]}$$

and

$$P_{R,\text{alt}} = P_{R,0} \left(\frac{\rho_0}{\rho}\right)^{1/2} \qquad \textbf{[6.39]}$$

Geometrically, these equations allow us to plot a point on the P_R curve at altitude from a given point on the sea-level curve. For example, consider point 1 on the sea-level P_R curve sketched in Fig. 6.22. By multiplying both the velocity and the power at point 1 by $(\rho_0/\rho)^{1/2}$, a new point is obtained—point 2 in Fig. 6.22. Point 2 is guaranteed to fall on the curve at altitude because of our analysis above. In this fashion, the complete P_R curve at altitude can be readily obtained from the sea-level curve. The results are qualitatively given in Fig. 6.23, where the altitude curves tend to experience an upward and rightward translation as well as a slight clockwise rotation.

With regard to P_A, the lower air density at altitude invariably causes a reduction in power for both the reciprocating and jet engines. In this book we assume P_A and T_A to be proportional to ambient density, as in Example 6.4. Reasons for this will be made clear in Chap. 9. For the reciprocating engine, the loss in power can be delayed by using a supercharger. Nevertheless, the impact on airplane performance due to altitude effects is illustrated in Fig. 6.24a and b for the propeller- and jet-powered

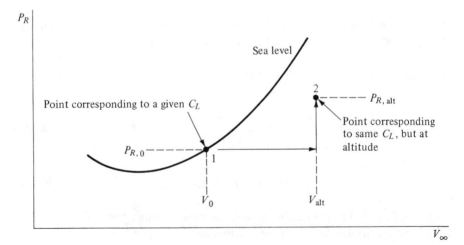

Figure 6.22   Correspondence of points on sea-level and altitude power-required curves.

Figure 6.23   Effect of altitude on power required.

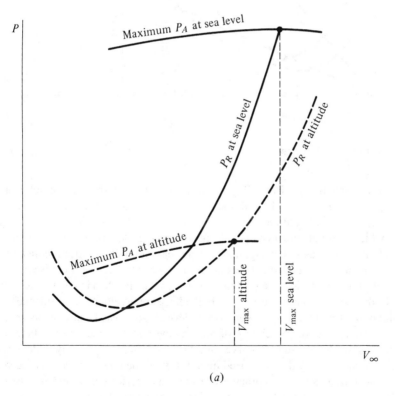

(a)

Figure 6.24   Effect of altitude on maximum velocity. (a) Propeller-driven airplane.
(continued)

(concluded)

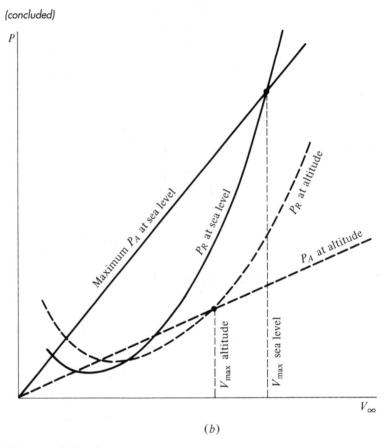

(b)

(b) Jet-propelled airplane.

airplanes, respectively. Both P_R and maximum P_A are shown; the solid curves correspond to sea level and the dashed curves to altitude. From these curves, note that V_{max} varies with altitude. Also note that at high enough altitude, the low-speed limit, which is usually dictated by V_{stall}, may instead be determined by maximum P_A. This effect is emphasized in Fig. 6.25, where maximum P_A has been reduced to the extent that, at velocities just above stalling, P_R exceeds P_A. For this case, we make the interesting conclusion that stalling speed cannot be reached in level, steady flight.

To this point in our discussion, only the horizontal velocity performance—both maximum and minimum speeds in steady, level flight—has been emphasized. We have seen that maximum velocity of an airplane is determined by the high-speed intersection of the P_A and P_R curves and that the minimum velocity is determined either by stalling or by the low-speed intersection of the power curves. These velocity considerations are an important part of airplane performance; indeed, for some airplanes, such as many military fighter planes, squeezing the maximum velocity out of the aircraft is the pivotal design feature. However, this is just the beginning of the performance story; we examine other important characteristics in the remaining sections of this chapter.

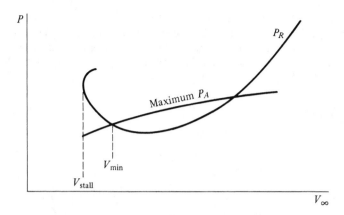

Figure 6.25 Situation when minimum velocity at altitude is greater than stalling velocity.

Using the method of this section, from the CJ-1 power-required curve at 22,000 ft in Example 6.4, obtain the CJ-1 power-required curve at sea level. Compare the maximum velocities at both altitudes.

Example 6.5

Solution

From Eqs. (6.38) and (6.39), corresponding points on the power-required curves for sea level and altitude are, respectively,

$$V_0 = V_{\text{alt}} \left(\frac{\rho}{\rho_0} \right)^{1/2}$$

and

$$\text{hp}_{R,0} = \text{hp}_{R,\text{alt}} \left(\frac{\rho}{\rho_0} \right)^{1/2}$$

We are given V_{alt} and $\text{hp}_{R,\text{alt}}$ for 22,000 ft from the CJ-1 curve in Example 6.4. Using the above formulas, we can generate V_0 and $\text{hp}_{R,0}$ as in the following table, noting that

$$\left(\frac{\rho}{\rho_0} \right)^{1/2} = \left(\frac{0.001184}{0.002377} \right)^{1/2} = 0.706$$

Given Points		$\left(\dfrac{\rho}{\rho_0} \right)^{1/2}$	Generated Points	
V_{alt}, ft/s	$\text{hp}_{R,\text{alt}}$		V_0, ft/s	$\text{hp}_{R,0}$
200	889	0.706	141	628
300	741		212	523
500	1190		353	840
800	3713		565	2621
1000	7012		706	4950

These results, along with the hp$_A$ curves for sea level and 22,000 ft, are plotted in Fig. 6.26. Looking closely at Fig. 6.26, note that point 1 on the hp$_R$ curve at 22,000 ft is used to generate point 2 on the hp$_R$ curve at sea level. This illustrates the idea of this section. Also, note that $V_{\max}$ at sea level is 975 ft/s = 665 mi/h. This is slightly larger than $V_{\max}$ at 22,000 ft, which is 965 ft/s = 658 mi/h.

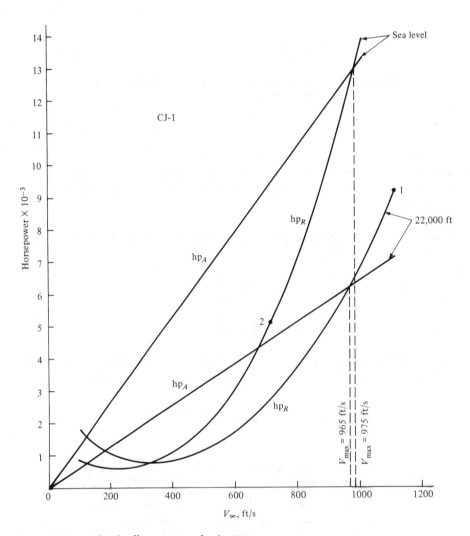

Figure 6.26 Altitude effects on $V_{\max}$ for the CJ-1.

DESIGN BOX

Maximum velocity at a given altitude is frequently a part of the set of specifications for a new airplane design. To design an airplane for a given V_{max}, what characteristics of the airplane would you, the airplane designer, be concerned with? That is, what design aspects of the airplane dictate the maximum velocity? The answer to this question reveals several critical design parameters that are important not only for V_{max} but also for other performance aspects of the airplane. Let us answer this question by obtaining an equation for V_{max} and examining the parameters in the equation. Combining Eqs. (6.1c) and (6.13), we have

$$T = q_\infty S C_D = q_\infty S \left(C_{D,0} + \frac{C_L^2}{\pi e \text{AR}} \right) \quad \textbf{[6.40]}$$

From Eq. (6.14), we obtain for steady, level flight

$$C_L = \frac{W}{q_\infty S} \quad \textbf{[6.41]}$$

Inserting Eq. (6.41) into (6.40) yields

$$T = q_\infty S \left(C_{D,0} + \frac{W^2}{q_\infty^2 S^2 \pi e \text{AR}} \right) \quad \textbf{[6.42]}$$

$$= q_\infty S C_{D,0} + \frac{W^2}{q_\infty S \pi e \text{AR}}$$

Multiply Eq. (6.42) by q_∞ and rearrange.

$$q_\infty S C_{D,0} - q_\infty T + \frac{W^2}{S \pi e \text{AR}} = 0 \quad \textbf{[6.43]}$$

Eq. (6.43) is a quadratic equation in terms of q_∞. Solving Eq. (6.43) for q_∞ by use of the quadratic formula, recalling that $q_\infty = \frac{1}{2} \rho_\infty V_\infty^2$, and setting T in Eq. (6.43) equal to the maximum thrust available (full-throttle thrust) $(T_A)_{max}$, we obtain for the maximum velocity (the details are left to you as a homework problem)

$$V_{max} = \left[\frac{\left(\frac{T_A}{W} \right)_{max} \left(\frac{W}{S} \right) + \left(\frac{W}{S} \right) \sqrt{\left(\frac{T_A}{W} \right)_{max}^2 - \frac{4 C_{D,0}}{\pi e \text{AR}}}}{\rho_\infty C_{D,0}} \right]^{1/2}$$

$$\textbf{[6.44]}$$

Examine Eq. (6.44) carefully. Note that $(T_A)_{max}$ does not appear alone, but appears only in the *ratio* $(T_A/W)_{max}$. Also note that the wing planform area S does not appear alone, but only in the *ratio* W/S. Hence, V_{max} does not depend on thrust alone, or weight alone, or wing area alone, but rather only on certain *ratios* of these quantities, namely,

$$\left(\frac{T_A}{W} \right)_{max} \quad : \text{maximum thrust-to-weight ratio}$$

$$\frac{W}{S} \quad : \text{wing loading}$$

We have just identified two of the most important airplane design parameters, namely, thrust-to-weight ratio and wing loading. In addition, from Eq. (6.44) we see that V_{max} depends on ρ_∞ (altitude), the zero-lift drag coefficient $C_{D,0}$, and the product eAR. Later in Sect. 6.15, we show that the product πeAR is equal to $4 C_{D,0}(L/D)_{max}^2$, where $(L/D)_{max}$ is the maximum value of the lift-to-drag ratio for the airplane. Hence $(L/D)_{max}$ is also an important design parameter. From Eq. (6.44) we conclude that V_{max} can be increased by

1. Increasing the maximum thrust-to-weight ratio $(T_A/W)_{max}$

2. Increasing the wing loading W/S

3. Decreasing the zero-lift drag coefficient $C_{D,0}$

These trends are almost intuitively obvious, even without looking at Eq. (6.44), except possibly for the benefit of increasing the wind loading. To help understand the advantage of a high wing loading in this case, imagine that W/S is increased by decreasing S. If the planform area is made smaller, the total skin friction drag on the wing is reduced (less surface for the shear stress to act on), and hence V_{max} is increased.

The results discussed here are important to other aspects of airplane performance. The design parameters T/W and W/S have a strong effect on other performance quantities in addition to V_{max}, as we will see in subsequent sections.

6.8   RATE OF CLIMB

Visualize a Boeing 777 transport (see Fig. 6.27) powering itself to takeoff speed on an airport runway. It gently lifts off at about 180 mi/h, the nose rotates upward, and the airplane rapidly climbs out of sight. In a matter of minutes, it is cruising at 30,000 ft. This picture prompts the following questions: How fast can the airplane climb? How long does it take to reach a certain altitude? The next two sections provide some answers.

Consider an airplane in steady, unaccelerated, climbing flight, as shown in Fig. 6.28. The velocity along the flight path is V_∞, and the flight path itself is inclined to the horizontal at angle θ. As always, lift and drag are perpendicular and parallel to V_∞, and the weight is perpendicular to the horizontal. Thrust T is assumed to be aligned with the flight path. Here, the physical difference from our previous discussion of level flight is that T is not only working to overcome the drag, but for climbing flight it is also supporting a component of weight. Summing forces parallel to the flight path, we get

$$T = D + W \sin \theta \qquad\qquad \textbf{[6.45]}$$

and perpendicular to the flight path, we have

$$L = W \cos \theta \qquad\qquad \textbf{[6.46]}$$

Note from Eq. (6.46) that the lift is now smaller than the weight. Equations (6.45) and (6.46) represent the equations of motion for steady, climbing flight and are analogous to Eqs. (6.11) and (6.12) obtained earlier for steady, horizontal flight.

Figure 6.27   Three-view of the Boeing 777-200 twin-turbofan high-capacity commercial airliner.

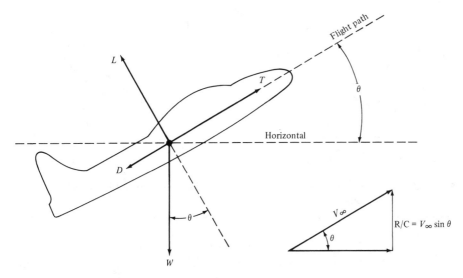

Figure 6.28   Airplane in climbing flight.

Multiply Eq. (6.38) by V_∞:

$$TV_\infty = DV_\infty + WV_\infty \sin\theta$$

$$\frac{TV_\infty - DV_\infty}{W} = V_\infty \sin\theta \qquad \textbf{[6.47]}$$

Examine Eq. (6.47) closely. The right-hand side, $V_\infty \sin\theta$, is the airplane's *vertical* velocity, as illustrated in Fig. 6.28. This vertical velocity is called the *rate of climb* R/C:

$$\boxed{R/C \equiv V_\infty \sin\theta} \qquad \textbf{[6.48]}$$

On the left-hand side of Eq. (6.47), TV_∞ is the power available, from Eq. (6.33), and is represented by the P_A curves in Fig. 6.20. The second term on the left-hand side of Eq. (6.47) is DV_∞, which for level flight is the power required, as represented by the P_R curve in Fig. 6.15. For climbing flight, however, DV_∞ is no longer precisely the power required, because power must be applied to overcome a component of weight as well as drag. Nevertheless, for small angles of climb, say $\theta < 20°$, it is reasonable to neglect this fact and to assume that the DV_∞ term in Eq. (6.47) is given by the level-flight P_R curve in Fig. 6.15. With this,

$$\boxed{TV_\infty - DV_\infty = \text{excess power}} \qquad \textbf{[6.49]}$$

where the excess power is the difference between power available and power required, as shown in Fig. 6.29a and 6.29b, for propeller-driven and jet-powered aircraft, respectively. Combining Eqs. (6.47) to (6.49), we obtain

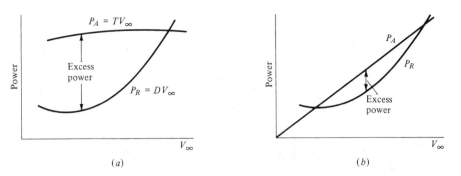

Figure 6.29   Illustration of excess power. (*a*) Propeller-driven airplane. (*b*) Jet-propelled airplane.

$$R/C = \frac{\text{excess power}}{W} \qquad\qquad \textbf{[6.50]}$$

where the excess power is clearly illustrated in Fig. 6.29.

Again, we emphasize that the P_R curves in Fig. 6.29*a* and 6.29*b* are taken, for convenience, as those already calculated for level flight. Hence, *in conjunction with these curves*, Eq. (6.50) is an *approximation* to the rate of climb, good only for small θ. To be more specific, a plot of DV_∞ versus V_∞ for climbing flight [which is exactly called for in Eq. (6.47)] is different from a plot of DV_∞ versus V_∞ for level flight [which is the curve assumed in Fig. 6.29 and used in Eq. (6.50)] simply because D *is smaller for climbing than for level flight at the same* V_∞. To see this more clearly, consider an airplane with $W = 5000$ lb, $S = 100$ ft^2, $C_{D,0} = 0.015$, $e = 0.6$, and AR $= 6$. If the velocity is $V_\infty = 500$ ft/s at sea level, and if the airplane is in *level* flight, then $C_L = L/(q_\infty S) = W/(\frac{1}{2}\rho_\infty V_\infty^2 S) = 0.168$. In turn,

$$C_D = C_{D,0} + \frac{C_L^2}{\pi e \text{AR}} = 0.015 + 0.0025 = 0.0175$$

Now consider the same airplane in a 30° climb at sea level, with the same velocity $V_\infty = 500$ ft/s. Here the lift is smaller than the weight, $L = W \cos\theta$, and therefore $C_L = W \cos 30°/(\frac{1}{2}\rho_\infty V_\infty^2 S) = 0.145$. In turn, $C_D = C_{D,0} + C_L^2/(\pi e \text{AR}) = 0.015 + 0.0019 = 0.0169$. This should be compared with the higher value of 0.0175 obtained above for level flight. As seen in this example, for steady climbing flight, L (hence C_L) is smaller, and thus induced drag is smaller. Consequently, total drag for climbing flight is smaller than that for level flight at the same velocity.

Return again to Fig. 6.29, which corresponds to a given altitude. Note that the excess power is different at different values of V_∞. Indeed, for both the propeller- and jet-powered aircraft there is some V_∞ at which the excess power is maximum. At this point, from Eq. (6.50), R/C will be maximum.

$$\text{max R/C} = \frac{\text{maximum excess power}}{W} \qquad\qquad \textbf{[6.51]}$$

This situation is sketched in Fig. 6.30*a*, where the power available is that at full throttle, that is, maximum P_A. The maximum excess power, shown in Fig. 6.30*a*, via Eq. (6.51) yields the maximum rate of climb that can be generated by the airplane at the given altitude. A convenient graphical method of determining maximum R/C is to plot R/C versus V_∞, as shown in Fig. 6.30*b*. A horizontal tangent defines the point of maximum R/C. Another useful construction is the *hodograph* diagram, which is a plot of the airplane's vertical velocity V_v versus its horizontal velocity V_h. Such a hodograph is sketched in Fig. 6.31. Remember that R/C is defined as the vertical velocity, R/C $\equiv V_v$; hence a horizontal tangent to the hodograph defines the point of maximum R/C (point 1 in Fig. 6.31). Also, any line through the origin and intersecting the hodograph (say, at point 2) has the slope V_v/V_h; hence, from the geometry of the velocity components, such a line makes the climb angle θ with respect to the horizontal axis, as shown in Fig. 6.31. Moreover, the length of the line is equal to V_∞. As this line is rotated counterclockwise, R/C first increases, then goes through its maximum, and finally decreases. Finally, the line becomes tangent to the hodograph at point 3. This tangent line gives the maximum climb angle for which the airplane can maintain steady flight, shown as θ_{max} in Fig. 6.31. It is interesting that maximum R/C does *not* occur at the maximum climb angle.

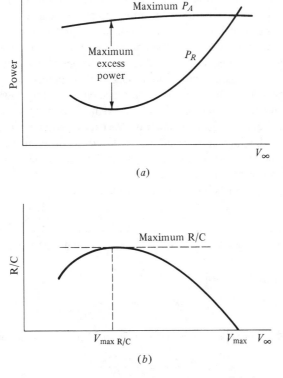

(a)

(b)

Figure 6.30   Determination of maximum rate of climb for a given altitude.

Figure 6.31   Hodograph for climb performance at a given altitude.

The large excess power and high thrust available in modern aircraft allow climbing flight at virtually any angle. For large climb angles, the previous analysis is not valid. Instead, to deal with large θ, the original equations of motion [Eqs. (6.45) and (6.46)] must be solved algebraically, leading to an exact solution valid for any value of θ. The details of this approach can be found in the books by Dommasch et al. and by Perkins and Hage (see Bibliography at the end of this chapter).

Returning briefly to Fig. 6.29*a* and *b* for the propeller-driven and jet-powered aircraft, respectively, we can see an important difference in the low-speed rate-of-climb performance between the two types. Due to the power-available characteristics of a piston engine–propeller combination, large excess powers are available at low values of V_∞, just above the stall. For an airplane on its landing approach, this gives a comfortable margin of safety in case of a sudden wave-off (particularly important for landings on aircraft carriers). In contrast, the excess power available to jet aircraft at low V_∞ is small, with a correspondingly reduced rate-of-climb capability.

Figures 6.30*b* and 6.31 give R/C at a given altitude. In the next section we will ask how R/C varies with altitude. In pursuit of an answer, we will also find the answer to another question, namely, how high the airplane can fly.

Example 6.6 | Calculate the rate of climb versus velocity at sea level for (*a*) the CP-1 and (*b*) the CJ-1.

Solution

a. For the CP-1, from Eq. (6.50)

$$R/C = \frac{\text{excess power}}{W} = \frac{P_A - P_R}{W}$$

With power in foot-pounds per second and W in pounds, for the CP-1, this equation becomes

$$R/C = \frac{P_A - P_R}{2950}$$

From Example 6.3, at $V_\infty = 150$ ft/s, $P_R = 0.326 \times 10^5$ ft · lb/s. From Example 6.4,

$P_A = 550(\text{hp}_A) = 550(184) = 1.012 \times 10^5$ ft · lb/s. Hence,

$$R/C = \frac{(1.012 - 0.326) \times 10^5}{2950} = 23.3 \text{ ft/s}$$

In terms of feet per minute,

$$R/C = 23.3(60) = \boxed{1395 \text{ ft/min}} \quad \text{at } V_\infty = 150 \text{ ft/s}$$

This calculation can be repeated at different velocities, with the following results:

V_∞, ft/s	R/C, ft/min
100	1492
130	1472
180	1189
220	729
260	32.6

These results are plotted in Fig. 6.32.
b. For the CJ-1, from Eq. (6.50),

$$R/C = \frac{P_A - P_R}{W} = \frac{550(\text{hp}_A - \text{hp}_R)}{19,815}$$

From the results and curves of Example 6.5, at $V_\infty = 500$ ft/s, $\text{hp}_R = 1884$, and $\text{hp}_A = 6636$.

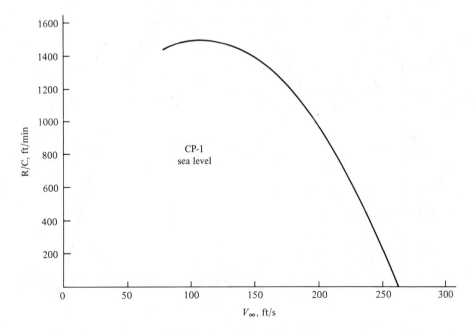

Figure 6.32 Sea-level rate of climb for the CP-1.

Hence
$$R/C = 550 \left(\frac{6636 - 1884}{19{,}815} \right) = 132 \text{ ft/s}$$

or
$$R/C = 132(60) = \boxed{7914 \text{ ft/min}} \qquad \text{at } V_\infty = 500 \text{ ft/s}$$

Again, a short tabulation for other velocities is given below for the reader to check.

V_∞, ft/s	R/C, ft/min
200	3546
400	7031
600	8088
800	5792
950	1230

These results are plotted in Fig. 6.33.

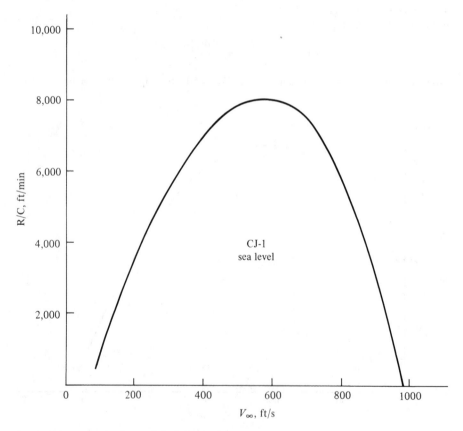

Figure 6.33   Sea-level rate of climb for the CJ-1.

DESIGN BOX

What airplane design parameters dictate maximum rate of climb? The answer is not explicitly clear from our graphical analysis carried out in this section. However, the answer can be obtained explicitly by deriving an equation for maximum rate of climb and identifying the design parameters that appear in the equation. The derivation is lengthy, and we are only interested in the final result here. For a detailed derivation, see Anderson, *Aircraft Performance and Design*, McGraw-Hill, New York, 1999. Denoting maximum rate of climb by $(R/C)_{max}$, and for compactness identifying the symbol Z as

$$Z = 1 + \sqrt{1 + \frac{3}{(L/D)_{max}^2 (T/W)_{max}^2}}$$

where $(L/D)_{max}$ is the maximum value of the lift-to-drag ratio for the given airplane, we can show that for a *jet-propelled* airplane

$$(R/C)_{max} = \left[\frac{(W/S)Z}{3\rho_\infty C_{D,0}}\right]^{1/2} \left(\frac{T}{W}\right)_{max}^{3/2}$$
$$\times \left[1 - \frac{Z}{6} - \frac{3}{2(T/W)_{max}^2 (L/D)_{max}^2 Z}\right]$$

[6.52]

and for a *propeller-driven* airplane

$$(R/C)_{max} = \left(\frac{\eta P}{W}\right)_{max} - 0.8776 \sqrt{\frac{W/S}{\rho_\infty C_{D,0}}} \frac{1}{(L/D)_{max}^{3/2}}$$

[6.53]

where η is the propeller efficiency defined by Eq. (6.31) and P is the shaft brake power from the engine (or engines, for a multiengine airplane).

Examining Eq. (6.52), we see once again that W, S, and T do not appear alone, but rather in ratios. From Eq. (6.52), the design parameters that dictate $(R/C)_{max}$ for a jet-propelled airplane are

- Wing loading W/S
- Maximum thrust-to-weight ratio $(T/W)_{max}$
- Zero-lift drag coefficient $C_{D,0}$
- Maximum lift-to-drag ratio $(L/D)_{max}$

These are the same design parameters that dictate V_{max} from Eq. (6.44). We also note, looking ahead to Sec. 6.14, that $(L/D)_{max}$ is determined by $C_{D,0}$, e, and AR, namely, that $(L/D)_{max}^2 = \pi e \text{AR}/(4 C_{D,0})$, as we will see. So identifying $(L/D)_{max}$ as a design parameter is the same as identifying a certain combination of e, AR, and $C_{D,0}$ as a design parameter. We will have more to say about the importance of $(L/D)_{max}$ in airplane design in subsequent sections.

Recall that for a propeller-driven airplane, the rating of the engine-propeller combination in terms of *power* is more germane than that in terms of thrust. Hence, Eq. (6.53) gives maximum rate of climb for a propeller-driven airplane in terms of the *power-to-weight* ratio $\eta P/W$. [Recall from Eq. (6.31) that ηP is the power available P_A for a propeller-driven airplane.] Therefore, for a propeller-driven airplane, an important design parameter that dictates $(R/C)_{max}$ is the power-to-weight ratio.

6.9 GLIDING FLIGHT

Consider an airplane in a power-off glide, as sketched in Fig. 6.34. The forces acting on this aircraft are lift, drag, and weight; the thrust is zero because the power is off. The glide flight path makes an angle θ below the horizontal. For an equilibrium unaccelerated glide, the sum of the forces must be zero. Summing forces along the flight path, we have

$$D = W \sin \theta$$

[6.54]

Figure 6.34   Airplane in power-off gliding flight.

and perpendicular to the flight path

$$L = W \cos \theta \qquad\qquad \textbf{[6.55]}$$

The equilibrium glide angle can be calculated by dividing Eq. (6.54) by (6.55), yielding

$$\frac{\sin \theta}{\cos \theta} = \frac{D}{L}$$

or

$$\boxed{\tan \theta = \frac{1}{L/D}} \qquad\qquad \textbf{[6.56]}$$

Clearly, the glide angle is strictly a function of the lift-to-drag ratio; the higher the L/D, the shallower the glide angle. From this, the smallest equilibrium glide angle occurs at $(L/D)_{\max}$, which corresponds to the maximum range for the glide.

Example 6.7 | The maximum lift-to-drag ratio for the CP-1 is 13.6. Calculate the minimum glide angle and the maximum range measured along the ground covered by the CP-1 in a power-off glide that starts at an altitude of 10,000 ft.

Figure 6.35 Range covered in an equilibrium glide.

Solution

The minimum glide angle is obtained from Eq. (6.56) as

$$\tan \theta_{min} = \frac{1}{(L/D)_{max}} = \frac{1}{13.6}$$

$$\boxed{\theta_{min} = 4.2°}$$

The distance covered along the ground is R, as shown in Fig. 6.35. If h is the altitude at the start of the glide, then

$$R = \frac{h}{\tan \theta} = h \frac{L}{D}$$

Hence,

$$R_{max} = h \left(\frac{L}{D}\right)_{max} = 10,000(13.6)$$

$$\boxed{R_{max} = 136,000 \text{ ft} = 25.6 \text{ mi}}$$

Repeat Example 6.7 for the CJ-1, for which the value of $(L/D)_{max}$ is 16.9.

Example 6.8

Solution

$$\tan \theta_{min} = \frac{1}{(L/D)_{max}} = \frac{1}{16.9}$$

$$\boxed{\theta_{min} = 3.39°}$$

$$R_{max} = h \left(\frac{L}{D}\right)_{max} = 10,000(16.9)$$

$$\boxed{R_{max} = 169,000 \text{ ft} = 32 \text{ mi}}$$

Note the obvious fact that the CJ-1, with its higher value of $(L/D)_{max}$, is capable of a larger glide range than the CP-1.

Example 6.9

For the CP-1, calculate the equilibrium glide velocities at altitudes of 10,000 and 2000 ft, each corresponding to the minimum glide angle.

Solution

$$L = \tfrac{1}{2}\rho_\infty V_\infty^2 S C_L$$

Combining this with Eq. (6.55) gives

$$W \cos\theta = \tfrac{1}{2}\rho_\infty V_\infty^2 S C_L$$

or

$$V_\infty = \sqrt{\frac{2\cos\theta}{\rho_\infty C_L}\frac{W}{S}}$$

where W/S is the by now familiar *wing loading*. From this equation we see that the higher the wing loading, the higher the glide velocity. This makes sense: A heavier airplane with a smaller wing area is going to glide to the earth's surface at a greater velocity. Note, however, that the glide angle, and hence range, depends not on the weight of the airplane and not on its wing loading but exclusively on the value of $(L/D)_{max}$, which is an aerodynamic property of the airframe design. A higher wing loading simply means that the airplane will have a *faster* glide and will reach the earth's surface sooner. From Example 6.1 we have for the CP-1,

$$\frac{W}{S} = \frac{2950}{174} = 16.95 \text{ lb/ft}^2$$

Also from the tabulation in Example 6.1, we see that $(L/D)_{max} = 13.6$ corresponds to a lift coefficient $C_L = 0.634$. (Note that both L/D and C_L are functions of the angle of attack of the airplane; these are aerodynamic data associated with the airframe and are not influenced by the flight conditions. Hence, $C_L = 0.634$ at maximum L/D, no matter whether the airplane is in level flight, is climbing, or is in a glide.) Therefore, at 10,000 ft, where $\rho_\infty = 0.0017556$ slug/ft^3, we have

$$V_\infty = \sqrt{\frac{(2\cos 4.2°)(16.95)}{0.0017556(0.634)}}$$

$$\boxed{V_\infty = 174.3 \text{ ft/s}} \qquad \text{at } h = 10,000 \text{ ft}$$

At 2000 ft, $\rho_\infty = 0.0022409$ slug/ft^3. Hence

$$V_\infty = \sqrt{\frac{(2\cos 4.2°)(16.95)}{0.0022409(0.634)}}$$

$$\boxed{V_\infty = 154.3 \text{ ft/s}} \qquad \text{at } h = 2000 \text{ ft}$$

Note that the equilibrium glide velocity decreases as altitude decreases.

6.10 ABSOLUTE AND SERVICE CEILINGS

The effects of altitude on P_A and P_R were discussed in Sec. 6.7 and illustrated in Fig. 6.24a and b. For the sake of discussion, consider a propeller-driven airplane: The results of this section will be qualitatively the same for a jet. As altitude increases, the maximum excess power decreases, as shown in Fig. 6.36. In turn, maximum R/C decreases. This is illustrated by Fig. 6.37, which is a plot of maximum R/C versus altitude, but with R/C as the abscissa.

There is some altitude high enough at which the P_A curve becomes tangent to the P_R curve (point 1 in Fig. 6.38). The velocity at this point is the only value at which steady, level flight is possible; moreover, there is zero excess power, hence zero maximum rate of climb, at this point. The altitude at which maximum R/C $= 0$ is defined as the *absolute ceiling* of the airplane. A more useful quantity is the *service ceiling*, defined as that altitude where maximum R/C $= 100$ ft/min. The service ceiling represents the practical upper limit of steady, level flight.

The absolute and service ceilings can be determined as follows:

1. Using the technique of Sec. 6.8, calculate values of maximum R/C for a number of different altitudes.

2. Plot maximum rate of climb versus altitude, as shown in Fig. 6.37.

3. Extrapolate the curve to 100 ft/min and 0 ft/min to find the service and absolute ceilings, respectively, as also shown in Fig. 6.37.

Calculate the absolute and service ceilings for (a) the CP-1 and (b) the CJ-1. | **Example 6.10**

Solution

a. For the CP-1, as stated in Example 6.1, all the results presented in all the examples of this chapter are taken from a computer program which deals with 100 different velocities, each at different altitudes, beginning at sea level and increasing in 2000-ft increments. In modern engineering, using the computer to take the drudgery out of extensive and repeated calculations

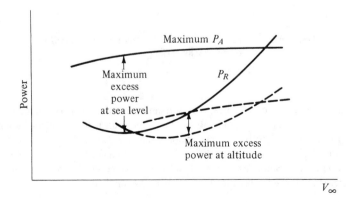

Figure 6.36   Variation of excess power with altitude.

Figure 6.37   Determination of absolute and service ceilings for the CP-1.

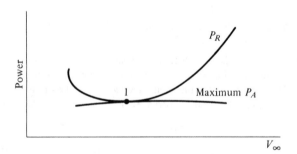

Figure 6.38   Power-required and power-available curves at the absolute ceiling.

is an everyday practice. For example, note from Example 6.6 that the maximum rate of climb at sea level for the CP-1 is 1500 ft/min. In essence, this result is the product of all the work performed in Examples 6.1 to 6.6. Now, to obtain the absolute and service ceilings, these calculations must be repeated at several different altitudes in order to find where R/C = 0 and 100 ft/min, respectively. Some results are tabulated and plotted in the table at the top of page 397; the reader should take the time to check a few of the numbers.

Altitude, ft	Maximum R/C, ft/min
0	1500
4,000	1234
8,000	987
12,000	755
16,000	537
20,000	331
24,000	135
26,000	40

These results are plotted in Fig. 6.37. From these numbers, we find

Absolute ceiling (R/C = 0) is ⎮ 27,000 ft ⎮

Service ceiling (R/C = 100 ft/min) is ⎮ 25,000 ft ⎮

b. For the CJ-1, utilizing the results from Examples 6.1 to 6.6 and making similar calculations at various altitudes, we tabulate the following results:

Altitude, ft	Maximum R/C, ft/min
0	8118
6,000	6699
12,000	5448
18,000	4344
24,000	3369
30,000	2502
36,000	1718

These results are plotted in Fig. 6.39.
From these results, we find

Absolute ceiling (R/C = 0) is ⎮ 49,000 ft ⎮

Service ceiling (R/C = 100 ft/min) is ⎮ 48,000 ft ⎮

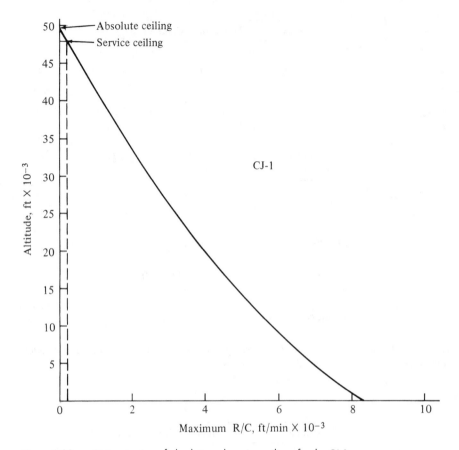

Figure 6.39   Determination of absolute and service ceilings for the CJ-1.

6.11   TIME TO CLIMB

To carry out its defensive role adequately, a fighter airplane must be able to climb from sea level to the altitude of advancing enemy aircraft in as short a time as possible. In another case, a commercial aircraft must be able to rapidly climb to high altitudes to minimize the discomfort and risks of inclement weather and to minimize air traffic problems. As a result, the time for an airplane to climb to a given altitude can become an important design consideration. The calculation of the time to climb follows directly from our previous discussions, as described below.

The rate of climb was defined in Sec. 6.8 as the vertical velocity of the airplane. Velocity is simply the time rate of change of distance, the distance here being the altitude h. Hence, $R/C = dh/dt$. Therefore,

$$dt = \frac{dh}{R/C} \qquad \textbf{[6.57]}$$

In Eq. (6.57), dt is the small increment in time required to climb a small increment dh in altitude. Therefore, from calculus, the time to climb from one altitude h_1 to another h_2 is obtained by integrating Eq. (6.57):

$$t = \int_{h_1}^{h_2} \frac{dh}{R/C}$$

Normally, time to climb is considered from sea level, where $h_1 = 0$. Hence, the time to climb to any given altitude h_2 is

$$t = \int_0^{h_2} \frac{dh}{R/C} \qquad \textbf{[6.58]}$$

To calculate t graphically, first plot $(R/C)^{-1}$ versus h, as shown in Fig. 6.40. The area under the curve from $h = 0$ to $h = h_2$ is the time to climb to altitude h_2.

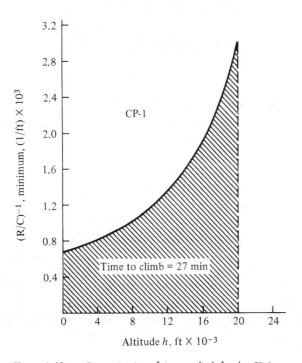

Figure 6.40   Determination of time to climb for the CP-1.

Example 6.11

Calculate and compare the time required for (*a*) the CP-1 and (*b*) the CJ-1 to climb to 20,000 ft.

Solution

a. For the CP-1, from Eq. (6.58), the time to climb is equal to the shaded area under the curve shown in Figure 6.40. The resulting area gives time to climb as $\boxed{27.0 \text{ min.}}$

b. For the CJ-1, Eq. (6.58) is plotted in Fig. 6.41. The resulting area gives time to climb as $\boxed{3.5 \text{ min.}}$

> Note that the CJ-1 climbs to 20,000 ft in one-eighth of the time required by the CP-1; this is to be expected for a high-performance executive jet transport in comparison to its propeller-driven piston engine counterpart.

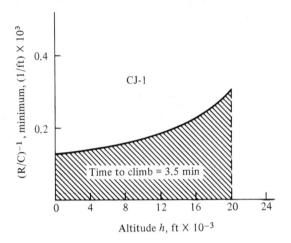

Figure 6.41   Determination of time to climb for the CJ-1.

6.12 RANGE AND ENDURANCE—PROPELLER-DRIVEN AIRPLANE

When Charles Lindbergh made his spectacular solo flight across the Atlantic Ocean on May 20–21, 1927, he could not have cared less about maximum velocity, rate of climb, or time to climb. Uppermost in his mind was the maximum distance he could fly on the fuel supply carried by the *Spirit of St. Louis*. Therefore, *range* was the all-pervasive consideration during the design and construction of Lindbergh's airplane. Indeed, throughout all 20th-century aviation, range has been an important design feature, especially for transcontinental and transoceanic transports and for strategic bombers for the military.

   Range is technically defined as the total distance (measured with respect to the ground) traversed by the airplane on a tank of fuel. A related quantity is *endurance*, which is defined as the total time that an airplane stays in the air on a tank of fuel. In different applications, it may be desirable to maximize one or the other of these characteristics. The parameters which maximize range are different from those which maximize endurance; they are also different for propeller- and jet-powered aircraft. The purpose of this section is to discuss these variations for the case of a propeller-driven airplane; jet airplanes are considered in the next section.

6.12.1 Physical Considerations

One of the critical factors influencing range and endurance is the *specific fuel consumption*, a characteristic of the engine. For a reciprocating engine, specific fuel consumption (commonly abbreviated SFC) is defined as the *weight of fuel consumed per unit power per unit time*. As mentioned earlier, reciprocating engines are rated in terms of horsepower, and the common units (although inconsistent) of specific fuel consumption are

$$\text{SFC} = \frac{\text{lb of fuel}}{(\text{bhp})(\text{h})}$$

where bhp signifies shaft brake horsepower, discussed in Sec. 6.6.

First, consider endurance. On a qualitative basis, to stay in the air for the longest time, common sense says that we must use the *minimum* number of pounds of fuel per hour. On a dimensional basis this quantity is proportional to the horsepower required by the airplane and to the SFC:

$$\frac{\text{lb of fuel}}{\text{h}} \propto (\text{SFC})(\text{hp}_R)$$

Therefore, minimum pounds of fuel per hour are obtained with minimum hp_R. Since minimum pounds of fuel per hour give maximum endurance, we quickly conclude that

Maximum endurance for a propeller-driven airplane occurs when the airplane is flying at minimum power required.

This condition is sketched in Fig. 6.42. Furthermore, in Sec. 6.5 we have already proved that minimum power required corresponds to a maximum value of $C_L^{3/2}/C_D$ [see Eq. (6.27)]. Thus

Maximum endurance for a propeller-driven airplane occurs when the airplane is flying at a velocity such that $C_L^{3/2}/C_D$ is maximum.

Now, consider range. To cover the longest distance (say, in miles), common sense says that we must use the minimum number of pounds of fuel per mile. On a dimensional basis, we can state the proportionality

$$\frac{\text{lb of fuel}}{\text{mi}} \propto \frac{(\text{SFC})(\text{hp}_R)}{V_\infty}$$

Figure 6.42   Points of maximum range and endurance on the power-required curve for a propeller-driven airplane.

(Check the units yourself, assuming V_∞ is in miles per hour.) As a result, minimum pounds of fuel per mile are obtained with a minimum hp_R/V_∞. This minimum value of hp_R/V_∞ precisely corresponds to the tangent point in Fig. 6.17, which also corresponds to maximum L/D, as proved in Sec. 6.5. Thus

> Maximum range for a propeller-driven airplane occurs when the airplane is flying at a velocity such that C_L/C_D is maximum.

This condition is also sketched in Fig. 6.42.

6.12.2 Quantitative Formulation

The important conclusions displayed above were obtained from purely physical reasoning. We will develop quantitative formulas which substantiate these conclusions and which allow the direct calculation of range and endurance for given conditions.

In this development, the specific fuel consumption is couched in units that are consistent, that is,

$$\frac{\text{lb of fuel}}{(\text{ft} \cdot \text{lb/s})(\text{s})} \quad \text{or} \quad \frac{\text{N of fuel}}{(\text{J/s})(\text{s})}$$

For convenience and clarification, c will designate the specific fuel consumption with consistent units.

Consider the product $cP\,dt$, where P is engine power and dt is a small increment of time. The units of this product are (in the English engineering system)

$$cP\,dt = \frac{\text{lb of fuel}}{(\text{ft}\cdot\text{lb/s})(\text{s})}\,\frac{\text{ft}\cdot\text{lb}}{\text{s}}(\text{s}) = \text{lb of fuel}$$

Therefore, $cP\,dt$ represents the differential change in weight of the fuel due to consumption over the short time period dt. The total weight of the airplane W is the sum of the fixed structural and payload weights, along with the changing fuel weight. Hence, any change in W is assumed to be due to the change in fuel weight. Recall that W denotes the weight of the airplane at any instant. Also, let $W_0 =$ gross weight of the airplane (weight with full fuel and payload), $W_f =$ weight of the fuel load, and $W_1 =$ weight of the airplane *without* fuel. With these considerations, we have

$$W_1 = W_0 - W_f$$

and

$$dW_f = dW = -cP\,dt$$

or

$$dt = -\frac{dW}{cP} \qquad \textbf{[6.59]}$$

The minus sign in Eq. (6.59) is necessary because dt is physically positive (time cannot move backward, except in science fiction novels), while at the same time W is decreasing (hence dW is negative). Integrating Eq. (6.59) between time $t = 0$, where $W = W_0$ (fuel tanks full), and time $t = E$, where $W = W_1$ (fuel tanks empty), we find

$$\int_0^E dt = -\int_{W_0}^{W_1} \frac{dW}{cP}$$

$$\boxed{E = \int_{W_1}^{W_0} \frac{dW}{cP}} \qquad \textbf{[6.60]}$$

In Eq. (6.60), E is the endurance in seconds.

To obtain an analogous expression for range, multiply Eq. (6.59) by V_∞:

$$V_\infty\,dt = -\frac{V_\infty\,dW}{cP} \qquad \textbf{[6.61]}$$

In Eq. (6.61), $V_\infty\,dt$ is the incremental distance ds covered in time dt.

$$ds = -\frac{V_\infty\,dW}{cP} \qquad \textbf{[6.62]}$$

The total distance covered throughout the flight is equal to the integral of Eq. (6.62) from $s = 0$, where $W = W_0$ (full fuel tank), to $s = R$, where $W = W_1$ (empty fuel tank):

$$\int_0^R ds = -\int_{W_0}^{W_1} \frac{V_\infty\,dW}{cP}$$

or

$$\boxed{R = \int_{W_1}^{W_0} \frac{V_\infty\,dW}{cP}} \qquad \textbf{[6.63]}$$

In Eq. (6.63), R is the range in consistent units, such as feet or meters.

Equations (6.60) and (6.63) can be evaluated graphically, as shown in Fig. 6.43a and b for range and endurance, respectively. Range can be calculated accurately by plotting $V_\infty/(cP)$ versus W and taking the area under the curve from W_1 to W_0, as shown in Fig. 6.43a. Analogously, the endurance can be calculated accurately by plotting $(cP)^{-1}$ versus W and taking the area under the curve from W_1 to W_0, as shown in Fig. 6.43b.

Equations (6.60) and (6.63) are accurate formulations for endurance and range. In principle, they can include the entire flight—takeoff, climb, cruise, and landing— as long as the instantaneous values of W, V_∞, c, and P are known at each point along the flight path. However, Eqs. (6.60) and (6.63), although accurate, are also long and tedious to evaluate by the method discussed above. Therefore, simpler but approximate analytic expressions for R and E are useful. Such formulas are developed below.

6.12.3   Breguet Formulas (Propeller-Driven Airplane)

For level, unaccelerated flight, we demonstrated in Sec. 6.5 that $P_R = DV_\infty$. Also, to maintain steady conditions, the pilot has adjusted the throttle such that power available from the engine-propeller combination is just equal to the power required: $P_A = P_R = DV_\infty$. In Eq. (6.59), P is the brake power output of the engine itself. Recall from Eq. (6.31) that $P_A = \eta P$, where η is the propeller efficiency. Thus

$$P = \frac{P_A}{\eta} = \frac{DV_\infty}{\eta} \qquad \textbf{[6.64]}$$

Substitute Eq. (6.64) into (6.63):

$$R = \int_{W_1}^{W_0} \frac{V_\infty\,dW}{cP} = \int_{W_1}^{W_0} \frac{V_\infty\eta\,dW}{cDV_\infty} = \int_{W_1}^{W_0} \frac{\eta\,dW}{cD} \qquad \textbf{[6.65]}$$

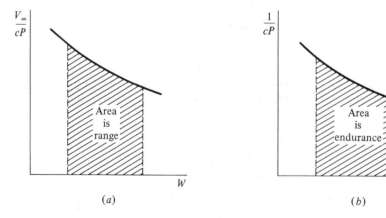

Figure 6.43   Determination of range and endurance.

Multiplying Eq. (6.65) by W/W and noting that for steady, level flight $W = L$, we obtain

$$R = \int_{W_1}^{W_0} \frac{\eta}{cD} \frac{W}{W} \, dW = \int_{W_1}^{W_0} \frac{\eta}{c} \frac{L}{D} \frac{dW}{W} \qquad \textbf{[6.66]}$$

Unlike Eq. (6.63), which is exact, Eq. (6.66) now contains the direct assumption of level, unaccelerated flight. However, for practical use, it will be further simplified by assuming that η, $L/D = C_L/C_D$, and c are constant throughout the flight. This is a reasonable approximation for cruising flight conditions. Thus, Eq. (6.66) becomes

$$R = \frac{\eta}{c} \frac{C_L}{C_D} \int_{W_1}^{W_0} \frac{dW}{W}$$

$$\boxed{R = \frac{\eta}{c} \frac{C_L}{C_D} \ln \frac{W_0}{W_1}} \qquad \textbf{[6.67]}$$

Equation (6.67) is a classic formula in aeronautical engineering; it is called the *Breguet range formula*, and it gives a quick, practical estimate for range which is generally accurate to within 10 to 20 percent. Also, keep in mind that, as with all proper physical derivations, Eq. (6.67) deals with consistent units. Hence, R is in feet or meters when c is in consumption of fuel in lb/(ft · lb/s)(s) or N/(J/s)(s), respectively, as discussed in Sec. 6.12.2. If c is given in terms of brake horsepower and if R is desired in miles, the proper conversions to consistent units should be made before using Eq. (6.67).

Look at Eq. (6.67). It says all the things that common sense would expect; namely, to maximize range for a reciprocating-engine, propeller-driven airplane, we want the following:

1. The largest possible propeller efficiency η.
2. The lowest possible specific fuel consumption c.
3. The highest ratio of W_0/W_1, which is obtained with the largest fuel weight W_F.
4. Most importantly, flight at *maximum L/D*. This confirms our above argument in Sec. 6.12.1 that for *maximum range*, we must fly at *maximum L/D*. Indeed, the Breguet range formula shows that range is directly proportional to L/D. This clearly explains why high values of L/D (high aerodynamic efficiency) have always been of importance in the design of airplanes. This importance was underscored in the 1970s due to the increasing awareness of the need to conserve energy (hence fuel).

A similar formula can be obtained for endurance. If we recall that $P = DV_\infty/\eta$ and that $W = L$, Eq. (6.60) becomes

$$E = \int_{W_1}^{W_0} \frac{dW}{cP} = \int_{W_1}^{W_0} \frac{\eta}{c} \frac{dW}{DV_\infty} = \int_{W_1}^{W_0} \frac{\eta}{c} \frac{L}{DV_\infty} \frac{dW}{W}$$

Since $L = W = \frac{1}{2}\rho_\infty V_\infty^2 S C_L$, then $V_\infty = \sqrt{2W/(\rho_\infty S C_L)}$. Thus

$$E = \int_{W_1}^{W_0} \frac{\eta}{c}\frac{C_L}{C_D}\sqrt{\frac{\rho_\infty S C_L}{2}}\,\frac{dW}{W^{3/2}}$$

Assuming that C_L, C_D, η, c, and ρ_∞ (constant altitude) are all constant, this equation becomes

$$E = -2\frac{\eta}{c}\frac{C_L^{3/2}}{C_D}\left(\frac{\rho_\infty S}{2}\right)^{1/2}[W^{-1/2}]_{W_1}^{W_0}$$

or
$$E = \frac{\eta}{c}\frac{C_L^{3/2}}{C_D}(2\rho_\infty S)^{1/2}(W_1^{-1/2} - W_0^{-1/2})$$ [6.68]

Equation (6.68) is the *Breguet endurance formula*, where E is in seconds (consistent units).

Look at Eq. (6.68). It says that to maximize endurance for a reciprocating-engine, propeller-driven airplane, we want

1. The highest propeller efficiency η.
2. The lowest specific fuel consumption c.
3. The highest fuel weight W_f, where $W_0 = W_1 + W_f$.
4. Flight at maximum $C_L^{3/2}/C_D$. This confirms our above argument in Sec. 6.12.1 that for *maximum endurance*, we must fly at maximum $C_L^{3/2}/C_D$.
5. Flight at sea level, because $E \propto \rho_\infty^{1/2}$, and ρ_∞ is largest at sea level.

It is interesting to note that, subject to our approximations, endurance depends on altitude, whereas range [see Eq. (6.67)] is independent of altitude.

Remember that the discussion in this section pertains only to a combination of piston engine and propeller. For a jet-powered airplane, the picture changes, as discussed in the next section.

Example 6.12

Estimate the maximum range and maximum endurance for the CP-1.

Solution

The Breguet range formula is given by Eq. (6.67) for a propeller-driven airplane. This equation is

$$R = \frac{\eta}{c}\frac{C_L}{C_D}\ln\frac{W_0}{W_1}$$

with the specific fuel consumption c in consistent units, say (lb fuel)/(ft · lb/s)(s) or simply per foot. However, in Example 6.1, the SFC is given as 0.45 lb of fuel/(hp)(h). This can be changed to consistent units, as

$$c = 0.45\frac{\text{lb}}{(\text{hp})(\text{h})}\frac{1\,\text{hp}}{550\,\text{ft}\cdot\text{lb/s}}\frac{1\,\text{h}}{3600\,\text{s}} = 2.27\times10^{-7}\,\text{ft}^{-1}$$

In Example 6.1, the variation of $C_L/C_D = L/D$ was calculated versus velocity. The variation of $C_L^{3/2}/C_D$ can be obtained in the same fashion. The results are plotted in Fig. 6.44.
From these curves,

$$\max\left(\frac{C_L}{C_D}\right) = 13.62 \qquad \max\left(\frac{C_L^{3/2}}{C_D}\right) = 12.81$$

These are results pertaining to the aerodynamics of the airplane; even though the above plots were calculated at sea level (from Example 6.1), the *maximum* values of C_L/C_D and $C_L^{3/2}/C_D$ are independent of altitude, velocity, etc. They depend only on the aerodynamic design of the aircraft.

The gross weight of the CP-1 is $W_0 = 2950$ lb. The fuel capacity given in Example 6.1 is 65 gal of aviation gasoline, which weighs 5.64 lb/gal. Hence, the weight of the fuel $W_p = 65(5.64) = 367$ lb. Thus, the empty weight $W_1 = 2950 - 367 = 2583$ lb.

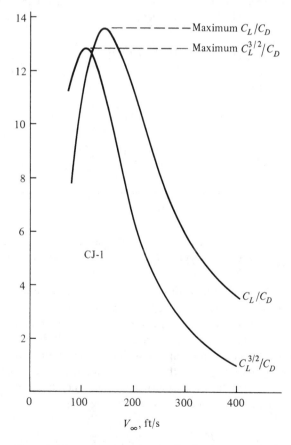

Figure 6.44 Aerodynamic ratios for the CP-1 at sea level.

Returning to Eq. (6.67), we have

$$R = \frac{\eta}{c} \frac{C_L}{C_D} \ln \frac{W_0}{W_1} = \frac{0.8}{2.27 \times 10^{-7}} (13.62) \left(\ln \frac{2950}{2583} \right)$$

$$\boxed{R = 6.38 \times 10^6 \text{ ft}}$$

Since 1 mi = 5280 ft,

$$R = \frac{6.38 \times 10^6}{5280} = \boxed{1207 \text{ mi}}$$

The endurance is given by Eq. (6.68):

$$E = \frac{\eta}{c} \frac{C_L^{3/2}}{C_D} (2\rho_\infty S)^{1/2} (W_1^{-1/2} - W_0^{-1/2})$$

Because of the explicit appearance of ρ_∞ in the endurance equation, maximum endurance will occur at sea level, $\rho_\infty = 0.002377$ slug/ft^3. Hence,

$$E = \frac{0.8}{2.27 \times 10^{-7}} (12.81)[2(0.002377)(174)]^{1/2} \left(\frac{1}{2583^{1/2}} - \frac{1}{2950^{1/2}} \right)$$

$$\boxed{E = 5.19 \times 10^4 \text{ s}}$$

Since 3600 s = 1 h,

$$E = \frac{5.19 \times 10^4}{3600} = \boxed{14.4 \text{ h}}$$

6.13   RANGE AND ENDURANCE—JET AIRPLANE

For a jet airplane, the specific fuel consumption is defined as the *weight of fuel consumed per unit thrust per unit time*. Note that *thrust* is used here, in contradistinction to *power*, as in the previous case for a reciprocating engine–propeller combination. The fuel consumption of a jet engine physically depends on the thrust produced by the engine, whereas the fuel consumption of a reciprocating engine physically depends on the brake power produced. It is this simple difference which leads to different range and endurance formulas for a jet airplane. In the literature, *thrust-specific fuel consumption (TSFC)* for jet engines is commonly given as

$$\text{TSFC} = \frac{\text{lb of fuel}}{(\text{lb of thrust})(\text{h})}$$

(Note the inconsistent unit of time).

6.13.1 Physical Considerations

The maximum endurance of a jet airplane occurs for minimum pounds of fuel per hour, the same as for propeller-driven aircraft. However, for a jet,

$$\frac{\text{lb of fuel}}{\text{h}} = (\text{TSFC})(T_A)$$

where T_A is the thrust available produced by the jet engine. Recall that in steady, level, unaccelerated flight, the pilot has adjusted the throttle such that thrust available T_A just equals the thrust required T_R: $T_A = T_R$. Therefore, minimum pounds of fuel per hour correspond to minimum thrust required. Hence, we conclude that

> Maximum endurance for a jet airplane occurs when the airplane is flying at minimum thrust required.

This condition is sketched in Fig. 6.45. Furthermore, in Sec. 6.3, minimum thrust required was shown to correspond to maximum L/D. Thus

> Maximum endurance for a jet airplane occurs when the airplane is flying at a velocity such that C_L/C_D is maximum.

Now, consider range. As before, maximum range occurs for a minimum number of pounds of fuel per mile. For a jet, on a dimensional basis,

$$\frac{\text{lb of fuel}}{\text{mi}} = \frac{(\text{TSFC})(T_A)}{V_\infty}$$

Recalling that for steady, level flight, $T_A = T_R$, we note that minimum pounds of fuel per mile correspond to a minimum T_R/V_∞. In turn, T_R/V_∞ is the slope of a line through the origin and intersecting the thrust-required curve; its minimum value

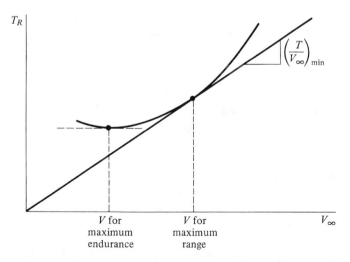

Figure 6.45 Points of maximum range and endurance on the thrust-required curve.

occurs when the line becomes tangent to the thrust-required curve, as sketched in Fig. 6.45. The aerodynamic condition holding at this tangent point is obtained as follows. Recall that for steady, level flight, $T_R = D$. Then

$$\frac{T_R}{V_\infty} = \frac{D}{V_\infty} = \frac{\frac{1}{2}\rho_\infty V_\infty^2 SC_D}{V_\infty} = \frac{1}{2}\rho_\infty V_\infty SC_D$$

Since $V_\infty = \sqrt{2W/(\rho_\infty SC_L)}$, we have

$$\frac{T_R}{V_\infty} = \frac{1}{2}\rho_\infty S\sqrt{\frac{2W}{\rho_\infty SC_L}}C_D \propto \frac{1}{C_L^{1/2}/C_D}$$

Hence, minimum T_R/V_∞ corresponds to maximum $C_L^{1/2}/C_D$. In turn, we conclude that

Maximum range for a jet airplane occurs when the airplane is flying at a velocity such that $C_L^{1/2}/C_D$ is maximum.

6.13.2 Quantitative Formulation

Let c_t be the thrust-specific fuel consumption in consistent units, for example,

$$\frac{\text{lb of fuel}}{\text{(lb of thrust)(s)}} \quad \text{or} \quad \frac{\text{N of fuel}}{\text{(N of thrust)(s)}}$$

Let dW be the elemental change in weight of the airplane due to fuel consumption over a time increment dt. Then

$$dW = -c_t T_A \, dt$$

or

$$dt = \frac{-dW}{c_t T_A} \qquad [6.69]$$

Integrating Eq. (6.69) between $t = 0$, where $W = W_0$, and $t = E$, where $W = W_1$, we obtain

$$E = -\int_{W_0}^{W_1} \frac{dW}{c_t T_A} \qquad [6.70]$$

Recalling that $T_A = T_R = D$ and $W = L$, we have

$$E = \int_{W_1}^{W_0} \frac{1}{c_t} \frac{L}{D} \frac{dW}{W} \qquad [6.71]$$

With the assumption of constant c_t and $C_L/C_D = L/D$, Eq. (6.71) becomes

$$\boxed{E = \frac{1}{c_t}\frac{C_L}{C_D}\ln\frac{W_0}{W_1}} \qquad [6.72]$$

Note from Eq. (6.72) that for maximum endurance for a jet airplane, we want

1. Minimum thrust-specific fuel consumption c_t.
2. Maximum fuel weight W_f.

3. Flight at maximum L/D. This confirms our above argument in Sec. 6.13.1 that for maximum endurance for a jet, we must fly such that L/D is maximum.

Note that, subject to our assumptions, E for a jet does not depend on ρ_∞; that is, E is independent of altitude.

Now consider range. Returning to Eq. (6.69) and multiplying by V_∞, we get

$$ds = V_\infty \, dt = -\frac{V_\infty \, dW}{c_t T_A} \qquad \text{[6.73]}$$

where ds is the increment in distance traversed by the jet over the time increment dt. Integrating Eq. (6.73) from $s = 0$, where $W = W_0$, to $s = R$, where $W = W_1$, we have

$$R = \int_0^R ds = -\int_{W_0}^{W_1} \frac{V_\infty \, dW}{c_t T_A} \qquad \text{[6.74]}$$

However, again noting that for steady, level flight, the engine throttle has been adjusted such that $T_A = T_R$ and recalling from Eq. (6.16) that $T_R = W/(C_L/C_D)$, we rewrite Eq. (6.74) as

$$R = \int_{W_1}^{W_0} \frac{V_\infty}{c_t} \frac{C_L}{C_D} \frac{dW}{W} \qquad \text{[6.75]}$$

Since $V_\infty = \sqrt{2W/(\rho_\infty S C_L)}$, Eq. (6.75) becomes

$$R = \int_{W_1}^{W_0} \sqrt{\frac{2}{\rho_\infty S}} \frac{C_L^{1/2}/C_D}{c_t} \frac{dW}{W^{1/2}} \qquad \text{[6.76]}$$

Again, assuming constant c_t, C_L, C_D, and ρ_∞ (constant altitude), we rewrite Eq. (6.76) as

$$R = \sqrt{\frac{2}{\rho_\infty S}} \frac{C_L^{1/2}}{C_D} \frac{1}{c_t} \int_{W_1}^{W_0} \frac{dW}{W^{1/2}}$$

$$\boxed{R = 2\sqrt{\frac{2}{\rho_\infty S}} \frac{1}{c_t} \frac{C_L^{1/2}}{C_D} (W_0^{1/2} - W_1^{1/2})} \qquad \text{[6.77]}$$

Note from Eq. (6.77) that to obtain maximum range for a jet airplane, we want the following:

1. Minimum thrust-specific fuel consumption c_t.

2. Maximum fuel weight W_f.

3. Flight at maximum $C_L^{1/2}/C_D$. This confirms our above argument in Sec. 6.13.1 that for maximum range, a jet must fly at a velocity such that $C_L^{1/2}/C_D$ is maximum.

4. Flight at high altitudes, that is, low ρ_∞. Of course, Eq. (6.77) says that R becomes infinite as ρ_∞ decreases to zero, that is, as we approach outer space. This is physically ridiculous, however, because an airplane requires the atmosphere to generate lift and thrust. Long before outer space is reached, the

assumptions behind Eq. (6.77) break down. Moreover, at extremely high altitudes, ordinary turbojet performance deteriorates and c_t begins to increase. All we can conclude from Eq. (6.77) is that range for a jet is poorest at sea level and increases with altitude, up to a point. Typical cruising altitudes for subsonic commercial jet transports are from 30,000 to 40,000 ft; for supersonic transports they are from 50,000 to 60,000 ft.

Example 6.13

Estimate the maximum range and endurance for the CJ-1.

Solution

From the calculations of Example 6.1, the variation of C_L/C_D and $C_L^{1/2}/C_D$ can be plotted versus velocity, as given in Fig. 6.46. From these curves, for the CJ-1.

$$\max\left(\frac{C_L^{1/2}}{C_D}\right) = 23.4$$

$$\max\left(\frac{C_L}{C_D}\right) = 16.9$$

In Example 6.1, the specific fuel consumption is given as TSFC = 0.6 (lb fuel)/(lb thrust)(h). In consistent units,

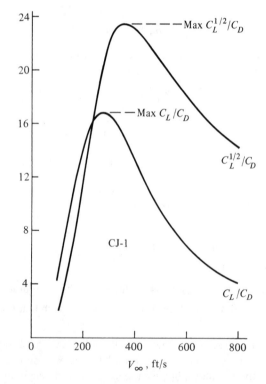

Figure 6.46 Aerodynamic ratios for the CJ-1 at sea level.

$$c_t = 0.6\frac{\text{lb}}{\text{(lb)(h)}}\frac{1\,\text{h}}{3600\,\text{s}} = 1.667 \times 10^{-4}\text{s}^{-1}$$

Also the gross weight is 19,815 lb. The fuel capacity is 1119 gal of kerosene, where 1 gal of kerosene weighs 6.67 lb. Thus, $W_f = 1119(6.67) = 7463$ lb. Hence, the empty weight is $W_1 = W_0 - W_f = 19,815 - 7463 = 12,352$ lb.

The range of a jet depends on altitude, as shown by Eq. (6.77). Assume the cruising altitude is 22,000 ft, where $\rho_\infty = 0.00184$ slug/ft^3. From Eq. (6.77), using information from Example 6.1, we obtain

$$R = 2\sqrt{\frac{2}{\rho_\infty S}}\frac{1}{c_t}\frac{C_L^{1/2}}{C_D}(W_0^{1/2} - W_1^{1/2})$$

$$= 2\sqrt{\frac{2}{0.001184(318)}}\left(\frac{1}{1.667 \times 10^{-4}}\right)(23.4)(19,815^{1/2} - 12,352^{1/2})$$

$$\boxed{R = 19.2 \times 10^6 \text{ ft}}$$

In miles,

$$R = \frac{19.2 \times 10^6}{5280} = \boxed{3630 \text{ mi}}$$

The endurance can be found from Eq. (6.72):

$$E = \frac{1}{c_t}\frac{C_L}{C_D}\ln\frac{W_0}{W_1} = \frac{1}{1.667 \times 10^{-4}}(16.9)\left(\ln\frac{19,815}{12,352}\right)$$

$$\boxed{E = 4.79 \times 10^4 \text{ s}}$$

or in hours

$$E = \frac{4.79 \times 10^4}{3600} = \boxed{13.3 \text{ h}}$$

6.14 RELATIONS BETWEEN $C_{D,0}$ AND $C_{D,i}$

In the previous sections, we have observed that various aspects of the performance of different types of airplanes depend on the aerodynamic ratios $C_L^{1/2}/C_D$, C_L/C_D, or $C_L^{3/2}/C_D$. Moreover, in Sec. 6.3, we proved that at minimum T_R, drag due to lift equals zero-lift drag, that is, $C_{D,0} = C_{D,i}$. Analogously, for minimum P_R, we proved in Sec. 6.5 that $C_{D,0} = \frac{1}{3}C_{D,i}$. In this section, such results are obtained strictly from aerodynamic considerations. The relations between $C_{D,0}$ and $C_{D,i}$ depend purely on the conditions for maximum $C_L^{1/2}/C_D$, C_L/C_D, or $C_L^{3/2}/C_D$; their derivations do not have to be associated with minimum T_R or P_R, as they were in Secs. 6.3 and 6.5.

For example, consider maximum L/D. Recalling that $C_D = C_{D,0} + C_L^2/(\pi e\text{AR})$, we can write

$$\frac{C_L}{C_D} = \frac{C_L}{C_{D,0} + C_L^2/(\pi e\text{AR})} \tag{6.78}$$

For maximum C_L/C_D, differentiate Eq. (6.78) with respect to C_L and set the result equal to 0:

$$\frac{d(C_L/C_D)}{dC_L} = \frac{C_{D,0} + C_L^2/(\pi e\text{AR}) - C_L\left[2C_L/(\pi e\text{AR})\right]}{\left[C_{D,0} + C_L^2/(\pi e\text{AR})\right]^2} = 0$$

Thus,

$$C_{D,0} + \frac{C_L^2}{\pi e\text{AR}} - \frac{2C_L^2}{\pi e\text{AR}} = 0$$

or

$$C_{D,0} = \frac{C_L^2}{\pi e\text{AR}}$$

$$\boxed{C_{D,0} = C_{D,i}} \quad \text{for } \left(\frac{C_L}{C_D}\right)_{\max} \qquad \textbf{[6.79]}$$

Hence, Eq. (6.79), which is identical to Eq. (6.22), simply stems from the fact that L/D is maximum. The fact that it also corresponds to minimum T_R is only because T_R happens to be minimum when L/D is maximum.

Now consider maximum $C_L^{3/2}/C_D$. By setting $d(C_L^{3/2})/dC_l = 0$, a derivation similar to that above yields

$$\boxed{C_{D,0} = \tfrac{1}{3}C_{D,i}} \quad \text{for } \left(\frac{C_L^{3/2}}{C_D}\right)_{\max} \qquad \textbf{[6.80]}$$

Again, Eq. (6.80), which is identical to Eq. (6.30), simply stems from the fact that $C_L^{3/2}/C_D$ is maximum. The fact that it also corresponds to minimum P_R is only because P_R happens to be minimum when $C_L^{3/2}/C_D$ is maximum.

Similarly, when $C_L^{1/2}/C_D$ is maximum, setting $d(C_L^{1/2})/dC_L = 0$ yields

$$\boxed{C_{D,0} = 3C_{D,i}} \quad \text{for } \left(\frac{C_L^{1/2}}{C_D}\right)_{\max} \qquad \textbf{[6.81]}$$

You should not take Eqs. (6.80) and (6.81) for granted; derive them yourself.

We stated in Example 6.12 that the *maximum values* of $C_L^{1/2}/C_D$, C_L/C_D, and $C_L^{3/2}/C_D$ are independent of altitude, velocity, etc.; rather, they depend only on the aerodynamic design of the aircraft. The results of this section allow us to prove this statement, as follows.

First, consider again the case of maximum C_L/C_D. From Eq. (6.79),

$$C_{D,0} = C_{D,i} = \frac{C_L^2}{\pi e\text{AR}} \qquad \textbf{[6.82]}$$

Thus,

$$C_L = \sqrt{\pi e\text{AR}C_{D,0}} \qquad \textbf{[6.83]}$$

Substituting Eqs. (6.82) and (6.83) into Eq. (6.78), we obtain

$$\frac{C_L}{C_D} = \frac{C_L}{2C_L^2/(\pi e\text{AR})} = \frac{\pi e\text{AR}}{2C_L} = \frac{\pi e\text{AR}}{2\sqrt{\pi e\text{AR}C_{D,0}}} \qquad \textbf{[6.84]}$$

Hence, the value of the maximum C_L/C_D is obtained from Eq. (6.84) as

$$\left(\frac{C_L}{C_D}\right)_{max} = \frac{(C_{D,0}\pi e \mathrm{AR})^{1/2}}{2C_{D,0}} \qquad \textbf{[6.85]}$$

Note from Eq. (6.85) that $(C_L/C_D)_{max}$ depends only on e, AR, and $C_{D,0}$, which are aerodynamic design parameters of the airplane. In particular, $(C_L/C_D)_{max}$ does not depend on altitude. However, note from Figs. 6.44 and 6.46 that maximum C_L/C_D occurs at a certain velocity, and the velocity at which $(C_L/C_D)_{max}$ is obtained *does* change with altitude.

In the same vein, it is easily shown that

$$\left(\frac{C_L^{1/2}}{C_D}\right)_{max} = \frac{(\frac{1}{3}C_{D,0}\pi e \mathrm{AR})^{1/4}}{\frac{4}{3}C_{D,0}} \qquad \textbf{[6.86]}$$

and

$$\left(\frac{C_L^{3/2}}{C_D}\right)_{max} = \frac{(3C_{D,0}\pi e \mathrm{AR})^{3/4}}{4C_{D,0}} \qquad \textbf{[6.87]}$$

Prove this yourself.

Example 6.14

From the equations given in this section, directly calculate $(C_L/C_D)_{max}$ and $(C_L^{3/2}/C_D)_{max}$ for the CP-1.

Solution

From Eq. (6.85),

$$\left(\frac{C_L}{C_D}\right)_{max} = \frac{(C_{D,0}\pi e \mathrm{AR})^{1/2}}{2C_{D,0}} = \frac{[0.025\pi(0.8)(7.37)]^{1/2}}{2(0.025)} = \boxed{13.6}$$

From Eq. (6.87),

$$\left(\frac{C_L^{3/2}}{C_D}\right)_{max} = \frac{(3C_{D,0}\pi e \mathrm{AR})^{3/4}}{4C_{D,0}} = \frac{[(3)(0.025)\pi(0.8)(7.37)]^{3/4}}{4(0.025)} = \boxed{12.8}$$

Return to Example 6.12, where the values of $(C_L/C_D)_{max}$ and $(C_L^{3/2}/C_D)_{max}$ were obtained graphically, that is, by plotting C_L/C_D and $C_L^{3/2}/C_D$ and finding their peak values. Note that the results obtained from Eqs. (6.85) and (6.87) agree with the graphical values obtained in Example 6.12 (as they should); however, the use of Eqs. (6.85) and (6.87) is much easier and quicker than plotting a series of numbers and finding the maximum.

Example 6.15

From the equations given in this section, directly calculate $(C_L^{1/2}/C_D)_{max}$ and $(C_L/C_D)_{max}$ for the CJ-1.

Solution

From Eq. (6.77),

$$\left(\frac{C_L^{1/2}}{C_D}\right)_{max} = \frac{(\frac{1}{3}C_{D,0}\pi e AR)^{1/4}}{\frac{4}{3}C_{D,0}} = \frac{[\frac{1}{3}(0.02)\pi(0.81)(8.93)]^{1/4}}{\frac{4}{3}(0.02)} = \boxed{23.4}$$

From Eq. (6.76)

$$\left(\frac{C_L}{C_D}\right)_{max} = \frac{(\pi e AR C_{D,0})^{1/2}}{2C_{D,0}} = \frac{[\pi(0.81)(8.93)(0.02)]^{1/2}}{2(0.02)} = \boxed{16.9}$$

These values agree with the graphically obtained maximums in Example 6.13.

Example 6.16

Using the result from this section, and Eqs. (6.44), (6.52), and (6.53), analytically calculate
a. V_{max} for the CP-1 at sea level.
b. $(R/C)_{max}$ for the CP-1 at sea level.
c. V_{max} for the CJ-1 at sea level.
d. $(R/C)_{max}$ for the CJ-1 at sea level.
Compare with the graphical solutions obtained in Examples 6.2, 6.4, and 6.6.

Solution

a. The maximum velocity is given by Eq. (6.44), repeated here:

$$V_{max} = \left[\frac{\left(\frac{T_A}{W}\right)_{max}\left(\frac{W}{S}\right) + \left(\frac{W}{S}\right)\sqrt{\left(\frac{T_A}{W}\right)_{max}^2 - \frac{4C_{D,0}}{\pi e AR}}}{\rho_\infty C_{D,0}}\right]^{1/2}$$

For the CP-1, from the data given in Example 6.1,

$$\frac{W}{S} = \frac{2950}{174} = 16.95 \text{ lb/ft}^2$$

$$\eta P = 0.8(230)(550) = 1.012 \times 10^5 \frac{\text{ft} \cdot \text{lb}}{\text{s}}$$

From Eq. (6.85) and the result from Example 6.14,

$$\frac{4C_{D,0}}{\pi e AR} = \frac{1}{(L/D)_{max}^2} = \frac{1}{13.6^2} = 5.4066 \times 10^{-3}$$

Also $\rho_\infty C_{D,0} = 0.002377(0.025) = 5.9425 \times 10^{-5} \text{slug/ft}^3$

Power available and thrust available are related by

$$T_A V_\infty = P_A = \eta P$$

For maximum T_A and P_A, $V_\infty = V_{max}$. Hence

$$(T_A)_{max} = \frac{\eta P}{V_{max}}$$

or

$$\left(\frac{T_A}{W}\right)_{max} = \frac{\eta P}{W}\frac{1}{V_{max}} = \frac{1.012 \times 10^5}{2950}\frac{1}{V_{max}}$$

or

$$\left(\frac{T_A}{W}\right)_{max} = \frac{34.305}{V_{max}} \qquad \text{[E6.1]}$$

Inserting the above data into Eq. (6.44), we have

$$V_{\text{max}} = \left[\frac{(T_A/W)_{\text{max}}(16.95) + 16.95\sqrt{(T_A/W)^2_{\text{max}} - 5.4066 \times 10^{-3}}}{5.9425 \times 10^{-5}} \right]^{1/2}$$

[E6.2]

$$\text{or} \quad V_{\text{max}} = 558.97 \left[\left(\frac{T_A}{W} \right)_{\text{max}} + \sqrt{\left(\frac{T_A}{W} \right)^2_{\text{max}} - 5.4066 \times 10^{-3}} \right]^{1/2}$$

Equations (E6.1) and (E6.2) must be solved for V_{max} by trial and error. Assume V_{max}, calculate $(T_A/W)_{\text{max}}$ from Eq. (E6.1), insert this into Eq. (E6.2), calculate V_{max} from Eq. (E6.2), and see if this matches the originally assumed V_{max}. If not, assume another value of V_{max}, and try again. A few iterations are tabulated below.

V_{max} (ft/s) (assumed)	$\left(\dfrac{T_A}{W} \right)_{\text{max}}$ [from Eq. (E6.1)]	V_{max} (ft/s) [from Eq. (E6.2)]
265	0.1295	271.6
270	0.12706	268.5
269	0.1275	269.1

From this, we have calculated for the CP-1 at sea level,

$$V_{\text{max}} = \boxed{269 \text{ ft/s}}$$

This is to be compared with $V_{\text{max}} = 265$ ft/s as obtained from the graphical solution in Example 6.4, which is limited by "graphical accuracy." The analytical solution of $V_{\text{max}} = 269$ ft/s obtained here is inherently more accurate.

b. The maximum rate of climb for a propeller-driven airplane is given by Eq. (6.53), repeated here:

$$(\text{R/C})_{\text{max}} = \left(\frac{\eta P}{W} \right)_{\text{max}} - 0.8776 \sqrt{\frac{W/S}{\rho_\infty C_{D,0}}} \frac{1}{(L/D)^{3/2}_{\text{max}}}$$

We have already obtained the following data:

$$\left(\frac{\eta P}{W} \right)_{\text{max}} = \frac{1.012 \times 10^5}{2950} = 34.305 \text{ ft/s}$$

$$\rho_\infty C_{D,0} = 5.9425 \times 10^{-5} \text{ slug/ft}^3$$

$$\frac{W}{S} = 16.95 \text{ lb/ft}^2$$

$$\left(\frac{L}{D} \right)_{\text{max}} = 13.6$$

Hence, Eq. (6.53) becomes

$$(R/C)_{max} = 34.305 - 0.8776 \sqrt{\frac{16.95}{5.9425 \times 10^{-5}} \frac{1}{(13.6)^{3/2}}}$$

or $\qquad (R/C)_{max} = 34.305 - 9.345 = 24.96 \text{ ft/s}$

Thus, $\qquad (R/C)_{max} = 24.96(60) = \boxed{1497.6 \text{ ft/min}}$

This is to be compared with $(R/C)_{max} = 1500$ ft/min as read from the peak of the graph in Fig. 6.22 from Example 6.6.

c. From the data given about the CJ-1 in Example 6.1, we have

$$\left(\frac{T_A}{W}\right)_{max} = \frac{7300}{19,815} = 0.3684$$

$$\frac{W}{S} = \frac{19,815}{318} = 62.31 \text{ lb/ft}^2$$

$$\rho_\infty C_{D,0} = 0.002377(0.02) = 4.754 \times 10^{-5} \text{ slug/ft}^3$$

Also, from Example 6.15, we have

$$\frac{4C_{D,0}}{\pi e AR} = \frac{1}{(L/D)_{max}^2} = \frac{1}{(16.9)^2} = 3.501 \times 10^{-3}$$

Substituting these data in Eq. (6.44), we obtain

$$V_{max} = \left[\frac{\left(\frac{T_A}{W}\right)_{max}\left(\frac{W}{S}\right) + \left(\frac{W}{S}\right)\sqrt{\left(\frac{T_A}{W}\right)_{max}^2 - \frac{4C_{D,0}}{\pi e AR}}}{\rho_\infty C_{D,0}} \right]^{1/2}$$

$$= \left[\frac{0.3684(62.31) + 62.31\sqrt{(0.3684)^2 - 3.501 \times 10^{-3}}}{4.754 \times 10^{-5}} \right]^{1/2}$$

or $\qquad V_{max} = \boxed{979.5 \text{ ft/s}}$

This is to be compared with $V_{max} = 975$ ft/s obtained by graphical means in Example 6.2.

d. The maximum rate of climb for a jet airplane is given by Eq. (6.52), repeated here:

$$(R/C)_{max} = \left[\frac{(W/S)Z}{3\rho_\infty C_{D,0}}\right]^{1/2}\left(\frac{T}{W}\right)_{max}^{3/2}\left[1 - \frac{Z}{6} - \frac{3}{2(T/W)_{max}^2(L/D)_{max}^2 Z}\right]$$

where $\qquad Z = 1 + \sqrt{1 + \frac{3}{(L/D)_{max}^2(T/W)_{max}^2}}$

Putting in the data for the CJ-1, we have

$$Z = 1 + \sqrt{1 + \frac{3}{(16.9)^2(0.3684)^2}} = 2.038$$

and

$$(R/C)_{max} = \left[\frac{(62.31)(2.038)}{3(4.754 \times 10^{-5})}\right]^{1/2} (0.3684)^{3/2} \left[1 - \frac{2.038}{6} - \frac{3}{2(0.3684)^2(16.9)^2(2.038)}\right]$$

$$= 135.28 \text{ ft/s}$$

or

$$(R/C)_{max} = 135.28(60) = \boxed{8117 \text{ ft/min}}$$

This is to be compared with $(R/C)_{max} = 8100$ ft/min read from the peak of the graph in Fig. 6.33 from Example 6.6.

DESIGN BOX

The ratio of lift to drag is a direct measure of the aerodynamic efficiency of a given airplane. For example, if for a given airplane $(L/D)_{max} = 15$, this means that the airplane can lift 15 lb of weight at a cost of only 1 lb of drag—quite a leverage. Indeed, for atmospheric flight, the wing of an airplane (usually its strongest lifting component by far) can be loosely likened to a lever that allows us to lift far more weight than we have to expend in thrust from the engine (to counterbalance the drag). The evolution of the airplane in the 20th century has been characterized by a steady increase in $(L/D)_{max}$; this evolution is discussed at length in Sec. 6.24. Some values of $(L/D)_{max}$ for typical airplanes are tabulated here.

Airplane	$\left(\dfrac{L}{D}\right)_{max}$
Wright Flyer (1903)	5.7
French SPAD XIII (World War I)	7.4
Douglas DC-3 (1930's)	14.7
Boeing 747 (contemporary)	20

The importance of $(L/D)_{max}$ as a parameter in airplane design cannot be overstated—it is one of the driving aspects that dictate the configuration of the airplane. Airplane designers usually try to squeeze as much $(L/D)_{max}$ into a new airplane as they can, subject to compromises with other aspects of the design. We have already seen that $(L/D)_{max}$ plays a role in dictating V_{max}, $(R/C)_{max}$, and especially range and endurance. Indeed, historically the quest for greater range has been the primary factor that has driven up the design value of $(L/D)_{max}$. (See Anderson, *A History of Aerodynamics and Its Impact on Flying Machines*, Cambridge University Press, New York, 1997.)

Strictly speaking, we have seen in Secs. 6.12 and 6.13 that the value of $(L/D)_{max} = (C_L/C_D)_{max}$ dictates maximum range for a propeller-driven airplane and maximum endurance for a jet airplane, whereas $(C_L^{3/2}/C_D)_{max}$ dictates maximum endurance for a propeller-driven airplane and $(C_L^{1/2}/C_D)_{max}$ dictates maximum range for a jet airplane. However, the geometric and aerodynamic features of an airplane that maximize C_L/C_D will also maximize $C_L^{1/2}/C_D$ and $C_L^{3/2}/C_D$, as seen in Eqs. (6.85) through (6.87). To obtain maximum values of these aerodynamic ratios, Eqs. (6.85) through (6.87) clearly indicate that the airplane designer should, as much as possible,

(continued)

1. Reduce the zero-lift drag coefficient $C_{D,0}$.

2. Increase the Oswald efficiency factor e.

3. Increase the aspect ratio AR.

Of course, this last point—increasing the aspect ratio—makes sense only for subsonic flight. We have discussed previously that for transonic and supersonic airplanes, wave drag is dominant, and wave drag can be somewhat reduced by using low–aspect ratio wings. For high-speed airplanes designed for cruise at supersonic speeds, the design wing aspect ratio is driven by considerations other than those for maximum range in subsonic flight. The low–aspect ratio, Mach 2, Lockheed F-104 shown in Fig. 4.44 is a case in point.

The value of $(L/D)_{max}$ is fixed by the aerodynamics and geometry of the given airplane configuration via $C_{D,0}$, e and AR. Hence, $(L/D)_{max}$ does not change with altitude. However, the *velocity* at which the airplane must fly to achieve $(L/D)_{max}$ does vary with altitude. To explain why this is so, first recall that L/D is a function of the airplane's angle of attack. For example, the variation of L/D versus α for the special-purpose F-111 TACT aircraft (illustrated in Fig. 5.31) is shown in Fig. 6.47. Note that $(L/D)_{max}$ occurs at an angle of attack of 6°; at this angle of attack, $C_L = 0.44$. If the airplane is flying at sea level, in order to fly at $(L/D)_{max}$ it must be flying at $\alpha = 6°$ with $C_L = 0.44$. For the given weight, this condition fixes the velocity at which the airplane must fly via the relation $W = q_\infty S C_L$, or

$$V_\infty = \sqrt{\frac{2W}{\rho_\infty S C_L}} \qquad \textbf{[6.88]}$$

To fly at $(L/D)_{max}$ at higher altitude, the airplane must still fly at $\alpha = 6°$ with $C_L = 0.44$. However, since ρ_∞ has decreased, V_∞ must be larger, as given by Eq. (6.88). That is, V_∞ must be increased to just the right value so that the lift remains equal to the weight for the fixed C_L at $\alpha = 6°$. As a result, the velocity required to fly at $(L/D)_{max}$ increases with an increase in altitude.

Although the value of $(L/D)_{max}$ is very important in airplane design, *flight* at $(L/D)_{max}$ is not always the holy grail of aeronautical engineering that it may seem. As usual, the airplane designer is faced with a compromise, this time involved with V_{max} relative to the velocity for $(L/D)_{max}$. The velocity

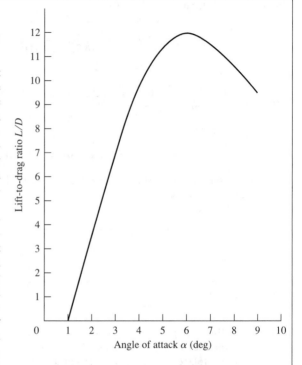

Figure 6.47 Flight data for lift-to-drag ratio versus angle of attack for the F-111 TACT airplane shown in Figure 5.31. $M_\infty = 0.7$. Wing sweep angle = 26°. Data from Baldwin et al, *Symposium on Transonic Aircraft Technology (TACT)*, Air Force Flight Dynamics Laboratory Technical Report AFFDL-TR-78-100, Wright-Patterson Air Force Base, Ohio, 1978.

for $(L/D)_{max}$ can be substantially *smaller* than the maximum velocity. For example, from Fig. 6.46, the velocity at sea level for $(L/D)_{max}$ for the CJ-1 is about 300 ft/s, whereas from Fig. 6.26, $V_{max} = 975$ ft/s—a considerable difference. For the CP-1 at sea level, from Fig. 6.44 the velocity for $(L/D)_{max}$ is about 150 ft/s, whereas from Fig. 6.21a, $V_{max} = 265$ ft/s at sea level. If the pilot of the CP-1 chooses to fly very efficiently by flying at $V_\infty = 150$ ft/s so that L/D is at its maximum value, then the flight will take almost 75 percent longer to go from point A to B compared to flying at V_{max}. Since time is valuable (indeed, the reason why most passengers fly is to save

(continued)

time), the design cruise speed for a given airplane may not correspond to $(L/D)_{max}$. The airplane designer must be ready to accept a higher-speed cruise with an L/D that is less than the value of $(L/D)_{max}$. However, this does not diminish the importance of $(L/D)_{max}$ as a design parameter. For example, an airplane with a high value of $(L/D)_{max}$ will still have comparatively high values of L/D while flying at velocities other than that for $(L/D)_{max}$. Also, Bernard Carson, a professor of aerospace engineering at the U.S. Naval Academy, has suggested a rational compromise that combines the concept of long range obtained by flying at the slower velocity for $(L/D)_{max}$ and the shorter flight times obtained by flying at higher speeds. His analysis leads to an optimum compromise for flight velocity called the *Carson speed*, which can be shown to be a factor of 1.32 higher than the velocity for $(L/D)_{max}$. The details can be found in Anderson, *Aircraft Performance and Design*, McGraw-Hill, New York, 1999.

6.15   TAKEOFF PERFORMANCE

Up to this point in our discussion of airplane performance, we have assumed that all accelerations are zero; that is, we have dealt with aspects of *static* performance as defined in Sec. 6.2. For the remainder of this chapter, we relax this restriction and consider several aspects of airplane performance that involve finite acceleration, such as takeoff and landing runs, turning flight, and accelerated rate of climb. With this, we move to the right-hand column on our chapter road map, shown in Fig. 6.4. We now take up the study of dynamic performance.

To begin, we ask the question, What is the running length along the ground required by an airplane, starting from zero velocity, to gain flight speed and lift from the ground? This length is defined as the *ground roll*, or *liftoff distance*, s_{LO}.

To address this question, let us first consider the accelerated rectilinear motion of a body of mass m experiencing a constant force F, as sketched in Fig. 6.48. From Newton's second law,

$$F = ma = m\frac{dV}{dt}$$

or

$$dV = \frac{F}{m}dt \qquad\qquad [6.89]$$

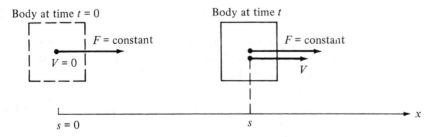

Figure 6.48   Sketch of a body moving under the influence of a constant force F, starting from rest $(V = 0)$ at $s = 0$ and accelerating to velocity V at distance s.

Assume that the body starts from rest ($V = 0$) at location $s = 0$ at time $t = 0$ and is accelerated to velocity V over distance s at time t. Integrating Eq. (6.89) between these two points, and remembering that both F and m are constant, we have

$$\int_0^V dV = \frac{F}{m} \int_0^t dt$$

or

$$V = \frac{F}{m} t \qquad \textbf{[6.90]}$$

Solving for t, we get

$$t = \frac{Vm}{F} \qquad \textbf{[6.91]}$$

Considering an instant when the velocity is V, the incremental distance ds covered during an incremental time dt is $ds = V\, dt$. From Eq. (6.90), we have

$$ds = V\, dt = \frac{F}{m} t\, dt \qquad \textbf{[6.92]}$$

Integrating Eq. (6.92) gives

$$\int_0^s ds = \frac{F}{m} \int_0^t t\, dt$$

or

$$s = \frac{F}{m} \frac{t^2}{2} \qquad \textbf{[6.93]}$$

Substituting Eq. (6.91) into (6.93), we obtain

$$\boxed{s = \frac{V^2 m}{2F}} \qquad \textbf{[6.94]}$$

Equation (6.94) gives the distance required for a body of mass m to accelerate to velocity V under the action of a constant force F.

Now consider the force diagram for an airplane during its ground roll, as illustrated in Fig. 6.49. In addition to the familiar forces of lift, drag, thrust, and weight, the airplane experiences a resistance force R due to rolling friction between the tires and the ground. This resistance force is given by

$$R = \mu_r (W - L) \qquad \textbf{[6.95]}$$

where $W - L$ is the net normal force exerted between the tires and the ground and μ_r is the coefficient of rolling friction. Summing forces parallel to the ground, and employing Newton's second law, we have

$$F = T - D - R = T - D - \mu_r (W - L) = m \frac{dV}{dt} \qquad \textbf{[6.96]}$$

Let us examine Eq. (6.96) more closely. It gives the local instantaneous acceleration of the airplane dV/dt as a function of T, D, W, and L. For takeoff, over most

Figure 6.49 Forces acting on an airplane during takeoff and landing.

of the ground roll, T is reasonably constant (this is particularly true for a jet-powered airplane). Also W is constant. However, both L and D vary with velocity, since

$$L = \tfrac{1}{2}\rho_\infty V_\infty^2 S C_L \tag{6.97}$$

and

$$D = \frac{1}{2}\rho_\infty V_\infty^2 S \left(C_{D,0} + \phi \frac{C_L^2}{\pi e \text{AR}} \right) \tag{6.98}$$

The quantity ϕ in Eq. (6.98) requires some explanation. When an airplane is flying close to the ground, the strength of the wing tip vortices is somewhat diminished because of interaction with the ground. Since these tip vortices induce downwash at the wing (see Sec. 5.13), which in turn generates induced drag (see Sec. 5.14), the downwash and hence induced drag are reduced when the airplane is flying close to the ground. This phenomenon is called *ground effect* and is the cause of the tendency for an airplane to flare, or "float," above the ground near the instant of landing. The reduced drag in the presence of ground effect is accounted for by ϕ in Eq. (6.98), where $\phi \leq 1$. An approximate expression for ϕ, based on aerodynamic theory, is given by McCormick (see Bibliography at the end of this chapter) as

$$\phi = \frac{(16h/b)^2}{1 + (16h/b)^2} \tag{6.99}$$

where h is the height of the wing above the ground and b is the wingspan.

In light of the above, to accurately calculate the variation of velocity with time during the ground roll, and ultimately the distance required for liftoff, Eq. (6.96) must be integrated numerically, taking into account the proper velocity variations of L and D from Eqs. (6.97) and (6.98), respectively, as well as any velocity effect on T. A typical variation of these forces with distance along the ground during takeoff is sketched in Fig. 6.50. Note from Eq. (6.94) that s is proportional to V^2, and hence the horizontal axis in Fig. 6.50 could just as well be V^2. Since both D and L are proportional to the dynamic pressure $q_\infty = \tfrac{1}{2}\rho_\infty V_\infty^2$, they appear as linear variations

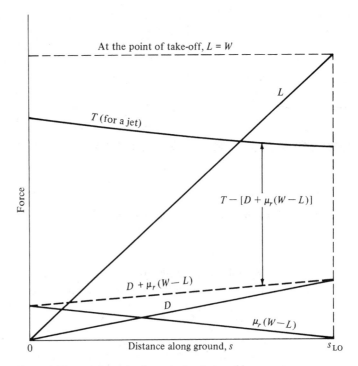

Figure 6.50   Schematic of a typical variation of forces acting on an airplane during takeoff.

in Fig. 6.50. Also, Fig. 6.50 is drawn for a jet-propelled airplane; hence T is relatively constant.

A simple but approximate expression for the liftoff distance s_{LO} can be obtained as follows. Assume that T is constant. Also, assume an *average value* for the sum of drag and resistance forces, $[D + \mu_r(W - L)]_{av}$ such that this average value, taken as a constant force, produces the proper liftoff distance s_{LO}. Then we consider an effective constant force acting on the airplane during its takeoff ground roll as

$$F_{eff} = T - [D + \mu_r(W - L)]_{av} = const \qquad \textbf{[6.100]}$$

These assumptions are fairly reasonable, as seen from Fig. 6.50. Note that the sum of $D + \mu_r(W - L)$ versus distance (or V^2) is reasonably constant, as shown by the dashed line in Fig. 6.50. Hence, the accelerating force $T - [D + \mu_r(W - L)]$, which is illustrated by the difference between the thrust curve and the dashed line in Fig. 6.50, is also reasonably constant. Now return to Eq. (6.94). Considering F given by Eq. (6.100), $V = V_{LO}$ (the liftoff velocity), and $m = W/g$, where g is the acceleration of gravity, Eq. (6.94) yields

$$s_{LO} = \frac{(V_{LO}^2)(W/g)}{2\{T - [D + \mu_r(W - L)]_{av}\}} \qquad \textbf{[6.101]}$$

To ensure a margin of safety during takeoff, the liftoff velocity is typically 20 percent higher than the stalling velocity. Hence, from Eq. (5.71), we have

$$V_{LO} = 1.2V_{stall} = 1.2\sqrt{\frac{2W}{\rho_\infty S C_{L,max}}} \qquad \textbf{[6.102]}$$

Substituting Eq. (6.102) into (6.101), we obtain

$$s_{LO} = \frac{1.44W^2}{g\rho_\infty S C_{L,max}\{T - [D + \mu_r(W - L)]_{av}\}} \qquad \textbf{[6.103]}$$

To make a calculation using Eq. (6.103), Shevell (see Bibliography at the end of this chapter) suggests that the average force in Eq. (6.103) be set equal to its instantaneous value at a velocity equal to $0.7V_{LO}$, that is,

$$[D + \mu_r(W - L)]_{av} = [D + \mu_r(W - L)]_{0.7V_{LO}}$$

Also, experience has shown that the coefficient of rolling friction μ_r in Eq. (6.103) varies from 0.02 for a relatively smooth, paved surface to 0.10 for a grass field.

A further simplification can be obtained by assuming that thrust is much larger than either D or R during takeoff. Refer to the case shown in Fig. 6.50; this simplification is not unreasonable. Hence, ignoring D and R compared to T, Eq. (6.103) becomes simply

$$s_{LO} = \frac{1.44W^2}{g\rho_\infty S C_{L,max} T} \qquad \textbf{[6.104]}$$

Equation (6.104) illustrates some important physical trends, as follows:

1. Liftoff distance is very sensitive to the weight of the airplane, varying directly as W^2. If the weight is doubled, the ground roll of the airplane is quadrupled.
2. Liftoff distance is dependent on the ambient density ρ_∞. If we assume that thrust is directly proportional to ρ_∞, as stated in Sec. 6.7, that is, $T \propto \rho_\infty$, then Eq. (6.104) demonstrates that

$$s_{LO} \propto \frac{1}{\rho_\infty^2}$$

This is why on hot summer days, when the air density is less than that on cooler days, a given airplane requires a longer ground roll to get off the ground. Also, longer liftoff distances are required at airports which are located at higher altitudes (such as at Denver, Colorado, a mile above sea level).

3. The liftoff distance can be decreased by increasing the wing area, increasing $C_{L,max}$, and increasing the thrust, all of which simply make common sense.

The total takeoff distance, as defined in the Federal Aviation Requirements (FAR), is the sum of the ground roll distance s_{LO} and the distance (measured along the ground) to clear a 35-ft height (for jet-powered civilian transports) or a 50-ft height (for all other airplanes). A discussion of these requirements, as well as more details regarding the total takeoff distance, can be found in Anderson, *Aircraft Performance and Design*, McGraw-Hill, New York, 1999. Also see the books by Shevell and by McCormick listed in the Bibliography at the end of this chapter for more information on this topic.

Example 6.17

Estimate the liftoff distance for the CJ-1 at sea level. Assume a paved runway, hence $\mu_r = 0.02$. Also, during the ground roll, the angle of attack of the airplane is restricted by the requirement that the tail not drag the ground; therefore assume that $C_{L,\max}$ during ground roll is limited to 1.0. Also, when the airplane is on the ground, the wings are 6 ft above the ground.

Solution

Use Eq. (6.103). To evaluate the average force in Eq. (6.103), first obtain the ground effect factor from Eq. (6.99), where $h/b = 6/53.3 = 0.113$.

$$\phi = \frac{(16h/b)^2}{1 + (16h/b)^2} = 0.764$$

Also, from Eq. (6.102),

$$V_{LO} = 1.2V_{\text{stall}} = 1.2\sqrt{\frac{2W}{\rho_\infty S C_{L,\max}}} = 1.2\sqrt{\frac{2(19,815)}{0.002377(318)(1.0)}} = 230 \text{ ft/s}$$

Hence, $0.7V_{LO} = 160.3$ ft/s. The average force in Eq. (6.103) should be evaluated at a velocity of 160.3 ft/s. To do this, from Eq. (6.97) we get

$$L = \tfrac{1}{2}\rho_\infty V_\infty^2 S C_L = \tfrac{1}{2}(0.002377)(160.3)^2(318)(1.0) = 9712 \text{ lb}$$

Equation (6.98) yields

$$D = \frac{1}{2}\rho_\infty V_\infty^2 S \left(C_{D,0} + \phi\frac{C_L^2}{\pi e AR} \right)$$

$$= \frac{1}{2}(0.002377)(160.3)^2(318)\left[0.02 + 0.764\frac{1.0^2}{\pi(0.81)(8.93)} \right] = 520.7 \text{ lb}$$

Finally, from Eq. (6.103),

$$s_{LO} = \frac{1.44W^2}{g\rho_\infty S C_{L,\max}\{T - [D + \mu_r(W - L)]_{\text{av}}\}}$$

$$= \frac{1.44(19,815)^2}{32.2(0.002377)(318)(1.0)\{7300 - [520.7 + (0.02)(19,815 - 9712)]\}} = \boxed{3532 \text{ ft}}$$

Note that $[D + \mu_r(W - L)]_{\text{av}} = 722.8$ lb, which is about 10 percent of the thrust. Hence, the assumption leading to Eq. (6.104) is fairly reasonable, that is, that D and R can sometimes be ignored compared with T.

6.16   LANDING PERFORMANCE

Consider an airplane during landing. After the airplane has touched the ground, the force diagram during the ground roll is exactly the same as that given in Fig. 6.49, and the instantaneous acceleration (negative in this case) is given by Eq. (6.96). However, we assume that in order to minimize the distance required to come to a complete stop, the pilot has decreased the thrust to zero at touchdown, and therefore the equation of motion for the landing ground roll is obtained from Eq. (6.96) with $T = 0$.

$$-D - \mu_r(W - L) = m\,\frac{dV}{dt} \qquad\qquad \textbf{[6.105]}$$

A typical variation of the forces on the airplane during landing is sketched in Fig. 6.51. Designate the ground roll distance between touchdown at velocity V_T and a complete stop by s_L. An accurate calculation of s_L can be obtained by numerically integrating Eq. (6.105) along with Eqs. (6.97) and (6.98).

However, let us develop an approximate expression for s_L which parallels the philosophy used in Sec. 6.15. Assume an average constant value for $D + \mu_r(W - L)$

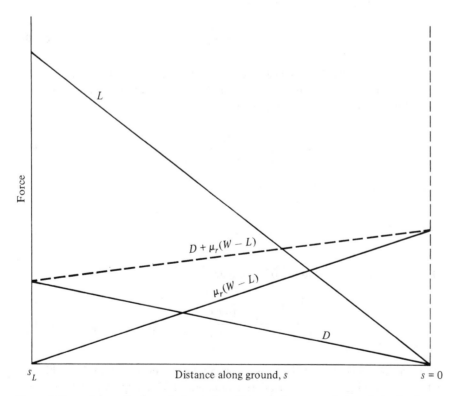

Figure 6.51   Schematic of a typical variation of forces acting on an airplane during landing.

which effectively yields the correct ground roll distance at landing s_L. Once again, we can assume that $[D + \mu_r(W - L)]_{av}$ is equal to its instantaneous value evaluated at $0.7V_T$.

$$F = -[D + \mu_r(W - L)]_{av} = -[D + \mu_r(W - L)]_{0.7V_T} \qquad \text{[6.106]}$$

[Note from Fig. 6.51 that the net decelerating force $D + \mu_r(W - L)$ can vary considerably with distance, as shown by the dashed line. Hence, our assumption here for landing is more tenuous than for takeoff.] Returning to Eq. (6.92), we integrate between the touchdown point, where $s = s_L$ and $t = 0$, and the point where the airplane's motion stops, where $s = 0$ and time equals t.

$$\int_{s_L}^{0} ds = \frac{F}{m} \int_{0}^{t} t \, dt$$

or

$$s_L = -\frac{F}{m} \frac{t^2}{2} \qquad \text{[6.107]}$$

Note that, from Eq. (6.106), F is a negative value; hence s_L in Eq. (6.107) is positive. Combining Eqs. (6.91) and (6.107), we obtain

$$s_L = -\frac{V^2 m}{2F} \qquad \text{[6.108]}$$

Equation (6.108) gives the distance required to decelerate from an initial velocity V to zero velocity under the action of a constant force F. In Eq. (6.108), F is given by Eq. (6.106), and V is V_T. Thus, Eq. (6.108) becomes

$$s_L = \frac{V_T^2(W/g)}{2[D + \mu_r(W - L)]_{0.7V_T}} \qquad \text{[6.109]}$$

To maintain a factor of safety,

$$V_T = 1.3V_{\text{stall}} = 1.3\sqrt{\frac{2W}{\rho_\infty SC_{L,\text{max}}}} \qquad \text{[6.110]}$$

Substituting Eq. (6.110) into (6.109), we obtain

$$\boxed{s_L = \frac{1.69W^2}{g\rho_\infty SC_{L,\text{max}}[D + \mu_r(W - L)]_{0.7V_T}}} \qquad \text{[6.111]}$$

During the landing ground roll, the pilot is applying brakes; hence, in Eq. (6.111) the coefficient of rolling friction is that during braking, which is approximately $\mu_r = 0.4$ for a paved surface.

Modern jet transports utilize thrust reversal during the landing ground roll. Thrust reversal is created by ducting air from the jet engines and blowing it in the upstream direction, opposite to the usual downstream direction when normal thrust is produced. As a result, with thrust reversal, the thrust vector in Fig. 6.49 is reversed and points in the drag direction, thus aiding the deceleration and shortening the ground roll. Designating the reversed thrust by T_R, we see that Eq. (6.105) becomes

$$-T_R - D - \mu_r(W - L) = m\frac{dV}{dt} \qquad \text{[6.112]}$$

Assuming that T_R is constant, Eq. (6.111) becomes

$$s_L = \frac{1.69W^2}{g\rho_\infty SC_{L,\max}\{T_R + [D + \mu_r(W - L)]_{0.7V_T}\}} \qquad \text{[6.113]}$$

Another ploy to shorten the ground roll is to decrease the lift to near zero, hence impose the full weight of the airplane between the tires and the ground and increase the resistance force due to friction. The lift on an airplane wing can be destroyed by spoilers, which are simply long, narrow surfaces along the span of the wing, deflected directly into the flow, thus causing massive flow separation and a striking decrease in lift.

The total landing distance, as defined in FAR, is the sum of the ground roll distance and the distance (measured along the ground) to achieve touchdown in a glide from a 50-ft height. Such details are beyond the scope of this book; see the books by Shevell and by McCormick (listed in the Bibliography at the end of this chapter) and by Anderson, *Aircraft Performance and Design*, McGraw-Hill, New York, 1999, for more information.

Example 6.18

Estimate the landing ground roll distance at sea level for the CJ-1. No thrust reversal is used; however, spoilers are employed such that $L = 0$. The spoilers increase the zero-lift, drag coefficient by 10 percent. The fuel tanks are essentially empty, so neglect the weight of any fuel carried by the airplane. The maximum lift coefficient, with flaps fully employed at touchdown, is 2.5.

Solution

The empty weight of the CJ-1 is 12,352 lb. Hence,

$$V_T = 1.3V_{\text{stall}} = 1.3\sqrt{\frac{2W}{\rho_\infty SC_{L,\max}}} = 1.3\sqrt{\frac{2(12,352)}{0.002377(318)(2.5)}} = 148.6 \text{ ft/s}$$

Thus, $0.7V_T = 104$ ft/s. Also, $C_{D,0} = 0.02 + 0.1(0.02) = 0.022$. From Eq. (6.98), with $C_L = 0$ (remember, spoilers are employed, destroying the lift),

$$D = \tfrac{1}{2}\rho_\infty V_\infty^2 SC_{D,0} = \tfrac{1}{2}(0.002377)(104)^2(318)(0.022) = 89.9 \text{ lb}$$

From Eq. (6.111), with $L = 0$,

$$s_L = \frac{1.69W^2}{g\rho_\infty S C_{L,\max}(D + \mu_r W)_{0.7V_T}}$$

$$= \frac{1.69(12,352)^2}{32.2(0.002377)(318)(2.5)[89.9 + 0.4(12,352)]} = \boxed{842 \text{ ft}}$$

6.17   TURNING FLIGHT AND THE *V-n* DIAGRAM

Up to this point in our discussion of airplane performance, we have considered rectilinear motion. Our static performance analyses dealt with zero acceleration leading to constant velocity along straight-line paths. Our discussion of takeoff and landing performance involved rectilinear acceleration, also leading to motion along a straight-line path. Let us now consider some cases involving *radial* acceleration, which leads to *curved* flight paths; that is, let us consider the turning flight of an airplane. In particular, we examine three specialized cases: a level turn, a pull-up, and a pull-down. A study of the generalized motion of an airplane along a three-dimensional flight path is beyond the scope of this book.

A level turn is illustrated in Fig. 6.52. Here, the wings of the airplane are banked through angle ϕ; hence the lift vector is inclined at angle ϕ to the vertical. The bank angle ϕ and the lift L are such that the component of the lift in the vertical direction exactly equals the weight:

$$L \cos \phi = W$$

and therefore the airplane maintains a constant altitude, moving in the same horizontal plane. However, the resultant of L and W leads to a resultant force F_r, which acts in the horizontal plane. This resultant force is perpendicular to the flight path, causing the airplane to turn in a circular path with a radius of curvature equal to R. We wish to study this turn radius R as well as the turn rate $d\theta/dt$.

From the force diagram in Fig. 6.52, the magnitude of the resultant force is

$$F_r = \sqrt{L^2 - W^2} \qquad [6.114]$$

We introduce a new term, the *load factor* n, defined as

$$n \equiv \frac{L}{W} \qquad [6.115]$$

The load factor is usually quoted in terms of "g's"; for example, an airplane with lift equal to 5 times the weight is said to be experiencing a load factor of 5 g's. Hence, Eq. (6.114) can be written as

$$F_r = W\sqrt{n^2 - 1} \qquad [6.116]$$

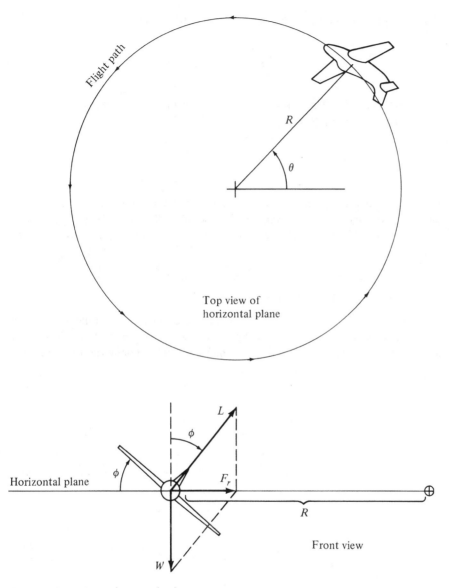

Figure 6.52 An airplane in a level turn.

The airplane is moving in a circular path at velocity V_∞. Therefore, the radial acceleration is given by V_∞^2/R. From Newton's second law

$$F_r = m \frac{V_\infty^2}{R} = \frac{W}{g} \frac{V_\infty^2}{R} \qquad \text{[6.117]}$$

Combining Eqs. (6.116) and (6.117) and solving for R, we have

$$R = \frac{V_\infty^2}{g\sqrt{n^2 - 1}}$$ [6.118]

The angular velocity, denoted by $\omega \equiv d\theta/dt$, is called the *turn rate* and is given by V_∞/R. Thus, from Eq. (6.118), we have

$$\omega = \frac{g\sqrt{n^2 - 1}}{V_\infty}$$ [6.119]

For the maneuvering performance of an airplane, both military and civil, it is frequently advantageous to have the smallest possible R and the largest possible ω. Equations (6.118) and (6.119) show that to obtain both a small turn radius and a large turn rate, we want

1. The highest possible load factor (i.e., the highest possible L/W)
2. The lowest possible velocity

Consider another case of turning flight, in which an airplane initially in straight, level flight (where $L = W$) suddenly experiences an increase in lift. Since $L > W$, the airplane will begin to turn upward, as sketched in Fig. 6.53. For this pull-up maneuver,

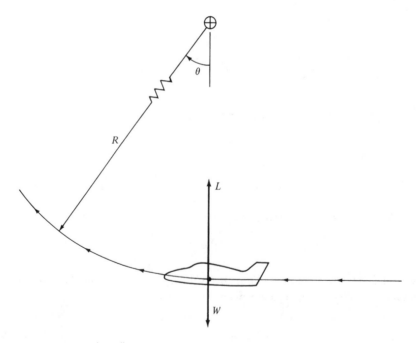

Figure 6.53   The pull-up maneuver.

the flight path becomes curved in the vertical plane, with a turn rate $\omega = d\theta/dt$. From the force diagram in Fig. 6.53, the resultant force F_r is vertical and is given by

$$F_r = L - W = W(n - 1) \qquad \textbf{[6.120]}$$

From Newton's second law

$$F_r = m\frac{V_\infty^2}{R} = \frac{W}{g}\frac{V_\infty^2}{R} \qquad \textbf{[6.121]}$$

Combining Eqs. (6.120) and (6.121) and solving for R give

$$R = \frac{V_\infty^2}{g(n - 1)} \qquad \textbf{[6.122]}$$

and since $\omega = V_\infty/R$,

$$\omega = \frac{g(n - 1)}{V_\infty} \qquad \textbf{[6.123]}$$

A related case is the pull-down maneuver, illustrated in Fig. 6.54. Here an airplane in initially level flight suddenly rolls to an inverted position, such that both L and W are pointing downward. The airplane will begin to turn downward in a circular flight path with a turn radius R and turn rate $\omega = d\theta/dt$. By an analysis similar to those above, the following results are easily obtained:

$$R = \frac{V_\infty^2}{g(n + 1)} \qquad \textbf{[6.124]}$$

Figure 6.54   The pull-down maneuver.

$$\omega = \frac{g(n+1)}{V_\infty}$$ [6.125]

Prove this to yourself.

Considerations of turn radius and turn rate are particularly important to military fighter aircraft; everything else being equal, those airplanes with the smallest R and largest ω will have definite advantages in air combat. High-performance fighter aircraft are designed to operate at high load factors, typically from 3 to 10. When n is large, then $n + 1 \approx n$ and $n - 1 \approx n$; for such cases, Eqs. (6.118), (6.119), and (6.122) to (6.125) reduce to

$$R = \frac{V_\infty^2}{gn}$$ [6.126]

and

$$\omega = \frac{gn}{V_\infty}$$ [6.127]

Let us work with these equations further. Since

$$L = \tfrac{1}{2}\rho_\infty V_\infty^2 S C_L$$

then

$$V_\infty^2 = \frac{2L}{\rho_\infty S C_L}$$ [6.128]

Substituting Eqs. (6.128) and (6.115) into Eqs. (6.126) and (6.127), we obtain

$$R = \frac{2L}{\rho_\infty S C_L g(L/W)} = \frac{2}{\rho_\infty C_L g}\frac{W}{S}$$ [6.129]

and

$$\omega = \frac{gn}{\sqrt{2L/(\rho_\infty S C_L)}}$$

$$= \frac{gn}{\sqrt{[2n/(\rho_\infty C_L)](W/S)}} = g\sqrt{\frac{\rho_\infty C_L n}{2(W/S)}}$$ [6.130]

Note that in Eqs. (6.129) and (6.130) the factor W/S appears. As we have discussed in previous sections, this factor occurs frequently in airplane performance analyses and is labeled

$$\frac{W}{S} \equiv \text{wing loading}$$

Equations (6.129) and (6.130) clearly show that airplanes with lower wing loadings will have smaller turn radii and larger turn rates, everything else being equal. However, the design wing loading of an airplane is usually determined by factors other than maneuvering, such as payload, range, and maximum velocity. As a result, wing loadings for light, general aviation aircraft are relatively low, but those for high-performance military aircraft are relatively large. Wing loadings for some typical airplanes are listed here.

Airplane	W/S, lb/ft^2
Wright Flyer (1903)	1.2
Beechcraft Bonanza	18.8
McDonnell Douglas F-15	66
General Dynamics F-16	74

From the above, we conclude that a small, light aircraft such as the Beechcraft Bonanza can outmaneuver a larger, heavier aircraft such as the F-16 because of smaller turn radius and larger turn rate. However, this is really comparing apples and oranges. Instead, let us examine Eqs. (6.129) and (6.130) for a *given* airplane with a given wing loading and ask the question, For this specific airplane, under what conditions will R be minimum and ω maximum? From these equations, clearly R will be minimum and ω will be maximum when both C_L and n are maximum. That is,

$$R_{\min} = \frac{2}{\rho_\infty g C_{L,\max}} \frac{W}{S} \qquad [6.131]$$

$$\omega_{\max} = g \sqrt{\frac{\rho_\infty C_{L,\max} n_{\max}}{2(W/S)}} \qquad [6.132]$$

Also note from Eqs. (6.131) and (6.132) that best performance will occur at sea level, where ρ_∞ is maximum.

There are some practical constraints on the above considerations. First, at low speeds, $n_{\max}$ is a function of $C_{L,\max}$ itself, because

$$n = \frac{L}{W} = \frac{\frac{1}{2}\rho_\infty V_\infty^2 S C_L}{W}$$

and hence
$$n_{\max} = \frac{1}{2}\rho_\infty V_\infty^2 \frac{C_{L,\max}}{W/S} \qquad [6.133]$$

At higher speeds, $n_{\max}$ is limited by the structural design of the airplane. These considerations are best understood by examining Fig. 6.55, which is a diagram showing load factor versus velocity for a given airplane—*the V-n diagram*. Here, curve AB is given by Eq. (6.133). Consider an airplane flying at velocity V_1, where V_1 is shown in Fig. 6.55. Assume that the airplane is at an angle of attack such that $C_L < C_{L,\max}$. This flight condition is represented by point 1 in Fig. 6.55. Now assume that the angle of attack is increased to that for obtaining $C_{L,\max}$, keeping the velocity constant at V_1. The lift increases to its maximum value for the given V_1, and hence the load factor $n = L/W$ reaches its maximum value $n_{\max}$ for the given V_1. This value of $n_{\max}$ is given by Eq. (6.133), and the corresponding flight condition is given by point 2 in Fig. 6.55. If the angle of attack is increased further, the wing stalls and the load factor drops. Therefore, point 3 in Fig. 6.55 is unobtainable in flight. Point 3 is in the *stall region* of the V-n diagram. Consequently, point 2 represents the highest possible load factor that can be obtained at the given velocity V_1. Now, as V_1 is increased, say, to a value of V_4, then the maximum possible load factor $n_{\max}$ also increases, as given

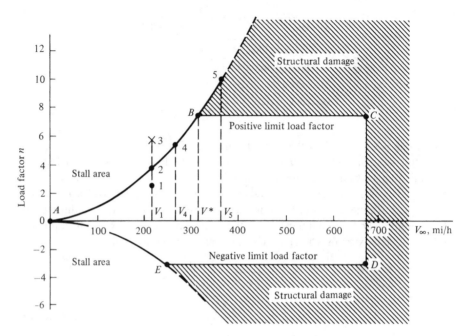

Figure 6.55   The *V*-*n* diagram for a typical jet trainer aircraft. (U.S. Air Force Academy.)

by point 4 in Fig. 6.55 and as calculated from Eq. (6.133). However, n_{max} cannot be allowed to increase indefinitely. Beyond a certain value of load factor, defined as the positive limit load factor and shown as the horizontal line BC in Fig. 6.55, structural damage may occur to the aircraft. The velocity corresponding to point B is designated as V^*. At velocities higher than V^*, say, V_5, the airplane must fly at values of C_L less than $C_{L,max}$ so that the positive limit load factor is not exceeded. If flight at $C_{L,max}$ is obtained at velocity V_5, corresponding to point 5 in Fig. 6.55, then structural damage will occur. The right-hand side of the V-n diagram, line CD, is a high-speed limit. At velocities greater than this, the dynamic pressure becomes so large that again structural damage may occur to the airplane. (This maximum velocity limit is, by design, much larger than the level-flight V_{max} calculated in Secs. 6.4 to 6.6. In fact, the structural design of most airplanes is such that the maximum velocity allowed by the V-n diagram is sufficiently greater than the maximum diving velocity for the airplane.) Finally, the bottom part of the V-n diagram, given by curves AE and ED in Fig. 6.55, corresponds to negative absolute angles of attack, that is, negative load factors. Curve AE defines the stall limit. (At absolute angles of attack less than zero, the lift is negative and acts in the downward direction. If the wing is pitched downward to a large enough negative angle of attack, the flow will separate from the bottom surface of the wing and the downward-acting lift will decrease in magnitude; that is, the wing *stalls*.) Line ED gives the negative limit load factor, beyond which structural damage will occur.

As a final note concerning the V-n diagram, consider point B in Fig. 6.55. This point is called the *maneuver point*. At this point, both C_L and n are simultaneously at their highest possible values that can be obtained anywhere throughout the allowable flight envelope of the aircraft. Consequently, from Eqs. (6.131) and (6.132), this point corresponds simultaneously to the smallest possible turn radius and the largest possible turn rate for the airplane. The velocity corresponding to point B is called the *corner velocity* and is designated by V^* in Fig. 6.55. The corner velocity can be obtained by solving Eq. (6.133) for velocity, yielding

$$V^* = \sqrt{\frac{2n_{\max}}{\rho_\infty C_{L,\max}} \frac{W}{S}} \qquad \textbf{[6.134]}$$

In Eq. (6.134), the value of $n_{\max}$ corresponds to that at point B in Fig. 6.55. The corner velocity is an interesting dividing line. At flight velocities less than V^*, it is not possible to structurally damage the airplane owing to the generation of too much lift. In contrast, at velocities greater than V^*, lift can be obtained that can structurally damage the aircraft (e.g., point 5 in Fig. 6.55), and the pilot must make certain to avoid such a case.

6.18 ACCELERATED RATE OF CLIMB (ENERGY METHOD)[1]

Modern high-performance airplanes, such as the supersonic General Dynamics F-16 shown in Fig. 6.56, are capable of highly accelerated rates of climb. Therefore, the performance analysis of such airplanes requires methods that go beyond the static rate-of-climb considerations given in Secs. 6.8 to 6.11. The purpose of this section is to introduce one such method, namely, a method dealing with the *energy* of an airplane. This is in contrast to our previous discussions that have dealt explicitly with forces on the airplane.

Consider an airplane of mass m in flight at some altitude h and with some velocity V. Due to its altitude, the airplane has *potential energy* PE equal to mgh. Due to its velocity, the airplane has *kinetic energy* KE equal to $\frac{1}{2}mV^2$. The total energy of the airplane is the sum of these energies;

$$\text{Total aircraft energy} = \text{PE} + \text{KE} = mgh + \tfrac{1}{2}mV^2 \qquad \textbf{[6.135]}$$

The energy per unit weight of the airplane is obtained by dividing Eq. (6.135) by $W = mg$. This yields the *specific energy*, denoted by H_e, as

$$H_e \equiv \frac{\text{PE} + \text{KE}}{W} = \frac{mgh + \tfrac{1}{2}mV^2}{mg}$$

[1] This section is based in part on material presented by the faculty of the department of aeronautics at the U.S. Air Force Academy at its annual aerodynamics workshop, held each July at Colorado Springs. This author has had the distinct privilege to participate in this workshop since its inception in 1979. Special thanks for this material go to Col. James D. Lang, Major Thomas Parrot, and Col. Daniel Daley.

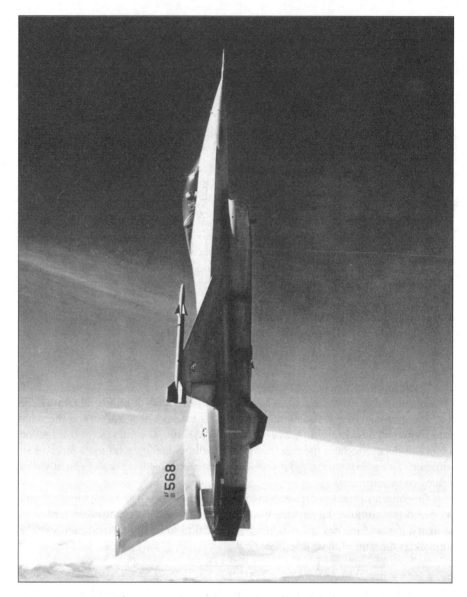

Figure 6.56   General Dynamics F-16 in 90° vertical accelerated climb. *(U.S. Air Force.)*

or

$$H_e = h + \frac{V^2}{2g}$$

[6.136]

The specific energy H_e has units of height and is therefore also called the *energy height* of the aircraft. Thus, let us become accustomed to quoting the energy of an airplane in terms of the energy height H_e, which is simply the sum of the potential and kinetic energies of the airplane per unit weight. Contours of constant H_e are illustrated in Fig. 6.57, which is an altitude-Mach number map. Here, the ordinate and abscissa are altitude h and Mach number M, respectively, and the dashed curves are lines of constant energy height.

To obtain a feeling for the significance of Fig. 6.57, consider two airplanes, one flying at an altitude of 30,000 ft at Mach 0.81 (point A in Fig. 6.57) and the other flying at an altitude of 10,000 ft at Mach 1.3 (point B). Both airplanes have the same energy height, 40,000 ft (check this yourself by calculation). However, airplane A has more potential energy and less kinetic energy (per unit weight) than airplane B. If both airplanes maintain their same states of total energy, then both are capable of "zooming" to an altitude of 40,000 ft at zero velocity (point C) simply by trading all their kinetic energy for potential energy. Consider another airplane, flying at an altitude of 50,000 ft at Mach 1.85, denoted by point D in Fig. 6.57. This airplane will have an energy height of 100,000 ft and is indeed capable of zooming to an actual altitude of 100,000 ft by trading all its kinetic energy for potential energy. Airplane D is in a much higher energy state ($H_e = 100,000$ ft) than airplanes A and B (which have $H_e = 40,000$ ft). Therefore, airplane D has a much greater capability for speed and altitude performance than airplanes A and B. In air combat, everything else being equal, it is advantageous to be in a higher energy state (have a larger H_e) than your adversary.

How does an airplane change its energy state; for example, in Fig. 6.57, how could airplanes A and B increase their energy heights to equal that of D? To answer

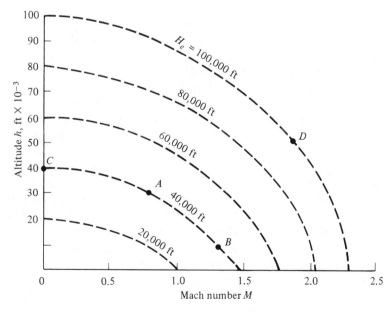

Figure 6.57 Altitude–Mach number map showing curves of constant energy height. These are universal curves that represent the variation of kinetic and potential energies per unit weight. They do not depend on the specific design factors of a given airplane.

this question, return to the force diagram in Fig. 6.5 and the resulting equation of motion along the flight path, given by Eq. (6.7). Assuming that α_T is small, Eq. (6.7) becomes

$$T - D - W \sin\theta = m \frac{dV}{dt} \qquad [6.137]$$

Recalling that $m = W/g$, we can rearrange Eq. (6.137) as

$$T - D = W \left(\sin\theta + \frac{1}{g} \frac{dV}{dt} \right)$$

Multiplying by V/W, we obtain

$$\frac{TV - DV}{W} = V \sin\theta + \frac{V}{g} \frac{dV}{dt} \qquad [6.138]$$

Examining Eq. (6.138) and recalling some of the definitions from Sec. 6.8, we observe that $V \sin\theta = R/C = dh/dt$ and that

$$\frac{TV - DV}{W} = \frac{\text{excess power}}{W} \equiv P_s$$

where the excess power per unit weight is defined as the *specific excess power* and is denoted by P_s. Hence, Eq. (6.138) can be written as

$$\boxed{P_s = \frac{dh}{dt} + \frac{V}{g} \frac{dV}{dt}} \qquad [6.139]$$

Equation (6.139) states that an airplane with excess power can use this excess for rate of climb (dh/dt) or to accelerate along its flight path (dV/dt) or for a combination of both. For example, consider an airplane in level flight at a velocity of 800 ft/s. Assume that when the pilot pushes the throttle all the way forward, an excess power is generated in the amount $P_s = 300$ ft/s. Equation (6.139) shows that the pilot can choose to use all this excess power to obtain a maximum unaccelerated rate of climb of 300 ft/s ($dV/dt = 0$, hence $P_s = dh/dt = $ R/C). In this case, the velocity along the flight path stays constant at 800 ft/s. Alternatively, the pilot may choose to maintain level flight ($dh/dt = 0$) and to use all this excess power to accelerate at the rate of $dV/dt = gP_s/V = 32.2(300)/800 = 12.1$ ft/s². On the other hand, some combination could be achieved, such as a rate of climb $dh/dt = 100$ ft/s along with an acceleration along the flight path of $dV/dt = 32.2(200)/800 = 8.1$ ft/s². [Note that Eqs. (6.138) and (6.139) are generalizations of Eq. (6.50). In Sec. 6.8, we assumed that $dV/dt = 0$, which resulted in Eq. (6.50) for a steady climb. In the present section, we are treating the more general case of climb with a finite acceleration.] Now return to Eq. (6.136) for the energy height. Differentiating with respect to time, we have

$$\frac{dH_e}{dt} = \frac{dh}{dt} + \frac{V}{g} \frac{dV}{dt} \qquad [6.140]$$

The right-hand sides of Eqs. (6.139) and (6.140) are identical, hence we see that

$$P_s = \frac{dH_e}{dt} \qquad [6.141]$$

That is, the *time rate of change of energy height is equal to the specific excess power*. This is the answer to the question at the beginning of this paragraph. An airplane can increase its energy state simply by the application of excess power. In Fig. 6.57, airplanes A and B can reach the energy state of airplane D *if* they have enough excess power to do so.

This immediately leads to the next question, namely, How can we ascertain whether a given airplane has enough P_s to reach a certain energy height? To address this question, recall the definition of excess power as illustrated in Fig. 6.29, that is, the difference between power available and power required. For a given altitude, say, h, the excess power (hence P_s) can be plotted versus velocity (or Mach number). For a subsonic airplane below the drag-divergence Mach number, the resulting curve will resemble the sketch shown in Fig. 6.58a. At a given altitude h_1, P_s will be an inverted, U-shaped curve. (This is essentially the same type of plot as shown in Figs. 6.32 and 6.33.) For progressively higher altitudes, such as h_2 and h_3, P_s becomes smaller, as also shown in Fig. 6.58a. Hence, Fig. 6.58a is simply a plot of P_s versus Mach number with altitude as a parameter. These results can be cross-plotted on an altitude–Mach number map using P_s as a parameter, as illustrated in Fig. 6.58b. For example, consider all the points on Figure 6.58a where $P_s = 0$; these correspond to points along a horizontal axis through $P_s = 0$, such as points a, b, c, d, e, and f in Fig. 6.58a. Now replot these points on the altitude–Mach number map in Fig. 6.58b. Here, points a, b, c, d, e, and f form a bell-shaped curve, along which $P_s = 0$. This curve is called the P_s *contour* for $P_s = 0$. Similarly, all points with $P_s = 200$ ft/s are on the horizontal line AB in Fig. 6.58a, and these points can be cross-plotted to generate the $P_s = 200$ ft/s contour in Fig. 6.58b. In this fashion, an entire series of P_s contours can be generated in the altitude–Mach number map. For a supersonic airplane, the P_s versus Mach number curves at different altitudes will appear as sketched in Fig. 6.59a. The "dent" in the U-shaped curves around Mach 1 is due to the large drag increase in the transonic flight regime (see Sec. 5.10). In turn, these curves can be cross-plotted on the altitude–Mach number map, producing the P_s contours as illustrated in Fig. 6.59b. Due to the double-humped shape of the P_s curves in Fig. 6.59a, the P_s contours in Fig. 6.59b have different shapes in the subsonic and supersonic regions. The shape of the P_s contours shown in Fig. 6.59b is characteristic of most supersonic aircraft. Now we are close to the answer to our question at the beginning of this paragraph. Let us overlay the P_s contours, say, from Fig. 6.59b, and the energy states illustrated in Fig. 6.57—all on an altitude–Mach number map. We obtain a diagram, as illustrated in Fig. 6.60. In this figure, note that the P_s contours always correspond to a given airplane at a given load factor, whereas the H_e lines are universal fundamental physical curves that have nothing to do with any given airplane. The usefulness of Fig. 6.60 is that it clearly establishes what energy states are obtainable by a given airplane. The regime of sustained flight for

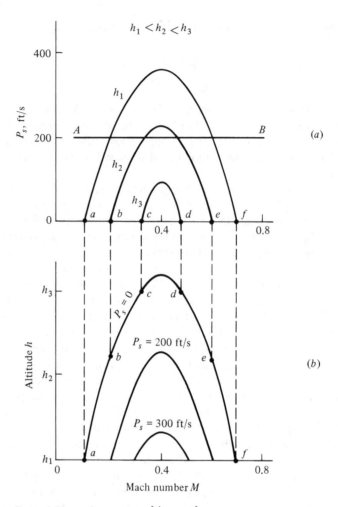

Figure 6.58   Construction of the specific excess-power contours in the altitude–Mach number map for a subsonic airplane below the drag-divergence Mach number. These contours are constructed for a fixed load factor; if the load factor is changed, the P_s contours will shift.

the airplane lies *inside* the envelope formed by the $P_s = 0$ contour. Hence, all values of H_e inside this envelope are obtainable by the airplane. A comparison of figures like Fig. 6.60 for different airplanes will clearly show in what regions of altitude and Mach number an airplane has maneuverability advantages over another.

Figure 6.60 is also useful for representing the proper flight path to achieve minimum time to climb. For example, consider two energy heights $H_{e,1}$ and $H_{e,2}$, where $H_{e,2} > H_{e,1}$. The time to move between these energy states can be obtained from

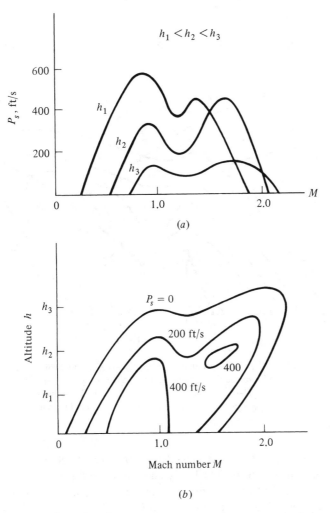

Figure 6.59 Specific excess-power contours for a supersonic airplane.

Eq. (6.141), written as

$$dt = \frac{dH_e}{P_s}$$

Integrating between $H_{e,1}$ and $H_{e,2}$, we have

$$t_2 - t_1 \int_{H_{e,1}}^{H_{e,2}} \frac{dH_e}{P_s} \qquad \textbf{[6.142]}$$

From Eq. (6.142), the time to climb will be a minimum when P_s is a maximum. Looking at Fig. 6.60, for each H_e curve, we see there is a point where P_s is a maximum.

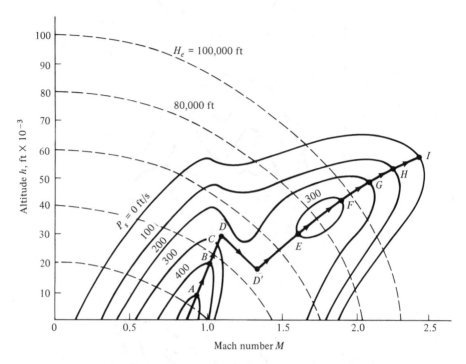

Figure 6.60   Overlay of P_s contours and specific energy states on an altitude–Mach number map. The P_s values shown here approximately correspond to a Lockheed F-104G supersonic fighter. Load factor $n = 1$. $W = 18,000$ lb. Airplane is at maximum thrust. The path given by points A through I is the flight path for minimum time to climb.

Indeed, at this point, the P_s curve is tangent to the H_e curve. Such points are illustrated by points A to I in Fig. 6.60. The heavy curve through these points illustrates the variation of altitude and Mach number along the flight path for minimum time to climb. The segment of the flight path between D and D' represents a constant-energy dive to accelerate through the drag-divergence region near Mach 1.

As a final note, analyses of modern high-performance airplanes make extensive use of energy concepts such as those described above. Indeed, military pilots fly with P_s diagrams in the cockpit. Our purpose here has been to simply introduce some of the definitions and basic ideas in involving these concepts. A more extensive treatment is beyond the scope of this book.

6.19   SPECIAL CONSIDERATIONS FOR SUPERSONIC AIRPLANES

The physical characteristics of subsonic flow and supersonic flow are totally different— a contrast as striking as that between day and night. We have already addressed some of these differences in Chaps. 4 and 5. However, these differences do not affect the airplane performance techniques discussed in this chapter. These techniques are general, and they apply to both subsonic and supersonic airplanes. The only way by which our performance analysis knows that the airplane is subsonic or supersonic is through the drag polar and the engine characteristics. Recall from our discussion in Sec. 5.3 that C_L and C_D are functions of free-stream Mach number; hence the drag polar is a function of M_∞. A given drag polar pertains to a specified Mach number; for example, the drag polar for the Lockheed C-141A shown in Fig. 6.2 pertains to low-speed flow $M_\infty \leq 0.3$. A generic comparison between the drag polars for a given subsonic Mach number and a given supersonic Mach number for the same airplane is sketched in Fig. 6.61. For a given C_L, C_D is much larger at supersonic speeds than at subsonic speeds because of the presence of supersonic wave drag. Hence the supersonic drag polar is displaced to the right of the subsonic drag polar and is a more tightly shaped parabola, as sketched in Fig. 6.61.

Consider an arbitrary point on the drag polar, such as point 1 shown in Fig. 6.61. A straight line O–1 drawn from the origin to point 1 will have a slope equal to $C_{L,1}/C_{D,1}$; that is, the slope is equal to the lift-to-drag ratio associated with flight at point 1. As we move point 1 up the drag polar, the slope of line O–1 will increase, associated with increased values of L/D. Let point A be the point where the straight line becomes tangent. Hence, the slope of the straight line OA is the maximum possible slope. This slope is equal to $(L/D)_{max}$, and point A corresponds to flight at maximum lift-to-drag ratio. This demonstrates the graphical construction from which $(L/D)_{max}$ can be obtained from the drag polar. Simply draw a straight line from the origin tangent to the drag polar—the slope of this line is equal to $(L/D)_{max}$.

With this in mind, let us compare the two drag polars in Fig. 6.61. Line OA is drawn tangent to the subsonic drag polar, and its slope gives $(L/D)_{max}$ at the given subsonic Mach number. Line OB is drawn tangent to the supersonic drag polar, and its slope gives $(L/D)_{max}$ at the given supersonic Mach number. Clearly, the slope of

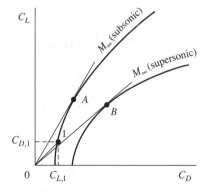

Figure 6.61   Generic comparison of a subsonic drag polar with a supersonic drag polar for the same airplane.

OB is smaller than the slope of OA. *The values of* $(L/D)_{max}$ *at supersonic speeds are smaller than at subsonic speeds.* This is dramatically shown in Fig. 6.62. As an airplane accelerates through Mach 1, there is a considerable drop in its $(L/D)_{max}$.

Perhaps the most severe effect on airplane performance associated with the decrease in $(L/D)_{max}$ at supersonic speeds is that on range. From Eq. (6.77) we saw that range for a jet airplane is proportional to $C_L^{1/2}/C_D$. If $(L/D)_{max}$ is smaller for a given supersonic Mach number, then so will be the value of $(C_L^{1/2}/C_D)_{max}$. This is the primary reason why the range of a given airplane cruising at supersonic speed is smaller than that at subsonic speed, everything else being equal.

Let us return to Eq. (6.75), repeated here:

$$R = \int_{W_1}^{W_0} \frac{V_\infty}{c_t} \frac{C_L}{C_D} \frac{dW}{W} \qquad [6.75]$$

This is the equation from which Eq. (6.77) was derived. Assuming flight at constant V_∞, c_t, and C_L/C_D, Eq. (6.75) becomes

$$R = \frac{V_\infty}{c_t} \frac{L}{D} \ln \frac{W_0}{W_1} \qquad [6.143]$$

You will frequently see Eq. (6.143) in the literature as the equation for range for a jet airplane. Note that Eq. (6.143) shows that maximum range is obtained *not* with

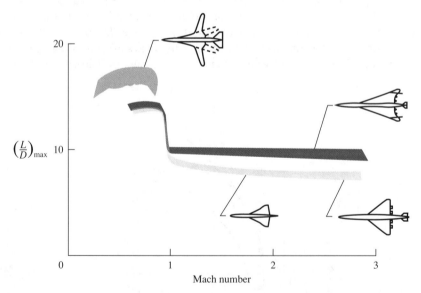

Figure 6.62   Variation of (L/D) max with Mach number for several generic airplane configurations. From M.R. Nichols, A.L. Keith, and W.E. Foss, "The Second-Generation Supersonic Transport," in *Vehicle Technology for Civil Aviation: The Seventies and Beyond.* NASA SP-292, pp. 409–428.

maximum L/D but rather with the maximum value of the *product*, $V_\infty(L/D)$. This product is maximum when $C_L^{1/2}/C_D$ is maximum, as shown through the derivation of Eq. (6.77). Nevertheless, Eq. (6.143) is a useful expression for the range for a jet airplane.

Equations (6.77) and (6.143) both indicate the obvious ways to compensate for the loss of $(L/D)_{max}$, hence $C_L^{1/2}/C_D$, on the range for a supersonic airplane, namely,

1. Decrease the thrust specific fuel consumption c_t.
2. Increase the fuel weight W_f, hence increasing the ratio W_0/W_1 in Eq. (6.143) and increasing the difference $W_0^{1/2} - W_1^{1/2}$ in Eq. (6.77).

Increasing the fuel weight is usually not a desirable design solution, because the additional fuel usually means a smaller useful payload for the airplane. Also, for turbojet and low–bypass ratio turbofans (see Chap. 9), the thrust specific fuel consumption increases with an increase in Mach number for supersonic speeds, further compounding the degradation of range.

DESIGN BOX

Based on the above discussion, the designer of a supersonic cruise airplane, such as a civil supersonic transport, must live with the realities embodied in Eq. (6.143). For example, during the 1990s, an extended study of a second-generation supersonic transport, labeled the *high-speed civil transport* (*HSCT*), was carried out by industry in the United States, supported by the high-speed research (HSR) program carried out by NASA. (By comparison, the Anglo-French Concorde designed in the 1960s, shown in Fig. 5.62, is a first-generation supersonic transport.) The baseline design specifications for the HSCT called for cruise at Mach 2.4 with a range of 5000 mi, carrying 300 passengers. This is an extreme design challenge, on the cutting edge of modern aeronautical technology. From Eq. (6.143), a few percent shortfall in L/D could prevent the achievement of the specified range. This underscores the importance of supersonic aerodynamic research aimed at improving supersonic L/D. The engine must produce the lowest possible thrust specific fuel consumption, while at the same time producing an environmentally acceptable low value of atmospheric pollutants in the jet exhaust, in order to protect the atmospheric ozone layer. Moreover, the engine noise must be an acceptably low value during takeoff and

landing—that is a major challenge for jet engines designed for supersonic flight, for which the exhaust jet velocities are large, hence very noisy. Hence, the design of engines for the HSCT is a massive challenge in itself. There are major structural and materials challenges as well. The design goal of the HSCT is a structural weight fraction (weight of the structure divided by the gross takeoff weight) of 0.2, which is considerably smaller than the more typical value of 0.25 and higher for conventional subsonic transports. With the smaller structural weight fraction, the HSCT can carry more fuel and/or more passengers to meet its other design specifications. And if this were not enough, the size of the baseline HSCT is so large, with a length longer than a football field, that there is a problem with elastic bending of the fuselage (in the longitudinal direction); as a result, stability and control are severely compromised. Indeed, this problem is compounded by the interaction of the aerodynamic force, the propulsive thrust, and the real-time control inputs. Called the APSE (aeropropulsiveservoelastic) effect, this is a problem that affects the HSCT in flight and on the ground. (For more details on the HSCT design challenges, see *U.S. Supersonic Commercial Aircraft: Assessing NASA's High Speed*

(continued)

Research Program, National Research Council Report, National Academy Press, Washington, DC, 1997.) We note that the sonic boom is not considered to be a problem for the HSCT because of the up-front decision that it would fly subsonically over land, the same restriction imposed on the Concorde SST.

In short, the design of an environmentally acceptable, economically viable supersonic transport is a ma-

jor aeronautical technological problem that has yet to be solved. It will be one of the most challenging aeronautical endeavors in the early 21st century, and perhaps many of the young readers of this book will have a hand in meeting this challenge.

6.20 A COMMENT, AND MORE ON THE ASPECT RATIO

We end the technical portion of this chapter by noting that detailed computer programs now exist within NASA and the aerospace industry for the accurate estimation of airplane performance. These programs are usually geared to specific types of airplanes, for example, general aviation aircraft (light single- or twin-engine private airplanes), military fighter aircraft, and commercial transports. Such considerations are beyond the scope of this book. However, the principles developed in this chapter are stepping-stones to more-advanced studies of airplane performance; the Bibliography at the end of this chapter provides some suggestions for such studies.

DESIGN BOX

In Chaps. 5 and 6 we have underscored the importance of the wing aspect ratio in airplane design. In particular, for subsonic flight we have noted that by increasing the aspect ratio, we can obtain a lower induced-drag coefficient, and hence a higher maximum L/D ratio. Now that we are at the end of our discussions of airplane aerodynamics and performance, it is worthwhile to expand this consideration by asking the question: For an airplane in steady, level flight, what design parameter dictates the induced *drag* itself (as contrasted with the induced drag *coefficient*)? Is it the aspect ratio, as intuition might indicate, or is it another design parameter? The answer is developed in the following discussion, which will help to expand our understanding of induced drag and will provide an enhanced physical understanding of the definition of aspect ratio.

From Eq. (6.1c), the *coefficient* of drag due to lift (which for subsonic flight at normal angles of attack is

mainly due to the induced-drag coefficient) is given by

$$C_{D,i} = \frac{C_L^2}{\pi e \mathrm{AR}} \qquad \textbf{[6.144]}$$

In turn, the *drag* due to lift is

$$D_i = q_\infty S C_{D,i} = q_\infty S \frac{C_L^2}{\pi e \mathrm{AR}} \qquad \textbf{[6.145]}$$

For steady, level flight, $L = W$. Hence

$$C_L^2 = \left(\frac{L}{q_\infty S}\right)^2 = \left(\frac{W}{q_\infty S}\right)^2 \qquad \textbf{[6.146]}$$

Substituting Eq. (6.146) into (6.145), we have

$$D_i = q_\infty S \left(\frac{W}{q_\infty S}\right)^2 \frac{1}{\pi e \mathrm{AR}} \qquad \textbf{[6.147]}$$

Since $\mathrm{AR} = b^2/S$, Eq. (6.147) can be written as

(continued)

$$D_i = \frac{1}{\pi e} q_\infty S \left(\frac{W}{q_\infty S} \right)^2 \left(\frac{S}{b^2} \right) \qquad \textbf{[6.148]}$$

Note that the wing area cancels out of Eq. (6.148), and we are left with

$$D_i = \frac{1}{\pi e q_\infty} \left(\frac{W}{b} \right)^2 \qquad \textbf{[6.149]}$$

This is a rather revealing result! The drag due to lift in steady, level flight—the force itself—does not depend explicitly on the aspect ratio, but rather on another design parameter, W/b, called the *span loading*.

$$\text{Span loading} \equiv \frac{W}{b} \qquad \textbf{[6.150]}$$

The drag due to lift varies with the square of the span loading.

From Eq. (6.149), we see that the drag due to lift, for a given weight airplane, can be reduced simply by increasing the wingspan. In so doing, the wing tip vortices (the physical source of induced drag) are simply moved farther away, hence lessening their effect on the rest of the wing and in turn reducing the induced drag. This makes good intuitive sense.

In light of the above, the span loading W/b takes its place as yet another design parameter that the airplane designers can adjust during the conceptual design process for a new airplane. Of course, the span loading and the aspect ratio are related via

$$\frac{W}{b} = \left(\frac{W}{S} \right) \frac{b}{\text{AR}} \qquad \textbf{[6.151]}$$

where W/S is the familiar wing loading.

Let us return to the concept of aspect ratio, which now takes on an enhanced significance, as follows. First, note that the zero-lift drag which we denoted by D_o is given by $q_\infty S C_{D,0}$ and hence is proportional to the wing area, whereas the drag due to lift for steady, level flight is proportional to the square of the span loading via Eq. (6.149). The *ratio* of these two drags is

$$\frac{D_i}{D_o} = \left[\frac{1}{\pi e q_\infty} \left(\frac{W}{b} \right)^2 \right] \frac{1}{q_\infty S C_{D,0}} \qquad \textbf{[6.152]}$$

In Eq. (6.152), the ratio $(W/b)^2/S$ can be cast as

$$\frac{(W/b)^2}{S} = \frac{(W/S)^2}{b^2/S} = \frac{(W/S)^2}{\text{AR}} \qquad \textbf{[6.153]}$$

Substituting Eq. (6.153) into (6.152), we have

$$\frac{D_i}{D_o} = \frac{1}{\pi e q_\infty^2 C_{D,0}} \frac{(W/S)^2}{\text{AR}} \qquad \textbf{[6.154]}$$

From Eq. (6.154), we can make the following statement: For specified values of the design parameters W/S and $C_{D,0}$, increasing the design aspect ratio will decrease the drag due to lift relative to the zero-lift drag. So the aspect ratio predominately controls the *ratio* of lift-induced drag to the zero-lift drag, whereas the span loading controls the actual *value* of the lift-induced drag.

6.21   HISTORICAL NOTE: DRAG REDUCTION—THE NACA COWLING AND THE FILLET

The radial piston engine came into wide use in aviation during and after World War I. As described in Chap. 9, a radial engine has its pistons arranged in a circular fashion about the crankshaft, and the cylinders themselves are cooled by airflow over the outer finned surfaces. Until 1927, these cylinders were generally directly exposed to the main airstream of the airplane, as sketched in Fig. 6.63. As a result, the drag on the engine-fuselage combination was inordinately high. The problem was severe enough that a group of aircraft manufacturers met at Langley Field on May

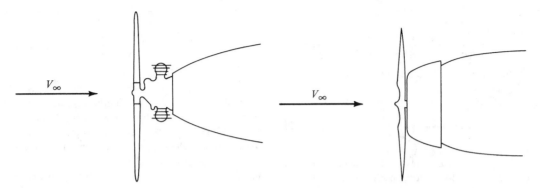

Figure 6.63   Engine mounted with no cowling.   **Figure 6.64**   Engine mounted with full cowling.

24, 1927, to urge NACA to undertake an investigation of means to reduce this drag. Subsequently, under the direction of Fred E. Weick, an extensive series of tests was conducted in the Langley 20-ft propeller research tunnel using a Wright Whirlwind J-5 radial engine mounted to a conventional fuselage. In these tests, various types of aerodynamic surfaces, called *cowlings*, were used to cover, partly or completely, the engine cylinders, directly guiding part of the airflow over these cylinders for cooling purposes but at the same time not interfering with the smooth primary aerodynamic flow over the fuselage. The best cowling, illustrated in Fig. 6.64, completely covered the engine. The results were dramatic! Compared with the unscowled fuselage, a full cowling reduced the drag by a stunning 60 percent! This is illustrated in Fig. 6.65,

Figure 6.65   Reduction in drag due to a full cowling.

Figure 6.66 Illustration of the wing fillet.

taken directly from Weick's report, entitled "Drag and Cooling with Various Forms of Cowling for a Whirlwind Radial Air-Cooled Engine," NACA Technical Report No. 313, published in 1928. After this work, virtually all radial engine–equipped airplanes since 1928 have been designed with a full NACA cowling. The development of this cowling was one of the most important aerodynamic advancements of the 1920s; it led the way to a major increase in aircraft speed and efficiency.

A few years later, a second major advancement was made, but by a completely different group and on a completely different part of the airplane. In the early 1930s, the California Institute of Technology at Pasadena, California, established a program in aeronautics under the direction of Theodore von Karman. Von Karman, a student of Ludwig Prandtl, became probably the leading aerodynamicist of the 1920–1960 time period. At Caltech, von Karman established an aeronautical laboratory of high quality, which included a large subsonic wind tunnel funded by a grant from the Guggenheim Foundation. The first major experimental program in this tunnel was a commercial project for Douglas Aircraft Company. Douglas was designing the DC-1, the forerunner of a series of highly successful transports (including the famous DC-3, which revolutionized commercial aviation in the 1930s). The DC-1 was plagued by unusual buffeting in the region where the wing joined the fuselage. The sharp corner at the juncture caused severe flow field separation, which resulted in high drag as well as shed vortices which buffeted the tail. The Caltech solution, which was new and pioneering, was to fair the trailing edge of the wing smoothly into the fuselage. These fairings, called *fillets*, were empirically designed and were modeled in clay on the DC-1 wind tunnel models. The best shape was found by trial and error. The addition of a fillet (see Fig. 6.66) solved the buffeting problem by smoothing out the separated flow and hence also reduced the interference drag. Since that time, fillets have become a standard airplane design feature. Moreover, the fillet is an excellent example of how university laboratory research in the 1930s contributed directly to the advancement of practical airplane design.

6.22 HISTORICAL NOTE: EARLY PREDICTIONS OF AIRPLANE PERFORMANCE

The airplane of today is a modern work of art and engineering. In turn, the prediction of airplane performance as described in this chapter is sometimes viewed as a relatively modern discipline. However, contrary to intuition, some of the basic concepts have roots deep in history; indeed, some of the very techniques detailed in previous sections were being used in practice only a few years after the Wright brothers' successful first flight in 1903. This section traces a few historic paths for some of the basic ideas of airplane performance, as follows:

1.   Some understanding of the *power required* P_R for an airplane was held by George Cayley. He understood that the rate of energy lost by an airplane in a steady glide under gravitational attraction must be essentially the power that must be supplied by an engine to maintain steady, level flight. In 1853, Cayley wrote:

The whole apparatus when loaded by a weight equal to that of the man intended ultimately to try the experiment, and with the horizontal rudder [the elevator] described
on the essay before sent, adjusted so as to regulate the oblique descent from some
elevated point, to its proper pitch, it may be expected to skim down, with no force but
its own gravitation, in an angle of about 11 degrees with the horizon; or possibly, if
well executed, as to direct resistance something less, at a speed of about 36 feet per
second, if loaded 1 pound to each square foot of surface. This having by repeated
experiments, in perfectly calm weather, been ascertained, for both the safety of the
man, and the datum required, let the wings be plied with the man's utmost strength;
and let the angle measured by the greater extent of horizontal range of flight be noted;
when this point, by repeated experiments, has been accurately found, we shall have
ascertained a sound practical basis for calculating what engine power is necessary
under the same circumstances as to weight and surface to produce horizontal flight....

2. The *drag polar*, a concept introduced in Secs. 5.14 and 6.1, sketched in
Figs. 5.35 and 6.1, and embodied in Eq. (6.1a), represents simply a plot of C_D
versus C_L, illustrating that C_D varies as the square of C_L. A knowledge of the drag
polar is essential to the calculation of airplane performance. It is interesting that the
first drag polars were drawn and published by Otto Lilienthal (see Sec. 1.5) in 1889,
although he did not call them such. The term *polar* for these diagrams was first
introduced by Gustave Eiffel in 1909. Eiffel, the designer of the Eiffel Tower in
Paris, built two wind tunnels and carried out extensive aerodynamic testing from
1909 to the time of his death in 1923.

3. Some understanding of the requirements for *rate of climb* existed as far back as
1913, when in an address by Granville E. Bradshaw before the Scottish
Aeronautical Society in Glasgow in December, the following comment was made:
"Among the essential features of all successful aeroplanes [is that] it shall climb
very quickly. This depends almost entirely on the weight efficiency of the engine.
The rate of climb varies directly as the power developed and indirectly as the weight
to be lifted." This is essentially a partial statement of Eq. (6.50).

4. No general understanding of the prediction of *airplane performance* existed
before the 20th century. The excellent summary of aeronautics written by Octave
Chanute in 1894, *Progress in Flying Machines*, does not contain any calculational
technique even remotely resembling the procedures set forth in this chapter. At best,
it was understood by that time that lift and drag varied as the first power of the area
and as the second power of velocity; but this does not constitute a performance
calculation. However, this picture radically changed in 1911. In that year, the
Frenchman Duchène received the Monthyon Prize from the Paris Academy of
Sciences for his book entitled *The Mechanics of the Airplane: A Study of the
Principles of Flight*. Captain Duchène was a French engineering officer, born in
Paris on December 27, 1869, educated at the famous Ecole Polytechnique, and later
assigned to the fortress at Toul, one of the centers of "aerostation" in France. It was
in this capacity that Captain Duchène wrote his book during 1910–1911. In this
book, the basic elements of airplane performance, as discussed in this chapter, are
put forth for the first time. Duchène gives curves of power required and power

available, as we illustrated in Fig. 6.21*a*; he discusses airplane maximum velocity; he also gives the same relation as Eq. (6.50) for rate of climb. Thus, some of our current concepts for the calculation of airplane performance date back as far as 1910–1911—four years before the beginning of World War I, and only seven years after the Wright brothers' first flight in 1903. Later, in 1917, Duchène's book was translated to English by John Ledeboer and T. O'B. Hubbard (see Bibliography at the end of this chapter). Finally, during 1918–1920, three additional books on airplane performance were written (again, see Bibliography), the most famous being the authoritative *Applied Aerodynamics* by Leonard Bairstow. By this time the foundations discussed in this chapter had been well set.

6.23   HISTORICAL NOTE: BREGUET AND THE RANGE FORMULA

Louis-Charles Breguet was a famous French aviator, airplane designer, and industrialist. Born in Paris on January 2, 1880, he was educated in electrical engineering at the Lycée Condorcet, the Lycée Carnot, and the École Superieure d'Electricité. After graduation, he joined the electrical engineering firm of his father, Maison Breguet. However, in 1909 Breguet built his first airplane and then plunged his life completely into aviation. During World War I, his airplanes were mass-produced for the French air force. In 1919, he founded a commercial airline company which later grew into Air France. His airplanes set several long-range records during the 1920s and 1930s. Indeed, Breguet was active in his own aircraft company until his death on May 4, 1955, in Paris. His name is associated with a substantial part of French aviation history.

The formula for range of a propeller-driven airplane given by Eq. (6.67) has also become associated with Breguet's name; indeed, it is commonly called the Breguet range equation. However, the reason for this association is historically obscure. In fact, the historical research of the present author can find no substance to Breguet's association with Eq. (6.67) until a presentation by Breguet to the Royal Aeronautical Society in London in 1922. On one hand, we find absolutely no reference to airplane range or endurance in any of the airplane performance literature before 1919, least of all a reference to Breguet. The authoritative books by Cowley and Levy (1918), Judge (1919), and Bairstow (1920) (see Bibliography at the end of this chapter) amazingly enough do not discuss this subject. On the other hand, in 1919, NACA Report No. 69, entitled "A Study of Airplane Ranges and Useful Loads," by J. G. Coffin, gives a complete derivation of the formulas for range, Eq. (6.67), and endurance, Eq. (6.68). However, Coffin, who was director of research for Curtiss Engineering Corporation at that time, gives absolutely no references to *anybody*. Coffin's work appears to be original and clearly seems to be the first presentation of the range and endurance formulas in the literature. However, to confuse matters, we find a few years later, in NACA Report No. 173, entitled "Reliable Formulae for Estimating Airplane Performance and the Effects of Changes in Weight, Wing Area or Power," by Walter

S. Diehl (we have met Diehl before, in Sec. 3.6), the following statement: "The common formula for range, usually credited to Breguet, is easily derived." Diehl's report then goes on to use Eq. (6.67), with no further reference to Breguet. This report was published in 1923, four years after Coffin's work.

Consequently, to say the least, the proprietorship of Eq. (6.67) is not clear. It appears to this author that, in the United States at least, there is plenty of documentation to justify calling Eq. (6.67) the Coffin-Breguet range equation. However, it has come down to us through the ages simply as Breguet's equation, apparently without documented substance.

6.24 HISTORICAL NOTE: AIRCRAFT DESIGN— EVOLUTION AND REVOLUTION

Sit back for a moment and think about the evolution of the airplane, beginning with Sir George Cayley's 1804 hand-launched glider. Indeed, Fig. 1.8 (Cayley's own sketch of this aircraft) shows the first airplane with a modern configuration. Now jump ahead a century in the design of the airplane to Fig. 1.2, the Wright brothers' historical photograph of their first successful flight in 1903; this is the true beginning of the practical airplane. Finally, jump another 80 years to Fig. 6.11, which shows a modern jet aircraft. Put these three aircraft side by side in your mind: Cayley's glider, the *Wright Flyer*, and the Cessna Citation 3. What a testimonial to the evolution of airplane design! Each machine is totally different, each being the product of three different worlds of scientific and engineering understanding and practice. One must marvel at the rapid technical progress, especially in the 20th century, that brings us to the present status of airplane design represented by the modern, fast, high-flying jet aircraft shown in Fig. 6.11. What were the major technical milestones in this progress? What were the evolutionary (and sometimes revolutionary) developments that swept us from Cayley's seminal concepts to the modern airplane? The eye-opening and exciting answers to these questions would require a separate book to relate, but in this section we highlight a few aspects of the technical progression of airplane design, using some of the technology we have covered in this chapter on airplane performance.

To provide a technical focus for our discussion, we chose two aerodynamic parameters as figures of merit to compare and evaluate different airplane designs. The first is the zero-lift drag coefficient $C_{D,0}$, an important characteristic of any airplane because it has a strong effect on the maximum flight speed. Recall that at V_{max} for an airplane since the angle of attack (and hence the induced drag) is small, the total drag given by the drag polar in Eq. (6.1c) is dominated by $C_{D,0}$ at high speeds. Everything else being equal, the lower the $C_{D,0}$, the faster the airplane. The other aerodynamic figure of merit highlighted here is the lift-to-drag ratio and especially its maximum value $(L/D)_{max}$. As we have already seen, L/D is a measure of the aerodynamic efficiency of an airplane, and it affects such flight characteristics as endurance and range. We will use both $C_{D,0}$ and $(L/D)_{\text{max}}$ to illustrate the historical progress in airplane design.

We start with the airplanes of Cayley early in the 19th century because they were the first designs to exemplify the fixed-wing heavier-than-air aircraft we know today. Return again to Fig. 1.8, showing the first airplane with a modern configuration, with a fixed wing for lift, a tail for stability, and a fuselage connecting the two. The mechanism of propulsion (in this case a hand launch) is separate from the mechanism of lift. The amount of technical knowledge Cayley was able to incorporate in his design is best reflected in his famous "triple paper" of 1809–1810 (see Sec. 1.3). The technical concepts of $C_{D,0}$ and L/D did not exist in Cayley's day, but he reflects a basic intuition about these quantities in his triple paper. For example, Cayley used a method called newtonian theory (which will be derived in Chap. 11) to estimate the aerodynamic force on an inclined plane (the wing). This theory takes into account only the pressure acting on the surface; surface shear stress (and hence friction drag) was not fully appreciated in Cayley's time; furthermore there were no methods for its prediction. The newtonian theory predicts a net force perpendicular to the inclined plane and therefore contains a component of drag. Cayley makes reference to this "retarding force" due to the component of the aerodynamic pressure force acting along the flow direction. In modern terms, we call this component of drag the *drag due to lift*. Cayley goes on to say (in discussing the flight of birds), "In addition to the retarding force thus received is the direct resistance, which the bulk of the bird opposes to the current. This is a matter to be entered into separately from the principle now under consideration." Here, Cayley is discussing what we would today call the *zero-lift drag* (sum of pressure drag due to separation and skin friction drag) due primarily to the body of the bird. Although Cayley was on the right track conceptually, he had no method of calculating the zero-lift drag, and measurements (made with a whirling arm such as sketched in Fig. 1.7) were wholly unreliable. Therefore, we have no value of $C_{D,0}$ for Cayley's 1804 glider in Fig. 1.8.

Although Cayley did not identify and use the concept of L/D directly, in his triple paper he refers to his glider sailing "majestically" from the top of a hill, descending at an angle of about 18° with the horizon. Using the results of Sec. 6.9 dealing with a power-off glide, we can today quickly calculate that the L/D ratio for the glider was 3.08, not a very impressive value. Typical values of L/D for modern airplanes are 15 to 20 and for modern gliders greater than 40. Cayley did not have an efficient airplane, nor did he know about aspect-ratio effects. Today we know that low–aspect ratio wings such as used by Cayley (aspect ratio about 1) are very inefficient because they produce large amounts of induced drag.

The technical evolution of airplane design after Cayley was gradual and evolutionary during the remainder of the 19th century. The change that occurred with the *Wright Flyer* (Figs. 1.1 and 1.2) was revolutionary (1) because the Wrights ultimately relied on virtually no previous data, doing everything themselves (see Sec. 1.8) and (2) because it was the first successful flying machine. The aerodynamic quality of the *Wright Flyer* is discussed by Culick and Jex, who report modern calculations and measurements of the drag polar for the *Wright Flyer* (Fig. 6.67). The experimental data were obtained on a model of the *Wright Flyer* mounted in a wind tunnel at the

Figure 6.67   Drag polar and lift curve for the 1903 *Wright Flyer*. Experimental data are from modern experiments using models of the *Wright Flyer* in modern wind tunnels. The vortex-lattice theory is a modern computer calculation. The values of $\hat{C}_L$, $\hat{C}_D$, and $\hat{\alpha}$ correspond to equilibrium trimmed-flight conditions (see Chap. 7), highlighted by the horizontal bar across the figure. *(From Culick and Jex.)*

California Institute of Technology. The theoretical data are supplied by a modern vortex-lattice computer program for calculating low-speed incompressible inviscid flow. (Since these methods do not include the effects of friction, they cannot be used to predict flow separation.) The data in Fig. 6.67 show that $C_{D,0}$ is about 0.10 and the maximum lift coefficient nearly 1.1. Moreover, drawing a straight line from the origin tangent to the drag polar curve, we see that the value of $(L/D)_{max}$ is about 5.7. By present standards, the *Wright Flyer* was not an aerodynamic masterpiece, but in 1903 it was the only successful flying machine in existence. Moreover, compared with Cayley's airplanes, the *Wright Flyer* was a *revolutionary* advancement in design.

After the *Wright Flyer*, advances in airplane design grew almost exponentially in the last half of the 20th century. Using our two figures of merit, $C_{D,0}$ and $(L/D)_{max}$, we can identify three general periods of progress in airplane design during the 20th century, as shown in Figs. 6.68 and 6.69. Values of $C_{D,0}$ (Fig. 6.68) and $(L/D)_{max}$ (Fig. 6.69) for representative airplanes are shown versus time in years. These data are obtained from Loftin, an authoritative publication which the interested reader is encouraged to examine; it contains detailed case studies of the technical designs of many famous aircraft. The data for $C_{D,0}$ in Fig. 6.68 suggest that airplane design has gone through three major evolutionary periods, distinguished from one another by a dramatic change. For example, the period of strut-and-wire biplanes (such as the SPAD XIII, shown in Fig. 6.70) extends from the *Wright Flyer* to the middle or end of the 1920s. Here, values of $C_{D,0}$ are typically on the order of 0.04, a high value

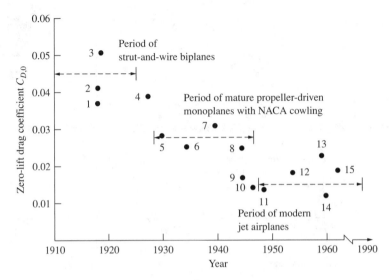

Figure 6.68   Use of zero-lift drag coefficient to illustrate three general periods of 20th-century airplane design. The numbered data points correspond to the following aircraft: (1) SPAD XIII, (2) Fokker D-VII, (3) Curtiss JN-4H Jenny, (4) Ryan NYP (*Spirit of St. Louis*), (5) Lockheed Vega, (6) Douglas DC-3, (7) Boeing B-17, (8) Boeing B-29, (9) North American P-51, (10) Lockheed P-80, (11) North American F-86, (12) Lockheed F-104, (13) McDonnell F-4E, (14) Boeing B-52, (15) General Dynamics F-111D.

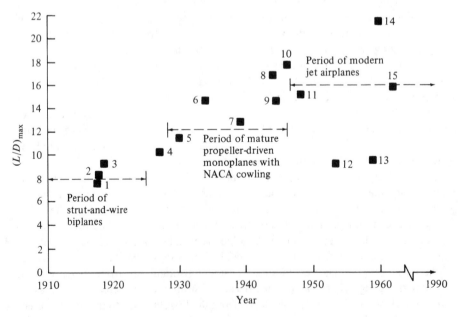

Figure 6.69   Use of lift-to-drag ratio to illustrate three general periods of 20th-century airplane design.

Figure 6.70 The French SPAD XIII, an example of the strut-and-wire biplane period. Captain Eddie Rickenbacker is shown at the front of the airplane. *(U.S. Air Force.)*

due to the large form drag (pressure drag due to flow separation) associated with the bracing struts and wires between the two wings of a biplane. In the late 1920s a revolution in design came with the adoption of the monoplane configuration coupled with the NACA cowl (see Sec. 6.21). The resulting second period of design evolution (exemplified by the DC-3 shown in Fig. 6.71) is characterized by $C_{D,0}$ values on the order of 0.027. In the mid-1940s, the major design revolution was the advent of the jet-propelled airplane. This period, which we are still in today (reflected in the famous F-86 of the Korean war era, shown in Fig. 6.72) is represented by $C_{D,0}$ values on the order of 0.015.

The use of $(L/D)_{max}$ as an aerodynamic figure of merit has been discussed in previous sections. As seen in Fig. 6.69, where $(L/D)_{max}$ is plotted versus years, the data points for the same airplanes as in Fig. 6.68 group themselves in the same three design periods as deduced from Fig. 6.68. Note that compared with the value of 5.7 for the *Wright Flyer*, the average value of $(L/D)_{max}$ for World War I airplanes was about 8—not a great improvement. After the introduction of the monoplane with the NACA cowling, typical $(L/D)_{max}$ values averaged substantially higher, on the order of 12 or sometimes considerably greater. [The Boeing B-29 bomber of World War II fame had an $(L/D)_{max}$ value of nearly 17, the highest for this period. This was in part due to the exceptionally large wing aspect ratio of 11.5 in a period when wing aspect ratios were averaging on the order of 6 to 8.] Today, $(L/D)_{max}$ values for modern

Figure 6.71   The Douglas DC-3, an example of the period of mature propeller-driven monoplanes with the NACA cowling and wing fillets. *(Douglas Aircraft Company.)*

Figure 6.72   The North American F-86, one of the most successful modern jet airplanes from the early 1950s. *(North American/Rockwell.)*

aircraft range over the whole scale, from 12 or 13 for high-performance military jet fighters to nearly 20 and above for large jet bombers and civilian transports such as the Boeing 747.

This section has given you the chance to think about the progress in aircraft design in terms of some of the aerodynamic performance parameters discussed in this chapter.

6.25 SUMMARY

A few of the important aspects of this chapter are listed here:

1. For a complete airplane, the drag polar is given as

$$C_D = C_{D,0} + \frac{C_L^2}{\pi e \text{AR}}$$ [6.1c]

where $C_{D,0}$ is the zero-lift drag coefficient and the term $C_L^2/(\pi e \text{AR})$ includes both induced drag and the contribution of parasite drag due to lift.

2. Thrust required for level, unaccelerated flight is

$$T_R = \frac{W}{L/D}$$ [6.16]

Thrust required is a minimum when L/D is maximum.

3. Power required for level, unaccelerated flight is

$$P_R = \sqrt{\frac{2W^3 C_D^2}{\rho_\infty S C_L^3}}$$ [6.27]

Power required is a minimum when $C_L^{3/2}/C_D$ is a maximum.

4. The rate of climb R/C $= dh/dt$ is given by

$$\frac{dh}{dt} = \frac{TV - DV}{W} - \frac{V}{g}\frac{dV}{dt}$$ [6.139]

where $(TV - DV)/W = P_s$, the specific excess power. For an unaccelerated climb, $dV/dt = 0$, and hence

$$\text{R/C} = \frac{dh}{dt} = \frac{TV - DV}{W}$$ [6.50]

5. In a power-off glide, the glide angle is given by

$$\tan \theta = \frac{1}{L/D}$$ [6.56]

6. The absolute ceiling is defined as that altitude where maximum R/C $= 0$. The service ceiling is that altitude where maximum R/C $= 100$ ft/min.

7. For a propeller-driven airplane, range R and endurance E are given by

$$R = \frac{\eta}{c}\frac{C_L}{C_D}\ln\frac{W_0}{W_1}$$ [6.67]

and
$$E = \frac{\eta}{c} \frac{C_L^{3/2}}{C_D} (2\rho_\infty S)^{1/2} (W_1^{-1/2} - W_0^{-1/2}) \qquad \textbf{[6.68]}$$

Maximum range occurs at maximum C_L/C_D. Maximum endurance occurs at sea level with maximum $C_L^{3/2}/C_D$.

8. For a jet-propelled airplane, range and endurance are given by

$$R = 2\sqrt{\frac{2}{\rho_\infty S}} \frac{1}{c_t} \frac{C_L^{1/2}}{C_D} (W_0^{1/2} - W_1^{1/2}) \qquad \textbf{[6.77]}$$

and
$$E = \frac{1}{c_t} \frac{C_L}{C_D} \ln \frac{W_0}{W_1} \qquad \textbf{[6.72]}$$

9. At *maximum* $C_L^{3/2}/C_D$, $C_{D,0} = \frac{1}{3}C_{D,i}$. For this case,

$$\left(\frac{C_L^{3/2}}{C_D}\right)_{\text{max}} = \frac{(3C_{D,0}\pi e\text{AR})^{3/4}}{4C_{D,0}} \qquad \textbf{[6.87]}$$

At *maximum* C_L/C_D, $C_{D,0} = C_{D,i}$. For this case,

$$\left(\frac{C_L}{C_D}\right)_{\text{max}} = \frac{(C_{D,0}\pi e\text{AR})^{1/2}}{2C_{D,0}} \qquad \textbf{[6.85]}$$

At *maximum* $C_L^{1/2}/C_D$, $C_{D,0} = 3C_{D,i}$. For this case

$$\left(\frac{C_L^{1/2}}{C_D}\right)_{\text{max}} = \frac{(\frac{1}{3}C_{D,0}\pi e\text{AR})^{1/4}}{\frac{4}{3}C_{D,0}} \qquad \textbf{[6.86]}$$

10. Takeoff ground roll is given by

$$s_{\text{LO}} = \frac{1.44W^2}{g\rho_\infty SC_{L,\text{max}}\{T - [D + \mu_r(W - L)]_{\text{av}}\}} \qquad \textbf{[6.103]}$$

11. The landing ground roll is

$$s_L = \frac{1.69W^2}{g\rho_\infty SC_{L,\text{max}}[D + \mu_r(W - L)]_{\text{av}}} \qquad \textbf{[6.111]}$$

12. The load factor is defined as

$$n = \frac{L}{W} \qquad \textbf{[6.115]}$$

13. In *level turning flight*, the turn radius is

$$R = \frac{V_\infty^2}{g\sqrt{n^2 - 1}} \qquad \textbf{[6.118]}$$

and the turn rate is
$$\omega = \frac{g\sqrt{n^2 - 1}}{V_\infty} \qquad \textbf{[6.119]}$$

14. The V-n diagram is illustrated in Fig. 6.55. It is a diagram showing load factor versus velocity for a given airplane, along with the constraints on both n and V due to structural limitations. The V-n diagram illustrates some particularly important aspects of overall airplane performance.

15. The energy height (specific energy) of an airplane is given by

$$H_e = h + \frac{V^2}{2g} \qquad \qquad \textbf{[6.136]}$$

This, in combination with the specific excess power

$$P_s = \frac{TV - DV}{W}$$

leads to the analysis of accelerated-climb performance using energy considerations only.

BIBLIOGRAPHY

Anderson, J. D., Jr.: *Aircraft Performance and Design*, McGraw-Hill, New York, 1999.

Bairstow, L.: *Applied Aerodynamics*, Longmans, London, 1920.

Cowley, W. L., and H. Levy: *Aeronautics in Theory and Experiment*, E. Arnold, London, 1918.

Culick, F. E. C., and H. R. Jex: "Aerodynamics, Stability and Control of the 1903 Wright Flyer," pp. 19–43 in Howard Wolko (ed.), *The Wright Flyer; An Engineering Perspective*, Smithsonian Press, Washington, 1987.

Dommasch, D. O., S. S. Sherbey, and T. F. Connolly, *Airplane Aerodynamics*, 3d ed., Pitman, New York, 1961.

Duchène, Captain: *The Mechanics of the Airplane: A Study of the Principles of Flight* (transl. by J. H. Ledeboer and T. O'B. Hubbard), Longmans, London, 1917.

Hale, F. J.: *Introduction to Aircraft Performance, Selection and Design*, Wiley, New York, 1984.

Judge, A. W.: *Handbook of Modern Aeronautics*, Appleton, London, 1919.

Loftin, Lawrence: *Quest for Performance: The Evolution of Modern Aircraft*, NASA SP-468, 1985.

McCormick, B. W.: *Aerodynamics, Aeronautics and Flight Mechanics*, Wiley, New York, 1979.

Perkins, C. D., and R. E. Hage: *Airplane Performance, Stability and Control*, Wiley, New York, 1949.

Shevell, R. S.: *Fundamentals of Flight*, Prentice-Hall, Englewood Cliffs, NJ, 1983.

PROBLEMS

6.1 Consider an airplane patterned after the twin-engine Beechcraft Queen Air executive transport. The airplane weight is 38,220 N, wing area is 27.3 m^2, aspect ratio is 7.5, Oswald efficiency factor is 0.9, and zero-lift drag coefficient $C_{D,0} = 0.03$. Calculate the thrust required to fly at a velocity of 350 km/h at (*a*) standard sea level and (*b*) an altitude of 4.5 km.

6.2 An airplane weighing 5000 lb is flying at standard sea level with a velocity of 200 mi/h. At this velocity, the L/D ratio is a maximum. The wing area and aspect ratio are 200 ft^2 and 8.5, respectively. The Oswald efficiency factor is 0.93. Calculate the total drag on the airplane.

6.3 Consider an airplane patterned after the Fairchild Republic A-10, a twin-jet attack aircraft. The airplane has the following characteristics: wing area $= 47$ m^2, aspect ratio $= 6.5$, Oswald efficiency factor $= 0.87$, weight $= 103,047$ N, and zero-lift drag coefficient $= 0.032$. The airplane is equipped with two jet engines with 40,298 N of static thrust *each* at sea level.

a. Calculate and plot the power-required curve at sea level.
b. Calculate the maximum velocity at sea level.
c. Calculate and plot the power-required curve at 5-km altitude.
d. Calculate the maximum velocity at 5-km altitude. (Assume the engine thrust varies directly with free-stream density.)

6.4 Consider an airplane patterned after the Beechcraft Bonanza V-tailed, single-engine light private airplane. The characteristics of the airplane are as follows: aspect ratio $= 6.2$, wing area $= 181$ ft^2, Oswald efficiency factor $= 0.91$, weight $= 3000$ lb, and zero-lift drag coefficient $= 0.027$. The airplane is powered by a single piston engine of 345 hp maximum at sea level. Assume the power of the engine is proportional to free-stream density. The two-blade propeller has an efficiency of 0.83.

a. Calculate the power required at sea level.
b. Calculate the maximum velocity at sea level.
c. Calculate the power required at 12,000-ft altitude.
d. Calculate the maximum velocity at 12,000-ft altitude.

6.5 From the information generated in Prob. 6.3, calculate the maximum rate of climb for the twin-jet aircraft at sea level and at an altitude of 5 km.

6.6 From the information generated in Prob. 6.4, calculate the maximum rate of climb for the single-engine light plane at sea level and at 12,000-ft altitude.

6.7 From the rate-of-climb information for the twin-jet aircraft in Prob. 6.5, estimate the absolute ceiling of the airplane. (*Note:* Assume maximum R/C varies linearly with altitude—not a precise assumption, but not bad either.)

6.8 From the rate-of-climb information for the single-engine light plane in Prob. 6.6, estimate the absolute ceiling of the airplane. (Again, make the linear assumption described in Prob. 6.7.)

6.9 The maximum lift-to-drag ratio of the World War I *Sopwith Camel* was 7.7. If the aircraft is in flight at 5000 ft when the engine fails, how far can it glide in terms of distance measured along the ground?

6.10 For the *Sopwith Camel* in Prob. 6.9, calculate the equilibrium glide velocity at 3000 ft, corresponding to the minimum glide angle. The aspect ratio of the airplane is 4.11, the Oswald efficiency factor is 0.7, the weight is 1400 lb, and the wing area is 231 ft^2.

6.11 Consider an airplane with a zero-lift drag coefficient of 0.025, an aspect ratio of 6.72, and an Oswald efficiency factor of 0.9. Calculate the value of $(L/D)_{\mathrm{max}}$.

6.12 Consider the single-engine light plane described in Prob. 6.4. If the specific fuel consumption is 0.42 lb of fuel per horsepower per hour, the fuel capacity is 44 gal, and the maximum gross weight is 3400 lb, calculate the range and endurance at standard sea level.

6.13 Consider the twin-jet airplane described in Prob. 6.3. The thrust specific fuel consumption is 1.0 N of fuel per newton of thrust per hour, the fuel capacity is 1900 gal, and the maximum gross weight is 136,960 N. Calculate the range and endurance at a standard altitude of 8 km.

6.14 Derive Eqs. (6.80) and (6.81).

6.15 Derive Eqs. (6.86) and (6.87).

6.16 Estimate the sea-level lift off distance for the airplane in Prob. 6.3. Assume a paved runway. Also, during the ground roll, the angle of attack is restricted by the requirement that the tail not drag the ground. Hence, assume $C_{L,\max}$ during the ground roll is limited to 0.8. Also, when the airplane is on the ground, the wings are 5 ft above the ground.

6.17 Estimate the sea-level liftoff distance for the airplane in Prob. 6.4. Assume a paved runway, and $C_{L,\max} = 1.1$ during the ground roll. When the airplane is on the ground, the wings are 4 ft above the ground.

6.18 Estimate the sea-level landing ground roll distance for the airplane in Prob. 6.3. Assume the airplane is landing at full gross weight. The maximum lift coefficient with flaps fully employed at touchdown is 2.8. After touchdown, assume zero lift.

6.19 Estimate the sea-level landing ground roll distance for the airplane in Prob. 6.4. Assume the airplane is landing with a weight of 2900 lb. The maximum lift coefficient with flaps at touchdown is 1.8. After touchdown, assume zero lift.

6.20 For the airplane in Prob. 6.3, the sea-level corner velocity is 250 mi/h, and the maximum lift coefficient with no flap deflection is 1.2. Calculate the minimum turn radius and maximum turn rate at sea level.

6.21 The airplane in Prob. 6.3 is flying at 15,000 ft with a velocity of 375 mi/h. Calculate its specific energy at this condition.

6.22 Derive Eq. (6.44).

6.23 From the data shown in Fig. 6.2, estimate the value of the Oswald efficiency factor for the Lockheed C-141A. The wing aspect ratio of the C-141A is 7.9.

6.24 Since the end of World War II, various claims have appeared in the popoular aviation literature of instances where powerful propeller-driven fighter airplanes from that period have broken the speed of sound in a vertical, power-on dive. The purpose of this problem is to show that such an event is technically not possible. Consider for example the Grumman F6F-3 Hellcat, a typical fighter from World War II. For this airplane, the zero-lift drag coefficient (at low speeds) is 0.0211, the wing planform area is 334 ft^2, and the gross weight is 12,441 lb. It is powered by a Pratt and Whitney R-2800 reciprocating engine which, with supercharging to an altitude of 17,500 ft, produces 1500 horsepower. Consider this airplane in a full-power vertical dive at (a) 30,000 ft, and then (b) 20,000 ft. Prove at these two altitudes that the airplane can not reach Mach 1.
Note: The aerodynamic characteristics of this airplane at Mach 1 have not been measured. So you will have to make some reasonable assumptions. For example,

what is the zero-lift drag coefficient at Mach 1? As an estimate, we can obtain from NACA TR 916 a zero-lift drag coefficient for the North American P-51 Mustang which, when extrapolated to Mach 1 shows an increase of 7.5 over its low-speed value. For the more blunt configuration of the F6F, let us assume $C_{D,0}$ (at $M = 1$) is 10 times larger than $C_{D,0}$ (low speed). Also, at Mach 1, the propeller efficiency would be almost zero (indeed, the propeller might even be producing an net drag rather than any thrust). To be conservative, let us assume the propeller efficiency at Mach 1 to be 0.3.

chapter

7

Principles of Stability and Control

An important problem to aviation is…improvement in the form of the aeroplane leading toward natural inherent stability to such a degree as to relieve largely the attention of the pilot while still retaining sufficient flexibility and control to maintain any desired path, without seriously impairing the efficiency of the design.

From the First Annual Report
of the NACA, 1915

7.1   INTRODUCTION

The scene: A French army drill field at Issy-les-Moulineaux just outside Paris. *The time:* The morning of January 13, 1908. *The character:* Henri Farman, a bearded, English-born but French-speaking aviator, who had flown for his first time just four months earlier. *The action:* A delicately constructed Voisin-Farman I-bis biplane (see Fig. 7.1) is poised, ready for the takeoff in the brisk Parisian wind, with Farman seated squarely in front of the 50-hp Antoinette engine. The winds ripple the fabric on the Voisin's box kite–shaped tail as Farman powers to a bumpy liftoff. Fighting against a head wind, he manipulates his aircraft to a marker 1000 m from his takeoff point. In a struggling circular turn, Farman deflects the rudder and mushes the biplane around the marker, the wings remaining essentially level to the ground. Continuing in its rather wide and tenuous circular arc, the airplane heads back. Finally, Farman lands at his original takeoff point, amid cheers from the crowd that had gathered for the occasion. Farman has been in the air for 1 min 28 s—the longest flight in Europe to that date—and has just performed the first circular flight of 1-km extent. For this, he is awarded the Grand Prix d'Aviation. (Coincidentally, in the crowd is a young Hungarian engineer, Theodore von Karman, who is present only due to the insistence of his female companion—waking at 5 AM in order to see history made. However,

467

Figure 7.1   The Voisin-Farman I-bis plane. *(National Air and Space Museum.)*

von Karman is mesmerized by the flight, and his interest in aeronautical science is catalyzed. Von Karman will go on to become a leading aerodynamic genius of the first half century of powered flight.)

The scene shifts to a small racetrack near Le Mans, France. *The time:* Just seven months later, August 8, 1908. *The character:* Wilbur Wright, intense, reserved, and fully confident. *The action:* A new Wright type A biplane (see Fig. 1.25), shipped to France in crates and assembled in a friend's factory near Le Mans, is ready for flight. A crowd is present, enticed to the field by much advance publicity and an intense curiosity to see if the "rumors" about the Wright brothers' reported success were really true. Wilbur takes off. Using the Wrights' patented concept of twisting the wing tips (*wing warping*), Wilbur is able to bank and turn at will. He makes two graceful circles and then effortlessly lands after 1 min 45 s of flight. The crowds cheer. The French press is almost speechless but then heralds the flight as epoch-making. European aviators who witness this demonstration gaze in amazement and then quickly admit that the Wrights' airplane is far advanced over the best European machines of that day. Wilbur goes on to make 104 flights in France before the end of the year and in the process transforms the direction of aviation in Europe.

The distinction between the two scenes above, and the reason for Wilbur's mastery of the air in comparison to Farman's struggling circular flight, involves stability and control. The Voisin aircraft of Farman, which represented the best European state of the art, had only rudder control and could make only a laborious, flat turn by simply swinging the tail around. In contrast, the Wright airplane's wing-twisting mechanism provided control of roll, which when combined with rudder control, allowed effortless turning and banking flight, figure-eights, etc. Indeed, the Wright brothers were *airmen* (see Chap. 1) who concentrated on designing total control into their aircraft before adding an engine for powered flight. Since those early days, airplane stability and control has been a dominant aspect of airplane design. This is the subject of this chapter.

Airplane performance, as discussed in Chap. 6, is governed by forces (along, and perpendicular to, the flight path), with the translational motion of the airplane as a response to these forces. In contrast, airplane stability and control, discussed in this chapter, are governed by moments about the center of gravity, with the rotational motion of the airplane as a response to these moments. Therefore, moments and rotational motion are the main focus of this chapter.

Consider an airplane in flight, as sketched in Fig. 7.2. The center of gravity (the point through which the weight of the complete airplane effectively acts) is denoted as cg. The xyz orthogonal axis system is fixed relative to the airplane; the x axis is along the fuselage, the y axis is along the wingspan perpendicular to the x axis, and the z axis is directed downward, perpendicular to the xy plane. The origin is at the center of gravity. The translational motion of the airplane is given by the velocity components U, V, and W along the x, y, and z directions, respectively. (Note that the resultant free-stream velocity V_∞ is the vector sum of U, V, and W.) The rotational

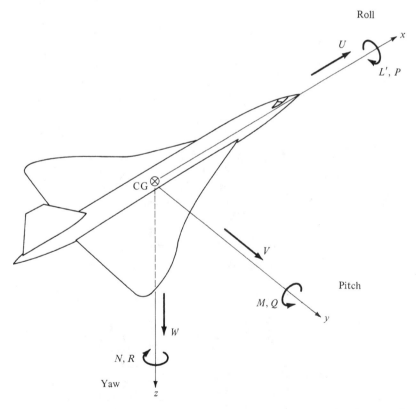

Figure 7.2 Definition of the airplane's axes along with the translational and rotational motion along and about these axes.

motion is given by the angular velocity components P, Q, and R about the x, y, z axes, respectively. These rotational velocities are due to the moments L', M, and N about the x, y, and z axes, respectively. (The prime is put over the symbol L so that the reader will not confuse it with lift.) Rotational motion about the x axis is called *roll*; L' and P are the *rolling* moment and velocity, respectively. Rotational motion about the y axis is called *pitch*; M and Q are the *pitching* moment and velocity, respectively. Rotational motion about the z axis is called *yaw*; N and R are the *yawing* moment and velocity, respectively.

There are three basic controls on an airplane—the ailerons, elevator, and rudder—which are designed to change and control the moments about the x, y, and z axes. These control surfaces are shown in Fig. 2.12 and repeated in Fig. 7.3; they are flaplike surfaces that can be deflected back and forth at the command of the pilot. The ailerons are mounted at the trailing edge of the wing, near the wing tips. The elevators are located on the horizontal stabilizer. In some modern aircraft, the complete horizontal stabilizer is rotated instead of just the elevator (so-called flying tails). The rudder is

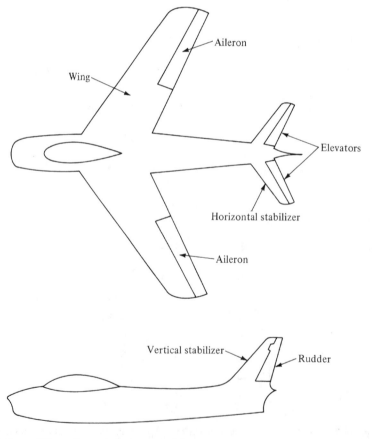

Figure 7.3   Some airplane nomenclature.

located on the vertical stabilizer, at the trailing edge. Just as in the case of wing flaps discussed in Sec. 5.17, a downward deflection of the control surface will increase the lift of the wing or tail. In turn, the moments will be changed, as sketched in Fig. 7.4. Consider Fig. 7.4a. One aileron is deflected up and the other down, creating a differential lifting force on the wings, thus contributing to the rolling moment L'. In Fig. 7.4b, the elevator is deflected upward, creating a negative lift at the tail, thus contributing to the pitching moment M. In Fig. 7.4c, the rudder is deflected to the

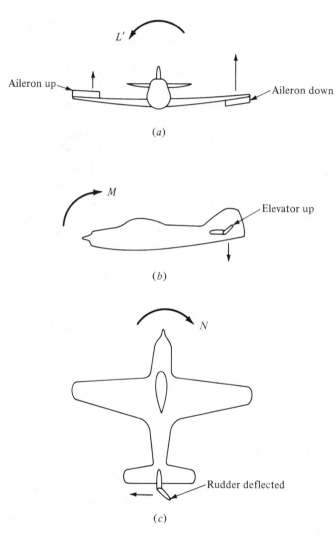

Aileron up

Aileron down

L'

(a)

M

Elevator up

(b)

N

Rudder deflected

(c)

Figure 7.4   Effect of control deflections on roll, pitch, and yaw. (a) Effect of aileron deflection; lateral control. (b) Effect of elevator deflection; longitudinal control. (c) Effect of rudder deflection; directional control.

right, creating a leftward aerodynamic force on the tail, thus contributing to the yawing moment N.

Rolling (about the x axis) is also called *lateral motion*. Referring to Fig. 7.4*a*, we see that ailerons control roll; hence they are known as *lateral controls*. Pitching (about the y axis) is also called *longitudinal motion*. In Fig. 7.4*b*, we see that elevators control pitch; hence they are known as *longitudinal controls*. Yawing (about the z axis) is also called *directional motion*. Figure 7.4*c* shows that the rudder controls yaw; hence it is known as the *directional control*.

All the above definitions and concepts are part of the basic language of airplane stability and control; they should be studied carefully. Also, in the process, the following question emerges: What is meant by the words *stability and control* themselves? This question is answered in the next section.

Return to the general road map for this book, shown in Fig. 2.1. With this chapter we are still dealing with the overall subject of flight mechanics, but now we are concentrating on the second subbox under flight mechanics, namely, stability and control. The road map for the present chapter is shown in Fig. 7.5. There are two general routes shown, that for stability in the left column and that for control in the right column. Both the subjects of stability and control can be subdivided into categories labeled *static* and *dynamic*, as shown in Fig. 7.5. We define the difference between these categories in the next section. In this chapter we concentrate primarily (though not exclusively) on longitudinal stability and control. We deal with such considerations of static longitudinal stability as the calculation of longitudinal moments about the center of gravity, equations that can be used to help us determine whether an airplane is stable or not; and we define two concepts used to describe the stability characteristics, namely, the neutral point and the static margin. For the latter part of this chapter, we run down the right side of the road map in Fig. 7.5, dealing primarily with static longitudinal control. Here we examine the concept of *trim* in greater detail, and we look at elevator deflections necessary to trim and the associated hinge moments for the elevator. We also look at the differences between *stick-fixed*

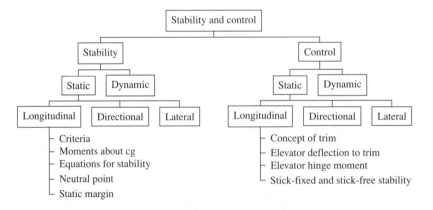

Figure 7.5   Road map for Chap. 7.

and *stick-free* stability. Many of the terms used above may seem unfamiliar and somewhat strange. However, we spend the rest of this chapter helping you to learn these concepts and making you more familiar with the language of airplane stability and control. It will be useful for you to frequently return to Fig. 7.5 as we proceed through this chapter, to help orient yourself about the details and where they fit into the bigger picture.

7.2 DEFINITION OF STABILITY AND CONTROL

There are two types of stability: static and dynamic. They can be visualized as follows.

7.2.1 Static Stability

Consider a marble on a curved surface, such as a bowl. Imagine that the bowl is upright and the marble is resting inside, as shown in Fig. 7.6a. The marble is stationary; it is in a state of *equilibrium*, which means that the moments acting on the marble are

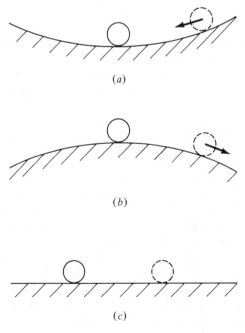

(a)

(b)

(c)

Figure 7.6   Illustration of static stability.
(a) Statically stable system. (b) Statically
unstable system. (c) Statically neutral system.

zero. If the marble is now disturbed (moved to one side, as shown by the dotted circle in Fig. 7.6a) and then released, it will roll back toward the bottom of the bowl, that is, toward its original equilibrium position. Such a system is *statically* stable. In general, we can state that

> If the forces and moments on the body caused by a disturbance tend initially to return the body toward its equilibrium position, the body is statically stable. The body has positive static stability.

Now, imagine the bowl is upside-down, with the marble at the crest, as shown in Fig. 7.6b. If the marble is placed precisely at the crest, the moments will be zero and the marble will be in equilibrium. However, if the marble is now disturbed (as shown by the dotted circle in Fig. 7.6b), it will tend to roll down the side, away from its equilibrium position. Such a system is *statically unstable*. In general, we can state that

> If the forces and moments are such that the body continues to move *away* from its equilibrium position after being disturbed, the body is *statically unstable.* The body has *negative* static stability.

Finally, imagine the marble on a flat horizontal surface, as shown in Fig. 7.6c. Its moments are zero; it is in equilibrium. If the marble is now disturbed to another location, the moments will still be zero, and it will still be in equilibrium. Such a system is *neutrally stable*. This situation is rare in flight vehicles, and we will not be concerned with it here.

We emphasize that static stability (or the lack of it) deals with the *initial* tendency of a vehicle to return to equilibrium (or to diverge from equilibrium) after being disturbed. It says nothing about whether it ever reaches its equilibrium position, or how it gets there. Such matters are the realm of dynamic stability, as follows.

7.2.2 Dynamic Stability

Dynamic stability deals with the *time history* of the vehicle's motion after it initially responds to its static stability. For example, consider an airplane flying at an angle of attack α_e such that its moments about the center of gravity are zero. The airplane is therefore in equilibrium at α_e; in this situation, it is *trimmed*, and α_e is called the *trim angle of attack*. Now assume that the airplane is disturbed (say, by encountering a wind gust) to a new angle of attack α, as shown in Fig. 7.7. The airplane has been pitched through a *displacement* $\alpha - \alpha_e$. Now, let us observe the subsequent pitching motion after the airplane has been disturbed by the gust. We can describe this motion by plotting the instantaneous displacement versus time, as shown in Fig. 7.8. Here $\alpha - \alpha_e$ is given as a function of time t. At $t = 0$, the displacement is equal to that produced by the gust. If the airplane is statically stable, it will *initially* tend to move back toward its equilibrium position, that is, $\alpha - \alpha_e$ will initially decrease. Over a lapse of time, the vehicle may monotonically "home in" to its equilibrium position, as shown in Fig. 7.8a. Such motion is called *aperiodic*. Alternately, it may first

Figure 7.7 Disturbance from the equilibrium angle of attack.

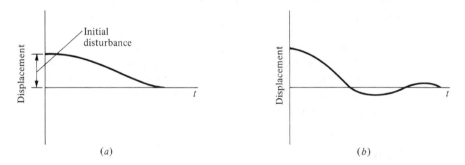

Figure 7.8 Examples of dynamic stability. (a) Aperiodic; (b) Damped oscillations.

overshoot the equilibrium position and approach α_e after a series of oscillations with decreasing amplitude, as shown in Fig. 7.8b. Such motion is described as *damped oscillations*. In both situations, Figs. 7.8a and 7.8b, the airplane eventually returns to its equilibrium position after some interval of time. These two situations are examples of *dynamic stability* in an airplane. Thus, we can state that

A body is dynamically stable if, out of its own accord, it eventually returns to and remains at its equilibrium position over a period of time.

On the other hand, after initially responding to its static stability, the airplane may oscillate with increasing amplitude, as shown in Fig. 7.9. Here, the equilibrium position is never maintained for any period of time, and the airplane eventually diverges completely; the airplane in this case is *dynamically unstable* (even though it is statically stable). Also, it is theoretically possible for the airplane to pitch back and forth with constant-amplitude oscillations. This is an example of a *dynamically neutral* body; such a case is of little practical interest here.

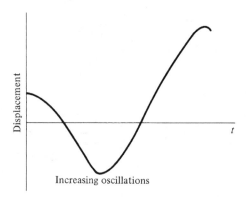

Figure 7.9   An example of dynamic instability.

It is important to observe from the above examples that a dynamically stable airplane must always be statically stable. However, static stability is *not* sufficient to ensure dynamic stability. Nevertheless, static stability is usually the first stability characteristic to be designed into an airplane. (There are some exceptions, to be discussed later.) Such considerations are of paramount importance in conventional airplanes, and therefore most of this chapter will deal with *static* stability and control. A study of dynamic stability, although of great importance, requires rather advanced analytical techniques beyond the scope of this book.

7.2.3   Control

The conventional control surfaces (elevators, ailerons, and rudder) on an airplane were discussed in Sec. 7.1 and sketched in Figs. 7.3 and 7.4. Their function is usually (1) to change the airplane from one equilibrium position to another and (2) to produce nonequilibrium accelerated motions such as maneuvers. The study of the *deflections* of the ailerons, elevators, and rudder necessary to make the airplane do what we want and of the amount of *force* that must be exerted by the pilot (or the hydraulic boost system) to deflect these controls is part of a discipline called *airplane control*, to be discussed later in this chapter.

7.2.4   Partial Derivative

Some physical definitions associated with stability and control have been given above. In addition, a mathematical definition, namely, that of the partial derivative, will be useful in the equations developed later, not only in this chapter but in our discussion of astronautics (Chap. 8) as well. For those readers having only a nodding acquaintance with calculus, hopefully this section will be self- explanatory; for those with a deeper calculus background, this should serve as a brief review.

Consider a function, say, $f(x)$, of a single variable x. The derivative of $f(x)$ is defined from elementary calculus as

$$\frac{df}{dx} \equiv \lim_{\Delta x \to 0} \left[\frac{f(x + \Delta x) - f(x)}{\Delta x} \right]$$

Physically, this limit represents the instantaneous rate of change of $f(x)$ with respect to x.

Now consider a function which depends on more than one variable, for example, the function $g(x, y, z)$ which depends on the three independent variables x, y, and z. Let x vary while y and z are held constant. Then the instantaneous rate of change of g with respect to x is given by

$$\frac{\partial g}{\partial x} \equiv \lim_{\Delta x \to 0} \left[\frac{g(x + \Delta x, y, z) - g(x, y, z)}{\Delta x} \right]$$

Here $\partial g/\partial x$ is the *partial derivative* of g with respect to x. Now let y vary while holding x and z constant. Then the instantaneous rate of change of g with respect to y is given by

$$\frac{\partial g}{\partial y} \equiv \lim_{\Delta y \to 0} \left[\frac{g(x, y + \Delta y, z) - g(x, y, z)}{\Delta y} \right]$$

Here, $\partial g/\partial y$ is the *partial derivative* of g with respect to y. An analogous definition holds for the partial derivative with respect to z, denoted by $\partial g/\partial z$.

In this book, we use the concept of the partial derivative as a definition only. The calculus of partial derivatives is essential to the advanced study of virtually any field of engineering, but such considerations are beyond the scope of this book.

If $g = x^2 + y^2 + z^2$, calculate $\partial g/\partial z$. | **Example 7.1**

Solution
From the definition given above, the partial derivative is taken with respect to z, holding x and y constant.

$$\frac{\partial g}{\partial z} = \frac{\partial(x^2 + y^2 + z^2)}{\partial z} = \frac{\partial x^2}{\partial z} + \frac{\partial y^2}{\partial z} + \frac{\partial z^2}{\partial z} = 0 + 0 + 2z = 2z$$

7.3 MOMENTS ON THE AIRPLANE

A study of stability and control is focused on moments: moments on the airplane and moments on the control surfaces. At this stage, it would be well for the reader to review the discussion of aerodynamically produced moments in Sec. 5.2. Recall that the pressure and shear stress distributions over a wing produce a pitching moment. This moment can be taken about any arbitrary point (the leading edge, the trailing edge, the quarter chord, etc.). However, there exists a particular point about which the moments are independent of the angle of attack. This point is defined as the *aerodynamic center* for the wing. The moment and its coefficient about the aerodynamic center are denoted by M_{ac} and $C_{M,ac}$, respectively, where $C_{M,ac} \equiv M_{ac}/(q_\infty S c)$.

Reflecting again on Sec. 5.2, consider the force diagram of Fig. 5.5. Assume the wing is flying at zero lift; hence F_1 and F_2 are equal and opposite forces. Thus, the moment established by these forces is a pure couple, which we know from elementary physics can be translated anywhere on the body at constant value. Therefore, at *zero lift*, $M_{ac} = M_{c/4} = M_{\text{any point}}$. In turn,

$$C_{M,ac} = (C_{M,c/4})_{L=0} = (C_{M,\text{any point}})_{L=0}$$

This says that the value of $C_{M,ac}$ (which is constant for angles of attack) can be obtained from the value of the moment coefficient about any point when the wing is at the zero-lift angle of attack $\alpha_{L=0}$. For this reason, M_{ac} is sometimes called the *zero-lift moment*.

The aerodynamic center is a useful concept for the study of stability and control. In fact, the force and moment system on a wing can be completely specified by the lift and drag acting through the aerodynamic center, plus the moment about the aerodynamic center, as sketched in Fig. 7.10. We adopt this convention for the remainder of this chapter.

Now consider the complete airplane, as sketched in Fig. 7.10. Here we are most concerned with the pitching moment about the center of gravity of the airplane M_{cg}. Clearly, by examination of Fig. 7.10, M_{cg} is created by (1) L, D, and M_{ac} of the wing; (2) lift of the tail; (3) thrust; and (4) aerodynamic forces and moments on other parts of the airplane, such as the fuselage and engine nacelles. (Note that weight does not contribute, since it acts through the center of gravity.) These contributions to M_{cg} will be treated in detail later. The purpose of Fig. 7.10 is simply to illustrate the important conclusion that a moment does exist about the center of gravity of an airplane, and it is this moment which is fundamental to the stability and control of the airplane.

The moment coefficient about the center of gravity is defined as

$$C_{M,cg} = \frac{M_{cg}}{q_\infty S c} \qquad \textbf{[7.1]}$$

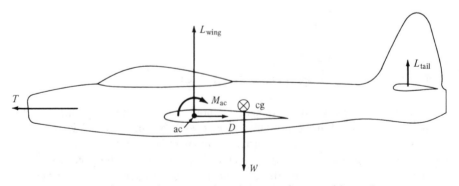

Figure 7.10   Contributions to the moment about the center of gravity of the airplane.

Combining the above concept with the discussion of Sec. 7.2, we find an airplane is in equilibrium (in pitch) when the moment about the center of gravity is zero; that is, when $M_{cg} = C_{M,cg} = 0$, the airplane is said to be *trimmed*.

7.4 ABSOLUTE ANGLE OF ATTACK

Continuing with our collection of tools to analyze stability and control, we consider a wing at an angle of attack such that lift is zero; that is, the wing is at the zero-lift angle of attack $\alpha_{L=0}$, as shown in Fig. 7.11a. With the wing in this orientation, draw a line through the trailing edge parallel to the relative wind V_∞. This line is defined as the *zero-lift line* for the airfoil. It is a fixed line; visualize it frozen into the geometry of the airfoil, as sketched in Fig. 7.11a. As discussed in Chap. 5, conventional cambered airfoils have slightly negative zero-lift angles; therefore the zero-lift line lies slightly above the chord line, as shown (with overemphasis) in Fig. 7.11a.

Now consider the wing pitched to the geometric angle of attack α such that lift is generated, as shown in Fig. 7.11b. (Recall from Chap. 5 that the geometric angle of attack is the angle between the free-stream relative wind and the chord line.) In the same configuration, Fig. 7.11b demonstrates that the angle between the zero-lift line and the relative wind is equal to the sum of α and the absolute value of $\alpha_{L=0}$. This angle is defined as the *absolute angle of attack* α_a. From Fig. 7.11b, $\alpha_a = \alpha + \alpha_{L=0}$ (using $\alpha_{L=0}$ in an absolute sense). The geometry of Fig. 7.11a and 7.11b should be studied carefully until the concept of α_a is clearly understood.

The definition of the absolute angle of attack has a major advantage. When $\alpha_a = 0$, then $L = 0$, no matter what the camber of the airfoil. To further illustrate, consider the lift curves sketched in Fig. 7.12. The conventional plot (as discussed

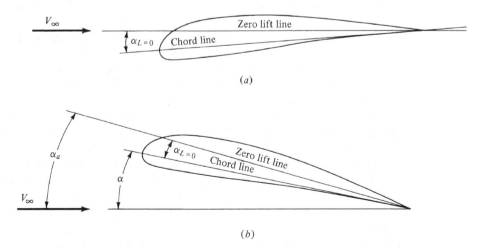

(a)

(b)

Figure 7.11   Illustration of the zero-lift line and absolute angle of attack. (a) No lift; (b) with lift.

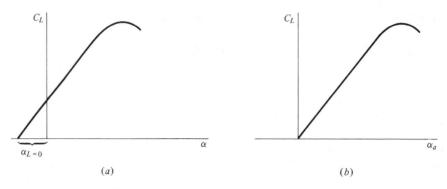

Figure 7.12   Lift coefficient versus (*a*) geometric angle of attack and (*b*) absolute angle of attack.

in detail in Chap. 5), C_L versus α, is shown in Fig. 7.12*a*. Here the lift curve does not go through the origin, and of course $\alpha_{L=0}$ is different for different airfoils. In contrast, when C_L is plotted versus α_a, as sketched in Fig. 7.12*b*, the curve always goes through the origin (by definition of α_a). The curve in Fig. 7.12*b* is identical to that in Fig. 7.12*a*; only the abscissa has been translated by the value $\alpha_{L=0}$.

The use of α_a in lieu of α is common in studies of stability and control. We adopt this convention for the remainder of this chapter.

7.5   CRITERIA FOR LONGITUDINAL STATIC STABILITY

Static stability and control about all three axes shown in Fig. 7.2 are usually a necessity in the design of conventional airplanes. However, a complete description of all three types—lateral, longitudinal, and directional static stability and control (see Fig. 7.4)—is beyond the scope of this book. Rather, the intent here is to provide only the flavor of stability and control concepts, and to this end only the airplane's longitudinal motion (pitching motion about the y axis) is considered in detail. This pitching motion is illustrated in Fig. 7.4*b*. It takes place in the plane of symmetry of the airplane. Longitudinal stability is also the most important static stability mode; in airplane design, wind tunnel testing, and flight research, it usually earns more attention than lateral or directional stability.

Consider a rigid airplane with fixed controls, for example, the elevator in some fixed position. Assume the airplane has been tested in a wind tunnel or free flight and that its variation of M_{cg} with angle of attack has been measured. This variation is illustrated in Fig. 7.13, where $C_{M,cg}$ is sketched versus α_a. For many conventional airplanes, the curve is nearly linear, as shown in Fig. 7.13. The value of $C_{M,cg}$ at zero lift (where $\alpha_a = 0$) is denoted by $C_{M,0}$. The value of α_a where $M_{cg} = 0$ is denoted by α_e; as stated in Sec. 7.3, this is the equilibrium, or trim, angle of attack.

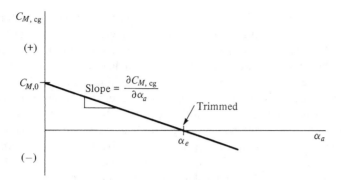

Figure 7.13 Moment coefficient curve with a negative slope.

Consider the airplane in steady, equilibrium flight at its trim angle of attack α_e, as shown in Fig. 7.14a. Suddenly, the airplane is disturbed by hitting a wind gust, and the angle of attack is momentarily changed. There are two possibilities: an increase or a decrease in α_a. If the airplane is pitched upward, as shown in Fig. 7.14b, then $\alpha_a > \alpha_e$. From Fig. 7.13, if $\alpha_a > \alpha_e$, the moment about the center

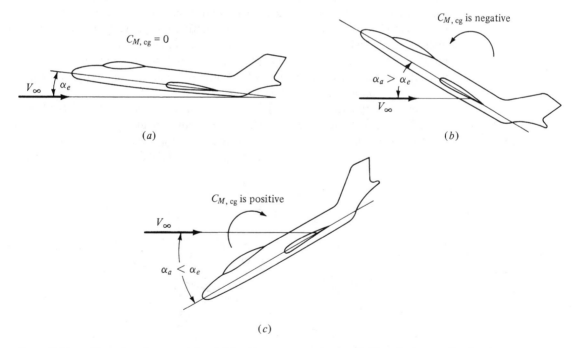

Figure 7.14 Illustration of static stability: (a) Equilibrium position (trimmed). (b) Pitched upward by disturbance. (c) Pitched downward by disturbance. In both (b) and (c), the airplane has the initial tendency to return to its equilibrium position.

of gravity is negative. As discussed in Sec. 5.4, a negative moment (by convention) is counterclockwise, tending to pitch the nose downward. Hence, in Fig. 7.14b, the airplane will initially tend to move back toward its equilibrium position after being disturbed. On the other hand, if the plane is pitched downward by the gust, as shown in Fig. 7.14c, then $\alpha_a < \alpha_e$. From Fig. 7.13, the resulting moment about the center of gravity will be positive (clockwise) and will tend to pitch the nose upward. Thus, again we have the situation in which the airplane will initially tend to move back toward its equilibrium position after being disturbed. From Sec. 7.2, this is precisely the definition of static stability. Therefore, we conclude that an airplane which has a $C_{M,\text{cg}}$-versus-α_a variation like that shown in Fig. 7.13 is *statically stable*. Note from Fig. 7.13 that $C_{M,0}$ is positive and that the slope of the curve $\partial C_{M,\text{cg}}/\partial \alpha_a$ is negative. Here, the partial derivative, defined in Sec. 7.2d, is used for the slope of the moment coefficient curve. This is because (as we shall see) $C_{M,\text{cg}}$ depends on a number of other variables in addition to α_a, and therefore it is mathematically proper to use $\partial C_{M,\text{cg}}/\partial \alpha_a$ rather than $dC_{M,\text{cg}}/d\alpha_a$ to represent the slope of the line in Fig. 7.13. As defined in Sec. 7.2d, $\partial C_{M,\text{cg}}/\partial \alpha_a$ symbolizes the instantaneous rate of change of $C_{M,\text{cg}}$ with respect to α_a, with all other variables held constant.

Consider now a different airplane, with a measured $C_{M,\text{cg}}$ variation as shown in Fig. 7.15. Imagine the airplane is flying at its trim angle of attack α_e as shown in Fig. 7.16a. If it is disturbed by a gust, pitching the nose upward, as shown in Fig. 7.16b, then $\alpha_a > \alpha_e$. From Fig. 7.15 this results in a positive (clockwise) moment, which tends to pitch the nose even farther away from its equilibrium position. Similarly, if the gust pitches the nose downward (Fig. 7.16c), a negative (counterclockwise) moment results, which also tends to pitch the nose farther away from its equilibrium position. Therefore, because the airplane always tends to diverge from equilibrium when disturbed, it is *statically unstable*. Note from Fig. 7.15 that $C_{M,0}$ is negative and $\partial C_{M,\text{cg}}/\partial \alpha_a$ is positive for this airplane.

For both airplanes, Figs. 7.13 and 7.15 show a positive value of α_e. Recall from Fig. 6.8 that an airplane moves through a range of angle of attack as it flies through its velocity range from V_{stall} (where α_a is the largest) to V_{max} (where α_a is the smallest). The value of α_e must fall within this flight range of angle of attack, or else the airplane

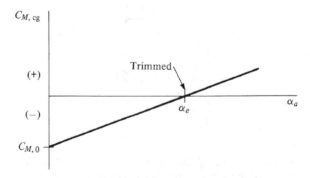

Figure 7.15   Moment coefficient curve with a positive slope.

$C_{M,\text{cg}}$ is positive

$C_{M,\text{cg}} = 0$

$\alpha_a > \alpha_e$

α_e

V_∞

(a)

(b)

$C_{M,\text{cg}}$ is negative

$\alpha_a < \alpha_e$

(c)

Figure 7.16 Illustration of static instability. (a) Equilibrium position (trimmed). (b) Pitched upward by disturbance. (c) Pitched downward by disturbance. In both (b) and (c), the airplane has the initial tendency to diverge farther away from its equilibrium position.

cannot be trimmed for steady flight. (Remember that we are assuming a fixed elevator position: We are discussing stick-fixed stability.) When α_e does fall within this range, the airplane is *longitudinally balanced*.

From the above considerations, we conclude the following.

The necessary criteria for longitudinal balance and static stability are:

1. $C_{M,0}$ must be positive.
2. $\partial C_{M,\text{cg}}/\partial \alpha_a$ must be negative.

That is, the $C_{M,\text{cg}}$ curve must look like Fig. 7.13.

Of course, implicit in the above criteria is that α_e must also fall within the flight range of angle of attack for the airplane.

We are now in a position to explain why a conventional airplane has a horizontal tail (the horizontal stabilizer shown in Fig. 7.3). First, consider an ordinary wing (by itself) with a conventional airfoil, say an NACA 2412 section. Note from the airfoil data in App. D that the moment coefficient about aerodynamic center is negative. This is characteristic of all airfoils with positive camber. Now assume that the wing is at zero lift. In this case, the only moment on the wing is a pure couple, as explained in Sec. 7.3; hence, at zero lift the moment about one point is equal to the moment about any other point. In particular,

$$C_{M,ac} = C_{M,cg} \quad \text{for zero lift (wing only)} \quad \textbf{[7.2]}$$

On the other hand, examination of Fig. 7.13 shows that $C_{M,0}$ is, by definition, the moment coefficient about the center of gravity at zero lift (when $\alpha_a = 0$). Hence, from Eq. (7.2)

$$C_{M,0} = C_{M,ac} \quad \text{wing only} \quad \textbf{[7.3]}$$

Equation (7.3) demonstrates that for a wing with positive camber ($C_{M,ac}$ negative), $C_{M,0}$ is also negative. Hence, such a wing by itself is *unbalanced*. To rectify this situation, a horizontal tail must be added to the airplane, as shown in Fig. 7.17a and 7.17b. If the tail is mounted behind the wing, as shown in Fig. 7.17a, and if it is inclined downward to produce a negative tail lift as shown, then a clockwise moment about the center of gravity will be created. If this clockwise moment is strong enough, then it will overcome the negative $C_{M,ac}$, and $C_{M,0}$ for the wing-tail combination will become positive. The airplane will then be balanced.

The arrangement shown in Fig. 7.17a is characteristic of most conventional airplanes. However, the tail can also be placed ahead of the wing, as shown in Fig. 7.17b; this is called a *canard configuration*. For a canard, the tail is inclined upward to produce a positive lift, hence creating a clockwise moment about the center of gravity. If this moment is strong enough, then $C_{M,0}$ for the wing-tail combination will become

Figure 7.17 (a) Conventional wing-tail combination. The tail is set at such an angle as to produce negative lift, thus providing a positive $C_{M,0}$. (b) Canard wing-tail combination. The tail is set at such an angle as to produce positive lift, thus providing a positive $C_{M,0}$.

positive and again the airplane will be balanced. Unfortunately, the forward-located tail of a canard interferes with the smooth aerodynamic flow over the wing. For this and other reasons, canard configurations have not been popular. Of course, a notable exception were the *Wright Flyers*, which were canards. In fact, it was not until 1910 that the Wright brothers went to a conventional arrangement. Using the word *rudder* to mean elevator, Orville wrote to Wilbur in 1909 that "the difficulty in handling our machine is due to the rudder being in front, which makes it hard to keep on a level course.... I do not think it is necessary to lengthen the machine, but to simply put the rudder behind instead of before." Originally, the Wrights thought the forward-located elevator would help to protect them from the type of fatal crash encountered by Lilienthal. This rationale persisted until the design of their model B in 1910. Finally, a modern example of a canard is the North American XB-70, an experimental supersonic bomber developed for the Air Force in the 1960s. The canard surfaces ahead of the wing are clearly evident in the photograph shown in Fig. 7.18. In recent years, canards have come back on the aeronautical scene, for some high-performance military airplanes and special general aviation designs. The X-29 shown in Fig. 5.60 is a canard.

In retrospect, using essentially qualitative arguments based on physical reasoning and without resort to complicated mathematical formulas, we have developed some fundamental results for longitudinal static stability. Indeed, it is somewhat amazing how far our discussion has progressed on such a qualitative basis. However, we now

Figure 7.18   The North American XB-70. Note the canard surfaces immediately behind the cockpit. *(Rockwell International Corp.)*

turn to some quantitative questions. For a given airplane, how far should the wing and tail be separated in order to obtain stability? How large should the tail be made? How do we design for a desired trim angle α_e? These and other such questions are addressed in the remainder of this chapter.

7.6 QUANTITATIVE DISCUSSION: CONTRIBUTION OF THE WING TO M_{cg}

The calculation of moments about the center of gravity of the airplane M_{cg} is critical to a study of longitudinal static stability. The previous sections have already underscored this fact. Therefore, we now proceed to consider individually the contributions of the wing, fuselage, and tail to moments about the center of gravity of the airplane, in the end combining them to obtain the total M_{cg}.

Consider the forces and moments on the wing only, as shown in Fig. 7.19. Here the zero-lift line is drawn horizontally for convenience; hence, the relative wind is inclined at the angle α_w with respect to the zero-lift line, where α_w is the absolute angle of attack of the wing. Let c denote the mean zero-lift chord of the wing (the chord measured along the zero-lift line). The difference between the zero-lift chord and the geometric chord (as defined in Chap. 5) is usually insignificant and will be ignored here. The center of gravity for the airplane is located a distance hc behind the leading edge, and zc above the zero-lift line, as shown. Hence, h and z are coordinates of the center of gravity in fractions of chord length. The aerodynamic center is a distance $h_{ac_w}c$ from the leading edge. The moment of the wing about the aerodynamic center of the wing is denoted by M_{ac_w}, and the wing lift and drag are L_w and D_w, respectively, as shown. As usual, L_w and D_w are perpendicular and parallel, respectively, to the relative wind.

We wish to take moments about the center of gravity with pitch-up moments positive as usual. Clearly, from Fig. 7.19, L_w, D_w, and M_{ac_w}, all contribute to moments

Figure 7.19   Airfoil nomenclature and geometry.

about the center of gravity. For convenience, split L_w and D_w into components perpendicular and parallel to the chord. Then, referring to Fig. 7.19, we find the moments about the center of gravity of the airplane due to the wing are

$$M_{\mathrm{cg}_w} = M_{\mathrm{ac}_w} + L_w \cos\alpha_w(hc - h_{\mathrm{ac}_w}c) + D_w \sin\alpha_w(hc - h_{\mathrm{ac}_w}c)$$
$$+ L_w \sin\alpha_w zc - D_w \cos\alpha_w zc \qquad \textbf{[7.4]}$$

[Study Eq. (7.4) and Fig. 7.19 carefully, and make certain that you understand each term before progressing further.] For the normal-flight range of a conventional airplane, α_w is small; hence, the approximation is made that $\cos\alpha_w \approx 1$ and $\sin\alpha_w \approx \alpha_w$ (where α_w is in radians). Then Eq. (7.4) becomes

$$M_{\mathrm{cg}_w} = M_{\mathrm{ac}_w} + (L_w + D_w\alpha_w)(h - h_{\mathrm{ac}_w})c + (L_w\alpha_w - D_w)zc \qquad \textbf{[7.5]}$$

Dividing Eq. (7.5) by $q_\infty Sc$ and recalling that $C_M = M/(q_\infty Sc)$, we obtain the moment coefficient about the center of gravity as

$$C_{M,\mathrm{cg}_w} = C_{M,\mathrm{ac}_w} + (C_{L,w} + C_{D,w}\alpha_w)(h - h_{\mathrm{ac}_w})$$
$$+ (C_{L,w}\alpha_w - C_{D,w})z \qquad \textbf{[7.6]}$$

For most airplanes, the center of gravity is located close to the zero-lift line; hence z is usually small ($z \approx 0$) and will be neglected. Furthermore, α_w (in radians) is usually much less than unity, and $C_{D,w}$ is usually less than $C_{L,w}$; hence, the product $C_{D,w}\alpha_w$ is small in comparison to $C_{L,w}$. With these assumptions, Eq. (7.6) simplifies to

$$\boxed{C_{M,\mathrm{cg}_w} = C_{M,\mathrm{ac}_w} + C_{L,w}(h - h_{\mathrm{ac}_w})} \qquad \textbf{[7.7]}$$

Referring to Fig. 7.12b, we find $C_{L,w} = (dC_{L,w}/d\alpha)\alpha_w = a_w\alpha_w$, where a_w is the *lift slope of the wing*. Thus, (Eq. (7.7) can be written as

$$C_{M,\mathrm{cg}_w} = C_{M,\mathrm{ac}_w} + a_w\alpha_w(h - h_{\mathrm{ac}_w}) \qquad \textbf{[7.8]}$$

Equations (7.7) and (7.8) give the contribution of the wing to moments about the center of gravity of the airplane, subject of course to the above assumptions. Closely examine Eqs. (7.7) and (7.8) along with Fig. 7.19. On a physical basis, they state that the wing's contribution to M_{cg} is essentially due to two factors: the moment about the aerodynamic center M_{ac_w} and the lift acting through the moment arm $(h - h_{\mathrm{ac}_w})c$.

The above results are slightly modified if a fuselage is added to the wing. Consider a cigar-shaped body at an angle of attack to an airstream. This fuselage-type body experiences a moment about its aerodynamic center, plus some lift and drag due to the airflow around it. Now consider the fuselage and wing joined together: a *wing-body combination*. The airflow about this wing-body combination is different from that over the wing and body separately; aerodynamic interference occurs where the flow over the wing affects the fuselage flow, and vice versa. Due to this interference, the moment due to the wing-body combination is *not* simply the sum of the separate wing and fuselage moments. Similarly, the lift and drag of the wing-body combination are affected by aerodynamic interference. Such interference effects are extremely difficult to predict theoretically. Consequently, the lift, drag, and moments of a wing-body combination are usually obtained from wind tunnel measurements. Let $C_{L_{wb}}$ and $C_{M,ac_{wb}}$ be the lift coefficient and moment coefficient about the aerodynamic center, respectively, for the wing-body combination. Analogous to Eqs. (7.7) and (7.8) for the wing only, the contribution of the wing-body combination to M_{cg} is

$$C_{M,cg_{wb}} = C_{M,ac_{wb}} + C_{L_{wb}}(h - h_{ac_{wb}})$$ [7.9]

$$C_{M,cg_{wb}} = C_{M,ac_{wb}} + a_{wb}\alpha_{wb}(h - h_{ac_{wb}})$$ [7.10]

where a_{wb} and α_{wb} are the slope of the lift curve and absolute angle of attack, respectively, for the wing-body combination. In general, adding a fuselage to a wing shifts the aerodynamic center forward, increases the lift curve slope, and contributes a negative increment to the moment about the aerodynamic center. We emphasize again that the aerodynamic coefficients in Eqs. (7.9) and (7.10) are almost always obtained from wind tunnel data.

Example 7.2

For a given wing-body combination, the aerodynamic center lies 0.05 chord length ahead of the center of gravity. The moment coefficient about the aerodynamic center is -0.016. If the lift coefficient is 0.45, calculate the moment coefficient about the center of gravity.

Solution

From Eq. (7.9),

$$C_{M,cg_{wb}} = C_{M,ac_{wb}} + C_{L_{wb}}(h - h_{ac_{wb}})$$

where
$$h - h_{ac_{wb}} = 0.05$$
$$C_{L_{wb}} = 0.45$$
$$C_{M,ac_{wb}} = -0.016$$

Thus
$$C_{M,cg_{wb}} = -0.016 + 0.45(0.05) = \boxed{0.0065}$$

Example 7.3

A wing-body model is tested in a subsonic wind tunnel. The lift is found to be zero at a geometric angle of attack $\alpha = -1.5°$. At $\alpha = 5°$, the lift coefficient is measured as 0.52. Also, at $\alpha = 1.0°$ and $7.88°$, the moment coefficients about the center of gravity are measured as -0.01 and 0.05, respectively. The center of gravity is located at $0.35c$. Calculate the location of the aerodynamic center and the value of $C_{M,ac_{wb}}$.

Solution

First, calculate the lift slope:

$$a_{wb} \equiv \frac{dC_L}{d\alpha} = \frac{0.52 - 0}{5 - (-1.5)} = \frac{0.52}{6.5} = 0.08 \text{ per degree}$$

Write Eq. (7.10),

$$C_{M,cg_{wb}} = C_{M,ac_{wb}} + a_{wb}\alpha_{wb}(h - h_{ac_{wb}})$$

evaluated at $\alpha = 1.0°$ [remember that α is the geometric angle of attack, whereas in Eq. (7.10) α_{wb} is the absolute angle of attack]:

$$-0.01 = C_{M,ac_{wb}} + 0.08(1 + 1.5)(h - h_{ac_{wb}})$$

Then evaluate it at $\alpha = 7.88°$:

$$0.05 = C_{M,ac_{wb}} + 0.08(7.88 + 1.5)(h - h_{ac_{wb}})$$

The above two equations have two unknowns, $C_{M,ac_{wb}}$ and $h - h_{ac_{wb}}$. They can be solved simultaneously.

Subtracting the second equation from the first, we get

$$-0.06 = 0 - 0.55(h - h_{ac_{wb}})$$

$$h - h_{ac_{wb}} = \frac{-0.06}{-0.55} = 0.11$$

The value of h is given: $h = 0.35$. Thus

$$h_{ac_{wb}} = 0.35 - 0.11 = \boxed{0.24}$$

In turn,

$$-0.01 = C_{M,ac_{wb}} + 0.08(1 + 1.5)(0.11)$$

$$C_{M,ac_{wb}} = \boxed{-0.032}$$

7.7 CONTRIBUTION OF THE TAIL TO M_{cg}

An analysis of moments due to an isolated tail taken independently of the airplane would be the same as that given for the isolated wing above. However, in real life the tail is obviously connected to the airplane itself; it is not isolated. Moreover, the tail is generally mounted behind the wing; hence it feels the wake of the airflow over the wing. As a result, there are two interference effects that influence the tail aerodynamics:

1. The airflow at the tail is deflected downward by the *downwash* due to the finite wing (see Secs. 5.13 and 5.14); that is, the relative wind seen by the tail is not in the same direction as the relative wind V_∞ seen by the wing.

2. Because of the retarding force of skin friction and pressure drag over the wing, the airflow reaching the tail has been slowed. Therefore, the velocity of the relative wind seen by the tail is less than V_∞. In turn, the dynamic pressure seen by the tail is less than q_∞.

These effects are illustrated in Fig. 7.20. Here V_∞ is the relative wind as seen by the wing, and V' is the relative wind at the tail, inclined below V_∞ by the downwash angle ε. The tail lift L_t and drag D_t are (by definition) perpendicular and parallel,

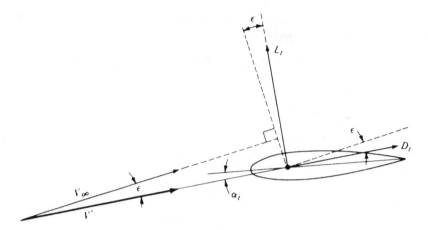

Figure 7.20   Flow and force diagram in the vicinity of the tail.

respectively, to V'. In contrast, the lift and drag of the complete airplane are always (by definition) perpendicular and parallel, respectively, to V_∞. Therefore, considering components of L_t and D_t perpendicular to V_∞, we demonstrate in Fig. 7.20 that the tail contribution to the total airplane lift is $L_t \cos \varepsilon - D_t \sin \varepsilon$. In many cases, ε is very small, and thus $L_t \cos \varepsilon - D_t \sin \varepsilon \approx L_t$. Hence, for all practical purposes, it is sufficient to add the tail lift directly to the wing-body lift to obtain the lift of the complete airplane.

Consider the tail in relation to the wing-body zero-lift line, as illustrated in Fig. 7.21. It is useful to pause and study this figure. The wing-body combination is at an absolute angle of attack α_{wb}. The tail is twisted downward to provide a positive $C_{M,0}$, as discussed at the end of Sec. 7.5. Thus the zero-lift line of the tail is intentionally inclined to the zero-lift line of the wing-body combination at the *tail-setting angle* i_t. (The airfoil section of the tail is generally symmetric, for which the tail zero-lift line and the tail chord line are the same.) The absolute angle of attack of the tail α_t is

Figure 7.21   Geometry of wing-tail combination.

measured between the local relative wind V' and the tail zero-lift line. The tail has an aerodynamic center, about which there is a moment M_{ac_t} and through which L_t and D_t act perpendicular and parallel, respectively, to V'. As before, V' is inclined below V_∞ by the downwash angle ε; hence L_t makes an angle $\alpha_{wb} - \varepsilon$ with the vertical. The tail aerodynamic center is located a distance l_t behind and z_t below the center of gravity of the airplane. Make certain to carefully study the geometry shown in Fig. 7.21; it is fundamental to the derivation which follows.

Split L_t and D_t into their vertical components $L_t \cos(\alpha_{wb} - \varepsilon)$ and $D_t \sin(\alpha_{wb} - \varepsilon)$ and their horizontal components $L_t \sin(\alpha_{wb} - \varepsilon)$ and $D_t \cos(\alpha_{wb} - \varepsilon)$. By inspection of Fig. 7.21, the sum of moments about the center of gravity due to L_t, D_t, and M_{ac_t} of the tail is

$$M_{cg_t} = -l_t[L_t \cos(\alpha_{wb} - \varepsilon) + D_t \sin(\alpha_{wb} - \varepsilon)]$$
$$+ z_t L_t \sin(\alpha_{wb} - \varepsilon) - z_t D_t \cos(\alpha_{wb} - \varepsilon) + M_{ac_t} \qquad \textbf{[7.11]}$$

Here, M_{cg_t} denotes the contribution to moments about the airplane's center of gravity due to the horizontal tail.

In Eq. (7.11), the first term on the right-hand side, $l_t L_t \cos(\alpha_{wb} - \varepsilon)$, is by far the largest in magnitude. In fact, for conventional airplanes, the following simplifications are reasonable:

1. $z_t \ll l_t$.
2. $D_t \ll L_t$.
3. The angle $\alpha_{wb} - \varepsilon$ is small, hence $\sin(\alpha_{wb} - \varepsilon) \approx 0$ and $\cos(\alpha_{wb} - \varepsilon) \approx 1$.
4. M_{ac_t} is small in magnitude.

With the above approximations, which are based on experience, Eq. (7.11) is dramatically simplified to

$$M_{cg_t} = -l_t L_t \qquad \textbf{[7.12]}$$

Define the *tail lift coefficient*, based on free-stream dynamic pressure $q_\infty = \frac{1}{2}\rho_\infty V_\infty^2$ and the tail planform area S_t, as

$$C_{L,t} = \frac{L_t}{q_\infty S_t} \qquad \textbf{[7.13]}$$

Combining Eqs. (7.12) and (7.13), we obtain

$$M_{cg_t} = -l_t q_\infty S_t C_{L,t} \qquad \textbf{[7.14]}$$

Dividing Eq. (7.14) by $q_\infty Sc$, where c is the wing chord and S is the wing planform area, gives

$$\frac{M_{cg_t}}{q_\infty Sc} \equiv C_{M,cg_t} = -\frac{l_t S_t}{cS} C_{L,t} \qquad \textbf{[7.15]}$$

Examining the right-hand side of Eq. (7.15), we note that $l_t S_t$ is a *volume* characteristic of the size and location of the tail and that cS is a *volume* characteristic of the size of the wing. The ratio of these two volumes is called the *tail volume ratio* V_H, where

$$V_H \equiv \frac{l_t S_t}{cS}$$ **[7.16]**

Thus, Eq. (7.15) becomes

$$\boxed{C_{M,\mathrm{cg}_t} = -V_H C_{L,t}}$$ **[7.17]**

The simple relation in Eq. (7.17) gives the total contribution of the tail to moments about the airplane's center of gravity. With the simplifications above and by referring to Fig. 7.21. Eqs. (7.12) and (7.17) say that the moment is equal to tail lift operating through the moment arm l_t.

It will be useful to couch Eq. (7.17) in terms of angle of attack, as was done in Eq. (7.10) for the wing-body combination. Keep in mind that the stability criterion in Fig. 7.13 involves $\partial C_{M,\mathrm{cg}}/\partial \alpha_a$; hence equations in terms of α_a are directly useful. Specifically, referring to the geometry of Fig. 7.21, we see that the angle of attack of the tail is

$$\alpha_t = \alpha_{\mathrm{wb}} - i_t - \varepsilon$$ **[7.18]**

Let a_t denote the lift slope of the tail. Thus, from Eq. (7.18),

$$C_{L,t} = a_t \alpha_t = a_t(\alpha_{\mathrm{wb}} - i_t - \varepsilon)$$ **[7.19]**

The downwash angle ε is difficult to predict theoretically and is usually obtained from experiment. It can be written as

$$\varepsilon = \varepsilon_0 + \frac{\partial \varepsilon}{\partial \alpha}\alpha_{\mathrm{wb}}$$ **[7.20]**

where ε_0 is the downwash angle when the wing-body combination is at zero lift. Both ε_0 and $\partial \varepsilon/\partial \alpha$ are usually obtained from wind tunnel data. Thus, combining Eqs. (7.19) and (7.20) yields

$$C_{L,t} = a_t \alpha_{\mathrm{wb}}\left(1 - \frac{\partial \varepsilon}{\partial \alpha}\right) - a_t(i_t + \varepsilon_0)$$ **[7.21]**

Substituting Eq. (7.21) into (7.17), we obtain

$$\boxed{C_{M,\mathrm{cg}_t} = -a_t V_H \alpha_{\mathrm{wb}}\left(1 - \frac{\partial \varepsilon}{\partial \alpha}\right) + a_t V_H(\varepsilon_0 + i_t)}$$ **[7.22]**

Equation (7.22), although lengthier than Eq. (7.17), contains the explicit dependence on angle of attack and will be useful for our subsequent discussions.

7.8 TOTAL PITCHING MOMENT ABOUT THE CENTER OF GRAVITY

Consider the airplane as a whole. The total M_{cg} is due to the contribution of the wing-body combination plus that of the tail:

$$C_{M,\text{cg}} = C_{M,\text{cg}_{\text{wb}}} + C_{M,\text{cg}_t}$$ **[7.23]**

Here, $C_{M,\text{cg}}$ is the total moment coefficient about the center of gravity for the complete airplane. Substituting Eqs. (7.9) and (7.17) into (7.23), we have

$$\boxed{C_{M,\text{cg}} = C_{M,\text{ac}_{\text{wb}}} + C_{L_{\text{wb}}}(h - h_{\text{ac}_{\text{wb}}}) - V_H C_{L,t}}$$ **[7.24]**

In terms of angle of attack, an alternate expression can be obtained by substituting Eqs. (7.10) and (7.22) into Eq. (7.23):

$$\boxed{C_{M,\text{cg}} = C_{M,\text{ac}_{\text{wb}}} + a_{\text{wb}}\alpha_{\text{wb}}\left[h - h_{\text{ac}_{\text{wb}}} - V_H \frac{a_t}{a_{\text{wb}}}\left(1 - \frac{\partial \varepsilon}{\partial \alpha}\right)\right] + V_H a_t(i_t + \varepsilon_0)}$$

[7.25]

The angle of attack needs further clarification. Referring again to Fig. 7.13, we find the moment coefficient curve is usually obtained from wind tunnel data, preferably on a model of the complete airplane. Hence, α_a in Fig. 7.13 should be interpreted as the absolute angle of attack referenced to the zero-lift line of the *complete airplane*, which is not necessarily the same as the zero-lift line for the wing-body combination. This comparison is sketched in Fig. 7.22. However, for many conventional aircraft, the difference is small. Therefore, in the remainder of this chapter, we assume the two zero-lift lines in Fig. 7.22 to be the same. Thus,

Figure 7.22 Zero-lift line of the wing-body combination compared with that of the complete airplane.

α_{wb} becomes the angle of attack of the complete airplane α_a. Consistent with this assumption, the total lift of the airplane is due to the wing-body combination, with the tail lift neglected. Hence, $C_{L_{wb}} = C_L$ and the lift slope $a_{wb} = a$, where C_L and a are for the complete airplane. With these interpretations, Eq. (7.25) can be rewritten as

$$C_{M,cg} = C_{M,ac_{wb}} + a\alpha_a\left[h - h_{ac_{wb}} - V_H\frac{a_t}{a}\left(1 - \frac{\partial\varepsilon}{\partial\alpha}\right)\right] + V_H a_t(i_t + \varepsilon_0)$$

[7.26]

Equation (7.26) is the same as Eq. (7.25), except that the subscript wb on some terms has been dropped in deference to properties for the whole airplane.

Example 7.4

Consider the wing-body model in Example 7.3. The area and chord of the wing are 0.1 m² and 0.1 m, respectively. Now assume that a horizontal tail is added to this model. The distance from the airplane's center of gravity to the tail's aerodynamic center is 0.17 m, the tail area is 0.02 m², the tail-setting angle is 2.7°, the tail lift slope is 0.1 per degree, and from experimental measurement, $\varepsilon_0 = 0$ and $\partial\varepsilon/\partial\alpha = 0.35$. If $\alpha = 7.88°$, calculate $C_{M,cg}$ for the airplane model.

Solution
From Eq. (7.26)

$$C_{M,cg} = C_{M,ac_{wb}} + a\alpha_a\left[h - h_{ac_{wb}} - V_H\frac{a_t}{a}\left(1 - \frac{\partial\varepsilon}{\partial\alpha}\right)\right] + V_H a_t(i_t + \varepsilon_0)$$

where $C_{M,ac_{wb}} = -0.032$ (from Example 7.3)
$a = 0.08$ (from Example 7.3)
$\alpha_a = 7.88 + 1.5 = 9.38°$ (from Example 7.3)
$h - h_{ac_{wb}} = 0.11$ (from Example 7.3)

$$V_H = \frac{l_t S_t}{cS} = \frac{0.17(0.02)}{0.1(0.1)} = 0.34$$

$a_t = 0.1$ per degree

$$\frac{\partial\varepsilon}{\partial\alpha} = 0.35$$

$i_t = 2.7°$
$\varepsilon_0 = 0$

Thus $$C_{M,cg} = -0.032 + 0.08(9.38)\left[0.11 - 0.34\left(\frac{0.1}{0.08}\right)(1 - 0.35)\right]$$
$$+ 0.34(0.1)(2.7 + 0)$$
$$= -0.032 - 0.125 + 0.092 = \boxed{-0.065}$$

7.9 EQUATIONS FOR LONGITUDINAL STATIC STABILITY

The criteria necessary for longitudinal balance and static stability were developed in Sec. 7.5. they are (1) $C_{M,0}$ must be positive and (2) $\partial C_{M,\text{cg}}/\partial \alpha_a$ must be negative, both conditions with the implicit assumption that α_e falls within the practical flight range of angle of attack; that is, the moment coefficient curve must be similar to that sketched in Fig. 7.13. In turn, the ensuing sections developed a quantitative formalism for static stability culminating in Eq. (7.26) for $C_{M,\text{cg}}$. The purpose of this section is to combine the above results to obtain formulas for the direct calculation of $C_{M,0}$ and $\partial C_{M,\text{cg}}/\partial \alpha_a$. In this manner, we will be able to make a quantitative assessment of the longitudinal static stability of a given airplane, as well as point out some basic philosophy of airplane design.

Recall that, by definition, $C_{M,0}$ is the value of $C_{M,\text{cg}}$ when $\alpha_a = 0$, that is, when the lift is zero. Substituting $\alpha_a = 0$ into Eq. (7.26), we directly obtain

$$C_{M,0} \equiv (C_{M,\text{cg}})_{L=0} = C_{M,\text{ac}_{\text{wb}}} + V_H a_t (i_t + \varepsilon_0) \qquad \textbf{[7.27]}$$

Examine Eq. (7.27). We know that $C_{M,0}$ must be positive in order to balance the airplane. However, the previous sections have pointed out that $C_{M,\text{ac}_{\text{wb}}}$ is negative for conventional airplanes. Therefore, $V_H a_t (i_t + \varepsilon_0)$ must be positive and large enough to more than counterbalance the negative $C_{M,\text{ac}}$. Both V_H and a_t are positive quantities, and ε_0 is usually so small that it exerts only a minor effect. Thus, i_t *must be a positive quantity*. This verifies our previous physical arguments that the tail must be set at an angle relative to the wing in the manner shown in Figs. 7.17*a* and 7.21. This allows the tail to generate enough negative lift to produce a positive $C_{M,0}$.

Consider now the slope of the moment coefficient curve. Differentiating Eq. (7.26) with respect to α_a, we obtain

$$\frac{\partial C_{M,\text{cg}}}{\partial \alpha_a} = a \left[h - h_{\text{ac}_{\text{wb}}} - V_H \frac{a_t}{a} \left(1 - \frac{\partial \varepsilon}{\partial \alpha} \right) \right] \qquad \textbf{[7.28]}$$

This equation clearly shows the powerful influence of the location h of the center of gravity and the tail volume ratio V_H in determining longitudinal static stability.

Equations (7.27) and (7.28) allow us to check the static stability of a given airplane, assuming we have some wind tunnel data for a, a_t, $C_{M,\text{ac}_{\text{wb}}}$, ε_0, and $\partial \varepsilon/\partial \alpha$. They also establish a certain philosophy in the design of an airplane. For example, consider an airplane where the location h of the center of gravity is essentially dictated by payload or other mission requirements. Then the desired amount of static stability can be obtained simply by designing V_H large enough, via Eq. (7.28). Once V_H is fixed in this manner, then the desired $C_{M,0}$ (or the desired α_e) can be obtained by designing i_t appropriately, via Eq. (7.27). Thus, the values of $C_{M,0}$ and $\partial C_{M,\text{cg}}/\partial \alpha_a$ basically dictate the design values of i_t and V_H, respectively (for a fixed center-of-gravity location).

Example 7.5 | Consider the wing-body-tail wind tunnel model of Example 7.4. Does this model have longitudinal static stability and balance?

Solution

From Eq. (7.28)

$$\frac{\partial C_{M,cg}}{\partial \alpha_a} = a\left[h - h_{ac_{wb}} - V_H \frac{a_t}{a}\left(1 - \frac{\partial \varepsilon}{\partial \alpha}\right)\right]$$

where, from Examples 7.3 and 7.4,

$$a = 0.08$$
$$h - h_{ac_{wb}} = 0.11$$
$$V_H = 0.34$$
$$a_t = 0.1 \text{ per degree}$$
$$\frac{\partial \varepsilon}{\partial \alpha} = 0.35$$

Thus $\dfrac{\partial C_{M,cg}}{\partial \alpha_a} = 0.08\left[0.11 - 0.34\dfrac{0.1}{0.08}(1 - 0.35)\right] = \boxed{-0.0133}$

The slope of the moment coefficient curve is negative, hence the airplane model is statically stable.

However, is the model longitudinally balanced? To answer this, we must find $C_{M,0}$, which in combination with the above result for $\partial C_{M,cg}/\partial \alpha$ will yield the equilibrium angle of attack α_e. From Eq. (7.27),

$$C_{M,0} = C_{M,ac_{wb}} + V_H a_t (i_t + \varepsilon_0)$$

where from Examples 7.3 and 7.4,

$$C_{M,ac_{wb}} = -0.032$$
$$i_t = 2.7°$$

Thus $C_{M,0} = -0.032 + 0.34(0.1)(2.7) = 0.06$

From Fig. 7.13, the equilibrium angle of attack is obtained from

$$0 = 0.06 - 0.0133\alpha_e$$

Thus $\alpha_e = 4.5°$

Clearly, this angle of attack falls within the reasonable flight range. Therefore, the airplane is longitudinally balanced as well as statically stable.

7.10   NEUTRAL POINT

Consider the situation where the location h of the center of gravity is allowed to move, with everything else remaining fixed. In fact, Eq. (7.28) indicates that static stability is a strong function of h. Indeed, the value of $\partial C_{M,\text{cg}}/\partial \alpha_a$ can always be made negative by properly locating the center of gravity. In the same vein, there is one specific location of the center of gravity such that $\partial C_{M,\text{cg}}/\partial \alpha_a = 0$. The value of h when this condition holds is defined as the *neutral point*, denoted by h_n. When $h = h_n$, the slope of the moment coefficient curve is zero, as illustrated in Fig. 7.23.

The location of the neutral point is readily obtained from Eq. (7.28) by setting $h = h_n$ and $\partial C_{M,\text{cg}}/\partial \alpha_a = 0$, as follows.

$$0 = a\left[h_n - h_{\text{ac}_{\text{wb}}} - V_H \frac{a_t}{a}\left(1 - \frac{\partial \varepsilon}{\partial \alpha}\right)\right] \qquad \textbf{[7.29]}$$

Solving Eq. (7.29) for h_n, we have

$$\boxed{h_n = h_{\text{ac}_{\text{wb}}} + V_H \frac{a_t}{a}\left(1 - \frac{\partial \varepsilon}{\partial \alpha}\right)} \qquad \textbf{[7.30]}$$

Examine Eq. (7.30). The quantities on the right-hand side are, for all practical purposes, established by the design configuration of the airplane. Thus, for a *given* airplane design, the neutral point is a *fixed quantity*, that is, a point that is frozen somewhere on the airplane. It is quite independent of the actual location h of the center of gravity.

The concept of the neutral point is introduced as an alternative stability criterion. For example, inspection of Eqs. (7.28) and (7.30) shows that $\partial C_{M,\text{cg}}/\partial \alpha_a$ is negative, zero, or positive depending on whether h is less than, equal to, or greater than h_n.

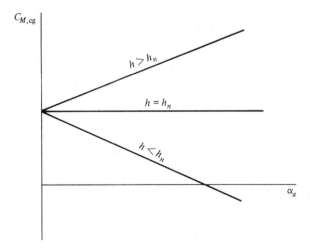

Figure 7.23 Effect of the location of the center of gravity, relative to the neutral point, on static stability.

These situations are sketched in Fig. 7.23. Remember that h is measured from the leading edge of the wing, as shown in Fig. 7.19. Hence, $h < h_n$ means that the center-of-gravity location is *forward* of the neutral point. Thus, an alternative stability criterion is as follows:

> For longitudinal static stability, the position of the center of gravity must always be forward of the neutral point.

Recall that the definition of the aerodynamic center for a wing is that point about which moments are independent of the angle of attack. This concept can now be extrapolated to the whole airplane by considering again Fig. 7.23. Clearly, when $h = h_n$, $C_{M,\text{cg}}$ is independent of the angle of attack. Therefore, the neutral point might be considered the aerodynamic center of the complete airplane.

Again examining Eq. (7.30), we see that the tail strongly influences the location of the neutral point. *By proper selection of the tail parameters, principally V_H, h_n can be located at will by the designer.*

Example 7.6	For the wind tunnel model of Examples 7.3 to 7.5, calculate the neutral point location.

Solution

From Eq. (7.30)

$$h_n = h_{\text{ac}_{\text{wb}}} + V_H \frac{a_t}{a} \left(1 - \frac{\partial \varepsilon}{\partial \alpha} \right)$$

where $h_{\text{ac}_{\text{wb}}} = 0.24$ (from Example 7.3). Thus

$$h_n = 0.24 + 0.34 \left(\frac{0.1}{0.08} \right) (1 - 0.35)$$

$$\boxed{h_n = 0.516}$$

Note from Example 7.3 that $h = 0.35$. Compare this center-of-gravity location with the neutral point location of 0.516. The center of gravity is comfortably *forward* of the neutral point; this again confirms the results of Example 7.5 that the airplane is statically stable.

7.11   STATIC MARGIN

A corollary to the above discussion can be obtained as follows. Solve Eq. (7.30) for $h_{\text{ac}_{\text{wb}}}$.

$$h_{\text{ac}_{\text{wb}}} = h_n - V_H \frac{a_t}{a} \left(1 - \frac{\partial \varepsilon}{\partial \alpha} \right) \qquad \textbf{[7.31]}$$

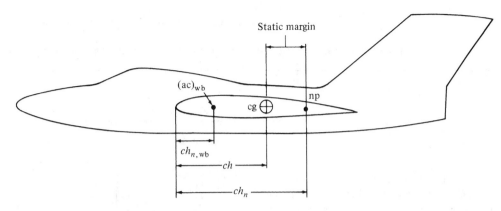

Figure 7.24   Illustration of the static margin.

Note that in Eqs. (7.29) to (7.31), the value of V_H is not precisely the same number as in Eq. (7.28). Indeed, in Eq. (7.28), V_H is based on the moment arm l_t measured from the center-of-gravity location, as shown in Fig. 7.21. In contrast, in Eq. (7.29), the center-of-gravity location has been moved to the neutral point, and V_H is therefore based on the moment arm measured from the neutral point location. However, the difference is usually small, and this effect will be ignored here. Therefore, substituting Eq. (7.31) into Eq. (7.28) and canceling the terms involving V_H, we obtain

$$\frac{\partial C_{M,\text{cg}}}{\partial \alpha_a} = a(h - h_n) \qquad \textbf{[7.32]}$$

The distance $h_n - h$ is defined as the *static margin* and is illustrated in Fig. 7.24. Thus, from Eq. (7.32),

$$\frac{\partial C_{M,\text{cg}}}{\partial \alpha_a} = -a(h_n - h) = -a \times \text{static margin} \qquad \textbf{[7.33]}$$

Equation (7.33) shows that the static margin is a direct measure of longitudinal static stability. For static stability, the static margin must be positive. Moreover, the larger the static margin, the more stable the airplane.

For the wind tunnel model of the previous examples, calculate the static margin.   **Example 7.7**

Solution

From Example 7.6, $h_n = 0.516$ and $h = 0.35$. Thus, by definition,

$$\text{Static margin} \equiv h_n - h = 0.516 - 0.35 = \boxed{0.166}$$

For a check on the consistency of our calculations, consider Eq. (7.33).

$$\frac{\partial C_{M,\text{cg}}}{\partial \alpha_0} = -a \times \text{static margin} = -0.08(0.166) = -0.0133 \text{ per degree}$$

This is the same value calculated in Example 7.5; our calculations are indeed consistent.

DESIGN BOX

Let us boil down all the above discussion to some plain speaking about the location of the lift force acting on the airplane relative to the center of gravity when the airplane is statically stable, and when it is trimmed. Such plain speaking helps the airplane designer to have a clearer concept of how to design for a specified amount of stability (or instability).

A diagram that is frequently shown for a *statically stable* airplane is sketched in Fig. 7.25a. Here we will see the lift acting through a point situated *behind* the center of gravity, and we say this is necessary for static stability. But what does this really mean? What is the real significance of Fig. 7.25a? Let us look at it more closely.

First, recall that the lift of the airplane is due to the component of the net integrated pressure distribution exerted over the external surface of the airplane—the wings, fuselage, tail, etc.—acting perpendicular to the relative wind. This pressure distribution exerts a *distributed* load over the whole airplane. However, as is frequently done, we can conceptualize the mechanical effect of this distributed load by replacing it with a single concentrated force acting through an arbitrary point plus the moments acting about the same point. This is what is shown in Fig. 7.25a; we show the lift as a single concentrated force acting through a point, and we also indicate the moments about this point. The point that is chosen in Fig. 7.25a is the aerodynamic center of the airplane (the neutral point). In Fig. 7.25a, the lift shown is the *total* lift of the airplane, including the contribution from the tail.

In Sec. 7.10 we demonstrated that the aerodynamic center (neutral point) must be located behind the center of gravity in order to have static stability. We now have a simple picture in Fig. 7.25b that easily

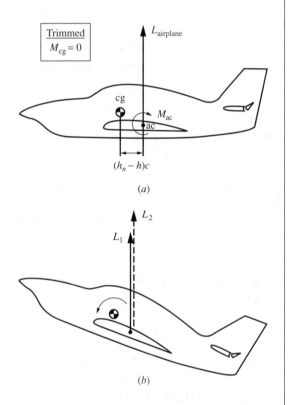

Figure 7.25   A diagram for static stability, with the lift acting behind the center of gravity.

proves this. In Fig. 7.25a the airplane is trimmed; that is, $M_{\text{cg}} = 0$. Imagine that the airplane encounters a gust such that its angle of attack is momentarily increased. In turn, the lift will momentarily increase, as shown in Fig. 7.25b. Here L_1 is the lift before the

(continued)

gust, and L_2 is the increased lift in response to the gust. Since the lift is acting through a point *behind* the center of gravity, the increased lift results in a negative (pitch-down) moment about the center of gravity, as shown in Fig. 7.25*b*. Hence the initial tendency after encountering the gust will be to pitch the nose down, reducing the angle of attack and restoring the airplane to its trimmed condition—the precise notion of static stability. It is clear from Fig. 7.25 that if the lift acting through the aerodynamic center is *behind* the center of gravity, the airplane will be statically stable.

We note in passing that for the airplane in Fig. 7.25*a* to be trimmed, $M_{cg} = 0$. The lift is shown acting through the moment arm $(h_n - h)c$, hence creating a pitch-down moment about the center of gravity equal to $-[(h_n - h)cL]$. In turn, the moment about the aerodynamic center of the airplane M_{ac} must be equal and opposite in order to have zero total moments about the center of gravity. That is, M_{ac} must be a positive (pitch-up) moment, as shown in Fig. 7.25*a*. Usually, much of this pitch-up moment is due to a download on the tail, similar to that illustrated in Fig. 7.17*a*. If the airplane configuration were a canard, then the positive M_{ac} would be due to an upload on the canard, similar to that illustrated in Fig. 7.17*b*. Indeed, when the airplane designer is trying to make a choice between the conventional rear-tail and the canard configurations, this situation is one of the advantages listed in favor of the canard. In Fig. 7.25, the lift shown is the *total* lift of the airplane, equal to the weight in steady, level flight. For a conventional rear tail configuration, the download on the tail requires the wing to produce more lift in order for the total lift to equal the weight. In contrast, with the canard configuration, the upload on the canard contributes to the overall lift, hence requiring less lift from the wing. In turn, this reduces the induced drag generated from the wing.

Figure 7.25 reflects a commonly shown diagram illustrating longitudinal static stability, with the lift shown acting behind the center of gravity. An alternative picture illustrating static stability is shown in Fig. 7.26. This picture is not so commonly seen, but it is perhaps a "purer" explanation of the nature of longitudinal static stability. Recall that the lift of the airplane is due to the net integrated effect of the pressure distribution acting over the entire surface

Figure 7.26 A diagram for static stability, with the lift acting at the center of pressure.

of the airplane. This pressure distribution has a *centroid* (analogous to the centroid of an area or a solid, which you calculate from differential calculus). The centroid of the pressure distribution is called the *center of pressure*. The center of pressure, being a centroid, is that point about which the net moment due to the distributed pressure is zero. Hence, when we simulate the mechanical effect of the distributed pressure distribution by a single concentrated force,

(continued)

it is most natural to locate this concentrated force at the center of pressure. Indeed, the center of pressure can be thought of as "that point on the airplane through which the lift effectively acts." To be more specific, we can simulate the mechanical effect of the distributed pressure loads on the airplane by first locating the center of pressure and then drawing the lift through this point, with zero moments about this point. This is the diagram shown in Fig. 7.26. Moreover, *when the airplane is trimmed, the center of pressure is precisely located at the center of gravity.* This is the case shown in Fig. 7.26a; with the lift acting through the center of pressure, and with the center of pressure at the center of gravity, there is no moment about the center of gravity, and by definition the airplane is trimmed. This is what nature does. When the airplane is trimmed, the pressure distribution over the airplane has been adjusted so that the center of pressure is precisely at the center of gravity.

When the angle of attack of the airplane changes, the pressure distribution over the surface changes, and

hence the center of pressure *shifts*—its location is a function of the angle of attack. For longitudinal static stability, the shift in the center of pressure must be in the direction shown in Fig. 7.26b. For static stability, the shift in the center of pressure must be *rearward*, to create a restoring moment about the center of gravity, as shown in Fig. 7.26b. Similarly, consider the originally trimmed airplane encountering a gust which *decreases* the angle of attack, as shown in Fig. 7.26c. For static stability, the shift in the center of pressure must be *forward* in order to create a restoring moment about the center of gravity, as shown in Fig. 7.26c. Hence, a statically stable airplane must be designed to have the shifts of the center of pressure in the directions shown in Fig. 7.26b and c.

In summary, Figs. 7.25 and 7.26 are alternative but equally effective diagrams to illustrate the necessary condition for longitudinal static stability. These figures supplement, and are totally consistent with, the more detailed mathematical descriptions in Secs. 7.6 to 7.11.

7.12   CONCEPT OF STATIC LONGITUDINAL CONTROL

A study of stability and control is double-barreled. The first aspect—that of stability itself—has been the subject of the preceding sections. However, for the remainder of this chapter, the focus will turn to the second aspect—control. In regard to our road map in Fig. 7.5, we are moving to the right-hand column.

Consider a statically stable airplane in trimmed (equilibrium) flight. Recalling Fig. 7.13, we see that the airplane must therefore be flying at the trim angle of attack α_e. In turn, this value of α_e corresponds to a definite value of lift coefficient, namely, the trim lift coefficient $C_{L_{\text{trim}}}$. For steady, level flight, this corresponds to a definite velocity, which from Eq. (6.26) is

$$V_{\text{trim}} = \sqrt{\frac{2W}{\rho_\infty S C_{L_{\text{trim}}}}}$$

[7.34]

Now assume that the pilot wishes to fly at a lower velocity $V_\infty < V_{\text{trim}}$. At a lower velocity, the lift coefficient, hence the angle of attack, must be increased to offset the decrease in dynamic pressure (remember from Chap. 6 that the lift must always balance the weight for steady, level flight). However, from Fig. 7.13, if α is increased, $C_{M,\text{cg}}$ becomes negative (i.e., the moment about the center of gravity is no longer zero), and the airplane is no longer trimmed. Consequently, if nothing else is changed about the airplane, it cannot achieve steady, level, equilibrium flight at any other velocity than V_{trim} or at any other angle of attack than α_e.

Obviously, this is an intolerable situation—an airplane must be able to change its velocity at the will of the pilot and still remain balanced. The only way to accomplish this is to effectively change the moment coefficient curve for the airplane. Say that the pilot wishes to fly at a *faster* velocity but still remain in steady, level, balanced flight. The lift coefficient must decrease, hence a new angle of attack α_n must be obtained, where $\alpha_n < \alpha_e$. At the same time, the moment coefficient curve must be changed such that $C_{M,\text{cg}} = 0$ at α_n. Figures 7.27 and 7.28 demonstrate two methods of achieving this change. In Fig. 7.27, the slope is made more negative, such that $C_{M,\text{cg}}$ goes through zero at α_n. From Eq. (7.28) or (7.32), the slope can be changed by shifting the center of gravity. In our example, the center of gravity must be shifted forward. Otto Lilienthal (see Sec. 1.5) used this method in his gliding flights. Figure 1.15 shows Lilienthal hanging loosely below his glider; by simply swinging his hips, he was able to shift the center of gravity and change the stability of the aircraft. This principle is carried over today to the modern hang gliders for sports use.

However, for a conventional airplane, shifting the center of gravity is highly impractical. Therefore, another method for changing the moment curve is employed, as shown in Fig. 7.28. Here, the slope remains the same, but $C_{M,0}$ is changed such that $C_{M,\text{cg}} = 0$ at α_n. This is accomplished by deflecting the elevator on the horizontal tail. Hence, we have arrived at a major concept of static, longitudinal control, namely, that the elevator deflection can be used to control the trim angle of attack, hence to control the equilibrium velocity of the airplane.

Consider Fig. 7.28. We stated above, without proof, that a translation of the moment curve without a change in slope can be obtained simply by deflecting the elevator. But *how* and *to what extent* does the elevator deflection change $C_{M,\text{cg}}$? To provide some answers, first consider the horizontal tail with the elevator fixed in the neutral position, that is, no elevator deflection, as shown in Fig. 7.29. The absolute angle of attack of the tail is α_t, as defined earlier. The variation of tail lift coefficient

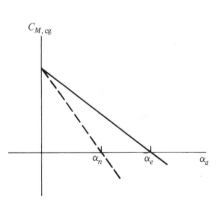

Figure 7.27 Change in trim angle of attack due to change in slope of moment coefficient curve.

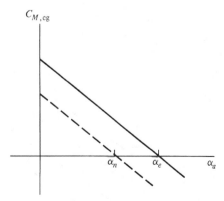

Figure 7.28 Change in trim angle of attack due to change in $C_{M,0}$.

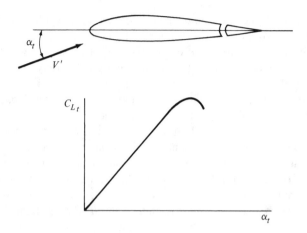

Figure 7.29   Tail-lift coefficient curve with no elevator deflection.

with α_t is also sketched in Fig. 7.29; note that it has the same general shape as the airfoil and wing lift curves discussed in Chap. 5. Now assume that the elevator is deflected downward through angle δ_e, as shown in Fig. 7.30. This is the same picture as a wing with a deflected flap, as discussed in Sec. 5.17. Consequently, just as in the case of a deflected flap, the deflected elevator causes the tail lift coefficient curve to shift to the left, as shown in Fig. 7.30. By convention (and for convenience later), a downward elevator deflection is positive. Therefore, if the elevator is deflected by an

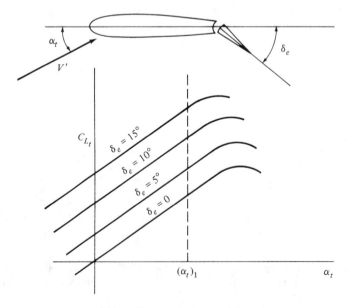

Figure 7.30   Tail lift coefficient with elevator deflection.

angle of, say, 5° and then held fixed as the complete tail is pitched through a range of α_t, then the tail lift curve is translated to the left. If the elevator is then deflected farther, say to 10°, the lift curve is shifted even farther to the left. This behavior is clearly illustrated in Fig. 7.30. Note that for all the lift curves, the slope $\partial C_{L,t}/\partial \alpha_t$ is the same.

With the above in mind, now consider the tail at a fixed angle of attack, say, $(\alpha_t)_1$. If the elevator is deflected from, say, 0 to 15°, then $C_{L,t}$ will increase along the vertical dashed line in Fig. 7.30. This variation can be cross-plotted as $C_{L,t}$ versus δ_e, as shown in Fig. 7.31. For most conventional airplanes, the curve in Fig. 7.31 is essentially linear, and its slope $\partial C_{L,t}/\partial \delta_e$ is called the *elevator control effectiveness*. This quantity is a direct measure of the "strength" of the elevator as a control; because δ_e has been defined as positive for downward deflections, $\partial C_{L,t}/\partial \delta_e$ is *always positive*.

Consequently, the tail lift coefficient is a function of *both* α_t and δ_e (hence, the partial derivative notation is used, as discussed earlier). Keep in mind that, physically, $\partial C_{L,t}/\partial \alpha_t$ is the rate of change of $C_{L,t}$ with respect to α_t, keeping δ_e constant; similarly, $\partial C_{L,t}/\partial \delta_e$ is the rate of change of $C_{L,t}$ with respect to δ_e, keeping α_t constant. Hence, on a physical basis,

$$C_{L,t} = \frac{\partial C_{L,t}}{\partial \alpha_t} \alpha_t + \frac{\partial C_{L,t}}{\partial \delta_e} \delta_e \qquad [7.35]$$

Recalling that the tail lift slope is $a_t = \partial C_{L,t}/\partial \alpha_t$, we see that Eq. (7.35) can be written as

$$C_{L,t} = a_t \alpha_t + \frac{\partial C_{L,t}}{\partial \delta_e} \delta_e \qquad [7.36]$$

Substituting Eq. (7.36) into (7.24), we have for the pitching moment about the center of gravity

$$C_{M,\text{cg}} = C_{M,\text{ac}_{\text{wb}}} + C_{L,\text{wb}}(h - h_{\text{ac}}) - V_H \left(a_t \alpha_t + \frac{\partial C_{L,t}}{\partial \delta_e} \delta_e \right) \qquad [7.37]$$

Equation (7.37) gives explicitly the effect of elevator deflection on moments about the center of gravity of the airplane.

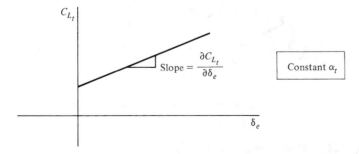

Figure 7.31 Tail lift coefficient versus elevator deflection at constant angle of attack; a cross-plot of Fig. 7.30.

The rate of change of $C_{M,\text{cg}}$ due *only* to elevator deflection is, by definition, $\partial C_{M,\text{cg}}/\partial\delta_e$. This partial derivative can be found by differentiating Eq. (7.37) with respect to δ_e, keeping everything else constant.

$$\frac{\partial C_{M,\text{cg}}}{\partial\delta_e} = -V_H\frac{\partial C_{L,t}}{\partial\delta_e} \qquad [7.38]$$

Note that, from Fig. 7.31, $\partial C_{L,t}/\partial\delta_e$ is constant; moreover, V_H is a specific value for the given airplane. Thus, the right-hand side of Eq. (7.38) is a constant. Therefore, on a physical basis, the increment in $C_{M,\text{cg}}$ due *only* to a given elevator deflection δ_e is

$$\Delta C_{M,\text{cg}} = -V_H\frac{\partial C_{L,t}}{\partial\delta_e}\delta_e \qquad [7.39]$$

Equation (7.39) now answers the questions asked earlier concerning how and to what extent the elevator deflection changes $C_{M,\text{cg}}$. Consider the moment curve labeled $\delta_e = 0$ in Fig. 7.32. This is the curve with the elevator fixed in the neutral position; it is the curve we originally introduced in Fig. 7.13. If the elevator is now deflected through a positive angle (downward), Eq. (7.39) states that all points on this curve will be shifted down by the constant amount $\Delta C_{M,\text{cg}}$. Hence, the slope of the moment curve is preserved; only the value of $C_{M,0}$ is changed by elevator deflection. This now proves our earlier statement made in conjunction with Fig. 7.28.

For emphasis, we repeat the main thrust of this section. The elevator can be used to change and control the trim of the airplane. In essence, this controls the equilibrium velocity of the airplane. For example, by a downward deflection of the elevator, a new trim angle α_n smaller than the original trim angle α_e can be obtained. (This is illustrated in Fig. 7.32.) This corresponds to an increase in velocity of the airplane.

As another example, consider the two velocity extremes—stalling velocity and maximum velocity. Figure 7.33 illustrates the elevator deflection necessary to trim the airplane at these two extremes. First consider Fig. 7.33a, which corresponds

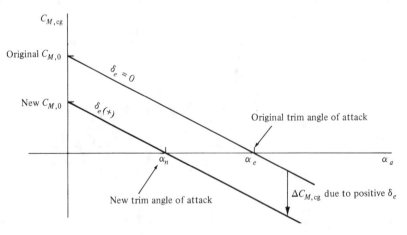

Figure 7.32   Effect of elevator deflection on moment coefficient.

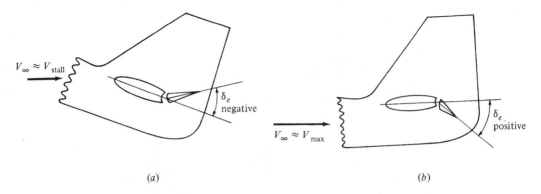

$V_\infty \approx V_{\text{stall}}$

δ_e negative

$V_\infty \approx V_{\text{max}}$

δ_e positive

(a)   (b)

Figure 7.33 Elevator deflection required for trim at (a) low flight velocity and (b) high flight velocity.

to an airplane flying at $V_\infty \approx V_{\text{stall}}$. This would be the situation on a landing approach, for example. The airplane is flying at $C_{L_{\text{max}}}$, hence the angle of attack is large. Therefore, from our previous discussion, the airplane must be trimmed by an *up-elevator* position, that is, by a negative δ_e. On the other hand, consider Fig. 7.33b, which corresponds to an airplane flying at $V_\infty \approx V_{\text{max}}$ (near full throttle). Because q_∞ is large, the airplane requires only a small C_L to generate the required lift force; hence the angle of attack is small. Thus, the airplane must be trimmed by a *down-elevator* position, that is, by a positive δ_e.

7.13 CALCULATION OF ELEVATOR ANGLE TO TRIM

The concepts and relations developed in Sec. 7.12 allow us now to calculate the precise elevator deflection necessary to trim the airplane at a given angle of attack. Consider an airplane with its moment coefficient curve given as in Fig. 7.34. The equilibrium angle of attack with no elevator deflection is α_e. We wish to trim the airplane at a new angle of attack α_n. What value of δ_e is required for this purpose?

To answer this question, first write the equation for the moment curve with $\delta_e = 0$ (the solid line in Fig. 7.34). This is a straight line with a constant slope equal to $\partial C_{M,\text{cg}}/\partial \alpha_a$ and intercepting the ordinate at $C_{M,0}$. Hence, from analytic geometry, the equation of this line is

$$C_{M,\text{cg}} = C_{M,0} + \frac{\partial C_{M,\text{cg}}}{\partial \alpha_a} \alpha_a \qquad \textbf{[7.40]}$$

Now assume the elevator is deflected through an angle δ_e. The value of $C_{M,\text{cg}}$ will change by the increment $\Delta C_{M,\text{cg}}$, and the moment equation given by Eq. (7.40) is now modified as

$$C_{M,\text{cg}} = C_{M,0} + \frac{\partial C_{M,\text{cg}}}{\partial \alpha_a} \alpha_a + \Delta C_{M,\text{cg}} \qquad \textbf{[7.41]}$$

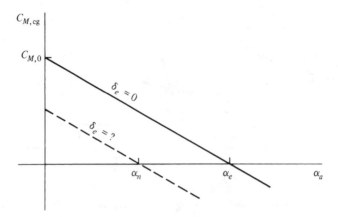

Figure 7.34   Given the equilibrium angle of attack at zero elevator deflection, what elevator deflection is necessary to establish a given new equilibrium angle of attack?

The value of $\Delta C_{M,\text{cg}}$ was obtained earlier as Eq. (7.39). Hence, substituting Eq. (7.39) into (7.41), we obtain

$$C_{M,\text{cg}} = C_{M,0} + \frac{\partial C_{M,\text{cg}}}{\partial \alpha_a} \alpha_a - V_H \frac{\partial C_{L,t}}{\partial \delta_e} \delta_e \qquad \textbf{[7.42]}$$

Equation (7.42) allows us to calculate $C_{M,\text{cg}}$ for any arbitrary angle of attack α_a and any arbitrary elevator deflection δ_e. However, we are interested in the specific situation where $C_{M,\text{cg}} = 0$ at $\alpha_a = \alpha_n$ and where the value of δ_e necessary to obtain this condition is $\delta_e = \delta_{\text{trim}}$. That is, we want to find the value of δ_e, which gives the dashed line in Fig. 7.34. Substituting the above values into Eq. (7.42), we have

$$0 = C_{M,0} + \frac{\partial C_{M,\text{cg}}}{\partial \alpha_a} \alpha_n - V_H \frac{\partial C_{L,t}}{\partial \delta_e} \delta_{\text{trim}}$$

and solving for δ_{trim}, we obtain

$$\delta_{\text{trim}} = \frac{C_{M,0} + (\delta C_{M,\text{cg}}/\partial \alpha_a)\alpha_n}{V_H(\partial C_{L,t}/\partial \delta_e)} \qquad \textbf{[7.43]}$$

Equation (7.43) is the desired result. It gives the elevator deflection necessary to trim the airplane at a given angle of attack α_n. In Eq. (7.43), V_H is a known value from the airplane design, and $C_{M,0}$, $\partial C_{M,\text{cg}}/\partial \alpha_a$, and $\partial C_{L,t}/\partial \delta_e$ are known values usually obtained from wind tunnel or free-flight data.

Example 7.8   Consider a full-size airplane with the same aerodynamic and design characteristics as the wind tunnel model of Examples 7.3 to 7.7. The airplane has a wing area of 19 m², a weight of

2.27×10^4 N, and an elevator control effectiveness of 0.04. Calculate the elevator deflection angle necessary to trim the airplane at a velocity of 61 m/s at sea level.

Solution

First, we must calculate the angle of attack for the airplane at $V_\infty = 61$ m/s. Recall that

$$C_L = \frac{2W}{\rho_\infty V_\infty^2 S} = \frac{2(2.27 \times 10^4)}{1.225(61)^2(19)} = 0.52$$

From Example 7.3, the lift slope is $a = 0.08$ per degree. Hence, the absolute angle of attack of the airplane is

$$\alpha_a = \frac{C_L}{a} = \frac{0.52}{0.08} = 6.5°$$

From Eq. (7.43), the elevator deflection angle required to trim the airplane at this angle of attack is

$$\delta_{\text{trim}} = \frac{C_{M,0} + (\delta C_{M,\text{cg}}/\partial\alpha_a)\alpha_n}{V_H(\partial C_{L,t}/\partial\delta_e)}$$

where

$$C_{M,0} = 0.06 \qquad \text{(from Example 7.5)}$$

$$\frac{C_{M,\text{cg}}}{\partial\alpha_a} = -0.0133 \qquad \text{(from Example 7.5)}$$

$$\alpha_n = 6.5° \qquad \text{(this is the } \alpha_a \text{ calculated above)}$$

$$V_H = 0.34 \qquad \text{(from Example 7.4)}$$

$$\frac{\partial C_{L,t}}{\partial\delta_e} = 0.04 \qquad \text{(given above)}$$

Thus, from Eq. (7.43),

$$\delta_{\text{trim}} = \frac{0.06 + (-0.0133)(6.5)}{0.34(0.04)} = \boxed{-1.94°}$$

Recall that positive δ is downward. Hence, to trim the airplane at an angle of attack of 6.5°, the elevator must be deflected *upward* by 1.94°.

7.14 STICK-FIXED VERSUS STICK-FREE STATIC STABILITY

The second paragraph of Sec. 7.5 initiated our study of a rigid airplane with *fixed controls*, for example, the elevator *fixed* at a given deflection angle. The ensuing sections developed the static stability for such a case, always assuming that the elevator can be deflected to a desired angle δ_e but held fixed at that angle. This is the situation when the pilot (human or automatic) moves the control stick to a given position and then rigidly holds it there. Consequently, the static stability that we have discussed to this point is called *stick-fixed static stability*. Modern high-performance airplanes designed to fly near or beyond the speed of sound have hydraulically assisted power controls; therefore a stick-fixed static stability analysis is appropriate for such airplanes.

However, consider a control stick connected to the elevator via wire cables without a power boost of any sort. This was characteristic of most early airplanes until the 1940s and is representative of many light, general aviation, private aircraft of today. In this case, to hold the stick fixed in a given position, the pilot must continually exert a manual force. This is uncomfortable and impractical. Consequently, in steady, level flight, the control stick is left essentially free; in turn, the elevator is left free to float under the influence of the natural aerodynamic forces and moments at the tail. The static stability of such an airplane is therefore called *stick-free static stability*. This is the subject of the next two sections.

7.15   ELEVATOR HINGE MOMENT

Consider a horizontal tail with an elevator which rotates about a hinge axis, as shown in Fig. 7.35. Assume the airfoil section of the tail is symmetric, which is almost always the case for both the horizontal and vertical tail. First, consider the tail at zero angle of attack, as shown in Fig. 7.35a. The aerodynamic pressure distribution on the top and bottom surfaces of the elevator will be the same, that is, symmetric about the chord. Hence, there will be no moment exerted on the elevator about the hinge line. Now assume that the tail is pitched to the angle of attack α_t, but the elevator is not deflected, that is, $\delta_e = 0$. This is illustrated in Fig. 7.35b. As discussed in Chap. 5, there will be a low pressure on the top surface of the airfoil and a high pressure on the bottom surface. Consequently, the aerodynamic force on the elevator will not be balanced, and there will be a moment about the hinge axis tending to deflect the elevator upward. Finally, consider the horizontal tail at zero angle of attack but with the elevator deflected downward and held fixed at the angle δ_e, as shown in Fig. 7.35c. Recall from Sec. 5.17 that a flap deflection effectively changes the camber of the airfoil and alters the pressure distribution. Therefore, in Fig. 7.35c, there will be low and high pressures on the top and bottom elevator surfaces, respectively. As a result, a moment will again be exerted about the hinge line, tending to rotate the elevator upward. Thus, we see that both the tail angle of attack α_t and the elevator deflection δ_e result in a moment about the elevator hinge line—such a moment is defined as the *elevator hinge moment*. It is the governing factor in stick-free static stability, as discussed in the next section.

Let H_e denote the elevator hinge moment. Also, referring to Fig. 7.36, we see the chord of the tail is c_t, the distance from the leading edge of the elevator to the hinge line is c_b, the distance from the hinge line to the trailing edge is c_e, and that portion of the elevator planform area that lies *behind* (aft of) the hinge line is S_e. The *elevator hinge moment* coefficient C_{h_e} is then defined as

$$C_{h_e} = \frac{H_e}{\frac{1}{2}\rho_\infty V_\infty^2 S_e c_e} \qquad\qquad \textbf{[7.44]}$$

where V_∞ is the free-stream velocity of the airplane.

(a)

(b)

(c)

Figure 7.35 Illustration of the aerodynamic generation of elevator hinge moment. (a) No hinge moment; (b) hinge moment due to angle of attack; (c) hinge moment due to elevator deflection.

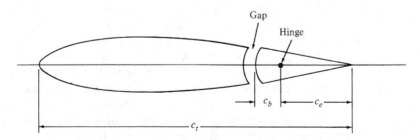

Figure 7.36 Nomenclature and geometry for hinge moment coefficient.

Recall that the elevator hinge moment is due to the tail angle of attack and the elevator deflection. Hence, C_{h_e} is a function of both α_t and δ_e. Moreover, experience has shown that at both subsonic and supersonic speeds, C_{h_e} is approximately a linear function of α_t and δ_e. Thus, recalling the definition of the partial derivative in Sec. 7.2d, we can write the hinge moment coefficient as

$$C_{h_e} = \frac{\partial C_{h_e}}{\partial \alpha_t} \alpha_t + \frac{\partial C_{h_e}}{\partial \delta_e} \delta_e \qquad \qquad \text{[7.45]}$$

where $\partial C_{h_e}/\partial \alpha_t$ and $\partial C_{h_e}/\partial \delta_e$ are approximately constant. However, the actual magnitudes of these constant values depend in a complicated way on c_e/c_t, c_b/c_e, the elevator nose shape, gap, trailing-edge angle, and planform. Moreover, H_e is very sensitive to local boundary layer separation. As a result, the values of the partial derivatives in Eq. (7.45) must almost always be obtained empirically, such as from wind tunnel tests, for a given design.

Consistent with the convention that downward elevator deflections are positive, hinge moments which tend to deflect the elevator downward are also defined as positive. Note from Fig. 7.35b that a positive α_t physically tends to produce a negative hinge moment (tending to deflect the elevator upward). Hence $\partial C_{h_e}/\partial \alpha_t$ is usually negative. (However, if the hinge axis is placed very far back, near the trailing edge, the sense of H_e may become positive. This is usually not done for conventional airplanes.) Also, note from Fig. 7.35c that a positive δ_e usually produces a negative H_e, hence $\partial C_{h_e}/\partial \delta_e$ is also negative.

7.16 STICK-FREE LONGITUDINAL STATIC STABILITY

Let us return to the concept of stick-free static stability introduced in Sec. 7.14. If the elevator is left free to float, it will always seek some equilibrium deflection angle such that the hinge moment is zero, that is, $H_e = 0$. This is obvious, because as long as there is a moment on the free elevator, it will always rotate. It will come to rest (equilibrium) only for that position where the moment is zero.

Recall our qualitative discussion of longitudinal static stability in Sec. 7.5. Imagine that an airplane is flying in steady, level flight at the equilibrium angle of attack. Now assume the airplane is disturbed by a wind gust and is momentarily pitched to another angle of attack, as sketched in Fig. 7.14. If the airplane is statically stable, it will initially tend to return toward its equilibrium position. In subsequent sections, we saw that the design of the horizontal tail was a powerful mechanism governing this static stability. However, until now, the elevator was always considered fixed. But if the elevator is allowed to float freely when the airplane is pitched by some disturbance, the elevator will seek some momentary equilibrium position different from its position before the disturbance. This deflection of the free elevator will change the static stability characteristics of the airplane. In fact, such stick-free stability is usually less than stick-fixed stability. For this reason, it is usually desirable to design an airplane such that the difference between stick-free and stick-fixed longitudinal stability is small.

With the above in mind, consider the equilibrium deflection angle of a free elevator. Denote this angle by δ_{free}, as sketched in Fig. 7.37. At this angle, $H_e = 0$. Thus, from Eq. (7.45),

$$C_{h_e} = 0 = \frac{\partial C_{h_e}}{\partial \alpha_t} \alpha_t + \frac{\partial C_{h_e}}{\partial \delta_e} \delta_{\text{free}} \qquad \qquad \textbf{[7.46]}$$

Solving Eq. (7.46) for δ_{free} gives

$$\boxed{\delta_{\text{free}} = -\frac{\partial C_{h_e}/\partial \alpha_t}{\partial C_{h_e}/\partial \delta_e} \alpha_t} \qquad \qquad \textbf{[7.47]}$$

Equation (7.47) gives the equilibrium, free-floating angle of the elevator as a function of tail angle of attack. As stated earlier, both partial derivatives in Eq. (7.47) are usually negative; hence, a positive α_t yields a negative δ_{free} (an upward deflection). This is intuitively correct, as verified by Fig. 7.37, which shows a negative δ_{free}.

Obviously, δ_{free} affects the tail lift coefficient, which in turn affects the static stability of the airplane. The tail lift coefficient for angle of attack α_t and fixed elevator deflection δ_e was given in Eq. (7.36), repeated here:

$$C_{L,t} = a_t \alpha_t + \frac{\partial C_{L,t}}{\partial \delta_e} \delta_e$$

However, for a free elevator, $\delta_e = \delta_{\text{free}}$. Denoting the tail lift coefficient for a free elevator as $C'_{L,t}$, we see that a substitution of Eq. (7.47) into (7.36) gives

$$C'_{L,t} = a_t \alpha_t + \frac{\partial C_{L,t}}{\partial \delta_e} \delta_{\text{free}}$$

$$C'_{L,t} = a_t \alpha_t - \frac{\partial C_{L,t}}{\partial \delta_e} \frac{\partial C_{h_e}/\partial \alpha_t}{\partial C_{h_e}/\partial \delta_e} \alpha_t$$

or

$$\boxed{C'_{L,t} = a_t \alpha_t F} \qquad \qquad \textbf{[7.48]}$$

Figure 7.37 Illustration of free elevator deflection.

where F is the *free elevator factor*, defined as

$$F = 1 - \frac{1}{a_t} \frac{\partial C_{L,t}}{\partial \delta_e} \frac{\partial C_{h_e}/\partial \alpha_t}{\partial C_{h_e}/\partial \delta_e}$$

The free elevator factor is a number usually less than unity and usually on the order of 0.7 to 0.8. It represents a reduction in the tail's contribution to static stability when the elevator is free. The magnitude of this reduction is developed below.

Consider now the moment about the center of gravity of the airplane. For a fixed elevator, the moment coefficient is given by Eq. (7.24),

$$C_{M,\text{cg}} = C_{M,\text{ac}_{wb}} + C_{L_{wb}}(h - h_{\text{ac}_{wb}}) - V_H C_{L,t}$$

For a free elevator, the tail lift coefficient is now changed to $C'_{L,t}$. Hence, the moment coefficient for a free elevator $C'_{M,\text{cg}}$ is

$$C'_{M,\text{cg}} = C_{M,\text{ac}_{wb}} + C_{L_{wb}}(h - h_{\text{ac}_{wb}}) - V_H C'_{L,t} \qquad \textbf{[7.49]}$$

Substituting Eq. (7.48) into (7.49), we get

$$\boxed{C'_{M,\text{cg}} = C_{M,\text{ac}_{wb}} + C_{L_{wb}}(h - h_{\text{ac}_{wb}}) - V_H a_t \alpha_t F} \qquad \textbf{[7.50]}$$

Equation (7.50) gives the final form of the moment coefficient about the center of gravity of the airplane with a free elevator.

By using Eq. (7.50), the same analyses as given in Sec. 7.9 can be used to obtain equations for stick-free longitudinal static stability. The results are as follows:

$$\boxed{\begin{array}{l} C'_{M,0} = C_{M,\text{ac}_{wb}} + F V_H a_t (i_t + \varepsilon_0) \qquad \textbf{[7.51]} \\[2mm] h'_n = h_{\text{ac}_{wb}} + F V_H \dfrac{a_t}{a}\left(1 - \dfrac{\partial \varepsilon}{\partial \alpha}\right) \qquad \textbf{[7.52]} \\[2mm] \dfrac{\partial C'_{M,\text{cg}}}{\partial \alpha} = -a(h'_n - h) \qquad \textbf{[7.53]} \end{array}}$$

Equations (7.51), (7.52), and (7.53) apply for stick-free conditions, denoted by the prime notation. They should be compared with Eqs. (7.27), (7.30), and (7.33), respectively, for stick-fixed stability. Note that $h'_n - h$ is the stick-free static margin; because $F < 1.0$, this is smaller than the stick-fixed static margin.

Hence, it is clear from Eqs. (7.51) to (7.53) that a free elevator usually decreases the static stability of the airplane.

Example 7.9 | Consider the airplane of Example 7.8. Its elevator hinge moment derivatives are $\partial C_{h_e}/\partial \alpha_t = -0.008$ and $\partial C_{h_e}/\partial \delta_e = -0.013$. Assess the *stick-free* static stability of this airplane.

Solution

First, obtain the free elevator factor F, defined from Eq. (7.48),

$$F = 1 - \frac{1}{a_t} \frac{\partial C_{L,t}}{\partial \delta_e} \frac{\partial C_{h_e}/\partial \alpha_t}{\partial C_{h_e}/\partial \delta_e}$$

where $\qquad a_t = 0.1 \qquad$ (from Example 7.4)

$$\frac{\partial C_{L,t}}{\partial \delta_e} = 0.04 \qquad \text{(from Example 7.8)}$$

$$F = 1 - \frac{1}{0.1}(0.04)\left(\frac{-0.008}{-0.013}\right) = 0.754$$

The stick-free static stability characteristics are given by Eqs. (7.51) to (7.53). First, from Eq. (7.51),

$$C'_{M,0} = C_{M,\text{ac}_{wb}} + FV_H a_t (i_t + \varepsilon_0)$$

where $\qquad C_{M,\text{ac}_{wb}} = -0.032 \qquad$ (from Example 7.3)

$$V_H = 0.34 \qquad \text{(from Example 7.4)}$$

$$i_t = 2.7° \qquad \text{(from Example 7.4)}$$

$$\varepsilon_o = 0 \qquad \text{(from Example 7.4)}$$

Thus, $\qquad C'_{M,0} = -0.032 + 0.754(0.34)(0.1)(2.7)$

$$\boxed{C'_{M,0} = 0.037}$$

This is to be compared with $C_{M,0} = 0.06$ obtained for stick-fixed conditions in Example 7.5. From Eq. (7.52),

$$h'_n = h_{\text{ac}_{wb}} + FV_H \frac{a_t}{a}\left(1 - \frac{\partial \varepsilon}{\partial \alpha}\right)$$

where $\qquad h_{\text{ac}_{wb}} = 0.24 \qquad$ (from Example 7.3)

$$\frac{\partial \varepsilon}{\partial \alpha} = 0.35 \qquad \text{(from Example 7.4)}$$

$$a = 0.08 \qquad \text{(from Example 7.4)}$$

$$h'_n = 0.24 + 0.754(0.34)\left(\frac{0.1}{0.08}\right)(1 - 0.35)$$

$$\boxed{h'_n = 0.448}$$

This is to be compared with $h_n = 0.516$ obtained for stick-fixed conditions in Example 7.6. Note that the neutral point has moved forward for stick-free conditions, hence decreasing the stability. In fact, the stick-free static margin is

$$h'_n - h = 0.448 - 0.35 = 0.098$$

This is a 41 percent decrease, in comparison with the stick-fixed static margin from

Example 7.7. Finally, from Eq. (7.53),

$$\frac{\partial C'_{M,\text{cg}}}{\partial \alpha} = -a(h'_n - h) = -0.08(0.098) = \boxed{-0.0078}$$

Thus, as expected, the slope of the stick-free moment coefficient curve, although still negative, is small in absolute value.

In conclusion, this problem indicates that stick-free conditions cut the static stability of our hypothetical airplane by nearly one-half. This helps to dramatize the differences between stick-fixed and stick-free considerations.

7.17   DIRECTIONAL STATIC STABILITY

Returning to Figure 7.2, we note that the preceding sections have dealt with longitudinal stability and control, which concerns angular motion about the y axis—pitching motion. In this section, we briefly examine the stability associated with angular motion about the z axis—yawing motion. Stability in yaw is called *directional stability*. In regard to our road map in Fig. 7.5, we are moving to the second box at the bottom of the left-hand column.

Examining Fig. 7.3, we see the vertical stabilizer (vertical fin, or vertical tail) is the conventional mechanism for directional stability. Its function is easily seen in Fig. 7.38. Consider an airplane in equilibrium flight with no yaw, as sketched in Fig. 7.38a. The vertical tail, which is designed with a symmetric airfoil section,

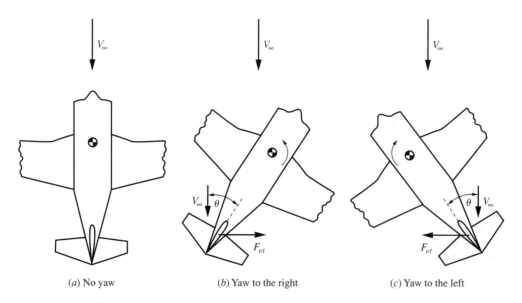

(a) No yaw   (b) Yaw to the right   (c) Yaw to the left

Figure 7.38   Effect of the vertical stabilizer on directional stability.

Figure 7.39 Moment arm of the vertical tail.

is at zero angle of attack to the free stream, and it experiences no net aerodynamic force perpendicular to V_∞. Assume the airplane is suddenly yawed to the right by a disturbance, as shown in Fig. 7.38*b*. The vertical tail is now at an angle of attack θ and experiences an aerodynamic force F_{vt} perpendicular to V_∞. This force creates a restoring yawing moment about the center of gravity which tends to rotate the airplane back toward its equilibrium position. The same situation prevails when the airplane is yawed to the left by a disturbance, as sketched in Fig. 7.38*c*.

The magnitude of the restoring moment in yaw is equal to $F_{vt}l_{vt}$, where l_{vt} is the moment arm from the aerodynamic center of the vertical tail to the airplane's center of gravity, as shown in Fig. 7.39. Since the aerodynamic force on the vertical tail F_{vt} is proportional to the area of the vertical tail S_{vt}, shown as the shaded area in Fig. 7.39, the design parameter governing directional stability can be shown to be the vertical tail volume ratio, defines as

$$\text{Vertical tail volume ratio} \equiv V_{vt} \equiv \frac{l_{vt} S_{vt}}{bS} \qquad \textbf{[7.54]}$$

where b is the wingspan and S is the wing planform area. The definition of V_{vt} in Eq. (7.54) is similar to the definition of the horizontal tail volume ratio V_H defined by Eq. (7.16), except that V_{vt} uses b rather than the chord c as the nondimensionalizing length in the denominator.

DESIGN BOX

For conventional airplanes, typical values of V_{vt} are given by Raymer (see Bibliography) as follows:

	V_{vt}
General aviation, single-engine	0.04
Twin turboprop	0.08
Jet fighter	0.07
Jet transport	0.09

These numbers are considerably smaller than typical values of V_H, which range from 0.4 to 1.0 (in Example 7.4, we used $V_H = 0.34$), because of the use of b rather than c in the definition of V_{vt}.

7.18   LATERAL STATIC STABILITY

Return to Fig. 7.2. In this section we briefly examine the stability associated with angular motion about the x axis—rolling motion. Stability in roll is called *lateral stability*. In regard to our road map in Fig. 7.5, we are moving to the third box at the bottom of the left-hand column.

Consider an airplane in steady, level flight. Let us take a view of this airplane from behind, looking in the direction of flight, as sketched in Fig. 7.40a. The lift equals the weight. They act equal and opposite to each other; there is no net side force. The airplane is suddenly perturbed by a gust which causes the right wing to dip; that is, a roll to the right ensues. This is sketched in Fig. 7.40b. The lift vector is now rotated from the vertical through angle ϕ, called the *bank angle*. The vector resolution of L and W results in a side force F, which causes the airplane to accelerate in the direction of F. This sidewise motion of the airplane is called a *slideslip*. Relative to the airplane, there appears a slideslip velocity V_S, shown in Fig. 7.40b.

Consider the effect of this slideslip velocity on the lift generated by the right and left wings. This is illustrated in Fig. 7.41. In Fig. 7.41a, the airplane is shown with the right and left wings in the same plane, perpendicular to the plane of symmetry of the fuselage. Let L_1 and L_2 be the lift generated by the right and left wings, respectively. The slideslip velocity V_S will affect the lift generated by each wing, but

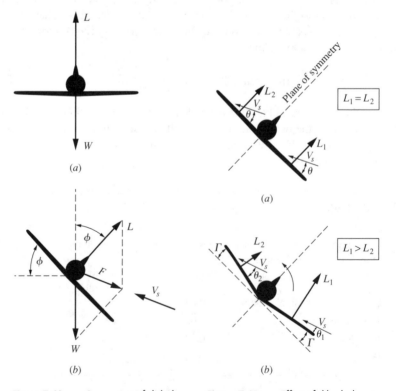

Figure 7.40   Generation of slideslip.   **Figure 7.41**   Effect of dihedral.

DESIGN BOX

For a given airplane design, the amount of dihedral depends on the location of the wing relative to the fuselage, that is, low-wing, midwing, or high-wing location. The schematics in Figs. 7.40 and 7.41 show a low-wing design. More dihedral is needed for a low-wing position than for a midwing or high-wing position. Also, a swept-back wing requires less dihedral than a straight wing. Some degree of lateral stability is usually necessary in conventional airplanes, but too much makes the airplane very sluggish to aileron control inputs. Indeed, the combination of mid- or high-wing location along with sweepback may have too much inherent lateral stability, and *anhedral* (negative dihedral) must be used to counteract some of this. Raymer (see Bibliography) gives the following typical values of dihedral (and anhedral) angle (in degrees) for various classes of airplanes:

	Wing Position		
	Low	**Middle**	**High**
Unswept (civil)	5 to 7	2 to 4	0 to 2
Subsonic swept wing	3 to 7	−2 to 2	−5 to −2
Supersonic swept wing	0 to 5	−5 to 0	−5 to 0

The amount of dihedral shown in Fig. 7.41b is greatly exaggerated for the purpose of illustration. The amount of dihedral (or anhedral) for some actual airplanes can be seen from the three views shown earlier in this book, that is, the F-86 (Fig. 2.13), the F4U Corsair (Fig. 2.14), the X-29 (Fig. 2.16), the F3F (Fig. 2.17), the F-104 (Fig. 4.4), the X-1 (Fig. 5.28), the U-2 (Fig. 5.48), the English Electric Lightning (Fig. 5.57), the Mirage C (Fig. 5.61), the Concorde (Fig. 5.62), and the P-38 (Fig. 7.42).

since the two wings are in the same plane, V_S makes the same angle θ with respect to both wings, and hence $L_1 = L_2$, as shown in Fig. 7.41a. As a result, there is no restoring moment to return the airplane to its original equilibrium position, shown in Fig. 7.40a. However, consider the case where both wings are bent upward through angle Γ, as shown in Fig. 7.41b; that is, the wings are designed with a V shape. This is called *dihedral*, and Γ is the dihedral angle. Here, the slideslip velocity makes an angle θ_1 with respect to the right wing and a larger angle θ_2 with respect to the left wing. As a result, the lift on the left wing L_2 is smaller than the lift on the right wing, and this creates a restoring rolling moment that tends to return the airplane to its equilibrium position, as shown in Fig. 7.41b. Hence, *dihedral is the design feature of the airplane which provides lateral stability.*

There are more sophisticated explanations of the dihedral effect. Also, there is always a coupling between yawing and rolling motion, so that one does not occur without the other. It is beyond the scope of this book to go into these matters further. You will examine these effects when you embark on a more advanced study of stability and control. The function of this section and the previous one has been only to introduce some of the most basic thoughts about directional and lateral stability.

7.19 A COMMENT

This brings to a close our technical discussion of stability and control. The preceding sections constitute an introduction to the subject; however, we have just scratched the surface. There are many other considerations: control forces, dynamic stability, etc. Such matters are the subject of more advanced studies of stability and control and are beyond the scope of this book. However, this subject is one of the fundamental pillars of aeronautical engineering, and the interested reader can find extensive presentations in books such as those of Perkins and Hage, and Etkin (see Bibliography at the end of this chapter).

7.20   HISTORICAL NOTE: THE WRIGHT BROTHERS VERSUS THE EUROPEAN PHILOSOPHY ON STABILITY AND CONTROL

The two contrasting scenes depicted in Sec. 7.1—the lumbering, belabored flight of Farman versus the relatively effortless maneuvering of Wilbur Wright—underscore two different schools of aeronautical thought during the first decade of powered flight. One school, consisting of virtually all early European and U.S. aeronautical engineers, espoused the concept of inherent stability (statically stable aircraft); the other, consisting solely of Wilbur and Orville Wright, practiced the design of statically unstable aircraft that had to be controlled every instant by the pilot. Both philosophies have their advantages and disadvantages, and because they have an impact on modern airplane design, we examine their background more closely.

The basic principles of airplane stability and control began to evolve at the time of George Cayley. His glider of 1804, sketched in Fig. 1.8, incorporated a vertical and horizontal tail that could be adjusted up and down. In this fashion, the complete tail unit acted as an elevator.

The next major advance in airplane stability was made by Alphonse Penaud, a brilliant French aeronautical engineer who committed suicide in 1880 at the age of 30. Penaud built small model airplanes powered by twisted rubber bands, a precursor of the flying balsa-and-tissue paper models of today. Penaud's design had a fixed wing and tail, like Cayley's, even though at the time Penaud was not aware of Cayley's work. Of particular note was Penaud's horizontal tail design, which was set at a negative $8°$ with respect to the wing chord line. Here we find the first true understanding of the role of the tail-setting angle i_t (see Secs. 7.5 and 7.7) on the static stability of an airplane. Penaud flew his model in the Tuileries Gardens in Paris on August 18, 1871, before members of the Société de Navigation Aérienne. The aircraft flew for 11 s, covering 131 ft. This event, along with Penaud's theory for stability, remained branded on future aeronautical designs right down to the present.

After Penaud's work, the attainment of "inherent" (static) stability became a dominant feature in aeronautical design. Lilienthal, Pilcher, Chanute, and Langley all strived for it. However, static stability has one disadvantage: The more stable the airplane, the harder it is to maneuver. An airplane that is highly stable is also sluggish in the air; its natural tendency to return to equilibrium somewhat defeats the purpose of the pilot to change its direction by means of control deflections. The Wright brothers recognized this problem in 1900. Since Wilbur and Orville were *airmen* in the strictest meaning of the word, they aspired for quick and easy maneuverability. Therefore, they discarded the idea of inherent stability that was entrenched by Cayley and Penaud. Wilbur wrote that "we . . . resolved to try a fundamentally different principle. We would arrange the machine so that it would not tend to right itself." The Wright brothers designed their aircraft to be statically unstable! This feature, along with their development of lateral control through wing warping, is primarily responsible for the fantastic aerial performance of all their airplanes from 1903 to 1912 (when Wilbur died). Of course, this design feature heavily taxed the pilot, who had to keep the airplane under control at every instant, continuously operating the controls to compensate for the unstable characteristics of the airplane. Thus, the Wright airplanes were difficult to fly, and long periods were required to train pilots for these aircraft. In the same vein, such unstable aircraft were more dangerous.

These undesirable characteristics were soon to be compelling. After Wilbur's dramatic public demonstrations in France in 1908 (see Sec. 1.8), the European designers quickly adopted the Wrights' patented concept of combined lateral and directional control by coordinated wing warping (or by ailerons) and rudder deflection. But they rejected the Wrights' philosophy of static instability. By 1910, the Europeans were designing and flying aircraft that properly mated the Wrights' control ideas with the long-established static stability principles. However, the Wrights stubbornly clung to their basic unstable design. As a result, by 1910 the European designs began to surpass the Wrights' machines, and the lead in aeronautical engineering established in the United States in 1903 now swung to France, England, and Germany, where it remained for almost 20 years. In the process, static stability became an unquestioned design feature in all successful aircraft up to the 1970s.

It is interesting that very modern airplane design has returned full circle to the Wright brothers' original philosophy, at least in some cases. Recent lightweight military fighter designs, such as the F-16 and F-18, are statically unstable in order to obtain dramatic increases in maneuverability. At the same time, the airplane is instantaneously kept under control by computer-calculated and electrically adjusted positions of the control surfaces—the *fly-by-wire* concept. In this fashion, the maneuverability advantages of static instability can be realized without heavily taxing the pilot: The work is done by electronics! Even when maneuverability is not a prime feature, such as in civil transport airplanes, static instability has some advantages. For example, the tail surfaces for an unstable airplane can be smaller, with a subsequent savings in structural weight and reductions in aerodynamic drag. Hence, with the advent of the fly-by-wire system, the cardinal airplane design principle of static stability may be somewhat relaxed in the future. The Wright brothers may indeed ride again!

7.21 HISTORICAL NOTE: THE DEVELOPMENT OF FLIGHT CONTROLS

Figure 7.3 illustrates the basic aerodynamic control surfaces on an airplane, the ailerons, elevator, and rudder. They have been an integral part of airplane designs for most of the 20th century, and we take them almost for granted. But where are their origins? When did such controls first come into practical use? Who had the first inspirations for such controls?

In the previous section, we already mentioned that by 1809 George Cayley employed a movable tail in his designs, the first effort at some type of longitudinal control. Cayley's idea of moving the complete horizontal tail to obtain such control persisted through the first decade of the 20th century. Henson, Stringfellow, Penaud, Lilienthal, and the Wright brothers all envisioned or utilized movement of the complete horizontal tail surface for longitudinal control. It was not until 1908 to 1909 that the first "modern" tail control configuration was put into practice. This was achieved by the French designer Levavasseur on his famous Antoinette airplanes, which had fixed vertical and horizontal tail surfaces with movable, flaplike rudder and elevator surfaces at the trailing edges. So the configuration for elevators and rudders shown in Fig. 7.3 dates back to 1908, five years after the dawn of powered flight.

The origin of ailerons (a French word for the extremity of a bird's wing) is steeped in more history and controversy. It is known that the Englishman M. P. W. Boulton patented a concept for lateral control by ailerons in 1868. Of course, at that time no practical aircraft existed, so the concept could not be demonstrated and verified, and Boulton's invention quickly retreated to the background and was forgotten. Ideas of warping the wings or inserting vertical surfaces (spoilers) at the wing tips cropped up several times in Europe during the late 19th century and into the first decade of the 20th century, but always in the context of a braking surface which would slow one wing down and pivot the airplane about a vertical axis. The true function of ailerons or wing warping, that for lateral control for banking and consequently turning an airplane, was not fully appreciated until Orville and Wilbur incorporated wing warping on their *Flyers* (see Chap. 1). The Wright brothers' claim that they were the first to invent wing warping may not be historically precise, but clearly they were the first to demonstrate its function and to obtain a legally enforced patent on its use (combined with simultaneous rudder action for total control in banking). The early European airplane designers did not appreciate the need for lateral control until Wilbur's dramatic public flights in France in 1908. This is in spite of the fact that Wilbur had fully described their wing warping concept in a paper at Chicago on September 1, 1901, and again on June 24, 1903; indeed, Octave Chanute clearly described the Wrights' concept in a lecture to the Aero Club de France in Paris in April 1903. Other aeronautical engineers at that time, if they listened, did not pay much heed. As a result, European aircraft before 1908, even though they were making some sustained flights, were awkward to control.

However, the picture changed after 1908, when in the face of the indisputable superiority of the Wrights' control system, virtually everybody turned to some type of lateral control. Wing warping was quickly copied and was employed on numerous different designs. Moreover, the idea was refined to include movable surfaces near the wing tips. These were first separate "winglets" mounted either above, below, or between the wings. But in 1909, Henri Farman (see Sec. 7.1) designed a biplane named the *Henri Farman III*, which included a flaplike aileron at the trailing edge of all four wing tips; this was the true ancestor of the conventional modern-day aileron, as sketched in Fig. 7.3. Farman's design was soon adopted by most designers, and wing warping quickly became passé. Only the Wright brothers clung to their old concept; a Wright airplane did not incorporate ailerons until 1915, six years after Farman's development.

7.22   HISTORICAL NOTE: THE "TUCK-UNDER" PROBLEM

A quick examination of Fig. 7.21, and the resulting stability equations such as Eqs. (7.26), (7.27), and (7.28), clearly underscores the importance of the downwash angle ε in determining longitudinal static stability. Downwash is a rather skittish aerodynamic phenomenon, very difficult to calculate accurately for real airplanes and therefore usually measured in wind tunnel tests or in free flight. A classic example of the stability problems that can be caused by downwash, and how wind tunnel testing can help, occurred during World War II, as described below.

In numerous flights during 1941 and 1942, the Lockheed P-38, a twin- engine, twin-boomed, high-performance fighter plane (see Fig. 7.42), went into sudden dives from which recovery was exceptionally difficult. Indeed, several pilots were killed in this fashion. The problem occurred at high subsonic speeds, usually in a dive, where the airplane had a tendency to nose over, putting the plane in yet a steeper dive. Occasionally, the airplane would become locked in this position, and even with maximum elevator deflection, a pullout could not be achieved. This "tuck-under" tendency could not be tolerated in a fighter aircraft which was earmarked for a major combat role.

Therefore, with great urgency, NACA was asked to investigate the problem. Since the effect occurred only at high speeds, usually above Mach 0.6, compressibility appeared to be the culprit. Tests in the Langley 30-ft by 60-ft low-speed tunnel and in the 8-ft high-speed tunnel (see Sec. 4.24) correlated the tuck-under tendency with the simultaneous formation of shock waves on the wing surface. Such compressibility effects were discussed in Secs. 5.9 and 5.10, where it was pointed out that beyond the critical Mach number for the wing, shock waves will form on the upper surface, encouraging flow separation far upstream of the trailing edge. The P-38 was apparently the first operational airplane to encounter this problem. The test engineers at Langley made several suggestions to rectify the situation, but all involved major modifications of the airplane. For a model already in production, a quicker fix was needed.

Next, the 16-ft high-speed wind tunnel at the NACA Ames Aeronautical Laboratory in California (see again Sec. 4.24) was pressed into service on the P-38 problem. Here, further tests indicated that the shock-induced separated flow over the wing was drastically reducing the lift. In turn, because the downwash is directly related to lift, as discussed in Secs. 5.13 and 5.14, the downwash angle ε was greatly reduced. Consequently (see Fig. 7.21), the tail angle of attack α_t was markedly increased. This caused a sharp increase in the positive lift on the tail, creating a strong pitching

Figure 7.42   The Lockheed P-38 of World War II fame.

moment, nosing the airplane into a steeper dive. After the series of Ames tests in April 1943, Al Erickson of NACA suggested the addition of flaps on the lower surface of the wing at the $0.33c$ point in order to increase the lift, hence increase the downwash. This was the quick fix that Lockheed was looking for, and it worked.

7.23   SUMMARY

Some of the important points of this chapter are given below.

1.   If the forces and moments on a body caused by a disturbance tend *initially* to return the body *toward* its equilibrium position, the body is statically stable. In

contrast, if these forces and moments tend *initially* to move the body *away from* its equilibrium position, the body is *statically unstable*.

2. The necessary criteria for longitudinal balance and static stability are (*a*) $C_{M,0}$ must be positive, (*b*) $\partial C_{M,\text{cg}}/\partial\alpha_a$ must be negative, and (*c*) the trim angle of attack α_e must fall within the flight range of angle of attack for the airplane. These criteria may be evaluated quantitatively for a given airplane from

$$C_{M,0} = C_{M,\text{ac}_{\text{wb}}} + V_H a_t (i_t + \varepsilon_0) \qquad \textbf{[7.27]}$$

and

$$\frac{\partial C_{M,\text{cg}}}{\partial\alpha_a} = a\left[h - h_{\text{ac}_{\text{wb}}} - V_H\frac{a_t}{a}\left(1 - \frac{\partial\varepsilon}{\partial\alpha}\right)\right] \qquad \textbf{[7.28]}$$

where the tail volume ratio is given by

$$V_H = \frac{l_t S_t}{cS}$$

3. The neutral point is that location of the center of gravity where $\partial C_{M,\text{cg}}/\partial\alpha_a = 0$. It can be calculated from

$$h_n = h_{\text{ac}_{\text{wb}}} + V_H\frac{a_t}{a}\left(1 - \frac{\partial\varepsilon}{\partial\alpha}\right) \qquad \textbf{[7.30]}$$

4. The static margin is defined as $h_n - h$. For static stability, the location of the center of gravity must be ahead of the neutral point; that is, the static margin must be positive.

5. The effect of elevator deflection δ_e on the pitching moment about the center of gravity is given by

$$C_{M,\text{cg}} = C_{M,\text{ac}_{\text{wb}}} + C_{L,\text{wb}}(h - h_{\text{ac}}) - V_H\left(a_t\alpha_t + \frac{\partial C_{L,t}}{\partial\delta_e}\delta_e\right) \qquad \textbf{[7.37]}$$

6. The elevator deflection necessary to trim an airplane at a given angle of attack α_n is

$$\delta_{\text{trim}} = \frac{C_{M,0} + (\delta C_{M,\text{cg}}/\partial\alpha_a)\alpha_n}{V_H(\partial C_{L,t}/\partial\delta_e)} \qquad \textbf{[7.43]}$$

BIBLIOGRAPHY

Etkin, B.: *Dynamics of Flight*, Wiley, New York, 1959.

Gibbs-Smith, C. H.: *Aviation: An Historical Survey from Its Origins to the End of World War II*, Her Majesty's Stationery Office, London, 1970.

Perkins, C. D., and R. E. Hage: *Airplane Performance, Stability, and Control*, Wiley, New York, 1949.

Raymer, D. P.: *Aircraft Design: A Conceptual Approach*, 2d ed., American Institute of Aeronautics and Astronautics, Reston, VA, 1992.

PROBLEMS

7.1   For a given wing-body combination, the aerodynamic center lies 0.03 chord length ahead of the center of gravity. The moment coefficient about the center of gravity is 0.0050, and the lift coefficient is 0.50. Calculate the moment coefficient about the aerodynamic center.

7.2   Consider a model of a wing-body shape mounted in a wind tunnel. The flow conditions in the test section are standard sea-level properties with a velocity of 100 m/s. The wing area and chord are 1.5 m^2 and 0.45 m, respectively. Using the wind tunnel force and moment-measuring balance, the moment about the center of gravity when the lift is zero is found to be -12.4 N · m. When the model is pitched to another angle of attack, the lift and moment about the center of gravity are measured to be 3675 N and 20.67 N · m, respectively. Calculate the value of the moment coefficient about the aerodynamic center and the location of the aerodynamic center.

7.3   Consider the model in Prob. 7.2. If a mass of lead is added to the rear of the model such that the center of gravity is shifted rearward by a length equal to 20 percent of the chord, calculate the moment about the center of gravity when the lift is 4000 N.

7.4   Consider the wing-body model in Prob. 7.2. Assume that a horizontal tail with no elevator is added to this model. The distance from the airplane's center of gravity to the tail's aerodynamic center is 1.0 m. The area of the tail is 0.4 m^2, and the tail-setting angle is 2.0°. The lift slope of the tail is 0.12 per degree. From experimental measurement, $\varepsilon_0 = 0$ and $\partial\varepsilon/\partial\alpha = 0.42$. If the absolute angle of attack of the model is 5° and the lift at this angle of attack is 4134 N, calculate the moment about the center of gravity.

7.5   Consider the wing-body-tail model of Prob. 7.4. Does this model have longitudinal static stability and balance?

7.6   For the configuration of Prob. 7.4, calculate the neutral point and static margin. $h = 0.26$.

7.7   Assume that an elevator is added to the horizontal tail of the configuration given in Prob. 7.4. The elevator control effectiveness is 0.04. Calculate the elevator deflection angle necessary to trim the configuration at an angle of attack of 8°.

7.8   Consider the configuration of Prob. 7.7. The elevator hinge moment derivatives are $\partial C_{h_e}/\partial\alpha_t = -0.007$ and $\partial C_{h_e}/\partial\delta_e = -0.012$. Assess the stick-free static stability of this configuration.

8

Space Flight (Astronautics)

It is difficult to say what is impossible, for the dream of yesterday is the hope of today and the reality of tomorrow.

> Robert H. Goddard, at his
> high school graduation, 1904

Houston, Tranquillity Base here. The Eagle has landed.

> Neil Armstrong, in a radio transmission
> to Mission Control, at the instant
> of the first manned landing
> on the moon, July 20, 1969

8.1 INTRODUCTION

Space—that last frontier, that limitless expanse which far outdistances the reach of our strongest telescopes, that region which may harbor other intelligent civilizations on countless planets; space—whose unknown secrets have attracted the imagination of humanity for centuries and whose technical conquest has labeled the latter half of the 20th century as the *space age*; space—that is the subject of this chapter.

To this point in our introduction of flight, we have emphasized aeronautics, the science and engineering of vehicles which are designed to move within the atmosphere and which depend on the atmosphere for their lift and propulsion. However, as presented in Sec. 1.8, the driving force behind the advancement of aviation has always been the desire to fly higher and faster. The ultimate, of course, is to fly so high and so fast that you find yourself in outer space, beyond the limits of the sensible atmosphere. Here, motion of the vehicle takes place only under the influence of gravity and possibly some type of propulsive force; however, the mode of propulsion must be entirely independent of the air for its thrust. Therefore, the physical fundamentals and engineering principles associated with space vehicles are somewhat different from those associated with airplanes. The purpose of this chapter is to introduce some of the basic concepts of space flight, that is, to introduce the discipline of *astronautics*. In particular, in the early sections of this chapter we emphasize the calculation and analysis of orbits and trajectories of space vehicles operating under the influence of gravitational forces only (such as in the vacuum of free space). In the later sections, we consider several aspects of the entry of a space vehicle into the earth's atmosphere, especially the entry trajectory and aerodynamic heating of the vehicle.

The space age formally began on October 4, 1957, when the Soviet Union launched *Sputnik I*, the first artificial satellite to go into orbit around the earth. Unlike the first flight of the Wright brothers in 1903, which took years to have any impact on society, the effect of *Sputnik I* on the world was immediate. Within 12 years, people had walked on the moon, and after another 7 years, unmanned probes were resting on the surfaces of Venus and Mars. A variety of different space vehicles designed for different missions have been launched since 1957. Most of these vehicles fall into three main categories:

1. Earth satellites, launched with enough velocity to go into orbit about the earth, as sketched in Fig. 8.1. As we will show later, velocities on the order of 26,000 ft/s (7.9 km/s) are necessary to place a vehicle in orbit about the earth, and these orbits are generally elliptical. Figure 8.2 shows a photograph of an artificial earth satellite.

2. Lunar and interplanetary vehicles, launched with enough velocity to overcome the gravitational attraction of the earth and to travel into deep space. Velocities of 36,000 ft/s (approximately 11 km/s) or larger are necessary for this purpose. Such trajectories are parabolic or hyperbolic. A typical path from the earth to

Figure 8.1   Earth orbit.

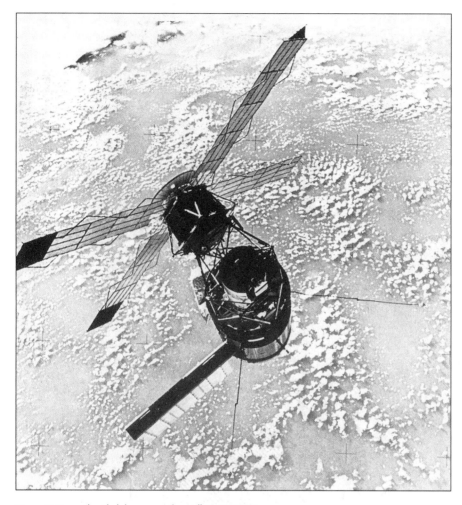

Figure 8.2   The Skylab—an earth satellite. (*NASA.*)

the moon is sketched in Fig. 8.3; here, the space vehicle is first placed in earth orbit, from which it is subsequently boosted by onboard rockets to an orbit about the moon, from which it finally makes a landing on the moon's surface. This is the mode employed by all the Apollo manned lunar missions, beginning with the historic first moon landing on July 20, 1969. A photograph of the Apollo spacecraft is shown in Fig. 8.4.

3. Space shuttles, designed to take off from the earth's surface, perform a mission in space, and then return and land on the earth's surface, all self-contained in the same vehicle. These are lifting reentry vehicles, designed with a reasonable L/D ratio to allow the pilot to land the craft just as an airplane. Earth orbit with

Figure 8.3   Earth-moon mission (not to scale).

a lifting reentry path is sketched in Fig. 8.5. The first successful flight of a space shuttle into space, with a subsequent lifting reentry and landing, was carried out by NASA's *Columbia* during the period April 12–14, 1981. A photograph of the space shuttle is given in Fig. 8.6.

Finally, a discussion of astronautics, even the present introductory one, requires slightly greater mathematical depth than just basic differential and integral calculus.

Figure 8.4 The Apollo spacecraft. (*Smithsonian National Air and Space Museum.*)

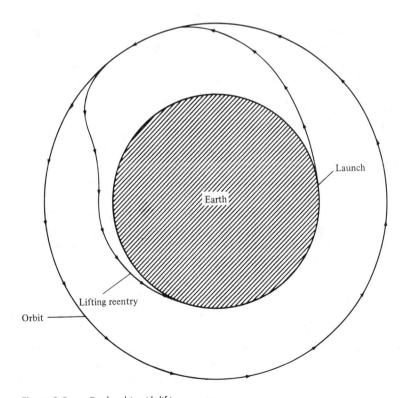

Figure 8.5 Earth orbit with lifting reentry.

Figure 8.6 The space shuttle. (*NASA.*)

Therefore, this chapter will incorporate more mathematical rigor than other parts of this book. In particular, some concepts from differential equations must be employed. However, it will be assumed that the reader has *not* had exposure to such mathematics, and therefore the necessary ideas will be introduced in a self-contained fashion.

The road map for this chapter is given in Fig. 8.7. Our study of astronautics is organized into three parts, following the three primary sequential phases of a space mission, as sketched in Fig. 8.8:

1. *Ascent through the atmosphere.* Most space vehicles initiate their mission by blasting off from the earth's surface, climbing out of the sensible atmosphere, and accelerating to orbital or escape velocity (we will define these velocities in Sec. 8.5). Called the *ascent phase*, this is mainly governed by the rocket engines which boost the vehicle into space. Rocket engines are discussed in Chap. 9 on propulsion. Hence, although for the sake of completeness the ascent phase is shown in our road map at the top left of Fig. 8.7, we defer the study of this phase until Chap. 9.

2. *Mission in space.* It is with this phase that we begin Chap. 8; namely, we study the motion of the space vehicle after is has been inserted into orbit or placed on a trajectory to carry it away from earth, that is, after the rocket engines have burned out. This is represented by the center box in Fig. 8.7.

3. *Planetary entry.* Some space vehicles continue on their paths indefinitely, moving into deep space without ever again encountering an atmosphere. Many others, especially those with human astronauts aboard, will eventually return to

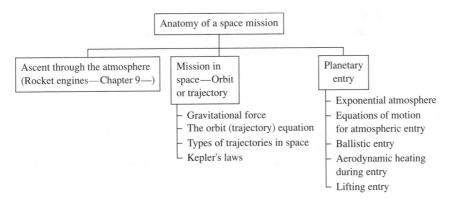

Figure 8.7 Road map for Chap. 8.

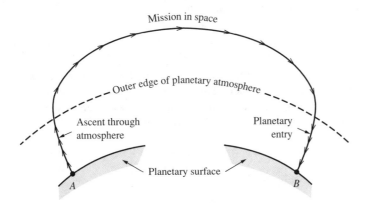

Figure 8.8 Anatomy of a space mission.

the earth's surface, and will have to come back through the earth's atmosphere at very high velocities. This is called *earth entry* (or sometimes by the misnomer *reentry*). Or, the space vehicle may enter the atmosphere of another planet in the solar system. This is called *planetary entry* (a more general term which includes earth entry). This phase is represented by the right-hand box in Fig. 8.7; it is the subject of the second half of Chap. 8.

8.2 DIFFERENTIAL EQUATIONS

Consider a dependent variable r which depends on an independent variable t. Thus, $r = f(t)$. The concept of the derivative of r with respect to t, denoted by dr/dt, has been used frequently in this book. The physical interpretation of dr/dt is simply the rate of change of r with respect to t. If r is a distance and t is time, then dr/dt is the rate of change of distance with respect to time, that is, velocity. The second derivative of r with respect to t is simply

$$\frac{d(dr/dt)}{dt} \equiv \frac{d^2r}{dt^2}$$

This is the rate of change of the derivative itself with respect to t. If r and t are distance and time, respectively, then d^2r/dt^2 is the rate of change of velocity with respect to time, that is, acceleration.

A differential equation is simply an equation which has derivatives in some of its terms. For example,

$$\frac{d^2r}{dt^2} + r\,\frac{dr}{dt} - 2t^3 = 2 \qquad \textbf{[8.1]}$$

is a differential equation; it contains derivatives along with the variables r and t themselves. By comparison, the equation

$$r + \frac{t^2}{r} = 0$$

is an algebraic equation; it contains only r and t without any derivatives.

To find a *solution* of the differential equation, Eq. (8.1), means to find a functional relation $r = f(t)$ which satisfies the equation. For example, assume that $r = t^2$. Then $dr/dt = 2t$ and $d^2r/dt^2 = 2$. Substitute into Eq. (8.1):

$$2 + t^2(2t) - 2t^3 = 2$$
$$2 + 2t^3 - 2t^3 = 2$$
$$2 = 2$$

Hence, $r = t^2$ does indeed satisfy the differential equation, Eq. (8.1). Thus, $r = t^2$ is called a solution of that equation.

Calculations of space vehicle trajectories involve distance r and time t. Some of the fundamental equations involve first and second derivatives of r with respect to t. To simplify the notation in these equations, we now introduce the *dot notation* for the time derivatives, that is,

$$\dot{r} \equiv \frac{dr}{dt}$$

$$\ddot{r} \equiv \frac{d^2r}{dt^2}$$

A single dot over the variable means the first time derivative of that variable; a double dot means the second derivative. For example, the differential equation Eq. (8.1) can be written as

$$\ddot{r} + r\dot{r} - 2t^3 = 2 \qquad \textbf{[8.2]}$$

Equations (8.1) and (8.2) are identical; only the notation is different. The dot notation for time derivatives is common in physical science; you will encounter it frequently in more advanced studies of science and engineering.

8.3 LAGRANGE'S EQUATION

In physical science, a study of the forces and motion of bodies is called *mechanics*. If the body is motionless, this study is further identified as *statics*; if the body is moving, the study is one of *dynamics*. In this chapter, we are concerned with the *dynamics* of space vehicles.

Problems in dynamics usually involve the use of Newton's second law, $F = ma$, where F is force, m is mass, and a is acceleration. Perhaps the reader is familiar with various applications of $F = ma$ from basic physics; indeed, we applied this law in Chap. 4 to obtain the momentum equation in aerodynamics and again in Chap. 6 to obtain the equations of motion for an airplane. However, in this section we introduce an equation, Lagrange's equation, which is essentially a corollary to Newton's second law. The use of Lagrange's equation represents an alternative approach to the solution of dynamics problems in lieu of $F = ma$; in the study of space vehicle orbits and trajectories, Lagrange's equation greatly simplifies the analysis. We do not derive Lagrange's equation; rather, we simply introduce it by way of an example and then, in the next section, apply it to obtain the orbit equation. A rigorous derivation of Lagrange's equation is left to more advanced studies of mechanics.

Consider the following example. A body of mass m is falling freely in the earth's gravitational field, as sketched in Fig. 8.9. Let x be the vertical distance of the body from the ground. If we ignore drag, the only force on the body is its weight w, directed downward. By definition, the weight of a body is equal to its mass m times the acceleration of gravity g; or $w = mg$. From Newton's second law,

$$F = ma \qquad \textbf{[8.3]}$$

The force is weight, directed downward. Since the direction of positive x is upward, then a downward-acting force is negative. Hence,

$$F = -w = -mg \qquad \textbf{[8.4]}$$

Figure 8.9 Falling body.

From the discussion in Sec. 8.2, the acceleration can be written as

$$a \equiv \frac{d^2x}{dt^2} \equiv \ddot{x} \qquad \textbf{[8.5]}$$

Substituting Eqs. (8.4) and (8.5) into Eq. (8.3) yields

$$-mg = m\ddot{x}$$

$$\boxed{\ddot{x} = -g} \qquad \textbf{[8.6]}$$

Equation (8.6) is the equation of motion for the body in our example. It is a differential equation whose solution will yield $x = f(t)$. Moreover, Eq. (8.6) was obtained by the application of Newton's second law.

Now consider an alternative formulation of this example, using Lagrange's equation. This will serve as an introduction to Lagrange's equation. Let T denote the *kinetic energy* of the body, where by definition

$$T = \tfrac{1}{2}mV^2 = \tfrac{1}{2}m(\dot{x})^2 \qquad \textbf{[8.7]}$$

Let Φ denote the potential energy of the body. By definition, the potential energy of a body referenced to the earth's surface is the weight of the body times the distance above the surface:

$$\Phi = wx = mgx \qquad \textbf{[8.8]}$$

Now define the *lagrangian function B* as the difference between kinetic and potential energy:

$$B \equiv T - \Phi \qquad \textbf{[8.9]}$$

For our example, combining Eqs. (8.7) to (8.9), we get

$$B = \tfrac{1}{2}m(\dot{x})^2 - mgx \qquad \textbf{[8.10]}$$

We now write *Lagrange's equation*, which will have to be accepted without proof; it is simply a corollary to Newton's second law:

$$\frac{d}{dt}\left(\frac{\partial B}{\partial \dot{x}}\right) - \frac{\partial B}{\partial x} = 0 \qquad \textbf{[8.11]}$$

In Lagrange's equation above, recall the definition of the partial derivative given in Sec. 7.2.4. For example, $\partial B/\partial \dot{x}$ means the derivative of B with respect to $\dot{x}$, holding everything else constant. Hence, from Eq. (8.10),

$$\frac{\partial B}{\partial \dot{x}} = m\dot{x} \qquad \textbf{[8.12]}$$

and
$$\frac{\partial B}{\partial x} = -mg \qquad \textbf{[8.13]}$$

Substituting Eqs. (8.12) and (8.13) into Eq. (8.11), we have

$$\frac{d}{dt}(m\dot{x}) - (-mg) = 0$$

or, because m is a constant,

$$m\frac{d}{dt}(\dot{x}) - (-mg) = 0$$

$$m\ddot{x} + mg = 0$$

$$\boxed{\ddot{x} = -g} \qquad \textbf{[8.14]}$$

Compare Eqs. (8.14) and (8.6); they are identical equations of motion. Therefore, we induce that Lagrange's equation and Newton's second law are equivalent mechanical relations and lead to the same equations of motion for a mechanical system. In the above example, the use of Lagrange's equation resulted in a slightly more complicated formulation than the direct use of $F = ma$. However, in the analysis of space vehicle orbits and trajectories, Lagrange's equation is the most expedient formulation, as will be detailed in the next section.

With the above example in mind, a more general formulation of Lagrange's equation can be given. Again, no direct proof is given; the reader must be content with the "cookbook" recipe given below, using the above example as a basis for induction. Consider a body moving in three-dimensional space, described by some generalized spatial coordinates q_1, q_2, and q_3. (These may be r, θ, and ϕ for a spherical coordinate system; x, y, and z for rectangular coordinate system; etc.) Set up the expression for the *kinetic energy* of the body, which may depend on the coordinates q_1, q_2, and q_3 themselves as well as the velocities $\dot{q}_1, \dot{q}_2$, and $\dot{q}_3$:

$$T = T(q_1, q_2, q_3, \dot{q}_1, \dot{q}_2, \dot{q}_3) \qquad \textbf{[8.15]}$$

Then set up the expression for the *potential energy* of the body, which depends only on spatial location:

$$\Phi = \Phi(q_1, q_2, q_3) \qquad \textbf{[8.16]}$$

Now form the *lagrangian function*

$$B = T - \Phi \qquad \textbf{[8.17]}$$

Finally, obtain three equations of motion (one along each coordinate direction) by writing Lagrange's equation for each coordinate:

$$
\begin{array}{ll}
q_1 \text{ coordinate:} & \dfrac{d}{dt}\left(\dfrac{\partial B}{\partial \dot{q}_1}\right) - \dfrac{\partial B}{\partial q_1} = 0 \\[3mm]
q_2 \text{ coordinate:} & \dfrac{d}{dt}\left(\dfrac{\partial B}{\partial \dot{q}_2}\right) - \dfrac{\partial B}{\partial q_2} = 0 \\[3mm]
q_3 \text{ coordinate:} & \dfrac{d}{dt}\left(\dfrac{\partial B}{\partial \dot{q}_3}\right) - \dfrac{\partial B}{\partial q_3} = 0
\end{array}
\qquad [8.18]
$$

Let us now apply this formalism to obtain the orbit or trajectory equations for a space vehicle.

8.4 ORBIT EQUATION

Space vehicles are launched from a planet's surface by rocket boosters. The rocket engines driving these boosters are discussed in Chap. 9. Here we are concerned with the motion of the vehicle after all stages of the booster have burned out and the satellite, interplanetary probe, etc., are smoothly moving through space under the influence of gravitational forces. At the instant the last booster stage burns out, the space vehicle is at a given distance from the center of the planet, moving in a specific direction at a specific velocity. Obviously, nature prescribes a specific path (a specific orbit about the planet or possibly a specific trajectory away from the planet) for these given conditions at burnout. The purpose of this section is to derive the equation which describes this path. Referring to our road map in Fig. 8.7, we begin with the center column.

8.4.1 Force and Energy

Consider a vehicle of mass m moving with velocity V in the vicinity of a planet of large mass M, as sketched in Fig. 8.10. The distance between the centers of the two

Figure 8.10   Movement of a small mass in the gravitational field of a large mass.

masses is r. In a stroke of genius during the last quarter of the 17th century, Isaac Newton uncovered the law of universal gravitation, which states that the gravitational force between two masses varies inversely as the square of the distance between their centers. In particular, this force is given by

$$F = \frac{GmM}{r^2} \qquad \text{[8.19]}$$

where G is the universal gravitational constant, $G = 6.67 \times 10^{-11} \text{ m}^3/(\text{kg})(\text{s})^2$.

Lagrange's equation deals with energy, both potential and kinetic. First consider the potential energy of the system shown in Fig. 8.10. Potential energy is always based on some reference point, and for gravitational problems in astronautics, it is conventional to establish the potential energy as zero at r equal to infinity. Hence, the potential energy at a distance r is defined as the work done in moving the mass m from infinity to the location r. Let Φ be the potential energy. If the distance between M and m is changed by a small increment dr, then the work done in producing this change is $F \, dr$. This is also the change in potential energy $d\Phi$. Using Eq. (8.19), we obtain

$$d\Phi = F \, dr = \frac{GmM}{r^2} \, dr$$

Integrating from r equals infinity, where Φ by definition is 0, to $r = r$, where the potential energy is $\Phi = \Phi$, we get

$$\int_0^{\Phi} d\Phi = \int_{\infty}^{r} \frac{-GmM}{r^2} \, dr$$

or

$$\boxed{\Phi = \frac{-GmM}{r}} \qquad \text{[8.20]}$$

Equation (8.20) gives the potential energy of small mass m in the gravitational field of large mass M at distance r. The potential energy at r is a negative value owing to our choice of $\Phi = 0$ at r going to infinity. However, if the idea of a negative energy is foreign to you, do not be concerned. In mechanical systems, we are usually concerned with *changes* in energy, and such changes are independent of our choice of reference for potential energy.

Now consider the kinetic energy. Here we need more precisely to establish our coordinate system. In more advanced studies of mechanics it can be proved that the motion of a body in a central force field (such as we are dealing with here) takes place in a plane. Hence, we need only two coordinates to designate the location of mass m. Polar coordinates are particularly useful in this case, as shown in Fig. 8.11. Here the origin is at the center of mass M, r is the distance between m and M, and θ is the angular orientation of r. The velocity of the vehicle of mass m is V. The velocity component parallel to r is $V_r = dr/dt = \dot{r}$. The velocity component perpendicular to r is equal to the radius vector r times the time rate of change of θ, that is, times the

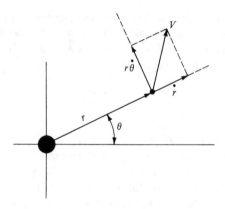

Figure 8.11   Polar coordinate system.

angular velocity; $V_\theta = r(d\theta/dt) = r\dot\theta$. Therefore, the kinetic energy of the vehicle is

$$T = \tfrac{1}{2}mV^2 = \tfrac{1}{2}[\dot r^2 + (r\dot\theta)^2]m \qquad\qquad \textbf{[8.21]}$$

8.4.2   Equation of Motion

From Eqs. (8.17), (8.20), and (8.21), the lagrangian function is

$$B = T - \Phi = \frac{1}{2}m[\dot r^2 + (r\dot\theta)^2] + \frac{GmM}{r} \qquad\qquad \textbf{[8.22]}$$

In orbital analysis, it is common to denote the product GM by k^2. If we are dealing with the earth, where $M = 5.98 \times 10^{24}$ kg, then

$$k^2 \equiv GM = 3.986 \times 10^{14} \text{ m}^3/\text{s}^2$$

Equation (8.22) then becomes

$$B = \frac{1}{2}m[\dot r^2 + (r\dot\theta)^2] + \frac{mk^2}{r} \qquad\qquad \textbf{[8.23]}$$

Now invoke Lagrange's equation, Eq. (8.18), where $q_1 = \theta$ and $q_2 = r$. First, the θ equation is

$$\frac{d}{dt}\left(\frac{\partial B}{\partial \dot\theta}\right) - \frac{\partial B}{\partial \theta} = 0 \qquad\qquad \textbf{[8.24]}$$

From Eq. (8.23)

$$\frac{\partial B}{\partial \dot\theta} = mr^2\dot\theta \qquad\qquad \textbf{[8.25]}$$

and
$$\frac{\partial B}{\partial \theta} = 0 \qquad\qquad \text{[8.26]}$$

Substituting Eqs. (8.25) and (8.26) into Eq. (8.24), we obtain

$$\frac{d}{dt}\left(mr^2\dot{\theta}\right) = 0 \qquad\qquad \text{[8.27]}$$

Equation (8.27) is the equation of motion of the space vehicle in the θ direction. It can be immediately integrated as

$$\boxed{mr^2\dot{\theta} = \text{const} = c_1} \qquad\qquad \text{[8.28]}$$

From elementary physics, linear momentum is defined as mass times velocity. Analogously, for angular motion, *angular momentum* is defined as $I\dot{\theta}$, where I is the moment of inertia and $\dot{\theta}$ is the angular velocity. For a point mass m, $I = mr^2$. Hence, the product $mr^2\dot{\theta}$ is the *angular momentum of the space vehicle*, and from Eq. (8.28),

$$\boxed{mr^2\dot{\theta} = \text{angular momentum} = \text{const}}$$

For a central force field, Eq. (8.28) demonstrates that the angular momentum is constant.

Now consider the r equation. From Eq. (8.18), where $q_2 = r$,

$$\frac{d}{dt}\frac{\partial B}{\partial \dot{r}} - \frac{\partial B}{\partial r} = 0 \qquad\qquad \text{[8.29]}$$

From Eq. (8.23)

$$\frac{\partial B}{\partial \dot{r}} = m\dot{r} \qquad\qquad \text{[8.30]}$$

$$\frac{\partial B}{\partial r} = -\frac{mk^2}{r^2} + mr\dot{\theta}^2 \qquad\qquad \text{[8.31]}$$

Substituting Eqs. (8.30) and (8.31) into Eq. (8.29), we get

$$\frac{d}{dt}m\dot{r} + \frac{mk^2}{r^2} - mr\dot{\theta}^2 = 0 \qquad\qquad \text{[8.32]}$$

or
$$m\ddot{r} - mr\dot{\theta}^2 + \frac{mk^2}{r^2} = 0 \qquad\qquad \text{[8.33]}$$

Equation (8.28) demonstrated that since m is constant, $r^2\dot{\theta}$ is constant. Denote this quantity by h.

$$r^2\dot{\theta} \equiv h = \text{angular momentum per unit mass}$$

Multiplying and dividing the second term of Eq. (8.33) by r^3 and canceling m yield

$$m\ddot{r} - m\frac{r^4\dot{\theta}^2}{r^3} + \frac{mk^2}{r^2} = 0$$

or
$$\boxed{\ddot{r} - \frac{h^2}{r^3} + \frac{k^2}{r^2} = 0}$$
[8.34]

Equation (8.34) is the equation of motion for the space vehicle in the r direction. Note that both h^2 and k^2 are constants. Recalling our discussion in Sec. 8.2, we see that Eq. (8.34) is a differential equation. Its solution will provide a relation for r as the function of time, that is, $r = f(t)$.

However, examine Fig. 8.11. The equation of the *path* of the vehicle in space should be geometrically given by $r = f(\theta)$, not $r = f(t)$. We are interested in this path; that is, we want the equation of the space vehicle motion in terms of its geometric coordinates r and θ. Therefore, Eq. (8.34) must be reworked, as follows.

Let us transform Eq. (8.34) to a new dependent variable u, where

$$r = \frac{1}{u}$$
[8.35]

Then
$$h \equiv r^2\dot{\theta} = \frac{\dot{\theta}}{u^2}$$
[8.36]

Hence
$$\dot{r} \equiv \frac{dr}{dt} = \frac{d(1/u)}{dt} = -\frac{1}{u^2}\frac{du}{dt}$$
[8.37]

$$= -\frac{1}{u^2}\frac{du}{d\theta}\frac{d\theta}{dt} = -\frac{\dot{\theta}}{u^2}\frac{du}{d\theta} = -h\frac{du}{d\theta}$$

Differentiating Eq. (8.37) with respect to t, we get

$$\ddot{r} = -h\frac{d}{dt}\frac{du}{d\theta} = -h\left(\frac{d}{d\theta}\frac{du}{d\theta}\right)\frac{d\theta}{dt}$$

$$= -h\left(\frac{d^2u}{d\theta^2}\right)\frac{d\theta}{du} = -h\frac{d^2u}{d\theta^2}\dot{\theta}$$
[8.38]

But from Eq. (8.36), $\dot{\theta} = u^2h$. Substituting into Eq. (8.38), we obtain

$$\ddot{r} = -h^2u^2\frac{d^2u}{d\theta^2}$$
[8.39]

Substituting Eqs. (8.39) and (8.35) into Eq. (8.34) yields

$$-h^2u^2\frac{d^2u}{d\theta^2} - h^2u^3 + k^2u^2 = 0$$

or, by dividing by h^2u^2,

$$\frac{d^2u}{d\theta^2} + u - \frac{k^2}{h^2} = 0$$
[8.40]

Equation (8.40) is just as valid an equation of motion as the original Eq. (8.34). Equation (8.40) is a differential equation, and its solution gives $u = f(\theta)$. Specifically, a solution of Eq. (8.40) is

$$u = \frac{k^2}{h^2} + A\cos(\theta - C) \qquad \text{[8.41]}$$

where A and C are constants (essentially, constants of integration). You should satisfy yourself that Eq. (8.41) is indeed a solution of Eq. (8.40) by substitution of (8.41) into (8.40).

Return to the original transformation, Eq. (8.35). Substituting $u = 1/r$ into Eq. (8.41) yields

$$r = \frac{1}{k^2/h^2 + A\cos(\theta - C)} \qquad \text{[8.42]}$$

Multiply and divide Eq. (8.42) by h^2/k^2:

$$\boxed{r = \frac{h^2/k^2}{1 + A(h^2/k^2)\cos(\theta - C)}} \qquad \text{[8.43]}$$

Equation (8.43) is the desired equation of the path (the orbit, or trajectory) of the space vehicle. It is an algebraic equation for $r = f(\theta)$; it gives the geometric coordinates r and θ for a given path. The *specific* path is dictated by the values of the constants h^2, A, and C in Eq. (8.43). In turn, refer to Fig. 8.12: these constants are fixed by conditions at the instant of burnout of the rocket booster. At burnout, the vehicle is a distance r_b from the center of the earth, and its velocity has a magnitude V_b in a direction β_b with respect to a perpendicular to r. These burnout conditions completely specify the vehicle's path; that is, they determine the values of h^2, A, and C for Eq. (8.43).

Equation (8.43) is sometimes generically called the *orbit equation*. However, it applies to the trajectory of a space vehicle escaping from the gravitational field of the earth as well as to an artificial satellite in orbit about the earth. In fact, what kind of orbit or trajectory is described by Eq. (8.43)? What type of mathematical curve is

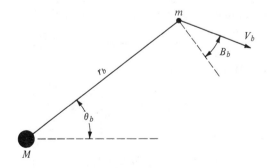

Figure 8.12   Conditions at the instance of burnout.

it? What physical conditions are necessary for a body to go into orbit or to escape from the earth? The answers can be found by further examination of Eq. (8.43), as discussed below.

8.5   SPACE VEHICLE TRAJECTORIES—SOME BASIC ASPECTS

Examine Eq. (8.43) closely. It has the general form

$$r = \frac{p}{1 + e\cos(\theta - C)} \qquad \textbf{[8.44]}$$

where $p = h^2/k^2$, $e = A(h^2/k^2)$, and C is simply a phase angle. From analytic geometry, Eq. (8.44) is recognized as the standard form of a *conic section* in polar coordinates; that is, Eq. (8.44) is the equation of a circle, ellipse, parabola, or hyperbola, depending on the value of e, where e is the *eccentricity* of the conic section. Specifically:

If $e = 0$, the path is a *circle*.

If $e < 1$, the path is an *ellipse*.

If $e = 1$, the path is a *parabola*.

If $e > 1$, the path is a *hyperbola*.

These possibilities are sketched in Fig. 8.13. Note that point b on these sketches denotes the point of burnout and that θ is referenced to the dashed line through b, that is, θ is arbitrarily chosen as zero at burnout. Then C is simply a phase angle which orients the x and y axes with respect to the burnout point, where the x axis is a line of symmetry for the conic section. From inspection of Fig. 8.13, circular and elliptical paths result in an orbit about the large mass M (the earth), whereas parabolic and hyperbolic paths result in escape from the earth.

On a physical basis, the eccentricity, hence the type of path for the space vehicle, is governed by the difference between the kinetic and potential energies of the vehicle. To prove this, consider first the kinetic energy $T = \frac{1}{2}mV^2$. From Eq. (8.21),

$$T = \tfrac{1}{2}m[\dot{r}^2 + (r\dot{\theta})^2]$$

Differentiate Eq. (8.44) with respect to t:

$$\frac{dr}{dt} = \dot{r} = \frac{[re\sin(\theta - C)]\dot{\theta}}{1 + e\cos(\theta - C)} \qquad \textbf{[8.45]}$$

Substitute Eq. (8.45) into (8.21):

$$T = \frac{1}{2}m\left\{ \frac{[r^2e^2\sin^2(\theta - C)]\dot{\theta}^2}{[1 + e\cos(\theta - C)]^2} + r^2\dot{\theta}^2 \right\} \qquad \textbf{[8.46]}$$

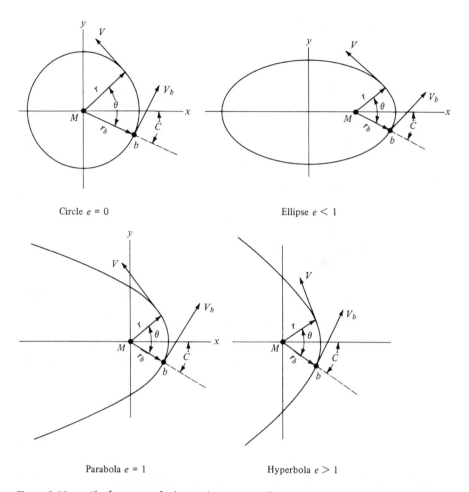

Circle $e = 0$

Ellipse $e < 1$

Parabola $e = 1$

Hyperbola $e > 1$

Figure 8.13 The four types of orbits and trajectories, illustrating the relation of the burnout point and phase angle with the axes of symmetry.

Recall that $r^2 \dot{\theta} \equiv h$, hence $\dot{\theta}^2 = h^2/r^4$. Thus, Eq. (8.46) becomes

$$T = \frac{1}{2} m \left\{ \frac{h^2 e^2 \sin^2(\theta - C)}{r^2 [1 + e \cos(\theta - C)]^2} + \frac{h^2}{r^2} \right\}$$ **[8.47]**

Putting the right-hand side of Eq. (8.47) under the same common denominator and remembering from Eq. (8.44) that

$$r^2 [1 + e \cos(\theta - C)]^2 = \left(\frac{h^2}{k^2} \right)^2$$

we transform Eq. (8.47) to

$$T = \frac{1}{2}m\frac{k^4}{h^2}[1 + 2e\cos(\theta - C) + e^2] \qquad \textbf{[8.48]}$$

The reader should fill in the few missing algebraic steps to obtain Eq. (8.48).

Consider now the absolute value of the potential energy, denoted as $|\Phi|$. From Eq. (8.20),

$$|\Phi| = \frac{GMm}{r} = \frac{k^2m}{r} \qquad \textbf{[8.49]}$$

Substitute Eq. (8.44) into Eq. (8.49):

$$|\Phi| = \frac{k^4m}{h^2}[1 + e\cos(\theta - C)] \qquad \textbf{[8.50]}$$

The difference between the kinetic and potential energies is obtained by subtracting Eq. (8.50) from Eq. (8.48):

$$T - |\Phi| = \frac{1}{2}m\frac{k^4}{h^2}[1 + 2e\cos(\theta - C) + e^2] - \frac{k^4m}{h^2}[1 + e\cos(\theta - C)] \qquad \textbf{[8.51]}$$

Let H denote $T - |\Phi|$. Then Eq. (8.51) becomes

$$H \equiv T - |\Phi| = -\frac{1}{2}m\frac{k^4}{h^2}(1 - e^2) \qquad \textbf{[8.52]}$$

Solving Eq. (8.52) for e, we get

$$\boxed{e = \sqrt{1 + \frac{2h^2H}{mk^4}}} \qquad \textbf{[8.53]}$$

Equation (8.53) is the desired result, giving the eccentricity e in terms of the difference between kinetic and potential energies H.

Examine Eq. (8.53). If the kinetic energy is smaller than the potential energy, H will be negative and hence $e < 1$. If the kinetic and potential energies are equal, $H = 0$ and $e = 1$. Similarly, if the kinetic energy is larger than the potential energy, H is positive and $e > 1$. Referring again to Fig. 8.13, we can make the following tabulation:

Type of Trajectory	e	Energy Relation
Ellipse	< 1	$\frac{1}{2}mV^2 < \frac{GMm}{r}$
Parabola	$= 1$	$\frac{1}{2}mV^2 = \frac{GMm}{r}$
Hyperbola	> 1	$\frac{1}{2}mV^2 > \frac{GMm}{r}$

From this we draw the important conclusion that a vehicle intended to escape the earth and travel into deep space (a parabolic or hyperbolic trajectory) must be launched such that its kinetic energy at burnout is equal to or greater than its potential energy, a conclusion that makes intuitive sense even without the above derivation.

Equation (8.53) tells us more. For example, what velocity is required for a circular orbit? To answer this question, recall that a circle has zero eccentricity. Putting $e = 0$ into Eq. (8.53), we get

$$0 = \sqrt{1 + \frac{2h^2 H}{mk^4}}$$

or
$$H = -\frac{mk^4}{2h^2} \qquad \textbf{[8.54]}$$

Recall that $H = T - |\Phi| = \frac{1}{2}mV^2 - GMm/r$. Hence, Eq. (8.54) becomes

$$\frac{1}{2}mV^2 = -\frac{mk^4}{2h^2} + \frac{GMm}{r} \qquad \textbf{[8.55]}$$

From Eq. (8.44), with $e = 0$,

$$r = \frac{h^2}{k^2} \qquad \textbf{[8.56]}$$

Substitute Eq. (8.56) into (8.55) and solve for V:

$$\frac{1}{2}mV^2 = -\frac{m}{2}\frac{k^2}{r} + \frac{k^2 m}{r} = \frac{k^2 m}{2r}$$

Thus
$$\boxed{V = \sqrt{\frac{k^2}{r}}} \quad \text{circular velocity} \qquad \textbf{[8.57]}$$

Equation (8.57) gives the velocity required to obtain a circular orbit. Recall from Sec. 8.4.2 that $k^2 = GM = 3.956 \times 10^{14}$ m^3/s^2. Assume that $r = 6.4 \times 10^6$ m, essentially the radius of the earth. Then

$$V = \sqrt{\frac{3.986 \times 10^{14}}{6.4 \times 10^6}} = 7.9 \times 10^3 \text{ m/s}$$

This is a convenient number to remember; *circular, or orbital, velocity is 7.9 km/s, or approximately 26,000 ft/s.*

The velocity required to escape the earth can be obtained in much the same fashion. We have demonstrated above that a vehicle will escape if it has a parabolic ($e = 1$) or a hyperbolic ($e > 1$) trajectory. Consider a parabolic trajectory. For this, we know that the kinetic and potential energies are equal; $T = |\Phi|$. Hence,

$$\frac{1}{2}mV^2 = \frac{GMm}{r} = \frac{k^2 m}{r}$$

Solving for V, we get

$$V = \sqrt{\frac{2k^2}{r}} \qquad \text{parabolic velocity} \qquad \text{[8.58]}$$

Equation (8.58) gives the velocity required to obtain a parabolic trajectory. This is called the *escape velocity*; note by comparing Eqs. (8.57) and (8.58) that the escape velocity is larger than the orbital velocity by a factor of $\sqrt{2}$. Again assuming r is the radius of the earth, $r = 6.4 \times 10^6$ m, then *escape velocity is 11.2 km/s, or approximately 36,000 ft/s*. Return to Fig. 8.12; if at burnout $V_b \geq 11.2$ km/s, then the vehicle will escape the earth, independent of the direction of motion β_b.

Example 8.1

At the end of a rocket launch of a space vehicle, the burnout velocity is 9 km/s in a direction due north and 3° above the local horizontal. The altitude above sea level is 500 mi. The burnout point is located at the 27th parallel (27°) above the equator. Calculate and plot the trajectory of the space vehicle.

Solution

The burnout conditions are sketched in Fig. 8.14. The altitude above sea level is

$$h_G = 500 \text{ mi} = 0.805 \times 10^6 \text{ m}$$

The distance from the center of the earth to the burnout point is (where the earth's radius is $r_e = 6.4 \times 10^6$ m)

$$r_b = r_e + h_G = 6.4 \times 10^6 + 0.805 \times 10^6 = 7.2 \times 10^6 \text{ m}$$

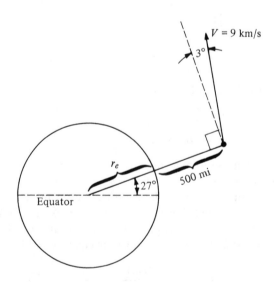

Figure 8.14   Burnout conditions for Example 8.1.

As given in Sec. 8.4.2,

$$k^2 \equiv GM = 3.986 \times 10^{14} \text{ m}^3/\text{s}^2$$

Also, as defined earlier,

$$h = r^2\dot{\theta} = r(r\dot{\theta}) = rV_\theta$$

where V_θ is the velocity component perpendicular to the radius vector r. Thus,

$$h = rV_\theta = r_b V \cos\beta_b = (7.2 \times 10^6)(9 \times 10^3)\cos 3° = 6.47 \times 10^{10} \text{ m}^2/\text{s}$$
$$h^2 = 4.188 \times 10^{21} \text{ m}^4/\text{s}^2$$

Hence $\quad p \equiv \dfrac{h^2}{k^2} = \dfrac{4.188 \times 10^{21}}{3.986 \times 10^{14}} = 1.0506 \times 10^7 \text{ m}$

The trajectory equation is given by Eq. (8.44), where the above value of p is the numerator of the right-hand side. To proceed further, we need the eccentricity e. This can be obtained from Eq. (8.53)

$$e = \sqrt{1 + \frac{2h^2 H}{mk^4}}$$

where $H/m = (T - |\Phi|)/m$.

$$\frac{T}{m} = \frac{V^2}{2} = \frac{(9 \times 10^3)^2}{2} = 4.05 \times 10^7 \text{ m}^2/\text{s}^2$$

$$\left|\frac{\Phi}{m}\right| = \frac{GM}{r_b} = \frac{k^2}{r_b} = \frac{3.986 \times 10^{14}}{7.2 \times 10^6} = 5.536 \times 10^7 \text{ m}^2/\text{s}^2$$

Hence $\qquad \dfrac{H}{m} = (4.05 - 5.536) \times 10^7 = -1.486 \times 10^7 \text{ m}^2/\text{s}^2$

Thus $\qquad e = \left[1 + \dfrac{2h^2}{k^4}\left(\dfrac{H}{m}\right)\right]^{1/2}$

$$= \left[1 + \frac{2(4.188 \times 10^{21})(-1.486 \times 10^7)}{(3.986 \times 10^{14})^2}\right]^{1/2} = \sqrt{0.2166} = 0.4654$$

Immediately we recognize that the trajectory is an elliptical orbit, because $e < 1$ and because $T < |\Phi|$. From Eq. (8.44),

$$r = \frac{p}{1 + e\cos(\theta - C)} = \frac{1.0506 \times 10^7}{1 + 0.4654\cos(\theta - C)}$$

To find the phase angle C, simply substitute the burnout location ($r_b = 7.2 \times 10^6$ m and $\theta = 0°$) into the above equation. (Note that $\theta = 0°$ at burnout, and hence θ is measured relative to the radius vector at burnout, with increasing θ taken in the direction of motion; this is sketched in Fig. 8.13.)

$$r_b = \frac{p}{1 + e\cos(-C)}$$

$$7.2 \times 10^6 = \frac{1.0506 \times 10^7}{1 + 0.4654\cos(-C)}$$

Solve for $\cos(-C)$:

$$\cos(-C) = 0.9878$$

Thus $$C = -8.96°$$

Finally, the complete equation of the orbit is

$$r = \frac{1.0506 \times 10^7}{1 + 0.4654\cos(\theta + 8.96°)}$$

where θ is in degrees and r is in meters.

The orbit is drawn to scale in Fig. 8.15. Note that b designates the burnout point, which is 27° above the equator. The x and y axes are the axes of symmetry for the elliptical orbit, and the phase angle orients the x axis at 8.96° *below* (because C is negative in this problem) the radius vector through point b. The angle θ is measured from the radius through b, with

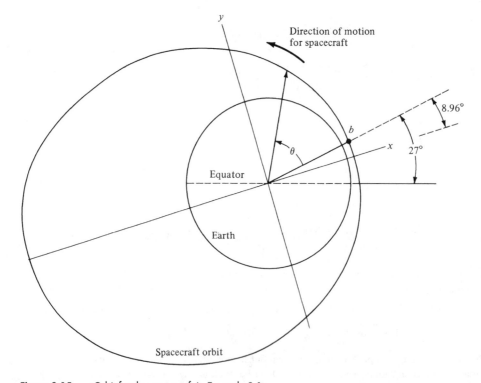

Figure 8.15   Orbit for the spacecraft in Example 8.1.

positive θ in the counterclockwise direction. The spacecraft is traveling in a counterclockwise direction, following an elliptical orbit. The perigee and apogee are 7.169×10^6 and 1.965×10^7 m, respectively.

8.6 KEPLER'S LAWS

To this point, our discussion has been couched in terms of an artificial space vehicle launched from the earth. However, most of the above analysis and results hold in general for orbits and trajectories of any mass in a central gravitational force field. The most familiar natural example of such motion is our solar system, that is, the orbits of the planets about the sun. Such motion has held people's attention since the early days of civilization. Early observations and mapping of planetary motion evolved over millennia, passing from the Babylonians to the Egyptians to the Greeks to the Romans, carried throughout the dark ages by the Arabians, and reaching the age of Copernicus in the 15th century (about the time Christopher Columbus was discovering America). However, by this time, astronomical observations were still inaccurate and uncertain. Then, during the period from 1576 to 1597, Tycho Brahe, a Danish noble, made a large number of precise astronomical observations which improved the accuracy of existing tables by a factor of 50. Near the end of his life, Brahe was joined by Johannes Kepler, a young German astronomer and mathematician, who further improved these observations. Moreover, Kepler made some pioneering conclusions about the geometry of planetary motion. From 1609 to 1618, Kepler induced and published three laws of planetary motion, obtained strictly from an exhaustive examination of the astronomical data. Kepler did not have the advantage of Newton's law of universal gravitational or newtonian mechanics, which came three-quarters of a century later. Nevertheless, Kepler's inductions were essentially correct, and his classical three laws are as important today for the understanding of artificial satellite motion as they were in the 17th century for the understanding of planetary motion. Therefore, his conclusions will be discussed in this section. Moreover, we will take advantage of our previous derivations of orbital motion to derive Kepler's laws, a luxury Kepler did not have himself.

Kepler's first major conclusion was as follows.

Kepler's first law: A satellite describes an *elliptical* path around its center of attraction.

We have already proved this fact in Secs. 8.4 and 8.5; therefore, nothing more needs to be said.

To prove Kepler's second law, recall from Eq. (8.28) that angular momentum is constant; that is, $mr^2\dot{\theta} = $ constant. Consider Fig. 8.16, which shows the radius vector r sweeping through an infinitesimally small angle $d\theta$. The area of the small triangle swept out is $dA = \frac{1}{2}r\, dh$. However, $dh = r\, d\theta$. Thus $dA = \frac{1}{2}r^2\, d\theta$. The time rate

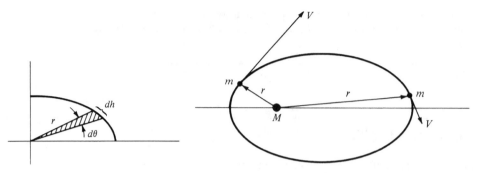

Figure 8.16   Area swept out by the
radius vector in mov-
ing through angle $d\theta$.

Figure 8.17   Illustration of the variation in velocity at
different points along the orbit.

of change of the area swept out by the radius is then

$$\frac{dA}{dt} = \frac{\frac{1}{2}r^2\,d\theta}{dt} = \frac{1}{2}r^2\dot{\theta} \qquad\qquad \textbf{[8.59]}$$

However, from Eq. (8.28), $r^2\dot{\theta}$ is a constant. Hence, Eq. (8.59) shows that

$$\frac{dA}{dt} = \text{const} \qquad\qquad \textbf{[8.60]}$$

which proves Kepler's second law.

> *Kepler's second law*: In equal times, the areas swept out by the radius vector of
> a satellite are the same.

An obvious qualitative conclusion follows from this law, as illustrated in Fig. 8.17.
Here, the elliptical orbit of a small mass m is shown about a large mass M. In order
for equal areas to be swept out in equal times, the satellite must have a larger velocity
when it is near M and a smaller velocity when it is far away. This is characteristic of
all satellite motion.

To derive Kepler's third law, consider the elliptical orbit shown in Fig. 8.18. The
point of closest approach, where r is minimum, is defined as the *perigee*; the point
farthest away, where r is maximum, is defined as the *apogee*. The mass M (say, that
of the earth or the sun) is at the focus of the ellipse. The major axis of the ellipse is
the distance from the perigee to the apogee, and one-half this distance is defined as
the *semimajor axis a*. The *semiminor axis b* is also shown in Fig. 8.18. Let us assume
for simplicity that the phase angle C of the orbit is zero. Thus, from Eq. (8.44), the
maximum and minimum radii are, respectively,

$$r_{\max} = \frac{h^2/k^2}{1-e} \qquad\qquad \textbf{[8.61]}$$

$$r_{\min} = \frac{h^2/k^2}{1+e} \qquad\qquad \textbf{[8.62]}$$

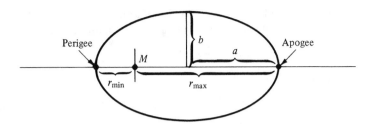

Figure 8.18 Illustration of apogee, perigee, and semimajor and semiminor axes.

From the definition of a, and using Eqs. (8.61) and (8.62), we obtain

$$a = \frac{1}{2}(r_{max} + r_{min}) = \frac{1}{2}\frac{h^2}{k^2}\left(\frac{1}{1-e} + \frac{1}{1+e}\right) = \frac{h^2/k^2}{1-e^2} \qquad [8.63]$$

The eccentricity e of the ellipse is geometrically related to the semimajor and semiminor axes; taking a result from analytic geometry, we get

$$e = \frac{(a^2 - b^2)^{1/2}}{a}$$

Solving for b gives

$$b = a(1 - e^2)^{1/2} \qquad [8.64]$$

If we lift another result from analytic geometry, we find the area of an ellipse is

$$A = \pi a b \qquad [8.65]$$

Substituting Eq. (8.64) into (8.65) yields

$$A = \pi a[a(1 - e^2)^{1/2}] = \pi a^2 (1 - e^2)^{1/2} \qquad [8.66]$$

Now return to Eq. (8.59):

$$dA = \tfrac{1}{2}r^2\dot{\theta}\,dt = \tfrac{1}{2}h\,dt \qquad [8.67]$$

Thus, the area of the ellipse can be obtained by integrating Eq. (8.67) around the complete orbit. That is, imagine the satellite starting at the perigee at time = 0. Now allow the satellite to move around one complete orbit, returning to the perigee. The area swept out by the radius vector is the whole area of the ellipse A. The time taken by the satellite in executing the complete orbit is defined as the *period* and is denoted by τ. Thus, integrating Eq. (8.67) around the complete orbit, we get

$$\int_0^A dA = \int_0^\tau \tfrac{1}{2}h\,dt$$

or

$$A = \tfrac{1}{2}h\tau \qquad [8.68]$$

We now have two independent results for A: Eq. (8.66) from analytic geometry and Eq. (8.68) from orbital mechanics. Equating these two relations, we have

$$\frac{1}{2}h\tau = \pi a^2(1 - e^2)^{1/2} \tag{8.69}$$

Solve Eq. (8.63) for h:

$$h = a^{1/2}k(1 - e^2)^{1/2} \tag{8.70}$$

Substitute Eq. (8.70) into Eq. (8.69):

$$\frac{1}{2}\tau a^{1/2}k(1 - e^2)^{1/2} = \pi a^2(1 - e^2)^{1/2}$$

or, squaring both sides,

$$\frac{1}{4}\tau^2 ak^2 = \pi^2 a^4$$

or

$$\tau^2 = \frac{4\pi^2}{k^2}a^3 \tag{8.71}$$

Examine Eq. (8.71). The factor $4\pi^2/k^2$ is a constant. Hence

$$\tau^2 = (\text{const})(a^3) \tag{8.72}$$

that is, the square of the period is proportional to the cube of the semimajor axis. If we have two satellites in orbit about the same planet, with values of τ_1, a_1 and τ_2, a_2, respectively, then Kepler's third law can be written as follows.

Kepler's third law: The periods of any two satellites about the same planet are related to their semimajor axes as

$$\frac{\tau_1^2}{\tau_2^2} = \frac{a_1^3}{a_2^3}$$

On this note, we conclude our discussion of space vehicle orbits and trajectories. We have finished the middle column in our road map, Fig. 8.7. Consistent with the scope of this book, we have provided only an introduction to the topic. Modern orbital and trajectory analysis is performed on high-speed digital computers, taking into account the gravitational attraction of several bodies simultaneously (e.g., gravitational attraction of the earth, sun, and moon on a lunar space vehicle), perturbations of the gravitational field due to the real nonspherical shape of the earth, trajectory corrections and orbit transfers due to in-flight propulsion, etc. Also, much attention is paid to satellite attitude control, that is, the proper orientation of the vehicle with respect to some given reference system. The reader is encouraged to look further into such matters in more advanced studies of astronautics.

Example 8.2

The period of revolution of the earth about the sun is 365.256 days. The semimajor axis of the earth's orbit is 1.49527×10^{11} m. In turn, the semimajor axis of the orbit of Mars is 2.2783×10^{11} m. Calculate the period of Mars.

Solution

From Kepler's third law, we have

$$\tau_2 = \tau_1 \left(\frac{a_2}{a_1}\right)^{3/2}$$

where

$$a_1 = 1.49527 \times 10^{11} \text{ m} \quad \text{earth}$$

$$\tau_1 = 365.256 \text{ days}$$

and

$$a_2 = 2.2783 \times 10^{11} \text{ m} \quad \text{Mars}$$

Hence

$$\tau_2 = 365.256 \left(\frac{2.2783}{1.49527}\right)^{3/2}$$

$$\boxed{\tau_2 = 686.96 \text{ days for Mars}}$$

8.7 INTRODUCTION TO EARTH AND PLANETARY ENTRY[1]

In all cases of contemporary manned space vehicles, and with many unmanned vehicles, it is necessary to terminate the orbit or trajectory at some time and return to earth. Obviously, this necessitates negotiating the atmosphere at high velocities. Recall from Sec. 8.5 that an orbital vehicle will enter the outer regions of the atmosphere at a velocity close to 26,000 ft/s; a vehicle returning from a moon mission (e.g., an Apollo vehicle) will enter at an even higher velocity, nearly 36,000 ft/s. These velocities correspond to flight Mach numbers of 30 or more! Such hypersonic flight conditions are associated with several, uniquely difficult aerodynamic problems—so unique and difficult that they dominated the research efforts of aerodynamicists during the late 1950s and throughout the 1960s. Subsequently, the successful manned entries of the Mercury, Gemini, and Apollo vehicles were striking testimonials to the success of this hypersonic research. Some aspects of hypersonic vehicles are discussed in Chap. 11.

Consider a space vehicle in orbit about earth, as shown in Fig. 8.19. We wish to terminate this orbit and land the vehicle somewhere on the earth's surface. First, the path of the vehicle is changed by firing a retrorocket, decreasing the vehicle's velocity. In terms of the orbit equation, Eq. (8.43) or (8.44), the retrorocket's firing effectively changes the values of h, e, and C such that the vehicle curves toward earth. When the vehicle encounters the outer region of the atmosphere (given by the dashed circle in Fig. 8.19), three types of entry paths are possible:

1. *Ballistic entry.* Here the vehicle has little or no aerodynamic lift. It falls through the atmosphere under the influence of drag and gravity, impacting the surface at point a in Fig. 8.19. The impact point is predetermined by the conditions at first entry to the atmosphere. The pilot has no control over his or her landing

---

[1] In much of the literature, you will find references to earth *reentry*, rather than earth *entry*. The word *reentry* implies that the space vehicle had entered the atmosphere before, and now it is doing so again. This is usually not true, except for the space shuttle. So we will use the word *entry* here; it seems grammatically more correct, and it is in keeping with modern use.

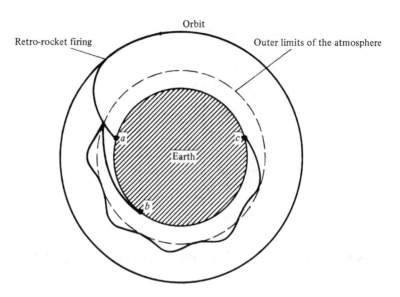

Figure 8.19   Three types of entry paths. (*a*) Ballistic; (*b*) glide; (*c*) skip.

position during this ballistic trajectory. It literally is the same as *falling* to the surface. Before the space shuttle, virtually all entries of existing space vehicles were ballistic. (A slight exception might be the Apollo capsule shown in Fig. 8.4, which at an angle of attack can generate a small lift-to-drag ratio, $L/D < 1$. However, for all practical purposes, this is still a ballistic entry vehicle.)

2.   *Skip entry.* Here the vehicle generates a value of L/D between 1 and 4 and uses this lifting ability to first graze the atmosphere, then slow down a bit, then pitch up such that the lift caries it back out of the atmosphere. This is repeated several times, much as a flat stone skipping over the surface of a pond, until finally the vehicle is slowed down appropriately and penetrates the atmosphere, landing at point *c* in Fig. 8.19. Unfortunately, the aerodynamic heating of a skip entry vehicle is inordinately large, and therefore such an entry mode has never been used and is not contemplated in the future.

3.   *Glide entry.* Here the vehicle is essentially an airplane, generating a lift-to-drag ratio of 4 or larger. The vehicle enters the atmosphere at a high angle of attack (30° or more) and flies to the surface, landing at point *b* in Fig. 8.19. An example of such a lifting entry vehicle is given in Fig. 8.6. The compelling advantages of the space shuttle are that the pilot can, in principle, choose the landing site and that the vehicle can be landed intact, to be used again.

In all the above entry modes, there are two overriding technical concerns—maximum deceleration and aerodynamic heating. For the safety of the occupants of a manned entry vehicle, the maximum deceleration should not exceed 10 times the acceleration of gravity, that is, 10 *g*'s. Furthermore, the aerodynamic heating of the vehicle should be low enough to maintain tolerable temperatures inside the capsule; if

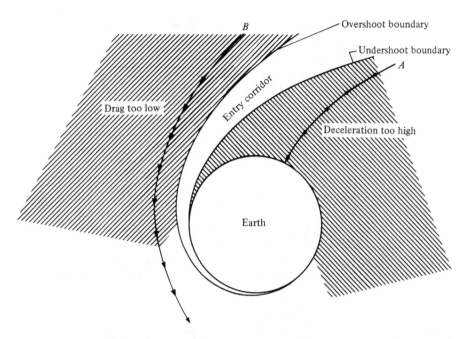

Figure 8.20   Illustration of the entry corridor.

the vehicle is unmanned, it still must be kept from burning up in the atmosphere. For these reasons, entry trajectories, maximum deceleration, and aerodynamic heating are the subject of the remainder of this chapter. With this, we move to the right-hand column in our road map, Fig. 8.7.

Finally, there is an extra consideration in regard to the entry of manned space vehicles returning from lunar or planetary missions. Such vehicles will approach the earth with parabolic or hyperbolic trajectories, as shown in Fig. 8.20. If the vehicle is traveling along path *A* in Fig. 8.20, penetration of the atmosphere will be too rapid and the maximum deceleration will be too large. On the other hand, if the vehicle is traveling along path *B*, it will not penetrate the atmosphere enough; the drag will be too low, the velocity will not be decreased enough for the vehicle to be captured by the earth, and it will go shooting past, back into outer space, never to return again. Consequently, there is a narrow *entry corridor* into which the vehicle must be guided for a successful return to the earth's surface. This entry corridor is shown in Fig. 8.20, bounded above by the *overshoot* boundary and below by the *undershoot* boundary.

8.8 EXPONENTIAL ATMOSPHERE

Since entry involves motion through the atmosphere, it is reasonable to expect entry performance to depend on the physical properties of the atmosphere. Such properties have been discussed in Chap. 3, where the atmospheric temperature distribution is given in Fig. 3.4. Detailed entry trajectory calculations made on computers take into account the precise variation of the standard atmosphere as given in Chap. 3. However, for a first approximation, a completely isothermal atmosphere with a constant temperature equal to some mean of the variation shown in Fig. 3.4, can be assumed. In this case, the density variation with altitude is a simple exponential, as given by Eq. (3.10). [At this point, the reader should review the derivation of Eq. (3.10).] Writing Eq. (3.10) with point 1 at sea level, we obtain

$$\frac{\rho}{\rho_0} = e^{-g_0 h/(RT)}$$ [8.73]

Equation (8.73) establishes the *exponential model atmosphere*. It agrees reasonably well with the actual density variation of the earth's standard atmosphere up to about 450,000 ft (about 140 km); above this height, the air is to thin that it has no meaningful influence on the entry trajectory. The exponential model atmosphere was used by NASA and other laboratories in the early studies of earth entry during the 1950s and early 1960s. We will adopt it here for the remainder of this chapter.

8.9   GENERAL EQUATIONS OF MOTION FOR ATMOSPHERIC ENTRY

Consider a space vehicle entering the atmosphere, as sketched in Fig. 8.21. At a given altitude h, the velocity of the vehicle is V, inclined at the angle θ below the local horizontal. The weight W is directed toward the center of the earth, and drag D and lift L are parallel and perpendicular, respectively, to the flight path, as usual. Summing

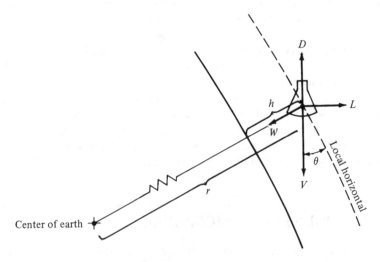

Figure 8.21   Geometry of entry vehicle forces and motion.

forces parallel and perpendicular to the flight path and using Newton's second law, we obtain, respectively,

$$-D + W \sin \theta = m \, \frac{dV}{dt} \qquad \textbf{[8.74]}$$

and

$$L - W \cos \theta = m \, \frac{V^2}{r_c} \qquad \textbf{[8.75]}$$

where r_c is the radius of curvature of the flight path. Equations (8.74) and (8.75) are identical to the equations of motion obtained in Chap. 6, specifically Eqs. (6.7) and (6.8), with $T = 0$ and θ measured below rather than above the horizontal.

We wish to establish an analysis which will yield velocity V as a function of altitude h. Dealing first with the drag equation, Eq. (8.74), we have

$$-D + W \sin \theta = m \, \frac{dV}{dt} = m \, \frac{dV}{ds} \frac{ds}{dt} = mV \, \frac{dV}{ds}$$

$$-D + W \sin \theta = \frac{1}{2} m \frac{dV^2}{ds} \qquad \textbf{[8.76]}$$

where s denotes distance along the flight path. From the definition of drag coefficient,

$$D = \tfrac{1}{2} \rho V^2 S C_D \qquad \textbf{[8.77]}$$

Also, from geometry shown in Fig. 8.22,

$$ds = -\frac{dh}{\sin \theta} \qquad \textbf{[8.78]}$$

Substitute Eqs. (8.77) and (8.78) into Eq. (8.76):

$$-\frac{1}{2} \rho V^2 S C_D + W \sin \theta = -\frac{1}{2} m \sin \theta \frac{dV^2}{dh} \qquad \textbf{[8.79]}$$

We are interested in obtaining V as a function of h. However, recall from Eq. (8.73) that $\rho = f(h)$, namely,

$$\frac{\rho}{\rho_0} = e^{-g_0 h/(RT)} = e^{-Zh} \qquad \textbf{[8.80]}$$

where $Z \equiv g_0/RT$ for simplicity of notation. Therefore, if we instead had a relation between velocity and density $V = f(\rho)$, we could still find the variation of V with h by using Eq. (8.80) as an intermediary. Let us take this approach and seek an equation relating V to ρ, as follows.

Figure 8.22   Flight path geometry.

Differentiating Eq. (8.80), we obtain

$$\frac{d\rho}{\rho_0} = e^{-Zh}(-Z\ dh) = \frac{\rho}{\rho_0}(-Z\ dh)$$

or
$$dh = -\frac{d\rho}{Z\rho} \qquad\qquad\qquad \textbf{[8.81]}$$

Substitute Eq. (8.81) into Eq. (8.79):

$$-\frac{1}{2}\rho V^2 S C_D + W\sin\theta = -\frac{1}{2}m\sin\theta\frac{dV^2}{d\rho}(-Z\rho) \qquad \textbf{[8.82]}$$

Divide Eq. (8.82) by $-\frac{1}{2}\rho Z m\sin\theta$:

$$\frac{V^2 S C_D}{Z m\sin\theta} - \frac{2mg}{Z\rho m} = -\frac{dV^2}{d\rho}$$

or
$$\boxed{\frac{dV^2}{d\rho} + \frac{1}{m/(C_D S)}\frac{V^2}{Z\sin\theta} = \frac{2g}{Z\rho}} \qquad \textbf{[8.83]}$$

Equation (8.83) is an exact equation of motion for a vehicle entering the atmosphere—the only approximation it contains is the exponential model atmosphere. Also note that the parameter $m/(C_D S)$, which appears in the second term in Eq. (8.83), is essentially a constant for a given space vehicle; it is identified as

$$\frac{m}{C_D S} \equiv \text{ballistic parameter}$$

The value of $m/(C_D S)$ strongly governs the entry trajectory, as will be demonstrated later.

Equation (8.83) is also a differential equation, and in principle it can be solved to obtain $V = f(\rho)$ and hence $V = f(h)$ through Eq. (8.80). However, in general, the angle θ in Eq. (8.83) also varies with altitude h, and this variation must be obtained before Eq. (8.83) can be solved. This is the role of our second equation of motion, Eq. (8.75)—the lift equation. Equation (8.75) can be reworked to obtain a differential equation in terms of $d\theta/d\rho$, which can then be solved simultaneously with Eq. (8.83) to obtain an explicit relation for V as a function of ρ for a vehicle with a given $m/(C_D S)$ and L/D. The details will not be given here; our intent has been to simply map out an approach to calculate a lifting entry path, as given above. The reader can obtain more details from the NACA and NASA reports given in the Bibliography at the end of this chapter.

After the above analysis is carried to completion, what does the actual entry path look like? An answer is given in Fig. 8.23, which illustrates the variation of velocity (the abscissa) with density (the ordinate). Since ρ is a function of altitude through Eq. (8.80), h is also given on the ordinate. Thus, Fig. 8.23 shows the entry path in terms of velocity versus altitude—a so-called *velocity-altitude map* for entry. Such velocity-altitude maps are frequently used in entry vehicle design and analysis. Examine Fig. 8.23 more closely. Imagine an entry vehicle just beginning to penetrate

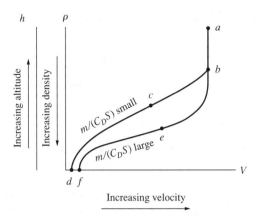

Figure 8.23   Entry trajectory on a velocity-altitude map.

the atmosphere. It is at a very high altitude and velocity, such as point a in Fig. 8.23. During the early portion of entry, the atmospheric density is so low that the drag is virtually insignificant; the vehicle penetrates the upper region of the atmosphere with only a small decrease in velocity, as shown from point a to point b in Fig. 8.23. However, below the altitude denoted by point b, the air density rapidly increases with an attendant marked increase in drag, causing the velocity to decrease rapidly. This is the situation at point c in Fig. 8.23. Finally, the vehicle impacts the surface at point d. In Fig. 8.23, the path a-b-c-d is for a given ballistic parameter. If $m/(C_D S)$ is made larger, the vehicle penetrates more deeply into the atmosphere before slowing down, as illustrated by path a-b-e-f. Thus, as suspected from an examination of Eq. (8.83), the ballistic parameter is an important design aspect of entry vehicles.

8.10   APPLICATION TO BALLISTIC ENTRY

A solution of the exact equations of motion, such as Eq. (8.83), must be performed numerically on a high-speed computer. That is, the curves in Fig. 8.23 are obtained from numbers generated by a computer; they are not given by simple, closed-form analytic equations. However, such an analytic solution can be obtained for a purely ballistic entry (no lift), with a few assumptions. This is the purpose of the present section.

Return to the picture of a vehicle entering the atmosphere, as shown in Fig. 8.21. If the path is purely ballistic, then $L = 0$ by definition. Also recall that the initial entry velocities are high—26,000 ft/s for circular orbits, 36,000 ft/s for parabolic space trajectories, etc. Therefore, the dynamic pressures associated with entry velocities throughout most of the velocity-altitude map are large. As a result, drag is large—much larger, in fact, than the vehicle's weight; $D \gg W$. With this in mind, W can be ignored, and the original drag equation, Eq. (8.74), becomes

$$-D = m \frac{dV}{dt} \tag{8.84}$$

Following Eq. (8.84) with the same derivation that led to Eq. (8.83), we obtain

$$\frac{dV^2}{d\rho} + \frac{1}{m/(C_D S)} \frac{V^2}{Z \sin \theta} = 0 \tag{8.85}$$

(The reader should carry through this derivation to satisfy her or his own curiosity.) Equation (8.85) is the same as Eq. (8.83), with the right side now zero because W has been neglected.

Furthermore, assume that θ is constant in Eq. (8.85). Referring to Fig. 8.21, we see this implies a straight-line entry path through the atmosphere. This is a reasonable approximation for many actual ballistic entry vehicles. If θ is constant, Eq. (8.85) can be integrated in closed form, as follows. First, rearrange Eq. (8.85):

$$\frac{dV^2}{V^2} = -\frac{d\rho}{[m/(C_D S)]Z \sin \theta} \tag{8.86}$$

Integrate Eq. (8.86) from the point of initial contact with the atmosphere, where $\rho = 0$ and $V = V_E$ (the initial entry velocity), to some point in the atmosphere, where the density is ρ and the vehicle velocity is V.

$$\int_{V_E}^{V} \frac{dV^2}{V^2} = -\frac{1}{[m/(C_D S)]Z \sin \theta} \int_{0}^{\rho} d\rho$$

or

$$\ln \frac{V^2}{V_E^2} = 2 \ln \frac{V}{V_E} = -\frac{\rho}{[m/(C_D S)]Z \sin \theta}$$

Thus,

$$\boxed{\frac{V}{V_E} = e^{-\rho/2[m/(C_D S)]Z \sin \theta}} \tag{8.87}$$

Equation (8.87) is a closed-form expression for the variation of V with ρ, hence of V with h via Eq. (8.80). It is an explicit equation for the entry trajectory on a velocity-altitude map, as sketched in Fig. 8.23, except Eq. (8.87) now tells us precisely how the velocity changes; before, we had to take the shapes of the curves in Fig. 8.23 on faith. For example, examine Eq. (8.87). As ρ increases (i.e., as the altitude decreases), V decreases. This confirms the shape of the curves shown in Fig. 8.23. Also, if $m/(C_D S)$ is made larger, the exponential term in Eq. (8.87) does not have as strong an effect until ρ becomes larger (i.e., until the altitude is smaller). Hence, a vehicle with a large $m/(C_D S)$ penetrates more deeply into the atmosphere with a high velocity, as shown in Fig. 8.23. Therefore, the variations shown in Fig. 8.23 are directly verified by the form of Eq. (8.87).

In Sec. 8.7, maximum deceleration was identified as an important entry consideration. We now have enough background to examine deceleration in greater detail. First, consider the equation of motion, Eq. (8.84), which neglects the vehicle's weight.

By definition, dV/dt in Eq. (8.84) is the acceleration, and from Eq. (8.84) it is negative for entry.

$$\frac{dV}{dt} = -\frac{D}{m}$$

Also, by definition, a negative value of acceleration is *deceleration*, denoted by $|dV/dt|$. From the above equation,

$$\text{Deceleration} = \left|\frac{dV}{dt}\right| = \frac{D}{m} \qquad \text{[8.88]}$$

From the definition of drag coefficient $D = \frac{1}{2}\rho V^2 S C_D$, Eq. (8.88) becomes

$$\left|\frac{dV}{dt}\right| = \frac{\rho V^2 S C_D}{2m} \qquad \text{[8.89]}$$

[In Eq. (8.89), the subscript ∞ has been dropped from ρ and V for convenience.] Note from Eq. (8.89) that $|dV/dt|$ increases as ρ increases, and decreases as V decreases. This allows us to qualitatively sketch the deceleration versus altitude curve shown in Fig. 8.24. At high altitudes, the velocity is large but relatively constant (see Fig. 8.23, from points a to b), whereas ρ is beginning to increase. Therefore, from Eq. (8.89), deceleration will first increase as the vehicle enters the atmosphere, as shown in Fig. 8.24 at high altitude. However, at lower altitudes, Fig. 8.23 shows that the velocity rapidly decreases. From Eq. (8.89), the velocity decrease now overshadows the increase in density, hence the deceleration will decrease in magnitude. This is shown in Fig. 8.24 at low altitude. Consequently, the deceleration experienced by a vehicle throughout entry first increases, then goes through a maximum, and finally decreases; this variation is clearly illustrated in Fig. 8.24.

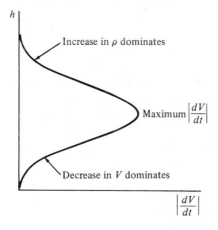

Figure 8.24   The variation of deceleration with altitude for ballistic entry.

The quantitative value of the *maximum* deceleration is of interest. It was stated in Sec. 8.7 that a manned entry vehicle should not exceed a maximum deceleration of 10 g's; furthermore, even unmanned vehicles have limitations dictated by structural failure of the vehicle itself or its components. Therefore, let us derive an equation for maximum deceleration. To begin, Eq. (8.89) gives an expression for deceleration which holds at any point along our straight-line ballistic trajectory. We wish to find the maximum deceleration. Therefore, from calculus, we wish to differentiate Eq. (8.89) and set the result equal to zero in order to find the conditions for maximum deceleration. Differentiating Eq. (8.89) with respect to time, and noting that both ρ and V vary along the trajectory, we have

$$\left| \frac{d^2 V}{dt^2} \right| = \frac{SC_D}{2m} \left(2\rho V \frac{dV}{dt} + V^2 \frac{d\rho}{dt} \right) \qquad \text{[8.90]}$$

From Eq. (8.84),

$$\frac{dV}{dt} = -\frac{D}{m} = -\frac{1}{m} \left(\frac{1}{2} \rho V^2 SC_D \right) \qquad \text{[8.91]}$$

Substitute Eq. (8.91) into Eq. (8.90):

$$\left| \frac{d^2 V}{dt^2} \right| = \frac{SC_D}{2m} \left[2\rho V \left(-\frac{\rho V^2 SC_D}{2m} \right) + V^2 \frac{d\rho}{dt} \right]$$

$$\left| \frac{d^2 V}{dt^2} \right| = \frac{SC_D V^2}{2m} \left(-\frac{\rho^2 V SC_D}{m} + \frac{d\rho}{dt} \right) \qquad \text{[8.92]}$$

Setting Eq. (8.92) equal to zero for conditions at maximum $|dV/dt|$, we find that

$$\frac{d\rho}{dt} = \frac{\rho^2 V SC_D}{m} \qquad \text{[8.93]}$$

From the exponential model atmosphere, differentiating Eq. (8.80) with respect to time gives

$$\frac{d\rho}{dt} = -\rho_0 Z e^{-Zh} \frac{dh}{dt} = -Z\rho \frac{dh}{dt} \qquad \text{[8.94]}$$

However, from the geometric construction of Fig. 8.22 and from Eq. (8.78),

$$\frac{dh}{dt} = -\frac{ds}{dt} \sin\theta = -V \sin\theta \qquad \text{[8.95]}$$

Substitute Eq. (8.95) into (8.94):

$$\frac{d\rho}{dt} = \rho Z V \sin\theta \qquad \text{[8.96]}$$

Substitute Eq. (8.96) into (8.93):

$$\rho Z V \sin\theta = \frac{\rho^2 V SC_D}{m} \qquad \text{[8.97]}$$

Solve Eq. (8.97) for ρ:

$$\rho = \frac{m}{C_D S} Z \sin \theta \qquad \textbf{[8.98]}$$

Equation (8.98) gives the value of density at the point of maximum deceleration. Substituting this into Eq. (8.89) in order to obtain maximum deceleration, we get

$$\left| \frac{dV}{dt} \right|_{\text{max}} = \frac{1}{2m} \frac{m}{C_D S} Z(\sin \theta) V^2 S C_D$$

$$\left| \frac{dV}{dt} \right|_{\text{max}} = \frac{1}{2} V^2 Z \sin \theta \qquad \textbf{[8.99]}$$

The velocity at the point of maximum deceleration is obtained by combining Eqs. (8.98) and (8.87), yielding

$$V = V_E e^{-1/2} \qquad \textbf{[8.100]}$$

Substituting Eq. (8.100) into (8.99), we find

$$\boxed{\left| \frac{dV}{dt} \right|_{\text{max}} = \frac{V_E^2 Z \sin \theta}{2e}} \qquad \textbf{[8.101]}$$

Equation (8.101) is the desired result. It gives us a closed-form expression from which we can quickly calculate the maximum deceleration for a straight-line ballistic entry trajectory. Note from Eq. (8.101) that

$$\left| \frac{dV}{dt} \right|_{\text{max}} \propto V_E^2 \quad \text{and} \quad \left| \frac{dV}{dt} \right|_{\text{max}} \propto \sin \theta$$

Hence, entry from a parabolic or hyperbolic trajectory ($V_E \geq 11.2$ km/s) is much more severe than from a near-circular orbit ($V_E = 7.9$ km/s). However, for entry, there is little we can do to adjust the value of V_E—it is primarily determined by the orbit or trajectory *before* entry, which in turn is dictated by the desired mission in space. Therefore, Eq. (8.101) tells us that maximum deceleration must be primarily adjusted by the entry angle θ. *In fact, we conclude from Eq. (8.101) that to have reasonably low values of deceleration during entry, the vehicle must enter the atmosphere at a shallow angle, that is, at a small θ.*

Finally, Eq. (8.101) yields a rather startling result. Maximum deceleration depends only on V_E and θ. Note that the design of the vehicle, that is, the ballistic parameter $m/(C_D S)$, does not influence the value of maximum deceleration. However, you might correctly suspect that $m/(C_D S)$ determines the altitude at which maximum deceleration occurs.

This concludes our discussion of deceleration, and of entry trajectories in general. In the next section, we examine the second major problem of entry as discussed in Sec. 8.7, namely, aerodynamic heating.

Example 8.3

Consider a solid iron sphere entering the earth's atmosphere at 13 km/s (slightly above escape velocity) and at an angle of 15° below the local horizontal. The sphere diameter is 1 m. The drag coefficient for a sphere at hypersonic speeds is approximately 1. The density of iron is 6963 kg/m³. Calculate (*a*) the altitude at which maximum deceleration occurs, (*b*) the value of the maximum deceleration, and (*c*) the velocity at which the sphere would impact the earth's surface.

Solution

First, calculate the ballistic parameter $m/(C_D S)$:

$$m = \rho v = \rho(\tfrac{4}{3}\pi r^3) \qquad S = \pi r^2$$

where r = radius of sphere. Hence,

$$\frac{m}{C_D S} = \frac{4}{3}\frac{r\rho}{C_D} = \frac{4}{3}\left[\frac{0.5(6963)}{1.0}\right] = 4642 \text{ kg/m}^2$$

Also, by definition, $Z = g_0/(RT)$. For our exponential atmosphere, assume a constant temperature of 288 K (recall from Sec. 8.8 that the exponential atmosphere is just an approximation of the detailed standard atmosphere discussed in Chap. 3). Hence,

$$Z = \frac{g_0}{RT} = \frac{9.8}{287(288)} = 0.000118 \text{ m}^{-1}$$

a. To obtain the altitude for maximum deceleration, calculate the corresponding density from Eq. (8.98):

$$\rho = \frac{m}{C_D S} Z \sin\theta = 4642(0.000118)(\sin 15°) = 0.1418 \text{ kg/m}^3$$

This can be translated to an altitude value via Eq. (8.73):

$$\frac{\rho}{\rho_0} = e^{-Zh}$$

or

$$h = -\frac{1}{Z}\ln\frac{\rho}{\rho_0} = \frac{1}{0.000118}\ln\frac{0.1418}{1.225} = 18{,}275 \text{ m}$$

Thus, the altitude for maximum deceleration is

$$\boxed{h = 18.275 \text{ km}}$$

b. The value of maximum deceleration is obtained from Eq. (8.101):

$$\left|\frac{dV}{dt}\right|_{max} = \frac{V_E^2 Z \sin\theta}{2e} = \frac{(13{,}000)^2(0.000118)(\sin 15°)}{2e} = 949.38 \text{ m/s}^2$$

Since 9.8 m/s² is the sea-level acceleration of gravity, the maximum deceleration in terms of *g*'s is

$$\left|\frac{dV}{dt}\right|_{max} = \frac{949.38}{9.8} = \boxed{96.87 \text{ } g\text{'s}}$$

This deceleration is very large; it is way beyond that which can be tolerated by humans.

c. The velocity at impact on the earth's surface is obtained from Eq. (8.87):

$$\frac{V}{V_E} = e^{-\rho/2[m/(C_D S)]Z \sin\theta}$$

where the value used for ρ is the standard sea-level value $\rho_0 = 1.225$ kg/m^3. Hence,

$$\frac{V}{V_E} = e^{-1.225/2(4642)(0.000118)\sin 15°} = 0.01329$$

Thus $V = 0.01329 V_E = 0.01329(13,000)$

$$\boxed{V = 172.8 \text{ m/s}}$$

It is interesting to note that the sphere has slowed down to subsonic velocity before impact. At sea level, $a_s = 340.9$ m/s, hence the Mach number at impact is

$$M = \frac{V}{a_s} = \frac{172.8}{340.9} = 0.507$$

Please note that, in reality, the iron sphere will encounter tremendous aerodynamic heating during entry, especially at the large velocity of 13 km/s. Hence, it is likely that the sphere would vaporize in the atmosphere and never impact the surface; this is the fate of most meteors that enter the atmosphere from outer space. Aerodynamic heating is the subject of the next section.

8.11   ENTRY HEATING

Imagine an entry body, say, the Apollo capsule, as it penetrates the atmosphere. For reasons to be developed later, this body has a very blunt nose, as shown in Fig. 8.25. The reentry velocities are extremely high, and the corresponding Mach numbers are hypersonic. From the aerodynamic discussions in Chap. 4, we know there will be a shock wave in front of the vehicle—the bow shock wave shown in Fig. 8.25. Because the entry velocities are so large, this shock wave will be very strong. Consequently, the temperature of the air behind the shock will be extraordinarily high. For example, during the 11.2 km/s entry of the Apollo, the air temperature behind the shock wave reaches 11,000 K—higher than the surface of the sun! At these temperatures, the air itself breaks down; the O_2 and N_2 molecules dissociate into O and N atoms, and ionize into O^+ and N^+ ions and electrons. The air becomes a chemically reacting gas. Of greater importance, however, is that such high temperatures result in large heat inputs to the entry vehicle itself. As shown in Fig. 8.25, the vehicle is sheathed in a layer of hot air, first from the hot shock layer at the nose, then from the hot boundary layer on the forward and rearward surfaces. These hot gases then flow downstream in the wake of the vehicle. A major objective of entry vehicle design is to shield the vehicle from this severe aerodynamic heating.

An alternate way of looking at this problem is to consider the combined kinetic and potential energies of the entry vehicle. At the beginning of entry, where V_E and h are large, this combined energy is large. At the end of entry, that is, at impact, V and

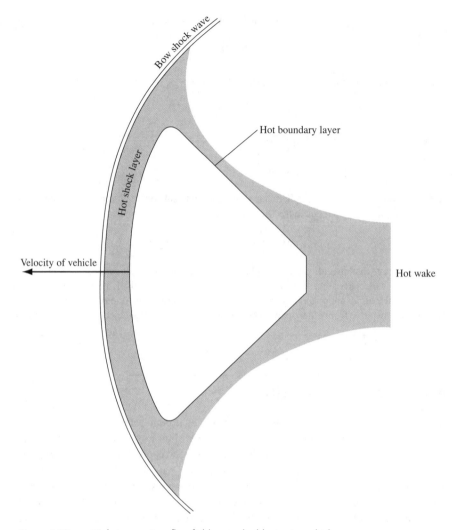

Figure 8.25   High-temperature flow field around a blunt entry vehicle.

h are essentially zero, and the vehicle has no kinetic or potential energy. However, energy is conserved, so where did it go? The answer is that the kinetic and potential energies of the vehicle are ultimately dissipated as *heat*. Returning to Fig. 8.25, we see that some of this heat goes into the vehicle itself, and the remainder goes into the air. The object of successful entry vehicle design is to minimize that heat which goes into the vehicle and maximize that which goes into the air.

The main physical mechanism of aerodynamic heating is related to the action of friction in the boundary layer, as discussed in reference to shear stress and drag in Chap. 4. Clearly, if you take the palm of your hand and rub it vigorously over the surface of a table, your skin will soon get hot. The same applies to the high-speed flow of a gas over an aerodynamic surface. The same frictional forces which create skin friction drag will also at the same time heat the air. The net result is heat transfer to the surface—*aerodynamic heating.*

Incidentally, aerodynamic heating becomes a problem at velocities far below entry velocity. For example, even at Mach 2 at sea level, the temperature behind a normal shock, and also deep within a boundary layer, can be as high as 520 K. Therefore, aerodynamic heating of the surfaces of supersonic airplanes such as the Concord Supersonic Transport and the F-15 is important and influences the type of materials used in their construction. For example, this is why titanium, rather than the more conventional aluminum, is extensively used on high-speed aircraft—titanium has greater strength at high temperatures. However, with the advent of hypervelocity entry vehicles in the space age, aerodynamic heating became an overriding problem with regard to the very survival of the vehicle itself. It even dictates the shape of the vehicle, as we will soon see.

For a quantitative analysis of aerodynamic heating, it is convenient to introduce a dimensionless heat transfer coefficient called the *Stanton number* C_H, defined as

$$C_H = \frac{dQ/dt}{\rho_\infty V_\infty (h_0 - h_w) S} \qquad [8.102]$$

where ρ_∞ and V_∞ are the free-stream density and velocity, respectively, h_0 is the total enthalpy (defined as the enthalpy of a fluid element which is slowed adiabatically to zero velocity, in the same spirit as the definition of T_0 in Chap. 4), h_w is the enthalpy at the aerodynamic surface (remember that the velocity is zero at the surface due to friction), S is a reference area (planform area of a wing, cross-sectional area of a spherical entry vehicle, etc.), and dQ/dt is the heating rate (energy per second) going into the surface. Let us use Eq. (8.102) to obtain a quantitative expression for entry vehicle heating.

Rewriting Eq. (8.102) gives

$$\frac{dQ}{dt} = \rho_\infty V_\infty (h_0 - h_w) S C_H \qquad [8.103]$$

Considering the energy equation, Eq. (4.41), and the definition of h_0, we obtain

$$h_0 = h_\infty + \frac{V_\infty^2}{2} \qquad [8.104]$$

For high-speed entry conditions, V_∞ is very large. Also the ambient air far ahead of the vehicle is relatively cool, hence $h_\infty = c_p T$ is relatively small. Thus, from Eq. (8.104)

$$h_0 \approx \frac{V_\infty^2}{2} \qquad [8.105]$$

Also the surface temperature, although hot by normal standards, still must remain less than a few thousand kelvins, below the melting or decomposition temperature of the surface. On the other hand, the temperatures associated with h_0 are large (11,000 K for the Apollo entry, as stated earlier). Hence, we can easily make the assumption that

$$h_0 \gg h_w \approx 0 \qquad \textbf{[8.106]}$$

Substituting Eqs. (8.106) and (8.105) into Eq. (8.103), we get

$$\frac{dQ}{dt} = \frac{1}{2}\rho_\infty V_\infty^3 S C_H \qquad \textbf{[8.107]}$$

Note that Eq. (8.107) states that the *aerodynamic heating rate varies as the cube of the velocity*. This is in contrast to aerodynamic drag, which varies only as the square of the velocity (as we have seen in Chaps. 4 and 5). For this reason, at very high velocities, aerodynamic heating becomes a dominant aspect and drag retreats into the background. Also recall the reasoning that led from Eq. (8.89) to the curve for deceleration versus altitude in Fig. 8.24. This same reasoning leads from Eq. (8.107) to the curve for heating rate versus altitude, sketched in Fig. 8.26. During the early part of entry, dQ/dt increases because of the increasing atmospheric density. In contrast, during the later portion of entry, dQ/dt decreases because of the rapidly decreasing velocity. Hence, dQ/dt goes through a maximum, as shown in Fig. 8.26.

In addition to the local heating rate dQ/dt, we are concerned with the *total heating Q*, that is, the total amount of energy transferred to the vehicle from beginning to end of entry. The result for Q will give us some vital information on the desired *shape* for entry vehicles. First, we draw on a relation between aerodynamic heating and skin friction, called *Reynold's analogy*. Indeed, it makes sense that aerodynamic heating and skin friction should somehow be connected because both are influenced by friction in the boundary layer. Based on experiment and theory, we approximate

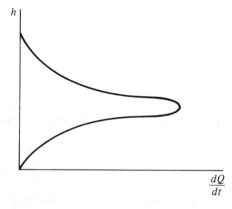

Figure 8.26   The variation of heat transfer rate during ballistic entry.

Reynold's analogy (without proof) as

$$C_H \approx \tfrac{1}{2} C_f \qquad \qquad \textbf{[8.108]}$$

where C_f is the mean skin friction coefficient averaged over the complete surface. Substituting Eq. (8.108) into (8.107), we obtain

$$\frac{dQ}{dt} = \frac{1}{4} \rho_\infty V_\infty^3 S C_f \qquad \qquad \textbf{[8.109]}$$

Returning to the equation of motion, Eq. (8.84), we have

$$\frac{dV_\infty}{dt} = -\frac{D}{m} = -\frac{1}{2m} \rho_\infty V_\infty^2 S C_D \qquad \qquad \textbf{[8.110]}$$

Mathematically, we can write dQ/dt as $(dQ/dV_\infty)(dV_\infty/dt)$, where dV_∞/dt is given by Eq. (8.110):

$$\frac{dQ}{dt} = \frac{dQ}{dV_\infty} \frac{dV_\infty}{dt} = \frac{dQ}{dV_\infty}\left(-\frac{1}{2m} \rho_\infty V_\infty^2 S C_D \right) \qquad \qquad \textbf{[8.111]}$$

Equating Eqs. (8.111) and (8.109),

$$\frac{dQ}{dV_\infty}\left(-\frac{1}{2m} \rho_\infty V_\infty^2 S C_D \right) = \frac{1}{4} \rho_\infty V_\infty^3 S C_f$$

or

$$\frac{dQ}{dV_\infty} = -\frac{1}{2} m V_\infty \frac{C_f}{C_D}$$

or

$$dQ = -\frac{1}{2} m \frac{C_f}{C_D} \frac{dV_\infty^2}{2} \qquad \qquad \textbf{[8.112]}$$

Integrate Eq. (8.112) from the beginning of entry, where $Q = 0$ and $V_\infty = V_E$, and the end of entry, where $Q = Q_{\text{total}}$ and $V_\infty = 0$:

$$\int_0^{Q_{\text{total}}} dQ = -\frac{1}{2} \frac{C_f}{C_D} \int_{V_E}^0 d\left(m \frac{V_\infty^2}{2} \right)$$

$$\boxed{Q_{\text{total}} = \frac{1}{2} \frac{C_f}{C_D} \left(\frac{1}{2} m V_E^2 \right)} \qquad \qquad \textbf{[8.113]}$$

Equation (8.113) is the desired result for total heat input to the entry vehicle. It is an important relation—examine it closely. It reflects two important conclusions:

1. The quantity $\frac{1}{2} m V_E^2$ is the initial kinetic energy of the vehicle as it first enters the atmosphere. Equation (8.113) says that total heat input is directly proportional to this initial kinetic energy.

2. Total heat input is directly proportional to the ratio of skin friction drag to total drag C_f/C_D.

The second conclusion is of particular importance. Recall from Chap. 5 that the total drag of a nonlifting body is pressure drag plus skin friction drag:

$$C_D = C_{D_p} + C_f$$

Equation (8.113) says that to minimize entry heating, we need to minimize the ratio

$$\frac{C_f}{C_{D_p} + C_f}$$

Now consider two extremes of aerodynamic configurations; a sharp-nosed, *slender body* such as the cone shown in Fig. 8.27a, and the *blunt body* shown in Fig. 8.27b. For a slender body, the skin friction drag is large in comparison to the pressure drag, hence $C_D \approx C_f$ and

$$\frac{C_f}{C_D} \approx 1 \qquad \text{slender body}$$

On the other hand, for a blunt body, the pressure drag is large in comparison to the skin friction drag, hence $C_D \approx C_{D_p}$ and

$$\frac{C_f}{C_D} \ll 1 \qquad \text{blunt body}$$

In light of Eq. (8.113), this leads to the following vital conclusion:

To minimize entry heating, the vehicle must have a blunt nose.

For this reason, all successful entry vehicles in practice, from intercontinental ballistic missiles (ICBMs) to the Apollo, have utilized rounded noses.

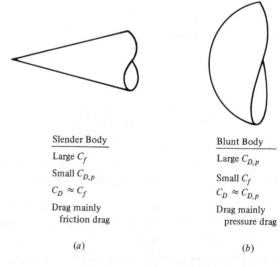

Slender Body	Blunt Body
Large C_f	Large $C_{D,p}$
Small $C_{D,p}$	Small C_f
$C_D \approx C_f$	$C_D \approx C_{D,p}$
Drag mainly friction drag	Drag mainly pressure drag
(a)	(b)

Figure 8.27   Comparison of blunt and slender bodies.

Returning to our qualitative discussion surrounding Fig. 8.25, we see that the advantage of a blunt body can also be reasoned on a purely physical basis. If the body is blunt, as shown in Fig. 8.25, the bow shock wave will be strong; that is, a substantial portion of the wave in the vicinity of the nose will be nearly normal. In this case, the temperature of extensive regions of the air will be high, and much of this high-temperature air will simply flow past the body without encountering the surface. Therefore, a blunt body will deposit much of its initial kinetic and potential energies into heating the air, and little into heating the body. In this fashion, a blunt body tends to minimize the total heat input to the vehicle, as proved quantitatively from Eq. (8.113).

The mechanism of aerodynamic heating discussed above is called *convective heating*. To conclude this section on entry heat transfer, another mechanism is mentioned—*radiative heating* from the shock layer. Consider Fig. 8.28, which shows a blunt entry body at high velocity. It was mentioned earlier that at speeds associated with lunar missions (11.2 km/s, or 36,000 ft/s), the air temperature behind the shock wave is as high as 11,000 K. At this high temperature, the shock layer literally *radiates* energy in all directions, as illustrated in Fig. 8.28, much as you feel the warmth radiated from a fireplace on a cold winter day. Some of this radiation is incident upon and absorbed by the vehicle itself, giving rise to an additional heat transfer component Q_R. This radiative heat transfer rate is proportional to a power of velocity ranging from V_∞^5 to V_∞^{12}, depending on the nose radius, density, and velocity. For ICBM and orbital vehicles, radiative heating is not significant. But as sketched in Fig. 8.29,

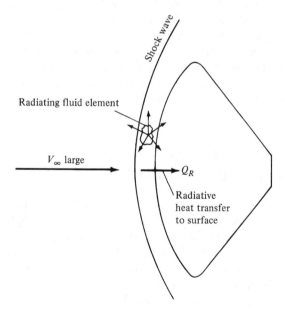

Figure 8.28 Mechanism of radiative heating from the high-temperature shock layer.

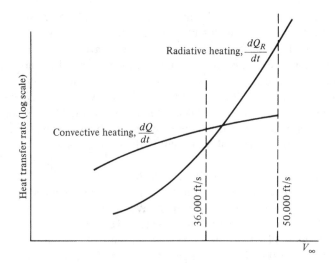

Figure 8.29   Comparison of convective and radiative heat transfer rates, illustrating dominance of radiative heating at high velocities.

because of its strong velocity dependence, radiative heating becomes dominant at very high velocities. For the Apollo mission from the moon ($V_E = 36{,}000$ ft/s), radiative heating was slightly less than convective heating. However, for future manned missions from the planets ($V_E \approx 50{,}000$ ft/s), radiative heating will swamp convecting heating. This is illustrated schematically in Fig. 8.29. Moreover, entry into the atmospheres of other large planets, especially Jupiter, will be overwhelmed by radiative heating. For these reasons, the designers of vehicles for advanced space missions will have to be vitally concerned about radiative heating from the shock layer during atmospheric entry. The interested reader can find more details on radiative heating in the AIAA paper by Anderson listed in the Bibliography at the end of this chapter.

8.12 LIFTING ENTRY, WITH APPLICATION TO THE SPACE SHUTTLE

On April 14, 1981, the space shuttle *Columbia* entered the atmosphere and successfully returned to the surface of the earth, ending the historic first flight of this unique space transportation system into space around the earth. A diagram of the space shuttle orbiter mounted on its rocket booster is shown in Fig. 8.30. The entry trajectory of the space shuttle differs considerably from the ballistic trajectories discussed in Sec. 8.10 because the shuttle is an aerodynamic vehicle that produces lift. Indeed, during the initial part of its entry, the space shuttle is flying at a very high angle of attack (on the order of 40°). For this angle of attack at the hypersonic speeds of reentry from low earth orbit (initially at Mach 25), the lift-to-drag ratio L/D for the blunt-nosed highly swept delta-wing configuration of the space shuttle (see Figs. 8.6 and 8.30) is on the order of 2—not a high value by conventional subsonic-airplane standards as seen

1/200TH SCALE

0   200   400   600   800   1000   cm

0   100   200   300   400   500   in

Booster nozzle gimbal travel

Main engine (3) gimbal travel

Figure 8.30   The space shuttle. (*Rockwell International.*)

in Chap. 6, but certainly high enough to produce substantial lift at such hypersonic velocities. Since the flight of the space shuttle during its return to earth is essentially an unpowered glide through the atmosphere with almost global range, the trajectory of the space shuttle on a velocity-altitude map should differ considerably from the ballistic trajectories sketched in Fig. 8.23. This section examines such matters further and obtains the flight trajectories for lifting entry vehicles on a velocity-altitude map.

Return to the general equations of motion for atmospheric entry, (8.74) and (8.75). In our previous study of ballistic entry in Sec. 8.10, we used Eq. (8.74) as the equation of motion parallel to the vehicle's flight path. It was dominated by aerodynamic drag, as expected for a ballistic vehicle. In contrast, for our present discussion of lifting entry, we use Eq. (8.75) as the equation of motion perpendicular to the flight path; as expected, it is dominated by aerodynamic lift. Equation (8.75) requires more interpretation than given it in Sec. 8.9. Specifically, the form of Eq. (8.75) with the right-hand side a positive term pertains to an upward-curved flight path, as shown by the dashed curve in Fig. 8.31; here, the lift is greater than the weight component, and the vehicle rises. On the other hand, when $L < W \cos \theta$, the vehicle is descending, as shown by the solid curve in Fig. 8.31. For this case, the right-hand side of

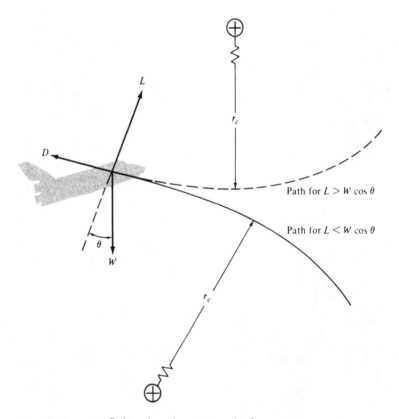

Figure 8.31   Two flight paths with opposite radii of curvature.

Eq. (8.75) must be negative (because the left-hand side is negative), and Eq. (8.75) must be written

$$L - W \cos \theta = -m \frac{V^2}{r_c} \qquad \textbf{[8.114]}$$

Equation (8.114) is the pertinent form for the lifting glide of the space shuttle, as sketched in Fig. 8.32. The vehicle is gliding at velocity V, and the flight path angle θ is measured below the local horizontal. Assume that the flight path is very shallow (θ is small and hence $\cos \theta \approx 1$). Furthermore, assume that the local radius of curvature r_c is approximately the radius of the earth r_e. Then Eq. (8.114) becomes

$$L - W = -\frac{mV^2}{r_e} \qquad \textbf{[8.115]}$$

Figure 8.32   Glide angle and velocity.

Since $L = \frac{1}{2}\rho V^2 S C_L$ and $W = mg$, Eq. (8.115) is written

$$\frac{1}{2}\rho V^2 S C_L + \frac{m V^2}{r_e} = mg \qquad \textbf{[8.116]}$$

Dividing Eq. (8.116) by m and factoring the V^2 give

$$V^2 \left(\frac{\rho}{2} \frac{C_L S}{m} + \frac{1}{r_e} \right) = g$$

or

$$V^2 = \frac{g}{(\rho/2)[m/(C_L S)]^{-1} + 1/r_e} \qquad \textbf{[8.117]}$$

Since both ρ and g are known functions of altitude, Eq. (8.117) gives the trajectory of a lifting entry vehicle on a velocity-altitude map. Moreover, Eq. (8.117) introduces a *lift parameter* $m/(C_L S)$ analogous to the ballistic parameter $m/(C_D S)$ defined in Sec. 8.9. Clearly, as we can see from Eq. (8.117), the value of $m/(C_L S)$ strongly governs the entry glide trajectory.

The influence of $m/C_L S$ is shown in Fig. 8.33; this velocity-altitude map illustrates lifting entry trajectories (A and B) for two different values of $m/C_L S$. Curve B pertains approximately to the space shuttle. Since higher values of $m/C_L S$ correspond to lower lift, the vehicle penetrates deeper into the atmosphere at higher velocity. For comparison Fig. 8.33 also shows the ballistic trajectories C, D, and E. Curve E, initiated at escape velocity, pertains approximately to the Apollo entry capsule. Although the Apollo generated a small amount of lift at the angle of attack during entry in order to modulate its flight path, it was essentially a ballistic reentry vehicle. Trajectories C and D represent earth entry from orbital velocity.

Figure 8.33 Comparison between lifting and ballistic entry paths on a velocity-altitude map.

8.13 HISTORICAL NOTE: KEPLER

The 16th century was a period of quandry for astronomy. The conservative line of scientific thought held the earth as the center of the universe, with the sun, planets, and stars revolving about it on various celestial spheres. This *geocentric* system was popular among the Greeks. Put into a somewhat rational form by Claudius Ptolemy in the second century A.D., this earth-centered system was adopted as the truth by the Church in western Europe and was carried through to the 16th century. However, about the time that Columbus was discovering America, a Polish scientist by the name of Nicolaus Copernicus was beginning to develop different ideas. Copernicus reasoned that the earth as well as all the other planets revolved around the sun, in a *heliocentric* system. He established his line of thought in a main work entitled *Six Books Concerning the Revolutions of the Heavenly Spheres*, published in the year of his death, 1543. Here Copernicus was diplomatic with regard to the Church's dogma. He stated that his heliocentric theory was not new, having been held by a few early Greek astronomers, and also that he was just "postulating and theorizing," not necessarily speaking the absolute truth. However, it was clear that Copernicus personally believed in what he wrote. Another astronomer, Giordano Bruno, who evangelized Copernicus's theory, was not so diplomatic and was burned at the stake in 1600. Galileo Galilei took up the heliocentric banner in 1632 and was ultimately exiled under guard for his heresy. Finally, the Danish astronomer Tycho Brahe, while shunning a direct association with the controversial heliocentric theory, spent virtually his complete life from 1546 to 1602 making astronomical observations of planet and star movements, resulting in spectacular improvements in the precision of existing knowledge.

Into this tenuous time, Johannes Kepler was born in Württemberg, Germany, on December 27, 1571. By winning scholarships, he was able to finish elementary school and to go on to the University of Tübingen. There he was converted to the heliocentric theory by Michael Mastlin, a professor of astronomy. Later, Kepler became a teacher of mathematics and an ardent astronomer. Through his writings on celestial motion, Kepler came to the attention of Tycho Brahe, who was now living in Prague. In 1599, Kepler went to Prague to work under Brahe, who died just two years later. Kepler stayed in Prague, extending and improving the existing tables of celestial movement. In 1627, he published his *Rudolphine Tables*, which were much more accurate than any existing tables at that time.

However, Kepler was also thinking and theorizing about his observations, attempting to bring some reason and order to the movement of the heavenly bodies. For example, the heliocentric system of Copernicus assumed circular orbits of the planets about the sun; but Kepler's accurate observations did not precisely fit circular motion. In 1609 he found that elliptical orbits fit his measurements exactly, giving rise to *Kepler's first law* (see Sec. 8.6). In the same year, he induced that a line drawn from the sun to a planet sweeps out equal areas in equal times—*Kepler's second law*. His first and second laws were published in his book *New Astronomy* in 1609. It was not until nine years later that he discovered that the square of the period of planetary orbits was proportional to the cube of the semimajor axis of the elliptical orbit—*Kepler's third law*. This was published in 1618 in his book *Epitome of the Copernican Astronomy*.

Kepler's impact on astronomy was massive; in fact, his work was the founding of modern astronomy. His contributions are all the more stunning because his laws were induced from empirical observation. Kepler did not have the tools developed later by Newton. Therefore, he could not derive his laws with the same finesse as we did in Sec. 8.6.

It is interesting to note that Kepler also wrote science fiction. In his book *Somnium* (Dream), Kepler describes a trip from the earth to the moon. Recognizing that the void of space would not support flight by wings, he had to resort to demons as a supernatural mode of propulsion. These demons would carry along humans, suitably anesthetized to survive the rigors of space travel. He described the moon in as much astronomical detail as was possible in that age, but imagined moon creatures that lived in caves. Modern historians of science fiction literature believe that Kepler's *Somnium* was really a vehicle to present his serious scientific ideas about the moon while attempting to avoid religious persecution. *Somnium* was published in 1634, four years after Kepler's death.

Kepler spent his later life as a professor of mathematics in Linz. He died in Regensburg on November 15, 1630, leaving a legacy that reaches across the centuries to the astronautics of the present day.

8.14   HISTORICAL NOTE: NEWTON AND THE LAW OF GRAVITATION

Newton's law of universal gravitation, Eq. (8.19), appears in every modern high school and college physics textbook; its existence is virtually taken for granted. Moreover, this equation is the very foundation for all modern astronautical calculations of motion through space, as discussed throughout this chapter. However, the disarming simplicity of Eq. (8.19) and its commonplace acceptance in classical physics belie the turmoil that swarmed about the concept of gravity before and during the 17th century, when Newton lived.

The earliest ideas on "gravity" were advanced by Aristotle during the period around 350 B.C. Believing that the four fundamental elements of the universe were earth, water, air, and fire, the aristolelian school held that everything in the universe had its appointed station and tended to return to this station if originally displaced. Objects made from "earth" held the lowest station, and thus heavy material objects would fall to the ground, seeking their proper status. In contrast, fire and air held a high station and would seek this status by rising toward the heavens. These ideas persisted until the age of Copernicus, when people began to look for more substantial explanations of gravity.

In 1600, the English scientist William Gilbert suggested that magnetism was the source of gravity and that the earth was nothing more than a gigantic lodestone. Kepler adopted these views, stating that gravity was "a mutual affection between cognate bodies tending towards union or conjunction, similar in kind to magnetism." Kepler used this idea in an attempt to prove his laws of planetary motion (see the previous section) but was not successful in obtaining a quantitative law for the force of gravity. About the same time, the French scientist and mathematician René Descartes (who introduced the cartesian coordinate system to the world of mathematics) proposed that gravity was the result of an astronomical fluid that was swirling in a vortex motion, pushing heavy objects toward the core of the vortex. Christian Huygens, a Dutch gentleman and amateur scientist, seemed to confirm Descartes's theory in the laboratory; he set up a whirlpool of water in a bowl and observed that pebbles "gravitated" to the center of the bowl.

Into this confused state of affairs was born Isaac Newton at Woolsthorpe near Grantham, Lincolnshire, England, on December 25, 1642. Newton's father died a few months before he was born, and Newton was raised by his grandmother. His education ultimately led to studies at Trinity College, Cambridge University, in 1661, where he quickly showed his genius for mathematics. In 1666, he left Cambridge for his home in Woolsthorpe Manor to avoid the Great Plague of 1665–1666. It was here, at the fresh age of 24, that Newton made some of his discoveries and conclusions that were to revolutionize science and mathematics, not the least of which was the development of differential calculus. Also Newton later maintained that during this stay in the country he deduced the law of centripetal force, that is, that a body in circular motion experiences a radial force that varies inversely with the distance from the center. (In today's language, the centripetal acceleration due to circular motion is equal to V^2/r, as shown in all elementary physics books.) From this result applied to Kepler's third law, Newton further deduced that the force of gravity between two objects varies inversely as the square of the distance separating them, which led to the universal law of gravitation, as given by Eq. (8.19). However, Newton did not bother to publish immediately or otherwise announce his findings. The public was kept in the dark for another 30 years!

Throughout the history of science and engineering, there are numerous examples of ideas whose "time had come" and which were conceived by several different people almost simultaneously. The same Christian Huygens made experiments with pendulums and circular moving bodies that led to his discovery of the law of centripetal force in 1673. With this, Robert Hooke (of Hooke's law fame), Christopher Wren (later to become an internationally famous architect), and Edmund Halley (of Halley's comet fame) all deduced the inverse-square law of gravity in 1679. Hooke wrote to Newton in the same year, telling him of the inverse-square discovery and asking Newton to use it to prove that a planet revolves in an elliptical orbit. Newton did not reply. In 1685, the problem was again posed to Newton, this time by Halley. Newton sent back such a proof. Halley was much impressed and strongly encouraged Newton to publish all his discoveries and thinking as soon as possible. This lead to Newton's *Philosophiae Naturales Principia Mathematica*—the famous *Principia*—which has become the foundation of classical physics. It is interesting to note that the *Principia* was originally to be published by the Royal Society. But Hooke, who laid claim to the prior discovery of the inverse-square law and who was the curator of the Royal

Society, apparently discouraged such publication. Instead, the *Principia*, the most important scientific document to that time in history, was published at the personal expense of Halley.

Hooke again put forward his claim to the inverse-square law during a meeting of the Royal Society in 1693. Shortly thereafter, Newton had a nervous breakdown, which lasted about a year. After his recovery, Newton finally announced that he had made the basic discoveries of both the centripetal force law and the inverse-square law of gravitation back in 1666. Because of his high standing and reputation of that time, as well as subsequently, Newton's claim has been generally accepted through the present time. However, the record shows that we have only his word. Therefore, the claim by Robert Hooke is certainly legitimate, at least in spirit. Equation (8.19), which comes down to us as Newton's law of universal gravitation, could legitimately be labeled (at the very least) the "Newton-Hooke law."

Of course, this is not to detract from Newton himself, who was *the* giant of science in the 17th century. During his later years, Newton entered public life, becoming warden of the British Mint in 1696, advancing to the chief post of master in 1699. In this capacity, he made many important contributions during Britain's massive recoinage program of that time. In 1703, he was elected president of the Royal Society, a post he held for the next 25 years. During this period, Newton was embroiled in another controversy, this time with the German mathematician Gottfried von Liebniz over the claim of the discovery of calculus. Also, during these later years, Newton's imposing prestige and authority via the Royal Society apparently tended to squelch certain ideas put forward by younger scientists. Because of this, some historians of science hint that Newton may have hindered the progress of science during the first 30 years of the 18th century.

Newton died in Kensington on March 20, 1727. He is buried at a prominent location at Westminster Abbey. Without Newton, and without Kepler before him, this chapter on astronautics might never have been written.

8.15   HISTORICAL NOTE: LAGRANGE

In Sec. 8.3, a corollary to Newton's second law is introduced, namely, Lagrange's equation. Lagrange came after Newton. He was one of the small group of European scientists and mathematicians who worked to develop and augment newtonian (classical) physics during the 18th century; he was a contemporary of Laplace and a friend of Leonhard Euler.

Joseph L. Lagrange was born of French parents at Turin, Italy, on January 25, 1736. His father was an officer in the French army; hence it is no surprise that at the age of 19, Lagrange was appointed as professor of mathematics at the Turin Artillery School. Active in scientific thought, he helped to found the Turin Academy of Sciences. In 1756, he wrote to Euler (see Sec. 4.20) with some original contributions to the calculus of variations. This helped to establish Lagrange's reputation. In fact, in 1766, he replaced Euler as director of the Berlin Academy at the invitation of Frederick II (Frederick the Great) of Prussia. For the next 20 years, Lagrange was extremely productive in the field of mechanics. His work was analytical, and he endeavored to reduce the many aspects of mechanics to a few general formulas. This is clearly reflected in the formalism discussed in Sec. 8.3. Lagrange's equations used in Sec. 8.3 were published in an important book by Lagrange entitled *Mécanique Analytique* in 1787. For these contributions, he is considered by some historians to be the greatest mathematician of the 18th century.

Lagrange moved to Paris in 1786. During the French Revolution, he was president of the committee for reforming weights and measures standards. At the time of his death in Paris on April 10, 1813, he was working on a revised version of his *Mécanique Analytique*.

8.16   HISTORICAL NOTE: UNMANNED SPACE FLIGHT

On the evening of October 4, 1957, the present author was a student of aeronautical engineering. The radio was on. Concentration on studies was suddenly interrupted by a new bulletin: The Soviet Union had just successfully launched the first artificial earth satellite in history. Labeled *Sputnik I* and shown in Fig. 8.34, this 184-lb sphere circled the earth in an elliptical orbit, with an apogee and perigee of 560 and 140 mi, respectively, and with a period of $1\frac{1}{2}$ h. The personal feeling of exhilaration that humanity had finally made the first great step toward space exploration was tempered by questions about the technical position of the United States in space flight. These feelings were to be reflected and amplified throughout the United States for weeks, months, and years to come. *Sputnik I* started a technological revolution which has influenced virtually all aspects of society, from education to business, from biology to philosophy. October 4, 1957, is a red-letter date in the history of humanity—the beginning of the space age.

Although the launching of *Sputnik I* came as a surprise to most of the general public, the technical community of the western world had been given some clear hints by Russian scientists. For example, on November 27, 1953, at the World Peace Council in Vienna, the Soviet academician A. N. Nesmeyanov stated that "science had reached such a stage that . . . the creation of an artificial satellite of the earth is a real possibility." Then, in April 1955, the U.S.S.R. Academy of Sciences announced the creation of the Permanent Interdepartmental Commission for Interplanetary Communications, with responsibility for developing artificial earth satellites for meteorological applications. In August of that year, the highly respected Russian scientist Leonid I. Sedov, at the Sixth International Astronautical Congress, in Copenhagen, said: "In my opinion, it will be possible to launch an artificial satellite of the Earth within the next two years, and there is the technological possibility of creating artificial satellites of various sizes and weights." Obviously, the Russian program kept to its schedule. Indeed, in June 1957, just four months before *Sputnik I*, the same A. N. Nesmeyanov blatantly stated that both the rocket launch vehicle and the satellite were ready and would be launched in a few months. Clear signs and clear words—yet the launching of *Sputnik I* still fell like a ton of bricks on the western world.

Figure 8.34   The first artificial earth satellite—*Sputnik I*—launched on October 4, 1957.

In 1957, the United States was not new to the idea of artificial satellites. Indeed, some far-sighted thinking and technical analyses of the prospects for launching such satellites were performed by the U.S. Navy and the U.S. Army Air Force beginning in 1945. Then in May 1946 (just one year after Germany had been defeated in World War II), a Project RAND report entitled "Preliminary Design of an Experimental World-Circling Spaceship" was submitted to Wright Field, Dayton, Ohio. This report showed the feasibility of putting a 500-lb satellite in orbit at around 300 mi high. Moreover, it outlined how this could be accomplished in a five-year time scale! The authors of this report made some seriously prophetic statements, as follows:

> Although the crystal ball is cloudy, two things seem clear—1. A satellite vehicle with appropriate instrumentation can be expected to be one of the most potent scientific tools of the Twentieth Century.

2. The achievement of a satellite craft by the United States would inflame the imagination of mankind, and would probably produce repercussions in the world comparable to the explosion of the atomic bomb....

Then the authors go on to state:

Since mastery of the elements is a reliable index of material progress, the nation which first makes significant achievements in space travel will be acknowledged as the world leader in both military and scientific techniques. To visualize the impact on the world, one can imagine the consternation and admiration that would be felt here if the U.S. were to discover suddenly that some other nation had already put up a successful satellite.

These were indeed prophetic words, written fully eleven years before *Sputnik I*.

The 1946 RAND report, along with several contemporary technical reports from the Jet Propulsion Laboratory at the California Institute of Technology, established some fundamental engineering principles and designs for rocket launch vehicles and satellites. However, these ideas were not seized upon by the U.S. government. The period after World War II was one of shrinking defense budgets, and money was simply not available for such a space venture. Of probably greater importance was the lack of a mission. What if a satellite were launched? What benefits would it bring, especially military benefits? Keep in mind this was in a period before miniaturized electronics and sophisticated sensing and telemetering equipment. Therefore, the first serious U.S. effort to establish a satellite program withered on the vine, and the idea lay essentially dormant for the next nine years.

Although upstaged by *Sputnik I*, the United States in 1957 finally did have an ongoing project to orbit an artificial satellite. On July 29, 1955, President Dwight D. Eisenhower announced that the United States would orbit a small earth satellite in conjunction with the International Geophysical Year. Making use of 10 years of high-altitude sounding-rocket technology, which started with a number of captured German V-2 rockets, the United States established the Vangard program, managed by the Office of Naval Research, to accomplish this goal. Martin Company in Baltimore, Maryland, was chosen as the prime contractor. During the next two years, a rocket booster was designed and built to launch a small, 3-lb experimental satellite. By government edict, the Vangard project was required not to draw upon or interface with the rapidly growing and high-priority ICBM program, which was developing large rocket engines for the military. Therefore, Dr. John P. Hagen, director of Project Vangard, and his small team of scientists and engineers had to struggle almost as second-class citizens to design the Vangard rocket in an atmosphere of relatively low priority. (This is in sharp contrast to the Russian space program, which from the very beginning utilized and benefited from the Soviet military ICBM developments. Since Russian atomic warheads of that day were heavier than comparable U.S. devices, the Soviet Union had to develop more powerful rocket boosters. Their space program correspondingly benefited, allowing *Sputnik I* and *II* to be the surprisingly large weights of 184 and 1120 lb, respectively.

By October 1957, two Vangard rockets had been successfully tested at Cape Canaveral; and the test program, which was aimed at putting a satellite into orbit before the end of 1958, was reasonably close to schedule. Then came *Sputnik I* on October 4. Not to be completely upstaged, the White House announced on October 11 that Project Vangard would launch a U.S. satellite "in the near future." Suddenly in the limelight of public attention, and now under intense political pressure, a third test rocket was successfully tested on October 23, carrying a 4000-lb dummy payload

Figure 8.35   *Explorer I,* the first U.S. artificial earth satellite, launched on January 31, 1958.

to an altitude of 109 mi and 335 mi downrange. Then, on December 6, 1957, in full view of the world's press, the first Vangard was prepared to put a small satellite into orbit. Unfortunately, the Vangard first-stage engine had its first (and last) failure of the program. With failing thrust, the rocket lifted a few feet off the launch pad and then fell back in a spectacular explosion. In Dr. Hagen's words: "Although we had three successful test launches in a row, the failure of TV-3 [the designation of that particular vehicle] was heard around the world."

Despite the original disadvantages of low priority, despite the emotional pressure after *Sputnik I*, and despite the inglorious failure of December 6, the Vangard project went on to be very successful. Vangards I, II, and III were put into orbit on March 17, 1958; February 17, 1959; and September 18, 1959; respectively, attributing to the fine efforts of Dr. Hagen and his group.

But Vangard I was not the first U.S. satellite. President Eisenhower's July 1955 announcement about U.S. plans to orbit a satellite was followed by much debate about whether military rocketry should be used. One proposal at the time was to make use of the rocket vehicles being developed at the Army's Redstone Arsenal at Huntsville, Alabama, under the technical direction of Dr. Wernher Von Braun. After the decision was made to go with the Vangard, the engineers at the Army Ballistic Missile Agency at Huntsville continued to propose a satellite program using the proven intermediate-range Jupiter C rocket. All such proposals were turned down. However, the picture changed after *Sputnik*. In later October 1957, Von Braun's group was given the green light to orbit a satellite—target date was January 30, 1958. A fourth stage was added to the Jupiter C rocket; this new configuration was labeled the Juno I. The target date was missed by only one day. On January 31, 1958, *Explorer I*, the first U.S. artificial satellite, was placed into orbit by Von Braun's team of scientists and engineers from Huntsville. *Explorer I*, shown in Fig. 8.35, weighed 18 lb, and its orbit had apogee and perigee of 957 and 212 mi, respectively; its period as 115 min. With the launchings of both *Sputnik I* and *Explorer I*, the two technological giants in the world—the United States and the Soviet Union—were now in competition in the arena of space.

It is not the purpose here to give an exhaustive survey of space exploration. For an authoritative presentation see the excellent book by Von Braun and Ordway, as well as others listed in the Bibliography at the end of this chapter.

8.17 HISTORICAL NOTE: MANNED SPACE FLIGHT

The previous section on unmanned space flight, the present section on manned space flight, and Sec. 9.16 on the early history of rocket engines are inexorably entwined—their division into three distinct sections in this book is purely artificial. Indeed, humanity's first imaginative thoughts about space flight involved the travel of human beings (not inanimate objects) to the moon. Later, during the technological revolution of the 19th and 20th centuries, it was correctly reasoned that manned space travel would have to be preceded by unmanned attempts just to learn about the problems that might be encountered. Also during this period, the rocket engine was recognized as the only feasible mechanism for propulsion through the void of space. In fact, the three early pioneers of rocket engines—Tsiolkovsky, Goddard, and Oberth (see Sec. 9.16)—were inspired in their work by the incentive of space travel rather than the military applications that ultimately produced the first successful large rockets. Clearly, the histories of unmanned and manned space flight and rocketry are overlapping and in many cases are indistinguishable.

Manned space flight really has its roots in science fiction and reaches as far back as the second century A.D., when the Greek writer Lucian of Samosata conceived a trip to the moon. In this book *Vera Historia*, Lucian's ship is caught in a storm, lifted into the sky by the higher winds, and after seven days and seven nights, is quite accidentally blown to the moon. There he finds a land that is "cultivated and full of inhabitants." Lucian's work was followed by other science fiction fantasies over the ensuing centuries, including Kepler's *Somnium*, mentioned in Sec. 8.13. These science fiction stories served a useful purpose in fueling the imaginative minds of some people and spurring them to deeper technological thought. Of particular note are books by Jules Verne and H. G. Wells in the 19th century, which were avidly read by many of the early rocket engineers. In particular, both Tsiolkovsky and Goddard avidly read Well's *War of the Worlds* and Verne's *From the Earth to the Moon*, and both have gone on record as being inspired by these works.

Considering that Wells and Verne wrote less than 100 years ago, and that just 40 years ago rockets were only the playthings of a few visionaries, it is astounding that manned space flight has now become a reality—and in the minds of the general public, a somewhat common reality. The ice was broken on April 12, 1961, when the Soviet Union orbited the 10,400-lb *Vostok I* spacecraft carrying Major Yuri A. Gagarin—the first human being to ride in space. Gagarin was a Russian air force major; his orbital flight lasted 1 h 48 min, with an apogee of 203 mi. Upon entry, *Vostok* was slowed first by retrorockets, and then by parachute, and came to rest on the solid ground somewhere deep within the interior of Russia. However, it is thought that, just before touchdown, Gagarin left the spacecraft and floated to earth with his own parachute. This entry mode was followed by several other Russian astronauts during subsequent years. Unfortunately, Gagarin was later killed in an airplane crash on March 27, 1968.

Humanity was now on its way in space! Less than a year later, the first American in space for a sustained period of time, Marine Colonel John H. Glenn, Jr., was orbited on February 20, 1962. Executing three orbits in a Mercury capsule with an apogee and perigee of 162.7 and 100.3 mi, respectively, Glenn's flight lasted 4 h 56 min from blastoff to touchdown. As with all subsequent U.S. manned spacecraft, Glenn rode the Mercury capsule all the way to the earth's surface, impacting at sea and being recovered by ship. Figure 8.36 shows a diagram of the single-seat Mercury space capsule and gives a clear picture of its size and shape relative to the astronaut himself. Glenn's successful flight in 1962 was a high point in Project Mercury, which was the United States' first manned space program. This project had its roots in an air Force study entitled "Manned Ballistic Rocket Research System" initiated in March 1956—a full year and a half *before Sputnik I*. Within two years under this project, the Air Force, NACA, and 11 private companies did much fundamental work on spacecraft

Hatch

Pitch and
yaw jets

Drogue chute

Horizon
scanners

Main and
reserve chutes

Periscope

Roll jets

Figure 8.36   The Mercury spacecraft. *(NASA.)*

design and life support systems. After *Sputnik I*, and after the formation of NASA in 1959, this work was centralized within NASA and designated Project Mercury. Thus, when Gagarin went into orbit in 1961, the United States was not far behind.

Indeed, the U.S. manned space flight program was galvanized when President John F. Kennedy, in a speech before Congress on May 25, 1961, declared: "I believe that this nation should commit itself to achieving the goal, before this decade is out, of landing [a person] on the Moon and returning him safely to Earth. . . ." In virtually a flash, the Apollo program was born. Over the next eight years, work on the Apollo manned lunar vehicle marshalled a substantial portion of the U.S. human and material aerospace resources. Then, almost like a page out of science fiction itself, at 4:18 P.M. (EDT) on July 20, 1969, a lunar descent vehicle named *Eagle*, carrying Neil A. Armstrong and Edwin E. Aldrin, Jr., came to rest on the moon's surface, with Michael Collins keeping watch in the Apollo Command Module orbiting above. President Kennedy's goal had been met; the dreams and aspirations of people over the centuries had been fullfilled; and the work of such minds as Copernicus, Kepler, Newton, and Lagrange had come to dramatic fruition.

The technical story of manned space flight is one of superhuman effort, fantastic advances in science and engineering, and unswerving dedication. It is still going on, albeit at a somewhat reduced frenzy after Apollo, and it will continue to progress as long as modern society exists. It is impossible to give justice to such a story in this short section; indeed, whole volumes have been written on this subject alone. Again, for a particularly authoritative and modern review, the reader is referred to the book by Von Braun and Ordway listed in the Bibliography.

8.18   SUMMARY

Some of the highlights of this chapter are summarized below:

1. The equation of the orbit or trajectory of a spacecraft under the influence of a central, inverse-square gravitational force field is

$$r = \frac{p}{1 + e\cos(\theta - C)} \qquad \text{[8.44]}$$

where e is the eccentricity and C is the phase angle. If $e = 0$, the orbit is a circle; if $e < 1$, the orbit is an ellipse; if $e = 1$, the trajectory is a parabola; if $e > 1$, the trajectory is a hyperbola.

2. The eccentricity depends on the difference between kinetic and potential energies of the spacecraft H.

$$e = \sqrt{1 + \frac{2h^2 H}{mk^4}} \qquad \text{[8.53]}$$

3. Circular velocity is given by

$$V = \sqrt{\frac{k^2}{r}} \qquad \text{[8.57]}$$

For earth satellites, circular, or orbital, velocity is 7.9 km/s, or approximately 26,000 ft/s (based on $r =$ earth's radius).

4. Escape velocity is given by

$$V = \sqrt{\frac{2k^2}{r}} \qquad \text{[8.58]}$$

For escape from the earth, based on the earth's radius, this velocity is 11.2 km/s, or approximately 36,000 ft/s.

5. Kepler's laws are: (1) a satellite describes an elliptical path around its center of attraction; (2) in equal times, the areas swept out by the radius vector of a satellite are the same; and (3) the periods of any two satellites about the same planet are related to their semimajor axes as

$$\left(\frac{\tau_1}{\tau_2}\right)^2 = \left(\frac{a_1}{a_2}\right)^3$$

6. The velocity variation of a ballistic entry vehicle through the atmosphere is given by

$$\frac{V}{V_E} = e^{-\rho/2[m/(C_D S)]Z\sin\theta} \qquad \text{[8.87]}$$

where ρ is a function of altitude, $m/(C_D S)$ is the ballistic parameter, θ is the entry angle, V_E is the initial entry velocity, and $Z = g_0/(RT)$. During entry, the maximum deceleration is given by

$$\left|\frac{dV}{dt}\right|_{max} = \frac{V_E^2 Z \sin\theta}{2e} \qquad \textbf{[8.101]}$$

7. Entry aerodynamic heating varies as the cube of the velocity:

$$\frac{dQ}{dt} = \frac{1}{2}\rho_\infty V_\infty^3 S C_H \qquad \textbf{[8.107]}$$

To minimize aerodynamic heating, the vehicle should have a blunt nose.

8. The lifting entry path depends on the lift parameter $m/(C_L S)$.

BIBLIOGRAPHY

Allen, H. J., and A. J. Eggers: *A Study of the Motion and Aerodynamic Heating of Missiles Entering the Earth's Atmosphere at High Supersonic Speeds*, NACA TR 1381, 1958.

Anderson, J. D., Jr.: "An Engineering Survey of Radiating Shock Layers," *AIAA Journal*, vol. 7, no. 9, Sept. 1969, pp. 1665–1675.

Chapman, D. R.: *An Approximate Analytical Method for Studying Entry into Planetary Atmospheres*, NASA TR R-11, 1959.

Emme, E. M.: *A History of Space Flight*, Holt, New York, 1965.

Hartman, E. P.: *Adventures in Research: A History of Ames Research Center 1940–1965*, NASA SP-4302, 1970.

Nelson, W. C., and E. E. Loft: *Space Mechanics*, Prentice-Hall, Englewood Cliffs, NJ, 1962.

Von Braun, W. and F. I. Ordway: *History of Rocketry and Space Travel*, 3d rev. ed., Crowell, New York, 1975.

PROBLEMS

8.1 At the end of a rocket launch of a space vehicle from earth, the burnout velocity is 13 km/s in a direction due south and 10° above the local horizontal. The burnout point is directly over the equator at an altitude of 400 mi above sea level. Calculate the trajectory of the space vehicle.

8.2 Calculate and compare the escape velocities from Venus, Earth, Mars, and Jupiter, given the information tabulated below.

	Venus	Earth	Mars	Jupiter
k^2, m^3/s^2	3.24×10^{14}	3.96×10^{14}	4.27×10^{13}	1.27×10^{17}
r, m	6.16×10^6	6.39×10^6	3.39×10^6	7.14×10^7

8.3 The mass and radius of the earth's moon are 7.35×10^{22} kg and 1.74×10^6 m, respectively. Calculate the orbital and escape velocities from the moon.

8.4   It is known that the period of revolution of the earth about the sun is 365.3 days and that the semimajor axis of the earth's orbit is 1.495×10^{11} m. An astronomer notes that the period of a distant planet is 29.7 earth years. What is the semimajor axis of the distant planet's orbit? Check in a reference source (encyclopedia, etc.) what planet of the solar system this might be.

8.5   Assume that you wish to place in orbit a satellite that always remains directly above the same point on the earth's equator. What velocity and altitude must the satellite have at the instant of burnout of the rocket booster?

8.6   Consider a solid iron sphere entering the earth's atmosphere at 8 km/s and at an angle of 30° below the local horizontal. The sphere diameter is 1.6 m. Calculate (*a*) the altitude at which maximum deceleration occurs, (*b*) the value of the maximum deceleration, and (*c*) the velocity at which the sphere would impact the earth's surface.

8.7   The aerodynamic heating rate of a given entry vehicle at 200,000 ft traveling at a velocity of 27,000 ft/s is 100 Btu/(ft^2)(s). What is the heating rate if the velocity is 36,000 ft/s at the same altitude?

chapter
9

Propulsion

We have sought power in the same fire which serves to keep the vessel aloft. The first which presented itself to our imagination is the power of reaction, which can be applied without any mechanism, and without expense: it consists solely in one or more openings in the vessel on the side opposite to that in which one wishes to be conveyed.

> Joseph Montgolfier, 1783—the
> first recorded technical statement
> in history on jet propulsion for
> a flight vehicle

I began to realize that there might be something after all to Newton's Laws.

> Robert H. Goddard, 1902

9.1 INTRODUCTION

The old saying that "you cannot get something for nothing" is particularly true in engineering. For example, the previous chapters have discussed the aerodynamic generation of lift and drag; the performance, stability, and control of airplanes; and the motion of spacecraft. All this takes the expenditure of *power*, or *energy*, which is supplied by an engine or propulsive mechanism of some type. The study of *propulsion* is the subject of this chapter. Here we examine what makes an airplane or space vehicle go.

Throughout Chap. 1 the dominant role played by propulsion in the advancement of manned flight is clearly evident. George Cayley was concerned in 1799, and he equipped his airplane designs with paddles. Henson and Stringfellow did better by considering "airscrews" powered by steam engines, although they were unsuccessful in their efforts. In 1874, Felix Du Temple momentarily hopped off the ground in a machine powered by an obscure type of hot-air engine; he was followed by Mozhaiski in 1884 who used a steam engine (see Figs. 1.13 and 1.14). By the late 19th century, the early aeronautical engineers clearly recognized that successful manned flight depended upon the development of a lightweight but powerful engine. Fortunately, the advent of the first practical internal combustion engine in 1860 paved the way for such success. However, in spite of the rapid development of these gasoline-powered engines and their role in the early automobile industry, such people as Langley (Sec. 1.7) and the Wright brothers (Sec. 1.8) still were forced to design their own engines in order to obtain the high horsepower-to-weight ratio necessary for flight. Such internal combustion reciprocating engines driving a propeller ultimately proved to be a winning combination and were the only practical means of airplane propulsion up to World War II. In the process, such engines grew in horsepower from the 12-hp Wright-designed engine of 1903 to the 2200-hp radial engines of 1945, correspondingly pushing flight velocities from 28 to more than 500 mi/h. Then a revolution in propulsion occurred. Frank Whittle took out a patent in Britain in 1930 for a jet-propelled engine and worked ceaselessly on its development for a decade. In 1939, the German Heinkel He 178 airplane flew with a turbojet engine developed by Dr. Hans von Ohain. It was the first successful jet-propelled test vehicle. This led to the German Me 262 jet fighter late in World War II. Suddenly, jet engines became the dominant power plants for high-performance airplanes, pushing flight velocities up to the speed of sound in the 1950s, and beyond in the 1960s and 1970s. Today, the airplane industry rides on jet propulsion, and jet-propelled supersonic flight for both commercial and military airplanes is a regular occurrence. Meanwhile, another revolution of even greater impact occurred: the advent of the successful rocket engine. Pioneered by Konstantin Tsiolkovsky (1857–1935) in Russia, Robert H. Goddard (1882–1945) in the United States, and Hermann Oberth (b. 1894) in Germany, the rocket engine first became operational in 1944 with the German V-2 missile. Being the only practical means of launching a vehicle into space, the rocket engine soon proved itself during the space age, allowing people to go to the moon and to probe the deep unknown regions of our solar system.

It is clear from the above brief historical sketch that propulsion has led the way for all major advancements in flight velocities. Propulsion is one of the major disciplines of aerospace engineering; therefore, in the following sections, some of the basic principles of propellers, reciprocating engines, turbojets, ramjets, and rockets will be examined. Such propulsion devices are highly aerodynamic. Therefore, a firm understanding of the aerodynamic and thermodynamic fundamentals presented in Chaps. 4 and 5 will help you grasp the propulsion concepts discussed in this chapter.

The road map for this chapter is shown in Fig. 9.1. Flight propulsion devices for aerospace vehicles can be categorized into the three main columns in Fig. 9.1—air-breathing engines, ricket engines, and advanced space propulsion devices. Air-breathing engines constitute the left-hand column in Fig. 9.1; from their name, they are designed to use the oxygen in the atmosphere for an oxidizer. We begin this chapter by discussing the propeller and the reciprocating engine, the combination of which was the main power plant combination for the first 50 years of successful powered flight. We then examine the principle of jet propulsion, and we derive the thrust

Figure 9.1 Road map for Chap. 9.

equation for jet propulsion devices (which include both jet and rocket engines). With this, we tour down the remainder of the left column and examine the three main types of air-breathing jet engines—the turbojet, turbofan, and ramjet. The middle column in Fig. 9.1 deals with rocket engines—propulsion devices that carry their own fuel and oxidizer and that therefore are independent of the atmosphere. We examine how the performance of a rocket engine (thrust and efficiency) can be calculated, and how we can predict what weight of payload can be accelerated to what velocity by a rocket (the rocket equation). The performance of a multistage rocket vehicle (as opposed to a single, large rocket) is calculated and discussed. Also the important aspects of chemical rocket engine propellants are mentioned. Finally, we examine some of the basic concepts for advanced space propulsion (the right-hand column in Fig. 9.1).

9.2 PROPELLER

Airplane wings and propellers have something in common: They are both made up of airfoil sections designed to generate an aerodynamic force. The wing force provides lift to sustain the airplane in the air; the propeller force provides thrust to push the airplane through the air. A sketch of a simple three-blade propeller is given in Fig. 9.2, illustrating that a cross section is indeed an airfoil shape. However, unlike a wing, where the chord lines of the airfoil sections are essentially all in the same direction, a propeller is twisted such that the chord line changes from being almost parallel to V_∞ at the root, to almost perpendicular at the tip. This is illustrated in Fig. 9.3, which shows a side view of the propeller, as well as two sectional views, one at the tip and the other at the root. This figure should be studied carefully. The angle between the chord line and the propeller's plane of rotation is defined as the *pitch angle* β. The distance from the root to a given section is r. Note that $\beta = \beta(r)$.

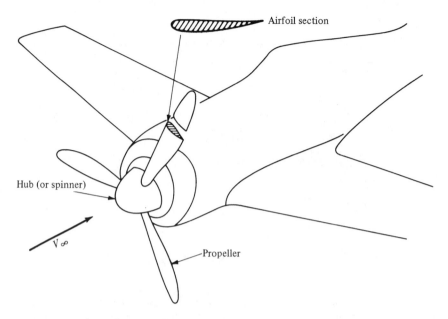

Figure 9.2   The airplane propeller, emphasizing that a propeller cross section is an airfoil shape.

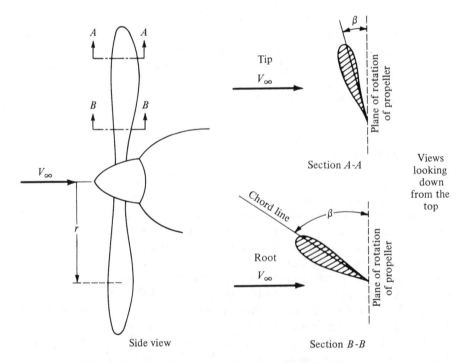

Figure 9.3   Illustration of propeller, showing variation of pitch along the blade.

The airflow seen by a given propeller section is a combination of the airplane's forward motion and the rotation of the propeller itself. This is sketched in Fig. 9.4a, where the airplane's relative wind is V_∞ and the speed of the blade section due to rotation of the propeller is $r\omega$. Here, ω denotes the angular velocity of the propeller in radians per second. Hence, *the relative wind seen by the propeller section is the vector sum of V_∞ and $r\omega$*, as shown in Fig. 9.4b.

Clearly, if the chord line of the airfoil section is at an angle of attack α with respect to the local relative wind V, then lift and drag (perpendicular and parallel to V, respectively) are generated. In turn, as shown in Fig. 9.5, the components of L and D in the direction of V_∞ produce a net *thrust* T:

$$T = L \cos\phi - D \sin\phi \qquad \textbf{[9.1]}$$

where $\phi = \beta - \alpha$. This thrust, when summed over the entire length of the propeller blades, yields the net thrust available (T_A, as defined in Chap. 6) which drives the airplane forward.

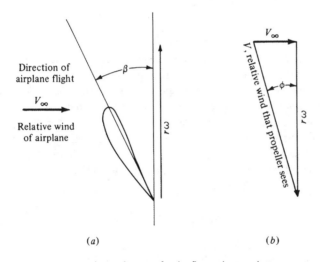

(a) **(b)**

Figure 9.4 Velocity diagram for the flow velocity relative to the propeller.

Figure 9.5 Generation of propeller thrust.

This simple picture is the essence of how a propeller works. However, the actual prediction of propeller performance is more complex. The propeller is analogous to a finite wing that has been twisted. Therefore, the aerodynamics of the propeller are influenced by the same induced flow due to tip vortices as was described for the finite wing in Secs. 5.13 and 5.14. Moreover, due to the propeller twist and rotational motion, the aerodynamic theory is even more complicated. However, propeller theory has been extensively developed, and more details can be found in the books by Dommasch et al. and Glauret (see Bibliography at the end of this chapter). Such theory is beyond the scope of this book.

Instead, let us concentrate on the understanding of propeller efficiency η introduced in Sec. 6.6. From Eq. (6.30) the propeller *efficiency* is defined as

$$\eta = \frac{P_A}{P} \qquad \text{[9.2]}$$

where P is the shaft brake power (the power delivered to the propeller by the shaft of the engine) and P_A is the power available from the propeller. As given in Eq. (6.31), $P_A = T_A V_\infty$. Hence, Eq. (9.2) becomes

$$\eta = \frac{T_A V_\infty}{P} \qquad \text{[9.3]}$$

As explained above, T_A in Eq. (9.3) is basically an aerodynamic phenomenon which is dependent on the angle of attack α in Fig. 9.5. In turn, α is dictated by the pitch angle β and ϕ, where ϕ itself depends on the magnitudes of V_∞ and $r\omega$. The angular velocity $\omega = 2\pi n$, where n is the number of propeller revolutions per second. Consequently, T_A must be a function of at least β, V_∞, and n. Finally, the thrust must also depend on the size of the propeller, characterized by the propeller diameter D. In turn, the propeller efficiency, from Eq. (9.3), must depend on β, V_∞, n, and D. Indeed, theory and experiment both show that for a fixed pitch angle β, η is a function of the dimensionless quantity

$$J = \frac{V_\infty}{nD} \qquad \text{advance ratio}$$

A typical variation of η with J for a fixed β is sketched in Fig. 9.6, three curves are shown corresponding to three different values of pitch. Figure 9.6 is important; it is from such curves that η is obtained for an airplane performance analysis, as described in Chap. 6.

Examine Fig. 9.6 more closely. Note that $\eta < 1$; this is because some of the power delivered by the shaft to the propeller is always lost, and hence $P_A < P$. These losses occur because of several different effects. First imagine that you are standing in an open field. The air is still; it has no velocity. Then assume a propeller-driven vehicle goes zooming by you. After the propeller has passed, you will feel a stiff breeze moving in the direction opposite that of the vehicle. This breeze is part of the slipstream from the propeller; that is, the air is set into both translational and rotational motion by the passage of the propeller. Consequently, you observe some translational and rotational *kinetic energy* of the air where before there was none. This kinetic energy has come from part of the power delivered by the shaft to the

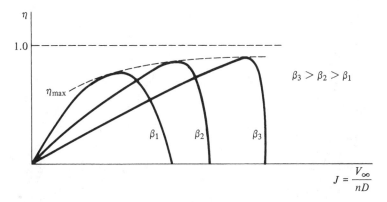

Figure 9.6 Propeller efficiency versus advance ratio. Note that D denotes propeller diameter.

propeller; it does no useful work and hence robs the propeller of some available power. In this fashion, the energy of the slipstream relative to the still air ahead of the vehicle is a source of power loss. Another source is frictional loss due to the skin friction and pressure drag (profile drag) on the propeller. Friction of any sort always reduces power. A third source is *compressibility* loss. The fastest-moving part of the propeller is the tip. For many high-performance engines, the propeller tip speeds result in a near-sonic relative wind. When this occurs, the same type of shock wave and boundary-layer separation losses which cause the drag-divergence increase for wings (see Sec. 5.10) now act to rob the propeller of available power. If the propeller tip speed is supersonic, η drops dramatically. This is the primary reason why propellers have not been used for transonic and supersonic airplanes. (After World War II, the NACA and other laboratories experimented with swept-back propellers, motivated by the success obtained with swept wings for high-speed flight, but nothing came of these efforts.) As a result of all the losses described above, the propeller efficiency is always less than unity.

Return again to Fig. 9.6. Note that for a fixed β, the efficiency is zero at $J = 0$, increases as J increases, goes through a maximum, and then rapidly decreases at higher J, finally again going to zero at some large finite value of J. Question: Why does η go to zero for the two different values of J? At the origin, the answer is simple. Consider a propeller with given values of n and D; hence, J depends only on V_∞. When $V_\infty = 0$, then $J = 0$. However, when $V_\infty = 0$, then $P_A = T_A V_\infty = 0$; consequently, $\eta = P_A/P = 0$. Thus, propeller efficiency is zero at $J = 0$ because there is no motion of the airplane, hence no power available. At the other extreme, when V_∞, hence J, is made large, the propeller loses lift owing to small angles of attack. This is seen in Fig. 9.7. Consider a given propeller airfoil section at a distance r from the center. Assume ω, hence $r\omega$, remains constant. If V_∞ is small, the relative wind will be as shown in Fig. 9.7a, where the airfoil section is at a reasonable angle of attack and hence produces a reasonable lift. Now, if V_∞ is increased, the relative

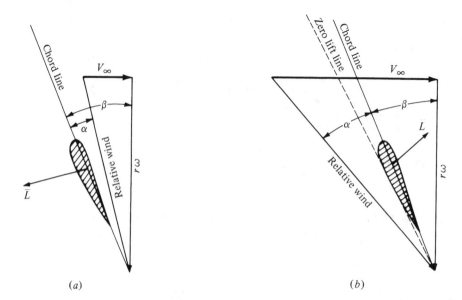

Figure 9.7   Explanation of the variation of propeller efficiency with advance ratio. (*a*) Velocity diagram for low V_∞. (*b*) Velocity diagram for high V_∞.

wind approaches the chord line; hence α, and therefore the lift coefficient, decreases. If the value of V_∞ is such that the relative wind corresponds to the zero-lift line, then the lift (hence thrust) is zero, and again $\eta = T_A V_\infty / P = 0$. In fact, if V_∞ is made even larger, the section will produce negative lift, hence reverse thrust, as shown in Fig. 9.7*b*.

A consideration of the relative wind also explains why a propeller blade is twisted, with a large β at the root and a small β at the tip, as was first sketched in Fig. 9.3. Near the root, r, and hence $r\omega$, is small. Thus, as shown in Fig. 9.8*a*, β must be large to have a reasonable α. In contrast, near the tip, r, hence $r\omega$, is large. Thus, as shown in Fig. 9.8*b*, β must be smaller in order to have a reasonable α.

Return again to Fig. 9.6. All early airplanes before 1930 had *fixed-pitch propellers*; that is, the values of β for all sections were geometrically fixed by the design and manufacture of the blades, and once the propeller was rigidly mounted on the engine shaft, the pilot could not change the blade angle. Thus from the curves in Fig. 9.6, maximum propeller efficiency could be obtained only at a specific value of the advance ratio J. At other flight velocities, the propeller always operated at efficiencies less than maximum. This characteristic severely limited airplane performance. Some improvement, although small, was attempted in 1916 at the Royal Aircraft Factory at Farnborough, England, by a design of a two-pitch propeller. However the ultimate solution was the *variable-pitch propeller*, patented in 1924 by Dr. H. S. Hele-Shaw and T. E. Beacham in England and first introduced into practical production in 1932 in the United States. The variable-pitch propeller is fixed to a mechanical mech-

$$\beta_a > \beta_b$$

(*a*)

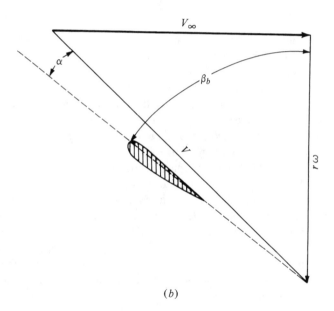

(*b*)

Figure 9.8   Difference in the relative wind along the propeller blade.
(*a*) Near the root; (*b*) near the tip.

anism in the hub; the mechanism rotates the entire blade about an axis along the length of the blade. In this fashion, the propeller pitch can be continuously varied to maintain maximum efficiency at all flight velocities. This can be visualized as riding along the peaks of the propeller efficiency curves in Fig. 9.6, as shown by the dotted η_{max} line. A further development of this concept was the introduction in 1935 of the *constant-speed propeller*, which allowed the pitch angle to be varied

NACA 16-009

NACA 16-509

NACA 16-1009

Figure 9.9   Typical high-speed airfoil sections for propellers.

continuously and automatically to maintain the proper torque on the engine such that the engine revolutions per minute were constant over the range of flight velocities. This is advantageous because the brake power output of aircraft piston engines is usually optimized at a given revolutions per minute. Nevertheless, the introduction of the variable-pitch and constant-speed propellers in the 1930s was one of the most important developments in the history of aeronautical engineering. As a result, values of η range from about 0.83 to 0.90 for most modern propellers.

A comment is in order concerning airfoil sections used for propellers. Early propellers from the World War I era typically utilized the RAF-6 airfoil; later, the venerable Clark Y shape was employed. During the late 1930s, some of the standard NACA sections were used. However, as aircraft speeds rapidly increased during World War II, special high-speed profiles were incorporated in propellers. The NACA developed a complete series, the 16 series, which found exclusive use in propellers. This series is different from the wing airfoil sections given in App. D; some typical shapes are sketched in Fig. 9.9. These are thin profiles, designed to minimize the transonic flow effects near the propeller tips. They should be compared with the more conventional shapes in App. D.

9.3   RECIPROCATING ENGINE

For the first 50 years of successful manned flight, the internal combustion, reciprocating, gasoline-burning engine was the mainstay of aircraft propulsion. It is still used today in airplanes designed to fly at speeds less than 300 mi/h, the range for the vast majority of light, private, general-aviation aircraft (such as the hypothetical CP-1 in the examples of Chap. 6). A photograph of a typical internal combustion reciprocating engine is shown in Fig. 9.10.

Figure 9.10   A large radial air-cooled internal combustion aircraft engine; the Pratt and Whitney Twin Wasp R-2000, produced from 1941 to 1959. (*Pratt and Whitney Aircraft, a Division of United Technologies.*)

The basic operation of these engines is a piston moving back and forth (reciprocating) inside a cylinder, with valves which open and close appropriately to let fresh fuel-air mixture in and burned exhaust gases out. The piston is connected to a shaft via a connecting rod which converts the reciprocating motion of the piston to rotational motion of the shaft. A typical four-stroke cycle is illustrated in Fig. 9.11. During the *intake* stroke (Fig. 9.11*a*), the piston moves downward, the intake valve is open, and a fresh charge of gasoline-air mixture is drawn into the cylinder. This process is sketched on the *p-V* diagram (a plot of pressure versus volume) in Fig. 9.11*a*. Here, point 1 corresponds to the beginning of the stroke (where the piston is at the top, called top dead center), and point 2 corresponds to the end of the stroke (where the piston is at the bottom, called bottom dead center). The volume *V* is the total mixture volume between the top of the cylinder and the face of the piston. The intake stroke takes place at essentially constant pressure, and the total mass of fuel-air mixture

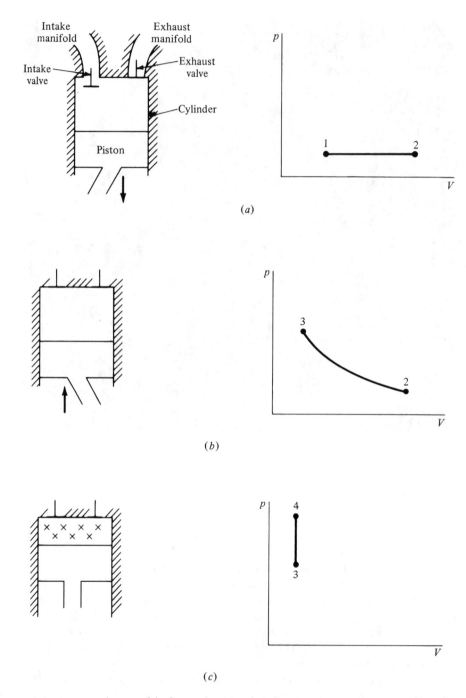

Figure 9.11   Elements of the four-stroke, internal combustion, reciprocating-engine cycle. (a) Intake stroke; (b) compression stroke; (c) constant-volume combustion. (continued)

(concluded)

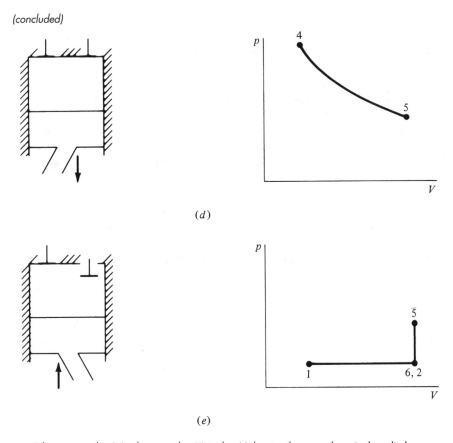

(d)

(e)

(d) power stroke; (e) exhaust stroke. Note that V denotes the gas volume in the cylinder.

inside the cylinder increases throughout the stroke. At the bottom of the intake stroke, the intake valve closes, and the *compression* stroke begins (Fig. 9.11b). Here, the piston compresses the now-constant mass of gas from a low pressure p_2 to a higher pressure p_3, as shown in the accompanying p-V diagram. If frictional effects are ignored, the compression takes place isentropically (see Sec. 4.6) because no heat is added or taken away. At the top of the compression stroke, the mixture is ignited, usually by an electric spark. *Combustion* takes place rapidly, before the piston has moved any meaningful distance. Hence, for all practical purposes, the combustion process is one of *constant volume* (Fig. 9.11c). Since energy is released, the temperature increases markedly; in turn, because the volume is constant, the equation of state, Eq. (2.9), dictates that pressure increases from p_3 to p_4. This high pressure exerted over the face of the piston generates a strong force which drives the piston downward on the *power stroke* (Fig. 9.11d). Again, assuming that frictional and heat transfer effects are negligible, the gas inside the cylinder expands isentropically to the pressure p_5. At the bottom of the power stroke, the exhaust valve opens. The pressure inside the cylinder instantly adjusts to the exhaust manifold pressure p_6, which is usually about the same value as p_2. Then during the *exhaust stroke*, Fig. 9.11e, the piston pushes the burned gases out of the cylinder, returning to conditions at point 1. Thus, the basic process of a conventional aircraft piston engine consists of a four-stroke cycle; intake, compression, power, and exhaust.

Because of the heat released during the constant-volume combustion, the cycle delivers a net amount of positive work to the shaft. This work can be calculated by using the complete p-V diagram for the cycle, as sketched in Fig. 9.12. Recall from Eq. (4.15) that the amount of work done on the gas due to a change in volume dV is $\delta w = -p\,dV$. In turn, the work done *by* the gas is simply

$$\delta w = p\,dV$$

For any part of the process, say, during the power stroke, this is equivalent to the small sliver of area of height p and base dV, as shown in Fig. 9.12. In turn, the work done by the gas on the piston during the whole power stroke is then

$$W_{\substack{\text{power} \\ \text{stroke}}} = \int_{V_4}^{V_5} p\,dV \qquad \textbf{[9.4]}$$

This is given by the area under the curve from point 4 to point 5 in Fig. 9.12. Analogously, the work done by the piston on the gas during the compression stroke is

$$W_{\substack{\text{compression} \\ \text{stroke}}} = \int_{V_3}^{V_2} p\,dV \qquad \textbf{[9.5]}$$

This is given by the area under the curve from point 2 to point 3. Consequently, the net work done during the complete cycle W is

$$W = W_{\substack{\text{power} \\ \text{stroke}}} - W_{\substack{\text{compression} \\ \text{stroke}}} \qquad \textbf{[9.6]}$$

Figure 9.12   The complete four-stroke cycle for a spark-ignition internal combustion engine (the Otto cycle).

This is equal to the shaded area of the *p-V* diagram shown in Fig. 9.12. Thus, we see the usefulness of *p-V* diagrams in analyzing thermodynamic processes in closed systems; namely, *the area bounded by the complete cycle on a p-V diagram is equal to the work done during the cycle*.

The power output of this arrangement is the *work done per unit time*. Consider the engine shaft rotating at *n* revolutions per second (r/s). The piston goes up and down once for each revolution of the shaft. Hence, the number of times the complete engine cycle is repeated in 1 s is $n/2$. The work output on each cycle is W, from Eq. (9.6). If the complete engine has N cylinders, then the power output of the engine is

$$\text{IP} = \frac{n}{2} NW \qquad \text{[9.7]}$$

The symbol IP is used to signify *indicated power*. This is the power that is generated by the thermodynamic and combustion processes inside the engine. However, transmission of this power to the shaft takes place through mechanical linkages, which always generate frictional losses due to moving parts in contact. As a result, the power delivered to the shaft is less than IP. If the *shaft brake power* is P (see Sec. 6.6), then

$$P = \eta_{\text{mech}}(\text{IP}) \qquad \text{[9.8]}$$

where η_{mech} is the mechanical efficiency which accounts for friction loss due to the moving engine parts. Then, from Eq. (6.30), the power available to propel the engine-propeller combination is

$$P_A = \eta\eta_{\text{mech}}(\text{IP}) \qquad \text{[9.9]}$$

or, from Eq. (9.7).

$$\boxed{P_A = \eta\eta_{\text{mech}}\frac{n}{2} NW} \qquad \text{[9.10]}$$

If rpm denotes the revolutions per minute of the engine, then $n = \text{rpm}/60$ and Eq. (9.10) becomes

$$P_A = \frac{\eta\eta_{\text{mech}}(\text{rpm})NW}{120} \qquad \text{[9.11]}$$

Equation (9.11) proves the intuitively obvious fact that the *power available for a propeller-driven airplane is directly proportional to the engine rpm*.

The work per cycle W in Eq. (9.10) can be expressed in more detailed terms. Consider the piston shown in Fig. 9.13. The length of the piston movement is called the *stroke s*; the diameter of the piston is called the *bore b*. The volume swept out by the piston is called the *displacement*, equal to $(\pi b^2/4)s$. Assume that a constant pressure p_e acts on the face of the piston during the power stroke; p_e is called the *mean effective pressure*. It is *not* the actual pressure acting on the piston, which in reality varies from p_4 to p_5 during the power stroke; rather, p_e is an artificially defined quantity which is related to the engine power output and which is an *average* representation of the actual pressure. Furthermore, assume that all the useful work

Figure 9.13   Illustration of bore, stroke, and mean effective pressure.

is done on the power stroke. Thus, W is equal to the force on the piston $(\pi b^2/4)p_e$ times the distance through which the force moves s; that is,

$$W = \frac{\pi b^2}{4} s p_e \tag{9.12}$$

Combining Eqs. (9.11) and (9.12), we obtain

$$P_A = \eta \eta_{\text{mech}}(\text{rpm}) N \frac{\pi b^2}{4} \frac{s p_e}{120} \tag{9.13}$$

The total displacement of the engine d is equal to the displacement of each cylinder times the number of cylinders:

$$d = \frac{\pi b^2}{4} s N \tag{9.14}$$

Combining Eqs. (9.13) and (9.14) yields

$$\boxed{P_A = \frac{\eta \eta_{\text{mech}}(\text{rpm}) d p_e}{120}} \tag{9.15}$$

Equation (9.15) indicates that power available is directly proportional to engine rpm, displacement, and mean effective pressure.

In Chap. 6, the altitude effect on P_A for a reciprocating engine–propeller combination was assumed to be governed by ambient density; that is, P_A was assumed to be directly proportional to ρ_∞. More credence can now be added to this earlier assumption in light of the above discussion. For example, Eq. (9.15) shows that P_A is proportional to p_e. However, p_e is representative of the mass of air originally obtained at ambient conditions, then mixed with a small amount of fuel in the intake manifold, and then sucked into the cylinder during the intake stroke. If this mass of air is reduced by flying at higher altitudes where ρ_∞ is lower, then p_e will be correspondingly lower. In turn, from Eq. (9.15), P_A will be correspondingly reduced. Therefore, the assumption that $P_A \propto \rho_\infty$ is reasonable.

The reduction of P_A with altitude can be delayed if a *supercharger* is used on the engine. This is basically a pump, driven from the engine crankshaft (a geared supercharger) or driven by a small turbine mounted in the engine exhaust jet (a turbosupercharger). The supercharger compresses the incoming air before it reaches the intake manifold, increasing its density and thereby avoiding a loss in P_A at altitude. Early work on superchargers was performed in the 1920s by the NACA at Langley. This was important research, because an unsupercharged airplane of that day was limited to altitudes on the order of 20,000 ft or less. However, on May 18, 1929, Navy Lt. Apollo Soueck, flying an Apache airplane powered by a supercharged Pratt & Whitney Wasp engine, reached 39,140 ft, an altitude record for that time. Subsequently, a substantial portion of the NACA propulsion research was channeled into superchargers, which led to the high-performance engines used in military aircraft during World War II. For modern general-aviation aircraft of today, supercharged engines are available as options on many designs and are fixed equipment on others.

A more extensive discussion of reciprocating, internal combustion engines is beyond the scope of this book. However, such engines are important to the general-aviation industry. In addition, their importance to the automobile industry goes without saying, especially in light of the modern demands of efficiency and low pollutant emissions. Therefore, the interested reader is strongly encouraged to study the subject more deeply; for example, more details can be found in the book by Obert (see Bibliography at the end of this chapter).

Example 9.1

Consider a six-cylinder internal combustion engine with a stroke of 9.5 cm and a bore of 9 cm. The compression ratio is 10. [Note that the compression ratio in internal combustion (IC) engine terminology is defined as the volume of the gas in the cylinder when the piston is at bottom dead center divided by the volume of the gas when the piston is at top dead center.] The pressure and temperature in the intake manifold are 0.8 atm and 250 K, respectively. The fuel-to-air ratio of the mixture is 0.06 (by mass). The mechanical efficiency of the engine is 0.75. If the crankshaft is connected to a propeller with an efficiency of 0.83, calculate the power available from the engine-propeller combination for 3000 rpm.

Solution

Consider the ideal cycle as sketched in Fig. 9.12. We want to calculate the work done per cycle in order ultimately to obtain the total power output. To do this, we first need to find p_3, p_4, p_5, $V_2 = V_5$, and $V_3 = V_4$. Since the compression stroke is isentropic, from Sec. 4.6,

$$\frac{p_3}{p_2} = \left(\frac{V_2}{V_3}\right)^\gamma = 10^{1.4} = 25.1$$

$$p_3 = 25.1(0.8) = 20.1 \text{ atm}$$

$$\frac{T_3}{T_2} = \left(\frac{V_2}{V_3}\right)^{\gamma-1} = (10)^{0.4} = 2.5$$

$$T_3 = 2.5(250) = 625 \text{ K}$$

Referring to Fig. 9.12, we see the combustion process from point 3 to point 4 is at constant volume. The chemical energy release in 1 kg of gasoline is approximately 4.29×10^7 J. Hence, the heat released per kilogram of fuel-air mixture is (recalling the fuel-to-air ratio is 0.06)

$$q = \frac{(4.29 \times 10^7)(0.06)}{1.06} = 2.43 \times 10^6 \text{J/kg}$$

From the first law of thermodynamics, Eq. (4.16), and from Eq. (4.23) for a constant-volume process

$$\delta q = de + p\,dv = de + 0 = c_v\,dT$$

Hence

$$q = c_v(T_4 - T_3)$$

or

$$T_4 = \frac{q}{c_v} + T_3$$

The value of c_v can be obtained from Eq. (4.68), recalling from Example 4.5 that $c_p = 1008$ J/(kg)(K) for air. Assume that the specific heats and gas constant for the fuel-air mixture are approximated by the air values alone; this is reasonable because only a small amount of fuel is present in the mixture. Hence $c_v = c_p - R = 1008 - 288 = 720$ J/(kg)(K). Thus

$$T_4 = \frac{q}{c_v} + T_3 = \frac{2.43 \times 10^6}{720} + 625 = 4000 \text{ K}$$

From the equation of state, noting that $V_4 = V_3$ and R is constant, we find $p_4/p_3 = T_4/T_3$. Thus

$$p_4 = p_3 \frac{T_4}{T_3} = 20.1 \frac{4000}{625} = 128.6 \text{ atm}$$

For the power stroke, the process is isentropic. Hence

$$\frac{p_5}{p_4} = \left(\frac{V_4}{V_5}\right)^\gamma = \left(\frac{1}{10}\right)^{1.4} = 0.0398$$

$$p_5 = 128.6(0.0398) = 5.12 \text{ atm}$$

We now have enough thermodynamic information to calculate the work done per cycle. From Eq. (9.5)

$$W_{\substack{\text{compression} \\ \text{stroke}}} = \int_{V_3}^{V_2} p\,dV$$

For an isentropic process, $pV^\gamma = c$, where c is a constant. Thus $p = cV^{-\gamma}$ and

$$W_{\text{compression}} = c \int_{V_3}^{V_2} V^{-\gamma}\,dV = \frac{c}{1-\gamma}(V_2^{1-\gamma} - V_3^{1-\gamma})$$

Since

$$c = p_2 V_2^\gamma = p_3 V_3^\gamma$$

$$W_{\text{compression}} = \frac{p_2 V_2 - p_3 V_3}{1 - \gamma}$$

We need volumes V_2 and V_3 to proceed further. Consider Fig. 9.13. The stroke of the piston is 9.5 cm, and the compression ratio is 10. If x denotes the distance from the top of the cylinder to the piston top dead center position, then from the definition of compression ratio,

$$\frac{x + 9.5}{x} = 10$$

$$x = 1.055 \text{ cm}$$

$$V_2 = \frac{\pi b^2}{4}(9.5 + 1.05) \qquad \text{where } b = \text{bore} = 9 \text{ cm}$$

$$= \pi(9)^2 \frac{9.5 + 1.05}{4} = 671.2 \text{ cm}^3$$

$$= 6.712 \times 10^{-4} m^3 \qquad \text{(remember consistent units)}$$

$$V_3 = \frac{V_2}{10} = 0.6712 \times 10^{-4} m^3$$

Thus
$$W_{\text{compression}} = \frac{p_2 V_2 - p_3 V_3}{1 - \gamma}$$

$$= \frac{[0.8(6.712 \times 10^{-4}) - 20.1(0.6712 \times 10^4)]1.01 \times 10^5}{-0.4} = 205 \text{ J}$$

Similarly, the work done by the power stroke from point 4 to point 5 (isentropic) is

$$W_{\substack{\text{power} \\ \text{stroke}}} = \int_{V_4}^{V_5} p \, dV$$

$$W_{\text{power}} = \frac{p_5 V_5 - p_4 V_4}{1 - \gamma}$$

$$= \frac{[5.12(6.712 \times 10^{-4}) - 128.6(0.6712 \times 10^{-4})]1.01 \times 10^5}{-0.4} = 1312 \text{ J}$$

Finally, from Eq. (9.6), the net work per cycle is

$$W = W_{\text{power}} - W_{\text{compression}} = 1312 - 205 = 1107 \text{ J}$$

The total power available from the engine-propeller combination is, from Eq. (9.11),

$$P_A = \frac{1}{120}\eta\eta_{\text{mech}}(\text{rpm}) \, NW = \frac{0.83(0.75)(3000)(6)(1107)}{120}$$

$$\boxed{P_A = 1.034 \times 10^5 \text{ J/s}}$$

From Sec. 6.6.1, we know that

$$1 \text{ hp} = 746 \text{ J/s}$$

Hence
$$\text{hp}_A = \frac{1.034 \times 10^5}{746} = \boxed{138.6 \text{ hp}}$$

Note: This example is rather long, with numerous calculations. However, it serves to illustrate many aspects of our discussion on IC engines, and the reader should examine it closely.

For the engine in Example 9.1, calculate the mean effective pressure. | **Example 9.2**

Solution

From Eq. (9.15)

$$P_4 = \frac{1}{120}\eta\eta_{\mathrm{mech}}(\mathrm{rpm})\,dp_e$$

where d is the displacement and p_e is the mean effective pressure. From Eq. (9.14),

$$d = \frac{\pi b^2}{4}sN = \frac{\pi(9)^2(9.5)(6)}{4} = 3626\ \mathrm{cm}^3 = 3.626 \times 10^{-3}\ \mathrm{m}^3$$

Hence, from Eq. (9.15) and the results of Example 9.1,

$$103,366 = \frac{1}{120}(0.83)(0.75)(3000)(3.626 \times 10^{-3})p_e$$

$$\boxed{p_e = 1.83 \times 10^6\ \mathrm{N/m}^2 = 18.1\ \mathrm{atm}}$$

9.4   JET PROPULSION—THRUST EQUATIONS

The previous two sections have discussed the production of thrust and power by a piston engine–propeller combination. Recall from Sec. 2.2 that the fundamental mechanisms by which nature communicates a force to a solid surface are by means of the surface pressure and shear stress distributions. The propeller is a case in point, where the net result of the pressure and shear stress distributions over the surface of the propeller blades yields an aerodynamic force, the thrust, which propels the vehicle forward. Also an effect of this thrust on the propeller is an equal and opposite reaction which yields a force on the air itself, pushing it backward in the opposite direction of the propeller thrust; that is, a change in momentum is imparted to the air by the propeller, and an alternate physical explanation of the production of thrust is that T is equal to the time rate of change of momentum of the airflow. For a propeller, this change in momentum is in the form of a large mass of air being given a small increase in velocity (about 10 m/s). However, keep in mind that the basic mechanism producing thrust is still the distribution of pressure and shear stress over the surface. Also, as in the case of lift produced by a wing, the thrust is primarily due to just the pressure distribution [see Eq. (9.1) and Fig. 9.5]; the shear stress is predominantly a drag-producing mechanism which affects the torque of the propeller.

These same principles carry over to jet propulsion. As sketched in Fig. 9.14a, the jet engine is a device which takes in air at essentially the free-stream velocity V_∞, heats it by combustion of fuel inside the duct, and then blasts the hot mixture of air and combustion products out the back end at a much higher velocity V_e. (Strictly speaking, the air velocity at the inlet to the engine is slightly larger than V_∞, but this is not important to the present discussion.) In contrast to a propeller, the jet engine creates a change in momentum of the gas by taking a small mass of air and giving it a large increase in velocity (hundreds of meters per second). By Newton's third law, the equal and opposite reaction produces a thrust. However, this reaction principle, which is commonly given as the basic mechanism for jet propulsion, is just an alternate explanation in the same vein as the discussion given above. The

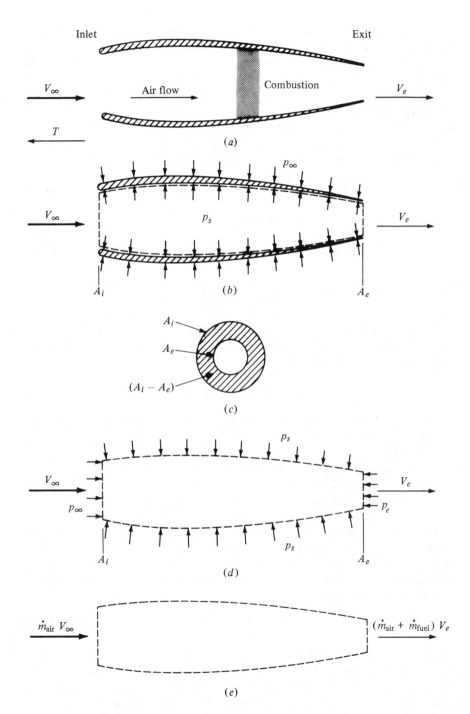

Figure 9.14 Illustration of the principle of jet propulsion (*a*) Jet propulsion engine. (*b*) Surface pressure on inside and outside surfaces of duct. (*c*) Front view, illustrating inlet and exit areas. (*d*) Control volume for flow through duct. (*e*) Change in momentum of the flow through the engine.

true fundamental source of the thrust of a jet engine is the net force produced by the pressure and shear stress distributions exerted over the surface of the engine. This is sketched in Fig. 9.14b, which illustrates the distribution of pressure p_s over the internal surface of the engine duct, and the ambient pressure, essentially p_∞, over the external engine surface. Shear stress, which is generally secondary in comparison to the magnitude of the pressures, is ignored here. Examining Fig. 9.14b, we let x denote the flight direction. The thrust of the engine in this direction is equal to the x component of p_s integrated over the complete internal surface, plus that of p_∞ integrated over the complete external surface. In mathematical symbols,

$$T = \int (p_s \, dS)_x + \int (p_\infty \, dS)_x \qquad [9.16]$$

Since p_∞ is constant, the last term becomes

$$\int (p_\infty \, dS)_x = p_\infty \int (dS)_x = p_\infty (A_i - A_e) \qquad [9.17]$$

where A_i and A_e are the inlet and exit areas, respectively, of the duct, as defined in Fig. 9.14b. In Eq. (9.17), the x component of the duct area, $\int (dS)_x$, is physically what you see by looking at the duct from the front, as shown in Fig. 9.14c. The x component of surface area is geometrically the projected frontal area shown by the crosshatched region in Fig. 9.14c. Thus, substituting Eq. (9.17) into (9.16), we obtain for the thrust T of the jet engine

$$T = \int (p_s \, dS)_x + p_\infty (A_i - A_e) \qquad [9.18]$$

The integral in Eq. (9.18) is not particularly easy to handle in its present form. Let us proceed to couch this integral in terms of the velocity and mass flow of gas through the duct. Consider the volume of gas bounded by the dashed lines in Fig. 9.14b. This is called a *control volume* in aerodynamics. This control volume is sketched again in Fig. 9.14d. The frontal area of the volume is A_i, on which p_∞ is exerted. The side of the control volume is the same as the internal area of the engine duct. Since the gas is exerting a pressure p_s, on the duct, as shown in Fig. 9.14b, by Newton's third law, the duct exerts an equal and opposite pressure p_s on the gas in the control volume, as shown in Fig. 9.14d. Finally, the rear area of the control volume is A_e, on which p_e is exerted. The pressure p_e is the gas static pressure at the exit of the duct. With the above in mind, and with Fig. 9.14d in view, the x component of the force on the gas inside the control volume is

$$F = p_\infty A_i + \int (p_s \, dS)_x - p_e A_e \qquad [9.19]$$

Now recall Newton's second law, namely, $F = ma$. This can also be written as $F = d(mV)/dt$; that is, the *force equals the time rate of change of momentum* (indeed, this is how Newton originally expressed his second law). Question: What is the time rate of change of momentum of the air flowing through the control volume? The answer can be obtained from Fig. 9.14e. The mass flow of air (kg/s or slug/s) entering the duct is $\dot{m}_{air}$; its momentum is $\dot{m}_{air} V_\infty$. The mass flow of gas leaving the duct (remember that fuel has been added and burned inside) is $\dot{m}_{air} + \dot{m}_{fuel}$; its momentum is $(\dot{m}_{air} + \dot{m}_{fuel})V_e$. Thus, the time rate of change of momentum of the airflow through the control volume is the difference between what comes out and what goes in: $(\dot{m}_{air} + \dot{m}_{fuel})V_e - \dot{m}_{air}V_\infty$. From Newton's second law, this is equal to the force on the control volume,

$$F = (\dot{m}_{air} + \dot{m}_{fuel})V_e - \dot{m}_{air}V_\infty \qquad \textbf{[9.20]}$$

Combining Eqs. (9.19) and (9.20) yields

$$(\dot{m}_{air} + \dot{m}_{fuel})V_e - \dot{m}_{air}V_\infty = p_\infty A_i + \int (p_s\,dS)_x - p_e A_e \qquad \textbf{[9.21]}$$

Solving Eq. (9.21) for the integral term, we obtain

$$\int (p_s\,dS)_x = (\dot{m}_{air} + \dot{m}_{fuel})V_e - \dot{m}_{air}V_\infty + p_e A_e - p_\infty A_i \qquad \textbf{[9.22]}$$

We now have the integral in the original thrust equation Eq. (9.18), in terms of velocity and mass flow, as originally desired. The final result for the engine thrust is obtained by substituting Eq. (9.22) into Eq. (9.18):

$$T = (\dot{m}_{air} + \dot{m}_{fuel})V_e - \dot{m}_{air}V_\infty + p_e A_e - p_\infty A_i + p_\infty(A_i - A_e) \qquad \textbf{[9.23]}$$

The terms involving A_i cancel, and we have

$$\boxed{T = (\dot{m}_{air} + \dot{m}_{fuel})V_e - \dot{m}_{air}V_\infty + (p_e - p_\infty)A_e} \qquad \textbf{[9.24]}$$

Equation (9.24) is the fundamental *thrust equation* for jet propulsion. It is an important result and will be examined in greater detail in subsequent sections. Keep in mind the reasoning that led to this result. First, the engine thrust was written down in purely mechanical terms; that is, the thrust is due to the pressure distribution acting over the internal and external surfaces of the duct; this is the essence of Eq. (9.18). Then the internal pressure distribution acting over the internal surface was couched in terms of the change of momentum of the gas flowing through the duct; this is the essence of Eq. (9.22). Finally, the two lines of thought were combined to yield Eq. (9.24). You should reread the concepts presented in this section several times until you feel comfortable with the ideas and results. The above derivation of the thrust equation using the control volume concept is an example of a general method commonly used for the solution of many aerodynamic problems. You will see it again in more advanced studies in aerodynamics and propulsion.

9.5 TURBOJET ENGINE

In 1944, the first operational jet fighter in the world was introduced by the German air force, the ME 262. By 1950, jet engines were the mainstay of all high-performance military aircraft, and by 1958 the commercial airlines were introducing the jet-powered Boeing 707 and McDonnell-Douglas DC-8. Today, the jet engine is the only practical propulsive mechanism for high-speed subsonic and supersonic flight. (Recall that our hypothetical CJ-1 in Chap. 6 was powered by two small jet engines.) A photograph of a typical turbojet engine is shown in Fig. 9.15.

The thrust of a turbojet engine is given directly by Eq. (9.24). The jet engine takes in a mass flow of cool air $\dot{m}_{air}$, at velocity essentially equal to V_∞ and exhausts a mass flow of hot air and combustion products $\dot{m}_{air} + \dot{m}_{fuel}$ at velocity V_e. This is illustrated in Fig. 9.16. The mass of fuel added is usually small compared to the mass of air; $\dot{m}_{fuel}/\dot{m}_{air} \approx 0.05$. Thus, Eq. (9.24) can be simplified by neglecting $\dot{m}_{fuel}$:

$$T = \dot{m}_{air}(V_e - V_\infty) + (p_e - p_\infty)A_e \qquad \textbf{[9.25]}$$

Equation (9.25) explicitly shows that T can be increased by increasing $V_e - V_\infty$. Thus, the function of a jet engine is to exhaust the gas out the back end faster than it comes in through the front end. The conventional turbojet engine performs this function by inducting a mass of air through the *inlet* (location 1 in Fig. 9.16). The flow is reduced to a low subsonic Mach number, $M \approx 0.2$ in a *diffuser* (point 1 to point 2 in Fig. 9.16). This diffuser is directly analogous to the wind tunnel diffusers discussed in Chap. 4. If V_∞ is subsonic, then the diffuser must increase the flow area to decelerate the flow; that is, the diffuser is a *divergent duct* [see Eq. (4.83)]. If V_∞ is supersonic, the diffuser must be a convergent-divergent duct, and the decrease in flow velocity is accomplished partly through shock waves, as shown in Fig. 9.16. For such supersonic inlets, a centerbody is sometimes employed to tailor the strength and location of the shock waves and to help form the convergent-divergent stream tube seen by the decelerating flow. In the diffusion process, the static pressure is increased from p_1 to p_2. After the diffuser, the flow is further compressed by a compressor (point 2 to point 3 in Fig. 9.16) from p_2 to p_3. The compressor is usually a series of alternating rotating and stationary blades. The rotating sections are called *rotors*, and the stationary sections are *stators*. The rotor and stator blades are nothing more than airfoil sections which alternately speed up and slow down the flow; the work supplied by the compressor serves to increase the total pressure of the flow. The compressor sketched in Fig. 9.16 allows the flow to pass essentially straight through the blades

Figure 9.15   The J52 turbojet engine. (*Pratt and Whitney Aircraft.*)

Figure 9.16 Turbojet engine and diffuser configurations.

without any major deviation in direction; thus, such devices are called *axial flow compressors*. This is in contrast to *centrifugal flow compressors* used in some early jet engines, where the air was turned sometimes more than 90°. After leaving the compressor, fuel is injected into the airstream and burned at essentially constant pressure in the *combustor* (point 3 to point 4 in Fig. 9.16), where the temperature is increased to about 2500°R. After combustion, the hot gas flows through the *turbine* (point 4 to point 5 in Fig. 9.16). The turbine is a series of rotating blades (again, basically airfoil sections) which extract work from the flowing gas. This work is then transmitted from the turbine through a shaft to the compressor; that is, the turbine drives the compressor. The flow through a turbine is an expansion process, and the pressure drops from p_4 to p_5. However, p_5 is still larger than the ambient pressure outside the engine. Thus, after leaving the turbine, the flow is expanded through a *nozzle* (point 5 to point 6 in Fig. 9.16) and is exhausted to the atmosphere at a high velocity V_e and at pressure $p_6 = p_e$. If the engine is designed for subsonic flight applications, the nozzle is usually convergent and V_e is subsonic, or at most sonic. However, if the engine is intended for supersonic aircraft, the exhaust nozzle is usually convergent-divergent and V_e is supersonic.

The thermodynamic process in an ideal turbojet engine is shown in the *p-v* diagram of Fig. 9.17. The ideal process ignores the effects of friction and heat losses. Here the air is isentropically compressed from p_1 to p_2 in the inlet diffuser, and the pressure is further isentropically increased to p_3 by the compressor. The process moves along the isentrope $pv^\gamma = c_1$, where c_1 is a constant (see Sec. 4.6). In the burner, the combustion process takes place at constant pressure (in contrast to the combustion process in an internal combustion reciprocating engine, which takes place at constant volume, as explained in Sec. 9.3). Since the temperature is increased by combustion and the pressure is constant, the equation of state $pv = RT$ dictates that v must increase from v_3 to v_4 in the burner. Expansion through the turbine

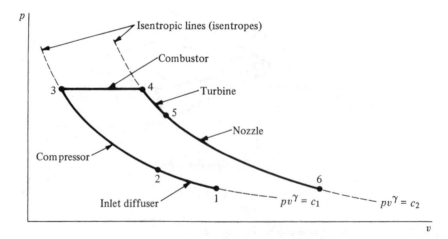

Figure 9.17   Pressure–specific volume diagram for an ideal turbojet.

isentropically drops the pressure to p_5, and further isentropic expansion through the nozzle decreases the pressure to p_6. The turbine and nozzle expansions follow the isentrope $pv^\gamma = c_2$, where c_2 is a constant different from c_1. The ideal engine process further assumes that the nozzle expands the gas to ambient pressure, such that $p_e = p_6 = p_1 = p_\infty$. In the real engine process, of course, there will be frictional and heat losses. The diffuser, compressor, turbine, and nozzle processes will not be exactly isentropic, the combustion process is not precisely at constant pressure; and the nozzle exit pressure p_e will be something different from p_∞. However, the ideal turbojet shown in Fig. 9.17 is a reasonable first approximation to the real case. The accounting of nonisentropic process in the engine is the subject for more advanced studies of propulsion.

Return again to the turbojet engine thrust equation, Eq. (9.25). We are now in a position to understand some of the assumptions made in Chap. 6 concerning thrust available T_A for a turbojet. In our performance analysis of the CJ-1, we assumed that (1) thrust did not vary with V_∞ and (2) the altitude effect on thrust was simply proportional to ρ_∞. From the continuity equation, Eq. (4.2), applied at the inlet, we find that $\dot{m}_{air} = \rho_\infty A_i V_\infty$. Hence, as V_∞ increases, $\dot{m}_{air}$ increases. From Eq. (9.25), this tends to increase T. However, as V_∞ increases, the factor $V_e - V_\infty$ decreases. This tends to decrease T. The two effects tend to cancel each other, and the net result is a relatively constant thrust at subsonic speeds. With regard to altitude effects, $\dot{m}_{air} = \rho_\infty A_i V_\infty$ decreases proportionately with a decrease in ρ_∞; the factor $V_e - V_\infty$ is relatively unaffected. The term $(p_e - p_\infty)A_e$ in Eq. (9.25) is usually much smaller than $\dot{m}_{air}(V_e - V_\infty)$; hence even though p_e and p_∞ change with altitude, this pressure term will not have a major effect on T. Consequently, the primary consequence of altitude is to decrease ρ_∞, which proportionately decreases $\dot{m}_{air}$, which proportionately decreases T. Hence, our assumption in Chap. 6 that $T \propto \rho_\infty$ is reasonable.

Consider a turbojet powered airplane flying at a standard altitude of 30,000 ft at a velocity of 500 mi/h. The turbojet engine itself has inlet and exit areas of 7 and 4.5 ft^2, respectively. The velocity and pressure of the exhaust gas at the exit are 1600 ft/s and 640 lb/ft^2, respectively. Calculate the thrust of the turbojet. | **Example 9.3**

Solution

At a standard altitude of 30,000 ft, from App. B, $p_\infty = 629.66$ lb/ft^2 and $\rho_\infty = 8.9068 \times 10^{-4}$ slug/ft^3. The free-stream velocity is $V_\infty = 500$ mi/h $= 500(88/60) = 733$ ft/s. Thus, the mass flow through the engine is

$$\dot{m}_{air} = \rho_\infty V_\infty A_i = (8.9068 \times 10^{-4})(733)(7) = 4.57 \text{ slugs/s}$$

From Eq. (9.25), the thrust is

$$T = \dot{m}_{air}(V_e - V_\infty) + (p_e - p_\infty)A_e$$
$$= 4.57(1600 - 733) + (640 - 629.66)(4.5)$$
$$= 3962 + 46.5 = \boxed{4008.5 \text{ lb}}$$

9.6 TURBOFAN ENGINE

Section 9.4 established the relation between thrust and rate of change of momentum of a mass of air. In the turbojet engine (see Sec. 9.5), all this mass flows through the engine itself, and all of it is accelerated to high velocity through the exhaust nozzle. Although this creates a large thrust, the *efficiency* of the process is adversely affected by the high exhaust velocities. Recall in Sec. 9.2 that one of the losses that reduces propeller efficiency is the kinetic energy remaining in the wake relative to the ambient air. In the case of the turbojet, the kinetic energy left in the jet exhaust is also a loss, and the high exhaust velocities produced by a jet engine just exacerbate the situation. This is why a piston engine–propeller combination is basically a more efficient device than a turbojet. (Remember, do not get efficiency and thrust confused—they are different things. A jet produces high thrust, but at a relatively low efficiency.) Therefore, the concepts of the pure turbojet and the propeller are combined in the *turbofan* engine. As sketched in Fig. 9.18, a turbofan engine is a turbojet engine that has a large ducted fan mounted on the shaft ahead of the compressor. The turbine drives both the fan and the compressor. The ducted fan accelerates a large mass of air that flows between the inner and outer shrouds; this unburned air then mixes with the jet exhaust downstream of the nozzle. The thrust of the turbofan is a combination of the thrust produced by the fan blades and jet from the exhaust nozzle. Consequently, the efficiency of a turbofan engine is better than that of a turbojet. This efficiency is denoted by the thrust-specific fuel consumption TSFC (see Sec. 6.12). For a typical turbojet, TSFC = 1.0 lb of fuel per pound of thrust per hour; for a typical turbofan, TSFC = 0.6 lb of fuel per pound of thrust per hour, a much better figure. This is why all modern commercial jet transports, such as the Boeing 747 and the McDonnell-Douglas MD-11, are equipped with turbofan engines. A photograph of a turbofan is given in Fig. 9.19*a*, and a cutaway view of the same engine is shown in Fig. 9.19*b*.

Of course, a further extension of this concept replaces the ducted fan and outer shroud with an out-and-out propeller, with the turbine driving both the compressor and the propeller. Such a combination is called a *turboprop*, where approximately 85 percent of the thrust comes from the propeller and the remaining 15 percent comes

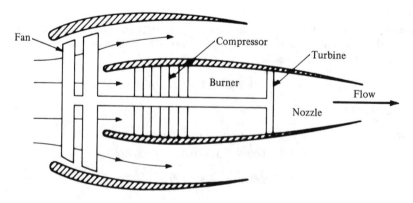

Figure 9.18   A turbofan engine.

(a)

(b)

Figure 9.19   (a) the Pratt and Whitney JT9D turbofan engine. (b) A cutaway view. (*Pratt and Whitney.*)

from the jet exhaust. Turboprops are efficient power plants that have found application in the range of 300 to 500 mi/h, one prime example is the Lockheed Electra transport of the 1950s.

9.7 RAMJET ENGINE

Let us now move in the opposite direction from the last section. Instead of adding fans and propellers to a turbojet, let us get rid of *all* rotating machinery; that is, consider the straight-through duct sketched in Fig. 9.20, where air is inducted through the inlet at velocity V_∞, decelerated in the diffuser (point 1 to point 2), burned in a region where fuel is injected (point 2 to point 3), and then blasted out the exhaust nozzle at very high velocity V_e (point 3 to point 4). Such a simple device is called a *ramjet engine*. A cutaway drawing of a ramjet engine is shown in Fig. 9.21. Because of their simplicity and high thrust, ramjets have always tickled the imaginations of aerospace engineers. However, because of some serious drawbacks, they have not yet been used as a prime propulsive mechanism on a manned aircraft. However, they are used on numerous guided missiles, and they appear to be the best choice for future hypersonic airplanes. For this reason, let us examine ramjets more closely.

The ideal ramjet process is shown in the p-v diagram of Fig. 9.22. All the compression from p_1 to p_2 takes place in the diffuser; that is, a ramjet compresses the air by simply "ramming" through the atmosphere. Obviously, the compression ratio p_2/p_1 is a function of flight Mach number. In fact, to enhance combustion, the airflow entering the combustion zone is at a low subsonic Mach number; hence, assuming that $M_2 \approx 0$, then $p_2 \approx p_0$ (total pressure), and from Eq. (4.74),

$$\frac{p_2}{p_1} \approx \left(1 + \frac{\gamma - 1}{2} M_\infty^2\right)^{\gamma/(\gamma-1)} \qquad \text{[9.26]}$$

The air decelerates isentropically in the diffuser, hence the compression from p_1 to p_2 follows the isentrope shown in Fig. 9.22. Fuel is injected into the air at the end of the diffuser, and combustion takes place, stabilized by mechanical flame holders. This combustion is at constant pressure, hence the specific volume increases from v_2

Figure 9.20   Ramjet engine.

Figure 9.21   A typical ramjet engine. (*Marquardt Aircraft Co.*)

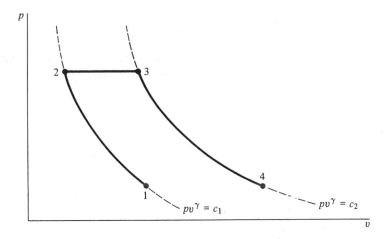

Figure 9.22   Pressure–specific volume diagram for an ideal ramjet.

to v_3. Then the hot, high-pressure gas is expanded isentropically through the exhaust nozzle, with the pressure dropping from p_3 to p_4.

One disadvantage of a ramjet is immediately obvious from the above discussion, and especially from Eq. (9.26): To start and operate, the ramjet must already be in motion. Otherwise, there would be no compression in the diffuser; that is, from Eq. (9.26), $p_2/p_1 = 1$ when $M_\infty = 0$. Therefore, all ramjet-powered vehicles must be launched by some independent mechanism (a catapult, or rockets) or must have a second engine of another type to develop enough flight speed to start the ramjet. At subsonic flight speeds, ramjets have another disadvantage. Although they produce high thrust, their subsonic efficiency is very low; typically TSFC $\approx$ 3 to 4 lb of fuel per pound of thrust per hour for ramjets at subsonic speeds. However, as shown in Fig. 9.23, TSFC decreases to 2 or less at supersonic speeds.

Indeed, Fig. 9.23 implicitly shows an advantage of ramjets for supersonic flight. At supersonic Mach numbers, TSFCs for turbojets and ramjets are somewhat comparable. Moreover, the curve for turbojets in Fig. 9.23 is terminated at Mach 3 for a specific reason. To operate at higher Mach numbers, the turbojet must increase its combustion temperature. However, there is a material limitation. If the gas temperature leaving the turbojet combustor and entering the turbine is too hot, the turbine blades will melt. This is a real problem: The high-temperature material properties of the turbine blades limit the conventional turbojet to comparatively low to moderate supersonic Mach numbers. However, a ramjet has no turbine; therefore, its combustion temperatures can be much higher, and a ramjet can zip right into the high-Mach-number regime. Therefore, for sustained and efficient atmospheric flight at Mach numbers above 3 or 4, a ramjet is virtually the only choice, given our present-day technology.

Starting with Fig. 9.20, we have described a conventional ramjet as a device which takes in the air at the inlet and diffuses it to a low subsonic Mach number before it enters the combustion zone. Consider this ramjet flying at $M_\infty = 6$. As a companion to Eq. (9.26), the temperature ratio T_2/T_1 can be estimated from Eq. (4.73) as

$$\frac{T_2}{T_1} \approx 1 + \frac{\gamma - 1}{2} M_\infty^2 \qquad \text{[9.27]}$$

(Note that the symbol T is used for both thrust and temperature; however, from the context, there should be no confusion.) If $M_\infty = 6$, Eq. (9.27) gives $T_2/T_1 \approx 7.9$. If

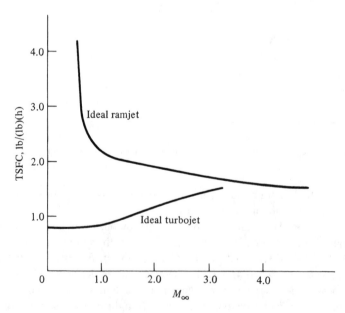

Figure 9.23   Comparison of thrust-specific fuel consumption for ideal ramjet and turbojet engines.

the ambient temperature $T_\infty = T_1 = 300$ K, then $T_2 = 2370$ K $= 4266°$R. At such high temperatures, the walls of the ramjet will tend to fail structurally. Thus, like turbojets, conventional ramjets are also limited by material problems, albeit at higher flight Mach numbers. Moreover, if the temperature of the air entering the combustor is too high, when the fuel is injected, it will be decomposed by the high temperatures rather than be burned; that is, the fuel will absorb rather than release energy, and the engine will become a drag machine rather than a thrust-producing device. Clearly, for hypersonic flight at very high Mach numbers, something else must be done. This problem has led to the concept of a *supersonic combustion ramjet*, the *SCRAMjet*. Here, the flow entering the diffuser is at high Mach number, say, $M_1 = M_\infty = 6$. However, the diffuser decelerates the airflow only enough to obtain a reasonable pressure ratio p_2/p_1; the flow is still supersonic upon entering the combustor. Fuel is added to the supersonic stream, where *supersonic combustion* takes place. In this way, the flow field throughout the SCRAMjet is completely supersonic; in turn, the static temperature remains relatively low, and the material and decomposition problems associated with the conventional ramjet are circumvented. Therefore, the power plant for a hypersonic transport in the future will most likely be a SCRAMjet. Research on such devices is now in process. Indeed, SCRAMjet research constitutes the very frontier of propulsion research today. One such example is the SCRAMjet design concept pioneered by the NASA Langley Research Center since the mid-1960s. Intended for application on a hypersonic transport, the Langley SCRAMjet consists of a series of side-by-side modules blended with the underside of the airplane, as sketched in the upper right corner of Fig. 9.24. The forward portion of the underside of the airplane acts as a compression surface; that is, the air flowing over the bottom

Figure 9.24 A concept for an airframe-integrated SCRAMjet engine, developed at NASA Langley Research Center. (*NASA.*)

surface is compressed (pressure is increased) when it passes through the shock wave from the nose of the vehicle. The configuration of an individual module is shown in the middle of Fig. 9.24. The compressed air from the bottom surface enters an inlet, where it is further compressed by additional shock waves from the leading edge of the inlet. This compressed air, still at supersonic velocity, subsequently flows over three struts, where H_2 is injected into the supersonic stream. A cross section of the struts is shown at the bottom left of Fig. 9.24. Combustion takes place downstream of the struts. The burned gas mixture is then expanded through a nozzle at the rear of each module. The flow is further expanded over the smooth underbody at the rear of the airplane, which is intentionally contoured to act as an extension of the engine nozzles. For all practical purposes, the entire undersurface of the complete airplane represents the whole SCRAMjet engine—hence the concept is called an *airframe-integrated* SCRAMjet.

9.8 ROCKET ENGINE

With the launching of *Sputnik I* on October 4, 1957, and with the subsequent massive space programs of the United States and the Soviet Union, the rocket engine came of age. The rocket is the ultimate high-thrust propulsive mechanism. With it, people have gone to the moon, and space vehicles weighing many tons have been orbited about the earth or sent to other planets in the solar system. Moreover, rockets have been used on experimental aircraft; the rocket-powered Bell X-1 was the first manned airplane to break the sound barrier (see Sec. 5.22), and the rocket-powered North American X-15 was the first manned hypersonic aircraft (see Sec. 5.23). Finally, almost all types of guided missiles, starting with the German V-2 in World War II, have been rocket-powered. With this in mind, let us examine the characteristics of a rocket engine.

All the propulsion engines discussed in previous sections have been air-breathing; the piston engine, turbojet, ramjet—all depend on the combustion of fuel with air, where the air is obtained directly from the atmosphere. In contrast, as sketched in Fig. 9.25, the rocket engine carries both its fuel and oxidizer and is completely independent of the atmosphere for its combustion. Thus, the rocket can operate in the vacuum of space, where obviously the air-breathing engines cannot. In Fig. 9.25, fuel and oxidizer are sprayed into the combustion chamber, where they burn, creating a high-pressure, high-temperature mixture of combustion products. The mixture velocity is low, essentially zero. Therefore, the combustion chamber in a rocket engine is directly analogous to the reservoir of a supersonic wind tunnel (see Sec. 4.13). Hence, the temperature and pressure in the combustion chamber are the total values T_0 and p_0, respectively. Also directly analogous to a supersonic wind tunnel, the products of combustion expand to supersonic speeds through the convergent-divergent rocket nozzle, leaving with an exit velocity V_e. This exit velocity is considerably higher than that for jet engines; hence, by comparison, rocket thrusts are higher, but efficiencies are lower. Figure 9.26 shows a typical rocket engine.

The thrust of a rocket engine is obtained from Eq. (9.24), where $\dot{m}_{air} = 0$ and $\dot{m}$ is the total mass flow of the products of combustion, $\dot{m} = \dot{m}_{fuel} + \dot{m}_{oxidizer}$. Hence,

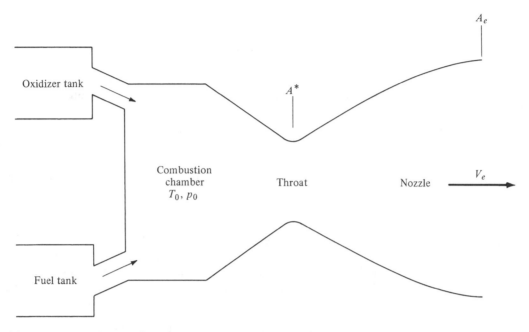

Figure 9.25 Schematic of a rocket engine.

Figure 9.26 The main rocket engine for the space shuttle. (*Rockwell International Corporation, Rocketdyne Corporation.*)

for a rocket engine,

$$T = \dot{m}V_e + (p_e - p_\infty)A_e \qquad \textbf{[9.28]}$$

The exit velocity V_e is readily obtained from the aerodynamic relations in Chap. 4. Write the energy equation, Eq. (4.41), between the combustion chamber and the nozzle exit:

$$h_0 = h_e + \frac{V_e^2}{2} \qquad \textbf{[9.29]}$$

$$c_p T_0 = c_p T_e + \frac{V_e^2}{2} \qquad \textbf{[9.30]}$$

Solve Eq. (9.30) for V_e^2:

$$V_e^2 = 2c_p(T_0 - T_e) = 2c_p T_0 \left(1 - \frac{T_e}{T_0}\right) \qquad \textbf{[9.31]}$$

The expansion through the nozzle is isentropic. Hence, from Eq. (4.36), $T_e/T_0 = (p_e/p_0)^{(\gamma-1)/\gamma}$. Also, from Eq. (4.69), $c_p = \gamma R/(\gamma - 1)$. Thus, Eq. (9.31) becomes

$$V_e = \left\{ \frac{2\gamma R T_0}{\gamma - 1} \left[1 - \left(\frac{p_e}{p_0}\right)^{(\gamma-1)/\gamma}\right] \right\}^{1/2} \qquad \textbf{[9.32]}$$

A comparative measure of the efficiency of different rocket engines can be obtained from the *specific impulse* I_{sp}, defined as the thrust per unit weight flow at sea level:

$$I_{\text{sp}} \equiv \frac{T}{\dot{w}} \qquad \textbf{[9.33]}$$

where $\dot{w} = g_0 \dot{m}$. (Recall that the weight is equal to the acceleration of gravity at sea level times the mass.) With this definition, the unit of I_{sp} in any consistent system of units is simply seconds. Furthermore, assume that the nozzle exit pressure is the same as the ambient pressure. Then combining Eqs. (9.28) and (9.33), we get

$$I_{\text{sp}} = \frac{T}{\dot{w}} = \frac{T}{g_0 \dot{m}} = \frac{\dot{m}V_e}{g_0 \dot{m}} = \frac{V_e}{g_0} \qquad \textbf{[9.34]}$$

Substitute Eq. (9.32) into Eq. (9.34), and note from chemistry that the specific gas constant R is equal to the universal gas constant $\overline{R}$ divided by the molecular weight $\overline{M}$; $R = \overline{R}/\overline{M}$.

$$I_{\text{sp}} = \frac{1}{g_0} \left\{ \frac{2\gamma R T_0}{\gamma - 1} \left[1 - \left(\frac{p_e}{p_0}\right)^{(\gamma-1)/\gamma}\right] \right\}^{1/2} \qquad \textbf{[9.35]}$$

Equation (9.35) is important. It tells what is necessary to have a high specific impulse; namely, the combustion temperature T_0 should be high, and the molecular weight $\overline{M}$

should be low. The combustion temperature is primarily dictated by the chemistry of the oxidizer and fuel; a given combination, say, oxygen and hydrogen, will burn at a specific T_0 called the adiabatic flame temperature, and this value of T_0 will be determined by the heat of reaction. The more highly reacting the propellants, the higher the T_0. Also $\overline{M}$ is a function of the chemistry. If lightweight propellants are used, then $\overline{M}$ will be small. Therefore, outside of adjusting the oxidizer-to-fuel ratio (the O/F ratio), there is not much the engineer can do to change radically the I_{sp} for a given propellant combination. It depends primarily on the propellants themselves. However, Eq. (9.35) certainly tells us to choose a very energetic combination of lightweight propellants, as dramatized by the following tabulation.

Fuel-Oxidizer Combination	Adiabatic Flame Temperature, K	Average Molecular Weight of Combustion Products	I_{sp}, s
Kerosine-oxygen	3144	22	240
Hydrogen-oxygen	3517	16	360
Hydrogen-fluorine	4756	10	390

The kerosine-oxygen combination was used in the first stage of the Saturn 5 launch vehicle, which sent the Apollo astronauts to the moon; hydrogen-oxygen was used for the Saturn 5 second and third stages. However, the best combination is hydrogen-fluorine, which gives a specific impulse of 390 s, about the most we can expect from any propellant combination. Unfortunately, fluorine is extremely poisonous and corrosive and is therefore difficult to handle. Nevertheless, rocket engines using hydrogen-fluorine have been built.

Consider again the rocket engine schematic in Fig. 9.25. We discussed above that T_0 in the combustion chamber is essentially a function of the heat of reaction of the propellants, a chemical phenomenon. But what governs the chamber pressure p_0? The answer is basically the mass flow of propellants being pumped into the chamber from the fuel and oxidizer tanks, and the area of the nozzle throat A^*. Moreover, we are in a position to prove this. From the continuity equation evaluated at the throat,

$$\dot{m} = \rho^* A^* V^* \qquad \textbf{[9.36]}$$

where the superscript * denotes conditions at the throat. Recall from Chap. 4 that the velocity is sonic at the throat of a convergent-divergent supersonic nozzle; that is, $M^* = 1$. Thus, V^* is the speed of sound, obtained from Eq. (4.54) as

$$V^* = \sqrt{\gamma R T^*} \qquad \textbf{[9.37]}$$

Also, from the equation of state,

$$\rho^* = \frac{p^*}{R T^*} \qquad \textbf{[9.38]}$$

Substitute Eqs. (9.37) and (9.38) into Eq. (9.36):

$$\dot{m} = \frac{p^*}{RT_*}A^*\sqrt{\gamma RT^*} = \frac{p^*A^*}{\sqrt{RT^*}}\sqrt{\gamma} \qquad [9.39]$$

Write Eqs. (4.73) and (4.74) between the combustion chamber and the throat:

$$\frac{T_0}{T^*} = 1 + \frac{\gamma-1}{2}M^{*2} = 1 + \frac{\gamma-1}{2} = \frac{\gamma+1}{2} \qquad [9.40]$$

$$\frac{p_0}{p^*} = \left(1 + \frac{\gamma-1}{2}M^{*2}\right)^{\gamma/(\gamma-1)} = \left(\frac{\gamma+1}{2}\right)^{\gamma/(\gamma-1)} \qquad [9.41]$$

Substitute Eqs. (9.40) and (9.41) into Eq. (9.39):

$$\dot{m} = \sqrt{\frac{\gamma}{R}}A^*\left(\frac{\gamma+1}{2}\right)^{-(\gamma+1)/[2(\gamma-1)]}\frac{p_0}{\sqrt{T_0}}$$

or

$$\dot{m} = \frac{p_0 A^*}{\sqrt{T_0}}\sqrt{\frac{\gamma}{R}\left(\frac{2}{\gamma+1}\right)^{(\gamma+1)/(\gamma-1)}} \qquad [9.42]$$

Equation (9.42) is important. It states that the mass flow through a nozzle that is choked, that is, when sonic flow is present at the throat, is *directly proportional to p_0 and A^* and inversely proportional to the square root of T_0*. Moreover, Eq. (9.42) answers the previous question about how p_0 is governed in a rocket engine combustion chamber. For a given combination of propellants, T_0 is fixed by the chemistry. For a fixed nozzle design, A^* is a given value. Hence, from Eq. (9.42),

$$p_0 = (\text{const})(\dot{m})$$

If $\dot{m}$ is doubled, then p_0 is doubled, etc. In turn, since mass is conserved, $\dot{m}$ through the nozzle is precisely equal to $\dot{m}_{\text{fuel}} + \dot{m}_{\text{oxidizer}}$ being fed into the chamber from the propellant tanks. Therefore, we repeat again the conclusion that p_0 is governed by the mass flow of propellants being pumped into the chamber from the fuel and oxidizer tanks and the area of the nozzle throat.

Before we leave this discussion of rocket engines, we note the very restrictive assumption incorporated in such equations as Eqs. (9.32), (9.35), and (9.42), namely, that γ is constant. The real flow through a rocket engine is chemically reacting and is changing its chemical composition throughout the nozzle expansion. Consequently, γ is really a variable, and the above equations are not strictly valid. However, they are frequently used for preliminary design estimates of rocket performance, and γ is chosen as some constant mean value, usually between 1.2 and 1.3, depending on the propellants used. A more accurate solution of rocket nozzle flows taking into account the variable specific heats and changing composition must be made numerically and is beyond the scope of this book.

Example 9.4 | Consider a rocket engine burning hydrogen and oxygen; the combustion chamber pressure and temperature are 25 atm and 3517 K, respectively. The area of the rocket nozzle throat is 0.1 m². The area of the exit is designed so that the exit pressure exactly equals ambient pressure at a standard altitude of 30 km. For the gas mixture, assume that $\gamma = 1.22$ and the molecular weight $\overline{M} = 16$. At a standard altitude of 30 km, calculate (a) specific impulse, (b) thrust, (c) area of the exit, and (d) flow Mach number at exit.

Solution

a. The universal gas constant, in SI units, is $\overline{R} = 8314$ J/(kg mol)(K). Hence, the specific gas constant is

$$R = \frac{\overline{R}}{\overline{M}} = \frac{8314}{16} = 519.6 \text{ J/(kg)(K)}$$

Thus, from Eq. (9.35)

$$I_{sp} = \frac{1}{g_0} \left\{ \frac{2\gamma \overline{R} T_0}{(\gamma - 1)\overline{M}} \left[1 - \left(\frac{p_e}{p_0} \right)^{(\gamma-1)/\gamma} \right] \right\}^{1/2}$$

$$= \frac{1}{9.8} \left\{ \frac{2(1.22)(8314)(3517)}{0.22(16)} \left[1 - \left(\frac{1.174 \times 10^{-2}}{25} \right)^{0.22/1.22} \right] \right\}^{1/2}$$

$$\boxed{I_{sp} = 397.9 \text{ s}}$$

Note that this value is slightly higher than the number tabulated in the previous discussion on specific impulse. The difference is that the tabulation gives I_{sp} for expansion to sea-level pressure, not the pressure at 30-km altitude as in this example.

b. From Eq. (9.28),

$$T = \dot{m} V_e + (p_e - p_\infty) A_e$$

In this equation, at 30 km, $p_e = p_\infty$. Hence,

$$T = m V_e \qquad \text{at 30-km altitude}$$

To obtain $\dot{m}$, use Eq. (9.42):

$$\dot{m} = \frac{p_0 A^*}{\sqrt{T_0}} \sqrt{\frac{\gamma}{R} \left(\frac{2}{\gamma + 1} \right)^{(\gamma+1)/(\gamma-1)}}$$

$$= \frac{25(1.01 \times 10^5)(0.1)}{\sqrt{3517}} \sqrt{\frac{1.22}{519.6} \left(\frac{2}{2.22} \right)^{2.22/0.22}} = 121.9 \text{ kg/s}$$

To obtain V_e, recall that the nozzle flow is isentropic. Hence,

$$\frac{T_e}{T_0} = \left(\frac{p_e}{p_0} \right)^{(\gamma-1)/\gamma}$$

$$T_e = 3517 \left(\frac{1.174 \times 10^{-2}}{25} \right)^{0.22/1.22} = 3517(0.2517) = 885.3 \text{ K}$$

Also, from Eq. (4.69),

$$c_p = \frac{\gamma R}{\gamma - 1} = \frac{1.22(519.6)}{0.22} = 2881.4 \text{ J/(kg)(K)}$$

From the energy equation, Eq. (4.42),

$$c_p T_0 = c_p T_e + \frac{V_e^2}{2}$$

$$V_e = \sqrt{2c_p(T_0 - T_e)} = \sqrt{2(2881.4)(3517 - 885.3)} = 3894 \text{ m/s}$$

Thus, the thrust becomes

$$T = \dot{m} V_e = 121.9(3894) = \boxed{4.75 \times 10^5 \text{ N}}$$

Note that 1 N = 0.2247 lb. Hence

$$T = (4.75 \times 10^5)(0.2247) = \boxed{106,700 \text{ lb}}$$

c. To obtain the exit area, use the continuity equation:

$$\dot{m} = \rho_e A_e V_e$$

To obtain the exit density, use the equation of state:

$$\rho_e = \frac{p_e}{RT_e} = \frac{1.1855 \times 10^3}{519.6(885.3)} = 2.577 \times 10^{-3} \text{ kg/m}^3$$

$$A_e = \frac{\dot{m}}{\rho_e V_e} = \frac{121.9}{(2.577 \times 10^{-3})(3894)}$$

$$\boxed{A_e = 12.14 \text{ m}^2}$$

d. To obtain the exit Mach number,

$$a_e = \sqrt{\gamma R T_e} = \sqrt{1.22(519.6)(885.3)} = 749 \text{ m/s}$$

$$M_e = \frac{V_e}{a_e} = \frac{3894}{749} = \boxed{5.2}$$

9.9   ROCKET PROPELLANTS—SOME CONSIDERATIONS

Recall from elementary chemistry that in order to produce a flame (such as on a gas stove) you need fuel (such as natural gas or propane) and an oxidizer (such as the oxygen in air). The burning process in the combustion chamber of a rocket engine is the same—it requires the burning of a fuel and oxidizer. The fuel and oxidizer together are called the *rocket propellants*. The choice of propellants is such a serious consideration in the design of rocket engines that we choose to devote this section to some basic discussion of rocket propellants. For example, the value of specific impulse for a rocket engine is mainly a function of the propellants used. The specific impulse, from its definition in Eq. (9.33), can be thought of the number of seconds after which one pound of propellants will produce one pound of thrust, and this number of seconds is critically dependent on the propellants themselves. The multiple choices of possible rocket propellants, and their combustion chemistry, are a subject in itself for which whole books have been written. Here, we just introduce some of the basic aspects of propellants.

In the most general sense, there are two different classifications of chemical propellants: liquid propellants and solid propellants. Let us examine each in turn.

9.9.1 Liquid Propellants

Here both the fuel and oxidizer are carried onboard the rocket in liquid form, and they are injected under pressure as a spray into the combustion chamber. This was illustrated schematically in Fig. 9.25, where the oxidizer and fuel are shown as separate sources being injected and mixed in the combustion chamber. The propellants are injected at high pressure. For example, in the space shuttle main engine (Fig. 9.26), the propellants are injected at a pressure of 440 atm—an extremely high pressure. Historically, the engineering design of the mechanisms to pressurize the propellants has been a challenge. There are two basic approaches to this problem as described below.

Mechanically the simplest is the *pressure-fed system*, shown schematically in Fig. 9.27a. Here, both the liquid fuel and oxidizer are placed under high pressure in their respective tanks by a high-pressure inert gas such as helium (He), which is

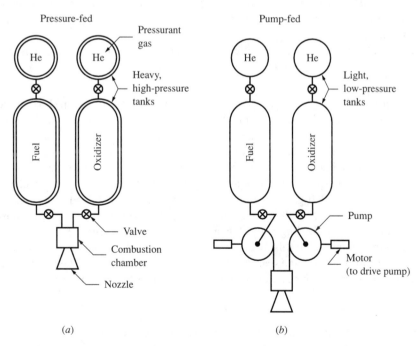

(a)   (b)

Figure 9.27   (a) Pressure fed rocket engine. (b) Pump fed rocket engine.

carried in separate (usually spherical) tanks. When valves connecting the propellant tanks to the combustion chamber are opened, the propellants, which are already under high pressure in their tanks, are forced into the combustion chamber. The advantage of this system is its relative simplicity. The disadvantage is that the propellant tanks must have thick walls to withstand the high pressure, and so the tanks are heavy. For this reason, pressure-fed systems are usually used for small rocket engines (thrust levels of 1000 lb or less) that operate for only short durations. Such engines are used as attitude control jets on spacecraft; they are usually not used as the primary rocket thrust-producing power plant. (An exception to this was the XLR11 rocket engine that powered the Bell X-1 to its historic first supersonic flight, as described in Sect. 5.22. Because no reliable fuel pump existed at the time of this flight, although one was feverishly being designed for the engine, the Bell and Reaction Motors engineers had to depend on a pressure-fed system for the rocket engine.)

The second type of mechanism is the *pump-fed system*, illustrated in Fig. 9.27*b*. Here the propellants are stored at relatively low pressure in thin-walled (hence lighter) tanks, and their pressure is increased by pumps before injection into the combustion chamber. In turn, the pumps can be driven by electric motors and batteries, or more usually by turbines that are themselves powered by burning a small amount of propellant. For the space shuttle main engine (Fig. 9.26), two low-pressure turbopumps boost the inlet pressure for two high-pressure turbopumps, which feed the propellants into the combustion chamber at a pressure of 440 atm or higher. Dual preburners generate the gases that power the high-pressure turbopumps. The combustion chamber pressure is about 210 atm; the difference between the 440-atm propellant injection pressure and the 210-atm combustion chamber pressure enhances the propellant spray and mixing process.

Liquid propellants come in different categories, some of which are itemized below.

Cryogenic Propellants   The space shuttle main engine utilizes hydrogen (H_2) for the fuel and oxygen (O_2) for the oxidizer. Because H_2 must be at or below 20 K (36°R, or -253°C, or -424°F) to be a liquid, and O_2 must be at or below 135 K (243 R, or -138°C, or -217°F), they are examples of cryogenic propellants—chemicals that must be stored at extremely low temperatures to remain in liquid form. At the launch pads for the space shuttle, liquid oxygen (Lox) is stored in a giant insulated sphere holding 900,000 gal, and liquid hydrogen (LH_2) is contained in a separate insulated sphere with an 850,000-gal capacity. These cryogenic temperatures must be maintained during fueling and the launch periods of the shuttle. However, it is worth all the trouble to do this, because the H_2–O_2 propellant combination yields a high specific impulse. For the space shuttle, the vacuum $I_{sp} = 455$ s. The combustion process in the rocket engine is started with an igniter, and the burning is self-sustaining after that.

Bipropellants and Monopropellants   The H_2–O_2 combination described above is an example of a *bipropellant* combination—two chemicals used for the combustion process. Other chemicals exist in which chemical energy can be released simply by decomposing the molecules—these are called *monopropellants*. Usually, a solid catalyst is used to promote the decomposition. Monopropellants usually have a smaller I_{sp} than bipropellant combinations do, but they are easier to deal with simply because only one chemical propellant is being used. This reduces weight, simplifies the fuel system, and usually increases reliability. Monopropellants find use in small rocket engines for spacecraft attitude control. Hydrazine (N_2H_4) is an extensively used monopropellant.

Hypergolic Propellants   As mentioned above, the H_2–O_2 system used for the space shuttle main engine requires the combustion process to be initiated by an igniter (a type of "spark plug"), after which combustion is self-sustaining. However, there are propellant combinations that ignite simply on contact with one another. These are called *hypergolic propellants*. Because of this feature, there is an added danger in handling the propellants. However, they have the advantage of eliminating the need for a separate ignition system. Fluorine (F_2) is hypergolic with most fuels, but F_2 is among the most dangerous of all rocket propellants and therefore is not frequently used. Hypergolic propellants are used on two propulsion subsystems on the space shuttle—the orbital maneuvering subsystem (OMS) used for orbital insertion and the reaction control subsystem (RCS) used for attitude control. The fuel is monomethylhydrazine (MMH), and the oxidizer is nitrogen tetroxide (N_2O_4). As one NASA space shuttle engineer has so glibly stated, "Because of the eagerness of these two propellants to ignite spontaneously, their storage facilities are widely separated on Complex 39's launch pads" (NASA Fact Sheet KSC 191-80, November 1980). This hypergolic system, MMH/N_2O_4 is not as energetic as the H_2–O_2 system used for the main engines; its I_{sp} ranges from 260 to 280 s in the RCS and 313 s in the OMS. The higher efficiency of the OMS is due to a higher expansion ratio in that rocket engine nozzle.

9.9.2 Solid Propellants

So far in this section, we have discussed *liquid* rocket propellants. These propellants usually require large tanks (especially H_2, which is a light, high-volume chemical). Return to Fig. 8.30, which is a four-view drawing of the space shuttle, and note the large, single tank on which the winged shuttle orbiter is mounted. This is the tank for the liquid propellants. But note the two smaller cylinders on each side of the big tank; these are the strap-on twin solid rocket boosters which help the main shuttle engines lift the entire shuttle system off the ground. These two rocket engines use *solid propellants* in contrast to the liquid propellants discussed earlier. Solid propellants are completely different from liquid propellants in both their nature and behavior. This is why the first and primary distinction made between rocket propellants is that of liquid versus solid propellants, as we are making here.

Historically, the first rockets used solid propellants (see Sec. 9.16 on the history of rockets), these were black powder rockets used more than 1300 years ago in China. In contrast, the first successful liquid-propellant rocket is a product of the 20th century, developed by Robert H. Goddard in 1926.

Solid rocket fuels are just that—the fuel and oxidizer are premixed and cast in solid form. The two solid rocket boosters of the space shuttle use a solid propellant consisting of atomized aluminum powder (16 percent) as a fuel and ammonium perchlorate (69.93 percent) as an oxidizer. The remainder is iron oxide powder (0.7 percent) as a catalyst and polybutadiene acrylic acid acrylonitrile (14 percent) as a rubber-based binder. The binder also burns as a fuel. The solid propellant is battleship gray and has the consistency of a hard rubber eraser.

Burning of a solid propellant is initiated by an igniter on the surface of the propellant grain. Then the surface burns and recedes away, much like a Fourth of July sparkler. Some propellant grains are designed to be an end burner (one end is ignited and burns away, as a cigarette does), as shown in Fig. 9.28a. Others have an inner cylindrical bore, where the inner surface is ignited, and the propellant grain burns outward toward the motor case, as shown in Fig. 9.28b. Such solid rockets are called *internal burners*. For the cylindrical bore shown in Fig. 9.28b, as the burning surface recedes, the burning surface area increases, hence increasing the mass flow of burned gases. In turn, because rocket thrust is proportional to mass flow [see Eq. (9.28)], the thrust will increase with time. Another internal burning configuration is a solid-propellant grain with a star-shaped internal hollow channel,

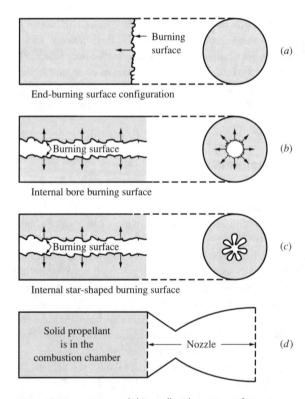

Figure 9.28   Some solid propellant burning configurations.

as sketched in Fig. 9.28c. With this configuration, ignition takes place on the star-shaped internal surface, and then the surface recedes away, becoming more circular in time. Because the star-shaped internal surface presents the maximum burning surface, which decreases with time, the thrust of this shaped grain is maximum at the beginning of burning and decreases with time. In essence, the timewise variation of the thrust of a solid rocket engine can be tailored by tailoring the shape of the solid-propellant grain. For the space shuttle's solid rocket booster, the internal cavity is an 11-point star which provides maximum thrust at liftoff. Please note that the solid-propellant grain configurations shown in Fig. 9.28 *a* to *c* are literally housed in the combustion chamber, and that the burned gases from these propellants are expanded through a convergent-divergent supersonic nozzle, the same as in a liquid-propellant rocket, as sketched in Fig. 9.28d.

One of the important physical characteristics of a solid propellant is the *linear burning rate r*, which is the time rate at which the burning surface of the propellant recedes normal to itself. The burning rate is a function mainly of the combustion chamber pressure p_0 and the initial temperature of the propellant. The pressure variation of r is given by

$$r = a p_0^n \qquad\qquad \textbf{[9.43]}$$

where r is the linear burning rate, p_0 is the combustion chamber pressure, and a and n are constants which are determined by experiment for a given propellant. Typically, for most propellants, n has a value between 0.4 and 0.8, where r is in units of inches per second and p_0 in pounds per square inch.

In comparison to liquid propellants, solid propellants have the following advantages and disadvantages.

Advantages

1. Solid rockets are simpler, safer, and more reliable. There is no need for pumps and complex propellant feed systems.

2. Solid propellants are more storable and stable. Some solid rockets can be stored for decades before use.

3. Solid propellants are dense, hence the overall volume of solid rockets is smaller. Compare the smaller size of the twin solid boosters on the space shuttle to the larger size of the main liquid propellant tank in Fig. 8.30.

Disadvantages

1. The specific impulse of solid propellants is considerably less than that of liquid propellants. For the space shuttle's solid rocket boosters $I_{sp} = 242$ s at sea level. In general, the specific impulse of solid rockets ranges from 200 to 300 s.

2. Once a solid rocket is ignited, it usually cannot be turned off. Also, it is difficult to throttle a solid rocket in order to vary the thrust at will. In contrast, liquid rockets are easily throttled, and the thrust can be cut off whenever desired just by manipulating the fuel and oxidizer valves.

9.9.3 A Comment

The choice of liquid versus solid propellants in the design of a new rocket engine depends on the design specifications, including engine performance, cost, reliability, maintainability, and so forth. However, the differences between liquid and solid propellants are so well defined that the engineering design choice is usually straightforward. We end this section by noting that the concept of hybrid rockets has been examined in recent years. Hybrid rockets are part solid, part liquid. The oxidizer may be solid and the fuel a liquid, or vice versa, in such hybrid rockets. Hybrid rockets are an attempt to combine the advantages of both solid and liquid propellants; but of course, as is true for any design compromise, hybrid rockets are just that—a compromise. At the time of writing, hybrid rockets are still mainly in the experimental stage.

9.10 ROCKET EQUATION

In Sec. 9.8 we developed some of the performance parameters for the rocket engine itself. In this section we will relate the rocket engine performance, as described by the specific impulse I_{sp}, to the velocity achieved by the complete rocket vehicle (such as the V-2 shown in Fig. 9.38).

The mass of a complete rocket vehicle consists of three parts: (1) the mass of the payload M_L (satellite, manned space capsule, etc.); (2) the mass of the structure of the vehicle M_s, including the rocket engine machinery, the propellant tanks, the structural beams, formers, and stringers; and (3) the mass of the propellants M_p. Hence, at any instant during the flight of the rocket vehicle, the total mass is

$$M = M_L + M_s + M_p \qquad \text{[9.44]}$$

Consider a rocket vehicle that blasts off from the surface of the earth and accelerates until all its propellants are exhausted. At the instant of liftoff, the vehicle velocity is zero; after the rocket engines have shut down because all the propellants have been consumed, the vehicle velocity is the burnout velocity V_b, which can be calculated from Newton's second law

$$F = M \frac{dV}{dt} \qquad \text{[9.45]}$$

The force on the vehicle is the net difference between the thrust of the rocket engine, the aerodynamic drag, and the weight of the vehicle. If we assume that the last two are small compared with the engine thrust, Eq. (9.45) can be written

$$T = M \frac{dV}{dt} \qquad \text{[9.46]}$$

The thrust is related to the specific impulse through Eq. (9.33), written as

$$T = \dot{w} I_{\text{sp}} = g_0 \dot{m} I_{\text{sp}} \qquad \text{[9.47]}$$

where $\dot{m}$ is the mass flow of the propellants. In Eq. (9.44), M is changing with time due to the decrease in M_p; indeed,

$$\dot{m} = -\frac{dM_p}{dt} = -\frac{dM}{dt} \qquad \textbf{[9.48]}$$

Combining Eqs. (9.47) and (9.48), we have

$$T = -g_0 I_{\text{sp}} \frac{dM}{dt} \qquad \textbf{[9.49]}$$

Substituting Eq. (9.49) into (9.46) gives

$$-g_0 I_{\text{sp}} \frac{dM}{dt} = M \frac{dV}{dt}$$

or

$$-\frac{dM}{M} = \frac{dV}{g_0 I_{\text{sp}}} \qquad \textbf{[9.50]}$$

integrating Eq. (9.50) between liftoff (where $V = 0$ and M is the initial mass M_i) and burnout (where $V = V_b$ and M is the final mass M_f), we have

$$-\int_{M_i}^{M_f} \frac{dM}{M} = \int_{M_f}^{M_i} \frac{dM}{M} = \frac{1}{g_0 I_{\text{sp}}} \int_0^{V_b} dV$$

or

$$\ln \frac{M_i}{M_f} = \frac{V_b}{g_0 I_{\text{sp}}}$$

or

$$V_b = g_0 I_{\text{sp}} \ln \frac{M_i}{M_f} \qquad \textbf{[9.51]}$$

This so-called *rocket equation* relates the burnout velocity of a rocket vehicle to the specific impulse associated with the engine and the mass ratio M_i/M_f. The equation can be turned inside out to relate the mass ratio necessary to achieve a given burnout velocity, as

$$\frac{M_i}{M_f} = e^{V_b/(g_0 I_{\text{sp}})} \qquad \textbf{[9.52]}$$

9.11 ROCKET STAGING

To the present, most space vehicles have been launched into space by *multistage rockets*—rocket boosters that are in reality two or more distinct rockets placed on top of each other (or beside each other, as in the case of the space shuttle, as shown in Fig. 8.30). Why do it this way? Why not have one large rocket booster that will do the job—why not have a single-stage-to-orbit vehicle? The answer is basically one of economics—which system will place a pound of payload in orbit for the least cost?

Until recently, the design choice was to use multistage rockets. This is the least-cost solution when the rockets are *expendable*, when the rocket stages are sequentially separated from the space vehicle (the payload) and are destroyed in the atmosphere while falling back to earth. However, at the time of writing, there is much discussion and technological development of totally *reusable* rocket boosters—rockets that are recovered and used again multiple times. (This is already partially achieved with the space shuttle. The expended solid rocket booster casings are recovered from the Atlantic Ocean after each launch. NASA then cleans and refurbishes these casings. They are returned to the manufacturer, which refills the casings with propellant. In this way, each casing is reused about 20 times. And, of course, the shuttle orbiter returns to earth with the main rocket engines intact, ready to be used again. Only the large liquid-propellant tank is lost on each launch.) If the rocket booster can be totally preserved after a launch, the hardware cost of replacing it is forgone. This sometimes swings the economic choice of least cost to a single-stage-to-orbit vehicle. To date, modern single-stage-to-orbit vehicles are only in the experimental phase.

In this section, we consider multistage rockets, which are currently the design choice for expendable rockets. Our purpose is simply to explain why a multistage booster is a cheaper solution to putting a given payload in space than one, larger, single-stage rocket.

In Sec. 9.10, we designated the payload mass by M_L, the mass of the structure by M_s, and mass of the propellants by M_p. In the rocket equation, M_i is the initial mass of the total vehicle before ignition, and M_f is the final mass at burnout. Let us first consider a single-stage rocket. The masses of the payload, structure, and propellant are represented schematically by the differently shaded areas in Fig. 9.29a, which is essentially a bar diagram for the mass breakdown. The burnout velocity for this

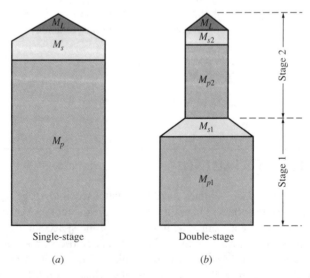

Single-stage   Double-stage

(a)   (b)

Figure 9.29   Schematic representation of the mass components of rockets. (a) Single stage. (b) Double stage.

single-stage rocket is given by Eq. (9.51), repeated here:

$$V_b = g_0 I_{sp} \ln \frac{M_i}{M_f} \qquad \textbf{[9.51]}$$

where
$$M_i = M_p + M_s + M_L$$

and
$$M_f = M_s + M_L$$

In contrast, consider the two-stage rocket is shown schematically in Fig. 9.29b. For the first stage, the propellant mass is M_{p1} and the structural mass is M_{s1}. The payload for the first stage is the *entire second stage*. For the second stage, the propellant mass is M_{p2}, the structural mass is M_{s2}, and the payload mass is M_L. Figure 9.29b is essentially a bar diagram showing the masses for both the first and second stages. The burnout velocity of the first stage (with the second stage attached) V_{b1} is given by Eq. (9.51):

$$V_{b1} = g_0 I_{sp} \ln \frac{M_i}{M_f} \qquad \textbf{[9.53]}$$

Here the initial mass is the sum of *all* the masses shown in Fig. 9.29b.

$$M_i = M_{p1} + M_{s1} + M_{p2} + M_{s2} + M_L \qquad \textbf{[9.54]}$$

and the final mass is the structural mass of the first stage plus the total mass of the second stage,

$$M_f = M_{s1} + M_{p2} + M_{s2} + M_L \qquad \textbf{[9.55]}$$

Substituting Eqs. (9.54) and (9.55) into Eq. (9.53), we have

$$V_{b1} = g_0 I_{sp} \ln \left[\frac{M_{p1} + M_{s1} + M_{p2} + M_{s2} + M_L}{M_{s1} + M_{p2} + M_{s2} + M_L} \right] \qquad \textbf{[9.56]}$$

The first stage at the instant of burnout separates from the second stage and drops away. The rocket engine of the second stage ignites, and boosts the second stage from its initial velocity V_{b1} to its final burnout velocity V_{b2}. The rocket equation, Eq. (9.51), when applied to the second stage which is already moving with the initial velocity V_{b1}, yields the *increase* in velocity, $V_{b2} - V_{b1}$ as

$$V_{b2} - V_{b1} = g_0 I_{sp} \ln \left(\frac{M_i}{M_f} \right)_2 \qquad \textbf{[9.57]}$$

where
$$M_i = M_{p2} + M_{s2} + M_L \qquad \textbf{[9.58]}$$

and
$$M_f = M_{s2} + M_L \qquad \textbf{[9.59]}$$

Substituting Eqs. (9.58) and (9.59) into Eq. (9.57), we have

$$V_{b2} - V_{b1} = g_0 I_{sp} \ln \frac{M_{p2} + M_{s2} + M_L}{M_{s2} + M_L} \qquad \textbf{[9.60]}$$

The advantage of a multistage rocket is illustrated by the following worked example.

Example 9.5 | Consider the single-stage rocket and the double-stage rocket sketched in Fig. 9.39a and b, respectively. Both rockets have the same total mass $M_{\text{total}} = 5000$ kg and the same specific impulse $I_{\text{sp}} = 350$ s. Both rockets have the same payload mass $M_L = 50$ kg. The total structural mass of the double-stage rocket is $M_{s1} + M_{s2} = 400$ kg $+ 100$ kg $= 500$ kg, which is the structural mass of the single-stage rocket. The total propellant mass of the double-stage rocket is $M_{p1} + M_{p2} = 3450 + 1000 = 4450$ kg, which is the propellant mass of the single-stage rocket. Both rockets are boosting the same payload mass of 50 kg into space. The breakdown between payload, structural, and propellant masses chosen in this example is purely arbitrary, but keeping the total masses in each category the same between the two rockets is intentional. In this way, the only difference between the rockets in Fig. 9.30a and b is that one is a single-stage rocket and the other is a double-stage rocket, but with the same total masses distributed over two stages. Calculate and compare the burnout velocities for the rockets in Fig. 9.30a and b.

Solution

For the single-stage rocket in Fig. 9.30a, the initial and final masses are

$$M_i = M_p + M_s + M_L = 4450 + 500 + 50 = 5000 \text{ kg}$$
$$M_f = M_s + M_L = 500 + 50 = 550 \text{ kg}$$

From Eq. (9.51),

$$V_b = g_0 I_{\text{sp}} \ln \frac{M_i}{M_f} = 9.8(350) \ln \frac{5000}{550} = 7570 \text{ m/s} = \boxed{7.57 \text{ km/s}}$$

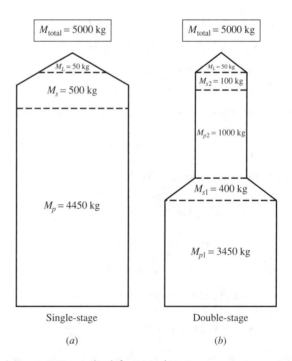

Single-stage

(a)

Double-stage

(b)

Figure 9.30   Sketch for Example 9.5.

For the double-stage rocket in Fig. 9.30b, we have for the burnout velocity of the first stage, from Eq. (9.56),

$$V_{b1} = g_0 I_{sp} \ln \frac{M_{p1} + M_{s1} + M_{p2} + M_{s2} + M_L}{M_{s1} + M_{p2} + M_{s2} + M_L}$$

$$= 9.8(350) \ln \frac{3450 + 400 + 1000 + 100 + 50}{400 + 1000 + 100 + 50}$$

$$= 9.8(350) \ln \frac{5000}{1550} = 4017 \text{ m/s}$$

The increase in velocity provided by the second stage is given by Eq. (9.60).

$$V_{b2} - V_{b1} = g_0 I_{sp} \ln \frac{M_{p2} + M_{s2} + M_L}{M_{s2} + M_L}$$

$$= 9.8(350) \ln \frac{1000 + 100 + 50}{100 + 50}$$

$$= 9.8(350) \ln \frac{1150}{150} = 6987 \text{ m/s}$$

Hence, the velocity at burnout of the second stage is

$$V_{b2} = 6987 + V_{b1} = 6987 + 4017 = 11{,}004 \text{ m/s} = \boxed{11 \text{ km/s}}$$

Comparison

From this example, we see that the payload of 50 kg is launched into space at a velocity of 11 km/s by the double-stage rocket, whereas for the same total expenditure of propellants, the single-stage rocket provides a velocity of only 7.57 km/s. Indeed, for this example, the single-stage rocket provides essentially orbital velocity for the payload, whereas the double-stage rocket gives the payload escape velocity, allowing the space vehicle to go into deep space.

9.12 ELECTRIC PROPULSION

The chemical rockets discussed in Secs. 9.8 through 9.11 are the "brute-force" propulsion devices for space vehicles—high thrust but relatively low I_{sp}. Their high thrust is absolutely necessary for ascent from the earth's surface to space. However, once in space, a space vehicle could take advantage of a propulsive device that produces much less thrust but has a much greater I_{sp}, and that could provide a sustained thrust for very long times, perhaps indefinitely. Unmanned missions to deep space would benefit from such devices. This has spawned a class of propulsion devices under the generic label of *advanced space propulsion*. In this section, we discuss only one type of advanced space propulsion—electric propulsion—and that only briefly. Our purpose is to give you only the flavor of such a device, so that you know that other propulsive mechanisms for space vehicles are feasible besides chemical rockets.

Electric propulsion describes the generic class of propulsion devices that use *electric power* to generate thrust. The idea is coupled with the fact that low–molecular weight propellants have high values of I_{sp} [recall the discussion surrounding Eq. (9.35) that I_{sp} varies inversely with molecular weight]. Electric propulsion concepts use electricity in various forms to accelerate a low–molecular weight gas, hence creating thrust and at the same time achieving a high I_{sp}. Some of the types of electric propulsion devices are discussed below.

9.12.1   Electron-Ion Thruster

This device produces thrust by accelerating positively charged ions in an electrostatic field. The basic concept is sketched in Fig. 9.31. A propellant (such as mercury, or an inert gas such as helium or argon) is fed into a chamber. Inside the chamber is an anode and a cathode. A beam of electrons is generated between the anode and cathode. The high-speed electrons collide with the atoms of the propellant, stripping off other electrons and leaving behind positively charged ions in the chamber. These ions then pass through a separately applied electrostatic field and are accelerated out of the device in the form of an ion beam. If nothing else were done, there would be a rapid buildup of negative charge in the chamber because of the flux of positively charged propellant leaving the device. In turn, the positively charged ion beam would be retarded by the massive negative charge in the chamber. Therefore, it is necessary to make the beam of particles exiting the chamber electrically neutral. This can be achieved by feeding electrons into the exhaust beam. Being neutral, the beam will not be retarded by the negative charge in the chamber.

Electron-ion thrusters have specific impulses from 3000 to 5000 s.

9.12.2   Magnetoplasmadynamic Thruster

The magnetoplasmadynamic (MPD) thruster uses a self-induced magnetic field to accelerate positive ions. The basic concept is sketched in Fig. 9.32. Here a powerful

Figure 9.31   Schematic of an electric-ion thruster.

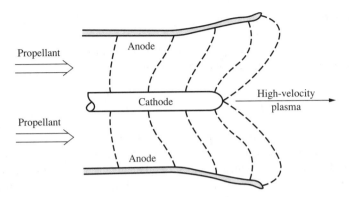

Figure 9.32   Schematic of a magnetoplasmadynamic (MPD) thruster.

pulse of electric current surges from a central cathode to the anode on the walls of a chamber. The propellant is ionized by the electric current. The current paths are illustrated by the dashed lines in Fig. 9.32. The electric current sets up an induced magnetic field in the chamber (recall that an electric current in a wire sets up an induced magnetic field about the wire) which then accelerates the plasma out the back end of the chamber.

Magnetoplasmadynamic thrusters can potentially create more thrust than an electric-ion thruster with approximately the same specific impulse.

9.12.3   Arc-Jet Thruster

This device is fundamentally simple, and it is more closely related to the chemical rockets than the other electric propulsion devices discussed above. In the arc-jet thruster, hydrogen is heated in a reservoir by an electric arc, and then the hot, low–molecular weight gas expands through a conventional convergent-divergent nozzle, as sketched in Fig. 9.33. There are no electromagnetic forces on the hot gas; the electric arc is simply a mechanism to create a hot gas in the reservoir, akin to the

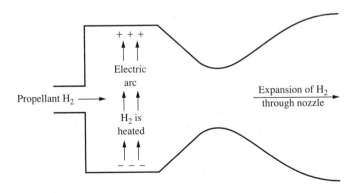

Figure 9.33   Arc-jet thruster.

energy release during combustion of chemical rocket propellants in the combustion chamber.

The arc-jet thruster has a specific impulse on the order of 800 to 1200 s, due mainly to the low molecular weight of hydrogen.

9.12.4   A Comment

All electric propulsion devices require a separate power source to drive the electromagnetic functions described above. The power supplies for electric propulsion devices can be solar cells, nuclear reactors, or other advanced energy sources that can be converted to electricity. Such matters are beyond the scope of our discussion.

For a more extensive but still fundamental discussion of advanced space propulsion, see the article by Frisbee listed in the Bibliography; this article is a primary reference source for this section.

The first ion engine to be employed on a deep space probe was launched on October 24, 1998, from Cape Canaveral. Appropriately named *Deep Space 1*, the space vehicles has the mission to test new, advanced technologies. On November 10, NASA engineers powered up the engine. It ran for 4.5 min before shutting itself off. On November 24, in response to commands sent to the spacecraft, the ion engine came to life again, and at the time of writing is continuing to run smoothly. At full throttle, the ion engine, which is powered by solar cells, consumes about 2500 W of electric power and produces 0.02 lb of thrust—a force equal to the weight of a sheet a paper in the palm of your hand.

9.13   HISTORICAL NOTE: EARLY PROPELLER DEVELOPMENT

The ancestry of the airplane propeller reaches as far back as the 12th century, when windmills began to dot the landscape of western Europe. The blades of these windmills, which were essentially large wood-and-cloth paddles, extracted energy from the wind to power mechanical grinding mills. Only a small intellectual adjustment was necessary to think of this process in reverse—to power mechanically the rotating paddles in order to add energy to the air and produce thrust. Indeed, Leonardo da Vinci developed a helical screw for a 16th century helicopter top. Later, a year after the first successful balloon flight in 1783 (see Chap. 1), a hand-driven propeller was mounted to a balloon by J. P. Blanchard. This was the first propeller to be truly airborne, but it did not succeed as a practical propulsive device. Nevertheless, numerous other efforts to power hot-air balloons with hand-driven propellers followed, all unsuccessful. It was not until 1852 that a propeller connected to a steam engine was successfully employed in an airship. This combination, designed by Henri Giffard, allowed him to guide his airship over Paris at a top speed of 5 mi/h.

As mentioned in Chap. 1, the parent of the modern airplane, George Cayley, eschewed the propeller and instead put his faith mistakenly in oarlike paddles for propulsion. However, Henson's aerial steam carriage (see Fig. 1.11) envisioned two pusher propellers for a driving force, and after that, propellers became the accepted propulsion concept for heavier-than-air vehicles. Concurrently, in a related fashion, the marine propeller was developed for use on steamships beginning in the early 19th century. Finally, toward the end of that century, the propeller was employed by Du Temple, Mozhaiski, Langley, and others in their faltering efforts to get off the ground (see, e.g., Figs. 1.13, 1.14, and 1.18).

However, a close examination of these 19th century aircraft reveals that the propellers were crude, wide, paddlelike blades which reflected virtually no understanding of propeller aerodynamics. Their efficiencies must have been exceedingly low, which certainly contributed to the universal failure of these machines. Even marine propellers, which had been extensively developed by 1900 for steamships, were strictly empirical in their design and at best had efficiencies on the order of 50 percent. There existed no rational hydrodynamic or aerodynamic theory for propeller design at the turn of the century.

This was the situation when Wilbur and Orville returned from Kill Devil Hills in the fall of 1902, flushed with success after more than 1000 flights of their number 3 glider (see Chap. 1) and ready to make the big step to a powered machine. Somewhat naively, Wibur originally expected this step to be straightforward; the engine could be ordered from existing automobile companies, and the propeller could be easily designed from existing marine technology. Neither proved to be the case. After spending several days in Dayton libraries, Wibur discovered that a theory for marine propellers did not exist and that even an appreciation for their true aerodynamic function had not been developed. So once again the Wright brothers, out of necessity, had to plunge into virgin engineering territory. Throughout the winter of 1902–1903, they wrestled with propeller concepts in order to provide accurate calculations for design. And once again they demonstrated that, without the benefit of a formal engineering education, they were the premier aeronautical engineers of history. For example, by early spring of 1903, they were the first to recognize that a propeller is basically a rotating wing, made up of airfoil sections that generated an aerodynamic force normal to the propeller's plane of rotation. Moreover, they made use of their wind tunnel data, obtained the previous year on several hundred different airfoil shapes, and chose a suitably cambered shape for the propeller section. They reasoned the necessity for twisting the blade in order to account for the varying relative airflow velocity from the hub to the tip. Indeed, in Orville's words:

It is hard to find even a point from which to start, for nothing about a propeller, or the medium in which it acts, stands still for a moment. The thrust depends upon the speed and the angle at which the blade strikes the air; the angle at which the blade strikes the air depends upon the speed at which the propeller is turning, the speed the machine is traveling forward, and the speed at which the air is slipping backward; the slip of the air backward depends upon the thrust exerted by the propeller and the amount of air acted upon. When any of these changes, it changes all the rest, as they are all interdependent upon one another. But these are only a few of the factors that must be considered....

By March of 1903, Wilbur had completed his theory to the extent that a propeller could be properly designed. Using a hatchet and drawknife, he carved two propellers out of laminated spruce and surfaced them with aluminum paint. Excited about their accomplishment, Orville wrote: "We had been unable to find anything of value in any of the works to which we had access, so we worked out a theory of our own on the subject, and soon discovered, as we usually do, that all the propellers built heretofore are *all wrong*, and then built a pair . . . based on our theory, which are *all right*!"

The propeller designed by the Wright brothers, principally by Wilbur, achieved the remarkably high efficiency of 70 percent and was instrumental in their successful flight on December 17, 1903, and in all flights thereafter. Moreover, their propellers remained the best in aviation for almost a decade. Indeed, until 1908, all other competitors clung to the older, paddlelike blades, both in the United States and in Europe. Then when Wilbur made his first dramatic public flight on August 8, 1908, at Hunaundières, France, the impact of his highly efficient propeller on the European engineers was almost as great as that of the Wrights' control system, which allowed smoothly maneuverable flight. As a result, subsequent airplanes in Europe and elsewhere adopted the type of aerodynamically designed propeller introduced by the Wrights.

Consequently, credit for the first properly designed propeller, along with the associated aerodynamic theory, must go to the Wright brothers. This fact is not often mentioned or widely recognized; however, this propeller research in 1903 represented a quantum jump in a vital area of aeronautical engineering, without which practical powered flight would have been substantially delayed.

The final early cornerstone in the engineering theory and design of airplane propellers was laid by William F. Durand about a decade after the Wright brothers' design was adopted. Durand was a charter member of NACA and became its chairman in 1916 (see Sec. 2.8). Durand was also the head of the mechanical engineering department at Stanford University at that time, and during 1916–1917 he supervised the construction of a large wind tunnel on campus designed purely for the purpose of experimenting with propellers. Then, in 1917, he published NACA Report No. 14 entitled "Experimental Research on Air Propellers." This report was the most extensive engineering publication on propellers to that date; it contained experimental data on numerous propellers of different blade shapes and airfoil sections. It is apparently the first technical report to give extensive plots of propeller efficiency versus advance ratio. Hence, the type of efficiency curve sketched in Fig. 9.6 dates back as far as 1917! Moreover, the values of maximum efficiency of most of Durand's model propellers were 75 to 80 percent, a creditable value for that point in history. It is interesting to note that, 60 years later, modern propeller efficiencies are not that much better, running between 85 and 90 percent. To Durand must also go the credit for the first dimensional analysis in propeller theory; in the same NACA reports, he shows by dimensional analysis that propeller efficiency must be a function of advance ratio, Reynolds number, and Mach number, and he uses these results to help correlate his experimental data. This early NACA report was an important milestone in the development of the airplane propeller. Indeed, a copy of the report itself is enshrined behind glass and is prominently displayed in the lobby of the new Durand Engineering Building on the Stanford campus.

9.14 HISTORICAL NOTE: EARLY DEVELOPMENT OF THE INTERNAL COMBUSTION ENGINE FOR AVIATION

The pivotal role of propulsion in the historical quest for powered flight was discussed in Chap. 1. The frustrating lack of a suitable prime mover was clearly stated as far back as 1852 by George Cayley, who wrote about his trials with a "governable parachute" (glider): "It need scarcely be further remarked, that were we in possession of a sufficiently light first mover to propel such vehicles by waftage, either on the screw principle or otherwise, with such power as to supply that force horizontally, which gravitation here supplies in the descent, mechanical aerial navigation would be at our command without further delay."

Indeed, Cayley devoted a great deal of thought to the propulsion problem. Before 1807, he had conceived the idea for a hot-air engine, in which air is drawn from the atmosphere, heated by passing it over a fire, and then expanded into a cylinder, doing work on a piston. This was to be an alternative to steam power. Considering his invention in a general sense, and not mentioning any possible application to flight, Cayley wrote in the October 1807 issue of Nicholson's *Journal* that "the steam engine has hitherto proved too weighty and cumbrous for most purposes of locomotion; whereas the expansion of air seems calculated to supply a mover free from these defects." Later, in 1843, Cayley summarized his work on aeronautical propulsion in a type of letter to the editor in *Mechanics' Magazine*:

> The real question rests now, as it did before, on the possibility of providing a sufficient power with the requisite lightness. I have tried many different engines as first movers, expressly for this purpose [flight]. Gun powder is too dangerous, but would, at considerable expense, effect the purpose: but who would take the double risk of breaking their neck or being blown to atoms? Sir Humphrey Davy's plan of using solid carbonic acid, when again expanded by heat, proved a failure in the hands of our most ingenious engineer, Sir M. Isambard Brunel.
>
> As all these processes require nearly the same quantity of caloric to generate the same degree of power, I have for some time turned my own attention to the use, as a power, of common atmospheric air expanded by heat, and with considerable success. A five-horse engine of this sort was shown at work to Mr. Babbage, Mr. Rennie, and many other persons capable of testing its efficiency, about three years ago. The engine was only an experimental one, and had some defects, but each horse power was steadily obtained by the combustion of about $6\frac{1}{2}$ pounds of coke per hour, and this was the whole consumption of the engine, no water being required. Another engine of this kind, calculated to avoid the defects of the former one, is now constructing, and may possibly come in aid of balloon navigation—for which it was chiefly designed—or the present project, if no better means be at hand.

Thus, in keeping with his remarkable and pioneering thinking on all aspects of aviation, George Cayley stated the impracticality of steam power for flight and clearly experimented with some forerunners of the IC engine. However, his thoughts were lost to subsequent aeronautical engineers of the 19th century, who almost universally attempted steam-powered flight (see Chap. 1).

The development of IC engines gained momentum with Lenoir's two-cycle gas-burning engine in 1860. Then, in 1876, Nikolaus August Otto designed and built the first successful four-stroke IC engine, the same type of engine discussed in Sec. 9.3. Indeed, the thermodynamic cycle illustrated in Fig. 9.12, consisting of isentropic compression and power strokes with constant-volume combustion, is called the *Otto cycle*. Although Otto worked in Germany, strangely enough, in 1877 he took out a U.S. patent on his engine. Otto's work was soon applied to land vehicle propulsion, heralding the birth of the automobile industry before 1900.

But automobiles and airplanes are obviously two different machines, and IC engines used in automobiles in 1900 were too heavy per horsepower for aeronautical use. One man who squarely faced this barrier was Samuel Pierpont Langley (see Sec. 1.7). He correctly recognized that the gasoline-burning IC engine was the appropriate power plant for an airplane. To power the newer versions of his Aerodromes, Langley contracted with Stephen M. Balzer of New York in 1898 for an engine of 12 hp weighing no more than 100 lb. Unfortunately, Balzer's delivered product, which was derived from the automobile engine, could produce only 8 hp. This was unacceptable, and Charles Manly, Langley's assistant, took the responsibility for a complete redesign of Balzer's engine in the laboratory of the Smithsonian Institution in Washington, District of Columbia. The net result was a power plant, finished in 1902, which could produce 52.4 hp while weighing only 208 lb. This was a remarkable achievement; it was not bettered until the advent of "high-performance" aircraft toward the end of World War I, 16 years later. Moreover, Manly's engine was a major departure from existing automobile engines of the time. It was a radial engine, with five cylinders equally spaced in a circular pattern around a central crankshaft. It appears to be the first aircraft radial engine in history, and certainly the first successful one. Unfortunately, the failure of Langley's Aerodromes in 1903 obscured the quality of Manly's engine, although the engine was in no way responsible for these failures.

Five hundred miles to the west, in Dayton, Ohio, the Wright brothers also originally planned to depend on a standard automobile engine for the power plant for their *Flyer*. In the fall of 1902, after their stunning success with their number 3 glider at Kill Devil Hills, the Wrights were rudely surprised to find that no automobile engine existed that was light enough to meet their requirement. Because Wilbur had taken the prime responsibility of developing a propeller (see Sec. 9.13) during this time, he assigned Orville the task of designing and building a suitable engine. It is interesting to note that Wilbur correctly considered the propeller to be a more serious problem than the engine. With the help of Charles Taylor, a mechanic who worked in the Wrights' bicycle shop, and using as a model the car engine of a Pope-Toledo (long since defunct), Orville expeditiously completed his engine design and construction in less than six weeks. In its first test in February 1903, the aluminum crankcase cracked. Two months later a local foundry finished casting a second case, and the engine was finally successfully tested in May.

The engine was a four-cylinder in-line design. It had only one speed, about 100 rpm, and could be stopped only by cutting off the supply of gasoline, which was fed to the cylinders by gravity. The engine produced 12 hp and weighed (without oil and fuel) about 100 lb. Although the Wrights' engine produced far less horsepower per pound of engine weight than Manly's design, it was nevertheless adequate for its purpose. The Wright brothers had little experience with IC engines before 1903, and their successful design is another testimonial to their unique engineering talents. In Orville's words: "Ignorant of what a motor this size ought to develop, we were greatly pleased with its performance. More experience showed us that we did not get one-half the power we should have had."

The Wright brothers' engine was obviously the first successful aircraft power plant to fly, by virtue of their history-making flight of December 17, 1903 (see Sec. 1.1). Subsequent development of the IC engine for airplanes came slowly. Indeed, nine years later, Captain H. B. Wild, speaking in Paris, gave the following pilot-oriented view of the aircraft engine:

> The comparatively crude and unreliable motor that we have at our disposal at the present time [1912] is no doubt the cause of many of the fatalities and accidents befalling the aeroplane. If one will look over the accessories attached to the aero engine of today, it will be noted that it is stripped clean of everything possible which would eliminate what he deems unnecessary parts in order to reduce the weight of the engine, and in doing so he often takes away the parts which help to strengthen the durability and reliability of the motor.

The eventual successful development of efficient, reliable, and long-endurance aircraft power plants is now a fact of history. However, it was accomplished only by an intensive and continuous engineering effort. Various reports on engine development—carburators, valves, radiators, etc.—perfuse the early NACA literature. Indeed, the recognition of the importance of propulsion was made clear in 1940 with the establishment of a complete laboratory for its research and development, the NACA Lewis Flight Propulsion Laboratory in Cleveland, Ohio.

The internal combustion reciprocating engine has now been supplanted by the gas-turbine jet engine as the main form of aeronautical propulsion. However, IC engines are still the most appropriate choice for general-aviation aircraft designed for speeds of 300 mi/h or less, and therefore their continued development and improvement will remain an important part of aerospace engineering.

9.15 HISTORICAL NOTE: INVENTORS OF EARLY JET ENGINES

By the late 1920s, the reciprocating engine–propeller combination was so totally accepted as *the* means of airplane propulsion that other concepts were generally discounted. In particular, jet propulsion was viewed as technically infeasible. For example, NACA reported in 1923 that jet propulsion was "impractical," but its studies were aimed at flight velocities of 250 mi/h or less, where jet propulsion is truly impractical. Eleven years later, the British government still held a similar opinion.

Into this environment came Frank Whittle (now Sir Frank Whittle). Whittle is an Englishman, born on June 1, 1907, in Coventry. As a young boy, he was interested in aviation, and in 1923 he enlisted in the Royal Air Force. Showing much intelligence and promise, he soon earned a coveted student's slot at the RAF technical college at Cranwell. It was here that Whittle became interested in the possibilities of gas-turbine engines for propelling airplanes. In 1928, he wrote a senior thesis at Cranwell entitled "Future Developments in Aircraft Design," in which he expounded the virtues of jet propulsion. It aroused little interest. Undaunted, Whittle went on to patent his design for a gas-turbine engine in January 1930. For the next five years, in the face of polite but staunch disinterest, Whittle concentrated on his career in the RAF and did little with his ideas on jet propulsion. However, in 1935, with the help of a Cranwell classmate, a firm of bankers agreed to finance a private company named Power Jets Ltd. specifically to develop the Whittle jet engine. So in June 1935, Frank Whittle and a small group of colleagues plunged into the detailed design of what they thought would be the first jet engine in the world. Indeed, the engine was finished in less than two years and was started up on a test stand on April 12, 1937—the first jet engine in the world to successfully operate in a practical fashion.

However, it was not the first to fly. Quite independently, and completely without the knowledge of Whittle's work, Hans von Ohain in Germany developed a similar gas-turbine engine. Working under the private support of the famous airplane designer Ernst Heinkel, von Ohain started his work in 1936. (As in the United States and England, the German government showed little initial interest in jet propulsion.) Three years after his work began, von Ohain's engine was mated with a specially designed Heinkel airplane. Then on August 28, 1939, the He 178 (see Fig. 9.34) became the first gas turbine–powered, jet-propelled airplane in history to fly. It was strictly an experimental aircraft, but von Ohain's engine of 838 lb of thrust pushed the He 178 to a maximum speed of 435 mi/h. Later, after the beginning of World War II, the German government reversed its lack of interest in jet propulsion, and soon Germany was to become the first nation in the world with operational military jet aircraft.

Figure 9.34   The German He 178—the first jet-propelled airplane in the world to fly successfully.

Meanwhile, in England, Whittle's success in operating a jet engine on a test stand finally overcame the Air Ministry's reluctance, and in 1938 a contract was let to Power Jets Ltd. to develop a revised power plant for installation in an airplane. Simultaneously, Gloster Aircraft received a contract to build a specially designed jet-propelled aircraft. Success was obtained when the Gloster E.28/39 airplane (see Fig. 9.35) took off from Cranwell on May 15, 1941, the first airplane to fly with a Whittle jet engine. The engine produced 860 lb of thrust and powered the Gloster airplane to a maximum speed of 338 mi/h. The Gloster E.28/39 now occupies a distinguished berth in the Science Museum in London, hanging prominently from the top-floor ceiling of the massive brick building in South Kensington, London. The technology gained with the Whittle engine was quickly exported to the United States and eventually fostered the birth of the highly successful Lockheed P-80 Shooting Star, the first U.S. production-line jet airplane.

In 1948, Frank Whittle retired from the RAF as an air commodore and was knighted for his contributions to British aviation. In 1976 he moved to the United States, where he worked and taught at the U.S. Naval Academy in Annapolis, Maryland. On August 8, 1996, he died at his home in Columbia, Maryland, where he was cremated.

Hans von Ohain was among the large group of German scientists and engineers who were brought to the United States at the end of World War II. He pursued a distinguished career at the Air Force's Aeronautical Research Laboratory at Wright-Patterson Air Force Base, Ohio, where he led a propulsion group doing research on advanced concepts. Indeed, the present author had the privilege of working for three years in the same laboratory with von Ohain and shared numerous invigorating conversations with this remarkable man. Later, von Ohain became affiliated with the U.S. Air Force Aeropropulsion Laboratory at Wright Field, from which he retired in 1980. He remained active after retirement as a tireless spokesman for aeronautics.

Figure 9.35   The Gloster E28—the first British airplane to fly with jet propulsion.

In 1984, he served a year at the National Air and Space Museum of the Smithsonian Institution in the prestigious Charles Lindbergh Chair (a chair that the present author was honored to occupy two years after von Ohain). Hans von Ohain died at his home in Melbourne, Florida, on March 13, 1998. He is buried in Dayton, Ohio. Within a span of two years, the world lost the two coinventors of the jet engine. History has already shown that these two gentlemen created a revolution in aeronautics—the jet revolution—perhaps on a par with the invention of the practical airplane by the Wright brothers.

9.16   HISTORICAL NOTE: EARLY HISTORY OF ROCKET ENGINES

"When it was lit, it made a noise that resembled thunder and extended 100 li [about 24 km]. The place where it fell was burned, and the fire extended more than 2000 feet.... These iron nozzles, the flying powder halberds that were hurled, were what the Mongols feared most." These words were written by Father Antonine Gaubil in 1739 in conjunction with his book on Genghis Khan; they describe how a Chinese town in 1232 successfully defended itself against 30,000 invading Mongols by means of rocket-propelled fire arrows. They are an example of the evidence used by most historians to show that rocketry was born and developed in the Orient many centuries ago. It is reasonably clear that the Chinese manufactured black powder at least as early as 600 A.D. and subsequently used this mixture of charcoal, sulfur, and saltpeter as a rocket propellant. Over the centuries, the rocket slowly spread to the West as a military weapon and was much improved as a barrage missile by Sir William Congreve in England in the early 1800s. (The "rocket's red glare" observed by Francis Scott Key in 1812 at Fort McHenry was produced by a Congreve rocket.) However, not until the end of the 19th century and the beginning of the 20th century was the rocket understood from a technical viewpoint and was its true engineering development begun.

The Soviet Union was first into space, both with an artificial satellite (*Sputnik I* on October 4, 1957) and with a human in orbit (Yuri Gagarin on April 12, 1961). Thus, in historical perspective, it is fitting that the first true rocket scientist was a Russian—Konstantin Eduardovitch Tsiolkovsky, born in September 1857 in the town of Izhevskoye. As a young student, he absorbed physics and mathematics and was tantalized by the idea of interplanetary space travel. In 1876, he became a school teacher in Borovsk, and in 1882 he moved to the village of Kaluga. There, in virtual obscurity, he worked on theories of space flight, hitting upon the idea of reactive propulsion in March 1883. Working without any institutional support, Tsiolkovsky gradually solved some of the theoretical problems of rocket engines. Fig. 9.36 shows his design of a rocket, fueled with liquid hydrogen (H_2) and liquid oxygen (O_2), which was published in the Russian magazine *Science Survey* in 1903 (the same year as the Wright brothers' successful first powered airplane flight). The fact that Tsiolkovsky knew to use the high–specific impulse combination of H_2–O_2 testifies to

Figure 9.36 Tsiolkovsky's rocket design of 1903, burning liquid hydrogen (H) and liquid oxygen (O).

the sophistication of his rocket theory. Tsiolkovsky was neither an experimentalist (it took money that he did not have to develop a laboratory) nor an engineer. Therefore, he conducted no practical experiments and generated no design data. Nevertheless, Tsiolkovsky was the first true rocket scientist, and he worked incessantly on his theories until his death on September 19, 1935, at the age of 78 years. In his later life, his contributions were finally recognized, and he became a member of the Socialist Academy (forerunner of the U.S.S.R. Academy of Science) in 1919, with a subsequent grant of a government pension.

At the turn of the century, progress in rocketry arrived in the United States in the form of Dr. Robert H. Goddard. Goddard was born at Worcester, Massachusetts, on October 5, 1882. His life had many parallels to Tsiolkovsky's: He, too, was an avid physicist and mathematician; he, too, was convinced that rockets were the key to space flight; and he, too, worked in virtual obscurity for most of his life. But there was one sharp difference. Whereas Tsiolkovsky's contributions were purely theoretical, Goddard successfully molded theory into practice and developed the world's first liquid-fueled rocket that worked.

Goddard was educated completely at Worcester, graduating from South High School in 1904, obtaining a bachelor's degree from Worcester Polytechnic Institute in 1908, and earning a doctorate in physics at Clark University in 1911. Subsequently, he became a professor of physics at Clark, where he began seriously to apply science and engineering to his childhood dreams of space flight. He, too, determined that liquid H_2 and O_2 would be very efficient rocket propellants, and he pursued these ideas during a leave of absence at Princeton University during 1912–1913. In July 1914, he was granted patents on rocket combustion chambers, nozzles, propellant feed systems, and multistage rockets. In 1917, he obtained a small grant ($5000) from the Smithsonian Institution in Washington, which permanently entrenched him in a career of rocketry. This grant led to one of the most historic documents of rocket engine history, a monograph entitled *A Method of Reaching Extreme Altitudes*, published as part of the Smithsonian Miscellaneous Collections in 1919. This book was a scholarly and authoritative exposition of rocket principles, but at that time few people seized upon Goddard's ideas.

Goddard increased his laboratory activities back at Worcester in the early 1920s. Here, after many tests and much engineering development, Goddard successfully launched the world's first liquid-fuel rocket on March 16, 1926. A picture of Goddard standing beside this rocket is shown in Fig. 9.37. The vehicle was 10 ft long; the motor itself was at the very top (far above Goddard's head in Fig. 9.37) and was fed liquid oxygen and gasoline through two long tubes which led from the propellant tanks at the rear of the vehicle (below Goddard's arm level in the figure). The conical nose on the fuel tanks was simply a deflector to protect the tanks from the rocket nozzle exhaust. This rocket reached a maximum speed of 60 mi/h and flew 184 ft. Although modest in performance, this flight on March 16, 1926 was to rocketry what the Wright brothers' December 17, 1903, flight was to aviation.

This work ultimately brought Goddard to the attention of Charles A. Lindbergh, who now had considerable stature because of his 1927 trans-Atlantic flight. Lindbergh was subsequently able to convince the Daniel Guggenheim Fund for the Promotion of Aeronautics to give Goddard a $50,000 grant to further pursue rocket engine development. Suddenly, Goddard's operation magnified, and in 1930 he and his wife moved to a more suitable testing location near Roswell, New Mexico. Here, for the next 11 years, Goddard made bigger and better rockets, although still in an atmosphere of obscurity. The government was simply not interested in any form of jet propulsion research during the 1930s. Also Goddard was cut somewhat from the same mold as the Wright brothers; he imposed a blanket of secrecy on his data for fear of others pirating his designs. However, at the beginning of World War II, the government's interest in Goddard's work turned from cold to hot, and his complete operation, personnel and facilities, was moved to the Naval Engineering Experiment Station at Annapolis, Maryland. Here, until July 1945, this group developed jet-assisted takeoff units for seaplanes and worked on a variable-thrust rocket engine.

On August 10, 1945, Dr. Robert H. Goddard died in Baltimore. Recognition for his contributions and realization of their importance to the development of modern rocketry came late. Indeed, only in the political heat of the post-*Sputnik* years did the United States really pay homage to Goddard. In 1959 he was honored by Congress; the same year he received the first Louis W. Hill Space Transportation Award of the Institute of Aeronautical Sciences (now the American Institute of Aeronautics and Astronautics). Also, on May 1, 1959, the new NASA Goddard Space Flight Center at Greenbelt, Maryland, was named in his honor. Finally, in 1960, the Guggenheim Foundation and Mrs. Goddard were given $1,000,000 by the government for use of hundreds of Goddard's patents.

During the 1930s, and completely independent of Goddard's operation, another small group in the United States developed rockets. This was the American Rocket Society (ARS), originally founded in March 1930 as the American Interplanetary Society and changing its name in 1934. This was a small group of scientists and engineers who believed in the eventual importance of rocketry. The society not only published technical papers, but also built and tested actual vehicles. Its first rocket, burning liquid oxygen and gasoline, was launched on May 14, 1933, at Staten Island, New York, and reached 250 ft. Following this, and up to World War II, the ARS was a public focal point for small rocket research and development, all without government support. After the beginning of the war, much of the ARS experimental activity was splintered and absorbed by other activities around the country. However, as

Figure 9.37 Robert H. Goddard and his first successful liquid-fuel rocket. This rocket made the world's first successful flight on March 16, 1926. (*National Air and Space Museum.*)

an information dissemination society, the ARS continued until 1963, publishing the highly respected *ARS Journal*. Then the American Rocket Society and the Institute of Aeronautical (by that time, Aerospace) Sciences were merged to form the present American Institute of Aeronautics and Astronautics.

As a brief example of how the threads of the history of flight are woven together, in 1941 members of the ARS formed a company, Reaction Motors, Inc., which went on to design and build the XLR-11 rocket engine. This engine powered the Bell X-1 and pilot Chuck Yeager to the first manned supersonic flight on October 14, 1947 (see Sec. 5.22 and Fig. 5.82).

The early history of rocket engines forms a geographic triangle, with one vertex in Russia (Tsiolkovsky), the second in the United States (Goddard), and the third in Germany. Representing this third vertex is Hermann Oberth, born in Transylvania on July 25, 1894, to later become a German citizen. Like Tsiolkovsky and Goddard before him, Oberth found inspiration in the novels of Jules Verne and began a mental search for a practical means of reaching the moon. During World War I, Oberth became interested in rockets, suggesting long-range liquid-fueled missiles to the German war department. In 1922, he combined these thoughts and suggested rockets for space flight. Oberth was at that time ignorant of the work of both Tsiolkovsky and Goddard. However, shortly thereafter, Goddard's work was mentioned in the German newspapers, and Oberth quickly wrote for a copy of the 1919 Smithsonian monograph. In 1923, Oberth published his own work on the theory of rocket engines, entitled *The Rocket into Planetary Space*. This was a rigorous technical text, and it laid the basis for the development of rockets in Germany. To foster Oberth's ideas, the German Society for Space Travel was formed in 1927 and began experimental work in 1929. (The American Rocket Society was subsequently patterned after the German society.) Oberth's ideas had a catalytic effect, especially on some of his students, such as Wernher Von Braun, and the 1930s found an almost explosive development of rocketry in Germany. This work, with Von Braun as the technical director, culminated in the development of the German V-2 rocket of World War II. Although an instrument of war, the V-2 was the first practical long-range rocket in history. A sketch of the V-2 is shown in Fig. 9.38. Powered by liquid oxygen and alcohol, this rocket was 46.1 ft long, 65 in in diameter, and 27,000 lb in weight. It was the first vehicle made by humans to fly outside the sensible atmosphere (hence, in space), with altitudes above 50 mi and a range of 200 mi. The missile reached supersonic speeds during its flight within the atmosphere. During the closing phases of World War II, hundreds of production V-2s were captured by both Russian and U.S. forces and shipped back to their respective countries. As a result, all modern rockets today can trace their ancestry directly back to the V-2 and hence through Von Braun back to Hermann Oberth.

The development of modern rockets, culminating in the huge Saturn booster for the Apollo program, is a story in itself and is beyond the scope of this book. However, the early history sketched in this section is intended to add appreciation for the technical aspects of rocket engines discussed in Sec. 9.8. For an authoritative presentation on the history of modern rocketry, see the books by Von Braun and Ordway and by Emme (see Bibliography at the end of this chapter).

Figure 9.38 The German World War II V-2 rocket. (*NASA.*)

9.17 SUMMARY

A few important aspects of the chapter are itemized below:

1. The cross section of a propeller is an airfoil shape designed to produce an aerodynamic force in the direction of motion of the airplane, that is, thrust. The efficiency of a propeller depends on the pitch angle and the advance ratio $J = V_\infty/(nD)$.

2. The four strokes of an Otto-cycle reciprocating internal combustion engine are intake, compression, power, and exhaust. Combustion takes place essentially at constant volume. The power generated by such an engine along with a propeller is the power available, expressed as

$$P_A = \eta\eta_{\text{mech}}\frac{n}{2}NW \qquad \text{[9.10]}$$

where η = propeller efficiency, η_{mech} = mechanical efficiency, n = revolutions per second of the engine shaft, N = number of cylinders, and W = work produced during the complete four-stroke cycle. The power available can also be expressed as

$$P_A = \frac{\eta\eta_{\text{mech}}(\text{rpm})\,dp_e}{120} \qquad \text{[9.15]}$$

where rpm is the revolutions per minute of the engine shaft, d is the displacement, and p_e is the mean effective pressure.

3. The thrust equation for a jet propulsion device is

$$T = (\dot{m}_{\text{air}} + \dot{m}_{\text{fuel}})V_e - \dot{m}_{\text{air}}V_\infty + (p_e - p_\infty)A_e \qquad \text{[9.24]}$$

4. The turbojet engine process involves aerodynamic compression of the intake air in a diffuser, further compression in a rotating compressor, constant-pressure combustion in the burner, expansion through a turbine which drives the compressor, and further expansion through an exhaust nozzle. In a turbofan engine, a large ducted fan is mounted on the shaft ahead of the compressor, which accelerates a large mass of auxiliary air outside the core of the engine itself, thus producing more thrust with higher efficiency. The ramjet engine has no rotating machinery and produces its thrust by means of aerodynamic compression in an inlet diffuser of the incoming air, burned at constant pressure in the combustor and exhausted through a nozzle.

5. The thrust for a rocket engine is

$$T = \dot{m}V_e + (p_e - p_\infty)A_e \qquad \text{[9.28]}$$

A rocket carries its own fuel and oxidizer and is not dependent on atmospheric air for the generation of thrust.

6. The specific impulse is a direct measure of the efficiency of a rocket engine–propellant combination.

$$I_{\text{sp}} = \frac{T}{\dot{w}} = \frac{1}{g_0}\left\{\frac{2\gamma RT_0}{\gamma - 1}\left[1 - \left(\frac{p_e}{p_0}\right)^{(\gamma-1)/\gamma}\right]\right\}^{1/2} \qquad \text{[9.35]}$$

For a high specific impulse, the combustion temperature T_0 should be high and the molecular weight of the combustion gas should be low.

7. The rocket equation relates burnout velocity to the specific impulse and the initial-to-final mass ratio.

$$V_b = g_0 I_{sp} \ln \frac{M_i}{M_f} \qquad \textbf{[9.51]}$$

BIBLIOGRAPHY

Dommasch, D. O., S. S. Sherbey, and T. F. Connolly: *Airplane Aerodynamics*, 3d ed., Pitman, New York, 1961.

Emme, E. M.: *A History of Space Flight*, Holt, New York, 1965.

Frisbee, R. H.: "Spacecraft Propulsion Systems—What They Are and How They Work," *Foundation Astronautics Notebook-6*, World Space Foundation, Pasadena, CA, 1983, pp. 2–20.

Glauret, H.: *The Elements of Aerofoil and Airscrew Theory*, Macmillan, New York, 1943.

Gray, G. W.: *Frontiers of Flight*, Knopf, New York, 1948.

Hill, P. G., and C. R. Peterson: *Mechanics and Thermodynamics of Propulsion*, Addison-Wesley, Reading, MA, 1964.

Obert, E. F.: *Internal Combustion Engines and Air Pollution*, Intext, New York, 1973.

Sutton, G. P.: *Rocket Propulsion Elements*, 4th ed., Wiley, New York, 1976.

Von Braun, W., and F. I. Ordway: *History of Rocketry and Space Travel*, 3d rev. ed., Crowell, New York, 1975.

Walsh, J. E.: *One Day at Kitty Hawk*, Crowell, New York, 1975.

PROBLEMS

9.1 A reciprocating engine for light aircraft, modeled after the Avco Lycoming O-235 engine, has the following characteristics: bore = 11.1 cm, stroke = 9.84 cm, number of pistons = 4, compression ratio = 6.75, mechanical efficiency = 0.83. It is connected to a propeller with an efficiency of 0.85. If the fuel-to-air ratio is 0.06 and the pressure and temperature in the intake manifold are 1 atm and 285 K, respectively, calculate the power available from the engine-propeller combination at 2800 rpm.

9.2 For the engine in Prob. 9.1, calculate the mean effective pressure.

9.3 Consider a turbojet mounted on a stationary test stand at sea level. The inlet and exit areas are the same, both equal to 0.45 m^2. The velocity, pressure, and temperature of the exhaust gas are 400 m/s, 1.0 atm, and 750 K, respectively. Calculate the static thrust of the engine. (*Note:* Static thrust of a jet engine is that thrust produced when the engine has no forward motion.)

9.4 Consider a turbojet-powered airplane flying at a standard altitude of 40,000 ft at a velocity of 530 mi/h. The turbojet engine has inlet and exit areas of 13 and 10 ft^2, respectively. The velocity and pressure of the exhaust gas at the exit are 1500 ft/s and 450 lb/ft^2, respectively. Calculate the thrust of the turbojet.

9.5 Consider a turbojet in an airplane flying at standard sea level with a velocity of 800 ft/s. The pressure ratio across the compressor is 12.5:1. The fuel-to-air ratio (by mass) is 0.05. If the nozzle exhausts the flow to ambient pressure, calculate the gas temperature at the exit. (In solving this problem, assume the air in the diffuser is slowed to a very low velocity before entering the compressor. Also assume the heat released per pound of fuel is 1.4×10^7 ft $\cdot$ lb/lb$_m$.)

9.6 A small ramjet engine is to be designed for a maximum thrust of 1000 lb at sea level at a velocity of 950 ft/s. If the exit velocity and pressure are 2000 ft/s and 1.0 atm, respectively, how large should the inlet be designed?

9.7 The mass flow through a rocket engine is 25 kg/s. If the exit area, velocity, and pressure are 2 m^2, 4000 m/s, and 2×10^4 N/m^2, respectively, calculate the thrust at a standard altitude of 50 km.

9.8 Consider a rocket engine in which the combustion chamber pressure and temperature are 30 atm and 3756 K, respectively. The area of the rocket nozzle exit is 15 m^2 and is designed so that the exit pressure exactly equals ambient pressure at a standard altitude of 25 km. For the gas mixture, assume $\gamma = 1.18$ and the molecular weight is 20. At a standard altitude of 25 km, calculate (*a*) specific impulse, (*b*) exit velocity, (*c*) mass flow, (*d*) thrust, and (*e*) throat area.

9.9 In a given rocket engine, a mass flow of propellants equal to 87.6 lb$_m$/s is pumped into the combustion chamber, where the temperature after combustion is 6000°R. The combustion products have mixture values of $R = 2400$ ft $\cdot$ lb/(slug)(°R) and $\gamma = 1.21$. If the throat area is 0.5 ft^2, calculate the pressure in the combustion chamber.

9.10 Consider a rocket with kerosine-oxygen as the fuel-oxidizer combination. The ratio of initial weight before blastoff to the final weight at burnout is 5.5. Calculate the burnout velocity.

9.11 A rocket using hydrogen-oxygen as the fuel-oxidizer combination has a specific impulse of 360 s. Calculate the ratio of propellant mass to initial mass required to achieve a burnout velocity equal to the escape velocity from the earth.

9.12 Consider a solid propellant rocket engine with an end-burning configuration as shown in Fig. 9.28*a*. The solid propellant is ammonium nitrate. The burning characteristics of this propellant when the initial grain temperature is 60°F are given by the following measured data: linear burning rate is 0.04 in/sec at a combustion pressure of 500 lb/in^2, and 0.058 in/sec at a combustion pressure of 1000 lb/in^2. The rocket engine is operating at a combustion pressure of 1500 lb/in^2. Calculate the distance the burning surface will recede in 5 seconds.

9.13 Consider a two stage rocket with the following design characteristics. *First stage*: propellant mass = 7200 kg; structural mass = 800 kg. *Second stage*: propellant mass = 5400 kg; structural mass = 600 kg. The payload mass is 60 kg. The specific impulse for both stages is 275 sec. Calculate the final burnout velocity.

10

Flight Vehicle Structures and Materials

Forces cannot be seen, but the effects of forces can be seen or understood. It is the effect of a force that is of interest in structural analysis.

Frederick K. Teichmann, Professor of
Aeronautics and Astronautics,
New York University, 1968

10.1 INTRODUCTION

In the grand scheme of flight vehicles, the consideration of structural design and analysis plays a special role. No matter how good the aerodynamics, or how powerful the propulsion, or how spectacular the flight dynamics, if the vehicle does not structurally hold together, then all is for naught. Samuel Langley learned this the hard way, as the disaster shown in Fig. 1.21 attests. Return to the road map for this book shown in Fig. 2.1, and read again the discussion surrounding this road map. In our "introduction to flight" we are almost at the end of the map—at the box labeled *Structures*.

The subject of flight vehicle structures is wide and rich. In this chapter we only scratch the surface. The purpose of this chapter simply is to give you a perspective on structures relative to the other flight disciplines shown in Fig. 2.1. Also this chapter will be brief; the road map shown in Fig. 10.1 has only three destinations, but they are sufficient to introduce the subject. The book by Raymer listed in the Bibliography has a nice review of structures, and much of the following material is gleaned from this source.

Figure 10.1   Road map for Chap. 10.

10.2   SOME PHYSICS OF SOLID MATERIALS

Our previous discussions of aerodynamics were built upon the science of fluid dynamics. A study of flight vehicle structures is built upon the science of solid mechanics. In this section we introduce some of the basic physics associated with forces exerted on solid materials, and their response to these forces.

10.2.1   Stress

When a force is impressed on a solid material, it is transmitted through the material, much as an externally applied pressure is transmitted through a fluid. This behavior is a molecular effect. The molecules of the solid material, being closely packed together, experience strong intermolecular forces. When an external force is applied to the solid, the shape or size of the solid tends to change. The molecules of the solid material, being so closely locked together by the intermolecular force, resist this change in the form of an internal force. The molecules become slightly displaced, enough that the internal force builds up and reaches an equilibrium with the external force.

This internal force, per unit area, is called *stress*. There are three general classes of stress, as follows.

Compression   Consider a segment of a solid rod with an external force F imposed on one end, as shown in Fig. 10.2a. The force acts in a direction into the rod. For the rod segment to remain in equilibrium, there must be an equal and opposite force F imposed on the other end, as shown in Fig. 10.2a. This force is pressing into the segment; by definition, it is a *compressive force*. The cross-sectional area of the rod is A. By definition, the stress, denoted by σ, is

$$\sigma = \frac{F}{A} \qquad\qquad [10.1]$$

The dimensions of stress are force per unit area, such as pounds per square inch (psi or lb/in^2), pounds per square foot (lb/ft^2), and newtons per square meter (N/m^2). This compressive stress acts perpendicular to the cross-sectional area A.

Tension   Consider the segment of a rod shown in Fig. 10.2b. Here, the force F is acting away from the segment on both ends. This is called a *tensile force*. The stress

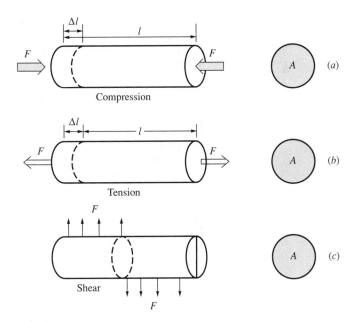

Figure 10.2 Three forms of stress.

in tension is also defined by Eq. (10.1). The tensile stress, as compressive stress does, acts perpendicular to the cross-sectional area A.

Shear Consider the segment of a rod shown in Fig. 10.2c. Here an external force F acts in the upward direction over the left part of the rod, and an equal and opposite force acts in the downward direction over the right part of the rod. These equal and opposite forces tend to slide the right part of the rod relative to the left part. This creates a *shear stress*, denoted by τ, defined as

$$\tau = \frac{F}{A} \qquad \textbf{[10.2]}$$

In contrast to compressive and tensile stress, which act perpendicular to the cross-sectional area A, shear stress acts *tangentially* to the cross-sectional area A. A classic example of shear is two plates fastened by a rivet, shown in Fig. 10.3. The force F acting to pull the plates apart creates a shear stress in the rivet.

10.2.2 Strain

When a compressive, tensile, or shear stress acts on a material, the material tends to change its shape and size. For example, return to Fig. 10.2a and b. When the compressive or tensile stress acts on the rod, the *length* of the rod changes. Let l denote the length of the rod before the stress acts on it, and let Δl be the change in

Figure 10.3   Rivet in shear.

length after the stress is imposed. By definition, the change in length per unit length is called *strain*, denoted by ϵ.

$$\epsilon \equiv \frac{\Delta l}{l} \qquad \textbf{[10.3]}$$

For most materials, up to a certain limiting value of the stress (the yield stress), the stress is directly proportional to the strain. The proportionality constant is defined as the *modulus of elasticity*, or *Young's modulus*, denoted by E. Hence

$$\boxed{\sigma = E\epsilon} \qquad \textbf{[10.4]}$$

Equation (10.4) is called *Hooke's law*. Since strain ϵ is a dimensionless quantity, the units of E are the same as the units of stress σ.

The deformation of the material due to shear is shown in Fig. 10.4, which is a greatly magnified view of the "kink" in the material that would be caused by the type of vertical shearing forces sketched in Fig. 10.2c. Note in Fig. 10.4 that the equal and opposite shear stresses acting on the vertical sides of the kinked segment set up a moment that acts in the counterclockwise direction on the segment. For the material to be in equilibrium, an equal and opposite moment must be set up by the induced shear stress τ_i acting on the top and bottom sides of the segment, as shown in Fig. 10.4. The measure of the deformation of the material is the angle θ, shown in Fig. 10.4. The angle θ is called the *shearing strain* and is given in units of radians. As in the case of compression and tension, up to a certain limit the shear stress is

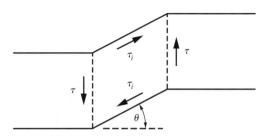

Figure 10.4 Deformation due to shear.

proportional to the strain. The proportionality constant is called the *shear modulus*, or *modulus of rigidity*, denoted by G. Hence,

$$\boxed{\tau = G\theta}$$

[10.5]

10.2.3 Other Cases

There are many types of loads exerted on a structure that cause compression, tension, and shear other than the simple cases illustrated in Fig. 10.2. For example, consider the cantilever beam shown in Fig. 10.5. In Fig. 10.5*a*, the beam is in its natural position—straight, and perpendicular to the wall. On the top and bottom surfaces, there are no compressive or tensile stresses. However, when the beam is bent upward by an applied load F, as shown in Fig. 10.5*b*, the top surface experiences a compressive stress and the bottom surface a tensile stress. These stresses are transferred to the juncture with the wall. This juncture must be able to handle these stresses as well as the shear stress at the wall caused by the upward-applied load. The bending case illustrated in Fig. 10.5 is not unlike that of an airplane wing bending upward under the influence of a lift force greater than the weight of the wing.

Another case is that of a cantilevered body being twisted by an applied torque, such as sketched in Fig. 10.6. The torque Q is applied to the body, and this results in a tangential shear stress induced in the body which resists the torque. This is not unlike the torsion created at the wing root of an airplane by the net aerodynamic moment exerted on the wing.

A third case is that of thermally induced stresses. Consider a body that is totally constrained, such as the beam supported between the two walls sketched in Fig. 10.7. Assume that heat is added to some location on the body. When the material gets warmer, its volume expands (thermal expansion). There is no room for it to move—it is constrained between two walls. As a result, a compressive stress is induced in the material of just the right value to produce a strain that cancels the thermal expansion. If the body is locally cooled instead, the thermal contraction is opposed by a tensile stress induced in the body. These are simple examples of *thermal stresses* induced in the material by local heating or cooling. Thermal stress is an important consideration

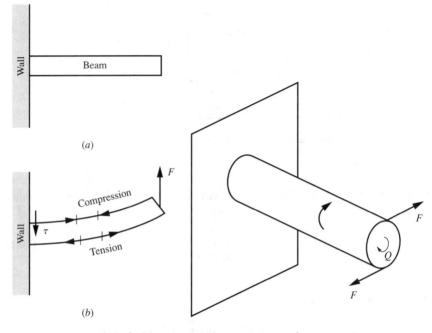

Figure 10.5 Bending of a beam. **Figure 10.6** Case of torsion.

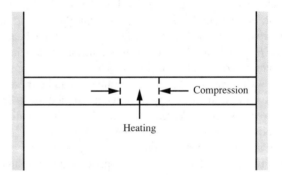

Figure 10.7 Example of thermal stress.

in the design of supersonic and hypersonic vehicles, where aerodynamic heating of the surface becomes a factor.

10.2.4 Stress-Strain Diagram

A generic diagram that plots stress versus strain for a given material is sketched in Fig. 10.8. As the stress is increased from zero, the strain follows in direct proportion as given by Eq. (10.4) or (10.5), up to a limiting value of stress called the *proportional*

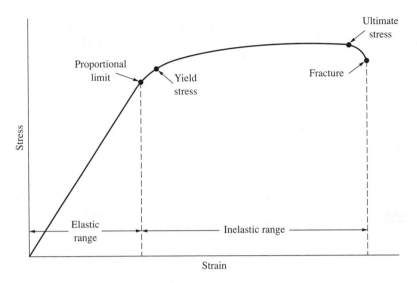

Figure 10.8 Stress-strain diagram.

limit. This regime in which stress varies linearly with strain is called the *elastic range*. When the stress at any point in this elastic range is relieved, the material returns to its original shape. There is no permanent structural damage. For stress levels above the proportional limit, there is a permanent set (a permanent deformation); this is called the *inelastic range*, as sketched in Fig. 10.8. In the inelastic range, when the stress is relieved, the material does not return to its original shape; this constitutes permanent structural damage. The point labeled *yield stress* in Fig. 10.8 is the shear stress at which the inelastic range effectively begins. The yield stress is rather arbitrarily defined as that stress which results in a permanent set of 0.002. Hooke's law, embodied in Eqs. (10.4) and (10.5), does not apply in the inelastic range. Finally, the *ultimate stress*, labeled in Fig. 10.8, is the highest stress the material can withstand. It is associated with a large permanent set, beyond which the material actually fractures. The fracture point is also labeled in Fig. 10.8.

DESIGN BOX

For aluminum alloys, the most common material used in aircraft structures, the ultimate stress is approximately 1.5 times the yield stress. Return to the *V-n* diagram shown in Fig. 6.55. Note the positive and negative limit load factors in the *V-n* diagram. These correspond to stress levels equal to the yield stress. An airplane designed not to exceed the limit load factor will not encounter the ultimate stress until the load factor is 1.5 times the limit factor. This provides a *safety factor* of 1.5—a much smaller value than that for earth-bound structures such as bridges, but typical for airplanes, where the saving of weight is paramount. Reflecting on the *V-n* diagram in Fig. 6.55, when an airplane exceeds its limit load factor, some elements of its structure will have a permanent deformation and must be repaired before the next flight.

The yield and ultimate stress, as well as the values of the modulus of elasticity E, and the shear modulus G, are tabulated below for several materials used in aircraft structures. In this tabulation, the following symbols are used:

σ_{tu} = ultimate tensile stress

σ_{ty} = yield tensile stress

σ_{cy} = yield compressive stress

σ_{su} = ultimate shear stress

Material	σ_{tu}	σ_{ty}	σ_{cy}	σ_{su}	E	G
	lb/in^2 × 10^{-3}				lb/in^2 × 10^{-6}	
Aircraft steel (5 Cr-M-V)	260	220	240	155	30	11
Stainless steel (AM-350)	185	150	158	120	29	11
Aluminum 2024	61	45	37	37	10.7	4
Titanium (Ti-6Al-4V)	160	145	154	100	16	6.2
Inconel X-750	155	100	100	101	31	11

Example 10.1

Consider a rod of stainless steel with a diameter of 0.75 in and a length of 100 in. The rod is in tension with a load of 12,000 lb. How much will the rod elongate under this load?

Solution

The cross-sectional area of the rod is

$$A = \frac{\pi d^2}{4} = \frac{\pi (0.75)^2}{4} = 0.442 \text{ in}^2$$

The tensile stress is

$$\sigma = \frac{F}{A} = \frac{12,000}{0.442} = 27,149 \text{ lb/in}^2$$

This is far below the tensile yield stress of 150,000 lb/in^2, as seen from the above tabulation. Hence, Hooke's law, Eq. (10.4), holds:

$$\sigma = E\epsilon$$

or

$$\epsilon = \frac{\sigma}{E} = \frac{27,149}{29 \times 10^6} = 9.36 \times 10^{-4}$$

Since $\epsilon = \Delta l/l$, the elongation is

$$\Delta l = \epsilon l = (9.36 \times 10^{-4})(100) = \boxed{0.0936 \text{ in}}$$

Consider an aluminum rod with a diameter of 1/4 in. A load of 2500 lb in tension is exerted on the rod. Will the rod experience a permanent set? | **Example 10.2**

Solution

The cross-sectional area of the rod is

$$A = \frac{\pi d^2}{4} = \frac{\pi (0.25)^2}{4} = 0.049 \text{ in}^2$$

The tensile stress in the rod is

$$\sigma = \frac{F}{A} = \frac{2500}{0.049} = 51{,}020 \text{ lb/in}^2$$

Comparing this stress with the tabulation above for aluminum 2024, we find that $\sigma_{ty} = 45{,}000$ lb/in^2 and $\sigma_{tu} = 61{,}000$ lb/in^2. The applied stress of 51,020 lb/in^2 is greater than the yield stress but less than the ultimate stress. Hence, the rod will experience a permanent set, but it will not fracture.

10.3   SOME ELEMENTS OF AN AIRCRAFT STRUCTURE

The structural design of an airplane is an intricate arrangement of various structural elements, such as wing spars, fuselage longerons, stringers, and stiffeners. This is certainly highlighted by the cutaway drawing of the B-17 shown in Fig. 2.15. Although a World War II vintage aircraft, the B-17 is a good example of a classical aircraft structure. Examine Fig. 2.15 closely, just to get a feel for what is meant by the term *aircraft structures*.

In this section, we identify the most basic elements of a conventional aircraft structure. Some of these are illustrated in Fig. 10.9, which shows a side view and top view of a generic airplane. The fuselage structure is characterized by *bulkheads*, which form the cross-sectional shape of the fuselage, and *longerons*, which are heavy strips that run the length of the fuselage and are attached to the outer edge of the bulkheads. The fuselage skin is attached to the longerons. The longerons carry to the wings the major loads exerted on the fuselage. They are usually I-shaped or H-shaped extrusions and are heavy elements of the fuselage structure. The primary wing structure is the *wing box*, shown as the crosshatched area in Fig. 10.9. The large moment created by the lift on the wing is transmitted by the wing box to the juncture with the fuselage, and this moment is carried across the fuselage by the *wing carry-through structure*.

The main element of the wing box is illustrated in Fig. 10.10, namely, the *wing spars*, which are large I-beams that run most of the span of the wing, with heights that reach from the bottom to the top surface of the wing. The spars are basically cantilever beams extending from the fuselage carry-through structure.

Three types of fuselage members are illustrated in Fig. 10.11. A keelson is shown in Fig. 10.11a; this is a strong beam placed at the bottom of the fuselage, much like the keel on a boat. The keelson is designed to carry the fuselage bending loads and is

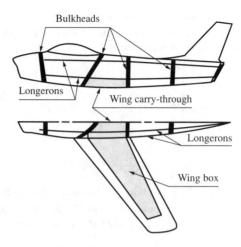

Figure 10.9   Some main structural elements for an airplane.

Figure 10.10   The wing box and carry-through structure.

frequently used in military fighter design. Another type of fuselage structural element is a stringer, illustrated in Fig. 10.11*b*. A stringer is a relatively light strip of small cross section that runs the length of the fuselage, and it is attached to the outer edge of the bulkhead. A large number of stringers are distributed over the circumference of the fuselage. Yet another fuselage element is a longeron, illustrated in Fig. 10.11*c* and mentioned earlier in conjunction with Fig. 10.9. Longerons are heavier than stringers, but lighter than keelsons.

Figure 10.12 shows a scale drawing of a cross section of the Concorde supersonic transport, illustrating one of the bulkheads, a section of the wing spar, and the carry-through structure. Figure 10.13 shows a detail of the wing structure for a small, general-aviation airplane, the Robin Hr100. Note the spars that run along the span of

Figure 10.11   Structural elements of the fuselage.

Figure 10.12   A fuselage bulkhead of the Concorde SST. From *The International Encyclopedia of Aviation*, ed. by David Mondey, Crown Publishers, New York, 1977.

Figure 10.13   Wing structure of the Robin Hr 100 general aviation airplane. From *The International Encyclopedia of Aviation*, ed. by David Mondey, Crown Publishers, New York, 1977.

the wing, and the airfoil sections that help to form the shape of the wing; these airfoil sections are called the *wing ribs*. Figures 10.12 and 10.13 give you an idea as to what some of these structural members actually look like. Note the cutout regions in some of these structures, such as the holes in the wing ribs—an effort to save weight.

10.4   MATERIALS

We are almost at the end of our brief journey through structures, as you can see from the short road map in Fig. 10.1. In this last section, we list and discuss some of the materials commonly used in flight structures.

Aluminum   Aluminum is the most widely used material in aircraft structures. Modern commercial transports such as the Boeing 747 use aluminum for about 80 percent of the structure. Aluminum is readily formed and machined, has reasonable cost, is corrosion-resistant, and has an excellent strength-to-weight ratio. In its pure form, aluminum is too soft for aircraft use. Therefore, alloys of aluminum are used, the most common being aluminum 2024, an alloy consisting of 93.5 percent aluminum, 4.4 percent copper, 1.5 percent manganese, and 0.6 percent magnesium. This alloy is also called *duralumin*. (Historical note: The first metal-covered airplanes were designed by Hugo Junkers, a professor at Aachen University in Germany, beginning in 1914. He first used all steel, which proved to be too heavy. In 1915, he turned to the use of duralumin and established a tradition that prevails to present day.)

Steel   For a typical commercial transport, steel makes up about 17 percent of the structure. It is used in those areas requiring very high strength, such as wing attachment fittings, landing gears, engine fittings, and flap tracks. Steel is an alloy of iron and carbon; typical steel alloys have about 1 percent carbon. Stainless steel is an alloy of steel and chromium that has good corrosion-resistant properties.

Titanium   Titanium has a better strength-to-weight ratio than aluminum and retains its strength at higher temperatures. However, it is hard to form and machine and is expensive, costing about 5 to 10 times more than aluminum. Titanium is rarely mined in the United States; most titanium comes from Russia. But some supersonic aircraft have to use titanium because of the high skin temperatures due to aerodynamic heating. The SR-71 is such a case. A cutaway drawing of the YF-12, an intercepter version of the SR-71, is shown in Fig. 10.14. This airplane cruises at Mach 3 and above, hence it was the first airplane to make extensive use of titanium. How to deal with titanium and how to obtain a sufficient supply were part of the challenge faced by the Lockheed Skunk Works engineers who designed the SF-71 and the YF-12. (See the historical case study of the SR-71 and YF-12 in the book by Anderson listed in the Bibliography at the end of this chapter.) Today, titanium is still a major consideration in the decision about the design Mach number of a second-generation supersonic transport. For an SST designed to cruise at Mach 2.2 or lower, aluminum can be used. But at a cruise Mach number of 2.4, the skin temperatures cross a threshold value such that titanium must be used.

High-Temperature Nickel Alloys   We note that hypersonic airplanes require advanced, high-temperature materials to withstand the high rates of aerodynamic heating at hypersonic speeds. Some nickel-based alloys are capable of withstanding the temperatures associated with moderate hypersonic speeds. The X-15, shown in Fig. 5.83, is

Figure 10.14   The stucture of the Lockheed YF-12.

such a case. This aircraft was designed to fly as fast as Mach 7; hence its structure made extensive use of inconel, a nickel-based alloy.

Composites   Composite materials are bringing about a revolution in aircraft structures because, for the same loads, a composite structure can yield at least a 25 percent reduction in weight. Composites are quite different from metals, in both their composition and physical properties, as discussed below.

From the *American Heritage Dictionary of the English Language*, the definition of the word *composite* is: "made up of distinct components." This is an apt descriptor of composite materials, which consist of a reinforcing material suspended in a matrix binder. A commonly used composite is graphite-epoxy, which uses carbon fibers as the reinforcing materials and an epoxy as the matrix binder. A schematic of a generic composite material is shown in Fig. 10.15, where the fibers are black and the matrix material is white. A structural part made of a composite material is usually molded, and then it can be cured at room or elevated temperatures and pressures to increase its strength. The arrangement of the fibers within the matrix can be of several types. Shown in Fig. 10.15 are alternate layers of fibers at 90° to each other. Some composites have the fibers running in just one direction, which is usually aligned with one of the principal axes to provide maximum strength in that direction. Other composites are made of alternate layers with the fibers oriented at 0°, 45°, and 90°; these alternate layers are called *plies*.

Composite materials usually follow a linear stress-strain relationship up to the proportional limit stress, as sketched in Fig. 10.16. However, composites have essentially no inelastic range compared to a metal, also shown in Fig. 10.16 for comparison. Rather, the composite will dramatically fracture at a stress very near the proportional stress. Hence, the factor of safety of 1.5 between yield stress and fracture which is built into a metal such as aluminum is not there for a composite material. Composite structures must have a safety factor explicitly included in the structural design.

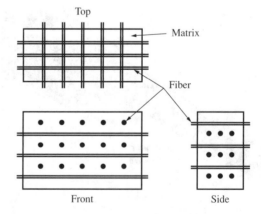

Figure 10.15   Generic sketch of a composite material.

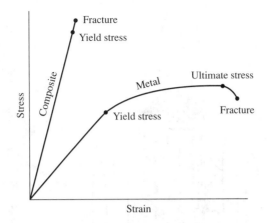

Figure 10.16   Stress-strain diagram comparing the behavior of composites with metals.

Modern airplanes are making extensive use of composite materials, and this trend will only grow in the future. For example, the Lockheed-Martin F-22, shown in Fig. 10.17, has 28 percent of its structure made up of composite material in the actual production aircraft, with the remaining being 33 percent aluminum, 24 percent titanium, 5 percent steel, and 10 percent miscellaneous. The Boeing 777, shown in Fig. 6.27, which is that company's latest production commercial transport at the time of writing, has 10 percent of its structural weight in composite material incorporated in such elements as the horizontal and vertical stabilizer torsion boxes; cabin floor beams; control surfaces such as the rudder, elevator, ailerons; the flaps, engine nacelles, landing gear doors, wing leading edge, and wing root fairing.

10.5   FATIGUE

Common experience shows that when you bend a piece of metal back and forth enough times, it will break. This is an example of *fatigue*. Various elements of an airplane structure are repeatedly being bent back and forth by the changing loads on the structure. The wing is bent upward by the high lift load at takeoff and is then bent downward upon landing by the weight of the wing after the lift is reduced to essentially zero. The structure goes through numerous load cycles in turbulent air. The fuselage experiences load cycles when the cabin pressure is increased and decreased. Military aircraft experience extreme load cycles during maneuvering in combat. So metal fatigue is a major concern in flight vehicles.

The nature of fatigue is shown in Fig. 10.18, which is a plot of the number of cycles that a metal structural element can experience before failure, as a function of the maximum stress imposed on it. Such information is determined empirically by *fatigue testing*. Figure 10.18 applies to fatigue tests of joints and splices using

Figure 10.17   The Lockheed-Martin F-22.

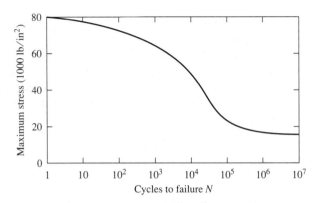

Figure 10.18   Fatigue test of joint and splices; 7075-76 aluminum alloy. After R. S. Shevell, *Fundamentals of Flight,* 2nd ed., Prentice Hall, New Jersey, 1989.

aluminum 7075-76. At the extreme left of Fig. 10.18, we see that a maximum stress of 80,000 lb/in² will break the material in just one cycle—this stress is, of course, the ultimate stress as defined in Sec.10.2. Fig. 10.18 shows that as the maximum stress is made smaller than the ultimate stress, then the metal can go through more load cycles before breaking. Note that by limiting the maximum stress levels in a given airplane structural design to much less than the ultimate stress, a large number of load cycles can be endured without structural failure. In airplane design, the structure is designed for certain extreme loads to be anticipated at the very edge of the airplane's operational envelope. Because these extreme loads are hardly encountered during routine operation of the airplane, the fatigue life of the structure is quite long.

The prediction of the fatigue life of various components of a flight vehicle structure is vital in the design and operation of an aircraft. This, in concert with regular airframe inspection and maintenance, will usually result in replacement of the component before it fails.

10.6   SOME COMMENTS

In this chapter we have just scratched the surface of the subjects of flight vehicle structural design and material selection. From our discussion, however, you can appreciate that an aircraft structure can be divided into different components, all integrated and working together. In a sense, an airplane fuselage is essentially a stiffened shell, and the wings are essentially a combination of stiffened plates and beams. In more advanced studies of flight vehicle structures, you will study in depth the structural nature of shells, plates, and beams. In further studies of materials, as applied to flight vehicles, you will be searching for materials with such favorable properties as high strength-to-weight ratio, ease of manufacture, low cost, good high-temperature performance, and long fatigue life.

And finally we mention the big picture. A flight vehicle is a finely tuned system in which aerodynamics, propulsion, flight dynamics, and structures are rolled into one "ball of wax," so to speak. The proper design and performance of the vehicle depend on all these disciplines working together. An airplane is analogous to a good stereo system, where the quality of the stereo is no better than the lowest quality of any one of its components. The quality of an airplane is no better than the lowest quality of its aerodynamics, propulsion, flight dynamics, and structures.

BIBLIOGRAPHY

Anderson, John D., Jr.: *Aircraft Performance and Design*, WCB/McGraw-Hill, New York, 1999.
Raymer, Daniel P.: *Aircraft Design: A Conceptual Approach*, 2d ed., American Institute of Aeronautics and Astronautics, Reston, VA, 1992.
Shevell, Richard S.: *Fundamentals of Flight*, 2d ed., Prentice-Hall, Englewood Cliffs, NJ, 1989, chap. 18.
Teichmann, Frederick, K.: *Fundamentals of Aircraft Structural Analysis*, Hayden Book Co., New York, 1968.

PROBLEMS

10.1   Consider two rods with identical length and diameter of 10 ft and 1 in respectively. One rod is AM-350 stainless steel, and the other is 2024 aluminum. Both rods are under the same tension load of 15,000 lb. Which rod will elongate the most, and by what amount compared to the other?

10.2   Consider a rod made from 2024 aluminum with a diameter of 0.5 in. What is the maximum load in tension that can be carried by this rod before it yields?

10.3   Consider an airplane with a tricycle landing gear stationary on the ground. The weight of the airplane is 5158 lb. Assume the nose wheel strut and the two main gear struts are perpendicular to the ground. The diameter of the nose wheel strut is 1 in; that of each main gear strut is 3 in. The nose wheel strut is located 5.62 ft ahead of the center of gravity of the airplane. The two main gear struts are located 1.12 ft behind the center of gravity. Calculate the compressive stresses in the nose wheel strut and the main gear struts.

10.4   Consider a kite flying in a wind, supported and restrained by a cord from the ground. The lift on the kite is 20 lb, and the cord makes an angle of 60° with the ground. The cord is taunt and straight. The diameter of the cord is 0.1 in. Calculate the tensile stress in the cord. Ignore the weight of the cord.

10.5   A one-foot square plate made of 2024 aluminum is subjected to a shear load along its edges. The displacement of one side relative to the opposite side is 0.1 in. Calculate the shear stress acting on the edges.

chapter

11

Hypersonic Vehicles

Within recent years the development of aircraft and guided missiles has brought a number of new aerodynamic problems into prominence. Most of these problems arise because of the extremely high flight velocities and are characteristically different in some way from the problems which arise in supersonic flight. The term "hypersonic" is used to distinguish flow fields, phenomena, and problems appearing at flight speeds far greater than the speed of sound from their counterparts appearing at flight speeds which are at most moderately supersonic. The appearance of new characteristic features in hypersonic flow fields justifies the use of a new term different from the well established term "supersonic."

Wallace D. Hayes and
Ronald F. Probstein, 1959

11.1 INTRODUCTION

The scene: A conventional airport for large jet aircraft, anywhere in the United States. *The time*: The 21st century. *The characters*: A flight crew, poised and ready for business. *The action*: The aircraft is ready, brimming with liquid hydrogen as fuel, and parked at the edge of the runway. The flight crew is notified and is rapidly transported from the airport terminal to the sleek, dartlike vehicle waiting on the runway. Within 30 minutes the aircraft takes off as a conventional airplane, but once in the air, the powerful air-breathing engines rapidly accelerate the vehicle through Mach 1. At Mach 5, supersonic combustion ramjet engines (Sec. 9.7 and Fig. 9.24) take over, and the aircraft continues to accelerate through the sensible atmosphere— Mach 10, Mach 15, Mach 20. When Mach 25 is reached, still within the sensible atmosphere at 200,000 ft, the vehicle has enough kinetic energy to coast into orbit

681

around the earth. It has done so strictly under the power of air-breathing propulsion and in a single stage after taking off from the airport. No rockets are used, and no intermediate propulsive stages were detached from the vehicle and dropped back to earth during the ascent. This airplane is simply a *single-stage-to-orbit vehicle.* A flight of fancy? A fantasy from the annals of science fiction? The author thinks not. The concept described above is that of a *transatmospheric vehicle.* Such a concept at various times at the end of the 20th century has been under active development in five different countries around the world, including a major program in the United States, where the vehicle was designated the *aerospace plane.* A general artist's concept of such an aerospace plane is shown in Fig. 11.1, and the flight trajectory of such a vehicle in Fig. 11.2 on a Mach number–altitude map (analogous to the velocity-altitude maps in Chap. 8). For comparison, the ascent and entry trajectories for the space shuttle are also shown in Fig. 11.2. Note that the ascent flight path for the aerospace plane takes place well below the shuttle ascent or entry, illustrating the need for the aerospace plane to stay within the sensible atmosphere so that the air-breathing engines can produce enough thrust to accelerate the vehicle to orbital velocity. This vehicle is a futuristic example of a *hypersonic* airplane, that is, an aircraft designed to fly faster than 5 times the speed of sound. Such hypersonic vehicles are the subject of this chapter.

Figure 11.1   Artist's concept of the National Aerospace Plane (NASP), a technology development program in the United States during 1985–1995 *(NASA).*

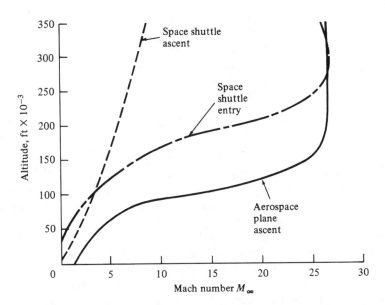

Figure 11.2 Flights paths for the ascent and entry of the space shuttle compared with the ascent path of an aerospace plane.

When such a transatmospheric vehicle flies successfully, it will by no means be the first hypersonic vehicle. The first time a piece of machinery ever flew faster than Mach 5 was on February 24, 1949, when a WAC Corporal second-stage rocket mounted on top of an old German V-2 rocket and launched from the White Sands proving ground in New Mexico achieved a top speed of 5150 m/h as it entered the atmosphere. (See Anderson, *Hypersonic and High-Temperature Gas Dynamics*, McGraw-Hill, 1988, for details.) By the 1950s, intercontinental ballistic missiles were flying at Mach 25 during entry tests of their nose cones. On April 12, 1961, the Russian astronaut Flight Major Yuri Gagarin became the first person to orbit the earth and hence to experience hypersonic flight at Mach 25 during entry. In the same year, on June 23, the X-15 hypersonic test aircraft (see Fig. 5.83) first exceeded Mach 5 in flight. In 1969 and the early 1970s, the Apollo lunar-return vehicles reached Mach 36 during entry into the earth's atmosphere. Thus we can clearly state that hypersonic flight is a reality and has been so since 1949.

Hypersonic aerodynamics and the impact it will have on the configuration of hypersonic vehicles are distinctly different from the lower supersonic regime, as noted in the passage quoted at the beginning of this chapter. Our purpose in this chapter is to describe briefly the physical aspects of hypersonic flow, to develop a simple but approximate aerodynamic theory for predicting pressure distributions on hypersonic vehicles, and to examine some of the performance and design aspects of such vehicles. By including this chapter in an introduction to aerospace engineering we are recognizing the importance of hypersonic flight in the past and assuming a continued growth of its importance in the future.

Figure 11.3   Road map for Chap. 11.

The road map for this chapter is shown in Fig. 11.3. It is a simple plan. First we examine some of the physical aspects of hypersonic flow, in keeping with the other chapters of this book where the fundamental physics of the given subject are emphasized. Then we examine a particular result from newtonian mechanics, based on Newton's study of fluid dynamics published in his *Principia* in 1687, called the newtonian sine-squared law. This law is useful for estimating pressure distributions on the surfaces of hypersonic vehicles. The chapter concludes with a discussion of some of the aerodynamic characteristics of hypersonic airplanes, including a presentation of a novel concept for the shape of such airplanes, called the wave rider.

11.2 PHYSICAL ASPECTS OF HYPERSONIC FLOW

Although it is generally accepted that hypersonic aerodynamics if defined as that part of the high-speed flight spectrum above Mach 5, this is no more than a rule of thumb; when a flow accelerates from $M = 4.99$ to $M = 5.01$, there is no clash of thunder or instant change of flow from green to red. No special flow phenomenon begins exactly at $M = 5.0$, in contrast to the distinct changes that occur when sonic flow, $M = 1.0$, is achieved. Instead hypersonic flow is best defined as that high–Mach number regime where certain physical flow phenomena become progressively more important as the Mach number is increased. In some cases, one or more of these phenomena become important above Mach 3, whereas in other cases they may not be compelling before Mach 7 or higher. Therefore, the designation of hypersonic flow as flow above Mach 5 is clearly just a convenient rule of thumb.

This section briefly describes the important physical aspects of hypersonic flow; in some sense this entire section will constitute a definition of hypersonic flow. Five main aspects that distinguish hypersonic flow from the lower-speed supersonic regime are described below.

11.2.1   Thin Shock Layers

Consider flow over a sharp wedge at two different Mach numbers: (1) a supersonic flow at $M_\infty = 2$ and (2) a hypersonic flow at $M_\infty = 20$. The shock waves and flow streamlines for these two cases are sketched in Fig. 11.4a and b, respectively. In both cases a straight oblique shock wave will emanate from the leading edge of the wedge,

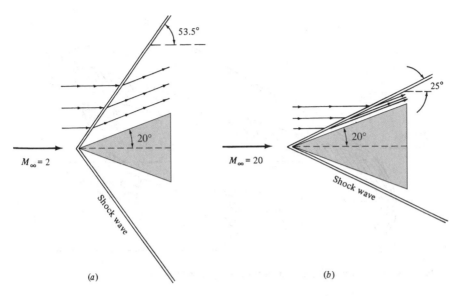

Figure 11.4 Shock waves and streamlines over a 20° half-angle wedge illustrating that hypersonic flows are characterized by thin shock layers.

as explained in Sec. 5.11. And in both cases the straight horizontal streamlines in the free stream ahead of the shock wave are discontinuously and uniformly bent in traversing the shock, the flow downstream consisting of straight uniform streamlines tangent to the wedge surface. However, at Mach 2 the shock wave angle is large (53.5°) whereas at Mach 20 the shock wave angle is much smaller (25°). The flow field between the shock wave and the body surface is called the *shock layer*, and we see from Fig. 11.4 that shock layers at hypersonic speeds are *thin*. A characteristic of hypersonic flow is that shock waves lie close to the surface, thus creating thin shock layers, which in turn can cause physical complications. For example, at low Reynolds number, the boundary layer on the body surface can grow quite thick, on the same order as the thickness of the thin shock layer itself. This leads to a merging of the shock wave with the boundary layer, constituting a *fully viscous shock layer*. However, the fact that the shock layer is thin allows the development of some simplified aerodynamic theories for the prediction of surface pressure at hypersonic speeds, one of which is described in Sec. 11.3.

11.2.2 Entropy Layer

Consider the hypersonic flow over a blunt-nosed body (Fig. 11.5). Consistent with the photographs in Fig. 4.23, the shock wave on a blunt-nosed body is slightly detached from the nose by the *shock detachment distance d* shown in Fig. 11.5. The shock wave curves downstream of the nose and at hypersonic speeds essentially wraps itself

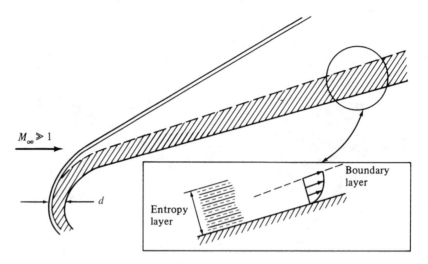

$M_\infty \gg 1$

d

Entropy
layer

Boundary
layer

Figure 11.5   Entropy layer on a blunt-nosed hypersonic body.

around the nose of the body. In the nose region the shock layer is very thin, and the shock wave is highly curved. This strong shock curvature induces large velocity gradients in the flow behind the shock in the nose region. These large velocity gradients are accompanied by strong thermodynamic changes in the flow. This region of strong gradients, called an *entropy layer*,[1] extends downstream close to the body surface. Downstream of the nose, the entropy layer interacts with the boundary layer growing along the surface; this interaction causes an increase in aerodynamic heating to the surface, above and beyond that which would be predicted without the entropy layer. At supersonic speeds the shock wave at the nose is also curved, but the magnitude of the curvature is far less than at hypersonic speeds. Since the strength of the entropy layer is related to shock curvature, the entropy-layer effect is primarily a hypersonic phenomenon.

11.2.3   Viscous Interaction

In Sec. 4.16 we stated that the thickness of the laminar boundary layer is inversely proportional to the square root of the Reynolds number. In addition, results for compressible-flow boundary layers show that the thickness is proportional to the Mach number squared. Hence,

$$\delta \propto \frac{M^2}{\sqrt{Re}}$$

---

[1] Entropy is a thermodynamic state variable alluded to in Sec. 4.6 but not defined in this book. Such matters are treated in the study of thermodynamics. Suffice it to say that the entropy varies greatly throughout the layer shown in Fig. 11.5.

As a result, at the high Mach numbers associated with hypersonic flows, δ can be very large. Indeed, for hypersonic vehicles flying at high altitudes and high Mach numbers (the upper right-hand portion of the map in Fig. 11.2), the boundary layer thickness can become so large that the flow outside the boundary layer, called the *inviscid flow*, is greatly affected. This creates a *viscous interaction*: The thick boundary layer flow affects the outer inviscid flow, and the changes in the inviscid flow feed back and influence the boundary layer growth. The practical consequence of viscous interaction on hypersonic vehicles is an increase in surface pressure and skin friction, leading to increased drag and increased aerodynamic heating. For example, consider the hypersonic flow over a sharp, flat plate sketched in Fig. 11.6. If the flow were inviscid, the pressure distribution over the flat plate would be constant and equal to the free-stream pressure p_∞, as shown by the dashed line in Fig. 11.6. However, in the real viscous flow over the flat plate, a boundary layer exists adjacent to the surface. At hypersonic speeds this boundary layer can be quite thick. In turn, the outer inviscid flow no longer sees a flat plate; instead it sees a body with some effective thickness induced by the thick boundary layer. (In boundary layer language, the effective thickness the boundary layer adds to a surface is called the *displacement thickness*.) In turn, the actual pressures exerted on the flat plate are higher than p_∞; the pressure on the surface induced by viscous interaction can be very high near the leading edge and then decreases downstream, as sketched by the solid curve in Fig. 11.6. The difference between the two curves in Fig. 11.6 is called the *induced-pressure increment*. Induced pressures near the leading edge of hypersonic vehicles generally tend to increase the drag.

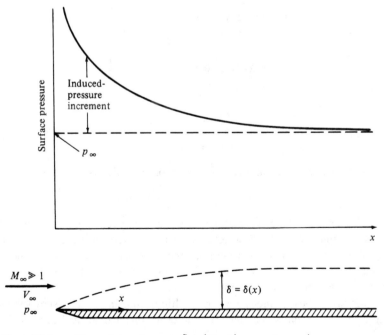

Figure 11.6   Viscous interaction on a flat plate at hypersonic speeds.

11.2.4 High-Temperature Effects

High–Mach number flows are high-energy flows; the ratio of kinetic energy to the gas internal energy increases as the square of the Mach number. This is easily seen by forming the ratio of kinetic energy to internal energy per unit mass of gas

$$\frac{V^2/2}{e} = \frac{V^2/2}{c_v T} = \frac{\gamma - 1}{2} \frac{V^2}{RT} = \frac{\gamma(\gamma - 1)}{2} \frac{V^2}{a^2} = \frac{\gamma(\gamma - 1)}{2} M^2 \qquad \textbf{[11.1]}$$

Therefore, a hypersonic free stream at $M_\infty = 20$ has a kinetic energy that is 112 times larger than its internal energy. However, when this flow enters a boundary layer (as in the flat plate in Fig. 11.6), it is slowed by the effects of friction. In such a case, the kinetic energy decreases rapidly and is converted in part to internal energy, which zooms in value. Since the gas temperature is proportional to internal energy, it also increases rapidly. Hence, hypersonic boundary layers are high-temperature regions of the flow, due to viscous dissipation of the flow kinetic energy. Another region of high-temperature flow is the shock layer behind the strong bow shock wave shown in Fig. 11.5. In this case, the flow velocity discontinuously decreases as it passes through the shock wave; once again, the lost kinetic energy reappears as an increase in internal energy and hence an increase in temperature behind the shock wave. Therefore, that portion of the shock layer behind a strong bow shock wave on a blunt-nosed body is a region of high-temperature flow.

High temperatures cause chemical reactions to occur in the flow. For example, in air when $T > 2000$ K, diatomic oxygen will dissociate

$$O_2 \rightarrow 2O$$

For $T > 4000$ K, diatomic nitrogen will dissociate

$$N_2 \rightarrow 2N$$

In this temperature range, nitric oxide will form

$$N_2 + O_2 \rightarrow 2NO$$

and will ionize

$$N + O \rightarrow NO^+ + e^-$$

At higher temperatures; the atoms will ionize; for example, for $T > 9000$ K,

$$O \rightarrow O^+ + e^- \qquad N \rightarrow N^+ + e^-$$

Clearly, a hypersonic flow can sometimes be a chemically reacting flow. In turn, these chemical reactions change the flow field properties and have an effect on aerodynamic heating of the surface. Since these high-temperature aspects are perhaps one of the dominant characteristics of hypersonic flow, any detailed study and analysis of a hypersonic flow should take them into account.

To emphasize these points, the velocity-altitude map in Fig. 11.7 shows the flight trajectories for lifting entry vehicles. Superimposed on the map are the regions where

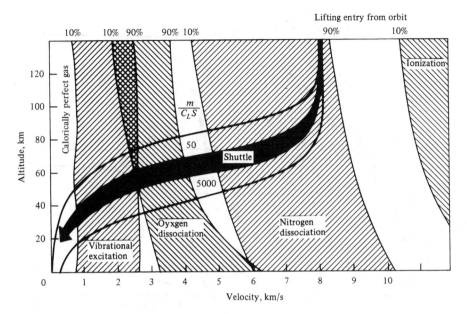

Figure 11.7 Velocity-altitude map showing where various chemical reactions are important in the blunt-nosed region of a hypersonic vehicle.

various chemical reactions occur around the nose of the vehicles. Clearly, much of the entry flight path is characterized by chemically reacting flow fields.

A typical variation of chemical species in the flow field around a blunt-nosed body is shown in Fig. 11.8. The body shape, shock wave shape, and two streamlines labeled A and B are shown in Fig. 11.8a for $V_\infty = 23,000$ ft/s at an altitude of 250,000 ft. The nose radius of the body is about 0.5 ft. In Fig. 11.8b the variation in concentration of atomic oxygen and atomic nitrogen along streamlines A and B is shown as a function of distance s along the streamlines. Note that dissociation occurs rapidly behind the shock wave and that large amounts of oxygen and nitrogen atoms are formed in the shock layer.

11.2.5 Low-Density Flow

Throughout this book we have treated air as a continuous medium. If you wave your hand through the air around you, the air feels like a continuous substance; but if you could go to an altitude of 300,000 ft and wave your hand about, the air would not feel so continuous. Your hand would begin to feel the influence of individual molecular impacts on the surface, and the air would seem to consist of distinct particles (molecules, atoms, ions, etc.) widely separated from one another. In this case, the air is no longer a continuous medium but is a gas at low density which exhibits certain special behavior. Let us examine this picture more closely.

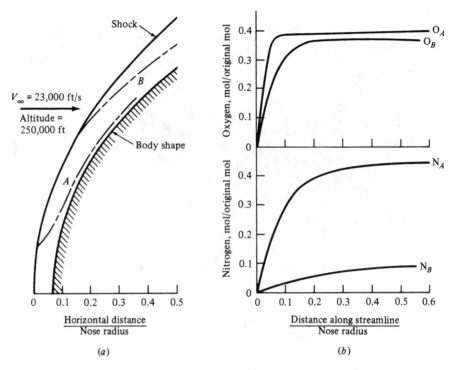

Figure 11.8   (a) Hypersonic flow over a blunt-nosed body, showing the shock wave, the body, and the shape of two streamlines labeled A and B. (b) Variation of concentrations of atomic oxygen and atomic nitrogen along the two streamlines in (a). Concentrations are given on the ordinates as moles of nitrogen or oxygen per original mole of air upstream of the shock wave. *(From J. G. Hall et al., " Blunt Nose Inviscid Airflows with Coupled Nonequilibrium Process," Journal of the Aeronautical Sciences, vol. 29, no. 9, September 1962, pp. 1038–1051.)*

The air around you is made up of individual molecules, principally oxygen and nitrogen, which are in random motion. Imagine that you can isolate one of these molecules and watch its motion. It will move a certain distance and then collide with one of its neighboring molecules. It will then move another distance and collide with another neighboring molecule, a process that will continue indefinitely. Although the distance between collisions is different, over time there will be some *average* distance that the molecule moves between successive collisions. This average distance is defined as the *mean free path* λ. At standard sea-level condition for air, $\lambda = 2.176 \times 10^{-7}$ ft, a very small distance. This implies that at sea level when you wave your hand through the air , the gas itself "feels" like a continuous medium, a so-called *continuum*. Imagine now that we raise ourselves to an altitude of 342,000 ft, where the air density is much lower and consequently the mean free path much larger than at sea level (at 342,000 ft, $\lambda = 1$ ft). Now, when you wave your hand through the air, you are more able to perceive individual molecular impacts; instead of a continuous substance, the air feels like an open region punctuated by individual widely spaced particles of matter. Under these conditions the aerodynamic concepts, equations, and results based on the assumption of a continuum begin to break down, and we must approach aerodynamics from a different point of view, using concepts from kinetic theory. This regime of aerodynamics is called *low-density flow*.

Whether or not low-density effects prevail for a given aerodynamic problem depends on the value of a nondimensional parameter called the *Knudson number K_n*, defined as

$$K_n = \frac{\lambda}{l}$$

where l is a characteristic dimension of the flow, for example, the length of a hypersonic vehicle or the diameter of a sphere. Continuum flow conditions will exist when $\lambda \ll l$, that is, when $K_n \ll 1$. Typically, $K_n < 0.03$ for continuum conditions at hold. At the other extreme when $\lambda \gg l$, we have *free-molecule flow*, that is, when $K_n > 10$. In free-molecule flow, a body surface feels only a small number of distinct molecular impacts. Moreover, the structure of the flow field becomes very blurred; for example, shock waves become very thick and essentially loose their identity. The aerodynamic force coefficients and surface heat transfer coefficients become strong functions of K_n (in addition to Mach number and Reynolds number), and the aerodynamic picture changes considerably. To illustrate such a change, recall from Example 8.3 that the drag coefficient for a sphere at hypersonic speeds is approximately 1. This is a continuum result, associated with $K_n \ll 1$. However, as K_n is increased C_D progressively increases, as shown in Fig. 11.9, approaching a value of 2 for free-molecule conditions, where $K_n > 10$.

Since low-density flows are not an inherent part of hypersonic flow, this discussion is not legitimately part of the definition of hypersonic flow. Nevertheless, hypersonic vehicles frequently fly at very high altitudes and therefore encounter low-density conditions. Hence, the design and analysis of hypersonic vehicles sometimes require the consideration of low-density flow. For example, the nose radius of the space shuttle is approximately 1 ft, and therefore at an altitude of 342,000 ft the value of the Knudsen number based on nose radius, $K_n = \lambda/R$ will be near unity. As a consequence, the flow in the nose region of the space shuttle encounters low-density effects at an altitude of approximately 300,000 ft, effects that are spread over the whole vehicle at higher altitudes. A glance at Fig. 11.2 shows that most of the important dynamics and aerodynamics for the entry of the space shuttle occur at altitudes below 300,000 ft, so that low-density effects are not a driving force in the performance of the shuttle. However, new generations of hypersonic airplanes may spend a considerable portion of their mission at high altitudes, and for these vehicles low-density effects will become more significant.

11.2.6 Recapitulation

To repeat, hypersonic flow is best defined as that regime where all or some of the physical phenomena discussed above become important as the Mach number is

Figure 11.9   Low-density effects on the drag coefficient of a sphere at hypersonic speeds; variation of C_D versus Knudsen number. *(The curve shown is from calculation made by Dr. James Moss at the NASA Langley Research Center.)*

increased to high values. For a vehicle of a given shape, some of these phenomena may begin to occur at Mach numbers below 5, whereas for other vehicles, the physical characteristics of hypersonic flow may not appear until Mach 7 or higher. We are therefore reminded once again that the definition of hypersonic flow as flight above Mach 5 is simply a convenient rule of thumb.

11.3   NEWTONIAN LAW FOR HYPERSONIC FLOW

In 1687 Newton published his famous *Principia*, which has formed the basis for all classical physics to the present. In the second book of the *Principia*, devoted to fluid mechanics, Newton postulated the following model of fluid flow. He considered a flow as a uniform rectilinear stream of particles, much like a cloud of pellets from a shotgun blast. As sketched in Fig. 11.10, Newton assumed that upon striking a surface inclined at a angle θ to the stream, the particles would transfer their normal momentum to the surface (thus exerting a force on it), but their tangential momentum would be preserved. Hence, the particles would move along the surface after colliding with it. For the inclined flat plate shown in Fig. 11.10, the force due to the loss of normal momentum by the impacting particles N is calculated as follows. The component

Figure 11.10   Model for the derivation of the newtonian sine-squared law.

of the free-stream velocity normal to the surface is $V_\infty \sin\theta$; according to Newton's model, this is the velocity lost by the particle upon impact with the surface. The area of the inclined surface A projects a cross-sectional area perpendicular to the flow equal to $A \sin\theta$ as show in Fig. 11.10. The mass flow across this area is the product of density, velocity, and projected area perpendicular to V_∞, as described in Sec. 4.1 and given by the product $\rho A V$ in Eq. (4.2). Therefore, we can write the following statement:

The time rate of change of momentum due to particles striking the surface is equal to

$$(\text{Mass flow}) \times (\text{change in normal component of velocity})$$

or $$(\rho_\infty V_\infty A \sin\theta)(V_\infty \sin\theta) = \rho_\infty V_\infty A \sin^2\theta$$

In turn, from Newton's second law, the force on the surface is equal to the time rate of change of momentum.

$$N = \rho_\infty V_\infty^2 A \sin^2\theta \qquad \qquad \textbf{[11.2]}$$

The force acts normal to the surface. From Eq. (11.2) the normal force per unit area is

$$\frac{N}{A} = \rho_\infty V_\infty^2 \sin^2\theta \qquad \qquad \textbf{[11.3]}$$

Let us now interpret the physical meaning of N/A, the normal force per unit area in Eq. (11.3) in terms of our modern knowledge of aerodynamics. Newton's model assumes a stream of individual particles all moving in straight parallel paths toward the surface; that is, the particles have a completely directed rectilinear motion. There is no random motion of the particles, which simply form a stream like pellets from a shotgun. In terms of our modern concepts, we know that a moving gas has molecular motion (composed of random motion of the molecules) as well as a directed motion. Moreover, as stated in Sec. 4.11, the static pressure (in this case, the static pressure of the free stream is p_∞) is simply a ramification of the purely *random* motion of the molecules. In Newton's model, there is no random motion, only directed motion.

Therefore, when the purely directed motion of the particles in Newton's model results in the normal force per unit area N/A in Eq. (11.3), this normal force per unit area must be construed as the *pressure difference* above p_∞, namely, $p - p_\infty$ on the surface. Hence, Eq. (11.3) becomes

$$p - p_\infty = \rho_\infty V_\infty^2 \sin^2 \theta \qquad \text{[11.4]}$$

This can be written in terms of the pressure coefficient C_ρ, defined by Eq. (5.27), as

$$\frac{p - p_\infty}{\frac{1}{2}\rho_\infty V_\infty^2} = 2 \sin^2 \theta$$

or

$$\boxed{C_p = 2 \sin^2 \theta} \qquad \text{[11.5]}$$

which is the famous *sine-squared law* of Newton. It allows us to calculate the pressure coefficient at a point on a surface where the angle between a tangent to the surface at that point and the free-stream direction is θ.

What does all this have to do with hypersonic flight or even with fluid mechanics in general? Equation (11.5) dates from the late 17th century, when hypersonic flight was not even a notion in anybody's mind. Newton's work on fluid mechanics was motivated by the need to calculate the resistance of bodies moving through fluids, such as a ship through water; and for that application, the newtonian sine-squared law was woefully inaccurate. The problem starts with the flow model itself (Fig. 11.10). In reality, in the low-speed flow of air or the flow of a liquid, the streamlines are not straight and parallel until they impact the body, as sketched in Fig. 11.10; actually the strealines begin to curve far ahead of the body and in general do not run into the surface of the body (usually only the single streamline through the stagnation point touches the body). Such real-flow phenomena are clear in the smoke photograph shown in Fig. 2.6. Therefore Eq. (11.5) is not expected to be an accurate result, and indeed our previous discussions of subsonic and supersonic aerodynamics have not used the newtonian sine-squared law. Now, however, let us return to the hypersonic flow pictured in Fig. 11.4b. If we look at it from across the room, the shock layer is so thin that it appears as if the straight parallel steamlines ahead of the shock wave were literally hitting the surface and then running tangentially along it. This is precisely the model used by Newton, as described earlier. Therefore, actual hypersonic flows come close to matching the newtonian model, with the result that the sine-squared law might be appropriate for estimating the pressure distributions over the surface of hypersonic vehicles. This indeed turns out to be the case, as shown in Fig. 11.11, where the surface-pressure distribution is given for a parabolically shaped axisymmetric body at Mach 4 in air. The solid line is from the exact numerical solution of the flow field obtained by the author on a high-speed digital computer, and the circles are from the sine-squared law, slightly modified from Eq. (11.5) as follows. In estimates of hypersonic pressure distributions, it is best to replace the pure number 2 in Eq. (11.5) with the value of the maximum pressure coefficient $C_{p,\max}$, which occurs at

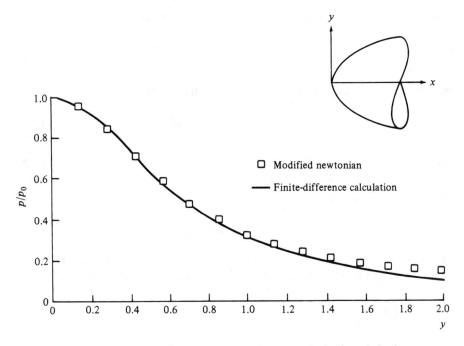

Figure 11.11 Surface pressure distribution on an axisymmetric body of parabolic shape, $M_\infty = 4$. Comparison between modified newtonian results and exact finite-difference calculations made on a high-speed digital computer. *(From Anderson, Modern Compressible Flow: With Historical Perspective, 2nd Edition McGraw-Hill, New York, 1990.)*

the stagnation point. That is, a *modified newtonian law* is

$$\boxed{C_p = C_{p,\text{max}} \sin^2 \theta} \qquad \text{[11.6]}$$

where

$$C_{p,\text{max}} = \frac{p_{0,2} - p_\infty}{\frac{1}{2} \rho_\infty V_\infty^2}$$

where $p_{0,2}$ is the total pressure behind a normal shock wave, given by the Rayleigh Pitot tube formula, Eq. (4.79). The squares in Fig. 11.11 are obtained from Eq. (11.6). Since excellent agreement is obtained with the exact results, the newtonian sine-squared law is useful for hypersonic applications.

Returning to Fig. 11.10, we calculate the lift and drag coefficients for the flat plate at an angle of attack α, using newtonian theory. For this case, since angle θ in Fig. 11.10 is the angle of attack, we will use α as usual to denote this angle, $\theta = \alpha$. From the geometry of Fig. 11.10

$$L = N \cos \alpha \qquad \text{[11.7]}$$

and

$$D = N \sin \alpha \qquad \text{[11.8]}$$

Substituting Eq. (11.2) in Eqs.(11.7) and (11.8), we find that

$$L = \rho_\infty V_\infty^2 A \sin^2 \alpha \cos \alpha \qquad \textbf{[11.9]}$$

and
$$D = \rho_\infty V_\infty^2 A \sin^3 \alpha \qquad \textbf{[11.10]}$$

In terms of lift and drag coefficients, Eqs.(11.9) and (11.10) become

$$C_L = \frac{L}{\frac{1}{2}\rho_\infty V_\infty^2 A} = 2 \sin^2 \alpha \cos \alpha \qquad \textbf{[11.11]}$$

and
$$C_D = \frac{D}{\frac{1}{2}\rho_\infty V_\infty^2 A} = 2 \sin^3 \alpha \qquad \textbf{[11.12]}$$

The lift-to-drag ratio become

$$\frac{L}{D} = \cot \alpha \qquad \textbf{[11.13]}$$

The results of Eqs. (11.11) to (11.13) are plotted in Fig. 11.12 as functions of the angle of attack. From this figure, note the following important characteristics:

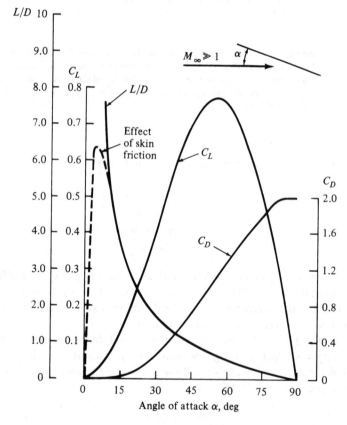

Figure 11.12   Newtonian results for lift and drag coefficient and lift-to-drag ratio for a flat plate as a function of angle of attack.

1. The lift coefficient increases gradually with angle of attack up to a high value of α. Indeed, maximum C_L occurs at $\alpha = 54.7°$, and C_L decreases for larger angles of attack. It is interesting to note that $\alpha \approx 55°$ for maximum lift is fairly realistic; the maximum lift coefficient for many practical hypersonic vehicles occurs at angles of attack in this neighborhood. The attainment of $C_{L,max}$ at such a high α at hypersonic speeds is certainly in contrast to our lower-speed experience discussed in Chap. 5, where it was seen that $C_{L,max}$ for subsonic airplanes occurs at values of α around 14 to 16°.

2. Another contrast between hypersonic conditions and our low-speed experience discussed earlier in this book is the variation of C_L versus α at low angle of attack, say, in the range of α from 0 to 15°. Note in Fig. 11.12 that the hypersonic C_L varies *nonlinearly* with α, in direct contrast to the linear variations seen at subsonic and supersonic speeds. From the point of view of theoretical aerodynamics, hypersonic flow is a very *nonlinear* phenomenon.

3. The value of L/D increases monotonically as α is decreases. Indeed, $L/D \rightarrow \infty$ as $\alpha \rightarrow 0$, but this is misleading; when skin friction is added to the picture, D becomes finite at $\alpha = 0$ and L/D reaches a maximum at some small angle of attack and then decreases to zero at $\alpha = 0$, as shown by the dashed line in Fig. 11.12, where laminar skin friction at a Reynolds number of 3×10^6 and a Mach number of 20 is assumed.

Consider the hypersonic flow over a sphere at Mach 25. Let s denote distance along the sphere surface, measured from the stagnation point, and let R denote the radius of the sphere. Point 1 is located a distance $s/R = 0.6$ from the stagnation point. Estimate the pressure coefficient at point 1.

Example 11.1

Solution

The location of point 1 is shown in Fig. 11.13; recalling that 1 rad is 57.3° and that ϕ in radians is given by s/R, we have, in degrees.

$$\phi = 57.3\frac{s}{R} = 57.3(0.6) = 34.38°$$

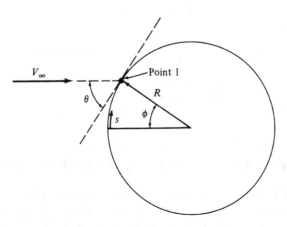

Figure 11.13 Geometry for Example 11.1.

In turn, the line tangent to the body at point 1 makes the angle θ with respect to the free-stream, where

$$\theta = 90° - \phi = 55.61°$$

From

$$\frac{p_{0,2}}{p_1} = \frac{p_{0,2}}{p_\infty}\left[\frac{(\gamma+1)^2 M_\infty^2}{4\gamma M_\infty^2 - 2(\gamma-1)}\right]^{\gamma/(\gamma-1)}\left[\frac{1-\gamma+2\gamma M_\infty^2}{\gamma+1}\right] \qquad \textbf{[4.79]}$$

where $p_{0,2}$ is the total pressure behind a normal shock wave (and hence the pressure at the stagnation point) and p_1 is the static pressure in the free stream ahead of the shock (that is, $p_1 = p_\infty$), we have for $\gamma = 1.4$ and $M_\infty = 25$

$$\frac{p_{0,2}}{p_\infty} = \left[\frac{2.4^2(25)^2}{4(1.4)(25)^2 - 2(0.4)}\right]^{1.4/0.4}\left[\frac{-0.4+2(1.4)(25)^2}{2.4}\right]$$

or

$$\frac{p_{0,2}}{p_\infty} = 1.1045(729) = 805.18$$

To convert the above ratio to a pressure coefficient, first note that the dynamic pressure can be written, using Eq. (4.53), as

$$q_\infty = \frac{1}{2}\rho_\infty V_\infty^2 = \frac{\gamma}{2}p_\infty\frac{\rho_\infty}{\gamma p_\infty}V_\infty^2 = \frac{\gamma}{2}p_\infty\frac{V_\infty^2}{a_\infty^2} = \frac{\gamma}{2}p_\infty M_\infty^2$$

Thus, from the definition of pressure coefficient

$$C_p = \frac{p - p_\infty}{q_\infty} = \frac{2}{\gamma M_\infty^2}\left(\frac{p}{p_\infty} - 1\right)$$

At the stagnation point $p = p_{0,2}$, and

$$C_{p,\text{max}} = \frac{2}{\gamma M_\infty^2}\left(\frac{p_{0,2}}{p_\infty} - 1\right) = \frac{2}{1.4(25)^2}(805.18 - 1)$$

or

$$C_{p,\text{max}} = 1.838$$

From the modified newtonian law, Eq. (11.6),

$$C_p = C_{p,\text{max}}\sin^2\theta$$

Evaluated at point 1,

$$C_p = C_{p,\text{max}}\sin^2\theta = 1.838\sin^2 55.62°$$

$$\boxed{C_p = 1.25}$$

11.4   SOME COMMENTS ON HYPERSONIC AIRPLANES

The infinitely thin, flat plate discussed in Sec. 11.3 is the most effective lifting surface at hypersonic speeds; the ratio of L/D from such a flat plate is the highest that can be expected at hypersonic flight conditions but the least effective in terms of volume capacity. It goes without saying that all practical flight vehicles must have a finite volume to carry fuel, payload, people, etc. Hence the flat-plate results, although instructive, are primarily of academic interest. This section briefly examines the characteristics of some more realistic hypersonic airplane configurations.

Figure 11.14 shows a three-view diagram of a conceptual hypersonic cruise aircraft, for example, a hypersonic transport. This NASA concept, in existence since the early 1970s, is a typical example of a hypersonic airplane configuration. The solid lines show the wing-body combination, which was tested in a hypersonic wind tunnel; the dashed lines show the propulsion module and vertical tail surface, which are part of the airplane design but were not included in the wind tunnel model.

The variation of lift coefficient with angle of attack for this aircraft is shown in Fig. 11.15 for $M_\infty = 8.0$. The solid lines are theoretical results calculated at two different Reynolds numbers, and the symbols are wind tunnel data. Note that

1. The lift coefficient varies *nonlinearly* with angle of attack, exhibiting concave curvature, a trend consistent with the flat-plate results shown in Fig. 11.12.

2. The lift coefficient is very insensitive to Reynolds number, a fact consistant with the low-speed experience discussed throughout this book.

Figure 11.14   A generic hypersonic transport configuration. The dimensions pertain to a wind tunnel model, the data for which are given in Figs. 11.15 to 11.17. Solid lines = wing-body model used for the wind tunnel tests; dashed lines = tail and propulsion modules for the complete configuration. *(From J. A. Penlund et al., Wall Temperature Effects on the Aerodynamics of a Hydrogen-Fueled Transport Concept in Mach 8 Blowdown and Shock Tunnels, NASA TP 2159, July 1983).*

Figure 11.15   Lift curve for the hypersonic transport configuration shown in Fig. 11.14. $M_\infty =$ 8. *(Wind tunnel data and theoretical curved from J. A. Penlund et al., Wall Temperature Effects on the Aerodynamics of a Hydrogen-Fueled Transport Concept in Mach 8 Blowdown and Shock Tunnels, NASA TP 2159, July 1983.)*

The lift-to-drag ratio versus angle of attack is given in Fig. 11.16. The two solid curves are theoretical results obtained assumed (1) turbulent flow at a high Reynolds number of 24.32×10^6 and (2) laminar flow at a low Reynolds number of 1.68×10^6. The wind tunnel data are partially bracketed by the theoretical curves and indicate that the actual flow on the model was transitional; that is, the flow near the nose and leading edges was laminar, followed by transition to turbulent flow. At the lower Reynolds numbers the flow was mainly laminar, whereas at the higher Reynolds numbers the flow was mainly turbulent. At high Mach numbers, transition to turbulent flow is usually delayed; hence, hypersonic flight vehicles frequently experience much larger regions of laminar flow than those expected at low speeds at the same Reynolds

Figure 11.16 Lift-to-drag ratio for the hypersonic transport configuration show in Fig. 11.14. $M_\infty = 8$. (Wind tunnel data and theoretical curves from J. A. Penlund et al., Wall Temperature Effects on the Aerodynamics of a Hydrogen-Fueled Transport Concept in Mach 8 Blowdown and Shock Tunnels, NASA TP 2159, July 1983.)

number. Note from Fig. 11.16 that the value of $(L/D)_{\max}$ is higher for turbulent flow than for laminar. At first this seems wrong; since in Chap. 4 we said that skin friction drag for a turbulent flow is much larger than for a laminar flow, L/D for turbulent flow should be much less. This would be true at the same Reynolds number. However, the two solid curves in Fig. 11.16 pertain to different Reynolds numbers. In Secs. 4.16 and 4.17, we saw that the skin friction coefficient decreases as the Reynolds number increases for both laminar and turbulent flows. Therefore, in Fig. 11.16 the turbulent curve corresponds to a lower skin friction drag coefficient C_F because the Reynolds number is so high (Re $= 24.32 \times 10^6$) whereas the laminar curve is given for a much lower Reynolds number (Re $= 1.68 \times 10^6$). In turn, the $(L/D)_{max}$ value is higher for the turbulent than for the laminar case. Note from Fig. 11.16 that

1. The L/D value is greatly affected by the Reynolds number.
2. Maximum L/D occurs in the angle-of-attack range of 3 to 5°.
3. The values of $(L/D)_{max}$ range from 4.5 to about 6, depending on the Reynolds number.

A drag polar is given in Fig. 11.17, plotted in the less conventional form of C_D versus C_L^2, in which the experimental data are almost linear, indicating that the drag

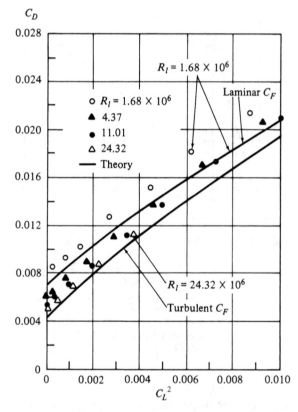

Figure 11.17   Drag polar for the hypersonic transport configuration shown in Fig. 11.14. $M_\infty = 8$. (Wind tunnel data and theoretical curver from J. A. Penlund et al., *Wall Temperature Effects on the Aerodynamics of a Hydrogen-Fueled Transport Concept in Mach 8 Blowdown and Shock Tunnels*, NASA TP 2159, July 1983.)

polar equation given by Eq. (6.1c) in the form of

$$C_D = C_{D,0} + rC_L^2 \qquad \textbf{[11.14]}$$

is reasonable valid at hypersonic speeds as well.

Let us return to a consideration of the lift-to-drag ratio L/D at hypersonic speeds. Note from Fig. 11.16 that the value of $(L/D)_{max}$ for the hypersonic airplane at Mach 8 is almost 6. Compare this with the typical subsonic airplane values of 14 to 17 (see, e.g., the L/D values shown in Figs.6.44 and 6.46). This is further dramatized by Table 11.1. It is a general trend that as the Mach number increases through the supersonic and hypersonic flight regimes, $(L/D)_{max}$ decreases. In fact, for $M_\infty > 1$, there is a general correlation for $(L/D)_{max}$ based on actual flight vehicle experience:

$$(L/D)_{max} = \frac{4(M_\infty + 3)}{M_\infty} \qquad \textbf{[11.15]}$$

This equation, which was first advanced in England by the famous airplane designer and aerodynamicist D. Kuchemann in 1978, is shown as the solid curve in Fig. 11.18, a plot of $(L/D)_{max}$ versus free-stream Mach number across the supersonic and hypersonic regime. Figure 11.18 also shows a shotgunlike scatter of open-circle data points corresponding to a variety of hypersonic vehicle designs; $(L/D)_{max}$ was obtained from wind tunnel tests, actual flight data, or theory. Details on these data points can be obtained from the references by Bowcutt and Anderson and by Corda and Anderson, listed in the Bibliography at the end of this chapter. The message from the solid curve and the open-circle data points in Fig. 11.18 is that high L/D is difficult to obtain at hypersonic speeds and that L/D decreases as Mach number increases. This natural phenomenon is due to the high drag associated with the strong shock waves and strong viscous effects encountered at hypersonic speeds. In some sense, the solid curve in Fig. 11.18 might be construed as a type of L/D barrier, which is hard to break at hypersonic speeds. Also note that although L/D decreases with Mach number, at high Mach numbers the rate of decrease becomes small; that is, the curve plateaus, and the variation of L/D with M_∞ becomes very small. Thus at high Mach number, L/D becomes almost independent of M_∞. This *Mach number–independence principle*, a basic principle of hypersonic aerodynamics, describes the

Table 11.1 Maximum lift-to-drag ratio for subsonic and supersonic aircraft

Airplane	$(L/D)_{max}$
North American P-51	14.6
Grumman F6F Hellcat	12.6
Boeing B-29	16.8
Beech Bonanza	13.8
Grumman A-6E	15.2
North American F-86	15.1
General Dynamics F-111	15.8
Hypersonic transport	6.0

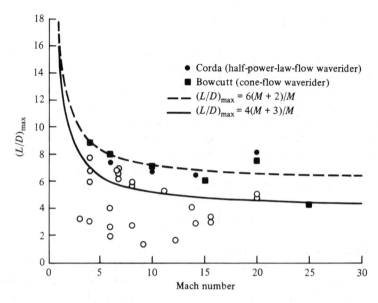

Figure 11.18   Comparison of the maximum lift-to-drag ratios for various hypersonic configurations. Solid symbols correspond to hypersonic wave riders generated by Corda, Bowcutt, and Anderson (see papers listed in the Bibliography).

fact that certain aerodynamic *coefficients*, for example, lift, drag, moment, and pressure coefficients, become relatively independent of Mach number when M_∞ is high enough (M_∞ greater than approximately 10). Mach number independence can be theoretically derived from the governing flow equations at hypersonic speeds; see the hypersonic text by Anderson listed in the Bibliography.

Returning to Fig. 11.18, we see that current research is aimed at breaking the L/D barrier discussed earlier. An example is a class of vehicles called *wave riders*, so named because they are designed to have an attached shock wave along their complete leading edge, so that it appears as if the vehicle were riding on top of its shock wave. An example of a modern wave rider shape is shown in Fig. 11.19, generated from the works of Bowcutt, Corda, and Anderson listed in the Bibliography. Although this rather complex and unusual shape is just an academic result today (no such wave riders have actually flown), the high predicted values of $(L/D)_{max}$ for wave riders are given in Fig. 11.18 by the solid symbols. Clearly, these theoretical results break the L/D barrier shown in Fig. 11.18. They are discussed here only as an example of the novel vehicle configurations that must be considered for efficient flight at hypersonic speeds.

As a final note in this section, we mention an important characteristic of any hypersonic airplane design, namely, the necessity to integrate the propulsion system fully with the airframe. For subsonic airplane design, some attention is always paid to the aerodynamic interaction between the engine nacelles and the rest of the air-

Figure 11.19 A typical wave rider configuration, designed for $M_\infty = 6$ (From Bowcutt and Anderson.)

frame. However, this is not a driving aspect of airplane design and in most subsonic airplanes the location of the engines is distinctly obvious; for example, in Fig. 6.11 the jet aircraft's engine nacelles are clearly evident and stand as a distinct component more or less by themselves. In contrast, at hypersonic speeds, extreme care must be taken to ensure that shock waves from one portion of the airplane, including the propulsion system, do not adversely impinge upon, and interact with, other portions of the airplane. Moreover, the flow that goes through the supersonic combustion ramjet engines has first passed through one or more shock wave systems from the forward portion of the vehicle, and it is necessary to tailor the aerodynamic properties of this air so as to encourage the most efficient engine performance. Therefore, for a hypersonic airplane, the propulsion system and the airframe must be highly integrated. An example appears in Fig. 9.24, which shows an airframe-integrated SCRAMjet engine. In the upper right-hand corner is a hypersonic airplane, and as explained at the end of Sec. 9.7, the entire undersurface of the complete airplane represents the whole SCRAMjet engine. In another example (Fig. 11.20) three typical generic configurations are compared: a Mach 3 supersonic transport, a Mach 6 hypersonic transport, and a Mach 12 hypersonic cruise vehicle. Note that the supersonic airplane (Fig. 11.20a) still has fairly distinct propulsion nacelles and that none of the engine exhaust is designed to touch the airframe. In contrast, the Mach 6 hypersonic transport

Figure 11.20   Comparison of high-speed airplane design from Mach 3 to 12. *(a)* Supersonic transport (M = 3), ram drag = 54,500 lb, gross thrust = 123,000 lb. *(b)* Hypersonic transport (M = 6), ram drag = 220,000 lb, gross thrust = 330,000 lb. *(c)* Hypersonic cruise vehicle (M = 12), ram drag = 1,950,000 lb, gross thrust = 2,100,000 lb. *(From Johnston et al.)*

(Fig. 11.20*b*) has a more fully integrated propulsion system, where the rear part of the airframe acts as part of the engine nozzle expansion. Also the wing and fuselage are more fully integrated, the wings being less distinct than for the Mach 3 airplane; that is, the Mach 6 airplane is more of a blended wing-body configuration than the supersonic airplane. At Mach 12 (Fig. 11.20*c*) these features are even more pronounced; the undersurface area of the airframe exposed to the engine exhaust is much greater, and the engine is much more an integral part of the airframe. Since the wings have become much smaller, the Mach 12 vehicle is more like a *lifting body* than a wing-body combination. Clearly, the design of hypersonic vehicles is a marked departure from conventional airplane design, and this will pose many interesting challenges for the aerospace engineers of the future.

11.5 SUMMARY

Hypersonic flow is that region of the high-speed flight spectrum in which the following physical phenomena become important as the Mach number increases to large values:

1. Thin shock layers
2. Entropy layers
3. Viscous interactions
4. High-temperature flow
5. Low-density flow

Depending on the vehicle size, shape, and altitude some of these hypersonic phenomena may occur at Mach numbers less than 5, while others may occur at Mach numbers greater than 5. As a rule of thumb only, hypersonic flow may be considered as flow where $M > 5$.

A convenient and sometimes reasonably accurate formula for predicting pressure distribution on the surface of hypersonic vehicles is the newtonian sine-squared law

$$C_p = 2 \sin^2 \theta \quad \text{original form} \qquad \textbf{[11.5]}$$
$$C_p = C_{p,\text{max}} \sin^2 \theta \quad \text{modified form} \qquad \textbf{[11.6]}$$

where $C_{p,\text{max}}$ is the pressure coefficient at a stagnation point and θ is the angle between a tangent a a given point on the surface and the free-stream direction.

Aerodynamic characteristics of hypersonic vehicles include the following:

1. Variation of C_L with angle of attack in nonlinear.
2. Maximum C_L usually occurs at a very high angle of attack, $\alpha \approx 55°$ or so.
3. Values of $(L/D)_{\text{max}}$ decrease as M_∞ increases. Hypersonic vehicles have lower values of $(L/D)_{\text{max}}$ than subsonic and supersonic vehicles.

BIBLIOGRAPHY

Anderson, John D., Jr.: *Hypersonic and High Temperature Gas Dynamics*, McGraw-Hill, New York, 1988.
Bowcutt, Kevin G., and John D. Anderson, Jr.: *Viscous Optimized Hypersonic Waveriders*, AIAA Paper 87-0272, 1987.
Corda, Stephen, and John D. Anderson, Jr.: *Viscous Optimized Hypersonic Waveriders Designed from Axisymmetric Flow Fields*, AIAA Paper 88-0369, 1988.
Hayes, Wallace D., and Ronald F. Probstein: *Hypersonic Flow Theory*, Academic Press, New York, 1959.
Johnston, Patrick J., and Allen H. Whitehead, Jr., and Gary T. Chapman: "Fitting Aerodynamic and Propulsion into the Puzzle," *Aerospace America*, vol. 25, no. 9, September 1987, pp. 32–37.

PROBLEMS

Consider a laminar boundary layer on a flat plate. At the trailing edge of the plate, with **11.1** a free-stream Mach number of 2, the boundary layer thickness in 0.3 in. Assuming the Reynolds number is held constant, calculate the boundary layer thickness for a Mach number of 20.

11.2   Consider a hypersonic vehicle flying at Mach 20 at a standard altitude of 59 km. Calculate the air temperature at a stagnation point on this vehicle. Comment on the accuracy of your answer.

11.3   Assume the nose of the space shuttle is spherical, with a nose radius of 1 ft. At Mach 18, calculate (*a*) the pressure coefficient at the stagnation point and (*b*) the pressure coefficient at a distance of 6 in away from the stagnation point measured along the surface.

11.4   Consider an infinitely thin, flat plate. Using newtonian theory, show that $C_{L,\max} = 0.77$ and that it occurs at $\alpha = 54.7°$.

11.5   Consider hypersonic flow over an infinitely thin, flat plate. The zero-lift drag coefficient is denoted by $C_{D,0}$. (Note that the zero-lift drag for a flat plate is entirely due to skin friction). Consider that the wave drag coefficient is given by the newtonian result for drag coefficient, that is, by Eq. (11.12). Also assume the lift coefficient is given by the newtonian result in Eq. (11.11). We wish to examine some results associated with $(L/D)_{\max}$ for this flat plate. Since $(L/D)_{\max}$ occurs at a small angle of attack, make the assumption of *small* α in Eqs. (11.11) and (11.12). Under these conditions show that at maximum L/D: (*a*) $(L/D)_{\max} = 0.67/C_{D,0}^{1/3}$ and occurs at $\alpha = C_{D,0}^{1/3}$; (*b*) wave drag coefficient $= 2C_{D,0}$.

appendix

A

Standard Atmosphere, SI Units

Altitude				
h_G, m	h, m	Temperature T, K	Pressure p, N/m^2	Density ρ, kg/m^3
−5,000	−5,004	320.69	1.7761 + 5	1.9296 + 0
−4,900	−4,904	320.03	1.7587	1.9145
−4,800	−4,804	319.38	1.7400	1.8980
−4,700	−4,703	318.73	1.7215	1.8816
−4,600	−4,603	318.08	1.7031	1.8653
−4,500	−4,503	317.43	1.6848	1.8491
−4,400	−4,403	316.78	1.6667	1.8330
−4,300	−4,303	316.13	1.6488	1.8171
−4,200	−4,203	315.48	1.6311	1.8012
−4,100	−4,103	314.83	1.6134	1.7854
−4,000	−4,003	314.18	1.5960 + 5	1.7698 + 0
−3,900	−3,902	313.53	1.5787	1.7542
−3,800	−3,802	312.87	1.5615	1.7388
−3,700	−3,702	212.22	1.5445	1.7234
−3,600	−3,602	311.57	1.5277	1.7082
−3,500	−3,502	310.92	1.5110	1.6931
−3,400	−3,402	310.27	1.4945	1.6780
−3,300	−3,302	309.62	1.4781	1.6631
−3,200	−3,202	308.97	1.4618	1.6483
−3,100	−3,102	308.32	1.4457	1.6336
−3,000	−3,001	307.67	1.4297 + 5	1.6189 + 0
−2,900	−2,901	307.02	1.4139	1.6044
−2,800	−2,801	306.37	1.3982	1.5900
−2,700	−2,701	305.72	1.3827	1.5757
−2,600	−2,601	305.07	1.3673	1.5615

Altitude		Temperature T, K	Pressure p, N/m^2	Density ρ, kg/m^3
h_G, m	h, m			
−2,500	−2,501	304.42	1.3521	1.5473
−2,400	−2,401	303.77	1.3369	1.5333
−2,300	−2,301	303.12	1.3220	1.5194
−2,200	−2,201	302.46	1.3071	1.5056
−2,100	−2,101	301.81	1.2924	1.4918
−2,000	−2,001	301.16	1.2778 + 5	1.4782 + 0
−1,900	−1,901	300.51	1.2634	1.4646
−1,800	−1,801	299.86	1.2491	1.4512
−1,700	−1,701	299.21	1.2349	1.4379
−1,600	−1,600	298.56	1.2209	1.4246
−1,500	−1,500	297.91	1.2070	1.4114
−1,400	−1,400	297.26	1.1932	1.3984
−1,300	−1,300	296.61	1.1795	1.3854
−1,200	−1,200	295.96	1.1660	1.3725
−1,100	−1,100	295.31	1.1526	1.3597
−1,000	−1,000	294.66	1.1393 + 5	1.3470 + 0
−900	−900	294.01	1.1262	1.3344
−800	−800	293.36	1.1131	1.3219
−700	−700	292.71	1.1002	1.3095
−600	−600	292.06	1.0874	1.2972
−500	−500	291.41	1.0748	1.2849
−400	−400	290.76	1.0622	1.2728
−300	−300	290.11	1.0498	1.2607
−200	−200	289.46	1.0375	1.2487
−100	−100	288.81	1.0253	1.2368
0	0	288.16	1.01325 + 5	1.2250 + 0
100	100	287.51	1.0013	1.2133
200	200	286.86	9.8945 + 4	1.2071
300	300	286.21	9.7773	1.1901
400	400	285.56	9.6611	1.1787
500	500	284.91	9.5461	1.1673
600	600	284.26	9.4322	1.1560
700	700	283.61	9.3194	1.1448
800	800	282.96	9.2077	1.1337
900	900	282.31	9.0971	1.1226

Altitude		Temperature T, K	Pressure p, N/m^2	Density ρ, kg/m^3
h_G, m	h, m			
1,000	1,000	281.66	8.9876 + 4	1.1117 + 0
1,100	1,100	281.01	8.8792	1.1008
1,200	1,200	280.36	8.7718	1.0900
1,300	1,300	279.71	8.6655	1.0793
1,400	1,400	279.06	8.5602	1.0687
1,500	1,500	278.41	8.4560	1.0581
1,600	1,600	277.76	8.3527	1.0476
1,700	1,700	277.11	8.2506	1.0373
1,800	1,799	276.46	8.1494	1.0269
1,900	1,899	275.81	8.0493	1.0167
2,000	1,999	275.16	7.9501 + 4	1.0066 + 0
2,100	2,099	274.51	7.8520	9.9649 − 1
2,200	2,199	273.86	7.7548	9.8649
2,300	2,299	273.22	7.6586	9.7657
2,400	2,399	272.57	7.5634	9.6673
2,500	2,499	271.92	7.4692	9.5696
2,600	2,599	271.27	7.3759	9.4727
2,700	2,699	270.62	7.2835	9.3765
2,800	2,799	269.97	7.1921	9.2811
2,900	2,899	269.32	7.1016	9.1865
3,000	2,999	268.67	7.0121 + 4	9.0926 − 1
3,100	3,098	268.02	6.9235	8.9994
3,200	3,198	267.37	6.8357	8.9070
3,300	3,298	266.72	6.7489	8.8153
3,400	3,398	266.07	6.6630	8.7243
3,500	3,498	265.42	6.5780	8.6341
3,600	3,598	264.77	6.4939	8.5445
3,700	3,698	264.12	6.4106	8.4557
3,800	3,798	263.47	6.3282	8.3676
3,900	3,898	262.83	6.2467	8.2802
4,000	3,997	262.18	6.1660 + 4	8.1935 − 1
4,100	4,097	261.53	6.0862	8.1075
4,200	4,197	260.88	6.0072	8.0222
4,300	4,297	260.23	5.9290	7.9376
4,400	4,397	259.58	5.8517	7.8536

Altitude		Temperature T, K	Pressure p, N/m^2	Density ρ, kg/m^3
h_G, m	h, m			
4,500	4,497	258.93	5.7752	7.7704
4,600	4,597	258.28	5.6995	7.6878
4,700	4,697	257.63	5.6247	7.6059
4,800	4,796	256.98	5.5506	7.5247
4,900	4,896	256.33	5.4773	7.4442
5,000	4,996	255.69	5.4048 + 4	7.3643 − 1
5,100	5,096	255.04	5.3331	7.2851
5,200	5,196	254.39	5.2621	7.2065
5,400	5,395	253.09	5.1226	7.0513
5,500	5,495	252.44	5.0539	6.9747
5,600	5,595	251.79	4.9860	6.8987
5,700	5,695	251.14	4.9188	6.8234
5,800	5,795	250.49	4.8524	6.7486
5,900	5,895	249.85	4.7867	6.6746
6,000	5,994	249.20	4.7217 + 4	6.6011 − 1
6,100	6,094	248.55	4.6575	6.5283
6,200	6,194	247.90	4.5939	6.4561
6,300	6,294	247.25	4.5311	6.3845
6,400	6,394	246.60	4.4690	6.3135
6,500	6,493	245.95	4.4075	6.2431
6,600	6,593	245.30	4.3468	6.1733
6,700	6,693	244.66	4.2867	6.1041
6,800	6,793	244.01	4.2273	6.0356
6,900	6,893	243.36	4.1686	5.9676
7,000	6,992	242.71	4.1105 + 4	5.9002 − 1
7,100	7,092	242.06	4.0531	5.8334
7,200	7,192	241.41	3.9963	5.7671
7,300	7,292	240.76	3.9402	5.7015
7,400	7,391	240.12	3.8848	5.6364
7,500	7,491	239.47	3.8299	5.5719
7,600	7,591	238.82	3.7757	5.5080
7,700	7,691	238.17	3.7222	5.4446
7,800	7,790	237.52	3.6692	5.3818
7,900	7,890	236.87	3.6169	5.3195
8,000	7,990	236.23	3.5651 + 4	5.2578 − 1
8,100	8,090	235.58	3.5140	5.1967

Altitude		Temperature T, K	Pressure p, N/m^2	Density ρ, kg/m^3
h_G, m	h, m			
8,200	8,189	234.93	3.4635	5.1361
8,300	8,289	234.28	3.4135	5.0760
8,400	8,389	233.63	3.3642	5.0165
8,500	8,489	232.98	3.3154	4.9575
8,600	8,588	232.34	3.2672	4.8991
8,700	8,688	231.69	3.2196	4.8412
8,800	8,788	231.04	3.1725	4.7838
8,900	8,888	230.39	3.1260	4.7269
9,000	8,987	229.74	3.0800 + 4	4.6706 − 1
9,100	9,087	229.09	3.0346	4.6148
9,200	9,187	228.45	2.9898	4.5595
9,300	9,286	227.80	2.9455	4.5047
9,400	9,386	227.15	2.9017	4.4504
9,500	9,486	226.50	2.8584	4.3966
9,600	9,586	225.85	2.8157	4.3433
9,700	9,685	225.21	2.7735	4.2905
9,800	9,785	224.56	2.7318	4.2382
9,900	9,885	223.91	2.6906	4.1864
10,000	9,984	223.26	2.6500 + 4	4.1351 − 1
10,100	10,084	222.61	2.6098	4.0842
10,200	10,184	221.97	2.5701	4.0339
10,300	10,283	221.32	2.5309	3.9840
10,400	10,383	220.67	2.4922	3.9346
10,500	10,483	220.02	2.4540	3.8857
10,600	10,582	219.37	2.4163	3.8372
10,700	10,682	218.73	2.3790	3.7892
10,800	10,782	218.08	2.3422	3.7417
10,900	10,881	217.43	2.3059	3.6946
11,000	10,981	216.78	2.2700 + 4	3.6480 − 1
11,100	11,081	216.66	2.2346	3.5932
11,200	11,180	216.66	2.1997	3.5371
11,300	11,280	216.66	2.1654	3.4820
11,400	11,380	216.66	2.1317	3.4277
11,500	11,479	216.66	2.0985	3.3743
11,600	11,579	216.66	2.0657	3.3217
11,700	11,679	216.66	2.0335	3.2699

Altitude		Temperature T, K	Pressure p, N/m^2	Density ρ, kg/m^3
h_G, m	h, m			
11,800	11,778	216.66	2.0018	3.2189
11,900	11,878	216.66	1.9706	3.1687
12,000	11,977	216.66	1.9399 + 4	3.1194 − 1
12,100	12,077	216.66	1.9097	3.0707
12,200	12,177	216.66	1.8799	3.0229
12,300	12,276	216.66	1.8506	2.9758
12,400	12,376	216.66	1.8218	2.9294
12,500	12,475	216.66	1.7934	2.8837
12,600	12,575	216.66	1.7654	2.8388
12,700	12,675	216.66	1.7379	2.7945
12,800	12,774	216.66	1.7108	2.7510
12,900	12,874	216.66	1.6842	2.7081
13,000	12,973	216.66	1.6579 + 4	2.6659 − 1
13,100	13,073	216.66	1.6321	2.6244
13,200	13,173	216.66	1.6067	2.5835
13,300	13,272	216.66	1.5816	2.5433
13,400	13,372	216.66	1.5570	2.5036
13,500	13,471	216.66	1.5327	2.4646
13,600	13,571	216.66	1.5089	2.4262
13,700	13,671	216.66	1.4854	2.3884
13,800	13,770	216.66	1.4622	2.3512
13,900	13,870	216.66	1.4394	2.3146
14,000	13,969	216.66	1.4170 + 4	2.2785 − 1
14,100	14,069	216.66	1.3950	2.2430
14,200	14,168	216.66	1.3732	2.2081
14,300	14,268	216.66	1.3518	2.1737
14,400	14,367	216.66	1.3308	2.1399
14,500	14,467	216.66	1.3101	2.1065
14,600	14,567	216.66	1.2896	2.0737
14,700	14,666	216.66	1.2696	2.0414
14,800	14,766	216.66	1.2498	2.0096
14,900	14,865	216.66	1.2303	1.9783
15,000	14,965	216.66	1.2112 + 4	1.9475 − 1
15,100	15,064	216.66	1.1923	1.9172
15,200	15,164	216.66	1.1737	1.8874
15,300	15,263	216.66	1.1555	1.8580

Altitude		Temperature T, K	Pressure p, N/m^2	Density ρ, kg/m^3
h_G, m	h, m			
15,400	15,363	216.66	1.1375	1.8290
15,500	15,462	216.66	1.1198	1.8006
15,600	15,562	216.66	1.1023	1.7725
15,700	15,661	216.66	1.0852	1.7449
15,800	15,761	216.66	1.0683	1.7178
15,900	15,860	216.66	1.0516	1.6910
16,000	15,960	216.66	1.0353 + 4	1.6647 − 1
16,100	16,059	216.66	1.0192	1.6388
16,200	16,159	216.66	1.0033	1.6133
16,300	16,258	216.66	9.8767 + 3	1.5882
16,400	16,358	216.66	9.7230	1.5634
16,500	16,457	216.66	9.5717	1.5391
16,600	16,557	216.66	9.4227	1.5151
16,700	16,656	216.66	9.2760	1.4916
16,800	16,756	216.66	9.1317	1.4683
16,900	16,855	216.66	8.9895	1.4455
17,000	16,955	216.66	8.8496 + 3	1.4230 − 1
17,100	17,054	216.66	8.7119	1.4009
17,200	17,154	216.66	8.5763	1.3791
17,300	17,253	216.66	8.4429	1.3576
17,400	17,353	216.66	8.3115	1.3365
17,500	17,452	216.66	8.1822	1.3157
17,600	17,551	216.66	8.0549	1.2952
17,700	17,651	216.66	7.9295	1.2751
17,800	17,750	216.66	7.8062	1.2552
17,900	17,850	216.66	7.6847	1.2357
18,000	17,949	216.66	7.5652 + 3	1.2165 − 1
18,100	18,049	216.66	7.4475	1.1975
18,200	18,148	216.66	7.3316	1.1789
18,300	18,247	216.66	7.2175	1.1606
18,400	18,347	216.66	7.1053	1.1425
18,500	18,446	216.66	6.9947	1.1247
18,600	18,546	216.66	6.8859	1.1072
18,700	18,645	216.66	6.7788	1.0900
18,800	18,745	216.66	6.6734	1.0731
18,900	18,844	216.66	6.5696	1.0564

716

Appendix A

Altitude		Temperature T, K	Pressure p, N/m^2	Density ρ, kg/m^3
h_G, m	h, m			
19,000	18,943	216.66	6.4674 + 3	1.0399 − 1
19,100	19,043	216.66	6.3668	1.0238
19,200	19,142	216.66	6.2678	1.0079
19,300	19,242	216.66	6.1703	9.9218 − 2
19,400	19,341	216.66	6.0744	9.7675
19,500	19,440	216.66	5.9799	9.6156
19,600	19,540	216.66	5.8869	9.4661
19,700	19,639	216.66	5.7954	9.3189
19,800	19,739	216.66	5.7053	9.1740
19,900	19,838	216.66	5.6166	9.0313
20,000	19,937	216.66	5.5293 + 3	8.8909 − 2
20,200	20,136	216.66	5.3587	8.6166
20,400	20,335	216.66	5.1933	8.3508
20,600	20,533	216.66	5.0331	8.0931
20,800	20,732	216.66	4.8779	7.8435
21,000	20,931	216.66	4.7274	7.6015
21,200	21,130	216.66	4.5816	7.3671
21,400	21,328	216.66	4.4403	7.1399
21,600	21,527	216.66	4.3034	6.9197
21,800	21,725	216.66	4.1706	6.7063
22,000	21,924	216.66	4.0420 + 3	6.4995 − 2
22,200	22,123	216.66	3.9174	6.2991
22,400	22,321	216.66	3.7966	6.1049
22,600	22,520	216.66	3.6796	5.9167
22,800	22,719	216.66	3.5661	5.7343
23,000	22,917	216.66	3.4562	5.5575
23,200	23,116	216.66	3.3497	5.3862
23,400	23,314	216.66	3.2464	5.2202
23,600	23,513	216.66	3.1464	5.0593
23,800	23,711	216.66	3.0494	4.9034
24,000	23,910	216.66	2.9554 + 3	4.7522 − 2
24,200	24,108	216.66	2.8644	4.6058
24,400	24,307	216.66	2.7761	4.4639
24,600	24,505	216.66	2.6906	4.3263
24,800	24,704	216.66	2.6077	4.1931
25,000	24,902	216.66	2.5273	4.0639

Altitude		Temperature T, K	Pressure p, N/m^2	Density ρ, kg/m^3
h_G, m	h, m			
25,200	25,100	216.96	2.4495	3.9333
25,400	25,299	217.56	2.3742	3.8020
25,600	25,497	218.15	2.3015	3.6755
25,800	25,696	218.75	2.2312	3.5535
26,000	25,894	219.34	2.1632 + 3	3.4359 − 2
26,200	26,092	219.94	2.0975	3.3225
26,400	26,291	220.53	2.0339	3.2131
26,600	26,489	221.13	1.9725	3.1076
26,800	26,687	221.72	1.9130	3.0059
27,000	26,886	222.32	1.8555	2.9077
27,200	27,084	222.91	1.7999	2.8130
27,400	27,282	223.51	1.7461	2.7217
27,600	27,481	224.10	1.6940	2.6335
27,800	27,679	224.70	1.6437	2.5484
28,000	27,877	225.29	1.5949 + 3	2.4663 − 2
28,200	28,075	225.89	1.5477	2.3871
28,400	28,274	226.48	1.5021	2.3106
28,600	28,472	227.08	1.4579	2.2367
28,800	28,670	227.67	1.4151	2.1654
29,000	28,868	228.26	1.3737	2.0966
29,200	29,066	228.86	1.3336	2.0301
29,400	29,265	229.45	1.2948	1.9659
29,600	29,463	230.05	1.2572	1.9039
29,800	29,661	230.64	1.2208	1.8440
30,000	29,859	231.24	1.1855 + 3	1.7861 − 2
30,200	30,057	231.83	1.1514	1.7302
30,400	30,255	232.43	1.1183	1.6762
30,600	30,453	233.02	1.0862	1.6240
30,800	30,651	233.61	1.0552	1.5735
31,000	30,850	234.21	1.0251	1.5278
31,200	31,048	234.80	9.9592 + 2	1.4777
31,400	31,246	235.40	9.6766	1.4321
31,600	31,444	235.99	9.4028	1.3881
31,800	31,642	236.59	9.1374	1.3455
32,000	31,840	237.18	8.8802 + 2	1.3044 − 2
32,200	32,038	237.77	8.6308	1.2646

Altitude		Temperature T, K	Pressure p, N/m^2	Density ρ, kg/m^3
h_G, m	h, m			
32,400	32,236	238.78	8.3890	1.2261
32,600	32,434	238.96	8.1546	1.1889
32,800	32,632	239.55	7.9273	1.1529
33,000	32,830	240.15	7.7069	1.1180
33,200	33,028	240.74	7.4932	1.0844
33,400	33,225	214.34	7.2859	1.0518
33,600	33,423	241.93	7.0849	1.0202
33,800	33,621	242.52	6.8898	9.8972 − 3
34,000	33,819	243.12	6.7007 + 2	9.6020 − 3
34,200	34,017	243.71	6.5171	9.3162
34,400	34,215	244.30	6.3391	9.0396
34,600	34,413	244.90	6.1663	8.7720
34,800	34,611	245.49	5.9986	8.5128
35,000	34,808	246.09	5.8359	8.2620
35,200	35,006	246.68	5.6780	8.0191
35,400	35,204	247.27	5.5248	7.7839
35,600	35,402	247.87	5.3760	7.5562
35,800	35,600	248.46	5.2316	7.3357
36,000	35,797	249.05	5.0914 + 2	7.1221 − 3
36,200	35,995	249.65	4.9553	6.9152
36,400	36,193	250.24	4.8232	6.7149
36,600	36,390	250.83	4.6949	6.5208
36,800	36,588	251.42	4.5703	6.3328
37,000	36,786	252.02	4.4493	6.1506
37,200	36,984	252.61	4.3318	5.9741
37,400	37,181	253.20	4.2176	5.8030
37,600	37,379	253.80	4.1067	5.6373
37,800	37,577	254.39	3.9990	5.4767
38,000	37,774	254.98	3.8944 + 2	5.3210 − 3
38,200	37,972	255.58	3.7928	5.1701
38,400	38,169	256.17	3.6940	5.0238
38,600	38,367	256.76	3.5980	4.8820
38,800	38,565	257.35	3.5048	4.7445
39,000	38,762	257.95	3.4141	4.6112
39,200	38,960	258.54	3.3261	4.4819
39,400	39,157	259.13	3.2405	4.3566

Altitude		Temperature T, K	Pressure p, N/m^2	Density ρ, kg/m^3
h_G, m	h, m			
39,600	39,355	259.72	3.1572	4.2350
39,800	39,552	260.32	3.0764	4.1171
40,000	39,750	260.91	2.9977 + 2	4.0028 − 3
40,200	39,947	261.50	2.9213	3.8919
40,400	40,145	262.09	2.8470	3.7843
40,600	40,342	262.69	2.7747	3.6799
40,800	40,540	263.28	2.7044	3.5786
41,000	40,737	263.87	2.6361	3.4804
41,200	40,935	264.46	2.5696	3.3850
41,400	41,132	265.06	2.5050	3.2925
41,600	41,300	265.65	2.4421	3.2027
41,800	41,527	266.24	2.3810	3.1156
42,000	41,724	266.83	2.3215 + 2	3.0310 − 3
42,400	41,922	267.43	2.2636	2.9489
42,400	42,119	268.02	2.2073	2.8692
42,600	42,316	268.61	2.1525	2.7918
42,800	42,514	269.20	2.0992	2.7167
43,000	42,711	269.79	2.0474	2.6438
43,200	42,908	270.39	1.9969	2.5730
43,400	43,106	270.98	1.9478	2.5042
43,600	43,303	271.57	1.9000	2.4374
43,800	43,500	272.16	1.8535	2.3726
44,000	43,698	272.75	1.8082 + 2	2.3096 − 3
44,200	43,895	273.34	1.7641	2.2484
44,400	44,092	273.94	1.7212	2.1889
44,600	44,289	274.53	1.6794	2.1312
44,800	44,486	275.12	1.6387	2.0751
45,000	44,684	275.71	1.5991	2.0206
45,200	44,881	276.30	1.5606	1.9677
45,400	45,078	276.89	1.5230	1.9162
45,600	45,275	277.49	1.4865	1.8662
45,800	45,472	278.08	1.4508	1.8177
46,000	45,670	278.67	1.4162 + 2	1.7704 − 3
46,200	45,867	279.26	1.3824	1.7246
46,400	46,064	279.85	1.3495	1.6799
46,600	46,261	280.44	1.3174	1.6366

Altitude		Temperature T, K	Pressure p, N/m^2	Density ρ, kg/m^3
h_G, m	h, m			
46,800	46,458	281.03	1.2862	1.5944
47,000	46,655	281.63	1.2558	1.5535
47,200	46,852	282.22	1.2261	1.5136
47,400	47,049	282.66	1.1973	1.4757
47,600	47,246	282.66	1.1691	1.4409
47,800	47,443	282.66	1.1416	1.4070
48,000	47,640	282.66	1.1147 + 2	1.3739 − 3
48,200	47,837	282.66	1.0885	1.3416
48,400	48,034	282.66	1.0629	1.3100
48,600	48,231	282.66	1.0379	1.2792
48,800	48,428	282.66	1.0135	1.2491
49,000	48,625	282.66	9.8961 + 1	1.2197
49,200	48,822	282.66	9.6633	1.1910
49,400	49,019	282.66	9.4360	1.1630
49,600	49,216	282.66	9.2141	1.1357
49,800	49,413	282.66	8.9974	1.1089
50,000	49,610	282.66	8.7858 + 1	1.0829 − 3
50,500	50,102	282.66	8.2783	1.0203
51,000	50,594	282.66	7.8003	9.6140 − 4
51,500	51,086	282.66	7.3499	9.0589
52,000	51,578	282.66	6.9256	8.5360
52,500	52,070	282.66	6.5259	8.0433
53,000	52,562	282.66	6.1493	7.5791
53,500	53,053	282.42	5.7944	7.1478
54,000	53,545	280.21	5.4586	6.7867
54,500	54,037	277.99	5.1398	6.4412
55,000	54,528	275.78	4.8373 + 1	6.1108 − 4
55,500	55,020	273.57	4.5505	5.7949
56,000	55,511	271.36	4.2786	5.4931
56,500	56,002	269.15	4.0210	5.2047
57,000	56,493	266.94	3.7770	4.9293
57,500	56,985	264.73	3.5459	4.6664
58,000	57,476	262.52	3.3273	4.4156
58,500	57,967	260.31	3.1205	4.1763
59,000	58,457	258.10	2.9250	3.9482
59,500	58,948	255.89	2.7403	3.7307

Standard Atmosphere, English Engineering Units

Altitude		Temperature T, °R	Pressure p, lb/ft^2	Density ρ, slugs/ft^3
h_G, ft	h, ft			
−16,500	−16,513	577.58	3.6588 + 3	3.6905 − 3
−16,000	−16,012	575.79	3.6641	3.7074
−15,500	−15,512	574.00	3.6048	3.6587
−15,000	−15,011	572.22	3.5462	3.6105
−14,500	−14,510	570.43	3.4884	3.5628
−14,000	−14,009	568.65	3.4314	3.5155
−13,500	−13,509	566.86	3.3752	3.4688
−13,000	−13,008	565.08	3.3197	3.4225
−12,500	−12,507	563.29	3.2649	3.3768
−12,000	−12,007	561.51	3.2109	3.3314
−11,500	−11,506	559.72	3.1576 + 3	3.2866 − 3
−11,000	−11,006	557.94	3.1050	3.2422
−10,500	−10,505	556.15	3.0532	3.1983
−10,000	−10,005	554.37	3.0020	3.1548
−9,500	−9,504	552.58	2.9516	3.1118
−9,000	−9,004	550.80	2.9018	3.0693
−8,500	−8,503	549.01	2.8527	3.0272
−8,000	−8,003	547.23	2.8043	2.9855
−7,500	−7,503	545.44	2.7566	2.9443
−7,000	−7,002	543.66	2.7095	2.9035
−6,500	−6,502	541.88	2.6631 + 3	2.8632 − 3
−6,000	−6,002	540.09	2.6174	2.8233
−5,500	−5,501	538.31	2.5722	2.7838

Altitude		Temperature T, °R	Pressure p, lb/ft^2	Density ρ, slugs/ft^3
h_G, ft	h, ft			
−5,000	−5,001	536.52	2.5277	2.7448
−4,500	−4,501	534.74	2.4839	2.7061
−4,000	−4,001	532.96	2.4406	2.6679
−3,500	−3,501	531.17	2.3980	2.6301
−3,000	−3,000	529.39	2.3560	2.5927
−2,500	−2,500	527.60	2.3146	2.5558
−2,000	−2,000	525.82	2.2737	2.5192
−1,500	−1,500	524.04	2.2335 + 3	2.4830 − 3
−1,000	−1,000	522.25	2.1938	2.4473
−500	−500	520.47	2.1547	2.4119
0	0	518.69	2.1162	2.3769
500	500	516.90	2.0783	2.3423
1,000	1,000	515.12	2.0409	2.3081
1,500	1,500	513.34	2.0040	2.2743
2,000	2,000	511.56	1.9677	2.2409
2,500	2,500	509.77	1.9319	2.2079
3,000	3,000	507.99	1.8967	2.1752
3,500	3,499	506.21	1.8619 + 3	2.1429 − 3
4,000	3,999	504.43	1.8277	2.1110
4,500	4,499	502.64	1.7941	2.0794
5,000	4,999	500.86	1.7609	2.0482
5,500	5,499	499.08	1.7282	2.0174
6,000	5,998	497.30	1.6960	1.9869
6,500	6,498	495.52	1.6643	1.9567
7,000	6,998	493.73	1.6331	1.9270
7,500	7,497	491.95	1.6023	1.8975
8,000	7,997	490.17	1.5721	1.8685
8,500	8,497	488.39	1.5423 + 3	1.8397 − 3
9,000	8,996	486.61	1.5129	1.8113
9,500	9,496	484.82	1.4840	1.7833
10,000	9,995	483.04	1.4556	1.7556
10,500	10,495	481.26	1.4276	1.7282
11,000	10,994	479.48	1.4000	1.7011
11,500	11,494	477.70	1.3729	1.6744
12,000	11,993	475.92	1.3462	1.6480
12,500	12,493	474.14	1.3200	1.6219
13,000	12,992	472.36	1.2941	1.5961

Altitude		Temperature T, °R	Pressure p, lb/ft^2	Density ρ, slugs/ft^3
h_G, ft	h, ft			
13,500	13,491	470.58	1.2687 + 3	1.5707 − 3
14,000	13,991	468.80	1.2436	1.5455
14,500	14,490	467.01	1.2190	1.5207
15,000	14,989	465.23	1.1948	1.4962
15,500	15,488	463.45	1.1709	1.4719
16,000	15,988	461.67	1.1475	1.4480
16,500	16,487	459.89	1.1244	1.4244
17,000	16,986	458.11	1.1017	1.4011
17,500	17,485	456.33	1.0794	1.3781
18,000	17,984	454.55	1.0575	1.3553
18,500	18,484	452.77	1.0359 + 3	1.3329 − 3
19,000	18,983	450.99	1.0147	1.3107
19,500	19,482	449.21	9.9379 + 2	1.2889
20,000	19,981	447.43	9.7327	1.2673
20,500	20,480	445.65	9.5309	1.2459
21,000	20,979	443.87	9.3326	1.2249
21,500	21,478	442.09	9.1376	1.2041
22,000	21,977	440.32	8.9459	1.1836
22,500	22,476	438.54	8.7576	1.1634
23,000	22,975	436.76	8.5724	1.1435
23,500	23,474	434.98	8.3905 + 2	1.1238 − 3
24,000	23,972	433.20	8.2116	1.1043
24,500	24,471	431.42	8.0359	1.0852
25,000	24,970	429.64	7.8633	1.0663
25,500	25,469	427.86	7.6937	1.0476
26,000	25,968	426.08	7.5271	1.0292
26,500	26,466	424.30	7.3634	1.0110
27,000	26,965	422.53	7.2026	9.9311 − 4
27,500	27,464	420.75	7.0447	9.7544
28,000	27,962	418.97	6.8896	9.5801
28,500	28,461	417.19	6.7373 + 2	9.4082 − 4
29,000	28,960	415.41	6.5877	9.2387
29,500	29,458	413.63	6.4408	9.0716
30,000	29,957	411.86	6.2966	8.9068
30,500	30,455	410.08	6.1551	8.7443
31,000	30,954	408.30	6.0161	8.5841
31,500	31,452	406.52	5.8797	8.4261
32,000	31,951	404.75	5.7458	8.2704

Altitude		Temperature T, °R	Pressure p, lb/ft^2	Density ρ, slugs/ft^3
h_G, ft	h, ft			
32,500	32,449	402.97	5.6144	8.1169
33,000	32,948	401.19	5.4854	7.9656
33,500	33,446	399.41	5.3589 + 2	7.8165 − 4
34,000	33,945	397.64	5.2347	7.6696
34,500	34,443	395.86	5.1129	7.5247
35,000	34,941	394.08	4.9934	7.3820
35,500	35,440	392.30	4.8762	7.2413
36,000	35,938	390.53	4.7612	7.1028
36,500	36,436	389.99	4.6486	6.9443
37,000	36,934	389.99	4.5386	6.7800
37,500	37,433	389.99	4.4312	6.6196
38,000	37,931	389.99	4.3263	6.4629
38,500	38,429	389.99	4.2240 + 2	6.3100 − 4
39,000	38,927	389.99	4.1241	6.1608
39,500	39,425	389.99	4.0265	6.0150
40,000	39,923	389.99	3.9312	5.8727
40,500	40,422	389.99	3.8382	5.7338
41,000	40,920	389.99	3.7475	5.5982
41,500	41,418	389.99	3.6588	5.4658
42,000	41,916	389.99	3.5723	5.3365
42,500	42,414	389.99	3.4878	5.2103
43,000	42,912	389.99	3.4053	5.0871
43,500	43,409	389.99	3.3248 + 2	4.9668 − 4
44,000	43,907	389.99	3.2462	4.8493
44,500	44,405	389.99	3.1694	4.7346
45,000	44,903	389.99	3.0945	4.6227
45,500	45,401	389.99	3.0213	4.5134
46,000	45,899	389.99	2.9499	4.4067
46,500	46,397	389.99	2.8801	4.3025
47,000	46,894	389.99	2.8120	4.2008
47,500	47,392	389.99	2.7456	4.1015
48,000	47,890	389.99	2.6807	4.0045
48,500	48,387	389.99	2.2173 + 2	3.9099 − 4
49,000	48,885	389.99	2.5554	3.8175
49,500	49,383	389.99	2.4950	3.7272
50,000	49,880	389.99	2.4361	3.6391
50,500	50,378	389.99	2.3785	3.5531

Altitude		Temperature T, °R	Pressure p, lb/ft^2	Density ρ, slugs/ft^3
h_G, ft	h, ft			
51,000	50,876	389.99	2.3223	3.4692
51,500	51,373	389.99	2.2674	3.3872
52,000	51,871	389.99	2.2138	3.3072
52,500	52,368	389.99	2.1615	3.2290
53,000	52,866	389.99	2.1105	3.1527
53,500	53,363	389.99	2.0606 + 2	3.0782 − 4
54,000	53,861	389.99	2.0119	3.0055
54,500	54,358	389.99	1.9644	2.9345
55,000	54,855	389.99	1.9180	2.8652
55,500	55,353	389.99	1.8727	2.7975
56,000	55,850	389.99	1.8284	2.7314
56,500	56,347	389.99	1.7853	2.6669
57,000	56,845	389.99	1.7431	2.6039
57,500	57,342	389.99	1.7019	2.5424
58,000	57,839	389.99	1.6617	2.4824
58,500	58,336	389.99	1.6225 + 2	2.4238 − 4
59,000	58,834	389.99	1.5842	2.3665
59,500	59,331	389.99	1.5468	2.3107
60,000	59,828	389.99	1.5103	2.2561
60,500	60,325	389.99	1.4746	2.2028
61,000	60,822	389.99	1.4398	2.1508
61,500	61,319	389.99	1.4058	2.1001
62,000	61,816	389.99	1.3726	2.0505
62,500	62,313	389.99	1.3402	2.0021
63,000	62,810	389.99	1.3086	1.9548
63,500	63,307	389.99	1.2777 + 2	1.9087 − 4
64,000	63,804	389.99	1.2475	1.8636
64,500	64,301	389.99	1.2181	1.8196
65,000	64,798	389.99	1.1893	1.7767
65,500	65,295	389.99	1.1613	1.7348
66,000	65,792	389.99	1.1339	1.6938
66,500	66,289	389.99	1.1071	1.6539
67,000	66,785	389.99	1.0810	1.6148
67,500	67,282	389.99	1.0555	1.5767
68,000	67,779	389.99	1.0306	1.5395
68,500	68,276	389.99	1.0063 + 2	1.5032 − 4
69,000	68,772	389.99	9.8253 + 1	1.4678

Altitude		Temperature T, °R	Pressure p, lb/ft^2	Density ρ, slugs/ft^3
h_G, ft	h, ft			
69,500	69,269	389.99	9.5935	1.4331
70,000	69,766	389.99	9.3672	1.3993
70,500	70,262	389.99	9.1462	1.3663
71,000	70,759	389.99	8.9305	1.3341
71,500	74,256	389.99	8.7199	1.3026
72,000	71,752	389.99	8.5142	1.2719
72,500	72,249	389.99	8.3134	1.2419
73,000	72,745	389.99	8.1174	1.2126
73,500	73,242	389.99	7.9259 + 1	1.1840 − 4
74,000	73,738	389.99	7.7390	1.1561
74,500	74,235	389.99	7.5566	1.1288
75,000	74,731	389.99	7.3784	1.1022
75,500	75,228	389.99	7.2044	1.0762
76,000	75,724	389.99	7.0346	1.0509
76,500	76,220	389.99	6.8687	1.0261
77,000	76,717	389.99	6.7068	1.0019
77,500	77,213	389.99	6.5487	9.7829 − 5
78,000	77,709	389.99	6.3944	9.5523
78,500	78,206	389.99	6.2437 + 1	9.3271 − 5
79,000	78,702	389.99	6.0965	9.1073
79,500	79,198	389.99	5.9528	8.8927
80,000	79,694	389.99	5.8125	8.6831
80,500	80,190	389.99	5.6755	8.4785
81,000	80,687	389.99	5.5418	8.2787
81,500	81,183	389.99	5.4112	8.0836
82,000	81,679	389.99	5.2837	7.8931
82,500	82,175	390.24	5.1592	7.7022
83,000	82,671	391.06	5.0979	7.5053
83,500	83,167	391.87	4.9196 + 1	7.3139 − 5
84,000	83,663	392.69	4.8044	7.1277
84,500	84,159	393.51	4.6921	6.9467
85,000	84,655	394.32	4.5827	6.7706
85,500	85,151	395.14	4.4760	6.5994
86,000	85,647	395.96	4.3721	6.4328
86,500	86,143	396.77	4.2707	6.2708
87,000	86,639	397.59	4.1719	6.1132
87,500	87,134	398.40	4.0757	5.9598
88,000	87,630	399.22	3.9818	5.8106

Altitude		Temperature T, °R	Pressure p, lb/ft^2	Density ρ, slugs/ft^3
h_G, ft	h, ft			
88,500	88,126	400.04	3.8902 + 1	5.6655 − 5
89,000	88,622	400.85	3.8010	5.5243
89,500	89,118	401.67	3.7140	5.3868
90,000	89,613	402.48	3.6292	5.2531
90,500	90,109	403.30	3.5464	5.1230
91,000	90,605	404.12	3.4657	4.9963
91,500	91,100	404.93	3.3870	4.8730
92,000	91,596	405.75	3.3103	4.7530
92,500	92,092	406.56	3.2354	4.6362
93,000	92,587	407.38	3.1624	4.5525
93,500	93,083	408.19	3.0912 + 1	4.4118 − 5
94,000	93,578	409.01	3.0217	4.3041
94,500	94,074	409.83	2.9539	4.1992
95,000	94,569	410.64	2.8878	4.0970
95,500	95,065	411.46	2.8233	3.9976
96,000	95,560	412.27	2.7604	3.9007
96,500	96,056	413.09	2.6989	3.8064
97,000	96,551	413.90	2.6390	3.7145
97,500	97,046	414.72	2.5805	3.6251
98,000	97,542	415.53	2.5234	3.5379
98,500	98,037	416.35	2.4677 + 1	3.4530 − 5
99,000	98,532	417.16	2.4134	3.3704
99,500	99,028	417.98	2.3603	3.2898
100,000	99,523	418.79	2.3085	3.2114
100,500	100,018	419.61	2.2580	3.1350
101,000	100,513	420.42	2.2086	3.0605
101,500	101,008	421.24	2.1604	2.9879
102,000	101,504	422.05	2.1134	2.9172
102,500	101,999	422.87	2.0675	2.8484
103,000	102,494	423.68	2.0226	2.7812
103,500	102,989	424.50	1.9789 + 1	2.7158 − 5
104,000	103,484	425.31	1.9361	2.6520
104,500	103,979	426.13	1.8944	2.5899
105,000	104,474	426.94	1.8536	2.5293
106,000	105,464	428.57	1.7749	2.4128
107,000	106,454	430.20	1.6999	2.3050
108,000	107,444	431.83	1.6282	2.1967

| Altitude | | Temperature T, °R | Pressure p, lb/ft^2 | Density ρ, slugs/ft^3 |
h_G, ft	h, ft			
109,000	108,433	433.46	1.5599	2.0966
110,000	109,423	435.09	1.4947	2.0014
111,000	110,412	436.72	1.4324	1.9109
112,000	111,402	438.35	1.3730 + 1	1.8247 − 5
113,000	112,391	439.97	1.3162	1.7428
114,000	113,380	441.60	1.2620	1.6649
115,000	114,369	443.23	1.2102	1.5907
116,000	115,358	444.86	1.1607	1.5201
117,000	116,347	446.49	1.1134	1.4528
118,000	117,336	448.11	1.0682	1.3888
119,000	118,325	449.74	1.0250	1.3278
120,000	119,313	451.37	9.8372 + 0	1.2697
121,000	120,302	453.00	9.4422	1.2143
122,000	121,290	454.62	9.0645 + 0	1.1616 − 5
123,000	122,279	456.25	8.7032	1.1113
124,000	123,267	457.88	8.3575	1.0634
125,000	124,255	459.50	8.0267	1.0177
126,000	125,243	461.13	7.7102	9.7410 − 6
127,000	126,231	462.75	7.4072	9.3253
128,000	127,219	464.38	7.1172	8.9288
129,000	128,207	466.01	6.8395	8.5505
130,000	129,195	467.63	6.5735	8.1894
131,000	130,182	469.26	6.3188	7.8449
132,000	131,170	470.88	6.0748 + 0	7.5159 − 6
133,000	132,157	472.51	5.8411	7.2019
134,000	133,145	474.13	5.6171	6.9020
135,000	134,132	475.76	5.4025	6.6156
136,000	135,119	477.38	5.1967	6.3420
137,000	136,106	479.01	4.9995	6.0806
138,000	137,093	480.63	4.8104	5.8309
139,000	138,080	482.26	4.6291	5.5922
140,000	139,066	483.88	4.4552	5.3640
141,000	140,053	485.50	4.2884	5.1460
142,000	141,040	487.13	4.1284 + 0	4.9374 − 6
143,000	142,026	488.75	3.9749	4.7380
144,000	143,013	490.38	3.8276	4.5473
145,000	143,999	492.00	3.6862	4.3649

Altitude		Temperature T, °R	Pressure p, lb/ft^2	Density ρ, slugs/ft^3
h_G, ft	h, ft			
146,000	144,985	493.62	3.5505	4.1904
147,000	145,971	495.24	3.4202	4.0234
148,000	146,957	496.87	3.2951	3.8636
149,000	147,943	498.49	3.1750	3.7106
150,000	148,929	500.11	3.0597	3.5642
151,000	149,915	501.74	2.9489	3.4241
152,000	150,900	503.36	2.8424 + 0	3.2898 − 6
153,000	151,886	504.98	2.7402	3.1613
154,000	152,871	506.60	2.6419	3.0382
155,000	153,856	508.22	2.5475	2.9202
156,000	154,842	508.79	2.4566	2.8130
157,000	155,827	508.79	2.3691	2.7127
158,000	156,812	508.79	2.2846	2.6160
159,000	157,797	508.79	2.2032	2.5228
160,000	158,782	508.79	2.1247	2.4329
161,000	159,797	508.79	2.0490	2.3462

appendix

C

Symbols and Conversion Factors

SYMBOLS

meter, m
kilogram, kg
second, s
kelvin, K
foot, ft
pound force, lb or lb_f
pound mass, lb_m
degree rankine, °R
newton, N
atmosphere, atm

CONVERSION FACTORS

1 ft = 0.3048 m
1 slug = 14.594 kg
1 slug = 32.2 lb_m
1 lb_m = 0.4536 kg
1 lb = 4.448 N
1 atm = 2116 lb/ft^2 = 1.01×10^5 N/m^2
1 K = 1.8°R

D

Airfoil Data

NACA 1408 Wing Section

NACA 1412 Wing Section

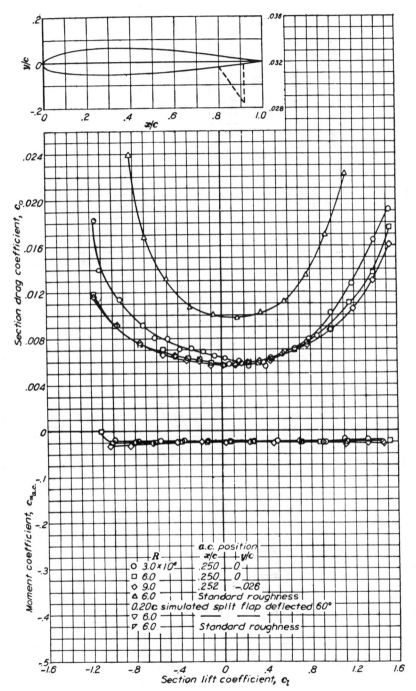

NACA 1412 Wing Section (*Continued*)

NACA 2412 Wing Section

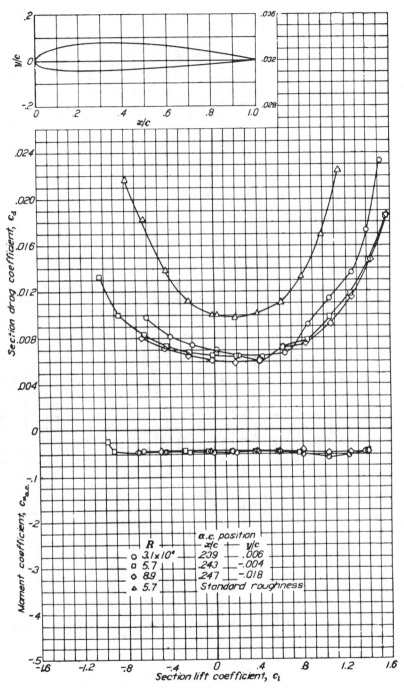

R	a.c. position x/c	y/c
○ 3.1x10⁶	.239	.006
□ 5.7	.243	-.004
◇ 8.9	.247	-.018
△ 5.7	Standard roughness	

NACA 2412 Wing Section (*Continued*)

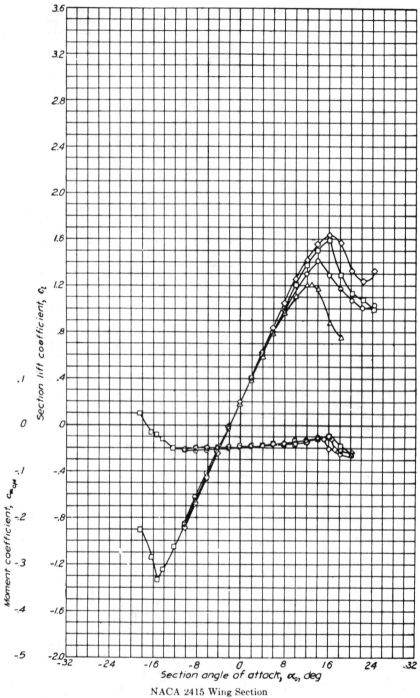

Section lift coefficient, c_l

Moment coefficient, $c_{m_{c/4}}$

Section angle of attack, α_0, deg

NACA 2415 Wing Section

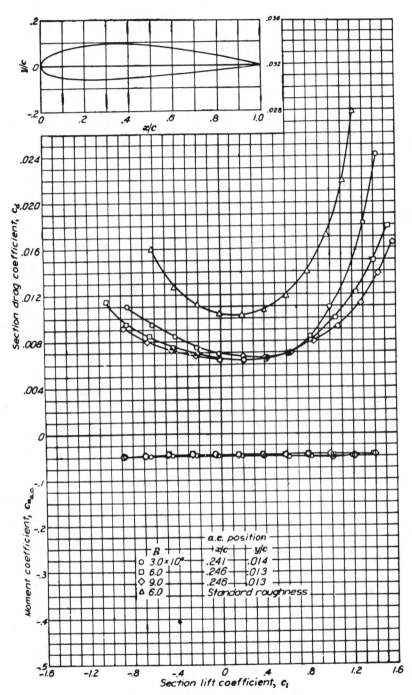

NACA 2415 Wing Section (*Continued*)

NACA 4412 Wing Section

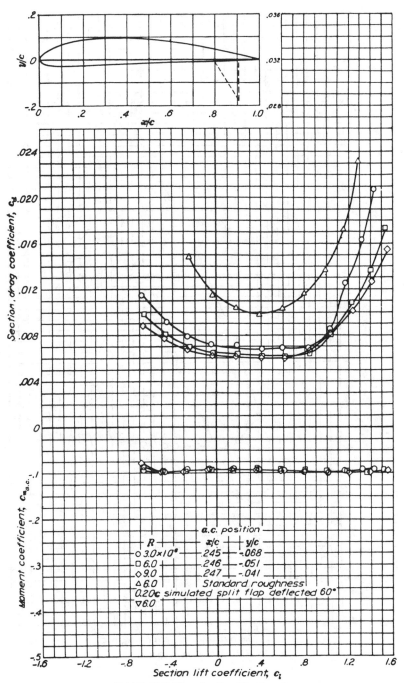

NACA 4412 Wing Section (*Continued*)

NACA 4415 Wing Section

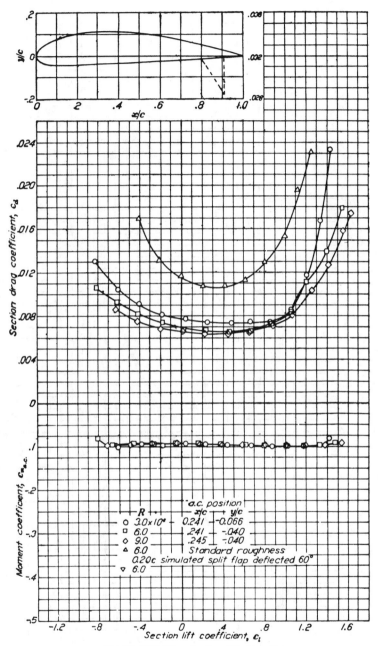

NACA 4415 Wing Section (*Continued*)

NACA 23012 Wing Section

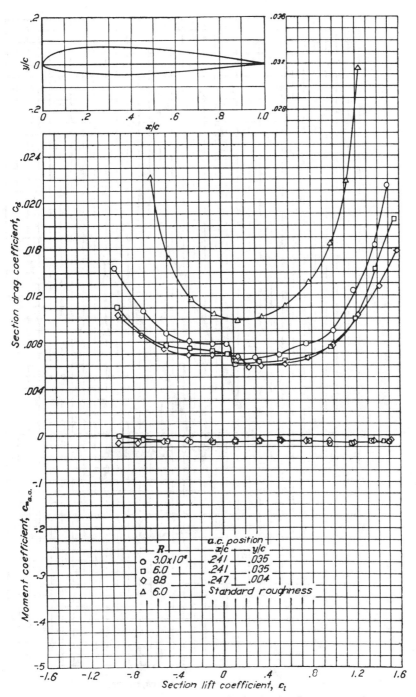

NACA 23012 Wing Section (*Continued*)

NACA 63-210 Wing Section

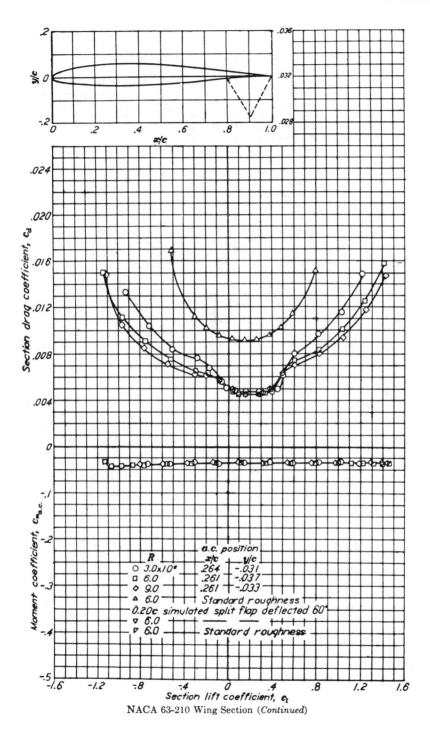

NACA 63-210 Wing Section (*Continued*)

NACA 64-210 Wing Section

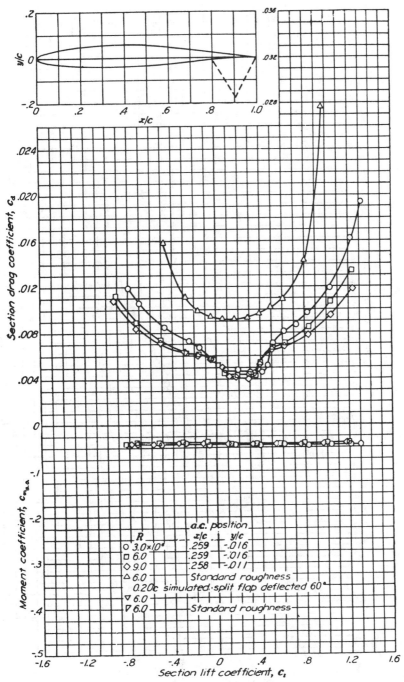

NACA 64-210 Wing Section (*Continued*)

NACA 65-210 Wing Section

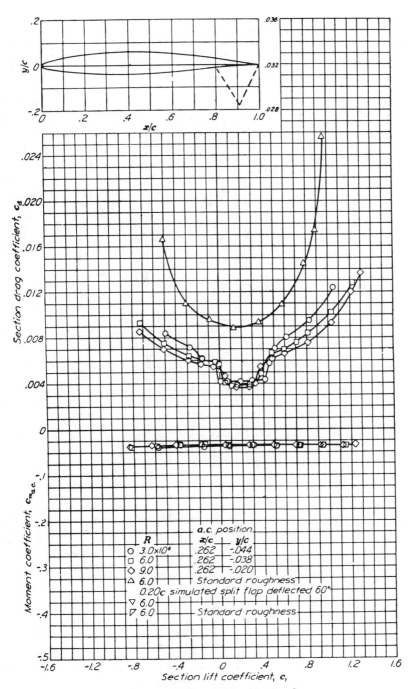

NACA 65–210 Wing Section (*Continued*)

NACA 0006 Wing Section

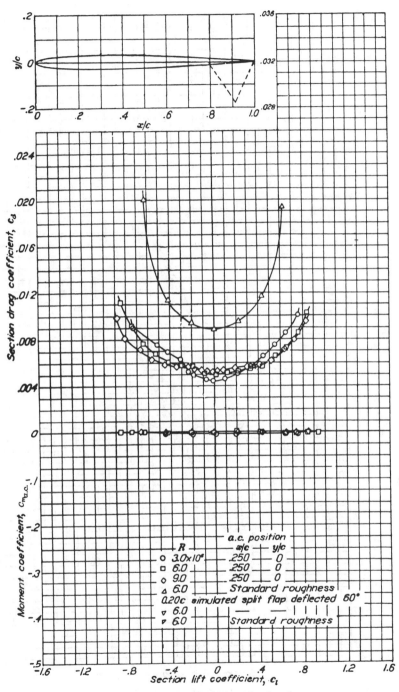

NACA 0006 Wing Section (*Continued*)

NACA 0009 Wing Section

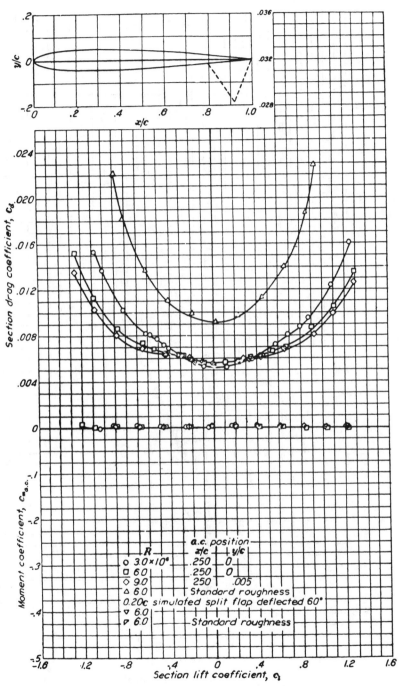

NACA 0009 Wing Section (*Continued*)

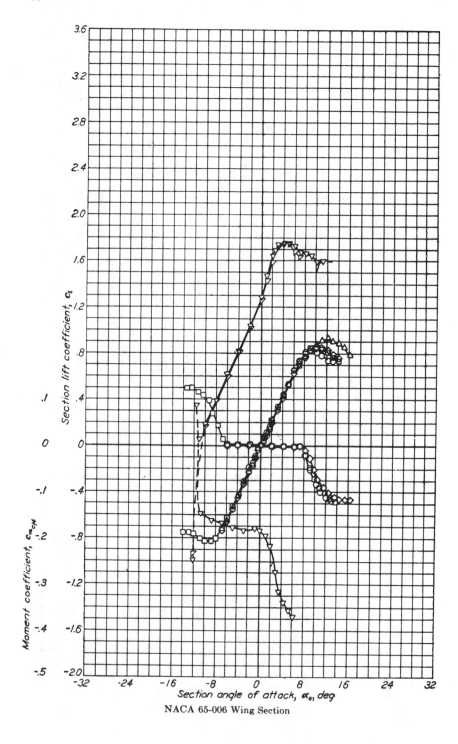

Section angle of attack, α_o, deg.

NACA 65-006 Wing Section

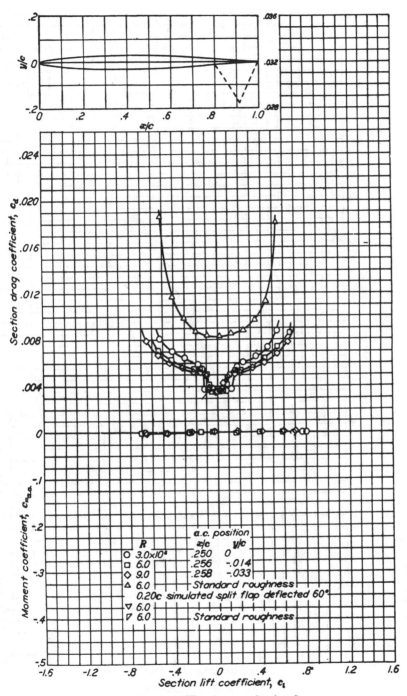

NACA 65-006 Wing Section (*Continued*)

NACA 65-009 Wing Section

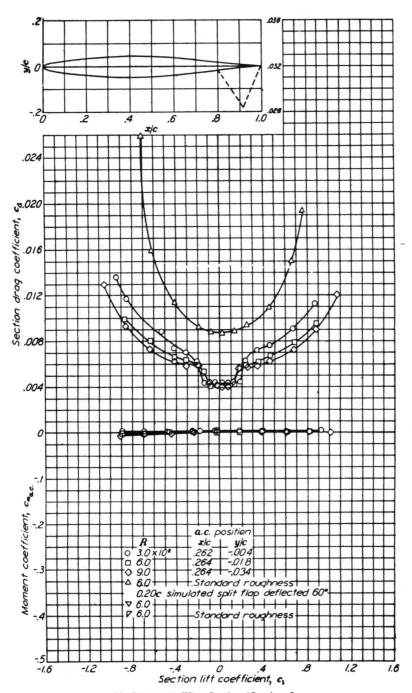

NACA 65-009 Wing Section (*Continued*)

INDEX